Innovation in the Food Sector Through the Valorization of Food and Agro-Food By-Products

Edited by Ana Novo de Barros and Irene Gouvinhas

Published in London, United Kingdom

IntechOpen

Supporting open minds since 2005

Innovation in the Food Sector Through the Valorization of Food and Agro-Food By-Products
http://dx.doi.org/10.5772/intechopen.91078
Edited by Ana Novo de Barros and Irene Gouvinhas

Contributors
Wei Xia, Ana Novo de Barros, Irene Gouvinhas, Amar Kaanane, Hind Mkadem, Manuel Pérez Nafarrate, Carlos Martín Enriquez Castro, Jesús Gerardo, Jirawat Yongsawatdigul, Ali Hamzeh, Maruf Ahmed, Abu Saeid, Ajita Tiwari, Roshna Khawas, Akeem Olayemi Raji, Renalda N. Munubi, Hieromin A. Lamtane, Marzieh Moosavi-Nasab, Armin Mirzapour-Kouhdasht, Najme Oliyaei, Jong Bang Eun, Jane Mara Block, Gabriela Polmann, Renan Danielski, Fartash Vasefi, Nicholas MacKinnon, Stanley Ng, John Chauvin, Ray Duran, Kouhyar Tavakolian, Thomas Burke, Kenneth Barton, Diego Aires da Silva, Giselle Cristine Melo Aires, Rosinelson Da Silva Pena, Fernando G. Fermoso, Juan Cubero-Cardoso, Antonio Serrano, Ángeles Trujillo-Reyes, Denys K. Villa-Gómez, Rafael Borja, Kusmiati Kusmiati, Fifi Afiati, Alireza Akhbardeh, Chrisna Widhiani, Alpinna Aditia, Destia D. Elviani, Atit Kanti

First published in London, United Kingdom, 2021 by IntechOpen
IntechOpen is the global imprint of INTECHOPEN LIMITED, registered in England and Wales, registration number: 11086078, 5 Princes Gate Court, London, SW7 2QJ, United Kingdom
Printed in Croatia

British Library Cataloguing-in-Publication Data
A catalogue record for this book is available from the British Library

Additional hard and PDF copies can be obtained from orders@intechopen.com

Innovation in the Food Sector Through the Valorization of Food and Agro-Food By-Products
Edited by Ana Novo de Barros and Irene Gouvinhas
p. cm.
Print ISBN 978-1-83880-682-8
Online ISBN 978-1-83880-683-5
eBook (PDF) ISBN 978-1-83880-690-3

For EU product safety concerns:
IN TECH d.o.o., Prolaz Marije Krucifikse Kozulić 3, 51000 Rijeka, Croatia, info@intechopen.com or visit our website at intechopen.com.

Meet the editors

Ana Novo de Barros is an assistant professor with habilitation in the Chemistry Department, University of Trás-os-Montes e Alto Douro, Portugal where she is also the director of the Centre for the Research and Technology of Agro-Environmental and Biological Sciences (CITAB). She is the author of more than eighty-five international peer-reviewed papers and holder of eight national and international patents. She has supervised eight Ph.D. theses and forty master's theses. Her main research area is the valorization of by-products of different industries, giving an added-value for the circular economy and industrial symbiosis. She has been working in the evaluation of antimicrobial activity against gastrointestinal and diabetic foot wound bacteria of hospital patients, in the concept of One Health.

Irene Gouvinhas is a full-time integrated researcher at the University of Trás-os-Montes e Alto Douro, Centre for the Research and Technology of Agro-Environmental and Biological Sciences (CITAB), Portugal, where she is collaborating on several R&D projects. She works in the areas of agrarian sciences and natural sciences with emphasis on the valorization of the major agrifood industrial by-products and waste for the recovery of compounds for several industrial applications. In her relatively short research career, she has already participated in several research projects funded by both national and European funds. She is the author of twenty-five JCR-indexed papers in specialized journals and two book chapters.

Contents

Preface

The progressive depletion of agro-diversity due to the emergence of new cultivars has led to a reduction of local varieties selected for millennia in traditional farming systems, which are now seriously threatened. The old varieties (landraces), now cultivated in traditional farming systems, are an important part of many production systems and are much appreciated for their taste and nutritional characteristics. A wide range of preharvest factors can modulate the appearance and nutritional value of the harvested products. These include biological factors (pathological, entomological, animal); physiological factors (physiological disorders, nutritional imbalances, maturity); environmental–cultural factors (climate, soils, water relations, light intensity); mechanical damage; and genetic variation. Creating and/or maintaining production conditions that minimize undesirable product appearance is essential. In recent years, consumers' perception of the roles played by the foods they ingest on their health status has evolved considerably. Foods are no longer seen only as a source of nutrients but are now viewed as a means to prevent disease and improve physical and mental well-being. This has changed consumer demands in the field of food production. Fruits, vegetables, and mushrooms are valuable sources of nutrients and contain many phytochemicals with biological activity. In addition, the use of genetic resources in the development of hypoallergenic and antioxidant products is an attractive solution. The identification/detection of antioxidants and antioxidant genes in different plant species is an interesting topic due to their importance for human health. A food traceability system is fundamentally based on product identification, data to trace, product routing, and traceability tools to ensure that consumers receive products that are safe and do not pose a threat to health. Another step in food valorization is the development of new products and new technologies for food processing.

This book contains fifteen chapters organized in three sections: "By-Products Valorization," "New Technologies," and "Innovation in the Food Sector."

Ana Novo de Barros and Irene Gouvinhas
CITAB – Centre for the Research and Technology of Agro-Environmental and Biological Sciences/UTAD – University of Trás-os-Montes e Alto Douro, Vila Real, Portugal

Section 1

By-Products Valorization

Chapter 1

Winery By-Products as Source of Bioactive Compounds for Pharmaceutical and Cosmetic Industries

Irene Gouvinhas and Ana Barros

Abstract

It is well established in the scientific community that agro-food wastes represent economic advantages and contribute to circular economy. For instance, wine industries of Região Demarcada do Douro involve the production of large quantities of by-products, such as stem, pomace, trimmed vine shoots, or wine lees, presenting a remarkable valuable composition in phytochemicals with putative health-promoting qualities. Nevertheless, the bioactive compounds obtained from these natural sources depends on the extraction process employed. In order to reduce production costs and optimize processes, new technologies—such as ultrasound-assisted extraction (UAE)—have been employed to decrease energy consumption and increase the product or process safety/control and quality. This work aims to characterize the phenolic compounds extracted from winery by-products (WBPs), namely grape stems, grape pomace, and wine lees of two grape (*Vitis vinifera* L.) varieties (Sousão and Tinta Barroca) from the same geographical site, as well as the antioxidant capacity. Wine lees and grape stems presented the highest concentration of phenolic compounds and the highest antioxidant capacity for Tinta Barroca variety, while grape pomace presented the highest values of these parameters for Sousão variety, demonstrating the high influence of the variety studied. Furthermore, wine lees revealed to be the winery by-product with the lowest antioxidant capacity and content of phenolics. These by-products revealed to be a rich source of phenolic compounds with high antioxidant capacities reveling to be of interest for pharmaceutical and cosmetic industries.

Keywords: Winery by-products, Ultrasound-assisted extraction, Bioactive compounds, Antioxidant capacity, Valorization

1. Introduction

The main strategies for the valorization of food wastes are related to their biotechnological transformation into chemicals or even the recovery of important substances, such as polyphenols that typically appear in the winery by-products (WBPs). Currently, the implemented alternatives for reducing the environmental impact of agronomic residues involves the development of new feeds and their use as soils amendments. Actually, these are the primary alternatives considered. Given their low added-value there is a need to search for new valorization alternatives.

Based on these premises, the content of bioactive phytochemicals of agro-food materials in general has allowed envisaging their use as donors of these kind of molecules to obtain materials that could contribute to enhance medical/nursing treatments.

Given the relevance of the winemaking companies, particularly at Douro region, and the high amount of underexploited wastes produced, the development of innovative applications for these materials urges [1]. On these materials ongoing research (also relevant studies developed by the research group) has revealed the valuable quantitative profile of bioactive compounds in WBPs, namely a variety of (poly) phenols and stilbenes that could be responsible for remarkable biological activities, such as anti-inflammatory, antioxidant, and antibacterial, among others [2–5].

However, for envisaging new applications for these materials, it is important to be aware about the close dependency of the phytochemical composition and therefore the biological power on an array of factors, namely the geographical growing conditions [3], the cultivar studied [4] and, most important, the extraction methodology employed [6]. In fact, the extraction methodology no just condition the phytochemical compounds obtained from a given plant material, but also is associated to environmental constraints, as well as to economic and toxicological issues depending on the solvent used [7]. To overcome these limitations, special attention has been paid to the extraction methods for bioactive compounds [8]. So, the use of eco-friendly techniques blended with reusable and non-toxic solvents is gaining a wide acceptance, due to its contribution to minimizing costs, heath related risks, and environmental impacts. As a valuable alternative to the traditional extraction methods, UAE arises as exceptional option to extract (poly)phenols, revealing to be an environment-friendly technology that offers several advantages over the conventional and non-conventional ones, such as a lower cost, versatility, and easily scale-up [7]. This technique has been already employed in diverse plant matrices and the outcomes reported have revealed it as one of the best alternatives to extract phenolic compounds from winery wastes [7, 9].

Based on these compositional features, potential applications for these materials, and specifically for grape stems, have been described by the research team, such as the spirits production, leading to an industrial alternative to traditional distilled spirits [10]. Beyond this application, recently, a preliminary study developed by the team demonstrated the stem extracts capacity to inhibit the growth of foot wound ulcers multidrug resistance bacteria (*S. aureus* and *Enterobacter aerogenes*) through disc diffusion and minimum inhibitory concentration assays [5, 11]. Whereby, WBPs are valuable candidates as wound healing agents for instance. Additional studies of our group also revealed that the quantitative (poly)phenolic profile of grape stems remains almost constant during storage for months, leading to the possibility to access this by-product all year-round, due to the preservation of the phytochemical composition [2], and thus, the biological activity expected.

In this work, we intend to generate new knowledge on the potential ability of WBPs (wine lees, grape pomace, and grape stems) bioactive compounds to be further used in pharmaceutical and cosmetic industries, using a sustainable and green extraction way, namely ultrasound-assisted extraction (UAE), enhancing the regional and circular economy.

2. Material and methods

2.1 Chemicals

Folin–Ciocalteu's reagent, 3,4,5-trihydroxybenzoic acid (gallic acid), acetic acid, both extra pure (>99%), and sodium hydroxide were purchased from Panreac

(Panreac Química S.L.U., Barcelona, Spain). Sodium nitrate, aluminum chloride, and sodium carbonate, all extra pure (>99%), and methanol were acquired from Merck (Merck, Darmstadt, Germany). Sodium molybdate (99.5%) was purchased from Chem-Lab (Chem-Lab N.V., Zedelgem, Belgium). The compounds 2,2-diphenyl-1-picrylhidrazyl radical (DPPH$^{\bullet}$), 2,2-azino-bis(3-ethylbenzothiazoline-6-sulphonic acid)diammonium salt (ABTS$^{\bullet+}$), and potassium phosphate were obtained from Sigma-Aldrich (Steinheim, Germany), as well as the standards compounds for the chromatographic separation. Additionally, 6-hydroxy-2,5,7,8-tetramethylchroman-2-carboxylic acid (Trolox) was purchased from Fluka Chemika (Neu-Ulm, Switzerland).

Ultrapure water was obtained using a Millipore water purification system. Chromatography solvents were of HPLC grade according to the analysis performed.

2.2 Sampling

The present work was carried on WBPs, namely, grape stems, grape pomace and wine lees of two varieties of *Vitis vinifera* L. (Sousão and Tinta Barroca), which are traditionally cultivated in the *Região Demarcada do Douro*, in northern Portugal. Plant material came from a farm located in Cima Corgo sub-region (Upper Corgo) (Sanfins do Douro - GPS: 41.1656, −7.2912, Average Altitude.: 730 m, vineyard altitude from 690 to 730 m), as demonstrated in **Figure 1**, where geology is essentially characterized by schist formations with occasional outcrops of granite in Mediterranean-like climatic conditions [12]. No irrigation was applied in the field trial of this investigation. WBPs were collected in 2020 growing season, at the wine company which possesses this vineyard. Once collected, samples were lyophilized, grounded to a fine powder, and stored, protected from light, at room temperature until analysis.

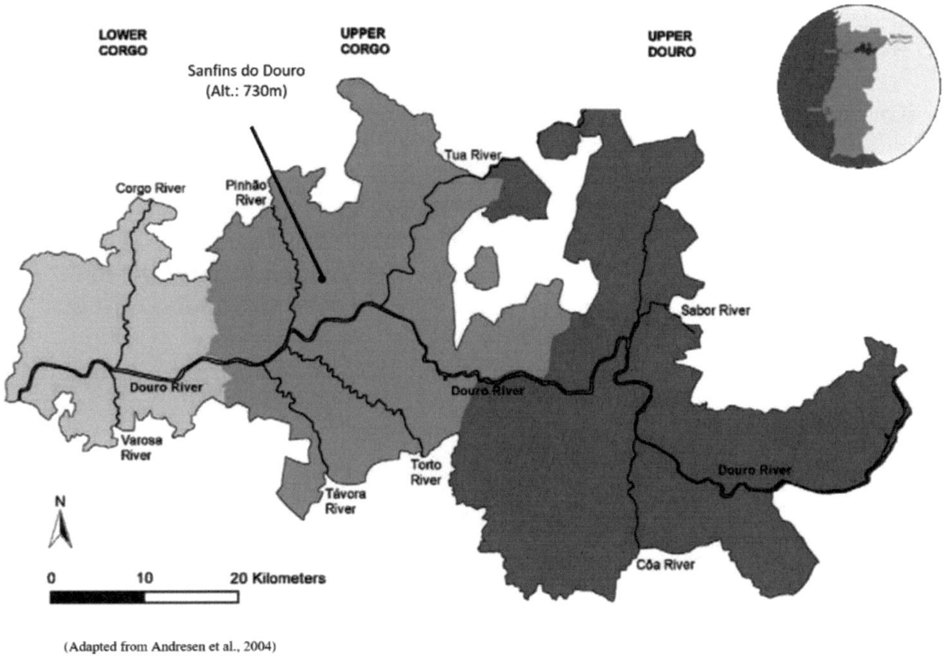

Figure 1.
Geographical origin of WBPs from the Região Demarcada do Douro *(Portugal).*

2.3 Ultrasound-assisted extraction (UAE) of bioactive compounds from olive seeds

For the phenolic compounds extraction, the protocol used was previously described by Lameirão et al. with some modifications [8]. The UAE was performed with an ultrasonic apparatus (VCX 500 Vibra-Cell™, Newtown, Connecticut, USA), using a 13 mm diameter tip with amplitude, temperature and time controller. The amplitude was employed at 50%. The powdered samples (2.5 g) were extracted with 50 mL of methanol:water (70:30, v/v) into the ultrasonic apparatus during 40 min and at 70°C. After ultrasonic extraction, the methanolic extracts were centrifuged (13,000 rpm, 4°C) for 15 min (Sigma Centrifuges 2–16 K, Germany) and filtered. Samples were stored at 4°C until analysis.

2.4 Phenolic content

The content in total phenols, flavonoids, and *ortho*-diphenols was determined according to spectrophotometric methodologies previously reported [13].

Briefly, the content of total phenolics in olive seed extracts was evaluated by the Folin–Ciocalteu spectrophotometric method, using gallic acid as standard, being the results expressed as mg of gallic acid per gram of dry weight (mg GA g^{-1} DW).

The content of *ortho*-diphenols in olive seeds was determined by adding Na_2MoO_4 (50 g L^{-1}) to the samples appropriately diluted, reading the absorbance at 375 nm. For the quantification, the gallic acid was used as standard. Results were expressed as mg GA g^{-1} DW.

For the assessment of flavonoid content, the aluminum complex method was performed, using catechin as standard. Results were expressed as mg of catechin per gram of dry weight (mg CAT g^{-1} DW).

All the assays were performed using 96-well micro plates (Nunc, Roskilde, Denmark) and an Infinite M200 microplate reader (Tecan, Grödig, Austria). For all analyses, three replicates (n = 3) of each sample were assessed.

2.5 Antioxidant capacity assays

The free radical scavenging capacity was determined by ABTS and DPPH spectrophotometric methods, according to the method described by [14]. FRAP methodology was also applied to measure ferric antioxidant power of WBPs extracts.

These assays were also performed using 96-well micro plates (Nunc, Roskilde, Denmark) and an Infinite M200 microplate reader (Tecan, Grödig, Austria), being the results expressed in mmol Trolox per gram of dried sample (mmol Trolox g^{-1} DW). All the analyses were made in triplicate (n = 3) for each sample [15].

2.6 Identification and quantification of phenolic compounds by RP–HPLC–DAD

The polyphenolic profile of WBPs extracts was assessed by Reverse Phase - High Performance Liquid Chromatography - Diode Array Detector (RP-HPLC-DAD), in an Agilent HPLC 1100 series equipped with a photodiode array detector and a mass detector in series (Agilent Technologies, Waldbronn, Germany), in accordance with the method previously described [13]. The equipment consisted of a photodiode array detector (model G1315B), an autosampler (model G1313A), a binary pump (model G1312A), and a degasser (model G1322A). The HPLC system was controlled by Xcalibur software (Agilent, version 08.03). A C18 column (250 x 4.6 mm, 5 μm particle size; ACE, Aberdeen, Scotland) was used, being the reverse phase HPLC method based on a polar mobile phase with the mixture of solvent A: H_2O/HCOOH (99.9:0.1, v/v), and solvent B: CH_3CN/HCOOH (99.9:0.1, v/v). The following linear gradient scheme was used (t in min; %B): (0; 5%), (15; 15%), (30; 30%), (40; 50%), (45; 95%), (50; 95%) and (55; 5%). At this last time (55 min), return to 5% of B to stabilize and prepare the column for the next sample. The analysis was performed at 25°C, with a flow rate of 1.0 mL/min and a sample injection volume of 20 μL. All samples were injected in triplicate. For the quantification of the identified compounds, the respective standards were used at 280 nm. Concentrations were expressed in mg g^{-1} of dry weight (mg g^{-1} DW).

2.7 Statistical analysis

The results are presented as mean (n = 3) ± standard deviation (SD). The data obtained were subjected to variance analysis (ANOVA) and a multiple range test (Tukey's test) for a p value <0.05, using IBM SPSS statistics 21.0 software (SPSS Inc., Chicago, IL, USA).

3. Results and discussion

3.1 Phenolic content of wine lees, grape pomace and grape stems

In the present work, the determination of the phenolic composition and the antioxidant capacity of wine lees, grape stems, and grape pomace extracts of two grape (*Vitis vinvifera* L.) varieties (Sousão and Tinta Barroca) were performed. The phenolic content of these samples collected from Douro region (Northern Portugal) was presented in **Table 1**. As it can be observed, in general, grape stems were the WPBs with the highest content of total phenols (168.75 mg GA g^{-1} DW, on average), *ortho*-diphenols (166.39 mg GA g^{-1} DW, on average), and flavonoids (152.31 mg CAT g^{-1} DW, on average), followed by grape pomace and wine lees. Concerning this last winery by-product, it can be stated that the samples from Sousão variety showed the lowest content of these three studied parameters, being significantly different from the other variety and WBPs ($p < 0.05$). In fact, wine lees and grape stems presented the highest values of phenolic content in Tinta Barroca samples which has not been observed for the grape pomace extracts, which can be explained by the different phenolic compounds present in these WBPs.

Romero et al. obtained similar values in wine lees of total phenols (38–254 mg CAT g^{-1}) and flavonoids (16–146 mg CAT g^{-1}) content from the Tempranillo variety, with these ranges caused by the extraction solvent employed by these authors [16]. However, Pérez-Serradilha et al. [17] obtained higher values in this WBP of total phenols content (364 mg g^{-1}) than those obtained in this work after a microwave-assisted extraction optimized. Our research group have analyzed the phenolic content of grape stem extracts prepared with conventional extraction methods (hydro-methanolic solvents) [3, 4, 11, 18].

The values ranged between 32 and 123 mg GA g^{-1} DW for total phenols, between 35 and 116 mg GA g^{-1} DW for *ortho*-diphenols, and from 34 to 106 mg CAT g^{-1} DW for flavonoids, depending on cultivar, geographical localization, crop season, among other factors [4, 18, 19]. The values of the present work were slightly higher than the values obtained in those studies, maybe due to the new efficient extraction method performed in this work (UAE). Grape pomace has been also analyzed by other authors concerning its phenolic content, obtaining values around 40 mg GA g^{-1} DW for total phenols and around 14 mg CAT g^{-1} DW for flavonoids which are significantly lower than those obtained in this work.

	Phenolic content						Antioxidant capacity					
	Total phenols		*Ortho*-diphenols		Flavonoids		ABTS		DPPH		FRAP	
	Sousão	Tinta Barroca	Sousão	Tinta Barroca	Sousão	Tinta Barroca	Sousão	Tinta Barroca	Sousão	Tinta Barroca	Sousão	Tinta Barroca
Wine lees	^{Z}15.44 ± 1.25^{a}	125.39 ± 1.12^{b}	118.91 ± 0.95^{a}	136.03 ± 1.10^{b}	18.50 ± 0.89^{a}	128.34 ± 1.07^{b}	1.71 ± 0.01^{a}	3.28 ± 0.03^{b}	1.24 ± 0.03^{a}	1.58 ± 0.03^{c}	1.54 ± 0.01^{a}	1.96 ± 0.03^{b}
Grape pomace	153.70 ± 0.53^{d}	135.32 ± 2.76^{c}	151.78 ± 1.89^{c}	138.70 ± 1.42^{b}	144.81 ± 1.75^{c}	129.93 ± 0.93^{b}	5.54 ± 0.09^{d}	4.01 ± 0.03^{c}	1.64 ± 0.03^{d}	1.59 ± 0.01^{c}	1.75 ± 0.01^{d}	1.61 ± 0.02^{b}
Grape stems	156.81 ± 1.29^{e}	180.68 ± 2.77^{f}	162.53 ± 1.01^{d}	170.24 ± 1.88^{e}	143.90 ± 2.13^{c}	160.71 ± 1.44^{d}	5.62 ± 0.02^{d}	8.02 ± 0.02^{e}	1.49 ± 0.07^{e}	1.85 ± 0.02^{c}	1.69 ± 0.01^{c}	2.02 ± 0.01^{e}
P-value	^{Y}***		***		***		**		**		**	

^{Z}*Data presented as Mean ($n = 3$) ± SD values for the same parameter evaluated followed by different superscript lowercase letters are significantly different at $p < 0.001$, according to Tukey's test.*
^{Y}*Level of significance: N.s.: not significant ($p > 0.05$);*
**significant at $p < 0.05$;*
***significant at $p < 0.01$;*
****significant at $p < 0.001$.*

Table 1.
Total phenols (mg GA g^{-1} DW), ortho-*diphenols (mg GA g^{-1} DW), and flavonoids (mg CAT g^{-1} DW) content and antioxidant capacity (mmol Trolox g^{-1} DW) of WBPs from Sousão and Tinta Barroca varieties.*

In fact, these contents are particularly dependent of several factors, such as the agronomic conditions [3], cultivar [4] and extraction methods employed [6]. In this sense, in order to reduce production costs and optimize processes, new technologies—such as ultrasound-assisted extraction (UAE) or microwave-assisted extraction (MAE)—have been employed to decrease energy consumption and increase the product or process safety/control and quality. These techniques, already used at large scale, emerged as efficient, energy/time-saving and clean extraction methodologies, providing higher recoveries of bioactive compounds using low amounts of solvent, with particular advantageous for natural sources [8, 20].

3.2 Antioxidant capacity of wine lees, grape pomace and grape stems

The antioxidant capacity of the WBPs investigated in this work was performed by three methods (ABTS, DPPH, and FRAP), being the results presented in **Table 1**. As expected, due to the correlation of most phenolic compounds with antioxidant capacity, Tinta Barroca samples presented the highest values for all the methods concerning wine lees and grape pomace samples. In contrast, grape pomace samples from Sousão variety showed the highest antioxidant capacity for ABTS (5.54 ± 0.09 mmol Trolox g^{-1} DW), DPPH (1.64 ± 0.03 mmol Trolox g^{-1} DW), and FRAP (1.75 ± 0.01 mmol Trolox g^{-1} DW) methodologies, being significantly different from the Tinta Barroca samples ($p < 0.05$).

Romero et al. [16] also determined the antioxidant capacity of wine lees extracts from Tempranillo variety, obtaining values ranged between 0.46 and 2.197 mmol Trolox g^{-1}. The values obtained in the present work are in the range of those presented by these authors, concerning FRAP method. In literature, grape stem extracts have been also analyzed concerning this biological property, which present antioxidant capacities of 0.35–0.84 mmol Trolox g^{-1} DW, 0.15–0.76 mmol Trolox g^{-1} DW, and 0.33–1.03 mmol Trolox g^{-1} DW for ABTS, DPPH, and FRAP methodologies [4, 18, 19], which were lower than those presented in the present study essentially due to several factors, such as different extraction methods, extraction solvents or protocols, varieties, among others.

3.3 Phenolic profile of wine lees, grape pomace and grape stems

The phenolic profile of wine lees, grape pomace and grape stems was performed by RP-HPLC-DAD, being the results presented in **Table 2**. Fifteen compounds were identified, being grape pomace samples the ones with more phenolic compounds identified, including phenolic acids, flavanols, and anthocyanins. In this study, it was possible to observe the same behavior referred above, namely the significant highest content of the phenolic compounds identified in grape pomace extracts from Sousão variety ($p < 0.05$). Similar compounds were identified in wine lees, namely gallic acid, catechin, epicathechin, and malvidin-3-*O*-gluside, beside others which were found only in this by-product, namely protocatechuic acid (0.337 mg g^{-1} DW, on average), delphinidin-3-*O*-glucoside (0.190 mg g^{-1} DW, on average), and petunidin-3-*O*-glucoside (0.268 mg g^{-1} DW, on average). All these compounds were also determined in higher concentrations in Tinta Barroca samples which is in agreement with the previous results reported above. Grape stem extracts presented three phenolic compounds which were not identified in wine lees and grape pomace, namely isorhammetin-3-*O*-(6-*O*-feruloyl)-glucoside, caftaric acid, and ε-viniferin, which were also present in higher concentration in Tinta Barroca samples, being significantly different from Sousão samples ($p < 0.05$).

Several compounds identified in this work have been also identified by other authors in these WBPs, namely gallic acid, catechin, delphinidin-3-*O*-glucoside,

	Wine lees		Grape pomace		Grape stems		*P*-value
	Sousão	Tinta Barroca	Sousão	Tinta Barroca	Sousão	Tinta Barroca	
1. Gallic acid	[Z]0.422 ± 0.001^{a}	0.503 ± 0.003^{b}	0.750 ± 0.002^{d}	0.641 ± 0.002^{c}	nd	nd	[Y]***
2. Isorhammetin-3-*O*-(6-*O*-feruloyl)-glucoside	nd	nd	nd	nd	0.203 ± 0.001^{a}	0.338 ± 0.001^{b}	***
3. Caftaric acid	nd	nd	nd	nd	0.171 ± 0.003^{a}	0.225 ± 0.008^{b}	***
4. Protocatechuic acid	0.216 ± 0.001^{a}	0.458 ± 0.001^{b}	nd	nd	nd	nd	**
5. *p*-coumaric acid	nd	nd	0.434 ± 0.005^{b}	0.125 ± 0.002^{a}	nd	nd	***
6. Delphinidin-3-*O*-glucoside	0.123 ± 0.002^{a}	0.258 ± 0.001^{b}	nd	nd	nd	nd	**
7. Catechin	0.559 ± 0.010^{a}	0.798 ± 0.018^{b}	1.501 ± 0.025^{d}	0.993 ± 0.031^{c}	nd	nd	***
8. Epicatechin	0.498 ± 0.001^{b}	0.660 ± 0.005^{d}	0.601 ± 0.007^{c}	0.412 ± 0.003^{a}	nd	nd	***
9. Petunidin-3-*O*-glucoside	0.215 ± 0.005^{a}	0.321 ± 0.007^{b}	nd	nd	nd	nd	**
10. Malvidin-3-*O*-glucoside	0.875 ± 0.023^{a}	0.968 ± 0.012^{c}	1.002 ± 0.024^{d}	0.934 ± 0.017^{b}	nd	nd	***
11. Quercetin-3-*O*-rutinoside	nd	nd	$0{,}450 \pm 0{,}002^{c}$	0.214 ± 0.012^{a}	$0.369 \pm 0{,}003^{b}$	0.605 ± 0.010^{d}	**
12. Quercetin-3-*O*-glucoside	nd	nd	$0.458 \pm 0{,}011^{c}$	0.385 ± 0.062^{a}	$0.374 \pm 0{,}013^{a}$	0.423 ± 0.019^{b}	**
13. Kaempferol-3-*O*-rutinoside	nd	nd	0.207 ± 0.001^{d}	0.110 ± 0.001^{c}	$0.071 \pm 0{,}001^{a}$	0.102 ± 0.003^{b}	**
14. Kaempferol-3-*O*-glucoside	nd	nd	0.305 ± 0.003^{c}	0.401 ± 0.001^{d}	0.211 ± 0.001^{a}	0.286 ± 0.006^{b}	***
15. ε-viniferin	nd	nd	nd	nd	0.087 ± 0.002^{c}	0.109 ± 0.004^{d}	**

[Z]*Data presented as Mean ($n = 3$) ± SD values in the same row followed by different superscript lowercase letters are significantly different at $p < 0.001$, according to Tukey's test.*
[Y]*Level of significance: N.s.: not significant ($p > 0.05$);*
**significant at $p < 0.05$;*
***significant at $p < 0.01$;*
****significant at $p < 0.001$. nd: not detected.*

Table 2.
Content of individual phenolics (mg g^{-1} dw) of WBPs from Sousão and Tinta Barroca varieties. Statistical treatment notes: Data were subjected to analysis of variance (ANOVA) and multiple range test (Tukey's test) with a significance of $p < 0.05$.

epicatechin, *p*-coumaric acid, petunidin-3-*O*-glucoside, malvidin-3-*O*-glucoside, among others [16, 17, 21, 22]. However, it is well known that the extraction method, the cultivars, the geographical conditions, the growing season, the plant diseases, among others, affect the chemical composition of WBPs, namely the secondary metabolites which are highly present in these matrices.

4. Conclusions

Nowadays, it is well established in the scientific community that wine has an important role in the prevention of some cardiovascular diseases, resulting from their content in bioactive phytochemicals with antioxidant capacity. Many of these compounds are derived from the solid parts of the grape cluster (stem, pomace, trimmed vine shoots, and wine lees). However, during the winemaking process, a complete extraction of these compounds to the juice/wine does not occur, and they may remain at high concentrations in certain wastes, such as in the stems. Indeed, the wine industry involves the production of large quantities of by-products, characterized by a valuable composition in phytochemicals with putative health-promoting qualities. Additionally, in light of the biological properties revealed recently, the search for natural bioactive compounds has paid attention on these materials as promising alternatives.

In this work, it was possible to observe the high content of phenolic compounds and the high antioxidant capacities demonstrated by several winery by-products, namely wine lees, grape pomace, and grape stems which were subjected to an ultrasound assisted extraction, obtaining higher values than those obtained by conventional extraction methods employed by the research group.

In this sense, the phenolics present in winery by-products may have an added-value to be used as an alternative to synthetic substances employed in distinct industries, giving rise to sustainable agro-industrial activities.

Acknowledgements

This work is supported by National Funds by FCT - Portuguese Foundation for Science and Technology, under the project UIDB/04033/2020 (CITAB). We also acknowledge Eng. Manuel Henrique from ROZÉS who kindly provided the samples.

Author details

Irene Gouvinhas and Ana Barros*
Centre for the Research and Technology of Agro-Environmental and Biological Sciences, University of Trás-os-Montes and Alto Douro (CITAB-UTAD), Inov4Agro - Institute for Innovation, Capacity Building and Sustainability of Agri-Food Production, Vila Real, Portugal

*Address all correspondence to: abarros@utad.pt

References

[1] Gouvinhas, I.; Queiroz, M.; Rodrigues, M.; Barros, A.I.R.N.A. *Evaluation of the Phytochemistry and Biological Activity of Grape (Vitis vinifera L.) Stems: Toward a Sustainable Winery Industry*; 2nd ed.; Elsevier Inc., 2019; ISBN 9780128137680.

[2] Gouvinhas, I.; Santos, R.A.; Queiroz, M.; Leal, C.; Saavedra, M.J.; Domínguez-Perles, R.; Rodrigues, M.; Barros, A.I.R.N.A. Monitoring the antioxidant and antimicrobial power of grape (*Vitis vinifera* L.) stems phenolics over long-term storage. *Ind. Crops Prod.* **2018**, *126*, doi:10.1016/j.indcrop.2018.10.006.

[3] Gouvinhas, I.; Pinto, R.; Santos, R.; José, M.; Isabel, A. Scientia Horticulturae Enhanced phytochemical composition and biological activities of grape (*Vitis vinifera* L.) Stems growing in low altitude regions. *Sci. Hortic. (Amsterdam).* **2020**, *265*, 109248, doi:10.1016/j.scienta.2020.109248.

[4] Leal, C.; Santos, R.A.; Pinto, R.; Queiroz, M.; Rodrigues, M.; José Saavedra, M.; Barros, A.; Gouvinhas, I. Recovery of bioactive compounds from white grape (Vitis vinifera L.) stems as potential antimicrobial agents for human health. *Saudi J. Biol.* Sci. **2020**, *27*, 1009-1015, doi:10.1016/j.sjbs.2020.02.013.

[5] Dias, C.; Domínguez-Perles, R.; Aires, A.; Teixeira, A.; Rosa, E.; Barros, A.; Saavedra, M.J. Phytochemistry and activity against digestive pathogens of grape (Vitis vinifera L.) stem's (poly) phenolic extracts. *LWT - Food Sci.* Technol. **2015**, *61*, 25-32, doi:10.1016/j.lwt.2014.11.033.

[6] Monteiro, M.; Santos, R.A.; Iglesias, P.; Couto, A.; Serra, C.R.; Gouvinhas, I.; Barros, A.; Oliva-Teles, A.; Enes, P.; Díaz-Rosales, P. Effect of extraction method and solvent system on the phenolic content and antioxidant activity of selected macro- and microalgae extracts. J. Appl. Phycol. **2020**, *32*, doi:10.1007/s10811-019-01927-1.

[7] Makris, D.P. Green extraction processes for the efficient recovery of bioactive polyphenols from wine industry solid wastes – Recent progress. Curr. Opin. Green Sustain. Chem. **2018**, *13*, 50-55, doi:10.1016/j.cogsc.2018.03.013.

[8] Lameirão, F.; Pinto, D.; F. Vieira, E.; F. Peixoto, A.; Freire, C.; Sut, S.; Dall'Acqua, S.; Costa, P.; Delerue-Matos, C.; Rodrigues, F. Green-Sustainable Recovery of Phenolic and Antioxidant Compounds from Industrial Chestnut Shells Using Ultrasound-Assisted Extraction: Optimization and Evaluation of Biological Activities In Vitro. Antioxidants **2020**, *9*, 267, doi:10.3390/antiox9030267.

[9] Barba, F.J.; Zhu, Z.; Koubaa, M.; Sant'Ana, A.S.; Orlien, V. Green alternative methods for the extraction of antioxidant bioactive compounds from winery wastes and by-products: A review. Trends Food Sci. Technol. **2016**, *49*, 96-109, doi:10.1016/j.tifs.2016.01.006.

[10] Barros, A.; Gouvinhas, I.; Machado, N.; Pinto, J.; Cunha, M.; Rosa, E.; Domínguez-Perles, R. New grape stems-based liqueur: Physicochemical and phytochemical evaluation. Food Chem. **2016**, *190*, 896-903, doi:10.1016/j.foodchem.2015.06.047.

[11] Leal, C.; Gouvinhas, I.; Santos, R.A.; Rosa, E.; Silva, A.M.; Saavedra, M.J.; Barros, A.I.R.N.A. Potential application of grape (*Vitis vinifera* L.) stem extracts in the cosmetic and

pharmaceutical industries: Valorization of a by-product. *Ind. Crops Prod.* **2020**, *154*, 112675, doi:10.1016/j.indcrop.2020.112675.

[12] Santos, J.A.; Malheiro, A.C.; Karremann, M.K.; Pinto, J.G. Statistical modelling of grapevine yield in the Port Wine region under present and future climate conditions. Int. J. Biometeorol. **2011**, *55*, 119-131, doi:10.1007/s00484-010-0318-0.

[13] Martins, R.B.; Gouvinhas, I.; Nunes, M.C.; Peres, J.A.; Raymundo, A.; Barros, A.I.R.N.A. Acorn flour as a source of bioactive compounds in gluten-free bread. Molecules **2020**, *25*, 1-19, doi:10.3390/molecules25163568.

[14] Monteiro, M.; Santos, R.A.; Iglesias, P.; Couto, A.; Serra, C.R.; Gouvinhas, I.; Barros, A.; Oliva-Teles, A.; Enes, P.; Díaz-Rosales, P. Effect of extraction method and solvent system on the phenolic content and antioxidant activity of selected macro- and microalgae extracts. J. Appl. Phycol. **2020**, *32*, 349-362, doi:10.1007/s10811-019-01927-1.

[15] Queiroz, M.; Oppolzer, D.; Gouvinhas, I.; Silva, A.M.; Barros, A.I.R.N.A.; Domínguez-Perles, R. New grape stems' isolated phenolic compounds modulate reactive oxygen species, glutathione, and lipid peroxidation in vitro : Combined formulations with vitamins C and E. Fitoterapia **2017**, *120*, 146-157, doi:10.1016/j.fitote.2017.06.010.

[16] Romero-Díez, R.; Rodríguez-Rojo, S.; Cocero, M.J.; Duarte, C.M.M.; Matias, A.A.; Bronze, M.R. Phenolic characterization of aging wine lees: Correlation with antioxidant activities. Food Chem. **2018**, *259*, 188-195, doi:10.1016/j.foodchem.2018.03.119.

[17] Pérez-Serradilla, J.A.; Luque de Castro, M.D. Microwave-assisted extraction of phenolic compounds from wine lees and spray-drying of the extract. *Food Chem.* **2011**, *124*, 1652-1659, doi:10.1016/j.foodchem.2010.07.046.

[18] Leal, C.; Costa, C.M.; Barros, A.I.R.N.A.; Gouvinhas, I. Assessing the Relationship Between the Phenolic Content and Elemental Composition of Grape (*Vitis vinifera* L.) Stems. *Waste and Biomass Valorization* **2020**, doi:10.1007/s12649-020-01090-6.

[19] Gouvinhas, I.; Pinto, R.; Santos, R.; Saavedra, M.J.; Barros, A.I. Enhanced phytochemical composition and biological activities of grape (*Vitis vinifera* L.) Stems growing in low altitude regions. *Sci. Hortic. (Amsterdam).* **2020**, *265*, doi:10.1016/j.scienta.2020.109248.

[20] Bruno, S.F.; Ekorong, F.J.A.A.; Karkal, S.S.; Cathrine, M.S.B.; Kudre, T.G. Green and innovative techniques for recovery of valuable compounds from seafood by-products and discards: A review. Trends Food Sci. Technol. **2019**, *85*, 10-22, doi:10.1016/j.tifs.2018.12.004.

[21] Luo, L.; Cui, Y.; Zhang, S.; Li, L.; Suo, H.; Sun, B. Detailed phenolic composition of Vidal grape pomace by ultrahigh-performance liquid chromatography–tandem mass spectrometry. J. Chromatogr. B Anal. Technol. Biomed. Life Sci. **2017**, *1068-1069*, 201-209, doi:10.1016/j.jchromb.2017.10.031.

[22] Milinčić, D.D.; Stanisavljević, N.S.; Kostić, A.Ž.; Soković Bajić, S.; Kojić, M.O.; Gašić, U.M.; Barać, M.B.; Stanojević, S.P.; Lj Tešić, Ž.; Pešić, M.B. Phenolic compounds and biopotential of grape pomace extracts from Prokupac red grape variety. LWT **2021**, *138*, 110739, doi:10.1016/j.lwt.2020.110739.

Chapter 2

Food Waste and Agro By-Products: A Step towards Food Sustainability

Ajita Tiwari and Roshna Khawas

Abstract

Utilization of food/agricultural waste is having challenge and necessity in day to day life. It's an important aspect for all the industries (food) for the process of modification and recovery. The main aim is to minimize deterioration and maximize utilization of food which will lead to less problems in waste management and environment pollution. In some of the meat packaging and food processing industries, waste utilization treatment has been implemented for successful and substantial processing. In need of growing demands of high nutritive and cheap price foods, requirements are getting high simultaneously with increasing world population. So, there is urgent need of nutrient recovery from wasted utilization and sources of food/feed will help to reduce the shortage of world food supplies to the coming generation.

Keywords: Innovation, agro by products, food waste utilization

1. Introduction

The organic waste generated during different unit operation from various sources including commercial and domestic kitchens, food processing plants, restaurants and cafeterias. According to FAO [1], in the food supply chain approximately 1.3 billion tonnes of vegetables, meat, fruits bakery, dairy and other food products are lost [2]. In Asian countries nearly every year amount of food waste (FW) could rise from 278 to 416 million tonnes from 2005 to 2025.

Kitchen and yard waste are the primary source of municipal solid waste (MSW). This waste can be further utilized for converting into useful products/energy generation at very low-cost rather than dumping and landfilling [3].

The process may attribute towards environmental and economic factors, such as capacity of municipal landfill; price associated with transportation of waste materials and landfilling; rules and regulation adoption for environment protection; less utilization of commercial fertilizers; more recycling of household waste and quality improvement of compost products [4]. Composting FW reduces the waste volume, kills pathogens, decreases weed germination in agricultural fields, and destroys malodorous compounds [5]. In concept of agriculture (organic), organic-grade agricultural waste compost gaining popularity due of its advantage on physical, biological, and chemical soil properties of soil [6]. Various food industries produced a large number of by-products or wastes which cause a serious disposable problem with the environment. Approximately 1.3 billion tonnes of food per year [7] for human consumption is lost or wasted globally. As the food production is resource-intensive, food wastes and losses are indirectly accompanied by impacts of environment, such as erosion of soil, deforestation, air and water pollution, as well as greenhouse gas emissions that occur in the processes of food production, storage, transportation, and waste management [8]. Domestic households generate the largest food-waste faction in the food supply chain [9]. As the food waste amount occurring high on the household level, in the food supply chain at the final stages, prevention must be taken at utmost importance to help prevent further climate change for food waste [10]. Recognition of appropriate management of waste has been implemented as essential prerequisite for sustainable development [11]. Historically, in the context of urban, removing potentially harmful substances or materials away from human settlements was the main focus of public waste management [12]. Increase in waste generation due to environmental, social and financial implications of unsustainable use of raw materials in the short and long term [13] waste management began to shift from a mere pollution prevention and control exercise, towards a more holistic approach.

This chapter presents the reasons for consumer food waste in a systematic, causes, replicable, systematic and transparent way. In this chapter it is reviewed and analyzed the observation collected from different studies which is carried out on the factors promoting or impending on consumer food waste. Food waste is generated when the unprocessed food is converted to suitable form for the human consumption, during the period of conversion of food it may lost, contaminated, discharged and degraded leading to the production of food waste. Nowadays waste management and its control are a great challenge from collection point to disposal unit and identifying of sustainable approach to solve the problem of waste management caused by the agricultural and industrial sectors, food supply chains and as well as retailers and final consumers. Some useful products for industrial purposes like biofuels or biopolymers are produced from the food waste. Fixation of carbon by composting and nutrients recovery can also be achieved and the final left out waste should be used as minimum desirable options for incineration and landfilling. The chapter reviews to provide an overview on food waste definitions, generation and reduction strategies, and conversion technologies emerging.

2. Food waste and agro by-products

Food wastes are usually organic residues produced by the processing of raw materials into food. Waste is characterized as a product that do not add value to a product whereas by-product is a secondary product obtained as a result of manufacturing of the main product, often with a market value. Many by-products require further processing before sale [14].

So, the wastes could be considered valuable by-product if appropriate technical means are available to generate value which exceeds the cost of reprocessing. Residues in this case cannot be considered as wastes but becomes a product of higher value. Utilization of food processing residues offer potential of converting these by-products to beneficial uses [15].

3. Agro by-products

Agro by-products or agro residues are mainly obtained from agricultural production, harvesting, and processing in farm areas and from agricultural processing industries such as oilseed extraction, brewery, malt production, cereal grain milling, fruit and vegetable processing. These by-products hold tremendous potential source of protein supply for animal feed and can also be converted to biofuels, bioenergy and other products in a way that produces economic value.

3.1 Types of agro by-products

The agro by-products derived from various crops play a significant role in animal nourishment. These by-products are of various types and can be classified into different groups, such as by-products from fruit and vegetable processing industry, crop waste and residue, by-products from sugar, starch and confectionary industry, by-products from distilleries and breweries, by-products from grain and legume milling industry, and oil industry. The handling and technologies used for processing of by-products are generally based on their type [16].

3.2 Value addition of agro by-products

Adding value to agricultural by-product makes it more desirable and enhances their economic value. Crop residue or agro by-products usually represent relatively high amounts of cellulosic material that could be returned to the soil for its future enrichment in carbon and nutrients or could be made available for further conversion to biofuels, bioenergy and other products. Such agricultural by-products can play an important role in triggering the transition of sustainable energy. There are many economic benefits that can be obtained through value-addition to agricultural by-products. These benefits include enhancing the resource use efficiency of agricultural production, increasing farm incomes and reducing the costs of production and thus increasing the profitability of farming, producing novel products, creating jobs, minimizing the disposal of the by-products into the environment to ensure improvement in environment quality [17].

3.3 Utilization of agro by-products

Earlier these agro residues were treated as waste by agriculturists and used to disposed into the surrounding environment causing pollution. However, they realized the significance of these by-products and the invulnerable costs of animal feed and fertilizers, and harmful impact to the environment and started to utilize it as animal feed. The use of crop residues is a good way of discarding materials that could otherwise be a potential health and environmental hazard. Agro by-products plays an important role in improving the nutritional status of various forms of rations and feeds of livestock as these by-products contain numerous amounts of macro and micro nutrients that are necessary for body growth and productivity [18]. There are several ways of utilizing agricultural by-products as feed for livestock, a source of fuel, and as inputs into agricultural production and rural industries [17]. The increased utilization of agricultural by-products can provide a sustainable basis for small and medium industries in a rural area and stimulate rural economic development. Harnessing crop residues as manure and bio-fertilizers, and as raw materials for producing energy and consumer products, can expand the profitability of agricultural enterprises, improve the quality of the environment and enhance energy security [19].

4. The global food supply chain: food losses and waste

The global food waste challenge concerns about over escalating emissions of Green House Gas and other impacts of environment associated with food waste [20],

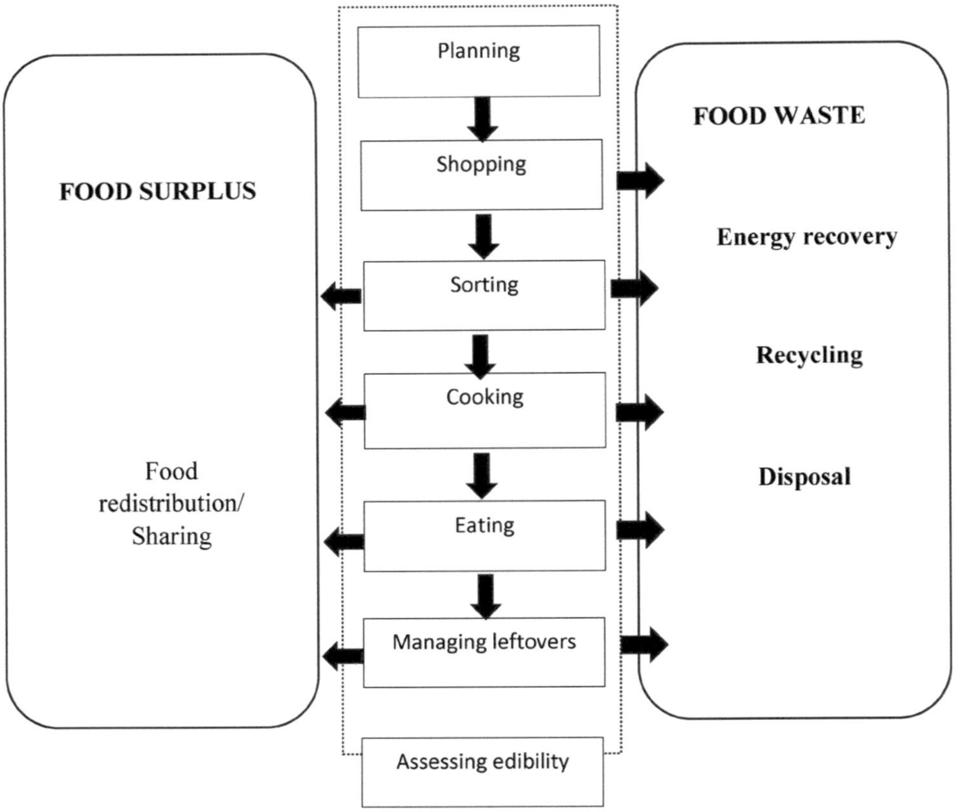

Figure 1.
Waste prevention and waste management.

as a priority waste stream, food waste is identified by growing number of policies (national and regional) [21–23]. In high range the food amount wasted in the global food supply chain (FSC) that could have otherwise been used to feed people food security is constantly pressing global issue and raises questions about it [24].

4.1 The waste hierarchy

The aim of the waste hierarchy is to recognize the problem and to generate the suitable environment friendly product from the waste. As shown in **Figure 1**, 'prevention' is the most favorable and disposal is the least favorable activity in the waste management 'pyramid'. It has been advised to consider the social and economic impacts as well as the environmental, the waste hierarchy, as a framework, primarily focuses on delivering the best environmental option by the European Waste Framework Directive (European Parliament Council, 2008).

5. Concepts in waste management and sustainability

An overview of these concepts is provided in the sections below:

5.1 Sustainable production and consumption

According to the United Nations Environmental Program [25] Sustainable Consumption and Production (SCP) defines "production and use of goods and services that respond to basic needs and bring a better quality of life, while minimizing the use of natural resources, toxic materials and emissions of waste and pollutants over the life cycle, so as not to jeopardize the needs of future generations". In this context, the SCP is seen to approach strategy implementing for achieving sustainable development, economy encompassing, environment and society with the use of both social innovation and technologies.

Framework of SCP policies includes growth of economy without damaging environment, fulfilling basic human requirements, and avert the rebound effect. SCP illustrate the phenomenon of negative impacts from growing consumption outweigh and the benefits of efficient and improved technologies. It is an integrated approach, involves both the supply and demand (goods and services), by reducing the adverse situation of production and consumption [26]. On the sustainable production side, some examples are prevention from pollution, cleaner production, efficient for ecosystem, and productivity towards greenery [27]. SCP on the consumption side, connects product and the producer with the consumer, allowing more sustainable choices to be made. Some traditional examples are eco-labeling, management in supply chain, minimization in waste, recycling, sustainable procurement and resource efficiency measures [28].

5.2 Avoidable and unavoidable food waste

WRAP defines avoidable food waste as food no longer wanted, thrown away. Avoidable food is composed of material that was, edible, at some point prior to disposal, even though a proportion is not edible at the time of disposal due to deterioration.

Avoidable food waste includes foods or parts of food that are considered edible by the vast majority of people. Unavoidable food waste is described as waste arising from food that is not, and has not been, edible under normal circumstances. This includes parts of foods such as fruit skin, apple cores and meat bones.

5.3 Waste prevention and waste management

Sinclair Knight Merz Enviros (SKM Enviros) explains waste prevention are the activities that avoids generation of waste, for instance, food surplus reduction, whereas waste management as shown in **Figure 1**, includes the activities which deals with food waste once it has been generated, such as composting and anaerobic digestion.

6. Fermentation processes

Chemical transformation of product into value added products is the process known as Fermentation which is one of the oldest methods used for product transformation through microorganisms. Fermentation processes are mostly done in three types/methods such as solid state, sub merged and liquid fermentation. Selections of the fermentation process are product specific. To obtained bioactive compounds of industrial interest from various substrates such as wastes, solid state and sub-merged fermentation processes are used [29].

6.1 Solid state fermentation

Solid-state fermentation (SSF) is the fermentation procedure in which development of microorganisms takes place on solid substrates in the absence of open liquid [30]. It focuses to attain the maximum nutrient attention from the substrate for fermentation by using the microbes such as fungi or bacteria. SSF further classified on the basis of seed culture used for fermentation is pure or mixed. In pure culture SSF, specific strains are used whereas, in with the mixed culture, different types of microorganisms are used for fermentation.

6.2 Sub merged/liquid fermentation

Submerged fermentation (SmF) is the type of fermentation in which the substrate is liquefied or put off in a water source. In industrial processes for high yield, low cost, and contamination SmF is mostly used. However, SmF has some disadvantages like physical space and energy or water requirements etc. [31]. Because of some advantageous SmF produced enzymes has been used over past of century as compared to SSF. Ease of process control and sterilization this fermentation process is easier to plan by researchers [32]. Pectinase, an enzyme production from fungi has been described by Favela-Tores et al. [33] using SmF. Pectinases are a gathering of related proteins engaged with the breakdown of pectin from an assortment of plants. Pectinases have various commercial as well as industrial importance.

6.3 Uses of fermentation for the production of bioactive/value added compounds

To elevate functional and nutritional values of the substrate to large extent SSF is a remarkable tool [34, 35]. For solid state fermentation several types of solid substrates generated from agro waste have been used which is contains of high nutritive value in terms of proteins, fibers, and minerals, respectively [36].

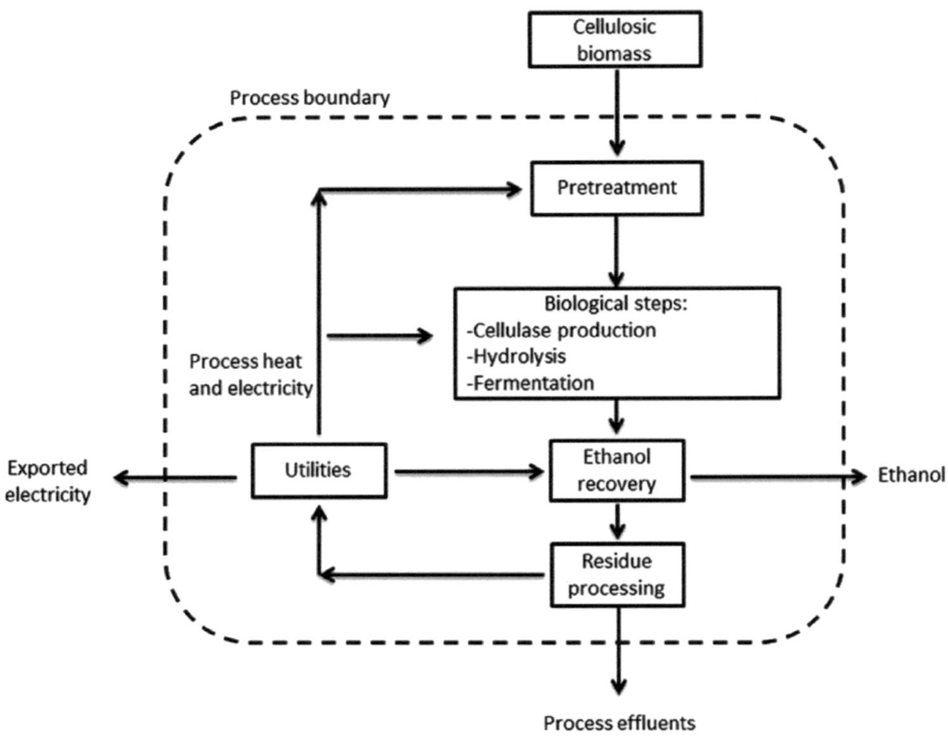

Figure 2.
Industrial waste and by-product streams via fermentation.

Figure 2 shows the outline of food industries waste through fermentation for production of various bioactive compounds. Real fact is that in human as well as animal diet these macro and micro molecules have tremendous value, therefore solid-state fermentation is an effective approach to improve their digestibility and bioavailability [37, 38]. Functional properties are the significant properties that define the pivotal phenomena of food, which are essentially used in food application [39, 40]. Also, the functional properties of food are always correlated with intrinsic components such as fats, proteins and starch, respectively.

From processing setups of the agro-industry especially the food industry produces a huge volume of wastes that are generally obtained [41]. The composition, quantity, and quality of wastes depend on the raw materials as well as the processing steps. Different type of wastes like date syrup, wheat straw, orange peel, and bran, rice straw and bran, banana, sugarcane bagasse and potato peel, soybean waste, apple pomace, oil press cakes, dairy waste, marine waste, brewery waste, food waste, etc. are produced by various types of food industries. Appropriate applications i.e., fermentations used for biotransformation of these wastes into valuable products having low cost and high nutritive value. Undeniably, use of wastes not only excludes the dumping problems but also resolves the pollution-related problems. Therefore, extra governing endorsements, as well as principal funds, are essential to bring these value-added products in the commercial market. The valorization of agro-industrial by-products to beneficial substances may not only provide future dimension to researchers but also decrease the existing environmental hazards.

7. Technologies for renewable energy generation from food/agricultural waste

Currently in many countries food waste are incinerated together with other combustible municipal wastes or landfilled for possible recovery of energy. However, due to these two approaches environment and economy of the countries are more stressed. Due to its composition of organic and nutrient-rich content, theoretically FW can be utilized as a useful resource for biofuel production through various processes of fermentation. It has attracted increasing interest in the production of biogas, hydrogen, ethanol and biodiesel as final products. Therefore, this section reviews all the food waste fermentation technologies for renewable energy generation.

7.1 Production of ethanol

The rapid global demand for the for the ethanol which has wide application in industries is increasing day by day. The main purpose of ethanol is to produce ethylene which is the main raw materials for the production of polyethylene and other plastics that is the reason for the high demand i.e. more than 140 million tonnes per year. Even the bioethanol has gained interest that is produced from cheap feedstocks [42, 43]. The source of bioethanol is the waste from starch and cellulose rich crops, e.g. sugar cane, rice and potato [44]. With the help fermentation in presence of *Saccharomyces cerevisiae* starch undergo breakdown resulted in the conversion of glucose by commercial enzymes and finally production of ethanol. In case of cellulose the breakdown due to hydrolysis is more difficult. If the FW contain large number of cellulose feedstocks than hydrolysis will become difficult, that is why, for the production of ethanol use of abundant and cheap wastes such as municipal, lignocellulosic and food waste has been explored as alternative substrates [45, 46].

7.2 Production of hydrogen

Hydrogen in the form of compressed gas gives high energy yield (142.35 kJ/g) which can also be produced from FW. The production of hydrogen is associate with the food waste containing higher amount of carbohydrate. The production rate of 0.9 to 8.35 mol H_2/mol hexose is generated from thee food waste according to recent studies [47]. The production of H_2 is influenced by many factors such as process configurations, pre-treatments and the composition of FW.

7.3 Production of methane

Methane is used as a fuel for ovens, homes, water heaters, automobiles, turbines, and other things. Because of its low cost, the production of methane via anaerobic processes is a good approach for management of waste, low production of residual waste and its utilization as a renewable energy source [48, 49]. In addition to biogas, a nutrient-rich digestate produced can also be used as soil conditioner or fertilizer. [50] investigated two-stage anaerobic digestion of fruit and vegetable wastes, in which 95.1% volatile solids (VS) conversion with a methane yield of 530 mL/g VS was achieved. [51] FW was converted to methane using a 5-L continuous digester fed with an organic loading rate (OLR) of 7.9 kg VS/m^3, resulting 70% VS conversion with a methane yield of 440 mL/g VS. [52] the methane production capacities of about 54 different fruit and vegetable wastes ranged from 180 to 732 mL/g VS depending on the origin of wastes.

7.4 Production of biodiesel

Biodiesel is synthesized through direct transesterification/acid catalyst using alkaline FW converted to fatty acids and biodiesel via various oleaginous microorganisms [53–56]. Many yeast strains produce microbial oil and then it can be used as the substitute of plant oils due to their similar fatty acid compositions. It also can be used as raw material for the production of biodiesel [57]. It has been found that the potential of FW hydrolyzate as culture medium and nutrient source in microalgae cultivation contributes for production of biodiesel [58].

In terms of prevention and concern towards economic and environment, management of FWs is utmost urgent and important to be implemented. The bioconversion of FW is economically viable for the conversion of biodiesel, ethanol, hydrogen, and methane. However, problems associated with FW in terms of transportation/collection should also be monitored. Nevertheless, the low or no cost of food waste along with the environmental benefits considering the waste disposal would balance the initial high capital costs of the biorefineries.

7.5 Production of bioactive compounds by fermentation of food waste

Bioactive and compounds are the two words which gives the term "Bioactive compounds". Scientifically, the meaning of this term is several molecules that have some biological activity.

These compounds are naturally present lesser content in plants and food stuffs, they are phytochemicals [59] and potentially able to growth in metabolism for the betterment of human health. Bioactive compounds are extremely heterogeneous class of compounds includes plant growth factors, alkaloids, mycotoxins, food-grade pigments, antibiotics, flavonoids and phenolic acids etc. with dissimilar chemical structures (hydrophilic or lipophilic), specific to ubiquitous distribution in nature, significant amount present in foods and in human body, efficient against oxidative species and possess the potential biological action [60, 61].

Through various food industries a large number of by-products or wastes are produced worldwide due to which it leads to environmental degradation. So, nowadays many approaches and new techniques are introduced for the use of the wastes, because these by-products are an excellent source of various bioactive components and beneficiary for human health. The composition of these wastes mainly depends on the waste source/type. Approximately half of the waste produced from food processing factories is lignocellulosic in nature. The dissimilar types of waste produced by food industries can be fortified by various processes. One of the oldest approaches is fermentation and carried out in three types of processes, that are carried out such as solid state, submerged and liquid fermentation used for product transformation into value added products through microorganisms [34].

8. Future perspectives

Agricultural industries generate a huge amount of wastes and by-products during production, handling and processing of agricultural products. Disposal of these wastes has a serious financial and ecological concern due to its detrimental environmental effects [62]. Therefore, to discover alternative methods of recycling and reprocessing of these wastes is a significant target taken into consideration globally. These wastes and by-products represent huge potential which have not been fully exploited, causing a loss of economic opportunity. There is thus need to identify the reasons for underutilization of agricultural by-products so that they can be addressed through suitable strategies and policy interventions. Part of the reason for the underutilization of agricultural by-products is due to lack of awareness about their properties and potential economic benefits. Proper research and studies need to be carried out on assimilating different value-added product manufacturing process. Value addition of by-products generates economic value as it facilitates the process of economic diversification by opening up new agricultural market and providing alternatives to low-cost commodity production, by offering new perspectives for the management of resources and by providing economic opportunities and environmental benefits. Markets for agricultural by-products are essential for their commercialization, value addition and efficient utilization. The lack of markets for the by-products restricts the use of crop residue to produce biofuels. So, there is a need to establish markets and to keep operational expenses of its value addition low enough to encourage the production and utilization of value-added products. These by-products also represent potential solutions to the problems of animal nutrition. Technologies needs to be developed for better utilization considering factors, such as characteristics of individual wastes and the environment in which they are produced, reprocessed and utilizied, such technologies need to convey products that are safe not only for animal feed use, but also from the point of view of human feeding.

The proper utilization of agricultural wastes and by-products has the potential to support entire industries, increase income and valuable employment opportunities, develop rural areas and solve the problem of waste and environmental pollution.

9. Conclusion

The world Population is increasing rapidly with the decreasing trend of natural resources are at the same time. Raising concerns over the security of global food due to the disparity between food wastage and food poverty, highlights the moral and social food waste dimensions. This chapter suggests that the first step towards a more sustainable resolution of the growing food waste issue is to adopt a sustainable production and consumption approach and tackle food surplus. The distinction between food surplus and food waste on one hand, and avoidable and unavoidable food waste on the other, are crucial in the process of identifying the most appropriate options for addressing the food waste challenge. This study proposes the food waste hierarchy as a framework to identify and prioritize the options for the minimization and management of food surplus and waste throughout the food supply chain. The proposed food waste hierarchy aims to challenge the current waste management approach to food waste, contribute to the debate about waste management and food security, and influence the current academic thinking and policies on waste and food to support more sustainable and holistic solutions.

Preventing food waste in agriculture and food processing requires improved infrastructure and technological solutions in harvesting, storage, transport and distribution, supported by large-scale investment and local policies. Waste management policies should be integrated and aligned with the wider policies on food, agriculture, food standards, food poverty alleviation and sustainable production and consumption.

Author details

Ajita Tiwari* and Roshna Khawas
Department of Agricultural Engineering, TSSOT, Assam University, Silchar, India

*Address all correspondence to: ajitatiwari@gmail.com

References

[1] FAO. Towards the future we want: End hunger and make the transition to sustainable agricultural and food systems. Food and agriculture organization of the United Nations Rome; 2012.

[2] Melikoglu, M., Lin, C. S. K. and Webb, C. 2013. Analyzing global food waste problem: pinpointing the facts and estimating the energy content. *Cent. Eur. J. Eng.*, 3(2), 157-64.

[3] Eriksen, G., Coale, F. and Bollero, G. 1999. Soil nitrogen dynamics and maize production in municipal solid waste amended soil. *Agron. J.* 91, 1009-1016.

[4] Hansen, T., Bhander, G., Christensen, T., Bruun, S. and Jensen, L. 2006. Life cycle modeling of environmental impacts of application of processed organic municipal solid waste on agricultural land (EASEWASTE). *Waste Manag. Res.*, 124, 153-166.

[5] Jakobsen, S. 1995. Aerobic decomposition of organic wastes 2. Value of compost as fertilizer. *Resour. Conserv. Recy.*, 13, 57-71.

[6] Iglesias-Jimenez, E. and Alvarez, C. 1993. Apparent availability of nitrogen in composted municipal refuse. *Biol. Fert. Soils,* 16, 313-318.

[7] Gustavsson, J., Cederberg, C., Sonesson, U., Van Otterdijk, R. and Meybeck, A. 2011. Global food losses and food waste. *Food and Agriculture Organization of the United Nations, Rom.* http://www.fao.org/docrep/014/mb060e/mb060e00.pdf (accessed 23.08.2017).

[8] Mourad, M. 2015. France moves toward a national policy against food waste. https://www.nrdc.org/sites/default/files/france-food-waste-policy-report.pdf (accessed 23.08.2017).

[9] BIOIS, 2010. Preparatory Study on Food Waste across EU 27. European Commission (DG ENV) Directorate C-Industry. Final Report. ISBN: 978-92-79-22138-5.

[10] Parfitt, J., Barthel, M. and Macnaughton, S. 2010. Food waste within food supply chains: quantification and potential for change to 2050. *Philosophical Transactions of the Royal Society B: Biological* Sciences 365, 3065-3081. http://rstb.royalsocietypublishing.org/content/365/1554/3065 (accessed 23.08.2017).

[11] UNEP, 2011. Decoupling Natural Resource Use and Environmental Impacts from Economic Growth. *United Nations Environment Programme, Paris*.

[12] Wilson, D.C., Rodic, L., Scheinberg, A., Velis, C. and Alabaster, G. 2012. Comparative analysis of solid waste management in 20 cities. *Waste Manag. Res.,* 30 (3), 237-254.

[13] Stern, N. 2006. Stern Review: The Economics of Climate Change. HM Treasury, London.

[14] Castro-Munoz, R., Barragan-Huerta, B. E., Fila, V., Denis, P. C. and Ruby-Figueroa, R. 2017. Current role of membrane technology: From the treatment of agro-industrial by-products up to the valorization of valuable compounds. *Waste biomass Valorization.*, 9, 513-529.

[15] Kathirvel, P. 2017. Bioconversion of fruit peel into value-added products. *International Conference on advances in biotechnology.*, 2(5), 71.

[16] Ajila, C. M., Brar, S. k., Verma, M., Tyagi, R. D., Godbout, S. and Valero, J. R. 2012. Bio-Processing of agro-byproducts to animal feed. *Critical Reviews in Biotechnology.*, 32(4), 382-400.

[17] Ayoo, C. and Bonti-Ankomah, S. 2020. Economic impacts of value addition to agricultural byproducts. Simpson, B. K., Aryee, A. N. A. and Toldra, F (Eds.) *Byproducts from agriculture and fisheries: adding value for food, feed, pharma, and fuels.* Chichester-UK: Johns Wiley & Sons Ltd. (Pp 685-698).

[18] Shamsi, I. H., Hussain, N. and Jiang, L. 2012. Agro-industrial by-products utilization in animal nutrition. Gupta, S. K (Ed) *Technological innovations in major world oil crops.* New York-USA: Springer. (Pp 209-220).

[19] Ayoo, C., Bonti-Ankomah, S. and Aryee, A. N. A. 2020. Constraints to value addition to agricultural byproducts. Simpson, B. K., Aryee, A. N. A. and Toldra, F (Eds.) *Byproducts from agriculture and fisheries: adding value for food, feed, pharma, and fuels.* Chichester-UK: Johns Wiley & Sons Ltd. (Pp 685-698).

[20] Garnett, T., Wilkes, A., 2014. Appetite for Change. Social, Economic and Environmental Transformations in China's Food System.

[21] EPA, 2012. Putting Surplus Food To Good Use. Washington DC

[22] Defra, 2011. Government Review of Waste Policy in England 2011. London.

[23] Government of South Australia, 2010. Valuing Our Food Waste. South Australia's Household Food Waste Recycling Pilot. Adelaide.

[24] Stuart, T., 2009. Waste. Uncovering the Global Food Scandal. Penguin, London.

[25] UNEP, 2008. Planning for Change. Guidelines for National Programmes, Paris.

[26] Greening, L.A., Greene, D.L. and Difiglio, C. 2000. Energy efficiency and consumption - the rebound effect e a survey. *Energy Policy,* 28, 389-401.

[27] Almeida, C.M.V.B., Bonilla, S.H., Giannetti, B.F. and Huisingh, D. 2013. Cleaner production initiatives and challenges for a sustainable world: an introduction to this special volume. *J. Clean. Prod,.* 47, 1-10.

[28] Tukker, A., Cohen, M.J., Hubacek, K. and Mont, O. 2010. The impacts of household consumption and options for change. *J. Industrial Ecol*, 14 (1): 13-30.

[29] Subramaniyam, R. and Vimala, R. 2012. Solid state and submerged fermentation for the production of bioactive substances: A comparative study. *Int. J. Sci. Nat.*, 3, 480-486.

[30] Bhargav, S., Panda, B. P., Ali, M. and Javed, S. 2008. Solid-state fermentation: An overview. *Chem. Biochem. Eng.*, 22, 49-70.

[31] Knob, A., Fortkamp, D., Prolo, T., Izidoro, S. C. and Almeida, J. M. 2014. Agro-residues as Alternative for Xylanase Production by Filamentous Fungi. *Biol. Res.,* 9, 5338-53773.

[32] Vidyalakshmi, R., Paranthaman, R. and Indhumathi, J. 2009. Amylase production on submerged fermentation by Bacillus spp. *World J. Chem.,* 4, 89-91.

[33] Favela-Torres, E., Volke-Sepulveda, T. and Viniegra-Gonzalez, G. 2006. Hydrolytic Depolymerising Pectinases. *Food Technol. Biotechnol.,* 44, 221-227.

[34] Chawla, P., Bhandari, L., Sadh, P. K. and Kaushik, R. 2017. Impact of Solid-State Fermentation (Aspergillus oryzae) on Functional Properties and Mineral Bioavailability of Black-Eyed Pea (*Vigna unguiculata*) Seed Flour. *Cereal Chem.,* 94, 437-442.

[35] Sadh, P. K., Duhan, S. and Duhan, J. S. 2018. Agro-industrial wastes and their

utilization using solid state fermentation: A review. *Bioresour. Bioprocess.*

[36] Sadh, P. K., Saharan, P. and Duhan, J. S. 2017. Bio-augmentation of antioxidants and phenolic content of *Lablab purpureus* by solid state fermentation with GRAS filamentous fungi. *Res. Effic. Technol.,* 3, 285-292.

[37] Oloyede, O. O., James, S., Ocheme, B. O., Chinma, C. E. and Akpa, V. E. 2016. Effects of fermentation time on the functional and pasting properties of defatted *Moringa oleifera* seed flour. *Food Sci. Nutr.,* 4, 89-95.

[38] Chi, C. H. and Cho, S. J. 2016. Improvement of bioactivity of soybean meal by solid-state fermentation with *Bacillus amyloliquefaciens* versus Lactobacillus spp. and *Saccharomyces cerevisiae*. *LWT Food Sci. Technol.*, 68, 619-625.

[39] Singh, H. 2011. Functional properties of milk proteins. *Ref. Modul. Food Sci.*

[40] Bhandari, L., Sodhi, N. S. and Chawla, P. 2016. Effect of acidified methanol modification on physico chemical properties of black-eyed pea (*Vigna unguiculata*) starch. *Int. J. Food Prop.*, 19, 2635-2648.

[41] Kumar, S. P., Kumar, S., Chawla, P. and Singh J. D. 2018. Fermentation: A Boon for Production of Bioactive Compounds by Processing of Food Industries Wastes (By-Products). *Molecules*, 23, 2560

[42] Lundgren, A. and Hjertberg, T. 2010. Ethylene from renewable resources. In: Surfactants Renewable Resources. John Wiley & Sons, Ltd. 109-26.

[43] International-Renewable-Energy-Agency. Production of bio-ethylene. IRENA- ETSAP; 2013.

[44] Thomsen, A. B., Medina, C. and Ahring, B. K. 2003. Biotechnology in ethanol production, in New and Emerging Bioenergy Technologies. In. *Riso National Laboratory, Denmark;* 2, 40-4.

[45] Jensen, J. W., Felby, C. and Jorgensen, H. 2011. Cellulase hydrolysis of unsorted MSW. *Appl Biochem Biotechnol,* 165(7-8), 1799-811.

[46] Kim, S. and Dale, B. E. 2004. Global potential bioethanol production from wasted crops and crop residues. *Biomass Bioenergy,* 26(4), 361-75.

[47] Patel, S. K. S., Kumar, P., and Kalia, V. C. 2012. Enhancing biological hydrogen production 707 through complementary microbial metabolisms. *Int. Jr. Hydrogen Energy*, 37(14), 10590-603.

[48] Morita, M. and Sasaki, K. 2012. Factors influencing the degradation of garbage in methanogenic bioreactors and impacts on biogas formation. *Appl, Microbiol Biotechnol,* 94(3), 575-82.

[49] Nasir, I. M., Ghazi, T. I. M. and Omar, R. 2012. Production of biogas from solid organic wastes through anaerobic digestion: a review. *Appl. Microbiol. Biotechnol.,* 95(2), 9-21.

[50] Viturtia, A. et al. 1989. Two-phase anaerobic digestion of a mixture of fruit and vegetable wastes. *Biolog. Wastes,* 29(3), 189-99.

[51] Lee, J. P., Lee, J. S. and Park, S. C. 1999. Two-phase methanization of food wastes in pilot scale. *Appl. Biochem Biotechnol – Part A Enzyme Eng Biotechnol*, 77– 79, 585-93.

[52] Gunaseelan, V. N. 2004. Biochemical methane potential of fruits and vegetable solid waste feedstocks. *Biomass Bioenergy,* 26(4), 389-99.

[53] Chen, Y. et al. 2009. Synthesis of biodiesel from waste cooking oil using immobilized lipase in fixed bed reactor. *Energy Convers Manage,* 50(3), 668-73.

[54] Mahmood, T. and Hussain, S. T. 2010. Nanobiotechnology for the production of biofuels from spent tea. *Afr J Biotechnol,* 9(6), 58-68.

[55] Papanikolaou, S. et al. 2011. Biotechnological conversion of waste cooking olive oil into lipid-rich biomass using Aspergillus and Penicillium strains. *J Appl Microbiol*, 110(5), 38-50.

[56] Yaakob, Z. et al. 2013. Overview of the production of biodiesel from waste cooking oil. *Renew Sustain Energy Rev*, 18, 84-93.

[57] Uckun Kiran, E., Trzcinski, A. and Webb, C. 2013. Microbial oil produced from biodiesel by-products could enhance overall production. *Bioresour Technol,* 29-50.

[58] Pleissner, D. 2013. Food waste as nutrient source in heterotrophic microalgae cultivation. *Bioresour Technol,* 137, 39-46.

[59] Martins, S., Mussatto, S. I., Martinez-Avila, G., Montanez-Saenz, J., Aguilar, C. N. and Teixeira, J. A. 2011. Bioactive phenolic compounds: Production and extraction by solid-state fermentation. A review. *Biotechnol. Adv.,* 29, 365-373.

[60] Carbonell-Capella, J. M., Buniowska, M., Barba, F. J., Esteve, M. J. and Frigola, A. 2014. Analytical methods for determining bioavailability and bioaccessibility of bioactive compounds from fruits and vegetables: A review. *Compr. Rev. Food Sci. Food Saf.*, 13, 155-171.

[61] Porrini, M. and Riso, P. 2008. Factors influencing the bioavailability of antioxidants in foods: A critical appraisal. *Nutr. Metab. Cardiovasc. Dis.,* 18, 647-650.

[62] The Government Office for Science, 2011. Foresight Project on Global Food and Farming Futures Synthesis Report C7: Reducing Waste. London.

Chapter 3

Valorization Technologies of Marine By-Products

Amar Kaanane and Hind Mkadem

Abstract

Generally, in different countries, strategies to improve food security have focused on increasing food production, which contributes to climate pollution and increases stress on scarce natural resources such as water and land. Due to the increase of world population (estimated to be 9 milliards in 2050), to the limited biological resources and to the increase of environmental pollution, there is a need in innovation in food industry. This can be done by improving food quality through new technologies for valorization of food and food by-products. According to Food and Agriculture Organization (FAO), one third of world food production is lost or wasted along the food supply chain. In the sector of fisheries and aquaculture, 35% of the world's harvest is lost or wasted each year. Thus, the valorization of marine by-products should be an obligation to assure the world food security and to satisfy the growing demand for fishery products. The objectives of this study are: First to review the sources of by-products and their characteristics and second to describe and evaluate the different technologies that are or can be used to valorize marine by-products in production of marine oils and concentrated fatty acids.

Keywords: by-products, valorization, technologies, fish, marine oil, n-3 fatty acids

1. Introduction

According to the FAO, world fishery production has reached around 179 million tons in 2018, of which 156.4 million tons were intended for direct human consumption. This is equivalent to an annual supply estimated at 20.5 kg per inhabitant. While world capture fisheries production stagnates at around 96.4 million tons, aquaculture is experiencing continuous growth in the supply of fish for human consumption, contributing with 46% of the total supply [1].

Unfortunately, despite the expanding demand for fishery and aquaculture products and their importance to the food security of many populations, a large part of the catch is wasted [2–4]. From the catch to the finished product, unused secondary products are generated. Today, some of these by-products are used but a great amount is wasted [5]. In 2016, Jackson & Newton estimated that 11.7 million tons of by-products produced in processing plants around the world are not collected for the production of marine ingredients [6].

In the past, marine by-products were often discarded as waste, used directly as feed for aquaculture, livestock, pets or used in silage and fertilizers [2, 7]. However, in past two decades, other uses of marine by-products have appeared based on their important characteristics and their contents of high value molecules. In some cases, compounds from by-products are identified higher in value than the starting material [8, 9]. Furthermore, with improved processing technologies, marine by-products can be now used differently and more efficiently.

Several reviews have been previously published discussing the possibilities of using marine by-products to produce high added value compounds [8, 10–15]. Different methods have allowed to produce useful molecules like proteins, gelatin, collagen, enzymes [7, 16, 17], biodiesel and biogas [18–24], natural pigments [25], minerals [26, 27], hydroxyapatite [28], chitin and chitosan [29, 30], creatine and taurine [31, 32].

The possible valorizations of marine by-products can be divided into three main categories: production of marine proteins (fishmeal, silage and hydrolysates), oils rich in polyunsaturated fatty acids (PUFAs) and preparation of high value compounds such as vitamins, enzymes, minerals, taurine and creatine, hydroxyapatite, biodiesel and biogas for human and animal nutrition, industrial or pharmaceutical uses. In this review, we mainly present the sources of marine by-products, their characteristics, and the possible technologies that can be used to produce marine oils and concentrated n-3 fatty acids.

2. Marine by-products and their characteristics

There is no one definition of marine by-products. In the past, marine by-products have been often considered as fish offal or waste [5, 7]. Actually, the term by-products designates all unused parts that can be recovered during production operations. They designate viscera, heads, trimmings, bones, cartilage, tails, skin, scales, blood, shells, carcasses, or damaged fish. Depending on the fishing period, reproductive elements such as eggs, milt or soft roe may be among these by-products [33].

In some works, the definition of by-products was reserved for feed. In others, the terms fish waste [34–37], waste streams [38], and rest raw material [5] have been used. In all cases, the biomass of by-products can be used to generate an added value unlike waste which has to be composted, burned or destroyed [5].

Generally, by-products can result from all aquatic food processing industries onshore or even during transformation on board. Marine by-products often constitute more than 50% of the body weight of processed fish [2, 4, 7, 39, 40]. However, this amount can reach up to 70% of the catch depending on catching species and area, postharvest conditions and industrial preparation processes [2, 7, 8, 11, 34, 41–44]. Processing operations like filleting, salting and smoking generate the most important amounts of by-products (50–75% of processed fish) [10], followed by the fish canning industry (30–65% of processed fish) and finally, the processing of crustaceans and mollusks [45]. It's estimated that the quantities of fish by-products generated by the processing industries will continue to increase due to the increasing demand for fishery products as source of valuable nutrients and a balanced diet for health [11, 45].

Knowledge of the properties of by-products allows their valorization into highly valuable products that could be higher in value than the fish fillets [8]. Analysis of the composition of by-products has revealed their richness in potentially valuable molecules such as proteins, essential fatty acids, oil, vitamins, minerals but also in bioactive compounds [4, 5, 34, 38, 46–48].

The by-products protein fraction is easily digestible and rich in essential amino acids. It can be used for production of peptides and amino acids, hydrolysates, gelatin and collagen, thermostable protein dispersions and protamine. While marine oils contain n-3 fatty acids [36, 37, 49–51], phospholipids, squalene, fat-soluble vitamins, and cholesterols. Additionally, other valuable components can be extracted from marine by-products including nucleic acids, calcium, phosphorous, and hydroxyapatite [14, 28, 35] and other bioactive compounds such as astaxanthin [8], chitin and chitosan [25, 30], creatine and taurine [10, 15].

There are significant compositional differences between parts composing by-products [38]. In cases where the separation between the different parts of marine by-products is possible, the valorization will be optimal. For example, prioritizing the extraction of protein derivatives from the skin of the fish or oil from viscera and/or heads. Fatty fish by-products present an important raw material for the fish oil extraction industries especially during the high fat season. Aidos and co-authors studied the possibility of oil extraction and quality of oil from salted by-products of the maatjes herring (heads, frames, skin, viscera, etc.) by demonstrating that salt does not prevent the production of an oil of good quality [47]. A recent study showed that sardine cooking condensate and cooked by-products have a high potential for the recovery of oil with yields that can reach 32.9% during the fatty season [52].

The greatest valorization of these by-products depends on their handling according to the hygiene rules applied for food production [33]. Special care must be taken to maintain the temperature low during storage and transport to avoid alteration and to preserve their nutritional qualities as marine by-products are highly sensitive to degradation (oxidation, microbial spoilage and enzymatic reactions) [53].

3. Main valorization technologies of marine by-products

3.1 Production of marine oils

Marine oils are rich in PUFAs, especially, eicosapentaenoic acid (EPA, 20: 5 n-3) and docosahexaenoic acid (DHA, 22: 6 n-3) [52, 54–59]. These n-3 fatty acids have valuable benefits and medicinal properties. Numerous articles have described the benefits of n-3 fatty acids in regard to blood pressure, prevention and treatment of coronary artery disease [60, 61], atherosclerosis and thrombosis [62–64], hyper-triglycemia [64, 65], schizophrenia and memory [66], stress and depression [67] and foetal development [57, 68–71]. However, the most widely discussed benefits relate to cardiovascular health [61, 65, 72–79] and the prevention and treatment of inflammatory diseases [57, 80–82].

These fatty acids are of marine origin, found mainly in fatty fish and seafood. They are obtained by consumption of algae, fungi and phytoplankton [83]. However, certain human groups, such as premature babies and ill people, are unable to synthesize them. Even in people not belonging to these groups, the amount of EPA and DHA synthesized by the body may not be enough because the biosynthesis of these two acids becomes slow with age as well as with bad habits such as smoking, alcohol intake and poor fitness habits [11, 84]. In this case, a diet based on marine lipids (fish and its derivatives) provides the needed intake of EPA and DHA [85, 86].

Marine oils are mainly composed of mixtures of fatty acids esterified with glycerol in triacylglycerides [11]. They are the main natural source of n-3 PUFAs particularly, EPA and DHA [37, 50, 51]. **Table 1** summarizes some variation intervals of EPA and DHA in certain oils extracted from fatty marine by-products. The variation depends on type of by-products used, the species, the catching season and the processing technology used for extraction and purification.

Species	By-products	EPA	DHA	Process	Ref
Sardinella maderensis	Liver	4.7	4.8	Solvent extraction	[87]
	skin	20.5	4.2		
Sardinella aurita	Liver	1.8	1.4		
	skin	10.4	2.5		
Cephalopholis taeniops	Liver	1.6	1.1		
	skin	3.1	6.9		
Cod (*Gadus morhua*)	Liver	8.6–11.4	11.8–16.2	Solvent extraction [88]	[89]
	Viscera	10.6–12.6	20.0–25.6		
	Trimming	14.2–16.5	30.4–33.8		
Saithe (*Pollachius virens*)	Liver	10.3–11.5	15.5–15.9		
	Viscera	7.3–13.6	9.5–23.2		
	Trimming	10.5–17.1	11.3–35.5		
Haddock (*Melanogrammus aeglefinus*)	Liver	13.1–14.8	15.2		
	Viscera	11.7–12.0	23.2–23.7		
	Trimming	14.6–16.6	33.3–35.7		
Tusk (*Brosme brosme*)	Liver	5.0	14.2		
	Viscera	7.0	25.5		
	Trimming	6.8	34.8		
Sardina pilchardus	cooked by-products	20.5–25.0	4.6–10.2	Batch hydraulic pressing	[52]
Tuna (*Thunnus obesus*)	Skins	4.2	23.6	CO_2 supercritical extraction	[90]
	Scales	4.8	23.5		
	Bones	5.1	21.6		
	Skins	3.6	21.8	Hexane Soxhlet extraction	
	Scales	4.5	21.5		
	Bones	4.7	20.0		
Sardina pilchardus	Heads, gut content, fins	14.20	18.59	Wet reduction method	[91]
Skipjack tuna	Precooked heads	0.1	25.5	Wet reduction method	[92]
	Non precooked heads	0.1	18.8		

Species	By-products	EPA	DHA	Process	Ref
Hake,	Offcuts	—	—	CO_2 supercritical extraction/Cold extraction/ Wet reduction/enzymatic extraction	[50]
Orange roughy,	Offcuts				
Salmon	Offcuts				
Jumbo squid	Livers				
Sardinella lemuru	Head Intestine liver	1.84 1.73 2.76	15.95 11.87 12.97	Solvent extraction	[37]
Sardina pilchardus	Heads	9.3	10.3	Enzymatic hydrolysis	[93]
Salmon	Frames without heads	9.3	11.3	Enzymatic hydrolysis	[94]
Black scabbardfish (*Aphanopus carbo*)	Heads, viscera, frames, skin, trimmings	2.7	6.2	Enzymatic hydrolysis	[95]
Sardine (Sardina pilchardus)	Heads, viscera, frames, trimmings)	—	—	Enzymatic hydrolysis	[96]
Salmon	Belly part	3.17	3.85	Pressing	[97]
	Belly part	4.45	3.62	CO_2 supercritical extraction	
	Trimmed muscle	3.53	3.46		
	Frame bone	4.27	3.60		
	Skin	3.87	3.26		
	Belly part	3.12	3.23	n- Hexane extraction	
	Trimmed muscle	3.22	3.98		
	Frame bone	3.85	4.32		
	Skin	2.79	3.09		
Indian mackerel (*Rastrelliger kanagurta*)	Skin	11.91–12.31	13.15–14.47	CO_2 supercritical extraction	[98]
		12.22	13.86	Soxhlet extraction	

Species	By-products	EPA	DHA	Process	Ref
Salmon	head, skin, viscera, backbone, frames, cuts off	2.46	2.97	Cold pressing	[99]
Salmon (*Salmo salar*)	Heads	7.7	11.9	Enzymatic hydrolysis	[100]
		8.4	12.1	Solvent extraction	
Rainbow trout (*Oncorhynchus mykiss*)	Roe	11.3	19.0	Enzymatic hydrolysis	[101]
		11.5	24.0	Solvent extraction [88]	
Rainbow Trout (Oncorhynchus mykiss)	Bones with leftover fish meat, skin, scales, fins	6.49–6.89	14.76–5.72	Isoelectric solubilization/precipitation	[102]
Nile perch (*Lates niloticus*)	Viscera	3.0	9.0	Enzymatic hydrolysis	[103]
Nile perch (Lates niloticus)	Heads	3.4	7.7	Enzymatic hydrolysis	[104]
Salmon (Salmo salar)	Heads	6.1	8.4		
Sardine	Heads	10.95	13.01	CO_2 supercritical extraction	[105]

Table 1.
EPA and DHA content (% of total fatty acid) in oil produced from marine by-products by different methods of extraction and purification.

Marine oils are produced from whole fish mainly small pelagic species, but also from by-products generated by the transformation industry. In 2018, it was estimated that between 25 and 35% of the total volume of fishmeal and fish oil produced came from by-products [2]. In production of marine oils, different techniques can be employed such as wet reduction process [52, 91, 92], solvent extraction [37, 87, 89, 106], supercritical fluid extraction [50, 90, 97, 98, 107, 108], urea complexation [108], cold pressing [99] and enzymatic treatment [9, 93–96, 101, 104, 109–112].

3.1.1 Wet reduction process

The traditional process for the production of marine oils is coupled with the production process of fishmeal. It is based on a heat treatment of the raw material which allows breaking the cell membrane to liberate the oil, pressing and separation [15, 106]. The principle of this process is based on the separation of the lipid phase from various fish compounds. The oil is separated by a decanter that separates insoluble compounds from the liquid phase, and a separator, separating oil and water. Another possibility is to use a tricanter, which separates solid, water phase and oil in one operation [8]. This process is mainly used in the treatment of fatty species containing high levels of fats such as anchovies, menhaden, sardines, Atlantic herring or their by-products [13].

On an industrial scale, the wet extraction technique is the most used method to obtain the oil and a substrate rich in proteins [106]. This process consists of 3 main stages:

Cooking, where the biomass is heated in a continuous screw cooker. Oil and water are separated from the solid protein. In order to recover an oil of high quality, the temperature and pressure must be adapted to the type and size of used biomass. Cooking can be done by direct injection of steam or by indirect steam heating in order to denature cell proteins and facilitate oil extraction [113, 114].

Pressing, a screw press squeezes the oil and water from the cooked biomass to separate the liquid phase containing the oil and the solid material. Press juice contains a considerable amount of suspended solids in the form of coagulated proteins, scales, edge fragments which have escaped the pressing. These particles are added to the fishmeal while the liquid phase undergoes centrifugation [115].

Centrifugation, which is now preferable compared to decantation, separates the oil from the aqueous phase. Draining and press water can be treated with live steam, allowing better separation of the oily phase on a horizontal centrifugal decanter [116]. In this step, water can be added also to wash the oil from any remaining impurities. At the end of this operation, the oil and stick water are obtained. The stickwater is evaporated to a concentrate before to add it to the pressing cake [36, 47, 117].

Quality and stability of the oil produced depend on process conditions. On one hand, the quality of oil depends on temperature, pumping speed and centrifugation. The marine oil is more stable if high pumping speed is used [118]. On the other hand, Chantachum and co-authors studied the influence of precooking on the separation and quality of tuna heads oil [92]. The results obtained have showed that cooking at 85°C for 30 min, followed by pressing at 140 tons/m^2 using a hydraulic press allows better release of the oily fractions which would be slowed down when cooked at a higher temperature due to the coagulation of proteins. The process conditions applied normally depend also on the type and quality of raw material used [117]. Comparison of quality of herring oils produced from three different types of by-products: heads, mixed by-products (heads, frames, skin, viscera, etc.), and headless byproducts showed that heads by-products and its oil presented the highest oxidation levels and the lowest R-tocopherol content. Heads contained the lowest PUFAs level and the highest amount of saturated fatty acids (SFAs) [55].

Oil recovery yield varies during the year depending on fat content of by-products. Extraction of oil from cooked by-products of *Sardina pilchardus* (skin, meat, bones and cooking condensate), using wet hydraulic pressing at 85°C for 30 min and centrifugation, gave an important oil yield varying between 6.0% and 32.9% depending on the fishing period [52]. Another study reported that the percentage of oil extracted from cooked by-products of *Sardinella gibbosa* was 8.96% [119].

Wet reduction process is more suitable to extract oil from fatty fish producing oil with improved quality and high level of n-3 PUFAs [50]. However, the main disadvantage of this technique is the use of high temperature during cooking that degrades oxidative quality of the oil and causes a loss of EPA and DHA contents due to hydrolysis and oxidation reactions [12, 111, 120].

In addition to the conventional extraction process called also wet reduction or hydraulic pressing, the extraction of marine oils could be obtained with several methods, such as enzymatic hydrolysis, physical fractionation, low-temperature solvent fractionation, supercritical fluid extraction and pH adjustement method to the isoelectric point (**Table 1**).

3.1.2 Enzymatic hydrolysis

The principle of enzymatic hydrolysis is based on the action of specific proteases at low temperature on protein tissue without use of solvents and high temperature [13, 50]. Firstly, the by-products are hydrolyzed by the use of commercial proteolytic enzymes and endogenous enzymes. These enzymes destroy the structure of cells walls. They broke down protein molecules to small peptides and amino acids which allows releasing the oil contained. After inactivation of the enzymes, the oil, hydrolysate and the insoluble residue are separated. The released oil can be centrifuged as previously described in the conventional process. For example, Batista and co-authors extracted oil from black scabbardfish (*Aphanopus carbo*) by-products using enzymatic hydrolysis with 1% Protamex [95]. The percentage of free oil released from the by-products has reached 36% of the total amount.

The enzymatic extraction is influenced by several operating factors, namely: nature of enzymes, temperature, pH, concentration of enzymes during hydrolysis, method and quality of grinding, and water content of the raw material [93, 110, 111, 120]. Optimal conditions of hydrolysis of sardine heads by Protamex were studied (temperature, hydrolysis time and enzyme-substrate ratio) [93]. Results have showed that optimum conditions were found to be similar to recover lipids and phospholipids (29 min, 31°C with 2.6 g.kg-1 enzyme). The same study has showed that hydrolysis could increase the extraction of lipids and phospholipids by 27% and 50%, respectively compared to classical extraction.

Šližyte and co-authors studied the enzymatic hydrolysis of mixtures of cod (*Gadus morhua*) by-products using Flavoenzyme and Neutrase [110]. The results demonstrated that the most important factor influencing the extraction yield is the added water regardless the type of enzyme. On the other hand, using an enzymatic process based on a proteolytic extraction of oil from crude tuna heads followed by a urea complexation (−5°C, 20 h) has allowed to obtain a mixture of DHA and EPA with a purity of 85.02% and a liquid recovery yield of 25.10% [121].

The enzymatic hydrolysis of salmon frames with Protamex, was able to separate the salmon frames into an aqueous fraction rich in soluble nitrogen (fish protein hydrolysate), an insoluble nitrogen fraction, an emulsion fraction, salmon oil and a bone fraction rich in protein and minerals. This process allows to separate oil from protein fraction that can be valorized to recover peptides, essential amino acids or other molecules [94].

Batista and co-authors studied the use of 0.5% of three commercial enzymes (Alcalase, Neutrase, and Protamex TM) and a water/fish ratio of 1:1 for production of protein hydrolysates and oil from raw and cooked sardine by-products from the canning industry [96]. Results have showed that the highest nitrogen solubilization and degree of hydrolysis were obtained with Alcalase and Protamex. The raw by-products were more easily hydrolyzed by these enzymes than the cooked sardine. The highest percentage of oil released was obtained from raw sardine, and Alcalase and Protamex were the most efficient. In another study, enzymatic extraction of oil from ground salmon heads at 55°C using different commercial enzymes (Alcalase, Neutrase, and Flavourzyme) showed that the highest oil recovery (17.4%) was obtained after 2 h by using Alcalase [111]. Same results were obtained later by other researchers [100, 101].

The enzymatic process can be chosen to extract the oil for many reasons. First, it is conducted under mild conditions protecting PUFAs from oxidation. Second, this technique does not use chemical solvents in addition to short time of hydrolysis [120, 122]. However, this process focuses more on the yield and production of protein hydrolysates than on the production of high quality oil. This disadvantage is due to the presence of lipase in the by-products which still active during the hydrolysis therefore it affects the oil quality [96]. Furthermore, this effect is more important if the hydrolysis is done in the presence of oxygen which increases significantly the free fatty acids in the oil, compared with the conventional method [8]. The main product from this process is the protein hydrolysates that are produced with higher yields compared with the traditional process [8].

3.1.3 Supercritical carbon dioxide

Extraction with supercritical carbon dioxide (SC-CO_2) is a promising technique, as CO_2 is nontoxic gas, nonflammable and clean solvent [123]. This technique can be carried out under mild operating conditions in an oxygen-free environment and at a moderate temperature, preventing degradation of PUFAs [12, 50, 90, 97, 98, 105]. CO_2 used is a green solvent. At or above critical temperature and pressure (31.1°C, 7.39 MPa), CO_2 is in a liquid state while at ambient temperature and pressure, CO_2 becomes a gas and evaporates [120].

This technique allows the extraction of lipids of low polarity, avoids the extraction of impurities and reduces the heavy metal content [12, 50, 90]. Rubio-Rodriguez and co-authors proposed to couple this technique with extraction-fractionation process to remove free fatty acids and improve fish oil quality, alternatively to physical and chemical refining [50]. Same authors compared different oil extraction methods from fish by-products, cold extraction or centrifuging, wet reduction, enzymatic extraction and supercritical fluid extraction. This study has showed that SC-CO_2 is an interesting method, operating conditions are suitable to prevent lipid oxidation and to reduce the amount of certain pollutants such as some arsenic products [50].

Another research has also compared 3 different oil extraction methods (supercritical carbon dioxide, n-hexane and traditional pressing) from Atlantic salmon by-products (belly part, trimmed muscle, frame bone and skin). The maximum oil yield was obtained by n-hexane extraction (total oil), followed by supercritical CO_2 extraction (highly selective technique extracting non-polar compounds) and the traditional pressing that has showed the lowest yield. Likewise, differences were noted in the oil quality parameters between the 3 studied techniques, the longer oxidative stability was obtained in the oil extracted by supercritical fluid CO_2 [97].

This technique was also compared to soxhlet extraction using hexane to produce oil from skins, bones and scales of bigeye tuna (*Thunnus obesus*). This study has confirmed improved quality parameters of oil obtained by supercritical CO_2 technique extraction (low heavy metal content in the oil) [90].

Using the ground skin of Indian mackerel (*Rastrelliger kanagurta*), various techniques of supercritical CO_2 were studied by varying pressure (20-35 MPa) and temperature (45–75°C). This study has showed that oil yield increased with pressure and temperature and the highest yields were 24.7, 53.2, 52.8, and 52.3/100 g sample (dry basis) for the continuous, cosolvent, soaking, and pressure swing techniques, respectively, at 35 MPa and 75°C [98].

Supercritical fluid technology coupled with membrane, enzymatic or adsorption process have been shown to produce high-quality oil with best reduction of levels of contaminants compared to traditional refining of oils [124, 125].

Faced with all these advantages, the main drawback of this method is the high cost of the application on an industrial scale [12, 50, 124]. In this contest, the effect of supercritical CO_2 techniques on CO_2 consumption was studied. The results have showed that the total amount of CO_2 consumption decreases significantly with temperature and increases with pressure in all extraction modes using supercritical CO_2 method. A higher amount of CO_2 was needed for the continuous technique, compared to the techniques of cosolvent, soaking and pressure swing regardless of levels of pressure and temperature. Consequently, the best extraction technique of the oil with least amount of CO_2 consumption was achieved with pressure swing mode at 35 MPa and 75 C [98].

3.1.4 Solvent extraction

Solvent extraction methods are numerous. They have been studied by several researchers applied on marine by-products [37, 87, 89, 106]. Unfortunately, these techniques have many disadvantages. Large amount of hazardous solvent and important energy are required. Besides, marine oils are oxidized exhibiting a strong red-brown or brown color when extraction of oil is done at high temperature and longtime [90]. Generally, extraction by solvent is only carried out on a laboratory scale for analytical purposes [124]. Among the widely used techniques, Bligh and Dyer method is the most recommended for the total extraction of lipids from biological tissues [126]. Most of the published data on lipid content are related to this method [105, 127–129]. The effectiveness of the Bligh and Dyer method was evaluated compared to Soxhlet method. The results have showed that the Bligh and Dyer extraction method is more effective in extracting polar and non-polar lipids from fish compared to the Soxhlet technique [130].

3.1.5 Cold pressing extraction

Other innovative processes are coming to the market such as cold pressing extraction, a patented process originating from the olive oil production industry [13]. This technique allows protection of PUFAs content, producing high quality marine oil from different types of by-products [50, 99]. This process is well known to produce a lower-yield, but higher quality of oil [131].

Due to its high degree in PUFAs, marine oils are very sensitive to oxidation. The degree of oxidation increases in the presence of air (oxygen), light and heat during extraction and storage [8, 47]. This phenomenon mainly reduces the shelf life of marine oils [53]. All techniques using high-temperature or toxic solvents can induce degradation and loss of nutritional qualities of marine oil.

For this, looking for gentle extraction of marine lipids or using non-heat processes might generate more stable lipid fractions [8]. It is also necessary to protect the oil by stopping or slowing down the oxidation process during the production and during storage. Adding antioxidants to the oil is one of the most used methods [132]. Using a modified atmosphere packaging [133], or encapsulation, which keeps marine oil away from oxygen and light [133, 134] can be also used. In addition to protecting the oil, the use of microencapsulation technology provides consumers with supplements n-3 fatty acids ready to consume.

3.2 Production of n-3 fatty acids concentrates

Another valorization of marine oils (produced from whole fish or from fish by-products) is their use in production of concentrated n-3 fatty acids in the form of free fatty acids, methyl and ethyl esters or acylglycerols [135]. Several processes can be used, the most important are urea complexation [108, 121, 136], molecular distillation [137], supercritical fluid extraction [98, 108], winterization [138, 139], fractionation by chromatography [120] and by enzymatic processes [111, 112, 140]. These techniques have been reviewed by many authors [13, 124, 135] and recently by [12]. The main challenge in the choice of concentration technique at industrial level is to reach higher yield and purity at lower cost [13, 124, 138]. **Table 2** outlines some methods to produce n-3 fatty acids concentrates with levels achieved of enrichment in EPA and DHA.

Concentration by winterization allows elimination of SFAs present in the oil, which crystallize at low temperatures [139]. Winterization is primarily designed for oils with a high content of SFAs. It allows elimination of stearic phase, by cooling the oil to 0–4°C. The degree of concentration of PUFAs by this process is evaluated as low as these interesting fatty acids could be lost in the stearic fraction [13]. However, this method produces n-3 PUFAs concentrate in natural form [138].

An alternative solvent winterization and enzymatic interesterification was studied to concentrate n-3 fatty acids in cod liver oil [138]. The optimization parameters considered were separation method, solvent, oil concentration, time and temperature of winterization. Likewise, enzymes used were examined for interesterification efficiency under different system air condition, time and temperature. Authors proposed the optimal conditions of the technique via winterization (0.1 g/mL oil/acetone, 24 h, −80°C, precooled Büchner filtration) and interesterification (Lipozyme TL IM, N_2 flow, 2.5 h, 40°C) improving n-3 fatty acid content to 43.20 mol%.

In another study, winterization was carried out on a bleached oil by a progressive cooling (30–5°C) in three phases. The effect of solvent type, solvent proportion, and agitation in the second cooling stage was studied. The results have demonstrated that using hexane has improved content of PUFA of 64.3% with 13% as a decrease percentage in level of SFAs compared to the fatty profile of bleached oil [139].

In addition to what is explained previously for CO_2 supercritical fluid extraction, this technique could be used to concentrate fatty acids. It's based on the use of CO_2 which, in the supercritical state, behaves like an extraction fluid and entrains fatty acids. In several passages, their concentrations therefore increase [13]. This method can achieve high concentration levels of n-3 PUFAs.

Supercritical carbon dioxide was used for simultaneous extraction and fractionation of fish oil from Tuna by-products [143]. The obtained oil was divided into six fractions based on molecular weight and the chain length of triglycerides in terms of fatty acid constituents. The results showed that the three first separated fractions were rich in SFAs followed by monounsaturated fatty acids (MUFAs), then PUFAs. While the three last fractions contained high levels of MUFAs and PUFAs.

Melgosa and co-authors studied the use of supercritical CO_2 as solvent in the lipase-catalyzed ethanolysis of fish oil. The effect of initial substrate ethanol/oil molar ratio (2–38), pressure (7.5–30 MPa), and temperature (323.15–353.15 K) on equilibrium conversion, reaction rate and oxidative status of the products were

Species	Type of oil	EPA	DHA	Process	Ref
Tuna (*Thunnus albacares*)	Tuna oil	EPA + DHA (85.02%) recovery 25.10%		Concentration by urea complexation	[121]
Rainbow sardine (*Dussumieria acuta*)	Oil from white muscle	15.39	17.45	Extraction [88]	[108]
	Concentrate oil	—	—	CO_2 Supercritical extraction	
		19.47	29.61	Urea complexation	
		17.74	25.51	Low-temperature crystallization with ethanol	
Pacific sardines (*Sardinops sagax*)	Crude oil from Skin-on fillets)	28.2	16.7	—	[140]
	Refined unhydrolyzed oil	26.86	13.62	Extraction and refining	
	Concentrate oil	33.74	29.94	Lipase-catalyzed hydrolysis	
Atlantic salmon (*Salmo salar*)	Crude Salmon by-products oil	3.71	9.02	—	[112]
	Salmon oil 40% hydrolysis	4.72	8.94	Lipolysis, filtration	
	Permeate	5.16	9.91		
	Esterified permeate	5.64	10.30	Enzymatic re-esterification	
Salmon	Oil from salmon heads	3.6	9.9	Enzymatic hydrolysis	[111]
	Permeate	5.6	11.6	Lipolysis	
	Permeate re-esterified	5.06	11.90	Re-esterification	
Arctic cod	Arctic cod liver oil	10.53	7.63	—	[138]
	Winterized oil	21.08	17.83	Alternate solvent winterization and enzymatic interesterification	
Tuna (*Thunnus thynnus*)	Tunafish oil	4.6	18.3	—	[141]
	Tunafish oil ethyl ester	5.3	23.7	Supercritical fluid chromatography	
Menhaden	Menhaden oil	13.5	12.6	—	[136]
	Menhaden oil after urea inclusion	29.4	41.8	Urea inclusion method	
Sardine (Sardinops sagax caeruleus)	Refined oil	14.51	12.55	—	[142]
	n-3 PUFAs concentrate	34.17	39.47	Chemical hydrolysis + urea complexation	
	n-3 PUFAs concentrate	46.26	40.32	Enzymatic hydrolysis + urea complexation	

Table 2.
Some methods of production of concentrated n-3 fatty acids in oil.

tested. The results have revealed the importance role of employing CO_2 in improving reaction kinetics by reduction of mass transfer limitations and prevention of n-3 PUFAs oxidation due to displacement of oxygen [144].

The results of supercritical fluid extraction applied on Rainbow sardine oil have showed that the highest decrease in SFAs and MUFAs were obtained at 50–60°C and 350 bars [108]. This technique was evaluated selective even in the fractionation of fish oil with lower content of EPA (4.6%) and DHA (6.7%) under conditions of lower pressure (100 and 200 bar) [145]. In another study, conditions (pressure, temperature and supercritical CO_2 flow rate) influencing concentration of fatty acids in fish oil by supercritical carbon dioxide were studied [146]. The results have demonstrated that fractionation by a supercritical fluid under optimal conditions: a pressure increase (at 5 kg/h flow rate) and flow rate increase (at 150 bar pressure), both determined a higher EPA + DHA concentration and decreased the EPA/DHA ratio. The same authors proposed to carry out a urea adduction during preparation to decrease the amount of SFAs in the starting oil before supercritical fractionation.

With regard to optimization conditions, numerous researches have worked on modeling, simulation, and optimization of the CO_2 supercritical fractionation of EPA and DHA esters in fish oil [147, 148].

Optimal conditions of application by using adjuvant material and modifying CO_2 volumetric density and temperature were also investigated in other studies to get the highest fractionation yield of EPA and DHA [149]. Likewise, Antunes-Corrêa and co-authors and Davarnejad and co-authors studied the optimal operating conditions (pressure and temperature) to fractionate fish oil. In the first study [150], the best results based on oil solubility were obtained using 7.8 MPa and 301.15 K. While in the second research [151], the maximal solubility of the fish oil (0.921 g of oil/100 g of CO_2) was obtained at optimum conditions of 40°C and 27.2 MPa. In both studies, EPA fractionation was recorded not possible and low, respectively.

The experience of CO_2 supercritical fluid chromatography was transferred to be used in laboratory in a pilot plant to produce EPA enriched mixtures. Fractionation was done on a silica adsorption column using CO_2 as supercritical solvent [152]. This allowed to achieve best purity of 93% in EPA ethyl ester fraction with a 24.6% yield. The study of the technical and economic feasibility to produce n-3 PUFAs ethyl ester concentrates from trans esterified fish oil using CO_2 supercritical fluid extraction has revealed that process cost is around 550 U.S $/kg DHA and EPA ethyl ester concentrate [141].

Other investigations have studied the use of enzymatic hydrolysis in production of n-3 PUFAs concentrates. This technique involves the use of specific enzymes (lipases), able to catalyze reactions such hydrolysis, ethanolysis or transesterification of triglycerides [124].

Concentration of Pacific sardines (*Sardinops sagax*) oil was carried out using lipase-catalyzed hydrolysis [140]. The results of this study have showed that hydrolysis with 250 U from *Candida rugosa* lipase has increased EPA concentration to a relatively constant level of 33.74% after 1.5 h. DHA levels were also significantly increased from 13.62% to 29.94% with 500 U after 9 h. This technique uses mild conditions (neutral pH and low temperatures), very important to preserve EPA and DHA from oxidation [140, 153].

Salmon oil produced from by-products of this species by controlled enzymatic procedure with Neutrase has followed a selective enzymatic hydrolysis under mild conditions, using Novozyme SP398 to enrich the n-3 PUFAs. The process used consist of a lipolysis, filtration in flat membrane device and enzymatic re-esterification with glycerol and Immobilized 1,3-specific lipase IM60 (Lipozym IM). This method induced a significant increase in the amount of PUFAs from 39.20 mol% of total fatty acids in the crude oil to 43.29 mol% in the re-esterified permeate [112].

The proteolytic extraction of oil from salmon heads using three different types of enzymes (Alcalase, Neutrase and flavourzyme) and the lipolysis of this oil to concentrate PUFAs were carried out. Lipolysis was done with Novozym SP398 to obtain a mixture of free fatty acids and glycerol (24 hours 45% hydrolysis). The mixture was then filtered. This process has allowed an increase of the PUFAs content from 41.6% in the crude oil to 46.5% in the permeate. Likewise, DHA and EPA percentages have increased from 9.9% to 11.6%, and from 3.6 to 5.6%, respectively [111]. The same authors used a re-esterification in the permeate with Lipozyme IM which permitted obtention of 5.06% and 11.90% in EPA and DHA contents, accordingly [111]. Moreover, other authors proposed combination of enzymatic or chemical hydrolysis with urea complexation to produce high concentrates of n-3 PUFAs. The enzymatic hydrolysis followed with urea complexation of refined sardine oil has increased the level of EPA and DHA from 14.51% to 46.26%, and from 12.55% to 40.32%, respectively [142].

Another technique, short path distillation was tested to purify Alaskan Walleye Pollock (*Gadus chalcogrammus*) and New Zealand Hoki (*Macruronus novaezelandiae*) liver oils [154]. Certainly, this process has reduced free fatty acids and lipid oxidation parameters, which is appreciated to produce purified oils. Consequently, the conduct of this operation at high temperatures may cause degradation of PUFAs or development of new undesirable compounds. The short path distillation was coupled to a previous enzymatic glycerolysis of sardine oil with glycerol [155]. This work showed that short path distillation is able to concentrate n-3 PUFAs in monoacylglycerols at suitable evaporator temperature (125°C) Same technique aided by a working fluid was evaluated efficient in removal of persistent organic pollutants in marine oils (PCDD/PCDF, dl-PCB and ndl-PCB) [156].

When comparing the effect of using urea complexation on the concentration yield compared with dry fractionation and low temperature solvent crystallization, results revealed that n-3 fatty acids were enriched in liquid fractions of all methods except by dry fractionation. The highest enrichment was achieved with the urea complexation method (83.00%) [157]. In the same context of valorization of marine by-products, application of urea crystallization on tuna oil recovered from liquid waste by-product from a tuna canning process allowed an increase in the concentration of n-3 PUFAs [158]. In another study conducted on concentration of fatty acids in sardine oil, the highest PUFA concentrations in low-temperature crystallization with ethanol were attained at −5°C, with EPA and DHA purities equivalent to 17.74 and 25.51%, respectively [108].

These authors also compared three different concentration techniques, supercritical fluid extraction (T = 40, 50, 60°C and 150, 250, 350 bar), Urea complexation (T = 1, −5, −10°C) and low-temperature crystallization with ethanol solvent (T = 10, 0, −5°C). The optimal conditions for each technique were determined. Nevertheless, the highest reduction of SFA and MUFA, the best increase in PUFA and the highest n-3 yield (47.53%), were obtained at −10°C in urea complexation method [108].

There are still several techniques used for the concentration of n-3 PUFAs, among which there is the use of polymeric membrane separation [159]. Optimal conditions of this method were found to be at the temperature of 36.19°C, pressure of 4.82 bar and stirring rate of 43.01 rpm with a desirability value of 0.99. With these conditions, a concentration of n-3 PUFAs of 34.98% was achieved.

Synthesized poly-vinylidene fluoride (PVDF) asymmetric membranes are also tested in concentration of n-3 PUFAs [160]. Conditions of preparation of PVDF membranes influences significantly results. In this work, PVDF membrane prepared at a coagulation bath temperature of 0°C resulted in the best n-3 PUFAs enrichment (40.4%) at 5 bar and 30°C.

4. Conclusion

Marine by-products (viscera, heads, trimmings, bones, cartilage, tails, skin, scales, blood, shells, carcasses, damaged fish, eggs, milt or soft roe), generated by marine transformation industries, constitute a good opportunity of valorization into highly valuable products. Their characterization determines the choice of the most suitable and efficient valorization method among all possibilities available, production of marine proteins (fishmeal, silage and hydrolysates), oils rich in polyunsaturated fatty acids (PUFAs) and preparation of high value compounds such as vitamins, enzymes, minerals, gelatin, collagen, chitin and chitosan, taurine and creatine, hydroxyapatite, natural pigments, biodiesel and biogas.

In this context, several studies have been carried out to explore possible technologies that can be used in the valorization of the marine by-products into marine oils and concentrated fatty acids. In addition to the conventional extraction process called also wet reduction process or hydraulic pressing, solvent extraction, supercritical fluid extraction, urea complexation, cold pressing or enzymatic hydrolysis processes could be used to transform these by-products into marine oils highly rich in PUFAs very demanded by food, nutraceutical and pharmaceutical industries.

For more advanced enhancement, the concentration of fatty acids in marine oils is also widely practiced. Several techniques can be used such as winterization, urea complexation, short path distillation, supercritical fluid extraction, low temperature solvent crystallization, fractionation by chromatography or by enzymatic processes. Combined methods were also tested like solvent winterization and enzymatic interesterification, urea adduction before a supercritical fractionation. Many studies have focused on comparison between these techniques to provide differences, advantages, disadvantages, or even optimal conditions of operating.

The main challenge in the choice of extraction and concentration techniques at industrial level is to reach higher yield, purity, quality, stability at lower cost and low unwanted environmental effects.

Conflict of interest

The authors declare that there is no conflict of interest.

Author details

Amar Kaanane* and Hind Mkadem
Institute of Agronomy and Veterinary Medicine Hassan II, Rabat, Morocco

*Address all correspondence to: a.kaanane@iav.ac.ma

References

[1] FAO. The State of Food and Agriculture 2019. Moving Forward on Food Loss and Waste Reduction. Rome; 2019

[2] FAO. 2020. State of the World Fisheries and Aquaculture. Sustainability in Action. Rome: Food and Agriculture Organization of the United Nations; 2020

[3] Barthomeuf DJ, Bergé C, J.P. How enzymes may be helpful for upgrading fish by-products: Enhancement of fat extraction. J Aquatic Food Product Technol. 2004;**13**(2):69-84

[4] Shahidi, F. 2007. Maximising the Value of Marine by-Products. Boca Raton, USA: Woodhead Publishing Ltd. ISBN: 9781845690137.

[5] Rustad T. Utilization of marine by-products. Electron. J. Environ. Agric. Food Chem. 2003;**2**(4):458-463 ISSN 1579-4377

[6] Jackson A, Newton RW. Project to Model the Use of Fisheries by-Products in the Production of Marine Ingredients with Special Reference to Omega-3 Fatty Acids EPA and DHA. Institute of Aquaculture: University of Stirling et IFFO; 2016

[7] Wasswa J, Tang J, Gu X. Utilization of fish processing by-products in the Gelatin industry. Food Reviews International. 2007;**23**:2,159-2,174. DOI: 10.1080/87559120701225029

[8] Rustad T, Storrø I, Slizyte R. Possibilities for the utilization of marine by-products. International Journal of Food Science and Technology. 2011;**2011**(46):2001-2014

[9] Dumay J, Donnay-Moreno C, Barnathan G, Jaouen P, Bergé JP. Improvement of lipid and phospholipid recoveries from sardine (Sardina pilchardus) viscera using industrial proteases. Process Biochemistry. 2006;**41**(11):2327-2332

[10] Ferraro V, Cruz IB, Jorge RF, Malcata FX, Pintado ME, Castro PML. Valorisation of natural extracts from marine source focused on marine by-products: A review. Food Research International. 2010;**43**:2221-2233

[11] Ferraro V, Carvalho A, Piccirillo C, Santos M, Castro P, Pintado M. Extraction of high added value biological compounds from sardine, sardine-type fish and mackerel canning residues - a review. Materials Science and Engineering, C. 2013;**33**:3111-3120

[12] Bonilla-Méndez JR, Hoyos-Concha JL. Methods of extraction, refining and concentration of fish oil as a source of omega-3 fatty acids. Corpoica Ciencia y Tecnología Agropecuaria. 2018;**19**(3):645-668

[13] Grosdemange, A.G. 2010. De l'huile de foie de morue aux huiles de poisson concentrées, riches en acides gras polyinsaturés Vue d'ensemble des techniques de production d'huile de poisson. Ocl vol. 17 n° 4 juillet-août 2010.

[14] Kim SK, Mendis E. Bioactive compounds from marine processing byproducts – A review. Food Research International. 2006;**39**(4):383-393

[15] Kristbergsson, K. Arason, S. 2007. Utilization of by-products and treatment of waste in the food industry. (Vol. 3, pp. 233-258). ISBN: 978-0-387-33511-7.

[16] Shahidi, F. Janak Kamil, Y.V.A. Enzymes from fish and aquatic invertebrates and their application in the food industry. Trends in Food Science and Technology. 2001;**12**:435-464

[17] Blanco M, Sotelo CG, Pérez-Martín RI. Hydrolysis as a valorization strategy for unused marine food biomass: Boarfish and small-spotted catshark discards and by-products. Journal of Food Biochemistry. 2015;**39**(2015):368-376

[18] Kato S, Kunisawa N, Kojima T, Murakami S. Evaluation of ozone treated fish waste oil as a fuel for transportation. Journal of Chemical Engineering of Japan. 2004;**37**:863-870

[19] Lopes da Silva T, Santos AR, Gomes R, Reis A. Valorizing fish canning industry by-products to produce ω-3 compounds and biodiesel. Environmental Technology and Innovation. 2017. DOI: 10.1016/j.eti.2017.11.002

[20] Makoure D, Arhaliass A, Echchelh A, et al. Valorization of fish by-products using reactive extrusion for biodiesel production and optimization. Waste Biomass Valor. 2019. DOI: 10.1007/s12649-019-00840-5

[21] Piccolo, T. 2008. Frame work analysis of fish waste to bio-diesel production – Aquafinca – case study, (Retrieved on September 12th, 2009). www.aquaticbiofuel.com.

[22] Leung DYC, Wu X, Leung MKH. A review on biodiesel production using catalyzed transesterification. Applied Energy. 2010;**87**(4):1083-1095. DOI: 10.1016/j.apenergy.2009.10.006

[23] Lin CY, Li RJ. Fuel properties of biodiesel produced from the crude fish oil from the soap stock of marine fish. Fuel Processing Technology. 2009;**90**(1):130-136

[24] El-Mashad HM, Zhang R, Avena-Bustillos RJ. A two-step process for biodiesel production from salmon oil. Biosysytems Engineering. 2008;**99**(2):220-227

[25] Sachindra NM, Bhaskar N, Mahendrakar NS. Recovery of carotenoids from shrimp waste in organic solvents. Waste Management. 2006;**26**:1092-1098

[26] Terzioğlu P, Öğüt H, Kalemtaş A. Natural calcium phosphates from fish bones and their potential biomedical applications. Materials Science and Engineering C. 2018;**91**(2018):899-911

[27] Ishikawa M, Kato M, Mihori T, Watanabe H, Sakai Y. Effect of vapor pressure on the rate of softening of fish bone by super-heated steam cooking. Nippon Suisan Gakkaishi. 1990;**56**:1687-1691

[28] Granito, R.N. Renno, A. C. M. Yamamura, H., Cruz de Almeida, M., Ruiz, P.L.M., Ribeiro, D.A. 2018. Hydroxyapatite from fish for bone tissue engineering: A promising approach. Int J Mol Cell Med 2018, 7(2): 80-90. DOI:10.22088/IJMCM.BUMS.7.2.80.

[29] Kurita K. Chitin and chitosan: Functional biopolymers from marine crustaceans. Marine Biotechnology. 2006;**8**:203-226

[30] Jang MK, Kong BG, Jeong YI, Lee CH, Nah JW. Physicochemical characterization of alpha-chitin, beta-chitin and Gama-chitin separated from natural resources. Journal of Polymer Science Part A: Polymer Chemistry. 2004;**42**:3423-3432. DOI: 10.1002/pola.20176

[31] Newsholme E, Hardy G. Creatine: A review on its nutritional applications in sport. Nutrition. 1998;**14**(3):322-324

[32] Omura Y, Inagaki M. Immunicytochemical localization of taurine in the fish retina under light and dark adaptation. Amino Acids. 2000;**19**:593-604

[33] Dumay J. Extraction de lipides en voie aqueuse par bioreacteur

enzymatique combiné à l'ultrafiltration: Application à la valorisation de co-produits de poisson (*Sardina pilchardus*). Thèse de doctorat de l'Université de Nantes; 2006 305 pp.

[34] Ghaly AE, Ramakrishnan VV, Brooks MS, Budge SM, Dave D. Fish processing wastes as a potential source of proteins, amino acids and oils: A critical review. J Microb Biochem Technol. 2013;**5**:107-129. DOI: 10.4172/1948-5948.1000110

[35] Esteban AJ, Garcıa P, Ramos MCM. Evaluation of fruit–vegetable and fish wastes as alternative feedstuffs in pig diets M.B. Waste Management. 2007;**27**(2007):193-200

[36] Skåra T, Cripps S. Upgrading of waste from salmon slaughteries using a three-phase separation process. In: Luten JB, Børresen T, Oehlenschläger J, editors. Seafood from Producer to Consumer, Integrated Approach to Quality. Amsterdam, The Netherlands: Elsevier Science; 1997. pp. 103-111

[37] Khoddami A, Ariffin, A.A. Bakar, J. Ghazali, H.M. 2009. Fatty acid profile of the oil extracted from fish waste (head, intestine and liver) (Sardinellale muru). World Allied Sciences Journal 7: 127-131.

[38] Bechtel PJ. Properties of different fish processing by-products from Pollock, cod and salmon. Journal of Food Processing Preservation. 2003;**27**(2003):101-116

[39] Babbit JK. Intrinsic quality and species of north specific fish. In: Keller S, editor. Making Profits out of Seafood by-Products. Alaska: Proceeding of the International conference on fish by-products; 1990. pp. 39-43

[40] Je JY, Qian ZJ, Byun HG, Kim SK. Purification and characterization of an antioxidant peptide obtained from tuna backbone protein by enzymatic hydrolysis. Process Biochemistry. 2007;**42**(5):840-846

[41] Pastoriza L, Sampedro G, Cabo ML, Herrera JJR, Bernardez M. Solubilisation of proteins from rayfish residues by endogenous and commercial enzymes. Journal of the Science of Food and Agriculture. 2004;**84**:83-88

[42] Torres J, Chen A, Rodrigo-Garcia YC, Jaczynski J, J. Recovery of byproducts from seafood processing streams. In: Shahidi F, editor. Maximising the Value of Marine by-Products. Boca Raton, USA: CRC Press; 2007. pp. 65-90

[43] Olsen RL, Toppe J, Karunasagar I. Challenges and realistic opportunities in the use of byproducts from processing of fish and shellfish. Trends in Food Science and Technology. 2014;**36**:144-151

[44] Anon. Assessment of Harvest and Post-Harvest Losses - Marine Fisheries. Cochin: National Agricultural Technology Project, CIFT; 2005 122 p.

[45] Lopes C, Antelo LT, Amaya Franco-Uría A, Alonso AA, Pérez-Martín R. Valorisation of fish by-products against waste management treatments – Comparison of environmental impacts. Waste Management Volume. 2015;**46**:103-112

[46] Gabrielsen BO, Austreng E. Growth, product quality and immune status of Atlantic salmon, Salmo salar L. fed wet feed with alginate. Aquaculture Research. 1998;**29**:397-401

[47] Aidos I, Luten JB, Boom RM, Padt AV. Upgrading of Maatjes herring byproducts: Production of crude fish oil. Journal of Agricultural Food Chemistry. 2001;**49**:3697-3704

[48] Archer, M. Watson, R. Denton, J.W. 2001. Fish waste production in the United Kingdom-the quantities

produced and opportunities for better utilisation. The sea fish industry authority sea fish technology, 1-57.

[49] Jacobsen CN, Horn NS, Sørensen AM, Bimbo AB. Food Enrichment with Omega-3 Fatty Acids. Cambridge: Woodhead Publishing Limited; 2013. p. 107

[50] Rubio-Rodriguez N, De Diego SM, Beltrán S, Jaime I, Sanz MT, Rovira J. Supercritical fluid extraction of fish oil from fish by-products: A comparison with other extraction methods. Journal of Food Engineering. 2012;**109**(2):238-248. DOI: 10.1016/j.jfoodeng.2011.10.011

[51] Hamilton MC, Hites RA, Schwager SJ, Foran JA, Knuth BA. Lipid composition and contaminants in farmed and wild salmon. Environmental Science & Technology. 2005;**39**:8622-8629

[52] Mkadem H, Kaanane A. Recovery and characterization of fish oil from by-products of sardine (Sardina pilchardus) in the canning process. Journal of Aquatic Food Product Technology. 2019. DOI: 10.1080/10498850.2019.1682733

[53] Falch E, Sandbakk M, Aursand M. Chapter 3: On-board handling of marine byproducts to prevent microbial spoilage, enzymatic reactions and lipid oxidation. In: Shahidi F, editor. Maximising the Value of Marine by-Products. Cambridge: Woodhead Publishing in Food Science, Technology and Nutrition; 2007. pp. 47-64 532 pp.

[54] Shahidi F. Bailey's Industrial Oil and Fat Products, 6 Volume Set. 6th ed. Hoboken, NJ: Wiley-Interscience; 2015

[55] Aidos I, Masbernat-Martinez S, Luten JB, Boom RM, Van Der Padt AJB. Composition and stability of herring oil recovered from sorted byproducts as compared to oil from mixed byproducts. Journal of Agricultural and Food Chemistry. 2002;**50**:2818-2824

[56] Zlatanos S, Sagredos AN. The fatty acid composition of some important Mediterranean fish species. Fat Sci Tech nol. 1993;**95**(29):6669

[57] Horrocks LA, Yeo YK. Health benefits of docosahexaenoic acid (DHA). Pharmacological Research. 1999;**40**:211-225

[58] Leaf A, Kang JX, Xiao YF, Billman GE, Voskuyl RA. The Antiarrythmic and anticonvulsant effects of dietary N-3 fatty acids. The Journal of Membrane Biology. 1999;**172**:1-11

[59] Holub DJ, Holub BJ. Omega-3 fatty acids from fish oils and cardiovascular disease. Molecular and Cellular Biochemistry, Kluwer Academic Publishers, Netherlands. 2004;**263**:217-225

[60] Sidhu KS. Health benefits and potential risks related to consumption of fish or fish oil. Regulatory Toxicology and Pharmacology. 2003;**38**:336-344

[61] Kris-Etherton PM, Harris WS, Appel LJ. American Heart Association nutrition committee. Fish consumption, fish oil, omega-3 fatty acids, and cardiovascular disease. Circulation. 2002;**106**:2747-2757

[62] Schacky CV. n-3 fatty acids and the prevention of coronary atherosclerosis. The American Journal of Clinical Nutrition. 2000;**71**:224-227

[63] Schmidt EB, Arnesen H, De Caterina R, Rasmussen LH, Kristensen SD. Marine n-3 polyunsaturated fatty acids and coronary heart disease: Part I. background, epidemiology, animal data, effects on risk factors and

safety. Thrombosis Research. 2005;**115**(3):163170

[64] Barnerjee I, Saha S, Dutta J. Comparison of the effect of dietary fish oils with different n-3 polyunsaturated fatty acid compositions on plasma and liver lipids in rats. Lipids. 1992;**27**:425-428

[65] Goodnight SHH, Jr WS, Connor WE, Illingworth DR. Polyunsaturated fatty acids, Hyperlipidemia, and thrombosis. Arteriosclerosis. 1982;**2**:87-112

[66] Mahadik SP, Evans D, Lal H. Oxidative stress and role of antioxidant and ω-3 essential fatty acid supplementation on schizophrenia. Prog. Neuro-psychopharmacol and Biol. Pshychiatry. 2001;**25**:463-493

[67] Bourre JM. Dietary omega-3 fatty acids and psychiatry: Mood, behavior, stress, depression, dementia and aging. The Journal of Nutrition, Health & Aging. 2005;**9**:31-38

[68] Weisinger HS, Armitage JA, Sinclair AJ, Vingrys AJ, Burns PL, Weisinger RS. Perinatal omega-3 fatty acid deficiency affetcs blood pressure laterin life. Nature Medecine. 2001;**7**:258-259

[69] Bourre J, Bonneil M, Clement M, Dumont O, Durand G, Lafront H, et al. Function of dietary polyunsaturated fatty acids in the nervous system. Prostaglandins, Leukotrienes and Essential Fatty Acids. 1993;**48**:5-15

[70] Makrides M, Neumann M, Simmer K, Pater J, Gibson R. Are long-chain polyunsaturated fatty-acids essential nutrients in infancy. Lancet. 1995;**345**(8963):1463-1468

[71] Hoffman DR, Uauy R. Essentiality of dietary omega-3 fatty acid for premature infants: Plasma and red blood cell fatty acid composition. Lipids. 1992;**27**:886-896

[72] Moyad MA. An introduction to dietary/supplemental omega-3 fatty acids for general health and prevention: Part II. Urologic Oncology: Seminars and Original Investigations. 2005;**23**:36-48

[73] Brouwer IA, Geelen A, Katan MB. n-3 fatty acids, cardiac arrhythmia and fatal coronary heart disease. Progress in Lipid Research. 2006;**45**:357-367

[74] Reiffel JA, McDonald A. Antiarrhythmic effects of Omega-3 fatty acids. The American Journal of Cardiology. 2006;**98**:50-60

[75] Psota TL, Gebauer SK, Kris-Etherton P. Dietary omega-3 fatty acid intake and cardiovascular risk. American Journal of Cardiology. 2006;**98**:3-18

[76] Jacobson TA. Secondary prevention of coronary artery disease with Omega-3 fatty acids. The American Journal of Cardiology. 2006;**98**:61-70

[77] Dagnelie PC, Rietveld T, Swart GR, Stijnen T, Van Den Berg JWO. Effect of dietary fish oil on blood levels of free fatty acids, ketone bodies and Triacylglycerolin humans. Lipids. 1994;**29**:41-45

[78] Gebauer SK, Psota TL, Harris WS, Kris-Etherton PM. n-3 fatty acid dietary recommendations and food sources to achieve essentiality and cardiovascular benefits. American Journal of Clinical Nutrition. 2006;**83**:1526-1535

[79] Middaugh JP. Cardiovascular deaths among Alaskan natives, 1980-1986. American Journal of Public Health. 1990;**80**:282-285

[80] Karmali RA, Marsh J, Fuchs C. Effect of omega-3 fatty acids on growth of a rat mammary tumor 2. Journal of the National Cancer Institute. 1984;**73**(2):457-461. DOI: 10.1093/jnci/73.2.457

[81] Rogers AE, Conner B, Boulanger C, Lee S. Mammary Tumorogenesis in rats fed diets high in lard. Lipids. 1986;**21**:275-280

[82] Wakimoto T, Kondo H, Nii H, Kimura K, Egami Y, Oka Y, et al. Furan fatty acid as an anti-inflammatory component from the green lipped mussel Perna canaliculus. Proceedings. National Academy of Sciences. United States of America. 2011;**108**:17533-17537

[83] Tang S, Qin C, Wang H, Li S, Tian S. Study on supercritical extraction of lipids and enrichment of DHA from oil-rich microalgae. The Journal of Supercritical Fluids. 2011;**57**(1):44-49

[84] Covington MB. Omega-3 fatty acid. American Family Physician. 2004;**70**(1):133-140

[85] Cho SY, Miyashita K, Miyazawa T, Fujimoto J, Kaneda T. Autoxidation of ethyl EPA and DHA. Journal of American Oil Chemists' Society. 1987;**64**:876-879

[86] Cunnane SC, Plourde M, Stewart K, Crawford MA. Docosahexaenoic acid and shore-based diets in hominin encephalization: A rebuttal. American Journal of Human Biology. 2007;**19**:578-581

[87] Njinkoue JM, Barnathan G, Miralles J, Gaydou EM, Samb A. Lipids and fatty acids in muscle, liver and skin of three edible fish from the Senegalese coast: Sardinellamaderensis, Sardinellaauritaand Cephalopholistaeniops. Comp. Biochem. Phys. Part B. 2002;**13**:395-402

[88] Bligh EG, Dyer WJ. A rapid method of total lipid extraction and purification. Canadian Journal of Biochemistry and Physiology. 1959;**37**:911-917

[89] Falch E, Rustad T, Aursand M. By-products from gadiform species as raw material for production of marine lipids as ingredients in food or feed. Process Biochemistry. 2006;**41**:666-674

[90] Raju A, Haq M, Cho YJ, Chun BS. Quality evaluation of oil recovered from by-products of bigeye tuna using supercritical carbon dioxide extraction. Turkish Journal of Fisheries and Aquatic Sciences. 2017;**17**:663-672

[91] Soldo B, Simat V, Vlahovi J, Skroza D, Ljubenkov I, Mekinic IG. High quality oil extracted from sardine by-products as an alternative to whole sardines: Production and refining. European Journal of Lipid Science and Technology. 2019;**121**:1800513

[92] Chantachum S, Benjakul S, Sriwirat N. Separation and quality of fish oil from precooked and non-precooked tuna heads. Food Chemistry. 2000;**69**:289-294

[93] Dumay J, Allery M, Donnay-Moreno C, Barnathan G, Jaouen P, Carbonneau ME, et al. Optimization of hydrolysis of sardine (Sardina pilchardus) heads with Protamex: Enhancemant of lipid and phospholipide extraction. Journal of the Science of Food and Agriculture. 2009;**89**:1599-1606

[94] Liaset B, Julshamn K, Espe M. Chemical composition and theoretical nutritional evaluation of the produced fractions from enzymatic hydrolysis of salmon frames with Protamex. Process Biochemistry. 2003;**38**:1747-1759

[95] Batista I, Ramos C, Coutinho J, Bandarra NM, Nunes ML. Characterization of protein hydrolysates and lipids obtained from black scabbardfish (Aphanopus carbo) by-products and antioxidative activity of the hydrolysates produced. Process Biochemistry. 2010;**45**(1):18-24. DOI: 10.1016/j.procbio.2009.07.019

[96] Batista I, Ramos C, Mendonça R, Nunes ML. Enzymatic hydrolysis

of sardine (Sardina pilchardus) by-products and lipid recovery. Journal of Aquatic Food Product Technology. 2009;**18**(1-2):120-134

[97] Haq M, Ahmed R, Cho YJ, Chun BS. Quality properties and bio-potentiality of edible oils from Atlantic salmon by-products extracted by supercritical carbon dioxide and conventional methods. Waste and Biomass Valorization. 2016;**8**(6):1953-1967. DOI: 10.1007/s12649-016-9710-2

[98] Sahena F, Zaidul IS, Jinap M, Jahurul S, Khatib MHA, Norulaini A, et al. Extraction of fish oil from the skin of Indian mackerel using supercritical fluids. Journal of Food Engineering. 2010;**99**(1):63-69. DOI: 10.1016/j.jfoodeng.2010.01.038

[99] Fouda T. Using green cold pressing to produce high quality fish oil from industrial Salmon waste. V Journal of Zoological Research. 2020;**2**(1)

[100] Gbogouri GA, Linder M, Fanni J, Parmentier M. Analysis of lipids extracted from salmon (Salmo salar) heads by commercial proteolytic enzymes. Euro J. Lipid Sci. Tech. 2006;**108**:766-775

[101] Al-Sayed M, Linder K, Fanni M, Parmentier J, M. Characterization of the lipid fractions obtained by proteolytic and chemical extraction from rainbow trout (Oncorhynchus mykiss) roe. Process Biochemistry. 2008;**43**:376-383

[102] Chen Y, Tou C, Jaczynski JC, J. Amino acid, fatty acid, and mineral profiles of materials recovered from rainbow trout (Oncorhynchus mykiss) processing by-products using isoelectric solubilization/precipitation. Journal of Food Science. 2007;**72**(9):C527-C535

[103] Mbatia B, Adlercreutz P, Mulaa F, Mattiasson B. Enzymatic enrichment of omega-3 polyunsaturated fatty acids in Nile perch (Lates niloticus) viscera oil. European Journal of Lipid Science and Technology. 2010b;**112**:977-984

[104] Mbatia B, Adlercreutz D, Adlercreutz P, Mahadhy A, Mulaa F, Mattiasson B. Enzymatic oil extraction and positional analysis of ω-3 fatty acids in Nile perch and salmon heads (2010a). Process Biochemistry. 2010a;**45**(5):815-819

[105] Létisse M, Rozières M, Hiol A, Sargent M, Comeau L. Enrichment of EPA and DHA from sardine by supercritical fluid extraction without organic modifier: I. optimization of extraction conditions. The Journal of Supercritical Fluids. 2006;**38**(1):27-36

[106] Adeniyi OD, Bawa AA. Mackerel (Scomber Scrombrus) oil extraction and evaluation as raw materials for industrial utilization. Leonardo Journal of Sciences. 2006;**8**:33-42

[107] Dunford N, Temelli T, LeBlanc F, F. Supercritical CO_2 extraction of oil and residual proteins from Atlantic mackerel (Scomber scombrus) as affected by moisture content. Journal of Food Science. 1997;**62**:289-294

[108] Homayooni B, Sahari MA, Barzegar M. Concentrations of omega-3 fatty acids from rainbow sardine fish oil by various methods. International Food Research Journal. 2014;**21**(2):743-748

[109] Aspmo SI, Horn SJ, Eijsink VGH. Enzymatic hydrolysis of Atlantic cod (Gadus morhua L.) viscera. Process Biochemistry. 2005;**40**:1957-1966

[110] Šližyte R, Rustad T, Storro I. Enzymatic hydrolysis of cod (Gadus morhua) by-products optimization of yield and properties of lipid and protein fractions. Process Biochemistry. 2005;**40**(2005):3680-3692

[111] Linder M, Fanni J, Parmentier M. Proteolytic extraction of salmon oil and

PUFA concentration by lipases. Marine Biotechnology. 2005;7:70-76

[112] Linder M, Matouba E, Fanni J, Parmentier M. n-3 PUFA enrichment of salmon oil as obtained by lipolysis, filtration and enzymatic re-esterifiaction. European Journal of Lipid Science and Technology. 2002;**8**(104):455-462

[113] Crowther, J.B. Booth, B.H. Blackwell, D.D. 2000. Procédé de production d'une huile de poisson de qualité alimentaire. Brevet WOUS0042535.

[114] Chandrasekaran, M. Shine, K. 2013. Oil Seeds. Valorization of Food Processing by-Products. 978-1-4398-4887-6.

[115] Gbogouri GA. Co-valorisation des protéines et des lipides riches en lécithine et en acides gras polyinsatures oméga 3 à partir de têtes de saumon (salmo salar) par hydrolyse enzymatique. Institut national polytechnique de lorraine. Ecole Nationale Supérieure d'Agronomie et des Industries Alimentaires Laboratoire de Science et Génie Alimentaires. Thesis. 2005;**2005**

[116] Ackman RG. Seafood lipids. In: Shahidi F, Botta JR, editors. Seafoods Chemistry, Processing Technology and Quality. New York: Blackie Academic and professional; 1994. pp. 34-48

[117] Aspevik T, Oterhals A, Rønning S, Altintzoglou B, Wubshet T, Gildberg SG, et al. Top Curr Chem (Z). 2017;**375**:53. DOI: 10.1007/s41061-017-0143-6

[118] Aidos, I.K.N. Boonman, M. Luten, J.B. Boom, R.M. Van Der, P.A. 2003. Influence of production parameters on fish oil quality in a pilot plant. J. of Food Sci., 68 (2): 581-587.

[119] Tengku, M. Tengku, R, Jeng, W.S. Amiza, M.A. 2018. Nutritional composition and thermal properties of Goldstripe Sardinella (Sardinella gibbosa) fillets and by-products, Journal of Aquatic Food Product Technology, 27:6, 667-679, DOI:10.1080/10498850.2018.1483991.

[120] Mbatia B. Valorisation of Fish Waste Biomass through Recovery of Nutritional Lipids and Biogas. (PhD Thesis). Sweden: Department of Biotechnology, Lund University; 2011

[121] Liu S, Zhang C, Hong P, Ji H. Concentration of docosahexaenoic acid (DHA) and eicosapentaenoic acid (EPA) of tuna oil by urea complexation: Optimization of process parameters. Journal of Food Engineering. 2006;**73**(3):203-209

[122] Rolle RS. Review: Enzyme applications for agro-processing in developing countries: An inventory of current and potential applications. World Journal of Microbiology and Biotechnology. 1998;**14**:611-619

[123] Lee JH, Asaduzzaman AKM, Yun JH, Yun JH, Chun BS. Characterization of the yellow croaker Larimichthys polyactis muscle oil extracted with supercritical carbon dioxide and an organic solvent. Fish Aquat. Sci. 2012;**15**:275-281. DOI: 10.5657/FAS.2012.0275

[124] Rubio-Rodriguez N, Beltrán S, Jaime, I. De Diego, S. M. Sanz, M. T. Carballido, J. R. Production of omega-3 polyunsaturated fatty acid concentrates: A review. Innovative Food Science and Emerging Technologies. 2010;**11**(1):1-12. DOI: 10.1016/j.ifset.2009.10.006

[125] Kawashima A, Iwakiri R, Honda K. Experimental study on the removal of dioxins and coplanar polychlorinated biphenyls (PCBs) from fish oil. Journal of Agriculture and Food Chemistry. 2006;**54**:10294-10299

[126] Smedes F, Askland TK. Revisiting the development of the Bligh and Dyer total lipid determination method. Marine Pollution Bulletin. 1999;**38**:193-201

[127] Mkadem H, Kaanane A. Seasonal changes in chemical composition and fatty acids of sardines (Sardina pilchardus) from the Dakhla coast (Morocco). Fisheries and Aquaculture. 2020;**1**(3)

[128] Chakraborty K, Joseph D. Cooking and pressing is an effective and eco-friendly technique for obtaining high quality oil from Sardinella longiceps. European Journal of Lipid Science and Technology. 2015;**117**(6):837-850. DOI: 10.1002/ejlt.201400539

[129] Immanuel G, Sathasivan S, Shankar V, Peter S, Palavesam MJP, A. Processing and characterisation of low cost Balistid fish Sufflamen capistratus liver oil for edible purpose. Food Chemistry. 2009;**115**(2):430-435. DOI: 10.1016/j.foodchem.2008.12.023

[130] Ozogul Y, Simşek A, Balikçi E, Kenar M. The effects of extraction methods on the contents of fatty acids, especially EPA and DHA in marine lipids. International Journal of Food Sciences and Nutrition. 2011

[131] Uitterhaegen E, Evon P. Twin–cold extrusion technology for vegetable oil extraction: A review. Journal of Food Engineering. 2017;**2017**(212):190-200

[132] Kerry J. Chapter 2: Physical and chemical properties of lipid by-products from seafood waste. In: Shahidi F, editor. Maximising the Value of Marine by-Products. Cambridge: Woodhead Publishing in Food Science, Technology and Nutrition; 2007. pp. 22-46 532 pp.

[133] Hosseini H, Ghorbani M, Jafari SM, Sadeghi A, Mahoonak AS. Encapsulation of EPA and DHA concentrate from Kilka fish oil by milk proteins and evaluation of its oxidative stability. Journal of Food Science and Technology. 2018. DOI: 10.1007/s13197018-3455-9

[134] Mehrad, B. Shabanpour, B. Jafari, S.M. Pourashouri, P. 2015. Characterization of dried fish oil from menhaden encapsulated by spray drying. AACL Bioflux, 2015, V 8, Issue 1.

[135] Shahidi F, Wanasundara UN. Omega-3 fatty acid concentrates: Nutritional aspects and production technologies. Trends in Food Science and Technology. 1998;**9**:230-240

[136] Lin T, Chen J, Chang SW, A. C. Enrichment of n-3 PUFA contents on triglycerides of fish oil by lipasecatalyzed trans-esterification under supercritical conditions. Biochemical Engineering Journal. 2006;**29**(1-2):27-34. DOI: 10.1016/j.bej.2005.02.035

[137] Cerón I, Cardona X, Toro CA, L. A. Simulación del proceso de concentración de aceite esencial de cidrón (Lippia citriodora) por destilación molecular de película descendente. Ingeniería y Competitividad. 2012;**14**(1):107-120

[138] Lei Q, Ba S, Zhang H, Wei Y, Lee JY, Li T. Enrichment of omega-3 fatty acids in cod liver oil via alternate solvent winterization and enzymatic interesterification. Food Chemistry. 2016;**199**:364-371. DOI: 10.1016/j.foodchem.2015.12.005

[139] Cunha D, Crexi C, De Almeida-Pinto VT, L. A. Winterization of fish oil with solvent. Food Science and Technology. 2009;**29**(1):207-213

[140] Okada T, Morrissey MT. Production of n-3 polyunsaturated fatty acid concentrates from sardine oil by lipase-catalyzed hydrolysis. Food Chemistry. 2007;**103**:1411-1419

[141] Alkio M, González C, Jäntti M, Aaltonen O. Purification of polyunsaturated fatty acid esters from tuna oil with supercritical fluid chromatography. JAOCS, Journal of the American Oil Chemists' Society. 2000;**77**(3):315-321

[142] Gamez-Meza N, Noriega-Rodrıguez JA, Medina-Juarez LA, Ortega-Garcia J, Monroy-Rivera J, Toro-Vazquez FJ, et al. Concentration of eicosapentaenoic acid and docosahexaenoic acid from fish oil by hydrolysis and urea complexation. Food Research International. 2003;**36**:721-727

[143] Ferdosh S, Sarker Z, Norulaini N, Akanda J, Ghafoor K, Kadir O. Simultaneous extraction and fractionation of fish oil from tuna by-product using supercritical carbon dioxide (SC-CO_2). Journal of Aquatic Food Product Technology. 2014;**25**(2):230-239. DOI: 10.1080/10498850.2013.843629

[144] Melgosa R, Sanz M, García T, Paz Á, Beltrán E, Lamas S, et al. Supercritical carbon dioxide as solvent in the lipase-catalyzed ethanolysis of fish oil: Kinetic study. Journal of CO_2 Utilization. 2017;**17**:170-179. DOI: 10.1016/j.jcou.2016.11.011

[145] Lopes BL, Sánchez-Camargo F, Ferreira AP, Grimaldi ALK, Paviani R, Cabral LC, et al. Selectivity of supercritical carbon dioxide in the fractionation of fish oil with a lower content of EPA+DHA. The Journal of Supercritical Fluids. 2012;**61**:78-85. DOI: 10.1016/j.supflu.2011.09.015

[146] Perretti G, Motori A, Bravi E, Favati F, Montanari L, Fantozzi P. Supercritical carbon dioxide fractionation of fish oil fatty acid ethyl esters. J. of Supercritical Fluids. 2007;**40**(2007):349-353

[147] Espinosa S, Diaz S, Brignole EA. Thermodynamic modeling and process optimization of supercritical fluid fractionation of fish oil fatty acid ethyl esters. Industrial and Engineering Chemistry Research. 2002;**41**(6):1516-1527

[148] Gironi F, Maschietti M. Separation of fish oils ethyl esters by means of supercritical carbon dioxide: Thermodynamic analysis and process modelling. Chemical Engineering Science. 2006;**61**:5114-5126

[149] Létisse M, Comeau L. Enrichment of eicosapentaenoic acid and docosahexaenoic acid from sardine by-products by supercritical fluid fractionation. Journal of Separation Science. 2008;**31**(8):1374-1380. DOI: 10.1002/jssc.200700501

[150] Antunes-Corrêa A, Arantes-Peixoto P, Guaraldo-Gonçalves C, Cabral LA, F. A. Fractionation of fish oil with supercritical carbon dioxide. Journal of Food Engineering. 2008;**88**:381-387

[151] Davarnejad R, Kassim K, Zainal M, Sata A, S. A. Extraction of fish oil by fractionation through supercritical carbon dioxide. Journal of Chemical and Engineering Data. 2008;**53**(9):2128-2132

[152] Pettinello G, Bertucco A, Pallado P, Stassi A. Production of EPA enriched mixtures by supercritical fluid chromatography: From the laboratory scale to the pilot plant. Journal of Supercritical Fluids. 2000;**19**:51-60

[153] Breivik H, Haraldsson GG, Kristinsson B. Preparation of highly purifiedconcentratesofeicosapentaenoic acid and docosahexaenoic acid. JAOCS. 1997;**74**(11):1425-1429

[154] Oliveira-Carvalho P, Bueno-Campos P, D'Addio-Noffs R, Oliveira M, TsuneziShimizu JG, Silva M,

et al. Aplicação de lipases microbianas na obtenção de concentrados de ácidos graxos poliinsaturados. Quimica Nova. 2003;**26**(1):75-80

[155] Solaesa Á, Sanz G, Falkeborg MT, Beltrán M, Guo S, Z. Production and concentration of monoacylglycerols rich in omega-3 polyunsaturated fatty acids by enzymatic glycerolysis and molecular distillation. Food Chemistry. 2016;**190**:960-967. DOI: 10.1016/j.foodchem.2015.06.061

[156] Olli J, Breivik J, Thorstad H, O. Removal of persistent organic pollutants in fish oils using short-path distillation with a working fluid. Chemosphere. 2013;**92**(3):273-278. DOI: 10.1016/j.chemosphere.2013.02.037

[157] Tengku, M. Tengku, R, Birch, E. J. 2013. Enrichment of omega-3 fatty acids of refined hoki oil. Journal of the American Oil Chemists' Society, 90(8), 1111-1119. doi:10.1007/s11746-013-2260.

[158] Suriani N, Lawalata W, Komansilan HJ, A. Urea crystallization on the concentrate making of omega-3 fatty acid from oil of tuna fish (Thunnus sp.) canning byproduct. International Journal of Pharm Tech Research. 2014;**6**(7):1981-1990

[159] Ghasemian S, Sahari M, Barzegar A, Gavlighi M, H. A. Concentration of Omega-3 polyunsaturated fatty acids by polymeric membrane. International Journal of Food Science and Technology. 2015;**50**(11):2411-2418. DOI: 10.1111/ijfs.12907

[160] Ghasemian S, Sahari M, Barzegar A, Ahmadi M, H. Omega-3 polyunsaturated fatty acids concentration using synthesized poly-vinylidene fluoride (PVDF) asymmetric membranes. Journal of the American Oil Chemists' Society. 2016;**93**(9):1201-1210. DOI: 10.1007/s11746-016-2876-8

Chapter 4

Animal Waste and Agro-by-Products: Valuable Resources for Producing Fish at Low Costs in Sub-Saharan Countries

Renalda N. Munubi and Hieromin A. Lamtane

Abstract

Animal and crop production throughout the world generate high amounts of wastes or by-products annually that may possess added value compounds with high functionality. These wastes and by-products may cause negative environmental impacts and significant expenses if not well managed and or controlled. Much of these wastes and by-products is valuable and cheaper source of potentially functional compounds such as proteins, lipids, starch, micronutrients, bioactive compounds, and dietary fibbers. In aquaculture, feed is expensive, and the existing body of literature has shown that animal manure and its extracts can be successfully incorporated into fishpond to increase fish production at a low cost. In addition, crop residues such as rice bran, maize bran, and seed cakes are commonly used as pond inputs in small-scale aquaculture. Animal waste and crop residues are added in a fishpond that filter-feeding fish can use directly as feed, and these may form a major proportion of the detritus in the pond. These resources also stimulate the growth of phytoplankton that are rich in protein and are the basis of the food web that can support the growth of a range of herbivorous and omnivorous fish. Therefore, technically, wastes are used as direct feed, a source of minerals for autotrophic production and a source of organic matter for heterotrophic production. In this context, animal manure and crop residues have been used to provide great opportunities to improve food security. The purpose of this review is to project the potential of animal waste and agro-by-products as a sustainable alternative as aquaculture inputs to reduce poverty, malnutrition, and hunger in developing countries.

Keywords: animal waste, fish farming, crop residues, farming systems, valorization

1. Introduction

Aquaculture is one of the world's fastest growing food production sectors with great potential for food supply, poverty alleviation, and enhanced trade and economic benefits, as targeted by sustainable development goals SDGs. The contribution of aquaculture to global fish supply increased from 3.9 percent in 1970 to over 41.3 percent in 2011 amounting to 63.7 million metric tonnes valued over USD

119 billion [1]. Its average growth rate of 8.8 percent has outpaced capture fisheries (1.2%) and terrestrial farmed meat production (2.8%) [1]. Aquaculture accounts for around 50 percent of seafood supply globally [2]. This quantity is expected to increase substantially as population increases (**Figure 1**). Aquaculture has gained much importance globally due to a decline in wild stock from natural water bodies; thus, aquaculture plays a key role in augmenting dwindling catch capture fisheries. It is well known that among other challenges facing the aquaculture sector, availability and quality of feeds affect its growth particularly in sub-Saharan (see for example [4–10]). Despite this challenge, aquaculture has been considered as one of the economic activities that contribute to poverty reduction, food security, and nutrition in the sub-Saharan Africa [4, 11, 12] and Asian countries [1, 13–17].

In order to realize the contribution of aquaculture in the alleviation of poverty and improvement of food security, development agencies should broaden their focus beyond poor/subsistence producers to include small and medium enterprises adopting a value chain perspective [18]. Bangladesh, which is among developing countries, has proven that aquaculture intervention in resource poor and marginalized group marked an increase in income savings and frequency of fish consumption [19]. Although small-scale fish farmers play a big role in poverty reduction and food security, the intensification from extensive to semi-intensive is essential [20]. However, for the intensification to take place, there should be an increase in investment in technological innovation and transfer through (i) Nutrition, feeds and feeding management, and (ii) low-impact production systems. This paper discusses

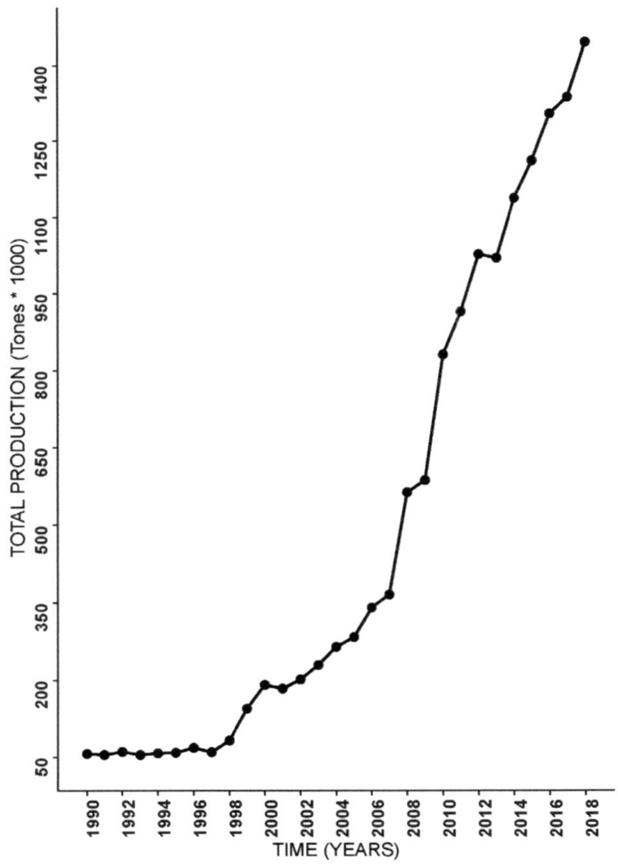

Figure 1.
Freshwater Aquaculture trend for African countries from 1990 to 2018 (data analyzed by this study see [3]).

the valorization of animal waste/by products and plants/crops-by-products to produce fish at low cost in order to increase nutrition and reduce food insecurity.

2. Aquaculture production in Africa

Africa's fisheries output is dominated by capture fisheries, but the contribution of aquaculture to the total amount of fish produced in the region has grown at a steady pace over the past decade (**Figure 1**). In these countries, fish is produced from capture fisheries and aquaculture. However, fish catches from wild sources have been declining, due to multiple anthropogenic pressures including climate change, overfishing, habitat destruction, invasion of non-native species, illegal, and unregulated fishing, and poor governance [21]. For example, consumption in the Eastern Africa region was projected to increase from 4.80 kg in 2013 to 5.49 kg by 2022 [22]. This implies that in order to meet the gap between fish production and the increasing demand for food fish, aquaculture production must double by 2050 to satisfy the Africa's fast-growing human population [23]. An appropriate way of keeping this sector growing constantly is the development of new researches aimed at determining the benefits of using different and cheap resources of feeds and determining how these strategies influence economic and productive parameters.

Country	Total production	Percent
Egypt	930,344	70.476
Nigeria	160,114	12.129
Ghana	70,628	5.350
Uganda	70,095	5.310
Zambia	17,500	1.326
Kenya	12,160	0.921
Tanzania	11,000	0.833
Zimbabwe	10,500	0.795
Malawi	5036	0.381
Rwanda	4526	0.343
Mali	3524	0.267
Congo	3185	0.241
Cã'te d'Ivoire	3000	0.227
Benin	2802	0.212
Lesotho	2500	0.189
Madagascar	2372	0.180
Algeria	2045	0.155
Angola	1752	0.133
South Africa	1503	0.114
Burundi	1455	0.110
Others	4045	0.306
Total	1,320,086.62	100

Table 1.
Freshwater aquaculture production (tones) in Africa by country in 2018. (data analyzed by this study see [3]).

In the aquaculture sector, Africa produced about 1400 tonnes of fish from freshwater aquaculture in 2018, but most of this came from Egypt, which contributed more than 70 percent of the total production (**Figure 1**, Data obtained in [3]). Major aquaculture producers in 2018 with more than 10,000 tonnes include Egypt, Nigeria, Ghana, Uganda, Zambia, Kenya, Tanzania, and Zimbabwe. Production has increased three times for the past ten tears from 563,000 in 2008 to 1,440,000 in 2018 (**Figure 1**, [3]). In general, African aquaculture production is overwhelmingly dominated by finfishes (99.3%), with only a small fraction of production from marine shrimps and mollusks [23]. Among the freshwater cultured finfishes, tilapia farming is the main product, which is also the most popular fish from a consumer perspective. Aquaculture production in Africa is also increasing as presented in **Table 1**.

3. Aquaculture production systems

In Africa, aquaculture systems are made up of extensive and semi-intensive systems. Small-scale earthen ponds (extensive systems) are characterized by low inputs and low yields. However, semi-intensive systems are characterized by human intervention where by fertilization is done to improve feed availability, hence, improved fish yield. In East Africa, semi intensive mainly used to produce *Oreochromis niloticus* and *Clarias gariepinus* in either monoculture or polyculture [24]. They consist mostly of earthen ponds, liner ponds and concrete ponds.

Other systems include cage particularly in areas with large water bodies including East Africa great lakes. Cage culture involves holding organisms under captivity within an enclosed space while maintaining free exchange of water. Cages use the existing water bodies, therefore, require comparatively low capital outlay and use simple technology, they can be used not only as a method for producing cheaply and high-quality protein but also for cleaning up eutrophicated waters through the culture and harvesting of caged planktivorous species. Although fish farming in cages in the existing water body is considered inexpensive relative to pond construction and its associated infrastructures [25], the feasibility and profitability of fish cage culture is influenced by the cost of input invested and revenue collected from output.

Although not common, re-circulating aquaculture system (RAS) has been used in some countries particularly South Africa. RAS refers to a fish farming technology that reuse wastewater from tanks/rearing premises [26]. Water reuse in RAS is supported by both inline and end pipe treatment using a series of mechanical filter for solid waste removal, bio-filter for dissolved nitrogenous waste removal and sludge pond to settle suspended solid [27]. RAS technology is termed as sustainable advanced production system that provides constant and independent production conditions and reduces water consumption compared with semi-intensive pond aquaculture, RAS technology provides high fish productivity with better effluent control of environmental conservation [28, 29]. Some of the sub Saharan countries have benefited from high temperature to which RAS performs efficiently [30]. The adoption of the system is low due to high cost of initial capital investment in tanks and high cost of electricity required in running the system and feeds. This has therefore called for sustainable aquaculture by integration of fish with livestock. Such integration involves the recycling of livestock wastes and processing by-products as manure and/or direct food for fish. Today, aquaculture in developing countries is mostly a small-scale activity and is usually not practiced as a stand-alone economic activity, but rather as subsistence farming integrated with agricultural activities such as horticulture and rearing of livestock.

4. Organic manure and fish growth

The production volume and market share of aquaculture products are advancing extremely rapidly. However, feed is usually recognized as the single largest cost to producers, hence, the best way of reducing the cost of fish production is using organic manure and supplementary feed when available. Animal manure is widely used in developing countries in fish production in earthen ponds. The quality of manure as a fertilizer varies depending on the source of animal and the quality of feed fed to the animal [31, 32]. Research showed that pig, chicken and duck manures increase fish production more than cow and sheep manure. For example, in Asia, fish farming is probably the only branch of animal husbandry in which the use of manures is a traditional management tool. In Sub-Saharan Africa, ponds are fertilized using organic manures such as cow dung, sheep, poultry or rabbit manure [33]. The use of animal manure to fertilize ponds has been widely practiced in many countries in order to increase plankton so that there is more natural food for fish to eat, hence, high fish production. Manuring is therefore considered a cheap and preferred source of nutrient to increase fish production.

Pond fertilization with animal manure stimulates production of bacteria, phytoplankton, zooplankton, and benthic organisms [34]. The use of animal waste (livestock) has been studied under integration systems in Africa [35–37] and extensively in Asian countries [38, 39]. Benefits of integrated Agro-aquaculture systems have been reported in resource poor areas particularly in developing countries [38, 40, 41]. Studies conducted in sub-Saharan countries on the integrated aquaculture and agricultural systems are presented in **Table 2**.

Several studies showed that organic supplements contributed to fish yields by supplying P, N and C for algal growth and by stimulating detritus production and heterotrophic utilization. It is well known that high fish yields can be achieved through abundance of plankton in the cultural system [46]. Africa has vast resources of livestock and poultry, which play a vital role in pond fertilization. Livestock wastes including animal manure and poultry by-products are valuable resources in fish farming [47]. Livestock manure contains protein content of about 15 percent, energy (1250) kilocalories per kilogram, manure, and soluble vitamins [48].

Name of IAA	Country	Fish spp.	Livestock	Author(s)
Fish-cum-vegetable	Kenya	Various	—	[42]
Unknown	Malawi	Tilapia	—	[41]
Fish-cum-vegetable	Tanzania	Tilapia	—	[43]
Fish-cum-vegetable	Tanzania	Tilapia & Catfish	—	[36]
Fish-poultry-vegetable	Tanzania	Tilapia	Poultry	[44]
Fish-poultry-vegetable	Tanzania	Tilapia & catfish	Poultry	[45]
Fish-cum-poultry	Tanzania	Tilapia	Poultry	[37]
Fish-cum-rabbit	Rwanda	Tilapia	Rabbit	[35]

Table 2.
Studies on the integrated agro-aquaculture in sub-Saharan countries.

5. Fish feed availability and the concept of valorization

One of the solutions of fish feed availability is to entice animal feeds producing industries to consider also the production of fish feed [49]. However, the big issue here will be affordability; of these industrial feed products; most of our farmers belong to subsistence income bracket, hence, they might not afford these feed products. The use of floating pellets needs higher investment [50], which in most cases is lacking among smallholders; and unless the government intervenes in addressing the problems through either credit facilities or the provision of subsidies, the situation is not likely to get any better. It has been established that profitability in aquaculture is influenced by the cost of feed [51]. In Sub-Saharan countries, justification for industrial scale production of fish feed is not a priority despite the availability of raw materials [4]. Therefore, in order to feed fish, farm-made feeds can be made using locally available ingredients including animal by-products and plant residues.

In Tanzania, more than 80 percent of fish farmers relied on locally available feed ingredients as a major feed supplement for their cultured fish [43]. These local feed ingredients are categorized into four groups, (i) animal by-products, (ii) agricultural by-products, (iii) plant leaves and weed, and (iv) industrial by-products. It has been reported that the early growth phase of tilapia in 1991–2000 was significantly contributed by the use of alternative sources of protein including fishery by-products, terrestrial animal by-products, and a wide range of plant by-products [52]. In this chapter, discussion is cantered on the valorization of two broad categories of ingredients, plant and animal based ingredients.

5.1 Plant based ingredients and by-products

In addition to fertilization, feeding in ponds is done using supplementary feeds formulated on farm or purchased from cottage fish feed production industries. In some cases, cereal bran such as grains as energy source (**Figure 2**) and soybeans as

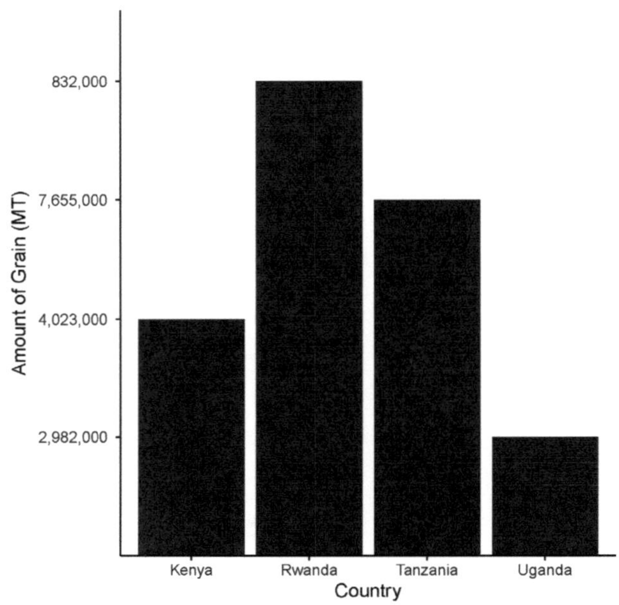

Figure 2.
Amount of grains required for fish feed compounding in East Africa. Source [53].

source of protein (**Figure 3**) are used in aquafeeds to increase pond productivity. The production from this system ranges from 1000 to 2500 kg/ha/year [33]. Most farmers prefer this system since it is less expensive in terms of feed inputs. Ten edible plant leaves were evaluated (see in [54]) as potential feed ingredients for aquatic animal, the results suggested that some of the plant leaves used contributed on growth performance, immune system, and disease resistance for the fish. Other important plant leaves which have been subjected to experiments to see whether they can be used as ingredients for fish feed formulation includes cassava leaves [55] and *Moringa* leaf [56]. In another study results showed that the integration of vegetables (*Brassica oleracea*) as pond inputs increased fish production and net yield than those reared under non-integrated systems [57]. In general, the amount of grain and soybean required in the four East African countries is given in **Figures 2** and **3**. Another experiment (see [58]), showed that when wheat bran, rice bran, and groundnut bran were used as agro-industrial by-products to examine their economic effectiveness in fish production, there were variability in growth rate and economic benefits, suggesting that variability of agro-by products reflects the growth rate of fish.

5.2 Animal based ingredients and by-products

According to the circular economy approach which focuses on the "reduce, reuse and recycle" of resources, waste from animal and food can be valorized leading to the production of proteins and other valuable compounds [59, 60]. For example, chicken, pig and cattle manures are substrates for production of housefly (*Musca domestica*) maggots which are in turn used as fish feed, or as supplement to fish meal in fish feed formulation [61]. Maggots are readily available and are accredited for having high nutritional value with an amino acid profile with biological value

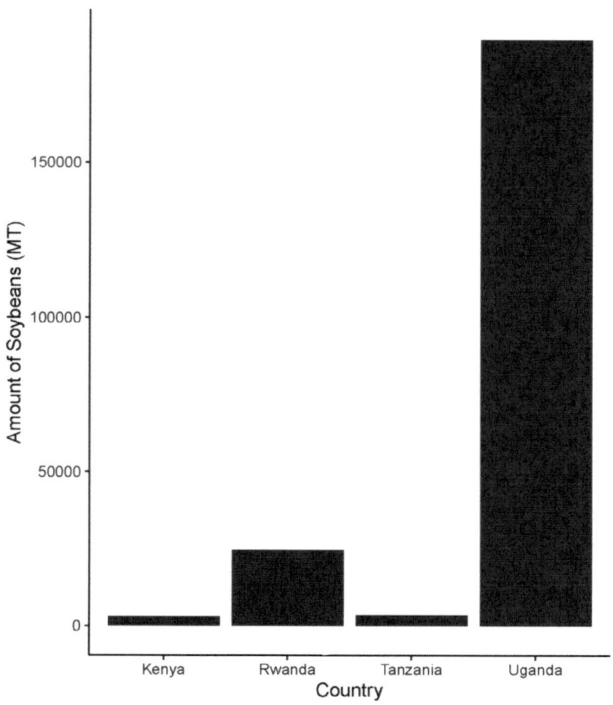

Figure 3.
Amount of soybean required for fish feed coumpound in four countries of East Africa. Source [53].

exceeding that of soybean and groundnut. Maggots can be harvested, processed into a meal that can be used to substitute or replace fish meal [61, 62]. Maggot grown on a mixture of cattle blood and wheat bran contained 92.7% dry matter, 47.6% crude protein, 25.3% fat, 7.5% crude fiber, 6.25% ash, and an amino acid profile comparable to fish meal [59] suggesting that animal wastes utilization can be used to produce insects which can be utilized as fish feed hence, reduce feed cost significantly, thus leading to a viable and sustainable aquaculture industry. The replacement of 25 percent fishmeal in catfish feed, culture with maggot gave high growth performance and profitability than fishmeal based diet [63, 64]. Several researches [65, 66] have been reporting on the use of red worms, black soldier fly, common housefly, and yellow mealworm as a source of protein to replace fish-meal. It is envisaged that the valorization of animal and animal by-products such as animal blood, offal of poultry, residues of traditional brewery waste, animal manure and fish wastes may contribute significantly on fish production hence, food nutrition and security.

6. Conclusion

It is clear that fish consumption in Sub-Saharan Africa is increasing. In order to maintain the present amount of fish consumption, considerable additional quantities of fish are required through aquaculture. In turn, aquaculture requires feed as a major input for increasing production. Since commercial fish feed production in most of the sub-Saharan countries is limited, considerable investments are required in local and low costs feed manufacturing. Raw materials of plant and animal origin are sufficiently available in the region albeit the possible competition from livestock and human consumption. Therefore, valorisation of animal and agro-products in the Sub-Saharan countries is imperative/inevitable for increasing fish production at low cost.

6.1 Recommendations

In order to increase food nutrition from aquaculture production through valorization of agro-by-products in the sub-Saharan countries, the following are recommended:

- Strengthen the use of Public Private partnering by putting more emphasis in services related to the collection of feed ingredients and preservation

- Public Private partnering must be embedded into an economic vision for aquaculture development

- Recognize small scale farmers as commercial ones and encourage small-scale farmers to work together by forming associations (work groups)

- Provide credit facilities for the private sector particularly for the small-scale holders

- Put emphasis on public private research partnerships and knowledge sharing on valorisation

- Provide capacity building and general education for small holders in order to improve their technological, managerial and commercial skills in handling agro by-products

DOI: http://dx.doi.org/10.5772/intechopen.95057

6.2 The way forward

With the ever-increasing human populations in sub-Saharan countries, the demand for food would increase and natural resources will become even scarcer. This situation will be more worsen with severe climate changes. These trends necessitate for a critical assessment of the situation to enable devise informed solutions in addressing issues pertaining to agro by-product processing and valorization.

Author details

Renalda N. Munubi* and Hieromin A. Lamtane
Department of Animal, Aquaculture and Range Sciences, Sokoine University of Agriculture, Morogoro, Tanzania

*Address all correspondence to: rmunubi@gmail.com; rmunubi@sua.ac.tz

References

[1] Moffitt CM, Cajas-Cano L. Blue Growth: The 2014 FAO State of the World Fisheries and Aquaculture. Fisheries. 2014;39(11):552-3.

[2] Little DC, Newton RW, Beveridge MC. Aquaculture: a rapidly growing and significant source of sustainable food? Status, transitions and potential. Proceedings of the Nutrition Society. 2016 Aug;75(3):274-86.

[3] Fisheries and aquaculture software. FishStatJ - software for fishery and Aquaculture statistical time series. In: *FAO Fisheries Division* [online]. Rome. Updated 14 September 2020. [Cited 16 November 2020]. http://www.fao.org/fishery/

[4] Brummett RE, Lazard J, Moehl J. African aquaculture: Realizing the potential. Food policy. 2008 Oct 1;33(5):371-85.

[5] FAO. Report of the FAO Expert Workshop on On-farm Feeding and Feed Management in Aquaculture.

[6] Hecht T, De Silva SS, Tacon AG. Study and analysis of feeds and fertilizers for sustainable aquaculture development. Hasan MR, editor. Rome: Food and Agriculture Organization of the United Nations; 2007.

[7] Huntington TC, Hasan MR. Fish as feed inputs for aquaculture–practices, sustainability and implications: a global synthesis. FAO Fisheries and Aquaculture Technical Paper. 2009 Jan;518:1-61.

[8] Pouomogne V, Brummett RE, Gatchouko M. Impacts of aquaculture development projects in western Cameroon. Journal of applied aquaculture. 2010 May 6;22(2):93-108.

[9] FAO. Aquaculture development. 5. Use of wild fish as feed in aquaculture. FAO. Technical guidelines for responsible fisheries. 2011.

[10] Tacon AG, Hasan MR, Metian M. Demand and supply of feed ingredients for farmed fish and crustaceans: trends and prospects. FAO Fisheries and Aquaculture technical paper. 2011(564):I.

[11] Kassam L, Dorward A. A comparative assessment of the poverty impacts of pond and cage aquaculture in Ghana. Aquaculture. 2017 Mar 1;470:110-22.

[12] Béné C, Arthur R, Norbury H, Allison EH, Beveridge M, Bush S, Campling L, Leschen W, Little D, Squires D, Thilsted SH. Contribution of fisheries and aquaculture to food security and poverty reduction: assessing the current evidence. World Development. 2016 Mar 1;79:177-96.

[13] De Silva SS, Davy FB. Aquaculture successes in Asia: contributing to sustained development and poverty alleviation. InSuccess stories in Asian aquaculture 2010 (pp. 1-14). Springer, Dordrecht.

[14] Béné C, Arthur R, Norbury H, Allison EH, Beveridge M, Bush S, Campling L, Leschen W, Little D, Squires D, Thilsted SH. Contribution of fisheries and aquaculture to food security and poverty reduction: assessing the current evidence. World Development. 2016 Mar 1;79:177-96.

[15] Bogard JR, Farook S, Marks GC, Waid J, Belton B, Ali M, Toufique K, Mamun A, Thilsted SH. Higher fish but lower micronutrient intakes: Temporal changes in fish consumption from capture fisheries and aquaculture in Bangladesh. PloS one. 2017 Apr 6;12(4):e0175098.

[16] Golden CD, Seto KL, Dey MM, Chen OL, Gephart JA, Myers SS, Smith M,

Vaitla B, Allison EH. Does aquaculture support the needs of nutritionally vulnerable nations? Frontiers in Marine Science. 2017 May 29;4:159.

[17] Teneva LT, Schemmel E, Kittinger JN. State of the plate: Assessing present and future contribution of fisheries and aquaculture to Hawai 'i's food security. Marine Policy. 2018 Aug 1;94:28-38.

[18] Beveridge M, Phillips M, Dugan P, Brummet R. Barriers to aquaculture development as a pathway to poverty alleviation and food security. (*Proceedings of the Advancing the Aquaculture Agenda, Paris, 2010*). Organization for Economic Cooperation and Development, 2010 Sep 14

[19] Pant J, Barman BK, Murshed-E-Jahan K, Belton B, Beveridge M. Can aquaculture benefit the extreme poor? A case study of landless and socially marginalized Adivasi (ethnic) communities in Bangladesh. Aquaculture. 2014 Jan 1;418:1-0.

[20] Waite R, Beveridge M, Brummett R, Castine S, Chaiyawannakarn N, Kaushik S, Mungkung R, Nawapakpilai SU, Phillips MI. Improving productivity and environmental performance of aquaculture. WorldFish; 2014 Sep 1.

[21] Chan CY, Tran N, Pethiyagoda S, Crissman CC, Sulser TB, Phillips MJ. Prospects and challenges of fish for food security in Africa. Global food security. 2019 Mar 1;20:17-25.

[22] Obiero K, Meulenbroek P, Drexler S, Dagne A, Akoll P, Odong R, Kaunda-Arara B, Waidbacher H. The contribution of fish to food and nutrition security in Eastern Africa: Emerging trends and future outlooks. Sustainability. 2019 Jan;11(6):1636.

[23] Cai J, Zhou X, Yan X, Lucente D, Lagana C. Top 10 species groups in global aquaculture 2017. Retrieved from Rome, Italy: http://www.fao.org/3/ca5224en/CA5224EN. pdf. 2019.

[24] Munguti JM, Musa S, Orina PS, Kyule DN, Opiyo MA, Charo-Karisa H, Ogello EO. An overview of current status of Kenyan fish feed industry and feed management practices, challenges and opportunities. International Journal of Fisheries and Aquatic Studies. 2014;1(6):128-37.

[25] Masser MP. Cage culture in freshwater and protected marine areas. Aquaculture production systems. 2012 May 11:50014-8300.

[26] Ebeling JM, Timmons MB. Recirculating aquaculture. Cayuga Aqua Ventures; 2010.

[27] Van Rijn J. Waste treatment in recirculating aquaculture systems. Aquacultural Engineering. 2013 Mar 1;53:49-

[28] Martins CI, Eding EH, Verdegem MC, Heinsbroek LT, Schneider O, Blancheton JP, d'Orbcastel ER, Verreth JA. New developments in recirculating aquaculture systems in Europe: A perspective on environmental sustainability. Aquacultural engineering. 2010 Nov 1;43(3):83-93.

[29] Bartelme RP, McLellan SL, Newton RJ. Freshwater recirculating aquaculture system operations drive biofilter bacterial community shifts around a stable nitrifying consortium of ammonia-oxidizing archaea and comammox Nitrospira. Frontiers in microbiology. 2017 Jan 30;8:101.

[30] Kinyage JP, Pedersen LF. Impact of temperature on ammonium and nitrite removal rates in RAS moving bed biofilters. Aquacultural Engineering. 2016 Nov 1;75:51-5.

[31] Boyd CE, Tucker CS. Pond aquaculture water quality management.

Springer Science & Business Media; 2012 Dec 6.

[32] Leip A, Ledgard S, Uwizeye A, Palhares JC, Aller MF, Amon B, Binder M, Cordovil CM, De Camillis C, Dong H, Fusi A. The value of manure-Manure as co-product in life cycle assessment. Journal of environmental management. 2019 Jul 1;241:293-304.

[33] Opiyo MA, Marijani E, Muendo P, Odede R, Leschen W, Charo-Karisa H. A review of aquaculture production and health management practices of farmed fish in Kenya. International Journal of Veterinary Science and Medicine. 2018 Dec 1;6(2):141-8.

[34] Kang'ombe J, Brown JA, Halfyard LC. Effect of using different types of organic animal manure on plankton abundance, and on growth and survival of *Tilapia rendalli* (Boulenger) in ponds. Aquaculture Research. 2006 Sep;37(13):1360-71.

[35] Tabaro SR, Mutanga O, Rugege D, Micha JC. Optimum rabbit density over fish ponds to optimize Nile tilapia production in an integrated rabbit–fish system in Rwanda. African Journal of Aquatic Science. 2012 Aug 1;37(2):165-74.

[36] Limbu SM, Shoko AP, Lamtane HA, Kishe-Machumu MA, Joram MC, Mbonde AS, Mgana HF, Mgaya YD. Fish polyculture system integrated with vegetable farming improves yield and economic benefits of small-scale farmers. Aquaculture Research. 2017 Jul;48(7):3631-44.

[37] Shoko AP, Limbu SM, Lamtane HA, Kishe-Machumu MA, Sekadende B, Ulotu EE, Masanja JC, Mgaya YD. The role of fish-poultry integration on fish growth performance, yields and economic benefits among smallholder farmers in sub-Saharan Africa, Tanzania. African Journal of Aquatic Science. 2019 May 2;44(1):15-24.

[38] Prein M. Integration of aquaculture into crop–animal systems in Asia. Agricultural systems. 2002 Jan 1;71(1-2):127-46.

[39] Bhatt BP, Bujarbaruah KM, Vinod K, Karunakaran M. Integrated fish farming for nutritional security in Eastern Himalayas, India. Journal of Applied Aquaculture. 2011 Apr 1;23(2):157-65.

[40] Cavalett O, De Queiroz JF, Ortega E. Emergy assessment of integrated production systems of grains, pig and fish in small farms in the South Brazil. Ecological Modelling. 2006 Mar 15;193(3-4):205-24.

[41] Dey MM, Paraguas FJ, Kambewa P, Pemsl DE. The impact of integrated aquaculture–agriculture on small-scale farms in Southern Malawi. Agricultural Economics. 2010 Jan;41(1):67-79.

[42] Kipkemboi J, van Dam AA, Denny P. Environmental impact of seasonal integrated aquaculture ponds ('fingerponds') in the wetlands of Lake Victoria, Kenya: an assessment, with the aid of Bayesian Networks. African Journal of Aquatic Science. 2007 Nov 1;32(3):219-34.

[43] Mmanda FP, Mulokozi DP, Lindberg JE, Norman Haldén A, Mtolera M, Kitula R, Lundh T. Fish farming in Tanzania: the availability and nutritive value of local feed ingredients. Journal of Applied Aquaculture. 2020 Jan 4:1-20.

[44] Mlelwa TI. *Growth performance and economic benefit of nile tilapia (Oreochromis niloticus) and chinese cabbage (Brassica rapachinensis) in aquaculture integration* (Doctoral dissertation, Sokoine University of Agriculture).

[45] Limbu SM, Shoko AP, Lamtane HA, Kishe-Machumu MA, Joram MC, Mbonde AS, Mgana HF,

Mgaya YD. Supplemental effects of mixed ingredients and rice bran on the growth performance, survival and yield of Nile tilapia, *Oreochromis niloticus* reared in fertilized earthen ponds. SpringerPlus. 2016 Dec;5(1):1-3.

[46] Jha P, Sarkar K, Barat S. Effect of different application rates of cow dung and poultry excreta on water quality and growth of ornamental carp, *Cyprinus carpio* vr. koi, in concrete tanks. Turkish Journal of fisheries and aquatic Sciences. 2004 Jan 1;4(1):17-22.

[47] Adewumi AA, Adewumi IK, Olaleye VF. Livestock wastes: fish-wealth solution. WIT Transactions on Ecology and the Environment. 2011 May 9;145:793-800.

[48] Kausar R. Review on recycling of animal wastes as a source of nutrients for freshwater fish culture within an integrated livestock system. Retreived April. 2009;24:2009.

[49] Brugère C, Ridler N, Haylor G, Macfadyen G, Hishamunda N. Aquaculture planning: policy formulation and implementation for sustainable development. FAO Fisheries and Aquaculture Technical Paper. 2010(542)

[50] Ansah YB, Frimpong EA. Impact of the adoption of BMPs on social welfare: A case study of commercial floating feeds for pond culture of tilapia in Ghana. Cogent Food & Agriculture. 2015 Dec 31;1(1):1048579.

[51] Neira I, Engle CR, Ngugi C. Economic and risk analysis of tilapia production in Kenya. Journal of Applied Aquaculture. 2009 May 19;21(2):73-95.

[52] Kumar G, Engle CR. Technological advances that led to growth of shrimp, salmon, and tilapia farming. Reviews in Fisheries Science & Aquaculture. 2016 Apr 2;24(2):136-52.

[53] Rothuis AJ, Turenhout MN, van Duijn A, Roem AJ, Rurangwa E, Katunzi EF, Shoko A, Kabagambe JB. Aquaculture in East Africa: a regional approach. IMARES/LEI; 2014.

[54] Olusola SE, Olaifa FE. Evaluation of some edible leaves as potential feed ingredients in aquatic animal nutrition and health. Journal of Fisheries. 2018 Mar 8;6(1):569-78.

[55] Madalla N. Novel. Feed Ingredients for Nile Tilapia (*Oreochromis niloticus* L.). Doctoral dissertation, University of Sterling

[56] Shigulu H.Effect of substituting sunflower seed cake meal and soya meal with *Moringa oleifera* leaf meal in Nile Tilapia *(Oreochromis niloticus)* diets (Doctoral dissertation, Dissertation for award of Msc. Degree at Sokoine University of Agriculture, Morogoro, Tanzania. 84pp).

[57] Shoko AP, Getabu A, Mwayuli G. Growth performance, yields and economic benefits of Nile tilapia *Oreochromis niloticus* and Kales *Brassica oleracea* cultured under vegetable-fish culture integration. Tanzania Journal of Science. 2011;37(1).

[58] Abarike ED, Attipoe FY, Alhassan EH. Effects of feeding fry of *Oreochromis niloticus* on different agro-industrial by-products. International Journal of Fisheries and Aquaculture. 2012 Sep 30;4(8):178-85.

[59] Gasco L, Acuti G, Bani P, Dalle Zotte A, Danieli PP, De Angelis A, Fortina R, Marino R, Parisi G, Piccolo G, Pinotti L. Insect and fish by-products as sustainable alternatives to conventional animal proteins in animal nutrition. Italian Journal of Animal Science. 2020 Dec 14;19(1):360-72.

[60] Stevens JR, Newton RW, Tlusty M, Little DC. The rise of aquaculture

by-products: Increasing food production, value, and sustainability through strategic utilisation. Marine Policy. 2018 Apr 1;90:115-24.

[61] Hezron L, Madalla N, Chenyambuga SW. Mass production of maggots for fish feed using naturally occurring adult houseflies (*Musca domestica*). Livestock Research for Rural Development. 2019;31(4).

[62] Hezron L, Madalla N, Chenyambuga SW. Alternate daily ration as a feeding strategy for optimum growth, nutrient utilization and reducing feed cost in Nile tilapia production. Livestock Research for Rural Development. 2019;31(7).

[63] Sogbesan OA. Nutritive potentials and utilization of garden snail (Limicolaria aurora) meat meal in the diet of *Clarias gariepinus* fingerlings. African Journal of Biotechnology. 2006;5(20).

[64] Djissou AS, Adjahouinou DC, Koshio S, Fiogbe ED. Complete replacement of fish meal by other animal protein sources on growth performance of *Clarias gariepinus* fingerlings. International Aquatic Research. 2016 Dec 1;8(4):333-41.

[65] Ghosh C. Integrated vermi-pisciculture–an alternative option for recycling of solid municipal waste in rural India. Bioresource Technology. 2004 May 1;93(1):71-5.

[66] Musyoka SN, Liti D, Ogello EO, Meulenbroek P, Waidbacher H. Earthworm, Eisenia fetida, bedding meal as potential cheap fishmeal replacement ingredient for semi-intensive farming of Nile Tilapia, (*Oreochromis niloticus*). Aquaculture Research. 2020 Jun;51(6):2359-68.

Section 2

New Technologies

Chapter 5

Innovation in Food Products Using Ozone Technology: Impact on Quality Assurance

Carlos Martín Enríquez-Castro, Manuel Pérez-Nafarrate and Jesús Enrique Gerardo Rodríguez

Abstract

Ozone application is a non-thermal technology used in food preservation, which is a powerful oxidant agent used in water and air treatment specially in disinfection processes for agriculture and food industry. The objective of this revision work is to publicize ozone applications in the growing, harvest, and postharvest handling of fruit and vegetables (F & V) across México. Ozonated water by foliar spraying and irrigation were proved to be effective in the control of pathogens, bacteria, and bugs. The use of Ozone was effective to heighten quality parameters of F & V, such as color, flavor, and soluble solids in mango, sugarcane, citric fruits, and nopal, increasing shelf life of fresh products up to 15 days after harvesting. Several protocols mentioned to fulfill the requirements of the producer were developed by TRIO3. The methodology proposed and the designed equipment by the company suggest a wider approach of this green technology in agriculture.

Keywords: ozone, climacteric fruits, ozonated water, foliar spraying, microbial growth control

1. Introduction

Ozone (O_3), also called trioxygen, is a gaseous substance whose molecule is formed by three oxygen atoms linked with an angular geometry. The ozone is formed by applying on the oxygen molecule enough energy to divide it forming various molecular structures [1]. This homonuclear molecule is the oxidant agent most powerful used for water and air treatment specially in disinfection processes for agriculture and food industry. Ozone is friendly with the environment and classified as Generally Recognized as Safe (GRAS) according to the Food and Drug Administration (FDA) [2].

Ozone is a universal disinfectant that reacts with the contaminant agents. It suppresses bad color and undesirable odor at the same time, destroying molds, bacteria, virus and algae [2, 3]. The deodorant action of ozone is due to the oxidation of chemical compounds such as ketones, hydrocarbons, acids, sulfides and nitrogenated derivatives [4]. Ozone oxidizes the cell wall, breaking its membrane and attacking to the DNA and RNA constituents. For this reason, the microorganisms are unable to develop immunity to the action of ozone as they do against other chemical compounds [1, 4, 5]. The use of ozone as sanitizer is effective without compromising health of consumers [6]. Traditional sanitizers are cheaper than ozone, however, sometimes they are not appropriate when a new outbreak emerges or when the presence of new food pathogens is detected [7, 8].

Ozone as a non-thermal technology is used in food preservation, which helps to improve the organoleptic characteristics of food following good manufacture practices [9]. Ozone can be pumped into a postharvest cold room. In water applications, ozone is drawn into a low negative pressure water stream using a Venturi injection system (Mazzei Company, LLT). The excess of not mixed ozone must be captured and destroyed to avoid the corrosion and personal injury [2]. A useful method of destroying the remanent ozone is combining UV light and catalytic agents such as granulated activated carbon [10]. The aim of this review is to describe several memories in the application of ozone in agriculture. Adhered to this, every project mentioned was adjusted to the state regulations achieving the expectations of the producer.

2. Biological factors involved in the deterioration of fruits and vegetables (F & V)

Some of the variables affecting the postharvest conditions of raw materials are related with their metabolic processes. Respiration as a process in living cells moderates the release of energy through the breakdown of carbon chains and the formation of new compounds, necessary for the maintenance and generation of synthetic reactions after harvest [11]. A low respiration rate in fresh materials increase postharvest life of perishable fruits. Ethylene concentration acts as a maturation hormone due to its biological activity in ripening climacteric fruits. Therefore, ethylene concentration can be monitored using precise instrumental methods [12]. Fundo et al. (2018) monitored the quality parameters of Cantaloupe melon juice using 30 min O_3 and 60 min O_3 treatments. They observed a diminution in color, vitamin C, total carotenoids, and antioxidant activity; but an increase in total phenolics registered in ozonated juices.

One of the most relevant stages of growth and development in F & V is probably the ripening or maturity. During ripeness, the enzymatic changes promote softening, hydrolytic conversion of storage material-starch to free sugars, and pigmentation changes, especially in climacteric fruits [13]. Flavor changes are considered an important topic in the metabolic pathway of F & V depending on sugars, acids and volatile compounds changes during ripening. As the fruit ripens, there is an increase in the synthesis of their volatile compounds [14].

2.1 Physical damage and pathological deterioration

In order to assure a good maturity index of raw materials, some factors such as water loss, changes in size, and changes in shape or surface must be considered, particularly when a marketable size is reached. Therefore, raw materials that presents physical damage must be separated to prevent postharvest losses. Measurement of texture, either manually or instrumentally can help to detect the maturity index of the fruit. Paciulli et al. [15] measured texture to frozen vegetables such as asparagus stems, zucchini and green beans. They observed a softening effect on the vegetable structure probably due to the blanching treatment affecting the inner cell wall. Depolymerization and/or solubilization reactions of the tissue occurred as a consequence of thermal treatment. In addition, it is advisable to avoid the improper handling or biodeterioration by microorganisms, insects, rodents or birds [16].

3. Applications of ozone

The harvesting of fresh F & V, its management during the freezing, packaging and processing involves the use of water. The presence of water on the agricultural products can increase the probability of contamination of plant pathogens and microorganisms of major concern in food safety. Inappropriate application of procedures related with prevention of contamination and disinfection of water can have severe consequences. Special attention on the microorganisms present in the water recirculation system must be taken because of a rapid reproduction and adaptability in adverse circumstances [16]. Ozone used for vegetable washing and disinfection either gas or dissolved in water has been effective in reducing several microorganisms such as *Escherichia coli*, Penicillium Italicum, and *P. digitatum* [17, 18]. In this way, damage and deterioration of the fruit can be prevented, and postharvest life increased [19]. Ozone is used to eliminate pests and insects without harming the food quality and the environment. Applied at proper doses, ozone diminishes the proliferation of bugs, and is an excellent choice to replace chemical products.

Feston et al. [20] used ozone and observed a reduction of bed bug (*Cimex lectularius* L.) in different life stages. They concluded that LC_{50} and LC_{90} values were higher in nymph and adults than the observed in eggs. The damage caused by insects it is not only restricted to what the vector physically damages and ingests, but also what is defecated over the plant. All these factors promote fungal development like Fusarium spp. and Aspergillus spp., and in most of the cases, they are responsible of promoting the development of mycotoxins [21].

3.1 Ozonated water by foliar spraying

Ozone applied in a foliar way allows the addition of nutrients in the plant and the elimination of pests and diseases. The effectiveness of this chemical agent in reducing microbial load of foods includes the presence of fungi, molds, spores, viruses, bacteria, eggs, and nymphs of insects as shown in **Figure 1**. Ozone has no unfavorable effects on the organoleptic characteristics of foods including textural and nutritional quality. Besides, ozone acts directly in the metabolism of the microorganism suppressing its development [19, 22]. During foliar spraying, plant diseases can be reduced due to the effect of nutrients and ozone added. Another benefit observed of using foliar spraying is that the dose of fertilizers applied in cultivars and the operating costs can be decreased. According to Ali [23] an appropriate foliar spraying improved the quality parameters of papaya using 2.5 ppm and 3.5 ppm ozone. Weight loss, firmness, ripening and soluble solids concentration were maintained and preserved for a longer storage period.

3.2 Ozonated water in the irrigation system

There are different ways to carry out an irrigation system: by gravity, sprinkling, dripping, micro sprinkling or sub-irrigation, and capillary diffusion as shown in **Figure 2**. The ozone irrigation system in cultivars allows the incorporation of nutrients and oxygen, the strengthening of roots, thickening of the tree trunk, and the increase of soluble solids and sugars. Ozonated water is effective to eliminate virus and bacteria and control in nopal-vegetable such as Fusarium spp. and Aspergillus spp. [24]. The nematodes present in their root are destroyed due to the sanitizing effect of ozonated water. The mature fruit (tuna or prickly pear fruit) can be sprayed with ozonated water to guarantee the innocuity and integrity of fruit [25].

Figure 1.
Ozonated water by foliar spraying in mango cultivars.

Figure 2.
Ozonated water in the irrigation system.

Ozonated water irrigation provides oxygen to the root and the stream that flows in its path reaching the root of the plant. The inner membrane of viruses, bacteria, fungi, algae, spores, and other microorganisms are destroyed using ozone dissolved

Commodity	Ozone treatment	Effects on quality	Reference
Broccoli	0.04 μL L^{-1}, 7 d, 10 °C	Delay of metabolic process & oxidative reactions	[2]
Cucumber	0.04 μL L^{-1}, 17 d, 3 °C	Enhancement of appearance. Higher values of Firmness (N) than the observed in control simples.	[2]
Mushroom	0.04 μL L^{-1}, 14 d, 4 °C	Blotch in cap surface was diminished.	[2]
Apple and pears	1.5 μL L^{-1}, 100 d, 20 °C	Total soluble solids and firmness were sustained.	[6]
Orange juice	Ozone variable dose	Control of decay, abatement of ethylene, removing of pesticides and residues.	[7]
Fresh cut lettuce	1 mg L^{-1}, 10 d, 4 °C	Storage life increased, decline of respiration rate & enzymatic browning.	[8]
Cantaloupe melon juice	7 ± 2.4 gL^{-1} for 30 & 60 min	Increase of polyphenols and vitamins.	[5]
Mango	10 μL L^{-1}, 3 d, 25 °C	Delay of ripening, Improvement of quality characteristics.	[7]
Papaya	2.5–3.5 ppm	Weight loss reduction, higher firmness values.	[8]
Soil and water deposits	Ozone variable dose	Bactericidal action on responsible for food delay (*P.aeruginosa*)	[6]

Table 1.
Uses of ozone in several fresh produces and its benefits.

in water, and promotes strength and productivity in the plant preventing diseases. This technique is widely applied in fruit trees, vineyards and nopal-vegetable among others [26]. **Table 1** shows uses of ozone in several materials and its benefits.

Several studies have been conducted to identify the vector disease in the plant as follows: leaf yellow curl virus in tomato [27], whitefly (*Bemisia tabici*) affects cotton [28], tropical fruits [29], and potato [30]. Ozonated water irrigation used in a regular way effectively eliminates plant disorders leaving no smell or trace [19]. Several researchers emphasize the effectiveness of ozone [30–34]. Contigiani et al. [35] applied ozonated water effectively in strawberries to reduce the risk of pathogen proliferation. They found significative differences ($P > 0.05$) in the fungal control (*B. cinedea*) using ozonated water for 5 min (2.73 mgL-1). Additionally, quality attributes were preserved. Ozone seems to be a very promising technology to reduce post-harvest losses.

4. Innovation project: processing of mangoes (*Mangifera indica*) using ozone technology

4.1 Introduction

Mango (Ataulfo variety) cultivated in the tropical areas of south of Mexico has a supreme quality but achieves a fast state of maturity and decomposes easily due to a high microbiological load. Local growers apply higher doses of fertilizers and pesticides several times per year to avoid the infestation of mango tree, specifically whitefly and their larvae. Effective post-harvest techniques to avoid major economic losses must be employed.

4.2 Methodology

The research team of Pérez-Nafarrate [36] designed processes using ozone gas and ozone water for disinfection and packaging of mango in an industrial plant in Guerrero, Mexico. The accomplishment of concise food safety regulations was according to studies realized by Tran et al. [37]. In order to preserve the physico-chemical characteristics of mango fruit, the next premises were followed:

Ozone gas used in a regular basis (one dose every third day) should eliminate gradually the presence of pests.

Gas monitoring was performed using an ozone analyzer (Model IN-2000, L2-LC, in USA Incorporated).

Ozone concentrations (1.5, 2.5, 3.5, 5 ppm) applied (96 h, 30 ± 3 °C) in the storage container should reduce the ethylene content generated as part of the fruit ripening process.

4.3 Results and discussion

The most effective ozone dose (3.5 ppm) preserved the organoleptic characteristics of the fruit and delayed the ripening for 15 days. According to producers, optimal shelf life of mango is 10 days [37]. Soluble solids, color, flavor and empiric firmness were improved (data not published) and a sensory evaluation of the fruit was developed by the operators. More and Rao [38, 39] used a combination of UV-C irradiation (210–280 nm), guar gum (2.5%), and ozone water (400 mg-h-1) to preserve organoleptic characteristics of mangoes at 32 ± 3 °C after 19 days.

Compliance with good manufacturing practices diminish the fruit deterioration, and the economic profit increased. Harvesting of climacteric fruits, including mangoes, papaya, and banana; requires storage temperature and relative humidity similar to that observed in subtropical environments. According to Pérez-Nafarrate [36], key factors to conduct an ozonated water experiment consist on reliability and accuracy of the measurement devices, type and growth stage of the microorganism, ozone concentration and time of contact with the product.

5. Innovation project: use of ozone in the harvesting and processing of nopal

5.1 Introduction

The nopal vegetable (*Opuntia ficus-indica*) grows in semi-arid zones around the world, it has a high dietary fiber content, and is usually consumed Mexico [40, 41]. Physical and morphological characteristics of nopal allow a quick adaptation to low precipitation and excessive heat (**Figure 3**). The nopal consist of two main parts: the edible part (84–90% water), and prickly pear (45–67% of reducing sugars) also called "tuna" [42]. Relative humidity is a crucial variable in nopal for development of pests and diseases. The most commonly microorganisms present in nopal are: cactus weevil *Cactophagus spinulae* (*Gyllenhal*), [43]; boll weevil (*Cylindrocopturus biradiatus Champs*), [40]; white worm (*Lanifera cyclades Druce*), zebra worn (*Olycella nephelepsa Dyar*), [44]; and cochineal (*Dactylopius coccus costa*), [45]. The most important include the bacterial stain (*Bacterium sp.*), rotting of the epidermis, anthracnose, prickly pear gold [46], the thickening of cladodes and the presence of bold among others [47]. The objective was to give added value to nopal-vegetable (raw and processed) using ozonated foliar spraying and ozonated water irrigation.

Figure 3.
Manual harvesting of nopal in México.

5.2 Methodology

Due to the nature of nopal, good manufacture practices in soil and water are required to control outbreaks of diseases. The experimental plantation was located in Matehuala, San Luis Potosi; México. The producers gave light watering to nopal seedlings every 8 or 15 days during the dry months as shown in **Figure 4**, and the irrigation program was monitored periodically to establish the ozone generator requirements. Salinity, hardness, pH and sedimented solids were registered for 15 days before installing the irrigation system. The ozone solubility control included the measurement of pressure and temperature, concentration of carbonates, nitrates and metals [6]. Once the nopal "pencas" (thick and fleshy central nerve that has leaves) were cut, they were placed in recyclable plastic boxes preserving the thorns to avoid the oxidation reactions [48].

5.3 Results and discussion

Ozonated water favored the organoleptic integrity of the nopal, reducing oxidation reactions and retaining green color in sliced raw nopal. Other goals achieved were the reduction in microbial load and the increase in shelf life (data not shown). The final product was packaged and sealed in plastic bags with 500 g and 1000 g. An investment proposal to install a rural cooperative company with capacity of processing 500 kg per day was presented to the cooperative.

Technical specifications and cost of a nopal sider, an immersion washing machine, a centrifugal dryer, and a vibrator dryer with cold were included in the investment project for $1,500,000.00 pesos.

Figure 4.
Sustainable management of the nopal irrigation system.

Good manufacturing practices are essential to accomplish nopal processing with ozone. Water quality and soil requirements monitoring were key factors to improve the postharvest life of nopal opuntia. An innovative option for rural population centers was presented to detonate the local economy. Almost 50% of inhabitants in Mexico are distributed in rural areas, therefore ozone technology should be affordable to most of the population.

6. Innovation project: use of ozone to counteract *Huanglongbing* in citrus trees

6.1 Introduction

The *Huanglongbing* (HLB) disease or "yellow dragon", affects the worldwide production of citrus (lemon, tangerine, orange). *Diaphorina citri* (Kuwayama), the insect vector of HLB, is the main responsible as shown in **Figure 5** [49, 50]. HLB is mainly distributed in Asia (Malaysia, Thailand and Vietnam) and Africa. Farmers invest up to 50 insecticide doses per year to control the vector and avoid the spread of the disease [51]. Symptoms of this plant disease include damage in tree and fruit, sparce yellow foliage, and stunting among others as shown in **Figure 6** [52], shortening of the productive life of the plant, and detriment of final quality product [53].

Figure 5.
HLB in citrus tree. Source: www.gob.mx/agricultura.

Figure 6.
HLB symptoms in Mexican lemon tree. Source: www.gob.mx/agricultura.

Since 2010, HLB has been spread to the American continent especially those areas with frequent rainfall and template temperatures. HLB in Mexico is a serious threat to the 526 thousand hectares of citrus distributed in twenty-three Federal Entities. This in turn represents a production risk of 6.7 million tons per year with a value of more than 400 million dollars. Around 67,000 producers engaged to citric industry generated approximately 70,000 direct jobs and 250,000 indirect jobs, all of them endangered by HLB [54]. Specific research in Mexico to counteract this pandemic is urgently recommended.

6.2 Methodology

In 2013, a collaboration protocol was signed between TRIO3 Food Technologies, the National Institute of Forestry, Agricultural and Livestock Research (INIFAP, for its acronym in spanish) and Los Limones Orchard, located in Tecoman; Colima (18°46′38.04′′ N, 103°48′56.74′′ W). Pérez-Nafarrate [55] evaluated the effect of ozone to eradicate the HLB disease in the productive region previously mentioned. Two main objectives were proposed: first, the early detection of infected trees in the growing season; and second to evaluate the damage spread by the vector in those trees with an advanced stage of the disease, and register its total loss [56]. Fifteen hectares planted with real lemon were micro-sprinkler irrigated with ozonated water, and nutrients were added to roots of trees every third day during six months. The sprinkler irrigation process was evaluated at the end of the experiment.

6.3 Results and discussion

Ozone applied in Los Limones Orchard considerably reduced the population of larvae and adult vectors increasing the citrus production (data not shown). After ozonated water process concluded in lemon trees, other benefits were observed by farmers and technical staff including the improvement of the foliage coloration, and strengthening of trunk and root trees, which are very promising. Showler [57] used an organic mixture composed of corn meal, humic acid, molasses and fish oil to suppress greasy spot in infected trees. An increase in soluble solids content and a weight loss reduction in treated trees, as a result of the healing treatment during three years, provided similar results as ozone did. According to Perez-Nafarrate [55], the insecticide and plaguicide action of ozonated water proved to be faster than traditional methods.

HLB, like many other plant diseases, be diminished effectively with ozonated water. Gas concentration used, soil quality and the effectiveness of the irrigation system were the critical parameters to observe in the experiment. This clean and affordable technology is efficient to control HLB, but more research is still needed to eradicate this bacterium from Mexico and its surroundings.

7. Innovation project: use of ozone to control diseases of sugarcane

7.1 Introduction

Sugarcane (*Saccharum spp hybrids*) is a tall perennial grass traditionally grown in tropical zones and used for obtaining sugar concentrates, molasses and other derivatives [58, 59]. Good manufacturing practices and biological control of diseases allow the correct development of this plant. It is common to find a considerable number of microorganisms in sugarcane such as bacteria, fungi and viruses, all these entities cause several diseases in the cultivar [60] as shown below:

Most common organisms found in sugarcane include: red rot (*Colletotrichum falcatum*) [61]; wilt (*Fusarium sacchari*) [62]; grassy shoot (*Exitanius indicus*) [63]; leaf scald (*Xanthomonas albilineans*) [64]; smut (*Sporisorium scitamineum*) [65, 66]; brown rust (*Puccinia melanocephala*) and orange rust (*Puccinia kuchnii*) [67].

The diseases of major economic importance in sugarcane are described as follows:

Mosaic disease. Is produced by Sugarcane mosaic virus and attacks young plants making areas appear on leaves pale green and yellowish within a normal green. The reeds become stunted causing production declines [68].

Eyespot. This name is derived from the powdery black mass of spores associated with this disease. The affected plants show a reddish elliptical lesion surrounded by a yellowish structure that varies in size. The spores of the fungus Ustilago scitaminea are transmitted by wind, rain, irrigation water, seeds or animals [69].

Rust. Produced by the fungus Puccinia melanocephala manifested by many elongated spots on the leaves, show no growth and the stem become thin. The yellowish spots look at the whole sheet [70].

Leaf scald. The causal organism Xanthomonas albineans is a bacterial disease first observed in Cuba in 1979. X. albineans causes sudden death of entire stems and seedlings affecting significantly sugarcane yields [64]. Leaf scald is also called in Latin America "gomosis", a plant disease characterized by an abundant production of gum. Gomosis caused by Xanthomona axonopodis involves widespread dwarfism in plant [70].

Ratoon stunting disease. RSD caused by Leifsonia xyli spp. was observed in Cuba for the first time in 1953. Several stunted stems very thin are present within the plant and short internodes. Damaged stems increase as the number of cuts in the field is higher [71].

The ozonated water irrigation provides a greater contribution of oxygen to the root. Ozonated water is free of viruses, bacteria, fungi, algae, spores and any other microorganism. Growth is achieved faster than usual with more liveliness and strength as well as more productivity. This irrigation method is beneficial for plants, and is currently used in fruit trees, vineyards and crops in general. Ozonated water achieves the prevention of plant diseases such as leaf scald and rust reducing the microbiological load. Chemical products such as pesticides and fungicides are reduced improving the shelf life of fresh products [60].

7.2 Methodology

On July 4 2008 was signed a collaborative project between the National Institute of Agrarian Innovation (INIA, for its acronym in spanish) and the National Institute of Research of Sugarcane in Havana-Pinar del Río. Red ferritic soil was used to grow up a 9-month agamic seed from 5 varieties identified as follows: C323–68, C86–56, C86–456, C88–380, and C90–530. The experimental design (divided plots 5 x 6 x 2) included 180 buds from each variety with two replicates, 5 varieties and 6 treatments. Finally, sugarcane certified quality seeds (P. Vista Florida cv. 115–2014) previously treated with ozonated water were introduced among the producers as shown in **Figure** 7.

The exposure time in ozonated water was 10, 20, and 30 min respectively. The treatments included a hot water-chemical treatment, a biological one, and a sample control [52]. The measurements were carried out one and six months after sowing. The main objective was to evaluate the effect of ozonated water on sugarcane seed growing. The viability and growth of agamic seeds of sugarcane used was also evaluated. The ozone equipment used is shown in **Figure 8**.

Figure 7.
Certified sugarcane seed P. Vista Florida. Cv. 115–2014.

Figure 8.
Equipment used for the ozonation of water.

7.2.1 Treatment with ozonated water

An irrigation and sprinkler system with ozonated water to promote the growth of sugarcane is shown in **Figures 9** and **10**. The fungicide was reduced progressively to control and eliminate fungus and pests as the amount of ozonated water increased. Previously, 90 buds were submerged in ozonated water during 24 h. Thirty buds per variety were immersed in ozonated water (1 ppm) for 10, 20, and 30 min respectively.

7.2.2 Witness without treatment

Thirty buds per variety were soaked on water for 24 h without applying any treatment.

Figure 9.
Ozonated water irrigation system as applied on sugarcane cultivation.

Figure 10.
Ozonized water by foliar spraying.

7.2.3 Hydrothermal + chemical treatment

Thirty buds per variety were submerged 24 h in water, then immersed in hot water at 51 °C for one hour. Immediately they were treated with propiconazole (Tilt EC 250) at 0.4% (w/w) for 15 min. This time was considered enough to eliminate leaf scald and red rot diseases.

7.2.4 Biological treatment

This treatment started on July 4 2008 with the immersion in water of 30 sugarcane buds per variety for 24 h, and then submerged for 30 min in a water solution of Nemacid (EDOCA, Amealco, Querétaro) at 2% (w/w). This organic insecticide is used in the control and inhibition of nemathodes. Quivican experimental station was used to plant the sprouted seeds treated with Nemacid. After 30 days, the sprouting count was carried out and analyzed in the laboratory. All the experiments were made by duplicate.

7.3 Results and discussion

All the treatments showed significant differences ($p < 0.0001$). Best results obtained on seeds immersed in ozonated water for different time intervals are shown in **Figure 11**. The sprouted seeds treated with 10 min-ozonated water immersion reached the highest germination level: 10 seeds in total. The sample control seeds and seeds treated with 30 min-ozonated water reached an acceptable germination rate: 9 and 8 respectively. Nemacid and hot water samples presented the lowest germination rate: 7 in total. Using ozonated water did not affect the viability of the seed. The experimental ozonated water immersion system proposed to producers of sugarcane reduced drastically eyespot and leaf scald. This treatment was beneficial for agamic seed viability showing a different response in every treatment, especially that with 10-min immersion time.

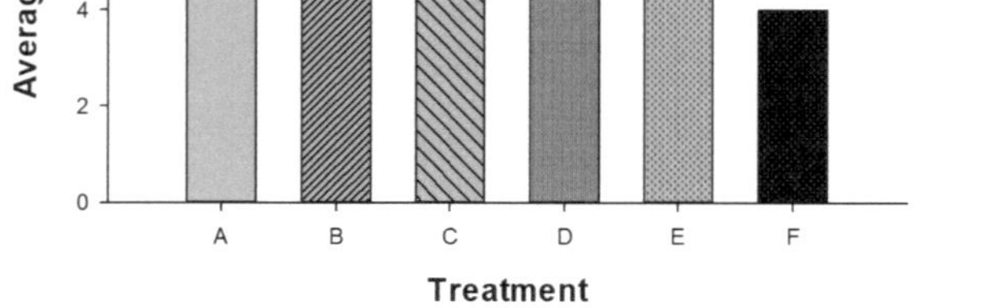

Figure 11.
Influence of ozonated water in sprouting of sugarcane agamic seed.

8. Conclusion

Ozone has demonstrated be useful for management of postharvest of climacteric fruits such as mangoes and papayas. This ecotechnology is cost effective and is very promising to use in developing countries of Latin América. Using ozone watering in Nopal opuntia crops, proved to be very effective to control diseases and increase plant health. An adequate ozone irrigation system prevents growth of bacteria and bugs in citrus trees, and other fruits, and is an alternative to reduce HLB infestation. Sugarcane and other important crops growing in tropical areas increase yield using ozonated water, and an early diagnose of diseases reduce losses and increase the economic benefit in productive areas. TRIO3 acknowledges all the producers and scientific investigators involved in the protocols developed during these last 10 years.

Acknowledgements

Thanks to TRIO3 Food Technology support and also thanks to Superior Technological Institute Zacatecas Norte (ITSZN-TecNM).

Conflict of interest

The authors declare no conflict of interest.

Author details

Carlos Martín Enríquez-Castro[1*], Manuel Pérez-Nafarrate[2] and Jesús Enrique Gerardo Rodríguez[3]

1 Superior Tecnological Institute Zacatecas Norte, Río Grande, Zacatecas, México

2 TRIO3 Food Technologies, Guadalajara, Jalisco, México

3 Department of Biochemistry and Molecular Biology and Robert M. Kerr Food and Agricultural Products Center, Oklahoma State University, Stillwater, OK, United States

*Address all correspondence to: carlos_enriquez_castro@hotmail.com

References

[1] Oner, ME, & Demirci A. Ozone for food decontamination: Theory and applications. In Handbook of hygiene control in the food industry. Woodhead Publishing. 2016. pp. 491-501

[2] Pandielvam, R., Subhashini, S., Banuu-Priya, E. P., Kothakota, A., Ramesh, S. V. and Shahir, S. Ozone based food preservation: a promising green technology for enhanced food safety. Ozone: Science & Engineering. 2019. **41**: 17-34

[3] Huyskens-Keil, S., Hassenberg, K. & Herppich, W. B. Impact of postharvest UV-C and ozone treatment on textural properties of white asparagus (*Asparagus officinalis L.).* Journal of Applied Botany and Food Quality. 2012. **84**, 229

[4] Wang, J., Wang, S., Sun, Y., Li, C., Li, Y., Zhang, Q., and Wu, Z. Reduction of Escherichia coli 0157: H7 and naturally present microbes on fresh-cut lettuce using lactic acid and aqueous ozone. 2019. RSC Advances. **9**, 22636-22643

[5] Fundo, J. F., Miller, F. A., Tremarin, A., García, E., Brandão, T. R. & Silva, C. L. Quality assesment of Cantaloupe melon juice under ozone processing. Innovative Food Science & Emerging Technologies. 2018. **47**, 461-466

[6] Giuliani, G., Ricevuti, G., Galoforo, A., and Franzini, M. Microbiological aspects of ozone: bactericidal activity and antibiotic/antimicrobial resistance in bacterial strains treated with ozone. 2018. Ozone therapy. **3**

[7] Prabha, V.I.T.H.U., Barma, R.D., Singh, R.A.N.J.I.T., and Madan, A. D. I. T.Y. A. Ozone technology in food processing: a review. 2015. Trends in Biosciences, **8**, 4031-4047

[8] Natha, A. Mukhimb, K., Swerb, T., Duttaa, D., Vermaa, N., Dekab, B.C., and Gangwara, B. A review on application of ozone in the food processing and packaging. 2014. Journal of Food Product Development and Packaging. **1**, 07-21

[9] Varga, L., ad Szigeti, J. Use of ozone in the dairy industry: A review. 2016. International Journal of Dairy Technology. **69**, 157-168

[10] Ghernaout, D., Alshammari Y., Alghamdi, A., Aichouini, M., Touahmia, M., and Ait Messaoudene, N. Water reuse: Extenuating membrane fouling in membrane processes. 2018. International Journal of Environmental Chemistry. **2**, 1-12

[11] Saltveit, M. E. Respiratory metabolism. In: Postharvest physiology and biochemistry of fruits and vegetables. Woodhead Publishing, 2019. pp. 73-91

[12] Martinelli, F., Scalenghe, R., Davino, S., Panno, S., Scuderi, G., Ruisi, P., Villa, P., Stroppiana, D., Boschetti, M., Goulart, L.R. and Davis, C. E. Advanced methods of plant disease detection: a review. Agronomy for Sustainable Development. 2015. **35**, 1-25

[13] Streif, J., Kittemann, D., Neuwald, D. A, McCormick, R. &Xuan, H. Pre and postharvest management of fruit quality, ripening and senescence. In VI International Postharvest Symposium. 2009. 877, 55-68

[14] Ramjan, M. and Ansari, M.T. Factors affecting of fruits, vegetables and its quality. Journal of Medicinal Plants. 2018. **6**, 16-18

[15] Paciuli, M., Ganino, T., Pellegrini, N., Rinaldi, M., Zaupa, M., Fabbri, A., and Chiavaro, E. Impact of the industrial freezing process on selected vegetables-Part I. Structure, texture and antioxidant capacity. 2015. Food Research International. **74**, 329-337

[16] Kiaya, V. Postharvest losses and strategies to reduce them. Technical Paper on Postharvest Losses, Action contre la Faim (ACF). 2014. **25**

[17] Luo, A., Bai, J., Li, R. Fang, Y., Li, L. Wang, D., Zhang L., Liang, J., Huang, T., and Kou, L. Effects of ozone treatment on the quality of kiwifruit during postharvest storage affected by Botritis cinerea and Penicillium expansum. Journal of Pythopathology. 2019. **167**, 470-478

[18] García-Martín, J.F., Olmo, M., and García, J.M. Effect of ozone treatment on post-harvest disease and quality of different citrus varieties at laboratory and at industrial facility. Postharvest Biology and Technology. 2018. **137**, 77-85

[19] Rodoni, L., Casadei, N., Concellon, A., Chaves,A. R. & Vicente, A. R. Effect of short term ozone treatments on tomato (*Solanum lycopersicum L.*) fruit quality and cell wall degradation. Journal of Agricultural and Food Chemistry. 2010. **58**, 594-599

[20] Feston, J. Gaire,S. Fardisi, M., Mason, L.J. & Gondhalekar, A. D. Determining baseline toxicity of ozone against an insecticide-susceptible strain of the common bed bug, Cimex lecturarius L. under laboratory conditions. Pest Management Science. 2020. 76, 3108-3116

[21] McDonough, M. X., Campabadal, C. A., Mason, L. J., Maier, D. E., Denvir, A & Woloshuk, C. Ozone application in a modified screw conveyor to treat grain for insect pests, fungal contaminants, and mycotoxins. Journal of Stored Products Research. 2011. **47**, 249-254

[22] Brodowska, A. J., Nowak, A. & Smigielski, K. Ozone in the food industry: principles of ozone treatment, mechanisms of action, and applications: an overview. Critical Reviews in Food Science and Nutrition. 2018. **58**, 2176-2201

[23] Ali, A., Ong, M. K. & Forney, C. F. Effect of ozone preconditioning on quality and antioxidant capacity of papaya fruit during ambient storage. Food Chemistry. 2014. **142**, 19-26

[24] Wang, L., Luo, Y., Wang, R., Li, Y., Li, Y., Chen, Z. Effect of deoxynivalenol detoxification by ozone treatment in wheat grains. 2016. Food Control. 66, 137-144

[25] Palmeri, R., Parafati, L., Arena, E., Grassenio,E., Restuccia, C., and Fallico, B. Antioxidant and antimicrobial properties of semi-processed frozen pricky pear juice as affected by cultivar and harvest time. 2020. Foods. **9**, 235

[26] Onopiuk, A., Póltorak, A., Wojtasik-Kalinowska, I., Szpicer, A., Marcinkowska-Lesiak, M., Pogorzelski, G., and Wierbicka, A. Impact of the storage atmosphere enriched with ozone on the quality of *Licopersicon esculentum* tomatoes. Journal of Food Processing and Preservation. 2019. **43**, 1-10

[27] Cui, H., Sun, Y., Zhao, Z., and Zhang, Y. The combined effect of elevated O3 levels and TYLCV infection increases the fitness of *Bemisia tabaci Meditarrean* on tomato plants. 2019. Environmental Entomology. **48**, 1425-1433

[28] Hong, Y., Yi, T., Tan, X., Zhao, Z., and Ge, F. High ozone (O3) affects the fitness associated with microbial composition and abundance of Q. biotype *Bemisia tabaci*. 2016. Frontiers in Microbiology. 7, 1593

[29] González Ramírez, J.E., Cabrera Jova, M., Robaina, A., Rodríguez Pérez, D., González Cadasio, A., and Portal, O. Water-dissolved ozone mediates potyvirus sanitation during in vitro propagation of *Dioscorea cayenensis subsp. rotundata* (*Poir*) Miége. 2020. Science and Engineering. **42**, 89-94

[30] Kim, J., Kil, E.J., Kim, S., Seo, H., Byun, H.S., Park, J., Lee, S. Seed

transmission of sweet potato leaf curl virus in sweet potato (Ipomoea batatas). 2015. Plant Pathology. **64**, 1284-1291

[31] Shezi, S., Magwaza, L.S., Mditshwa, A., and Tesfay, S.Z. Changes in biochemistry of fresh produce in response to ozone postharvest treatment. 2020. Scientia Horticulturae. 269, 109397

[32] Chen, C., Zhang, H., Zhang, X.,Dong, C., Xue, W., and Xu, W. The effect of different doses of ozone treatments on the postharvest quality and biodiversity of cantaloupes. 2020. Postharvest Biology and Technology. **163**, 11114

[33] Chitravathi, K., Chauhan, O., Raju, P.S., and Madhukar, N. Efficacy of aqueous ozone and chlorine in combination with passive modified packaging on the postharvest shelf-life extension of green chillies (Capsicum annum L.). 2015. Food and Bioprocess Technology. **8**, 1386-1392

[34] Tzortzakis, N. A powerful tool for the fresh produce preservation. In: Postharvest management approaches for maintaining quality of fresh produce. 2016. Switzerland, Springer. 175-180

[35] Contigiani, E. V., Jaramillo-Sánchez, G., Castro, M.A., Gómez, P.L., and Alzamora, S.M. Postharvest quality of strawberry fruit (*Fragaria x Ananassa Duch cv. Albion*) as affected by ozone washing: fungal spoilage, mechanical properties,and structure. 2018. Food and Bioprocess Technology. **11**, 1639-1650

[36] Pérez-Nafarrate, M. Sustainable ecological proposal for the implementation of a phytosanitary growing process of mango and postharvest treatment in the tropical zone of Guerrero: effect of ozonated water foliar spraying and use of an irrigation system. TRIO3 Food Technologies. Guadalajara, Jal. 2012

[37] Canut, A., Pascual, A., De Valencia, P. T., Franklin, C. B. & Valencia, E. P. Ozone Cip: ozone cleaning in place in food industries. In: IOA Conference and Exhibition Valencia, Spain. 2007. October 29-31

[38] More, S., and Rao, T.R. Elicitor-mediated sanitization in combination with edible coatings improve postharvest shelf life and antioxidant potential of mango fruit. 2019. Environmental and Experimental Biology. **17**, 107-114

[39] Tran, T. L., Aiamla-or, S., Srilaong, V., Jitareerat, P., Wongs-Aree, C. and Uthairatanakij, A. Ozone fumigation to delay ripening of mango "Nam Dok Mai N° 4". In: Southeast Asia Symposium on Quality Management in Postharvest Systems. 2013. **1088**, 103-106

[40] Vargas-Mendoza, A., Flores-Hernández, A. and Basaldua-Suárez, J. F. Dinámica de las principales plagas de nopal *Opuntia spp* en la zona semiárida de Querétaro. Revista Chapingo Serie Zonas Áridas. 2008. **7**, 21-27

[41] Kumar, K., Singh, D. & Singh, D. S. Cactus pear: cultivation and uses. Central Institute for Arid Horticulture (CIAH). Bikaner, Rajasthan, India. 2018. **73,** pp. 38

[42] Saroj, P. L, Singh, D. & Kumar, K. Culinary exploitation of nopal cactus pear in arid region. Indian Horticulture. 2017. **62**, 13-16

[43] López-Martínez, V., Pérez De la O, N. B., Ramírez-Bustos, I. I., Alia-Tejacal, I. & Jiménez-García, D. Current and potential distribution of the cactus weevil, *Cactophagus spinolae* (Gyllenhal) (Coleoptera: Curculionidae) in México. The Coleopterists Bulletin. 2016. **70**, 327-334

[44] Flores, R. G. R., Martínez, O. G., Quintanilla, J.A., V., and Peña, S. R. S. Especies de *Hemiptera-Heteroptera*

asociadas a *Opuntia spp*. y *Nopalea spp*. en el Desierto Chihuahuense Mexicano. 2017. Revista Mexicana de Ciencias Agrícolas. **8**, 1773-1784

[45] Abrha, H., Birhane, E., Zenebe, A., Hagos, H., Girma, A., Aynekulu, E. & Alemie, A. Modeling the impacts of climate change and cochineal (*Dactylopius coccus Costa*) invasion on the future distribution of cactus pear (*Opuntia ficus indica*) in Northern Ethiopia. Journal of the Professional Association for Cactus Development. 2018. *20*, 128-150

[46] Tafolla-Arellano, J.C., Baéz-Sañudo, R., and Tiznado-Hernández, M. E. The cuticle as a factor in the quality of horticultural crops. 2018. Scientia Horticulturae. **232**, 145-152

[47] Suaste-Dzul, A., Rojas-Martínez, R. I., Zavaleta-Mejía, E. & Pérez-Brito, D. Detección molecular de fitoplasmas en nopal tunero (*Opuntia ficus indica*) con síntomas de engrosamiento del cladodio. Revista Mexicana de Fitopatología. 2012. **30**, 81-85

[48] Pérez-Nafarrate, M. Ozone application for the implementation in the phytosanitary process of cultivation and postharvest of fresh vegetable-nopal in Zapopan, Jalisco. 2011. Unpublished data

[49] Doud, M. M., Wang, Y., Hoffman, M. T., Latza, C. L., Luo, W., Armstrong, C. M. & Duan, Y. Solar thermotherapy reduces the titer of *Candidatus Liberibacter asiaticus* and enhanced canopy growth by altering gene expression profiles in HLB-affected citrus plants. Horticulture Research. 2017. **4**, 1-9

[50] Gottwald, T. R. Current epidemiological understanding of citrus *Huanglongbing*. Annual review of phytopathology. 2010. **48**, 119-139

[51] Saénz-Pérez, C. A., Osorio-Hernández, E., Estrada-Drouaillet, B., Poot-Poot, W. A., Delgado-Martínez, R. & Rodríguez-Herrera, R. Principales enfermedades de los cítricos. Revista Mexicana de Ciencias Agrícolas. 2019. **10**, 1653-1665

[52] Batool, A., Iftikhar, Y., Mughal, S. M., Khan, M. M., Jaskani, M. J., Abbas, M. & Khan, I. A. Citrus greening disease: a major cause of citrus decline in the world. A review. Horticultural Science (Prague). 2007. **34**, 159-166

[53] Yi, T., Lei, L., He, l., Yi, J., Li, L., Dai, L., and Hong, Y. Symbiotic fungus affected the Asian citrus psyllid (ACP) resistance to Imidacloprid and Thiamethoxam. 2020. Frontiers in Microbiology. **11**, 3040

[54] SADER (Secretaría de Agricultura y Desarrollo Rural). Control frontal a la plaga del dragón amarillo (yellow dragon). 2014. http://www.gob.mx/agricultura. Accessed on line: 07/03/2020

[55] Pérez-Nafarrate, M. Phytosanitary support with the application of ozone for its implementation in the dispersal control of *Huanlongbing* in Tecoman, Colima. 2013. TRIO3 Food Technologies. Unpublished data

[56] Martinelly, F., Scalengue, R., Davino, S., Panno, S., Scuderi, G., Ruisi, P. & Davis, C. E. Advanced methods of plant disease detection. A review. Agronomy for Sustainable Development. 2015. **35**, 1-25

[57] Showler, A. T. Suppression of greasy spot disease caused by *Mycodphaerella citri whiteside* on grapefruit trees in an organic orchard using an aqueous organic mixture of composted cornmeal, humic acid, molasses, and fish oil versus vegetable oil. Crop Protection. 2017. **99**, 137-143

[58] Elsharif, A. A. & Abu-Naser, S. S. An expert system for diagnosing sugarcane diseases. International Journal of Academic Engineering Research. 2019. **3**, 19-27

[59] Kaur, L., Dharshini, S., Ram, B., and Appunu, C: Sugarcane genomics and transcriptomics. In Sugarcane Biotechnology: Challenges and Prospects. 2017. 13-32. Springer, Cham

[60] Pérez-Nafarrate, M., Fernández, L. A. & Bataller, M. Technological innovation in the germination, irrigation, and phytosanitary control of sugar cane cultivation. TRIO3 Food Technologies. 2015. Lambayeque, Perú

[61] Tariq, M., Khan, A., Tabassum B., Toufiq, N., Bhatti, M., Riaz, S. ... and Husnain, T. Antifungal activity of chitinase II against Colletotrichum falcatum went causing red hot disease in transgenic sugarcane. 2018.Turkish Journal of Biology. **42**, 45-53

[62] Poongothai, M., Wiswanathan, R., Malathi, P. & Sundar, A. R. Sugarcane wilt: pathogen recovery from different tissues and variation in cultural characters. Sugar Technology. 2014. **16**, 50-66

[63] Pratap, T., Singh, R., Pal, R., Yadaw, S. & Singh, V. Integrated weed management studies in sugarcane ratoon. Indian Journal of Weed Science. 2013. **45**, 257-259

[64] Mattielo, L., Riaño-Pachón, D.M., Martins, M.C.M., da Cruz, L. P., Bassi,D., Marchiori, P.E.R., Vasconcelos-Ribeiro, R., Venzian Labate M.T., Labate, C.A., and Meossi, M. Physiological and transcriptional analyses of developmental stages along sugarcane leaf. 2015. BMC Plant Biology. 15, 1-21

[65] Sundar, A. R., Barnabas, E. L., Malathi, P., Viswanathan, R., Sundar, A. R. & Barnabas, E. L. A mini-review on smult disease of sugarcane caused by *Sporisorium scitamineum.* Botany. 2012. **226,** 107-128

[66] Schaker, P. D., Palhares, A. C., Taniguti, L. M., Peters, L. P., Creste, S., Aitken, K. S., Van Sluys, M.A., Kitajima, J.P., Vieira, M.L.C., and Monteiro-Vitorello, C. B. RNAseq transcriptional profiling following whip development in sugarcane smut disease. Plos One. 2016. **11**, 1-21

[67] Elsayed, A.I., Komor, E., Boulila, M., Viswanathan, R., and Odero, D. C. Biology and management of sugarcane yellow leaf virus: an historical review. Archives of Virology. 2015. **160**, 2921-2934

[68] Xu, D. L., Park, J. W., Mirkov, T. E. & Zhou, G. H. Viruses causing mosaic disease in sugarcane and their genetic diversity in southern China. Archives of Virology. 2008. **153**, 1031

[69] Borrás-Hidalgo, O., Thomma, B. P., Carmona, E., Borroto, C. J, Pujol, M., Arenciba, A. & López, J. Identification of sugarcane genes induced in disease resistant somaclones upon inoculation with *Ustilago scitaminea* or *Bipolaris sacchari*. Plant Physiology and Biochemistry. 2005. **43**, 1115-1121

[70] Harrison, J. & Studholme, D. J. Draft genoma sequence of Xanthomonas axonopodis patovar vasculorum. Federation of European Microbiological Societes. 2014. **360**, 113-116

[71] Cia, M.C., de Carvalho, G., Azevedo, R.A., Monteiro-Vitorello, C.B., Souza, G.M., Nishiyama-Junior, M.Y., Gimiliani Lembke,C., da Cnha Antunes de Faria, R. S., Rodríguez-Marques, J. P., Meotto, M., and Camargo, L. E. A. Novel insights into the early stages of ratoon stunting disease of sugarcane inferred from transcript and protein analysis. 2018. Phytopathology. **108**, 1455-1466

Chapter 6

Advanced Optical Technologies in Food Quality and Waste Management

John Chauvin, Ray Duran, Stanley Ng, Thomas Burke, Kenneth Barton, Nicholas MacKinnon, Kouhyar Tavakolian, Alireza Akhbardeh and Fartash Vasefi

Abstract

Food waste is a global problem caused in large part by premature food spoilage. Seafood is especially prone to food waste because it spoils easily. Of the annual 4.7 billion pounds of seafood destined for U.S. markets between 2009 and 2013, 40 to 47 percent ended up as waste. This problem is due in large part to a lack of available technologies to enable rapid, accurate, and reliable valorization of food products from boat or farm to table. Fortunately, recent advancements in spectral sensing technologies and spectroscopic analyses show promise for addressing this problem. Not only could these advancements help to solve hunger issues in impoverished regions of the globe, but they could also benefit the average consumer by enabling intelligent pricing of food products based on projected shelf life. Additional technologies that enforce trust and compliance (e.g., blockchain) could further serve to prevent food fraud by maintaining records of spoilage conditions and other quality validation at all points along the food supply chain and provide improved transparency as regards contract performance and attribution of liability. In this chapter we discuss technologies that have enabled the development of hand-held spectroscopic devices for detecting food spoilage. We also discuss some of the analytical methods used to classify and quantify spoilage based on spectral measurements.

Keywords: Spoilage, valorization, spectroscopy, hyperspectral imaging, artificial intelligence

1. Introduction

Food waste is a significant problem in both developed and developing economies [1]. The global seafood industry faces unprecedented challenges as demand increases, consumer preferences change, and expectations of quality increase. Consumers are demanding more transparency and a commitment to sustainability while access to at-capacity or overfished fishery resources is strained and food production and retail profit margins are thin. Global food supply chains can be a cause of improper food storage, which leads to by-product waste. Whether through a lack of quality control or a hold in the distribution process, food spoilage can occur even before the product reaches markets. Approximately 35% of fish are lost to waste globally with between 30% and 35% loss in most regions of the world [2]. Seafood's perishability is largely to blame.

A major goal to improve the overall valorization of food and reduce agro-food waste or diversion to by-products is earlier and rapid detection of spoilage. Like medical evaluations for disease, early detection can lead to quicker response from manufacturers or consumers to increase the shelf life of food products. Microbiologists and food scientists have developed a variety of methods to detect surface microbials and pathogenic microorganisms including culturing and colony-counting methods, polymerase chain reaction (PCR)-based amplification for DNA analysis, immunoassay analysis, chromatography, and mass spectrometry [3, 4]. Unfortunately, these techniques have limited versatility and restrictive methodologies that are not practical with on-site and on-demand food quality and safety control [3–5]. However, spectroscopic technologies have shown great promise for enabling early detection of spoilage to help minimize food waste.

Another problem affecting consumers and contributing to global food waste is the lack of transparent pricing for food products as a function of shelf life. Alongside government and industry regulation, intelligent dynamic pricing based on projected shelf life at retail and other upstream points along the supply chain can encourage efforts to reduce waste. This requires new tools for tracking food products at all points along the supply chain. These tools must be easy to incorporate, objective, verifiable, and provide data on quality, provenance, and freshness.

A pioneer in the development of food quality and traceability technologies, SafetySpect is developing a new handheld quality, adulteration, and traceability (QAT) scanner to address many of these issues. Utilizing hyperspectral multi-mode technology to provide species identification and direct measurements of freshness/spoilage in a handheld device can address challenges of waste and mislabeling. In seafood and meat processing, distribution, and storage supply chains in developed and developing economies, this is likely to meaningfully decrease food waste and increase sustainable access to safe, healthy, and nutritious foods. It will also decrease costs and increase profit within supply chains by providing better attribution of liability and verification of supply contract performance. This transparency will provide incentives to upstream supply chain participants to improve operational methods that can result in the degradation of product or unnecessary, accelerated spoilage.

1.1 Current trends for examining fish quality

The main approach to improving valorization and by-product management is early detection of spoilage. A common method for detecting spoilage in fish is the Torry Freshness Score [6]. This systematic scoring method was developed in the UK to provide an objective assessment of fish quality. It uses the human senses to examine specific parts of the fish. For example, an evaluator will observe gill odors, skin tension, opaqueness of the eyes, and overall smell of the fish and provide a freshness rating between 0 (lowest) to 10 (highest). However, this manual approach to evaluating fish samples is time consuming and may be more susceptible to evaluator bias or human error. This motivates the development of technologies that enable rapid evaluation of fish quality with minimal human interpretation.

Spectroscopic approaches offer a robust, non-destructive means of detecting and evaluating the extent of food quality issues. In recent decades, advancements in micro-electro-mechanical systems (MEMS) and micro-electro-opto-mechanical systems (MEOMS) have enabled the development of miniaturized spectroscopic devices that can be used for analysis at all points along the food supply chain, from farm fields to distribution centers to retail markets. Hyperspectral imaging (HSI) combines spectroscopy and imaging to enable evaluation of an object's spectroscopic composition at a high spatial resolution, thus providing a more comprehensive evaluation tool for any given sample [7–9]. As the scale and complexity of food supply networks continues to grow, there is an ever increasing need for low-cost, portable, analytical devices to combat the corresponding growth in vulnerability of food products to adulteration, contamination, and fraud [10]. In the next section, we discuss a variety of technologies that have enabled the recent development of portable and handheld spectroscopic devices that can and have been used for evaluating the quality of food products.

2. Spectroscopy and hyperspectral imaging technologies

2.1 Infrared spectroscopy

One of the most common approaches used for quality control of food products involves the analysis of vibrational spectra via infrared spectroscopy. The spectral peaks and valleys formed by the fundamental vibrational modes (and their harmonics) of key structures within organic molecules can be used to detect the presence of abnormalities or measure the abundances of specific chemical components. The near infrared (NIR) and mid-infrared (MIR) spectral regions are of high interest in food analysis applications.

2.1.1 Infrared detectors

The rapid proliferation of visible digital camera technology over the past few decades is due in large part to the use of inexpensive silicon-based detectors which can sense wavelengths in the visible region and in the infrared region up to about 1050 nm. For longer wavelengths, however, detectors composed of different materials are required. Indium gallium arsenide (InGaAs) detectors have become the dominant technology for detectors on the market, surpassing germanium (Ge), lead sulfide (PbS), and lead selenide (PbSe) detectors [11]. Unfortunately, these detectors are generally more costly than silicon-based detectors. Furthermore, for wavelengths beyond 1700 nm, the noise in these detectors becomes so high that cooling is required to keep it to a manageable level [12]. To minimize the cost of these more expensive detectors, developers of handheld infrared spectrometers have sought to simplify detector designs by reducing the number of elements required.

2.1.1.1 Near-infrared spectroscopy (NIRS)

NIRS covers the approximate wavelength spectrum of 780 to 2500 nm. Within this range lie signals from the vibration of organic chemical bonds such as oxygen-hydrogen (O-H), carbon-hydrogen (C-H), nitrogen-hydrogen (N-H), and sulfur-hydrogen (S-H), as well as their overtones [13]. Instrument cost and robustness is generally better for NIR than for MIR [14]. However, NIR spectral peaks tend to be weak and broad with significant overlapping of absorption peaks because of a combination of vibrational spectra from multiple chemical bonds, making straightforward interpretation difficult, if not impossible [15]. Spectral preprocessing techniques (e.g., smoothing, detrending, and taking derivatives) and multivariate statistical methods (e.g., nonlinear partial least squares, Fisher determinant analysis, and artificial neural networks) are invoked to extract the information hidden in the spectra. Despite these disadvantages, the advantage in terms of lower cost, increased safety for the environment and operators, and superior chemical specificity and applicability to a broad range of sample types has made NIR spectroscopy a popular approach for food analysis [11, 15]. These advantages have in turn encouraged the development of numerous portable NIR spectrometers based on a variety of designs.

2.1.1.2 Dispersive NIR spectrometers

In conventional dispersive spectrometer designs, broadband light is passed through the sample and into an entrance slit to create a narrow line of light which is imaged onto a detector. A dispersive grating is inserted in the path causing the image of the narrow line to be spread out into a spectrum where the light is separated into its various wavelengths. These are then focused onto an array detector. **Figure 1** shows a basic dispersive spectrometer based on the Czerny-Turner design which uses mirrors to minimize the overall size of the design.

One major disadvantage of the design shown in **Figure 1** is the need for an array InGaAs detector which can be expensive. Alternative designs requiring only a single-element detector have been developed to help mitigate this expense. For example, instead of collecting all wavelengths simultaneously, spectrometers based on the Fabry-Perot interferometer design shown in **Figure 2** collect the wavelengths sequentially. This design features one fixed and one moveable half-silvered mirror aligned along the same optical axis. As the light bounces between the mirrors, constructive and destructive interference determines the spectrum of light that passes through the other side of the moveable mirror and onto the detector. When the spacing between the fixed and moveable mirrors equals an integer number of half wavelengths, maximum constructive interference occurs leading to a peak in the output spectrum at that wavelength. As with the Michelson interferometer, the spectral range of interest is thus examined by translating the moveable mirror over a specific spatial extent. An example of a compact instrument based on the Fabry-Perot design is Spectral Engines' MEMS Fabry-Perot spectrometer, the NIRONE [18].

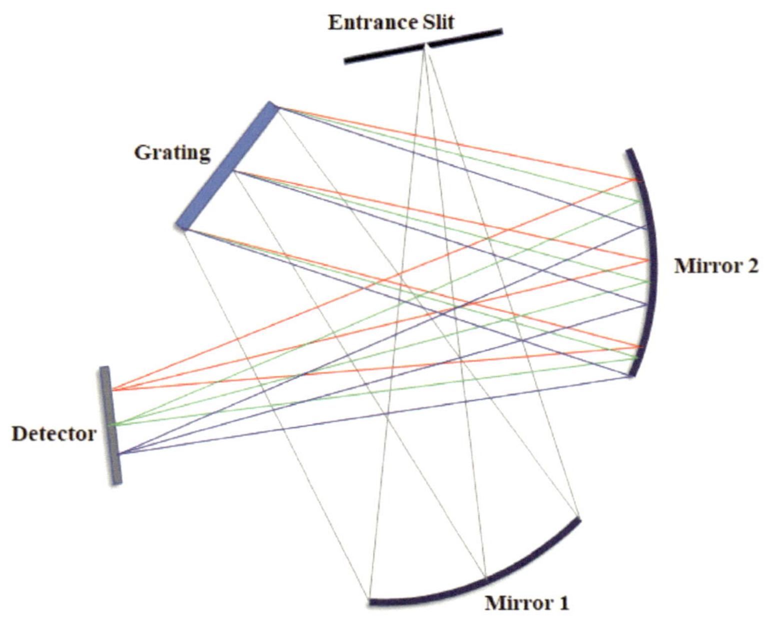

Figure 1.
Dispersive spectrometer based on the Czerny-Turner design [16].

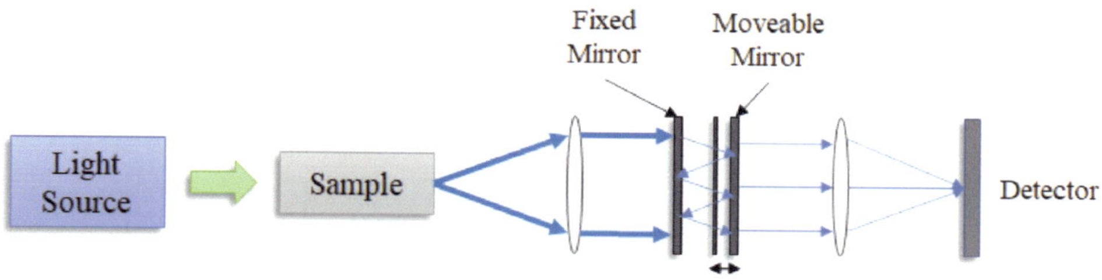

Figure 2.
Fabry-Perot spectrometer design [17].

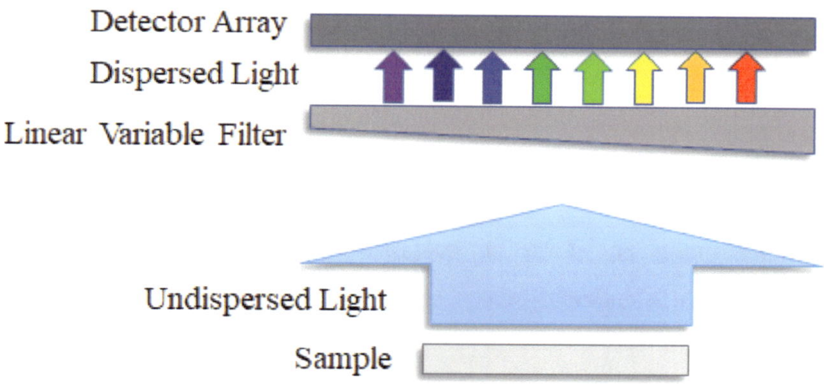

Figure 3.
Diagram showing the operating principle behind the LVF used in the MicroNIR OnSite [20].

2.1.1.3 Linear variable filter (LVF) NIR spectrometers

One problem common to all dispersive designs like the Czerny-Turner is that the light must be allowed to disperse over a given spatial extent such that the wavelengths can separate before reaching the detector. This causes limitations in designing for compactness [19]. One way to surmount this problem is with the use of LVFs which are generally formed from wedge-shaped optics and behave much like a Fabry-Perot interferometer but scan by lateral position along the filter instead of by movement of a mirror along the optical axis [12]. The LVF can be applied directly to a detector array, leading to a simple and compact mechanical design with no moving parts (see **Figure 3**). Viavi's MicroNIR OnSite features an LVF applied to a 128-pixel InGaAs array [21].

2.1.1.4 Hadamard spectrometers

The Hadamard spectrometer design has a couple of key advantages over the conventional dispersive design. First, it overcomes the slow scanning process of dispersive techniques where individual wavelengths must be collected one after another. This is often referred to as the multiplexing or Fellget advantage. Additionally, Hadamard spectrometers tend to be more sensitive and the sensors themselves have a higher optical throughput, resulting in what is termed the Jacquinot advantage [22]. **Figure 4** shows the basic layout for this type of spectrometer, the key component of which is the mask positioned just before the focusing lens. This mask blocks out a certain portion (usually ~50%) of the diffracted light at a time. The blocking elements are moved in discrete steps to form a binary matrix where the elements

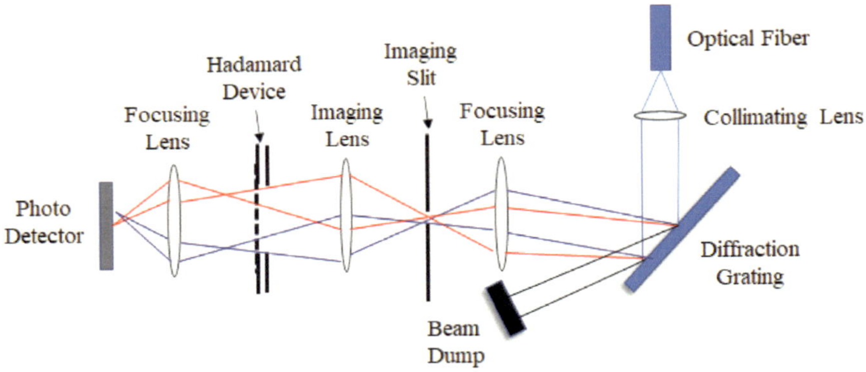

Figure 4.
Basic design layout for a Hadamard spectrometer [23].

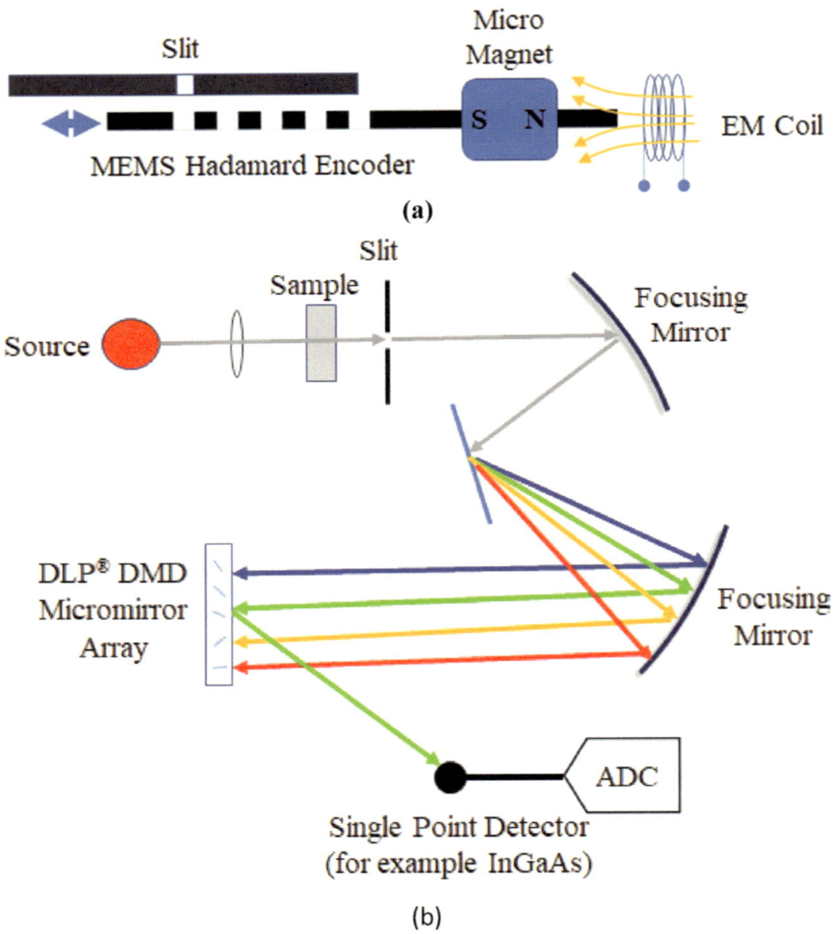

Figure 5.
Hadamard spectrometer designs and devices. (a) a coil and magnet [23] mask design; (b) layout of Texas Instruments' DLP® DMD-based NIRscan Nano optical engine [12].

of each successive row are shifted by one position to the left. The detector readings are recorded with each shift and pieced together to form a data vector. A Hadamard transformation using the binary matrix is applied to the data matrix to yield the measured spectrum [12].

The Hadamard mask itself can be implemented in a variety of ways. Early designs used a coil and magnet to move the mask linearly in front of a separate entrance slit (see **Figure 5(a)**). The microPHAZIR NIR spectrometer by Thermo Fisher Scientific uses a programmable MEMS diffraction grating as the Hadamard mask. Texas Instruments offers two NIR Hadamard spectrometer devices based on its Digital Light Projection (DLP®) digital micromirror device (DMD) technology, the DLP NIRscan and the DLP NIRscan Nano (see **Figure 5(b)**) [12]. The DMD contains an array of individually pivotable micromirrors which can be aligned and flipped in sequence to form the Hadamard mask [12].

2.1.1.5 Applications of handheld NIR spectrometers for food analysis

Advancement in MEMS technology and LVFs has led to the rapid miniaturization of NIR spectrometers, thus enabling the development of portable NIR spectrometers. Portable NIR spectrometers have been used to evaluate the quality of fruits and vegetables primarily during the pre-harvest stage while on the vine/tree. Such analyses focus on maturity parameters to enable determination of optimal harvest dates and include measurements of soluble solids content (SSC), titratable acidity, pH, weight, size, firmness, juice content, juice weight, pericarp thickness, and others [15].

NIR analyses of meat and fish are typically performed for shelf-life estimation and freshness evaluation. Examples include traceability analysis of pasture-fed lambs and stall-fed lambs, authenticity testing for pork and pork fat in veal sausages, moisture, protein, and fat analysis in Iberian pork muscles, fat characterization in Iberian ham, freshness evaluation in beef sirloin and beef eye of round, shelf life estimation of pork meat, and monitoring and control of the drying process in fermented sausages [13].

Portable NIR spectrometers have also been used to measure quality factors in milk and beverages. Components such as fat, protein, lactose, and moisture percentages have been measured to determine milk quality [15], and NIR spectral differences have been exploited to distinguish between organic and non-organic milk products [24]. Quality of rice wine, tea drinks, and beers have been evaluated via NIR measurement of alcohol, nitrogen, apparent extract, and non-sugar solids percentages, polyphenol and free amino acid concentrations, and bitterness and beer distinction factors [15].

2.2 Mid-infrared spectroscopy

The mid-infrared (MIR) spectrum covers a range of wavelengths from ~2500 nm to ~5000 nm and contains many of the fundamental absorption bands of organic components. Spectra in this range are very sensitive to chemical composition, leading to high specificity. Furthermore, organic functional groups produce well-delineated absorption bands in this region, a feature that can be exploited to individually separate different components present in a mixture by their unique fingerprints in the absorption spectrum [14]. Given the high cost of InGaAs detectors and the need for cooling to lower noise to a manageable level, MIR spectrometers generally feature single-element detector designs. Most exploit a technique based on the Fourier transform.

2.2.1 Fourier transform infrared spectrometer (FTIR)

One subset of FTIR spectrometers is based on the Michelson interferometer design that was used for Michelson and Morley's speed of light measurements.

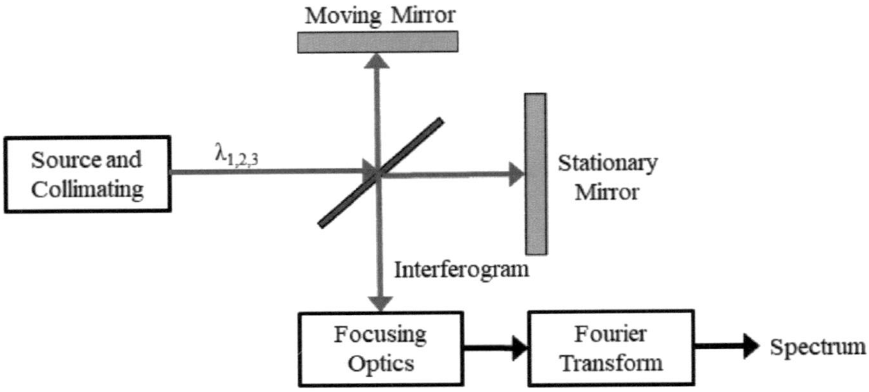

Figure 6.
Diagram of an FTIR spectrometer based on the Michelson interferometer [25].

This interferometer consists of two optical pathways oriented perpendicular to one another (see **Figure 6**). A collimated broadband light source enters from the left and strikes a half-silvered mirror (i.e., beam-splitter) oriented at 45°. Half of the beam then passes through this mirror and strikes a stationary mirror at the end of the pathway where it is reflected back toward the beam-splitter. The other half of the incident beam is directed toward a mirror that is allowed to move back and forth along this pathway. Upon reflection from this mirror and arrival at the beam-splitter, the two reflected beams produce an interference pattern which is focused on a single-element detector. The sample is typically inserted between the beam-splitter and the detector. After the moving mirror is swept through its full range of motion and the full interferogram recorded, these patterns are processed to produce the spectrum. Processing in this case is done with a Fourier transform which converts the sensor response as a function of spatial mirror position to a function of frequency. The Fourier transform accomplishes this by determining the optimal mixture of sine and cosine functions that can replicate the sensor response.

Like Hadamard spectrometry, FTIR spectrometry has several advantages over dispersive spectrometry such as that used in most NIR spectrometers. It enjoys both the multiplexing (Fellget) advantage and the throughput (Jacquinot) advantage. This latter characteristic serves to significantly reduce the noise in the sensor output. As this design includes only one moving component, the mirror in the path with the variable length, FTIR instruments have a mechanical design that it is highly robust to breakdown. Finally, many FTIR instruments include a HeNe laser that acts as an internal calibration standard, eliminating the need for calibration during operation (Connes advantage) [26]. SiWare Systems' NeoSpectra-Scanner is an FTIR NIR spectrometer with a MEMS-based Michelson interferometer design [27].

Another popular design for FTIR spectrometers is based on the property of attenuated total reflection (ATR). As shown in **Figure 7**, broadband infrared light is directed into a high refractive index crystal typically made of germanium, silicon, zinc sulfide, or diamond [28]. The ends of this crystal are cut such that the angle of incidence for the light will result in total internal reflection through the crystal. Although the light wave does not propagate outside of the crystal, an evanescent wave can still pass through the top of the crystal where the sample is placed. This evanescent wave interacts with the sample and absorbs portions of the infrared light, resulting in an attenuation of the light that reaches the detector. One of the primary advantages of this technique is that the light does not have to travel through the entire sample as it does for other designs, which often results in severe

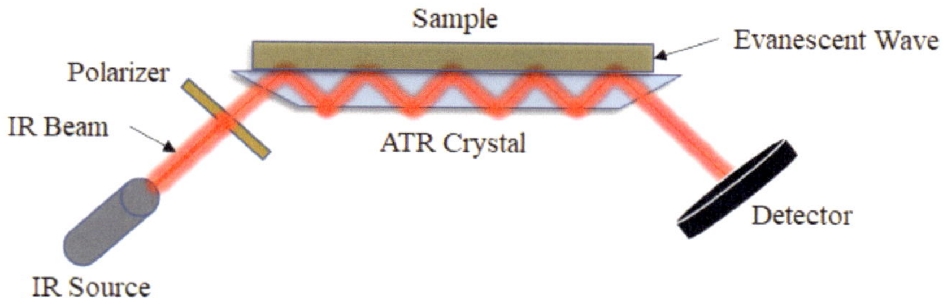

Figure 7.
Diagram illustrating the ATR concept [28].

attenuation and loss of signal. An example of a handheld FTIR spectrometer based on the ATR design is the Ocean MZ5, a miniature ATR-FTIR spectrometer produced by Ocean Optics [29].

2.2.2 Applications of handheld MIR spectrometers for food analysis

MIR spectrometers have long been used for food analysis, but most have been conducted in a laboratory setting. Examples include detection of food spoilage bacteria in meat and dairy produce, brand authentication of a range of Trappist beers, and adulteration of milk and of beef burgers [10]. More recently, portable MIR devices have been used for the simultaneous analysis of sugar and amino acid concentrations in raw potato tubers, the measurement of quality factors in tomato juices, and the measurement of fatty acid content of marine oil dietary supplements [30].

2.3 Raman spectroscopy

Raman spectroscopy is often seen as complimentary to infrared spectroscopy given the relative nature of the phenomena involved. While infrared spectroscopy measures the absorption of energy, Raman spectroscopy measures the exchange of energy with radiation provided by a monochromatic light source (usually a laser with a wavelength in the ultraviolet to NIR range). This exchange causes a shift in the source's wavelength. Molecules are infrared active only if the vibration induced by the source results in a change to the dipole moment, whereas the Raman shift is caused by changes in the molecules' polarization [10]. Thus, these two methods provide mutually exclusive information. Raman peaks tend to be much sharper than infrared peaks and data collection tends to be faster, but the Raman effect is inherently weaker. Furthermore, Raman spectrometers tend to be more expensive to manufacture than their infrared counterparts.

2.3.1 Raman spectrometers

Figure 8 shows an example design for a Raman spectrometer. Light from the laser is directed to the sample and the output is passed through a notch filter to separate out all but the Raman scattered light. A spectrograph grating then disperses this light into its constituent wavelengths and onto a detector. Metrohm's Mira M-1 is an example of a portable Raman spectrograph with a 785 nm laser [32]. Laser wavelengths for other Raman spectrometers can range from the ultraviolet (UV) to the NIR bands. Since spectral sensitivity and resolution increase with decreasing laser wavelength, UV lasers tend to be optimal for applications featuring bio-molecules [33].

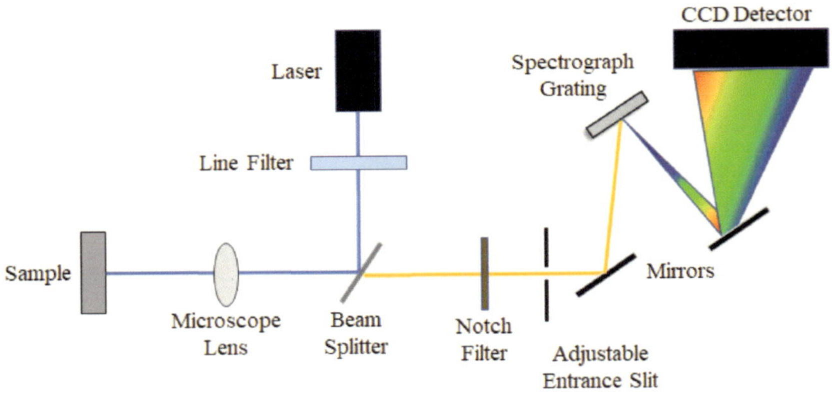

Figure 8.
Basic diagram of a Raman spectrometer [31].

2.3.2 Applications of handheld Raman spectrometers for food analysis

Recent applications of portable and handheld Raman spectrometers for food analysis include the detection of organophosphate and organothiophosphate pesticides on apple skins, the detection of fungicides and parasiticides on citrus fruits and bananas, authenticity and origin of vegetable and essential oils, detection of marker compounds for illegal alcoholic beverages, detection of adulteration in beef burgers, identification of rapid meat spoilage, and prediction of pork quality on a slaughterhouse line [10].

3. Artificial intelligence and machine learning techniques for spectral analysis

Once the spectroscopic data has been collected, sophisticated algorithms, and capable processors to host these algorithms, are needed to convert this data into useful information. The microchip revolution that started back in the 1960's has continued unabated [34] and silicon vendors continue to innovate with even mature technologies such as field programmable gate arrays (FPGAs) [35] with transistor counts that exceed one billion in number. This growth of computing power coupled with advances in spectroscopy have enabled modern machine learning algorithms to be implemented that can lead to significant positive changes to food safety, adulteration and fraud.

In this section, we discuss key machine learning algorithms that have been applied to spectroscopy in general and to food valorization applications specifically. We first examine methods used to extract from the data features that are non-redundant and information-rich and can be used for accurate classification and quantification of food spoilage and food quality.

3.1 Feature extraction

Most spectral datasets contain subsets of features that are highly redundant or subject to high amounts of noise. The inclusion of such features in a classification or regression algorithm generally leads to suboptimal performance. Feature extraction is the process by which redundant or noisy features are removed from the dataset, leaving a smaller set of features with a high amount of signal content. Here, we discuss popular methods for feature extraction that have been used for spectroscopic applications.

3.1.1 Principal component analysis (PCA)

PCA is a common method of feature extraction enabled through dimensionality reduction. PCA provides dimensionality reduction by representing the variance in the data within the smallest number of components possible. Each principal component is a linear combination of the original components and is calculated in an iterative fashion by identifying the weight vector that, when applied to the original data components, contains that largest amount of the remaining variance and is orthogonal to the previously calculated principal components. As a result, the majority of the variance (typically ~99% or more) is contained within the first few (typically 3–5) principal components, meaning the others can be safely ignored with negligible loss of information. These principal components are also eigenvectors of the data's covariance matrix and can be computed by eigendecomposition. The corresponding eigenvalues are proportional to the variance represented within each principal component and can be used to identify the principal components which are considered "significant." According to the Kaiser criterion [36], eigenvectors with eigenvalues less than 1 can be considered insignificant.

In additional to its dimensionality reduction benefit, PCA tends to yield principal components that provide good separability between data collected from different classes. It is this property that makes PCA such an effective tool for feature extraction. Principal components also provide qualitative clues to key underlying molecular constituent differences and relative abundances, since their spectral characteristics, often are the key features in the second, third and higher principal components.

PCA is widely used in food chemistry studies [37] and specifically for analysis of food spoilage. For example, in 2020 Saleem et al. [38] presented a new method for predicting microbial spoilage and detecting its location in bakery goods using HSI. HSI cameras monitored baked goods over a period of time as they were allowed to spoil. PCA was applied to difference images created by subtracting images collected at the beginning of the monitoring period, when the goods were fresh, and images collected at later times. The researchers then used PCA to separate pixels representing spoiled portions of the good from unspooled portions.

3.1.2 Sparse representation

Similar in concept to PCA, sparse representation methods are mathematical processes applied to data with the goal of transforming the data to a new representation containing as few non-zero elements as possible. This is achieved by conducting a trade-off between goodness-of-fit and sparsity. The transformation attempts to produce an accurate reproduction of the original data but is regulated with a cost penalty the increases with the number of non-zero components. Example sparse representation algorithms include basis pursuit, sparse dictionary learning, L1-regularization, and non-negative matrix factorization (NMF) [39].

With respect to HSI, sparsity can also be enforced through wavelength selection processes that identify a small number of information-rich wavelengths and discard all other wavelengths. Lei and Sun [40] developed a sparse coefficients wavelength selection and regression (SCWR) method for NIR spectral calibration to select the wavelengths that contributed most to the determination of the spectral response. They applied this method to a dataset if NIR spectra from potatoes with dehydration loss as the response variable. A model based on 23 selected wavelengths (from an original set of 200) predicted hydration loss with an accuracy that exceeded those yielded by common competing methods.

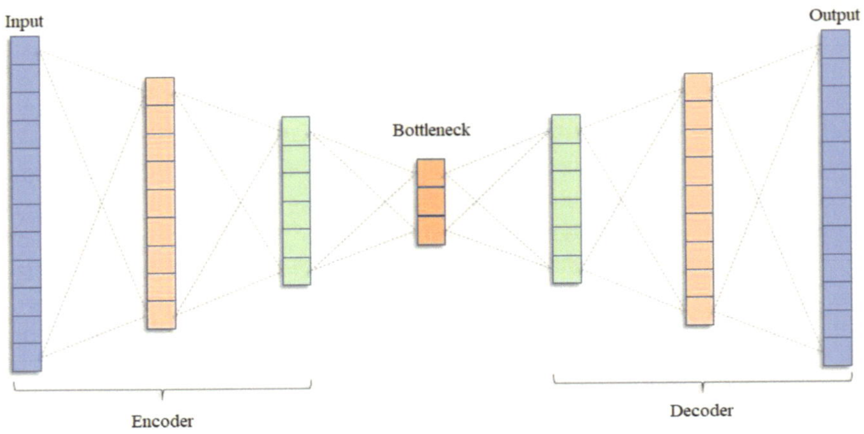

Figure 9.
Autoencoder neural network showing bottleneck which separates the encoder and decoder portions.

3.1.3 Autoencoders

Benefitting from recent advancements in both algorithms and processing technology, neural networks and their derivatives have experienced rapid development over the past decade. Another method for dimensionality reduction and feature extraction is based on a particular type of neural network called an autoencoder. An autoencoder network contains input (e,g., spectral measurement) and output layers of the same size but includes hidden layers in between with gradually decreasing numbers of nodes (see **Figure 9**). During training, the network weights are updated until the output is the same as the input within an acceptable tolerance. The layers from the input to the bottleneck center thus effectively encode a compressed version of the input signal. This set of layers is referred to as the "encoder" section. The compressed signal can then be uncompressed in the later layers (called the "decoder") to form a copy of the signal in the output layer. This process has the added benefit of removing noise in the input during encoding such that the decoded copy is more representative of the true response. In 2021, Vasafi et al. [41] made an initial application of an autoencoder in the field of food production process control by using it to detect anomalies such as changes in fat, temperature, added water, and cleaning solution during milk processing. Anomalies were found to result in significantly higher reconstruction error at the autoencoder output layer as compared with the control (i.e., "normal") data.

3.1.4 Partial least squares regression (PLSR)

PLSR is a well-known and often used means of conducting regression in the presence of noise. Regression provides a function that predicts a response from a data input (as opposed to classification which assigns the input to a class). While both PCA and PLSR are derived from experimental data, PCA is more qualitative by nature, often used in an exploratory manner, and is an unsupervised learning method. PLSR on the other hand is more quantitative and is a multi-dimensional evaluation that is linear. Both methods rely on computing a maximum covariance, PCA in the original data and PLSR in the data and response.

PLSR works best when the observed variables are highly correlated and noisy [42], which is a benefit in hyperspectral analysis where data at nearby wavelengths can be highly correlated. Also, PLSR assumes that the data set is linear and that that projection holds in a new subspace. However, if linearity does not hold up for the model, there are several ways to deal with this problem that include polynomials, splines or small neural nets [43]. There is also an easy way to deal with non-linearity [44], the basic idea being to expand the data matrix with square, cubic, or cross product terms.

PLSR continues to be a popular tool for food analysis and quality control applications. Jiang et al. [45] invoked PLSR to model the relationship between NIR spectra from hyperspectral images of chicken to total *Pseudomonas* spp. and *Enterobacteriaceae* counts (PEC) to predict PEC rapidly. Cavaglia et al. [46] similarly applied PLSR to predict density and pH in ATR-MIR spectra from alcoholic fermentation samples.

3.1.5 Wavelets

Wavelets are mathematical functions that, like Fourier analysis, transform data into its constituent spectral components. However, unlike the standard Fourier transform, wavelet transforms can provide frequency information for specific locations in the temporal or spatial domain. Wavelets of different shapes (called mother wavelets) focus on different portions of the frequency spectrum and are typically used in combination to analyze the full spectral bandwidth of concern. Each mother wavelet can also be rescaled to form daughter wavelets to change the resolution in the temporal or spatial domain and thus examine higher or lower portions of the frequency spectrum in more detail. Some wavelet examples are shown in **Figure 10**.

Wavelet analysis has been used in spectroscopic applications as a means of extracting useful features from specific regions of the spectra. For example, Qi et al. [47] applied a single wavelet form at seven different scales to extract features from shortwave infrared (SWIR) hyperspectral reflectance images from peanuts. Using these features, they were able to distinguish moldy portions of peanuts from healthy portions. Wavelet analysis can also be applied in the image domain to extract 2D features. Ji et al. [48] applied wavelet transforms to hyperspectral visible-NIR images of potatoes (after first applying PCA) to decompose the original images into sub-band images at different scales to extract textural features that would enable the identification of bruising in the potatoes.

3.1.6 Splines

Splines are piecewise linear or polynomial functions that are combined to approximate a given set of data. Splines are often used as smoothing functions to approximate data curves while eliminating the "roughness" caused by noise. Like the other techniques discussed above, splines benefit the feature extraction process by focusing on the true signal within the data.

One application of splines common to chemometric analysis is the regression analysis method of multivariate adaptive regression splines (MARS) [49]. MARS

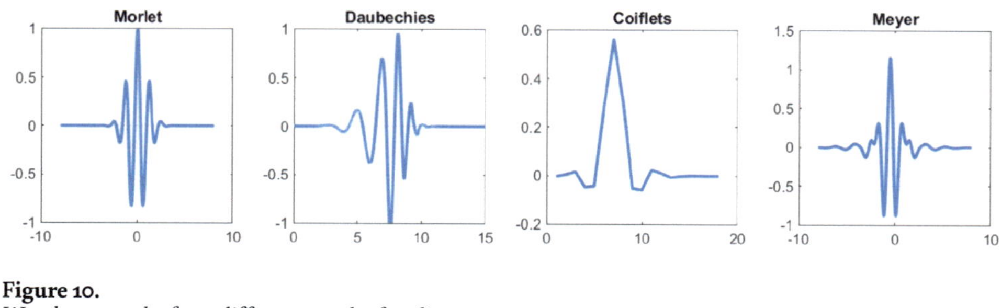

Figure 10.
Wavelet examples from different wavelet families.

models nonlinearities in data by fitting splines to specific regions of the input variable range. The regions are separated by "hinge" functions that have a value of zero for all locations except within the region of applicability. The transition points that link consecutive splines are called "knots." In a forward process knots or splines are added to yield a close fit to the data. In a backward process the least contributing terms are pruned to minimize overfitting. Garre et al. [50] compared a model developed using MARS to similar regression models developed to predict the amount of waste in food production and quantify model uncertainties. The MARS model achieved a precision comparable to that of more sophisticated machine learning models such as random forest methods developed to deal with the high variability in decision trees while maintaining low bias [51].

Closely related to spline regression is Savitzky–Golay filtering in which the data points are convolved with a set of filter weights, much like a weighted moving average. However, as the filter moves to each successive data point, a polynomial of degree p is fit to the data within the filter window, and the point in the center of the window is replaced by the polynomial value at that point [52]. One key benefit of Savitzky–Golay filtering is that it tends to preserve high frequency signal components while rejecting high frequency noise (often found in CCD or InGaAs arrays or photon-starved detection systems) whereas standard finite impulse response (FIR) filters tend to remove these signal components [53].

Savitzky–Golay filtering is a popular pre-processing technique that has been used extensively for food spectral analysis. Examples since 2020 include the use of Savitzky–Golay filtering in pre-processing NIR spectra to improve classification performance in the identification of allergens in powdered food materials [54], filter noise from FTIR spectra of instant freeze-dried coffee and MIR spectra of fruit puree samples [53], and NIR reflectance spectra of Indonesia rice flour-based food to enable accurate classification and level estimation of added sweeteners [55].

3.2 Classification

Automatic detection of food spoilage requires an algorithm that can successfully classify a food product (or part of a food product) as spoiled or healthy, either by detecting the presence of contaminants or by classifying physical changes to the product. A variety of sophisticated machine learning algorithms have been developed over the past few decades to provide accurate classification, and many of these have been used in spectroscopic and food quality applications. Here, we discuss two of the most popular classification algorithms, the support vector machine (SVM) and the artificial neural network, both of which take as input a set of features that are typically generated using the methods described in the previous section. We also discuss deep learning methods, which have advanced rapidly since 2012 and are being used in a wide variety of applications including food analysis. Unlike more conventional machine learning methods, deep learning methods include their own automatic feature extraction process [56].

3.2.1 Support vector machines (SVM)

An SVM is a supervised learning algorithm that seeks to find the separating hyperplane between data points of different classes that minimizes classification error. The position of the hyperplane is determined by the set of points (called "support vectors") that are closest to it. The basic concept of the SVM is intuitive when the hyperplane is linear and the classification is binary (see **Figure 11**).

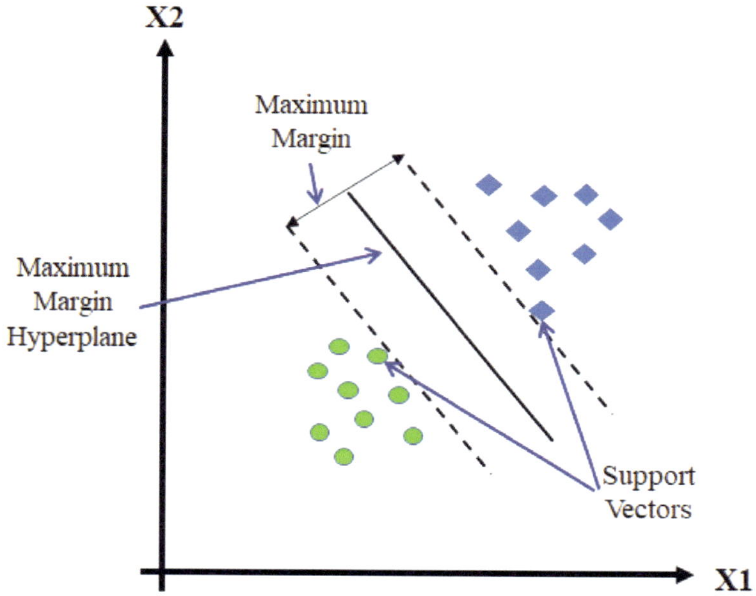

Figure 11.
Separating hyperplane determination for a Support Vector Machine. The hyperplane is positioned to maximize the margin between the support vectors.

However, SVMs can also be applied to data whose classes are not linearly separable by transforming the data from the original space into one in which they are linearly separable. This is often accomplished using the so-called "kernel trick" in which a kernel function compares vectors in the new space without performing the actual transformation, thus minimizing computation cost. Common kernels include linear, radial (i.e., Gaussian), and polynomial.

A reliable and robust machine learning classifier, the SVM has been used in many hyperspectral imaging food analysis applications. A few examples since 2020 include the detection of spoilage in visible-NIR imagery of baked goods [38], detection of bacterial foodborne pathogens in visible-NIR imagery [57], and detection of fish fillet substitution and mislabeling through accurate classification of fillet species from imagery collected from visible-NIR, fluorescence with UV excitation, SWIR, and Raman spectral bands [58].

3.2.2 Artificial neural networks

Artificial neural networks are another popular supervised classification method that has been surging in popularity with the advances in processing technology over the past few decades. Conventional artificial neural networks are based on the multilayer perceptron (MLP) architecture (see **Figure 12**) which was designed to resemble neurons in the brain. Such neurons accept some number of input values and remain dormant until the sum of inputs rises above a certain threshold value, at which point the neurons "fire." This nonlinear thresholding effect is enabled in artificial neural network nodes by nonlinear activation functions that determine each node's output value. Common activation functions include the sigmoid, hyperbolic tangent, and rectified linear unit functions.

Artificial neural networks are trained by initializing the network weights (usually with random values) and comparing the predicted results at the output layer to known target values. An error metric is calculated based on the difference between the prediction and target values, and the network weights are updated by

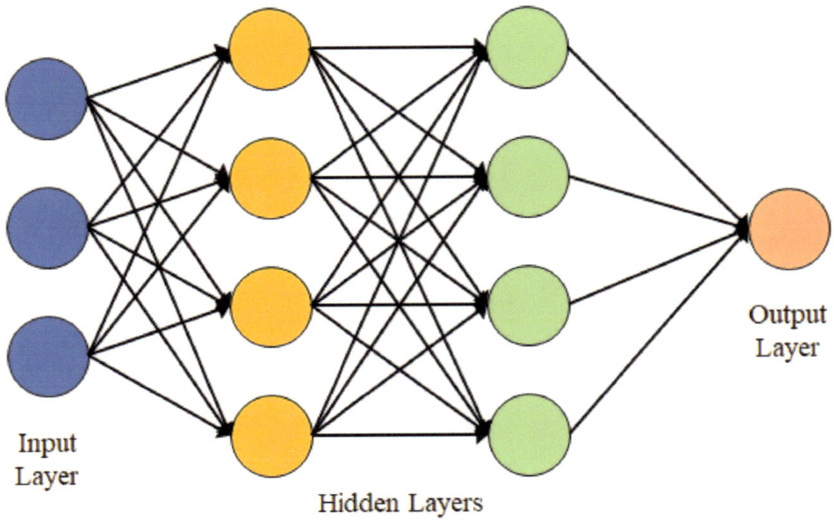

Figure 12.
A simple illustration of a multi-layer perceptron neural network architecture. Each circle represents a neural network node and each arrow represents the weight that connects a node in one layer to a node in the subsequent layer.

calculating partial derivatives of the error with respect to each weight, starting with the output layers and moving backward toward the input layer in a process known as backpropagation. This entire process is repeated until the error is brought below an acceptable tolerance or other stopping criteria are met.

Like SVMs, artificial neural networks have been widely used for spectral classification applications due to their ability to achieve high accuracy. In 2020, Balabanov et al. [59] developed a vison-based system with an artificial neural network to detect defects in apples passing on a conveyer belt by analyzing HSI in the visible and NIR spectral ranges. Although once popular, these MLP-based neural networks are rapidly being replaced by deep learning neural networks which not only offer superior performance, but also ease the data processing pipeline by eliminating the need for manual feature selection and extraction.

Prior to the advent of modern sophisticated processing technology, neural networks were limited in size due to their computational loads which grew with the number of layers and the number of nodes within each layer. As this processing technology advanced, more and more layers could be added to neural networks to improve their performance (although the theoretical reason for why this is the case is still poorly understood). Furthermore, as shown in **Figure 13**, layers could be added to perform different operations on the data, such as convolution and averaging (often called "pooling" in this context). The network then performs feature extraction by learning the weights in the convolutional layers which yield accurate classifications. In essence, the network learns which filters should be applied to the data to best extract the signal within. Pooling layers following the convolutional layers then apply averaging to help prevent overfitting. Following the successive convolutional and pooling (and possibly other) layers, the results are concatenated into a single-dimensional vector and fed into an MLP neural network to combine these features for classification.

In 2021 alone, deep learning neural networks have been used to classify beef freshness from visible-NIR reflectance spectra [60], to analyze NIR HSI to detect the presence of contamination during food packing [61], and to conduct a series of different food quality analyses from NIR spectra [62].

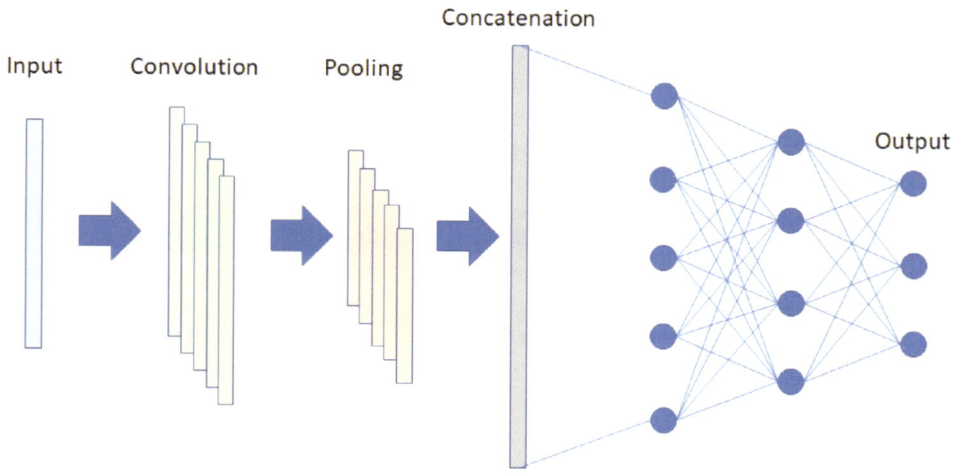

Figure 13.
Basic CNN architecture. Data at the input layer is passed through a convolutional layer which generates feature sets. These are then reduced in size through averaging in the pooling layer and the resulting features are concatenated. The final layers form an MLP-based neural network to yield the final classifications.

4. Food traceability and dynamic pricing

4.1 Inadequacies of existing traceability technologies

Traceability and real time analysis of food products that can help minimize waste will require new tools for quality assurance, authentication and digital supply chain management that can track products from harvest to market. Such tools must be objective, verifiable, provide data on quality, provenance, and freshness, and easy to incorporate at multiple nodes in the supply chain. Current technologies are inadequate to address most of these challenges and address only some components of the most difficult problems.

State of the art seafood traceability platforms provide tools for establishing chain of custody but lack dynamic pricing features or the verifiable and trusted freshness and authenticity data. These current platforms rely on estimates of shelf life based on catch date and storage conditions. These inputs are insufficiently verifiable and quantifiable for digital tools based on them to be broadly trusted and accepted for dynamic pricing, and they do not address authentication and quality metrics or capabilities at all.

As of 2021, dynamic pricing software solutions also lack higher quality verifiable and trusted freshness and authenticity data and are primarily designed for final retail discounting, often integrated only into broader retailer systems. This makes them less effective for application to upstream supply chain node tasks and adding value for each node in the supply chain.

Products for quantitative measurement of fish freshness rely primarily on destructive laboratory-based methods that are not capable of accurate spot checks of individual fish or fish portions or cannot be realistically and easily repeated at low cost at multiple points along the supply chain. Tools are available that measure tissue conductivity through fish skin, primarily to assess moisture, but they are not designed to address broader nutrient content, species, and traceability.

4.2 SafetySpect's quality, adulteration, and traceability (QAT) technology

One approach to addressing the problem of traceability and rapid detection of spoilage is SafetySpect's newly developed handheld QAT scanner that optically detects previously established chemical signatures of seafood freshness, quality, and species ID. This device integrates several types of spectroscopic data through its fusion-AI algorithm into simple, human readable reports. The handheld scanner will enable spot-checks of quality, species ID, and freshness all along the supply chain. Integration with a mobile device and app can couple the QAT output to blockchain-enabled, cloud-based, supply chain management platforms for tracking product quality, freshness, and species ID from harvest to market. When integrated with a digital platform using blockchain and dynamic pricing technology, these tools will support the modernization of the global fishing and seafood processing industries, supply chains and retail outlets, as well as provide accurate information to consumers about the sustainability, freshness, and quality of their seafood purchases. Specially designed apps can also integrate smallholder producers in developing economies into the broader, emerging digital supply chain platforms in these markets.

By providing trusted product and pricing data at any node of the supply chain, QAT fundamentally changes the business models of seafood processors, wholesalers and retailers, making it practical to (a) identify mislabeled product, and (b) dynamically price perishable seafood at multiple purchase decision points – beyond traditional final-discounting by retailers. The quantitative underlying data provides high confidence and additional visibility and trust in the freshness of such a highly perishable good, and makes intelligent pricing based on quality/freshness practical at all nodes along the supply chain before final retail sale.

With this capability, QAT will spur innovation in the execution of seafood supply chains by providing accountability at each node for maintaining quality. This will drive improvements in purchase decision making, traceability, authentication, and inventory planning. Retailers can use dynamic pricing in both their purchase and final sale decision making. Given the high proportion of seafood sales attributable to the largest retailers in both developed and developing markets, retailers can have significant economic incentive to adopt such technologies, and the power to encourage its adoption by upstream suppliers.

Technologies like SafetySpect QAT will have four major impacts on the global seafood industry: (1) Reduce waste by tracking fish freshness, thus enabling vastly improved freshness-based dynamic pricing tools at multiple supply chain nodes; (2) Increased visibility and trusted information; (3) Increased value of seafood across markets and positive impact on public health by providing assurance of quality, species, and freshness; (4) Positive impact on a number of sustainability goals including better management of at-risk wild fisheries, combatting illegal poaching of fish and wild animal species, improving food security, and providing better economic engagement and access for marginalized smallholder producers to broader digital supply chain systems and platforms, reducing resource consumption and combatting climate change.

5. Conclusion

Food waste is a global problem caused in large part by food spoilage that has gone undetected. This problem exacerbates world hunger issues and affects consumers by causing them to pay too high a price for food products whose shelf lives have been improperly or inadequately estimated. Several technologies have been applied to improve the detection of food spoilage and provide better valorization of food products at all points along the food supply chain, but many of these methods are either unreliable or damage the samples being evaluated. Spectroscopic techniques, on the other hand, offer a reliable and non-invasive means of detecting and quantifying spoilage. Recent developments in fundamental spectroscopic technologies have enabled the development and productization of portable and handheld devices for conducting analysis of food products in-situ. Furthermore, algorithmic advancements have improved our ability to extract the most relevant features from the spectroscopic data and yield highly accurate classifications and quantifications of spoilage. These technologies, in combination with advancements such as blockchain, used in conjunction with technologies like SafetySpect's QAT scanner, offer the promise to reduce food waste and extend shelf lives through detection of spoilage at earlier points along the food supply chain and will provide the ability to impose intelligent pricing and traceability tracking for the benefit of consumers around the globe.

Author details

John Chauvin[1], Ray Duran[1], Stanley Ng[1], Thomas Burke[2], Kenneth Barton[3], Nicholas MacKinnon[4], Kouhyar Tavakolian[1], Alireza Akhbardeh[4] and Fartash Vasefi[3]*

1 University of North Dakota, Grand Forks, ND, USA

2 SafetySpect, London, United Kingdom

3 SafetySpect, Los Angeles, CA, USA

4 Clinical Research, SafetySpect Inc., Los Angeles, CA, USA

*Address all correspondence to: fvasefi@safetyspect.com

References

[1] Love DC, Fry JP, Milli MC, Neff RA. Wasted seafood in the United States: Quantifying loss from production to consumption and moving toward solutions. Glob Environ Change. 2015 Nov 1;35:116-24.

[2] The State of World Fisheries and Aquaculture 2020 [Internet]. FAO; 2020 [cited 2020 Nov 14]. Available from: http://www.fao.org/documents/card/en/c/ca9229en

[3] Wang K, Pu H, Sun D-W. Emerging Spectroscopic and Spectral Imaging Techniques for the Rapid Detection of Microorganisms: An Overview. Compr Rev Food Sci Food Saf. 2018;17(2):256-73.

[4] Vasefi F, Booth N, Hafizi H, Farkas DL. Multimode Hyperspectral Imaging for Food Quality and Safety. Hyperspectral Imaging Agric Food Environ [Internet]. 2018 Nov 5 [cited 2020 Jun 14]; Available from: https://www.intechopen.com/books/hyperspectral-imaging-in-agriculture-food-and-environment/multimode-hyperspectral-imaging-for-food-quality-and-safety

[5] Feng Y-Z, Sun D-W. Application of hyperspectral imaging in food safety inspection and control: a review. Crit Rev Food Sci Nutr. 2012;52(11):1039-58.

[6] Sensory assessment scoresheets for fish and shellfish - Torry & QIM [Internet]. Seafish. [cited 2020 Nov 7]. Available from: https://www.seafish.org/document/?id=c551ce18-93d1-41c1-ae82-82185632cd98

[7] Liu D, Zeng X-A, Sun D-W. Recent developments and applications of hyperspectral imaging for quality evaluation of agricultural products: a review. Crit Rev Food Sci Nutr. 2015;55(12):1744-57.

[8] Elmasry G, Barbin DF, Sun D-W, Allen P. Meat quality evaluation by hyperspectral imaging technique: an overview. Crit Rev Food Sci Nutr. 2012;52(8):689-711.

[9] Foca G, Ferrari C, Ulrici A, Sciutto G, Prati S, Morandi S, et al. The potential of spectral and hyperspectral-imaging techniques for bacterial detection in food: A case study on lactic acid bacteria. Talanta. 2016 Jun 1;153:111-9.

[10] Ellis DI, Muhamadali H, Haughey SA, Elliott CT, Goodacre R. Point-and-shoot: rapid quantitative detection methods for on-site food fraud analysis – moving out of the laboratory and into the food supply chain. Anal Methods. 2015;7(22):9401-14.

[11] Yan H, Siesler HW. Hand-held near-infrared spectrometers: State-of-the-art instrumentation and practical applications. NIR News. 2018 Nov;29(7):8-12.

[12] Crocombe RA. Portable Spectroscopy. Appl Spectrosc. 2018 Dec 1;72(12):1701-51.

[13] Kademi HI, Ulusoy BH, Hecer C. Applications of miniaturized and portable near infrared spectroscopy (NIRS) for inspection and control of meat and meat products. Food Rev Int. 2019 Apr 3;35(3):201-20.

[14] Masna N, Paul SD, Chen C, Mandal S, Bhunia S. Eat, but Verify: Low-Cost Portable Devices for Food Safety Analysis. IEEE Consum Electron Mag. 2018 Dec 11;8.

[15] Santos CAT dos, Lopo M, Páscoa RNMJ, Lopes JA. A Review on the Applications of Portable Near-Infrared Spectrometers in the Agro-Food Industry. Appl Spectrosc. 2013 Nov 1;67(11):1215-33.

[16] Part 4: The Optical Bench - B&W Tek [Internet]. [cited 2020 Aug 29]. Available from: https://bwtek.com/spectrometer-part-4-the-optical-bench/

[17] Oliveres R. What is a Laser Spectrum Analyzer? [Internet]. My Laser Spectrum. 2016 [cited 2020 Aug 29]. Available from: http://mylaserspectrum.com/2016/08/laser-spectrum-analyzer/

[18] NIRONE SENSOR | Spectral Engines | High-performance, compact and reliable NIR spectral sensor | Acal BFi BE [Internet]. [cited 2020 Aug 29]. Available from: https://www.acalbfi.com/be/Photonics/Spectroscopy/MEMS-based-VIS-IR-spectrometers/p/High-performance--compact-and-reliable-NIR-spectral-sensor/0000008RR9

[19] INC TT ERIC BALTZ and ROGER KIRSCHNER, RESEARCH ELECTRO-OPTICS. For Compactness and Ruggedness, Linear Variable Filters Fit the Bill [Internet]. [cited 2020 Aug 29]. Available from: https://www.photonics.com/Articles/For_Compactness_and_Ruggedness_Linear_Variable/a60941

[20] Pederson CG, Friedrich DM, Hsiung C, von Gunten M, O'Brien NA, Ramaker H-J, et al. Pocket-size near-infrared spectrometer for narcotic materials identification. In: Druy MA, Crocombe RA, editors. Baltimore, Maryland, USA; 2014 [cited 2020 Aug 30]. p. 91010O. Available from: http://proceedings.spiedigitallibrary.org/proceeding.aspx?doi=10.1117/12.2050019

[21] MicroNIR OnSite-W [Internet]. 2017 [cited 2020 Aug 30]. Available from: https://www.viavisolutions.com/en-us/osp/products/micronir-onsite-w

[22] Gat N, Pottebaum T, Scriven G, Brandt R, Systems O-K. Hadamard Transform Imaging Spectrometry (HTIS) and Compressive Detection Techniques. :19.

[23] Du Y, Zhou G. A MEMS-driven Hadamard transform spectrometer. In: MOEMS and Miniaturized Systems XVII [Internet]. International Society for Optics and Photonics; 2018 [cited 2020 Aug 29]. p. 105450X. Available from: https://www.spiedigitallibrary.org/conference-proceedings-of-spie/10545/105450X/A-MEMS-driven-Hadamard-transform-spectrometer/10.1117/12.2292808.short

[24] Liu N, Parra HA, Pustjens A, Hettinga K, Mongondry P, van Ruth SM. Evaluation of portable near-infrared spectroscopy for organic milk authentication. Talanta. 2018 Jul 1;184:128-35.

[25] Fourier-Transform Spectrometers [Internet]. [cited 2020 Aug 28]. Available from: https://spie.org/publications/tt61_123_fourier_transform_spectrometers?SSO=1

[26] FTIR Spectroscopy Basics - US [Internet]. [cited 2020 Aug 28]. Available from: //www.thermofisher.com/us/en/home/industrial/spectroscopy-elemental-isotope-analysis/spectroscopy-elemental-isotope-analysis-learning-center/molecular-spectroscopy-information/ftir-information/ftir-basics.html

[27] Spectral Sensors [Internet]. NeoSpectra Sensors. 2018 [cited 2020 Aug 30]. Available from: https://www.neospectra.com/our-offerings/product-overview/

[28] Ausili A, Sánchez M, Gómez-Fernández JC. Attenuated total reflectance infrared spectroscopy: A powerful method for the simultaneous study of structure and spatial orientation of lipids and membrane proteins. Biomed Spectrosc Imaging. 2015 Jan 1;4(2):159-70.

[29] Ocean MZ5 - Mini ATR MIR Spectrometer Features | SpectrEcology [Internet]. Spectrecology - Spectroscopy

& Optical Sensing Solutions. [cited 2020 Aug 30]. Available from: https://www.spectrecology.com/mz5-attenuated-total-reflectance-mid-infrared-spectrometer/

[30] Bureau S, Cozzolino D, Clark CJ. Contributions of Fourier-transform mid infrared (FT-MIR) spectroscopy to the study of fruit and vegetables: A review. Postharvest Biol Technol. 2019 Feb;148:1-14.

[31] The Raman Spectrophotometer [Internet]. [cited 2020 Aug 30]. Available from: https://www.sas.upenn.edu/~crulli/TheRamanSpectrophotometer.html

[32] Mira M-1 Advanced Package [Internet]. [cited 2020 Aug 30]. Available from: https://www.metrohm.com/en-us/products-overview/spectroscopy/mira-handheld-raman-spectrometers/29230020

[33] Raman FAQs - What laser wavelengths are used for Raman spectroscopy? - HORIBA [Internet]. [cited 2021 Mar 28]. Available from: https://www.horiba.com/uk/scientific/products/raman-spectroscopy/raman-academy/raman-faqs/what-laser-wavelengths-are-used-for-raman-spectroscopy/

[34] Moore's Law Turns 50- How Much Longer Can it Last? : vTools Events [Internet]. [cited 2020 Nov 8]. Available from: https://meetings.vtools.ieee.org/m/35279

[35] FPGAs & 3D ICs [Internet]. Xilinx. [cited 2020 Nov 8]. Available from: https://www.xilinx.com/products/silicon-devices/fpga.html

[36] Kaiser HF. The application of electronic computers to factor analysis. Educ Psychol Meas. 1960;20:141-51.

[37] Granato D, Santos JS, Escher GB, Ferreira BL, Maggio RM. Use of principal component analysis (PCA) and hierarchical cluster analysis (HCA) for multivariate association between bioactive compounds and functional properties in foods: A critical perspective. Trends Food Sci Technol. 2018 Feb 1;72:83-90.

[38] Saleem Z, Khan MH, Ahmad M, Sohaib A, Ayaz H, Mazzara M. Prediction of Microbial Spoilage and Shelf-Life of Bakery Products Through Hyperspectral Imaging. IEEE Access. 2020;8:176986-96.

[39] Zhang Z, Xu Y, Yang J, Li X, Zhang D. A survey of sparse representation: algorithms and applications. IEEE Access. 2015;3:490-530.

[40] Lei T, Sun D-W. A novel NIR spectral calibration method: Sparse coefficients wavelength selection and regression (SCWR). Anal Chim Acta. 2020 May;1110:169-80.

[41] Vasafi PS, Paquet-Durand O, Brettschneider K, Hinrichs J, Hitzmann B. Anomaly detection during milk processing by autoencoder neural network based on near-infrared spectroscopy. J Food Eng. 2021 Jun;299:110510.

[42] Rosipal R. Nonlinear Partial Least Squares: An Overview. Chemoinformatics Adv Mach Learn Perspect Complex Comput Methods Collab Tech. 2010 Jan 1;

[43] Björk A. Chemometric and signal processing methods for real time monitoring and modeling: applications in the pulp and paper industry. 2007.

[44] Berglund A, Wold S. INLR, implicit non-linear latent variable regression. J Chemom. 1997;11(2):141-56.

[45] Jiang S, He H, Ma H, Chen F, Xu B, Liu H, et al. Quick assessment of chicken spoilage based on hyperspectral NIR spectra combined with partial least

squares regression [Internet]. 2021 [cited 2021 Mar 26]. Available from: / paper/Quick-assessment-of-chicken-spoilage-based-on-NIR-Jiang-He/ 94d06b673cd5e0eaf04 52144c9ee30d2e9e25b44

[46] Cavaglia J, Schorn-García D, Giussani B, Ferré J, Busto O, Aceña L, et al. ATR-MIR spectroscopy and multivariate analysis in alcoholic fermentation monitoring and lactic acid bacteria spoilage detection. Food Control. 2020 Mar;109:106947.

[47] Qi X, Jiang J, Cui X, Yuan D. Moldy Peanut Kernel Identification Using Wavelet Spectral Features Extracted from Hyperspectral Images. Food Anal Methods. 2020 Feb;13(2):445-56.

[48] Ji Y, Sun L, Li Y, Ye D. Detection of bruised potatoes using hyperspectral imaging technique based on discrete wavelet transform. Infrared Phys Technol. 2019 Dec;103:103054.

[49] Friedman JH. Multivariate Adaptive Regression Splines. Ann Stat. 1991 Mar;19(1):1-67.

[50] Garre A, Ruiz MC, Hontoria E. Application of Machine Learning to support production planning of a food industry in the context of waste generation under uncertainty. Oper Res Perspect. 2020;7:100147.

[51] Breiman L. Random Forests. Mach Learn. 2001 Oct 1;45(1):5-32.

[52] Press WH, Teukolsky SA. Savitzky-Golay Smoothing Filters. Comput Phys. 1990;4(6):669.

[53] Kew H. A model for spectroscopic food sample analysis using data sonification. Int J Speech Technol [Internet]. 2021 Jan 13 [cited 2021 Mar 27]; Available from: http://link.springer.com/10.1007/s10772-020-09794-9

[54] Rady A, Fischer J, Reeves S, Logan B, James Watson N. The Effect of Light Intensity, Sensor Height, and Spectral Pre-Processing Methods When Using NIR Spectroscopy to Identify Different Allergen-Containing Powdered Foods. Sensors. 2020 Jan;20(1):230.

[55] Masithoh RE, Rondonuwu FF, Setyabudi FMCS, Cho BK. Development of calibration model for determination of sweeteners additives in Indonesia rice flour-based food by FT-NIR spectroscopy. IOP Conf Ser Earth Environ Sci. 2020 Aug 7;542:012017.

[56] Hastie T, Tibshirani R, Friedman J. The Elements of Statistical Learning: Data Mining, Inference, and Prediction, Second Edition [Internet]. 2nd ed. New York: Springer-Verlag; 2009 [cited 2020 Nov 8]. (Springer Series in Statistics). Available from: https://www.springer.com/gp/book/9780387848570

[57] Bonah E, Huang X, Yi R, Aheto JH, Yu S. Vis-NIR hyperspectral imaging for the classification of bacterial foodborne pathogens based on pixel-wise analysis and a novel CARS-PSO-SVM model. Infrared Phys Technol. 2020 Mar;105:103220.

[58] Qin J, Vasefi F, Hellberg RS, Akhbardeh A, Isaacs RB, Yilmaz AG, et al. Detection of fish fillet substitution and mislabeling using multimode hyperspectral imaging techniques. Food Control. 2020 Aug 1;114:107234.

[59] Balabanov PV, Divin AG, Egorov AS, Yudaev VA, Lyubimova DA. Vision system for detection of defects on apples using hyperspectral imaging coupled with neural network and Haar cascade algorithm. IOP Conf Ser Mater Sci Eng. 2020 May 28;862:052058.

[60] Shin S, Lee Y, Kim S, Choi S, Kim JG, Lee K. Rapid and non-destructive spectroscopic method for classifying beef freshness using a deep spectral network fused with myoglobin

information. Food Chem. 2021 Aug;352:129329.

[61] Medus LD, Saban M, Francés-Víllora JV, Bataller-Mompeán M, Rosado-Muñoz A. Hyperspectral image classification using CNN: Application to industrial food packaging. Food Control. 2021 Jul;125:107962.

[62] Zeng J, Guo Y, Han Y, Li Z, Yang Z, Chai Q, et al. A Review of the Discriminant Analysis Methods for Food Quality Based on Near-Infrared Spectroscopy and Pattern Recognition. Mol Basel Switz. 2021 Feb 1;26(3).

Section 3

Innovation in the Food Sector

Chapter 7

Innovation in the Seafood Sector through the Valorization of By-Products

Marzieh Moosavi-Nasab, Najme Oliyaei, Jong-Bang Eun and Armin Mirzapour-Kouhdasht

Abstract

Aquatic, marine and algae, is reservoir of bioactive compounds, which have considerable potential to supply novel ingredients toward the development of commercial functional food products. Meanwhile, several valuable by-products generate during the manufacturing process. Seafood is still an intact reservoir of valuable compounds with significant potential to provide unique compounds applicable in functional food development. Seafood, as an important part of the diet all around the world, can be used as a source of functional components that are positively affecting the human health. Annually, 50–80 percent of the seafood processing is discarded as waste every year. Algae are also the novel natural resources for their biological and pharmacological properties. This chapter will be discussing the innovations in seafood and algae sector through the valorization of their by-products. Firstly, protein production, its characterization and the protein hydrolysates derived from seafood will be reviewed. Subsequently, bioactivity of the peptides obtained from these protein hydrolysates and other bioactive compounds such as carotenoid compounds derived from seafood including fish, shrimp, alga, and so on will be included. Finally, the main components of algae including sulfated polysaccharides, pigments and proteins will be surveyed.

Keywords: seafood by-products, algae by-products, bioactive compounds, protein, pigments, carotenoids, sulfated polysaccharides

1. Introduction

It is well-known that the seafood has been one of the most important parts of the human nutrition for a long time. According to reports obtained from FAO, the annual discard from global marine capture between 2010 and 2014 was 9.1 million tons. This huge amount of by-products represents 10.8% (10.1% –11.5%) of the annual average catch of 2010 to 2014 [1]. Utilizing this discarded part of the fishery industries could be environmentally and economically profitable.

Several value added products can be generated from seafood processing by-products depending on which kind of seafood is processed. Based on this, this chapter is divided into 3 major parts; (I) fish by-products, (II) crustaceans, and (III) seaweeds. This study has provided a review of use of fish by-products to produce some value added products including proteins, peptides, and oil. These products are the most important major products that have a promising future in global market. During last decades, different efforts have been done to utilize the seafood by-products to generate these value added products [2]. Obtaining proteins and peptides as functional and nutritional compounds from seafood by-products have been the objective of many researches [3–9].

Algae are an important renewable source of food, medicines and fertilizers and their utilization have increased in all around the world. They are considered to possess a high nutritional value and their metabolites, and associated biological activities, have particular significance for multiple nutraceutical, cosmetic and pharmaceutical applications [10, 11]. Seaweed consumption has a long tradition in Asian countries and has increased in European countries in over recent decades, due to increased awareness of their beneficial effects [12]. Thus, development of way for the utilization of marine algae for food, feed, and bioenergy is essential. One of the best way is conversion of biomass into a variety of valuable products which is known as biorefinery [13].

In recent years, numerous compounds with biological activities or pharmacological properties such as antibacterial, anti-inflammatory, anticancer, antiviral and anticoagulant are discovered in algae. Algae by-products can be used for human and animal as food, animal feed and ingredients of dietary supplements. Sulfated polysaccharides, pigments, proteins and lipid are the main by-products of algae [12].

This chapter focuses on important value added bioactive chemicals identified in seafood by products over the last years and describes the range of biological activities as well as industrial applications for which they are responsible.

2. Fish by-products

2.1 Proteins

Fish by-products obtained from seafood processing industries contain huge amounts of head, skin, scales, bones, fins, viscera, and dark muscle. The protein content of these by-products is approximately 15%, which is similar to that of fish fillets. The muscle which is attached to this by-product contains two distinct type of proteins including structural (myofibrillar) (approximately 70–80%) and sarcoplasmic proteins (approximately 20–30%). These high nutritional value proteins (even more than red meat and milk casein) indicate remarkable functional and technological properties like water holding capacity, emulsifying activity, film forming ability, foam forming capacity, and gel forming ability [14–17]. Commercial gelatins are mostly obtained from mammalian (porcine and bovine) skins and bones. As the researches confirm, the substitution of mammalian gelatin with fish gelatin is an appropriate and appealing due to increasing concerns of researchers and consumers about the risks of transmission of the pathogenic vectors such as prions. Albeit, number of committees like the Scientific Steering Committee of the European Union, have stated that consumption of bovine bone gelatin is safe [18], researchers are still debating on this.

Nowadays, researches have become to notice on a unique protein which can be easily extracted from fish by-products especially skin, scales, bones, and fins. This valuable protein is collagen/gelatin. Collagen is the most abundant protein in tissues including skin and bones (approximately 30% of the total protein). The structural investigates show that collagen is a triple helix with three identical polypeptide chains. The primary structure of this protein is continuous repeating of the Gly-X-Y-sequence. The positions of X and Y are mostly proline and hydroxyproline, respectively. Different types of collagen (29 distinct types) have been discovered

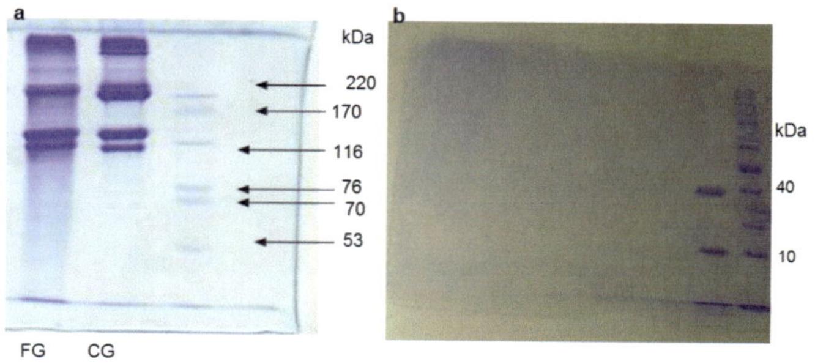

Figure 1.
Molecular weight distribution analysis by SDS-PAGE for gelatins. CG (commercial gelatin) and FG (fish wastes gelatin) (a) and for protease (b). Adapted from [23].

so far, which have right-handed triple helical conformation. The difference among these types is due to the variety in their amino acid sequences as a result of genetic variants [19–21]. Fish gelatin could be extracted from its by-products by a partially denaturation of collagen usually performed by hot water. Before extraction of fish by-products, some pretreatments are needed to ready them for being used as a gelatin source. The pretreatment step is ordinarily an alkaline and/or and acidic swelling process. The alkaline and/or acidic pretreatment is used to partial cleavage of rigid cross-links in the collagen and remove non-collagenous materials. The enzymatic aided chemical pretreatments are those which can be supplemented or replaced by enzymatic reaction. The "conditioning process" is the known name of this step by manufacturers of gelatin. Afterward, the gelatin (warm water soluble) will be extracted from collagen (not soluble) by hot water at a specific temperature and time. There are lots of studies performed in this research area. In a paper authored by Mirzapour-Kouhdasht, Moosavi-Nasab [22], gelatin was optimized at different levels of time and temperature using the response surface methodology (RSM). The responses including yield, protein content, gel strength, and viscosity indicated that the optimum conditions were 70.71°C and 5.85 h. Rheological, structural, and functional experiments showed that the gelatin characteristics were acceptable compared to the commercial bovine gelatin. The pretreatment in these experiments was performed by alkaline solution. In another study [23], gelatin was produced from Common carp wastes using alkaline protease from *Bacillus licheniformis* PTCC 1595. The enzymatic reaction was performed in 5, 10, 15, 20, and 25 units per gram of wastes. The molecular weight distribution of the gelatin (**Figure 1**) showed that this gelatin could be successively replace the commercial gelatin.

In some researches also fish gelatin is modified by some functional groups or chemical agents to improve the functional characteristics. In a study performed by [24], rheological, emulsifying, and structural properties of phosphorylated fish gelatin was investigated. The results of this study revealed that phosphorylation in a short time, enhances gel and rheological behavior of fish gelatin. Phosphorylation could improve the emulsions stability of fish gelatin as well. Authors stated that the structural properties of fish gelatin were significantly affected by this modification **Figure 2**.

2.2 Peptides

Peptides obtained from seafood processing by-products have been reported to have potent biological activities including antioxidant activity [25–31], antihypertensive, anticancer, anti-inflammatory, and anticoagulant properties [22, 32–37]. Among all these researches, the use of gelatin derived from fish by-products has

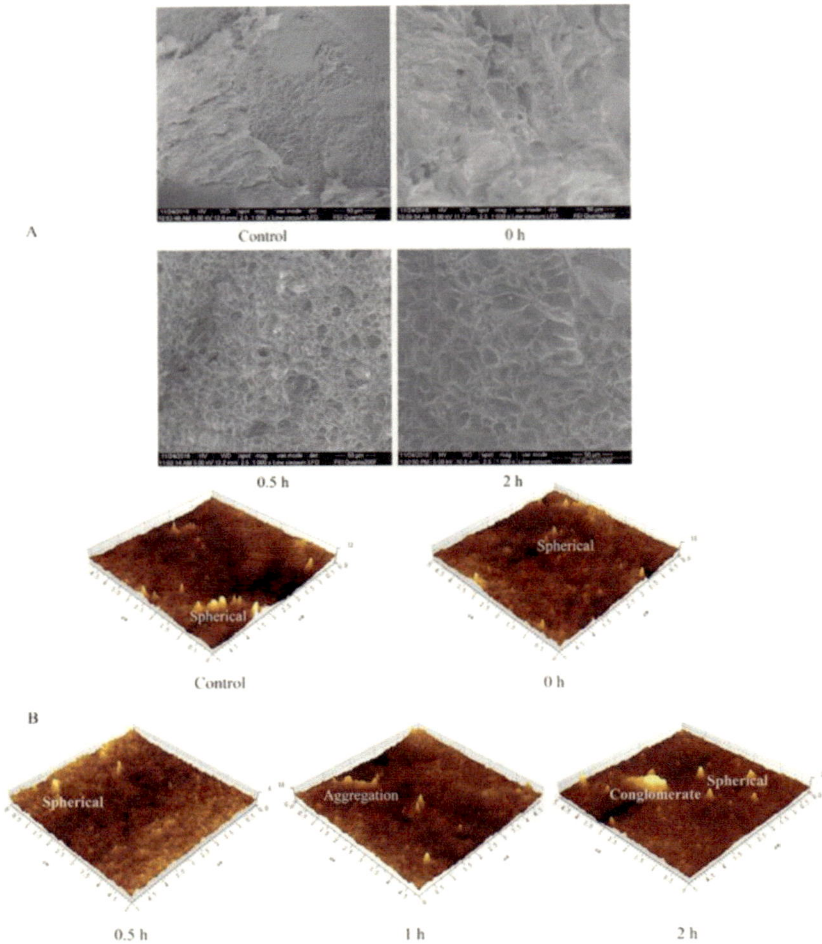

Figure 2.
Micrographs of control and phosphorylated fish gelatin. SEM (A) and AFM (B). Adapted from [24].

been well investigated as a source of bioactive peptides with various biological activities. In a study performed by Jin, Teng [38], salmon skin collagen was hydrolyzed by different proteolytic enzymes including pepsin, trypsin, papain, and Alcalase 2.4 L. Hydrolysates obtained from trypsin hydrolysis reaction indicated the highest dipeptidyl peptidase IV (DPP-IV) inhibitory activity (66.12%). After fractionation and identification processes, a bioactive peptide with sequence of LDKVFR for DPP-IV inhibitory activity was detected to be responsible for this activity (IC50 value of 0.1 ± 0.03 mg/mL). In another research conducted by Mirzapour-Kouhdasht and Moosavi-Nasab [39], gelatin extracted from *Scomberomorus commerson* skin in combination with its hydrolysates obtained by Actinidin from kiwifruit was used to extent the shelf-life of whole shrimp (*Penaeus merguiensis*). The results revealed that the gelatin hydrolysates can be applied as a preservative coating agent for whole shrimp.

2.3 Oil

Nowadays, of the most important nutritional substances which have gained much attention are Omega-3 long-chain polyunsaturated fatty acids (LCPUFA). These LCPUFA are necessary for human and animal physiology due to their structural and regulatory functions [40]. Fish by-products are a good natural source of LCPUFA, especially EPA (eicosapentaenoic acid) and DHA (docosahexaenoic acid). Fish oil is rich in vitamins (E, D, A). Due to these valuable components, fish oil consumption could be a promising way to impede some health risks such as inflammation, coronary heart diseases, obesity, arthritis, autoimmune disorders, and cancer [41–44].

Generally, the extraction of oils from fish by-products can be divided in two categories including conventional and modern methods. Generally, in conventional methods the raw material (fish by-products obtained from fish processing industries) are first cooked. After the cooking, the by-products are sieved followed by pressing for oil extraction. Subsequently the extracted slurry is decanted and the oil is stored in oil storing tanks [45].

In comparison with conventional extraction method, the modern extraction methods such as supercritical fluid extraction (SFE) could be useful for reducing the oxidation of LCPUFA. In a research performed by Rubio-Rodríguez and coworkers [46], SFE method with carbon dioxide under moderate conditions (25 MPa and 313 K) was used to extract oil from different fish by-products. They resulted that SFE is an advantageous method for oil extraction from fish by-products. The authors stated that the SFE can impede lipid oxidation and reduce extraction of impurities. In another study conducted by Sabzipour and others [47], quality of rainbow trout (*Oncorhynchus mykiss*) by-products oil was investigated. However, the aim of this study was to determine the effect of different postmortem processing times and blanching methods. The authors presented that the degradation of fish by-products oil occurs faster than the fish tissue oil. So they surveyed the effect of different treatments on the quality of the fish by-products oil. According to their report, salt blanching could decrease the effects of delayed processing and led to a higher quality.

However, the limitation of fish oil for utilization in food and pharmaceutical industries is related to the low stability and strong fishy flavor. The solution for this problem is to encapsulate the fish oil using different strategies to cover the off-flavor and also increase the stability. In a research performed by Drusch et al. [48], fish oil with was microencapsulated by spray-drying in a matrix of n-octenylsuccinate-derivatized starch and sugars. The results of this study indicated that this protocol can increase the oxidative stability of fish oil without any significant changes in physicochemical properties of the oil such as particle size, oil droplet size, and true density. Another study conducted by Chen et al. [49], the fish oil co-encapsulated with phytosterol ester and limonene, prepared by spray-drying and freeze-drying methods. The wall material used for encapsulation were whey protein isolate and soluble corn fiber. Sensory analysis of the encapsulated fish oil showed that the addition of limonene could cover the fishy flavor. The authors also reported that this procedure could significantly enhance the oxidative stability of the fish oil during 168 h of storage.

3. Crustaceans

3.1 Proteins and peptides

Tremendous amounts of shrimp processing by-products (head and body carapace) are discarded annually, which could be an important source of bioactive molecules. The amount of by-products generated during processing is about 48–56% of the whole shrimp depending on the species. The major composition of these by-products are protein (35–50%), polysaccharide (predominantly chitin) (15–25%), minerals (10–15%), and a few percent carotenoids [50]. Recently production of bioactive peptides from shrimp by-products has gained attentions. Several researchers found that this source of by-products could be a good one to generate bioactive peptides with especial activities such as angiotensin converting enzyme inhibitory (ACE inhibitory) [51, 52], antimicrobial activity [53], antioxidant activity [52, 54], etc. More investigations are required to characterize the biological and functional properties of these peptides.

3.2 Chitin

The major value added product obtained from crustaceans is chitin which has the second position among frequent and used biopolymers in the world after cellulose [55, 56]. In fact, chitin is a polymer of β-(1 → 4)- *N* -acetyl- D–glucosamine units which is extracted mainly from shrimp and crabs. This polysaccharide could be found in arthropods exoskeleton or in the cell walls of fungi and yeast as the major prominent structural component [57–65]. Chitosan is a linear polysaccharide derived from chitin deacetylation [66]. Chitin and chitosan have attained lots of attentions due to their non-toxicity, biocompatibility, biodegradability, and low cost [56, 67]. Chitosan is known as a biologically active component in many fields such as food and pharmaceutical applications. A number of activities of this polysaccharide such as making delivery systems [68], tissue engineering [69], food packaging and film forming [70, 71], and antimicrobial and wound healing [72] are investigated.

One of the most important characteristics of chitosan which can affect its pharmaceutical and functional properties is the degree of acetylation. In case of designing delivery systems, the molecular weight of this bioactive molecule becomes more important due to changing the encapsulation efficiency [73]. It is very important to know that chitosan has a higher solubility in lower pH values due to protonation of the amino groups of the molecule [74]. Permeation enhancers substances can increase the absorption of encapsulated biological active compounds in the gastrointestinal tract. One of the mechanisms of this action is opening the tight junctions of the epithelium cells [75, 76]. Chitosan has a mucoadhesive nature and capable to open epithelial connections (tight junctions) of the epithelium cells [77, 78]. **Figure 3** shows a schematically the action place of permeation enhancers to increase the absorbance of bioactive components in gastrointestinal tract.

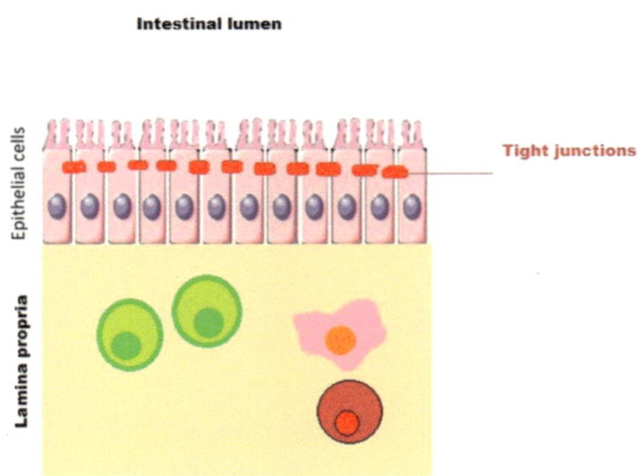

Figure 3.
The action place of permeation enhancers to increase the absorbance of bioactive components in gastrointestinal tract.

4. Algae

4.1 Sulfated polysaccharides

Phycocolloids or hydrocolloids are polysaccharides have been one of the most accessible and widely used in food industry as thickening and gel forming agent. Indeed, numerous sulfated polysaccharides from algae including agars, carrageenans and fucoidan (**Figure 4**) are the main bioactive components that have been determined to possess significant various biological activities [79].

Agar is polysaccharide comprised of two major components, agarose and agaropectin and has been extracted from seaweeds for industrial purposes in pharmaceutical, cosmetics and food industry as gelling and thickening agent [80]. The commercially used seaweeds for the extraction of agar are mainly *Gracilaria* and *Gelidium* species [81].

In addition, carrageenan is another linear sulfated polysaccharides that extracted from red seaweed and exhibits several applications in food industries as gelling, thickening, and emulsifying attributes, clarification of beer and wines. Carrageenan mainly obtain from two algae *Kappaphycus* and *Eucheuma* [82].

Fucoidans, a complex sulfated groups with fucose which found mainly in cell-wall matrix of brown macroalgae [83]. In addition to fucose, fucoidan contain other monosaccharides such as glucose, galactose, rhamnose, xylose, mannose and uronic acids [84]. Numerous brown seaweeds have been used for fucoidan extraction including *Sargassum* [85, 86], *Undaria* [87], *Laminaria* [88], *Cladosiphon* [89], *Fucus* [90], *Saccharina* [91] and *Ascophyllum* [92]. Several investigations have been confirmed the biological activities of fucoidan including antitumor, anticoagulant, antioxidant, immunomodulatory, anti-inflammatory, antiviral, antithrombotic, and hepatoprotective effects [93, 94]. This bioactivity of fucoidan is depend on its molecular weight, the monosaccharide composition, the sulfate content, the position of the sulfate ester group, the extraction technique, and fucoidan structure [94]. Thus, several extraction techniques are used such as conventional methods (hot water) [95] and non-conventional methods such as pressurized liquid extraction [84], ultrasound [96], enzyme assisted [90], microwave assisted [97] and subcritical water [91] extraction.

Subsequently, the green algae *Monostroma nitidum* is the commercial source of a sulfated polysaccharide named rhamnan sulfate [98]. Rhamnan sulfate found in

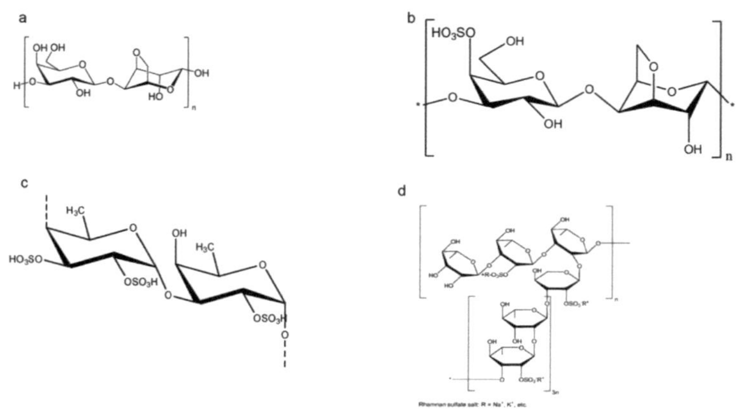

Figure 4.
The chemical structure of (a) agar; (b) carrageenan, (c) fucoidan and (d) Rhamnan sulfate.

cell wall of *M. nitidum* and structurally consists of rhamnose with a sulfate-group substituent that forms main chains with branched side chains [98, 99].

This polysaccharide is extracted by hot water, though is poorly water soluble [100]. Several studies exhibit its biological activities such as antiviral, anticoagulant, antitumor, anti-inflammatory, anti-hypercholesterolemic, anti-obesity and anti-hypertensive properties. Further, *M. nitidum*-derived rhamnan sulfate is considered to promote the human health [100].

Calcium spirulan (Ca-SP) is another novel sulfated polysaccharide isolated from blue-green alga Spirulina platensis. Ca-SP is an attractive candidate therapeutic agent for viral infectious diseases because of its antivirus and antitumor activities [101, 102].

4.2 Pigments

4.2.1 Carotenoids

Carotenoids and chlorophylls are generally wasted together with the residual biomass during the extraction of phycocyanin or sulfated polysaccharide, while can isolate as valuable product from algae [103].

Carotenoids are the most widespread class of pigments that are characterized as natural colorant and antioxidants with healthy effects including anti-cancer, anti-diabetic anti-obesity and eye diseases. The bio-functional properties of algal carotenoids make them potentially to use in nutraceuticals, cosmeceuticals and feed supplements in aquaculture sectors. Carotenoids divided into primary and secondary based on their metabolism and function. Primary carotenoids are structural and functional components in the photosynthetic apparatus, which take direct part in photosynthesis. Secondary carotenoids refer to extra-plastidic pigments produced in large quantities, through carotenogenesis, after exposure to specific environmental stimuli [104].

Microalgae are a potential renewable resource of primary and secondary carotenoids. α-carotene, β-carotene, lutein, fucoxanthin, violaxanthin, zeaxanthin, and neoxanthin, are characterized as primary carotenoids while astaxanthin, canthaxanthin, and echinenone are secondary carotenoids. Astaxanthin, zeaxanthin, fucoxanthin and lutein receive much attention as commercial carotenoids [104].

Seaweeds are the important sources of bioactive compounds which have several human health benefits. The most predominant seaweed carotenoids, such as fucoxanthin, lutein, β-carotene and siphonaxanthin have remarkable biological functions and applications [105]. Pigments are waste during the polysaccharide extraction process. Thus, carotenoids are recovered from microalgae and seaweeds by different approaches including conventional solvent extraction, non-conventional methods including pulsed electric field [106, 107], moderate electric field [108], supercritical fluid extraction [109], pressurized liquid extraction [110], microwave ssisted extraction [111, 112], ultrasound assisted extraction [113], high pressure homogenization [114].

Fucoxanthin ($C_{42}H_{58}O_6$) is the predominant carotenoid in brown algae (*Sargassum angustifolium*, *Laminaria japonica* and *Undaria pinnatifida*) and some microalgae (*Phaeodactylum tricornutum*, *Isochrysis galbana*, *Odontella aurita*) that accounting for more than 10% of the estimated total natural production of carotenoids. This yellowish-brown pigment exhibit remarkable biological properties, including anticancer, anti-inflammatory, antiobesity and neuroprotective activity [115–117]. Moreover, fucoxanthin extraction can be by-product of fucoidan extraction process as Yip et al., [118] obtained the fucoxanthin-rich extract from *S. binderi* with yield of 7.4 ± 0.4 mg/g.

Astaxanthin as king of antioxidant is found in microalgae such as *Haematococcus*. *H. pluvialis* is rich in astaxanthin and provide a natural and inexpensive source of astaxanthin [119]. The antioxidant activity of astaxanthin is 100 and 10 times greater than those of vitamin E and β-carotene. Moreover, astaxanthin has a superior preventive effect toward photo-oxidative compared with canthaxanthin, and β-carotene [120].

4.2.2 Phycobiliproteins

Phycobiliproteins are natural fluorescent dyes which participate in photosynthesis. These pigments are assembled large, distinct granules as phycobilisomes, which are attached to the thylakoid membrane of chloroplast. These pigment-protein complex plays an important role in light-harvesting in cyanobacteria, red algae cryptomonads, glaucophytes and some pyrrophyceae [121, 122]. Phycobiliproteins are divided into two main groups; phycoerythrin (PE –bright pink red), phycocyanin (PC –deep blue). The main components of phycocyanins are C-phycocyanin (C-PC), R-phycocyanin (R-PC), and allophycocyanin (AP – bluish green) [121, 122]. Moreover, there are differences between in their structural position. PE is at the tip of the rod-like phycobilisomes, PC is in the middle, while AP forms a core attached to the reaction and energy transfer proceeds successively from PE to PC to AP and to chlorophyll [123]. The other classification of phycobiliproteins is based on their spectral attributes which including phycoerythrobilin (PEB, A max 560 nm), phycocyanobilin (PCB, A max 620–650 nm), phycobiliviolin (PXB, A max 575 nm) and phycourobilin (PUB, A max 498 nm) [123]. These biopigments have attracted much attention in medicines, foods, cosmetics and fluorescent materials. The recent research has brought attention to the use of phycobiliproteins as food colorant, health drink and coloring agent in confectionary and cosmetics because they are hydrophilic and stable at low temperature with some preservative like citric acid, in acidic and basic solutions [121, 123]. Moreover, phycobiliproteins are used in diagnostic kits in immunology as fluorescent tracer of antibodies [123] and gel electrophoresis and gel exclusion chromatography as marker because of their high molecular absorptivity at visible wavelengths [122].

Phycocyanins have an apparent molecular mass of 140–210 kD and two subunits, α and β [124]. C-Phycocyanin is found in cyanobacteria strains such as *Spirulina* sp. (freshwater), *Phormidium* sp. (marine water) and *Lyngbya* sp. (marine water) [125]. However, the commercial source of this pigment is *Spirulina* which consists of about 20% of the dry weight of this algae [126]. Further, the other new source of phycocyanin is *Anabaena oryzae* SOS13 [124, 127].

Recent studies have demonstrated the role of C-PC as antioxidant, anti-inflammatory, hepatoprotective, and as well as free radical scavenger [128, 129]. Various techniques are used to extract phycocyanin from *Arthrospira platensis* (*Spirulina*) biomass including in various approaches such as physical (freeze–thaw) or an enzymatic (lysozyme) [124], supercritical fluid extraction [130] andsonication and microwave [131].

Phycoerythrin also have numerous health benefits, however, the absorption spectrum of cyanobacteria phycoerythrin is deferent from red algae. The cyanobacteria phycoerythrin exhibits a single peak at 565 nm in the visible wavelength region, while the absorption spectrum of red algae phycoerythrin includes three peaks in the visible wavelength region at 500, 550 and 565 nm (R-phycoerythrin) [123].

Allophycocyanin is a light-harvesting pigment protein complex found mainly in *A. platensis*. This water-soluble pigment is broadly used in biochemical techniques such as a fluorescent probe, especially for flow cytometry. Further, allophycocyanin has promising applications as antioxidative, anti-inflammatory, antitumor, anti-enterovirus and hepatoprotective [132]. Despite its potential biochemical and therapeutic benefits, there are some challenges in its downstream processing including difficulty in primary extraction and purification, containing lower proportion of phycobiliprotein rather than phycocyanin and the resistance of cell membrane to disruption cause extraction of 50–60% of A-PC by conventional methods. Moreover, the main objective of pigment extraction form *spirulina* is C-PC, consequently, remaining high content of A-PC (about 40–50%) in biomass after C-PC extraction [133].

4.3 Proteins

Algae protein waste is a by-product derived from water-extraction process of microalgae, during algae essence manufacturing. The underutilized algae wastes, containing above 50% protein, have low economical value to be used as animal feed. The pepsin hydrolysate from algae protein waste exhibited antioxidative activity in preliminary experiments, indicating that algae waste might become a new protein source for selection of novel antioxidative peptides [134].

Furthermore, protein hydrolysates from marine sources such as algae by-products, have generally been used to produce seafood flavorings. A high flavor quality is difficult to ensure for seafood flavoring that is produced from marine animal sources because of their high susceptibility to lipid oxidation and the high cost of removing excess fat. Seaweed by-products after agar extraction are good sources of plant protein and contain taste-active amino acids, such as aspartic acid, glutamic acid, arginine, and lysine, in addition to a low fat content [135].

A seaweed protein hydrolysate using 10% bromelain for 3 h, resulted in high level of arginine, lysine, and leucine as free amino acids. These amino acids exhibited an umami taste and a seaweed odor [135].

Most microalgae contain high level of protein which discarded or damaged during biofuels production, while are good candidate for protein extraction and consequently, obtain lipid-rich product as by-product as feedstock for biofuels production. Even though proteins are major algae biomass component, usually they are undervalued compared to minor components such as omega fatty acids, pigments or other possible valuable buy-products [136].

For instance, Garcia-Moscoso et al. [136] extracted more than 60 wt% of nitrogen content of *Scenedesmus* sp. by subcritical water medium then the lipid-rich residue used as suitable feedstock for biofuel production.

There are numerous investigations about algae protein waste and extraction of peptides or amino acids with functional properties. For instance, the antioxidative peptide of VECYGPNRPQF was isolated by pepsin from *Chlorella vulgaris*. This peptide had some bioactivity such as DNA protective effect against hydroxyl radicals, gastrointestinal enzyme-resistance, and strong antioxidant properties. Fractionation of proteins exhibited the high level of aspartic acid, glutamic acid, leucine and lysine [134]. This amino acid sequence (VECYGPNRPQF) can act as cheap and natural anticancer peptide because had antiproliferation and induced a post-G1 cell cycle arrest in AGS cells with no cytotoxicity effect in WI-38 lung fibroblasts cells [137].

Moreover, protein isolation, as valuable by-product, from defatted *Nannochloropsis*, can be obtained after lipid extraction during biofuel production. Defatted and non-defatted *Nannochloropsis* contained 56.9% and 40.5% protein respectively. The protein yields by alkaline (pH 11 and 60 C) extraction method were 16% and 30% respectively. These isolated proteins had a high molecular weight approximately 250 kDa [138].

Macroalgae are also a suitable protein source and rich in protein after extraction of their polysaccharide, lipid and polyphenols. Among three seaweed *Porphyra umbilicalis*, *Ulva lactuca*, and *Saccharina latissimi*, the highest protein isolated using pH- shift method (71%) was related to the *P. umbilicalis*. Furthermore, among different extraction methods including pH-shift method, accelerated solvent extraction and sonication in water and precipitation by ammonium sulfate, pH shift process is promising approach. However, the yield and extraction approach are influence by type and species of seaweed [139].

Brown algae such as *Laurencia filiformis*, *L. intricata*, *Gracilaria domingensis* and *Gracilaria. birdiae* can supply dietary proteins for human and animals because their protein content reported 18.3, 4.6, 6.2 and 7.1% respectively [140].

Combination of acid-alkaline process is another protein isolation from algae. First acid and then alkaline extraction is an alternative extraction by 59% protein recovery from brown seaweed *Ascophyllum nodosum*. The obtained protein had about 2–4 kDa molecular weight [141].

5. Conclusions

This chapter indicated that seafood by-products are one of the most important sources of value added products that can play an important role in the global market due to the increasing growth of demands for health beneficiary products. Through this opportunity and based on our research background for many years, we decided to provide important information about some value-added products obtained from seafood by-products. Proteins and peptides are a major part of the seafood by-products composition that can easily provide essential amino acids and bioactive peptides with health beneficent. Fish oil is another valuable product that could be extracted from seafood by-products. This source is rich in LCPUFA and decreases the risks of chronic diseases such as cardiovascular issues, thereby directly related to our health. Marine algae are a versatile, abundant, and valuable source of many compounds that have been widely used for many industries. The presence of bioactive compounds such as sulfated polysaccharide, carotenoid, and protein makes them a suitable candidate in biomedical applications. It seems, they will play an important role in human life because of their broad applications in food, pharmaceutical, and cosmetic industries.

Conflict of interest

The authors declare no conflict of interest.

Author details

Marzieh Moosavi-Nasab[1,2*], Najme Oliyaei[1,2], Jong-Bang Eun[3] and Armin Mirzapour-Kouhdasht[1,2,3*]

1 Seafood Processing Research Group, Shiraz University, Shiraz, Iran

2 Department of Food Science and Technology, Shiraz University, Shiraz, Iran

3 Department of Integrative Food, Bioscience and Biotechnology, Chonnam National University, Gwangju, South Korea

*Address all correspondence to: marzieh.moosavi-nasab@mail.mcgill.ca and armin.mirzapour@shirazu.ac.ir

References

[1] Roda, M.A.P., et al., *A third assessment of global marine fisheries discards*. 2019: Food and Agriculture Organization of the United Nations.

[2] Sila, A. and A. Bougatef, *Antioxidant peptides from marine by-products: Isolation, identification and application in food systems. A review.* Journal of Functional Foods, 2016. **21**: p. 10-26.

[3] Cahú, T.B., et al., *Recovery of protein, chitin, carotenoids and glycosaminoglycans from Pacific white shrimp (Litopenaeus vannamei) processing waste.* Process Biochemistry, 2012. **47**(4): p. 570-577.

[4] Sila, A., M. Nasri, and A. Bougatef, *Isolation and characterisation of carotenoproteins from deep-water pink shrimp processing waste.* International Journal of Biological Macromolecules, 2012. **51**(5): p. 953-959.

[5] Sila, A., et al., *Biochemical and antioxidant properties of peptidic fraction of carotenoproteins generated from shrimp by-products by enzymatic hydrolysis.* Food Chemistry, 2014. **148**: p. 445-452.

[6] Alasalvar, C., et al., *Handbook of seafood quality, safety and health applications*. 2011: John Wiley & Sons.

[7] Kleekayai, T., et al., *Extraction of antioxidant and ACE inhibitory peptides from Thai traditional fermented shrimp pastes.* Food Chemistry, 2015. **176**: p. 441-447.

[8] Mirzapour-Kouhdasht, A., et al., *Antioxidant mechanism, antibacterial activity, and functional characterization of peptide fractions obtained from barred mackerel gelatin with a focus on application in carbonated beverages.* Food Chemistry, 2020: p. 128339.

[9] Oliyaei, N., Ghorbani, M., Moosavi-Nasab, M., Sadeghimahoonak, A. R., & Maghsoudloo, Y. (2017). Effect of temperature and alkaline pH on the physicochemical properties of the protein isolates extracted from the whole ungutted Lanternfish (Benthosema pterotum). Journal of aquatic food product technology, **26**(10), 1134-1143.

[10] Nitschke, U. and D.B. Stengel, *A new HPLC method for the detection of iodine applied to natural samples of edible seaweeds and commercial seaweed food products.* Food Chemistry, 2015. **172**: p. 326-334.

[11] Moosavi-Nasab, M., Mirzapour-Kouhdasht, A., & Oliyaei, N. (2019). Application of essential oils for shelf-life extension of seafood products. In Essential Oils-Oils of Nature. IntechOpen.

[12] Ambrozova, J.V., et al., *Influence of extractive solvents on lipid and fatty acids content of edible freshwater algal and seaweed products, the green microalga Chlorella kessleri and the cyanobacterium Spirulina platensis.* Molecules, 2014. **19**(2): p. 2344-2360.

[13] Lange, L. and A.S. Meyer, *Potentials and possible safety issues of using biorefinery products in food value chains.* Trends in Food Science & Technology, 2019. **84**: p. 7-11.

[14] Ghaly, A., et al., *Fish Processing Wastes as a Potential Source of Proteins.* Amino acids and oils: a critical review. J Microb Biochem Technol, 2013. **5**(4): p. 107-129.

[15] Menon, V.V. and S.S. Lele, *Nutraceuticals and bioactive compounds from seafood processing waste*, in *Springer handbook of marine biotechnology*. 2015, Springer. p. 1405-1425.

[16] Halim, N., H. Yusof, and N. Sarbon, *Functional and bioactive properties of fish protein hydolysates and peptides: A comprehensive review.* Trends in Food

Science & Technology, 2016. **51**: p. 24-33.

[17] Oliyaei, N., M. Moosavi-Nasab, and M. Ghorbani, *Effect of salt and alkaline on the physicochemical properties of the protein isolates extracted from lanternfish (Benthosema pterotum).* Iranian Journal of Fisheries Sciences, 2019. **18**(2): p. 371-385.

[18] Karim, A. and R. Bhat, *Fish gelatin: properties, challenges, and prospects as an alternative to mammalian gelatins.* Food Hydrocolloids, 2009. **23**(3): p. 563-576.

[19] Mirzapour-Kouhdasht, A., et al., *Physicochemical, rheological, and molecular characterization of colloidal gelatin produced from Common carp by-products using microwave and ultrasound-assisted extraction.* Journal of Texture Studies, 2019. **50**(5): p. 416-425.

[20] Pal, G.K., T. Nidheesh, and P. Suresh, *Comparative study on characteristics and in vitro fibril formation ability of acid and pepsin soluble collagen from the skin of catla (Catla catla) and rohu (Labeo rohita).* Food Research International, 2015. **76**: p. 804-812.

[21] Gómez-Guillén, M., et al., *Functional and bioactive properties of collagen and gelatin from alternative sources: A review.* Food Hydrocolloids, 2011. **25**(8): p. 1813-1827.

[22] Mirzapour-Kouhdasht, A., et al., *Optimization of gelatin production from Barred mackerel by-products: Characterization and hydrolysis using native and commercial proteases.* Food Hydrocolloids, 2020: p. 105970.

[23] Kouhdasht, A.M., M. Moosavi-Nasab, and M. Aminlari, *Gelatin production using fish wastes by extracted alkaline protease from Bacillus licheniformis.* Journal of Food Science and Technology, 2018. **55**(12): p. 5175-5180.

[24] Huang, T., et al., *Rheological behavior, emulsifying properties and structural characterization of phosphorylated fish gelatin.* Food Chemistry, 2018. **246**: p. 428-436.

[25] Kim, S.-Y., J.-Y. Je, and S.-K. Kim, *Purification and characterization of antioxidant peptide from hoki (Johnius belengerii) frame protein by gastrointestinal digestion.* The Journal of Nutritional Biochemistry, 2007. **18**(1): p. 31-38.

[26] Qian, Z.-J., et al., *Protective effect of an antioxidative peptide purified from gastrointestinal digests of oyster, Crassostrea gigas against free radical induced DNA damage.* Bioresource Technology, 2008. **99**(9): p. 3365-3371.

[27] Rajapakse, N., et al., *Purification and in vitro antioxidative effects of giant squid muscle peptides on free radical-mediated oxidative systems.* The Journal of Nutritional Biochemistry, 2005. **16**(9): p. 562-569.

[28] Kumar, N.S., R. Nazeer, and R. Jaiganesh, *Purification and biochemical characterization of antioxidant peptide from horse mackerel (Magalaspis cordyla) viscera protein.* Peptides, 2011. **32**(7): p. 1496-1501.

[29] Yang, X.-R., et al., *Preparation and characterization of gelatin and antioxidant peptides from gelatin hydrolysate of skipjack tuna (Katsuwonus pelamis) bone stimulated by in vitro gastrointestinal digestion.* Marine Drugs, 2019. **17**(2): p. 78.

[30] Abuine, R., A.U. Rathnayake, and H.-G. Byun, *Biological activity of peptides purified from fish skin hydrolysates.* Fisheries and Aquatic Sciences, 2019. **22**(1): p. 10.

[31] Hemker, A.K., et al., *Effects of pressure-assisted enzymatic hydrolysis on functional and bioactive properties of tilapia (Oreochromis niloticus)*

by-product protein hydrolysates. LWT, 2020. **122**: p. 109003.

[32] Kim, S.-K. and I. Wijesekara, *Development and biological activities of marine-derived bioactive peptides: A review.* Journal of Functional Foods, 2010. **2**(1): p. 1-9.

[33] Najafian, L. and A.S. Babji, *A review of fish-derived antioxidant and antimicrobial peptides: their production, assessment, and applications.* Peptides, 2012. **33**(1): p. 178-185.

[34] Rajanbabu, V. and J.-Y. Chen, *Applications of antimicrobial peptides from fish and perspectives for the future.* Peptides, 2011. **32**(2): p. 415-420.

[35] Huang, C.-Y., et al., *Characterization and antioxidant and angiotensin I-converting enzyme (ACE)-inhibitory activities of gelatin hydrolysates prepared from extrusion-pretreated milkfish (Chanos chanos) scale.* Marine Drugs, 2018. **16**(10): p. 346.

[36] Ling, Y., S. Liping, and Z. Yongliang, *Preparation and identification of novel inhibitory angiotensin-I-converting enzyme peptides from tilapia skin gelatin hydrolysates: Inhibition kinetics and molecular docking.* Food & Function, 2018. **9**(10): p. 5251-5259.

[37] Kouhdasht, A.M. and M. Moosavi-Nasab, *Bioactive peptides derived from fish by-product collagen.* International Journal of Environmental Sciences & Nartural Resources, 2018. **13**(2): p. 555859.

[38] Jin, R., et al., *Identification of novel DPP–IV inhibitory peptides from Atlantic salmon (Salmo salar) skin.* Food Research International, 2020: p. 109161.

[39] Mirzapour-Kouhdasht, A. and M. Moosavi-Nasab, *Shelf-life extension of whole shrimp using an active coating containing fish skin gelatin hydrolysates produced by a natural protease.* Food Science & Nutrition, 2020. **8**(1): p. 214-223.

[40] Bannenberg, G., et al., *Omega-3 long-chain polyunsaturated fatty acid content and oxidation state of fish oil supplements in New Zealand.* Scientific Reports, 2017. **7**(1): p. 1-13.

[41] Mezzomo, N., et al., *Evidence of anti-obesity and mixed hypolipidemic effects of extracts from pink shrimp (Penaeus brasiliensis and Penaeus paulensis) processing residue.* The Journal of Supercritical Fluids, 2015. **96**: p. 252-261.

[42] Minihane, A.M., et al., *Consumption of fish oil providing amounts of eicosapentaenoic acid and docosahexaenoic acid that can be obtained from the diet reduces blood pressure in adults with systolic hypertension: a retrospective analysis.* The Journal of Nutrition, 2016. **146**(3): p. 516-523.

[43] von Schacky, C., *The role of omega-3 fatty acids in cardiovascular disease.* Current Atherosclerosis Reports, 2003. **5**(2): p. 139-145.

[44] Sveinsdottir, K., E. Martinsdottir, and A. Ramel, *Blood pressure-lowering effects of long chain n-3 fatty acids from meals enriched with liquid fish oil and from microencapsulated powder.* International Journal of Food Sciences and Nutrition, 2016. **67**(8): p. 1017-1023.

[45] Tacon, A.G. and M. Metian, *Global overview on the use of fish meal and fish oil in industrially compounded aquafeeds: Trends and future prospects.* Aquaculture, 2008. **285**(1-4): p. 146-158.

[46] Rubio-Rodríguez, N., et al., *Supercritical fluid extraction of fish oil from fish by-products: A comparison with other extraction methods.* Journal of Food Engineering, 2012. **109**(2): p. 238-248.

[47] Sabzipour, F., et al., *Effect of various postmortem processing times and*

blanching methods on quality of rainbow trout (Oncorhynchus mykiss) waste oil. Food Science & Nutrition, 2019. **7**(9): p. 3093-3102.

[48] Drusch, S., et al., *Physicochemical characterization and oxidative stability of fish oil encapsulated in an amorphous matrix containing trehalose.* Food Research International, 2006. **39**(7): p. 807-815.

[49] Chen, Q., et al., *Properties and stability of spray-dried and freeze-dried microcapsules co-encapsulated with fish oil, phytosterol esters, and limonene.* Drying Technology, 2013. **31**(6): p. 707-716.

[50] Sachindra, N. and N. Mahendrakar, *Process optimization for extraction of carotenoids from shrimp waste with vegetable oils.* Bioresource Technology, 2005. **96**(10): p. 1195-1200.

[51] Cheung, I.W. and E.C. Li-Chan, *Angiotensin-I-converting enzyme inhibitory activity and bitterness of enzymatically-produced hydrolysates of shrimp (Pandalopsis dispar) processing byproducts investigated by Taguchi design.* Food Chemistry, 2010. **122**(4): p. 1003-1012.

[52] He, H., et al., *Preparation and functional evaluation of oligopeptide-enriched hydrolysate from shrimp (Acetes chinensis) treated with crude protease from Bacillus sp. SM98011.* Bioresource Technology, 2006. **97**(3): p. 385-390.

[53] Bartlett, T.C., et al., *Crustins, homologues of an 11.5-kDa antibacterial peptide, from two species of penaeid shrimp, Litopenaeus vannamei and Litopenaeus setiferus.* Marine Biotechnology, 2002. **4**(3): p. 278-293.

[54] Huang, G., Z. Ren, and J. Jiang, *Separation of iron-binding peptides from shrimp processing by-products hydrolysates.* Food and Bioprocess Technology, 2011. **4**(8): p. 1527-1532.

[55] Kim, S.-K., et al., *Seafood processing by-products*. 2016: Springer.

[56] Marzieh, M.-N., et al., *Comparison of the physicochemical and structural characteristics of enzymatic produced chitin and commercial chitin.* International Journal of Biological Macromolecules, 2019. **139**: p. 270-276.

[57] Alishahi, A. and M. Aïder, *Applications of chitosan in the seafood industry and aquaculture: a review.* Food and Bioprocess Technology, 2012. **5**(3): p. 817-830.

[58] Andres, Y., et al., *Antibacterial effects of chitosan powder: mechanisms of action.* Environmental Technology, 2007. **28**(12): p. 1357-1363.

[59] Gerente, C., et al., *Removal of arsenic (V) onto chitosan: From sorption mechanism explanation to dynamic water treatment process.* Chemical Engineering Journal, 2010. **158**(3): p. 593-598.

[60] Goosey, M. and R. Kellner, *Recovery of copper from PCB manufacturing effluent using chitin and chitosan.* Circuit World, 2012.

[61] Kim, I.-Y., et al., *Chitosan and its derivatives for tissue engineering applications.* Biotechnology Advances, 2008. **26**(1): p. 1-21.

[62] Kwok, K.C. and G. McKay, *Novel batch reactor design for the adsorption of arsenate on chitosan.* Journal of Chemical Technology & Biotechnology, 2010. **85**(12): p. 1561-1568.

[63] Rong, C., et al., *Combined effect of ozonated water and chitosan on the shelf-life of Pacific oyster (Crassostrea gigas).* Innovative Food Science & Emerging Technologies, 2010. **11**(1): p. 108-112.

[64] Senevirathne, M. and S.-K. Kim, *Utilization of seafood processing by-products: medicinal applications*, in *Advances in Food and Nutrition Research*. 2012, Elsevier. p. 495-512.

[65] Wang, S.-L., T.-W. Liang, and Y.-H. Yen, *Bioconversion of chitin-containing wastes for the production of enzymes and bioactive materials.* Carbohydrate Polymers, 2011. **84**(2): p. 732-742.

[66] Shariatinia, Z., *Pharmaceutical applications of chitosan.* Advances in Colloid and Interface Science, 2019. **263**: p. 131-194.

[67] Khan, G., et al., *Tinidazole functionalized homogeneous electrospun chitosan/poly (ε-caprolactone) hybrid nanofiber membrane: development, optimization and its clinical implications.* International Journal of Biological Macromolecules, 2017. **103**: p. 1311-1326.

[68] Shariatinia, Z. and Z. Zahraee, *Controlled release of metformin from chitosan–based nanocomposite films containing mesoporous MCM-41 nanoparticles as novel drug delivery systems.* Journal of Colloid and Interface Science, 2017. **501**: p. 60-76.

[69] Ahsan, S.M., et al., *Chitosan as biomaterial in drug delivery and tissue engineering.* International Journal of Biological Macromolecules, 2018. **110**: p. 97-109.

[70] de Moraes Crizel, T., et al., *Active food packaging prepared with chitosan and olive pomace.* Food Hydrocolloids, 2018. **74**: p. 139-150.

[71] Shariatinia, Z. and M. Fazli, *Mechanical properties and antibacterial activities of novel nanobiocomposite films of chitosan and starch.* Food Hydrocolloids, 2015. **46**: p. 112-124.

[72] Ferreira, M.O.G., et al., *Chitosan Hydrogel in combination with Nerolidol for healing wounds.* Carbohydrate Polymers, 2016. **152**: p. 409-418.

[73] Sannan, T., K. Kurita, and Y. Iwakura, *Studies on chitin, 2. Effect of deacetylation on solubility.* Die Makromolekulare Chemie: Macromolecular Chemistry and Physics, 1976. **177**(12): p. 3589-3600.

[74] Kumar, M.R., et al., *Chitosan chemistry and pharmaceutical perspectives.* Chemical Reviews, 2004. **104**(12): p. 6017-6084.

[75] Cao, S.-j., et al., *Nanoparticles: oral delivery for protein and peptide drugs.* AAPS PharmSciTech, 2019. **20**(5): p. 190.

[76] Perry, S.L. and D.J. McClements, *Recent Advances in Encapsulation, Protection, and Oral Delivery of Bioactive Proteins and Peptides using Colloidal Systems.* Molecules, 2020. **25**(5): p. 1161.

[77] Fernández-Urrusuno, R., et al., *Enhancement of nasal absorption of insulin using chitosan nanoparticles.* Pharmaceutical Research, 1999. **16**(10): p. 1576-1581.

[78] Vila, A., et al., *Low molecular weight chitosan nanoparticles as new carriers for nasal vaccine delivery in mice.* European Journal of Pharmaceutics and Biopharmaceutics, 2004. **57**(1): p. 123-131.

[79] Rupérez, P., E. Gómez-Ordóñez, and A. Jiménez-Escrig, *Biological activity of algal sulfated and nonsulfated polysaccharides.* Bioactive compounds from marine foods: Plant and animal sources, 2013: p. 219-247.

[80] Vuai, S.A.H. and F. Mpatani, *Optimization of agar extraction from local seaweed species, Gracilaria salicornia in Tanzania.* Phycological Research, 2019. **67**(4): p. 261-266.

[81] Arvizu-Higuera, D.L., et al. *Effect of alkali treatment time and extraction time on agar from Gracilaria vermiculophylla.* in *Nineteenth International Seaweed Symposium*. 2007. Springer.

[82] Milledge, J.J., B.V. Nielsen, and D. Bailey, *High-value products from*

macroalgae: the potential uses of the invasive brown seaweed, Sargassum muticum. Reviews in Environmental Science and Bio/Technology, 2016. **15**(1): p. 67-88.

[83] Xin, L., et al., *Extraction, fractionation, and chemical characterisation of fucoidans from the brown seaweed Sargassum pallidum.* Czech Journal of Food Sciences, 2016. **34**(5): p. 406-413.

[84] Saravana, P.S., et al., *Structural, antioxidant, and emulsifying activities of fucoidan from Saccharina japonica using pressurized liquid extraction.* Carbohydrate Polymers, 2016. **153**: p. 518-525.

[85] Palanisamy, S., et al., *Isolation of fucoidan from Sargassum polycystum brown algae: Structural characterization, in vitro antioxidant and anticancer activity.* International Journal of Biological Macromolecules, 2017. **102**: p. 405-412.

[86] Huang, C.-Y., et al., *Antioxidant activities of crude extracts of fucoidan extracted from Sargassum glaucescens by a compressional-puffing-hydrothermal extraction process.* Food Chemistry, 2016. **197**: p. 1121-1129.

[87] Zhao, Y., et al., *Fucoidan extracted from Undaria pinnatifida: Source for nutraceuticals/functional foods.* Marine drugs, 2018. **16**(9): p. 321.

[88] Zhao, D., J. Xu, and X. Xu, *Bioactivity of fucoidan extracted from Laminaria japonica using a novel procedure with high yield.* Food Chemistry, 2018. **245**: p. 911-918.

[89] Nagamine, T., et al., *Intestinal absorption of fucoidan extracted from the brown seaweed, Cladosiphon okamuranus.* Marine Drugs, 2015. **13**(1): p. 48-64.

[90] Nguyen, T.T., et al., *Enzyme-Assisted Fucoidan Extraction from Brown Macroalgae Fucus distichus subsp. evanescens and Saccharina latissima.* Marine Drugs, 2020. **18**(6): p. 296.

[91] Saravana, P.S., et al., *Subcritical water extraction of fucoidan from Saccharina japonica: optimization, characterization and biological studies.* Journal of Applied Phycology, 2018. **30**(1): p. 579-590.

[92] Okolie, C.L., et al., *The comparative influence of novel extraction technologies on in vitro prebiotic-inducing chemical properties of fucoidan extracts from Ascophyllum nodosum.* Food Hydrocolloids, 2019. **90**: p. 462-471.

[93] Menshova, R.V., et al., *Fucoidans from brown alga Fucus evanescens: Structure and biological activity.* Frontiers in Marine Science, 2016. **3**: p. 129.

[94] Wang, C.-Y. and Y.-C. Chen, *Extraction and characterization of fucoidan from six brown macroalgae.* Journal of Marine Science and Technology, 2016. **24**(2): p. 319-328.

[95] January, G., et al., *Assessing methodologies for fucoidan extraction from South African brown algae.* Algal Research, 2019. **40**: p. 101517.

[96] Hmelkov, A.B., et al., *Ultrasound-assisted extraction of polysaccharides from brown alga Fucus evanescens. Structure and biological activity of the new fucoidan fractions.* Journal of Applied Phycology, 2018. **30**(3): p. 2039-2046.

[97] Yuan, Y. and D. Macquarrie, *Microwave assisted extraction of sulfated polysaccharides (fucoidan) from Ascophyllum nodosum and its antioxidant activity.* Carbohydrate Polymers, 2015. **129**: p. 101-107.

[98] Terasawa, M., et al., *Anti-influenza A virus activity of rhamnan sulfate from green algae Monostroma nitidum in mice with normal and compromised immunity.* Marine Drugs, 2020. **18**(5): p. 254.

[99] Tako, M., et al., *Structure-function relationship of rhamnan sulfate isolated from commercially cultured edible green seaweed, Monostroma nitidum.* Am. J. Appl. Chem, 2017. **5**: p. 38-44.

[100] Suzuki, K. and M. Terasawa, *Biological Activities of Rhamnan Sulfate Extract from the Green Algae Monostroma nitidum (Hitoegusa).* Marine Drugs, 2020. **18**(4): p. 228.

[101] Mader, J., et al., *Calcium spirulan derived from Spirulina platensis inhibits herpes simplex virus 1 attachment to human keratinocytes and protects against herpes labialis.* Journal of Allergy and Clinical Immunology, 2016. **137**(1): p. 197-203. e3.

[102] El-Baky, A.H., H.K. El Baz, and S. El-Latife, *Induction of sulfated polysaccharides in Spirulina platensis as response to nitrogen concentration and its biological evaluation.* Journal of Aquaculture Research & Development, 2014. **5**(1): p. 1.

[103] Marzorati, S., et al., *Carotenoids, chlorophylls and phycocyanin from Spirulina: supercritical CO 2 and water extraction methods for added value products cascade.* Green Chemistry, 2020. **22**(1): p. 187-196.

[104] Poojary, M.M., et al., *Innovative alternative technologies to extract carotenoids from microalgae and seaweeds.* Marine Drugs, 2016. **14**(11): p. 214.

[105] Ganesan, A.R., U. Tiwari, and G. Rajauria, *Seaweed nutraceuticals and their therapeutic role in disease prevention.* Food Science and Human Wellness, 2019. **8**(3): p. 252-263.

[106] Luengo, E., et al., *Effect of pulsed electric field treatments on permeabilization and extraction of pigments from Chlorella vulgaris.* The Journal of Membrane Biology, 2014. **247**(12): p. 1269-1277.

[107] Parniakov, O., et al., *Pulsed electric field assisted extraction of nutritionally valuable compounds from microalgae Nannochloropsis spp. using the binary mixture of organic solvents and water.* Innovative Food Science & Emerging Technologies, 2015. **27**: p. 79-85.

[108] Jaeschke, D.P., et al., *Carotenoid and lipid extraction from Heterochlorella luteoviridis using moderate electric field and ethanol.* Process Biochemistry, 2016. **51**(10): p. 1636-1643.

[109] Liau, B.-C., et al., *Supercritical fluids extraction and anti-solvent purification of carotenoids from microalgae and associated bioactivity.* The Journal of Supercritical Fluids, 2010. **55**(1): p. 169-175.

[110] Castro-Puyana, M., et al., *Pressurized liquid extraction of Neochloris oleoabundans for the recovery of bioactive carotenoids with anti-proliferative activity against human colon cancer cells.* Food Research International, 2017. **99**: p. 1048-1055.

[111] Zhao, L., et al., *Optimization of microwave-assisted extraction of astaxanthin from Haematococcus pluvialis by response surface methodology and antioxidant activities of the extracts.* Separation Science and Technology, 2009. **44**(1): p. 243-262.

[112] Tinoco, N.A., et al., *Generation of volatile compounds from carotenoids of Dunaliella bardawil algae by water bath heating and microwave irradiation.* Journal of the Brazilian Chemical Society, 2016. **27**(8): p. 1452-1458.

[113] Jaeschke, D.P., et al., *Ultrasound as an alternative technology to extract carotenoids and lipids from Heterochlorella luteoviridis.* Bioresource Technology, 2017. **224**: p. 753-757.

[114] Bernaerts, T.M., et al., *Cell disruption of Nannochloropsis sp. improves in vitro bioaccessibility of carotenoids and*

ω3-LC-PUFA. Journal of Functional Foods, 2020. **65**: p. 103770.

[115] Oliyaei, N., et al., *Encapsulation of fucoxanthin in binary matrices of porous starch and halloysite.* Food Hydrocolloids, 2020. **100**: p. 105458.

[116] Gómez-Loredo, A., J. Benavides, and M. Rito-Palomares, *Growth kinetics and fucoxanthin production of Phaeodactylum tricornutum and Isochrysis galbana cultures at different light and agitation conditions.* Journal of Applied Phycology, 2016. **28**(2): p. 849-860.

[117] Koo, S.Y., et al., *Anti-obesity effect of standardized extract of microalga Phaeodactylum tricornutum containing fucoxanthin.* Marine Drugs, 2019. **17**(5): p. 311.

[118] Yip, W.H., et al., *Characterisation and stability of pigments extracted from Sargassum binderi obtained from Semporna, Sabah.* Sains Malaysiana, 2014. **43**(9): p. 1345-1354.

[119] Shah, M., et al., *Astaxanthin-producing green microalga Haematococcus pluvialis: from single cell to high value commercial products.* Frontiers in Plant Science, 2016. 7: p. 531.

[120] Rao, A.R., et al., *Effective inhibition of skin cancer, tyrosinase, and antioxidative properties by astaxanthin and astaxanthin esters from the green alga Haematococcus pluvialis.* Journal of Agricultural and Food Chemistry, 2013. **61**(16): p. 3842-3851.

[121] İlter, I., et al., *Optimization of phycocyanin extraction from Spirulina platensis using different techniques.* Journal of Food Composition and Analysis, 2018. **70**: p. 78-88.

[122] Kuddus, M., et al., *Recent developments in production and biotechnological applications of C-phycocyanin.* BioMed Research International, 2013. **2013**.

[123] Mishra, S.K., A. Shrivastav, and S. Mishra, *Preparation of highly purified C-phycoerythrin from marine cyanobacterium Pseudanabaena sp.* Protein Expression and Purification, 2011. **80**(2): p. 234-238.

[124] Salama, A., et al., *Maximising phycocyanin extraction from a newly identified Egyptian cyanobacteria strain: Anabaena oryzae SOS13.* International Food Research Journal, 2015. **22**(2): p. 517.

[125] Patel, A., et al., *Purification and characterization of C-Phycocyanin from cyanobacterial species of marine and freshwater habitat.* Protein Expression and Purification, 2005. **40**(2): p. 248-255.

[126] Chaiklahan, R., N. Chirasuwan, and B. Bunnag, *Stability of phycocyanin extracted from Spirulina sp.: influence of temperature, pH and preservatives.* Process Biochemistry, 2012. **47**(4): p. 659-664.

[127] Abd El-Ghany, A.M., et al., *New approach for controlling snail host of Schistosoma mansoni, biomphalaria alexandrina with cyanobacterial strains-derived C-phycocyanin.* Vector-Borne and Zoonotic Diseases, 2018. **18**(9): p. 464-468.

[128] Martelli, G., et al., *Thermal stability improvement of blue colorant C-Phycocyanin from Spirulina platensis for food industry applications.* Process Biochemistry, 2014. **49**(1): p. 154-159.

[129] Wu, H.-L., et al., *Stability and antioxidant activity of food-grade phycocyanin isolated from Spirulina platensis.* International Journal of Food Properties, 2016. **19**(10): p. 2349-2362.

[130] Dejsungkranont, M., H.-H. Chen, and S. Sirisansaneeyakul, *Enhancement of antioxidant activity of C-phycocyanin of Spirulina powder treated with supercritical fluid carbon dioxide.*

Agriculture and Natural Resources, 2017. **51**(5): p. 347-354.

[131] Vali Aftari, R., et al., *The optimized concentration and purity of Spirulina platensis C-phycocyanin: a comparative study on microwave-assisted and ultrasound-assisted extraction methods.* Journal of Food Processing and Preservation, 2015. **39**(6): p. 3080-3091.

[132] Tavanandi, H.A., P. Vanjari, and K. Raghavarao, *Synergistic method for extraction of high purity Allophycocyanin from dry biomass of Arthrospira platensis and utilization of spent biomass for recovery of carotenoids.* Separation and Purification Technology, 2019. **225**: p. 97-111.

[133] Tavanandi, H.A., A.C. Devi, and K. Raghavarao, *A newer approach for the primary extraction of allophycocyanin with high purity and yield from dry biomass of Arthrospira platensis.* Separation and Purification Technology, 2018. **204**: p. 162-174.

[134] Sheih, I.-C., T.-K. Wu, and T.J. Fang, *Antioxidant properties of a new antioxidative peptide from algae protein waste hydrolysate in different oxidation systems.* Bioresource Technology, 2009. **100**(13): p. 3419-3425.

[135] Laohakunjit, N., O. Selamassakul, and O. Kerdchoechuen, *Seafood-like flavour obtained from the enzymatic hydrolysis of the protein by-products of seaweed (Gracilaria sp.).* Food Chemistry, 2014. **158**: p. 162-170.

[136] Garcia-Moscoso, J.L., et al., *Flash hydrolysis of microalgae (Scenedesmus sp.) for protein extraction and production of biofuels intermediates.* The Journal of Supercritical Fluids, 2013. **82**: p. 183-190.

[137] Sheih, I.-C., et al., *Anticancer and antioxidant activities of the peptide fraction from algae protein waste.* Journal of agricultural and food Chemistry, 2010. **58**(2): p. 1202-1207.

[138] Gerde, J.A., et al., *Optimizing protein isolation from defatted and non-defatted Nannochloropsis microalgae biomass.* Algal Research, 2013. **2**(2): p. 145-153.

[139] Harrysson, H., et al., *Production of protein extracts from Swedish red, green, and brown seaweeds, Porphyra umbilicalis Kützing, Ulva lactuca Linnaeus, and Saccharina latissima (Linnaeus) JV Lamouroux using three different methods.* Journal of Applied Phycology, 2018. **30**(6): p. 3565-3580.

[140] Gressler, V., et al., *Lipid, fatty acid, protein, amino acid and ash contents in four Brazilian red algae species.* Food Chemistry, 2010. **120**(2): p. 585-590.

[141] Kadam, S.U., et al., *Extraction and characterization of protein from Irish brown seaweed Ascophyllum nodosum.* Food Research International, 2017. **99**: p. 1021-1027.

Chapter 8

Valorization of Native Nuts from Brazil and Their Coproducts

Renan Danielski, Gabriela Polmann and Jane Mara Block

Abstract

The consumption of nuts as part of a healthy diet and active lifestyle has long been associated with chronic disease prevention. Nuts and their coproducts such as oil, cake, plant-based milk substitutes, flour, and shell are rich in lipids, proteins, phenolics, and other bioactive compounds. Nut flour also presents interesting physical properties, such as water or oil holding capacity, foam properties, emulsifying activity, and emulsion stability. These biological and physical properties make these products commercially attractive as organic ingredients in several foods such as spreads, bakery products, and cereal bars. In this chapter, the nutritional and bioactive profiles, as well as the evidenced health-promoting effects of nuts originating from Brazil, will be discussed. The focus will be on commercial nuts such as cashews, pecan, and Brazil nuts, along with some underexplored and relatively unknown indigenous species, such as sapucaia, chichá, monguba, and pracaxi. The knowledge of these Brazilian native nuts and their coproducts is important for stimulating their consumption among the population and their large-scale commercialization.

Keywords: oilseeds, biodiversity, bioactive compounds, lipids

1. Introduction

The consumption of nuts as part of a healthy diet and active lifestyle has long been associated with chronic disease prevention such as cardiovascular diseases, type 2 diabetes, cognitive function impairment, inflammatory disorders, among others. A closer look into the composition of nuts may help understand their health-promoting effects. They are rich in unsaturated acids such as oleic and linoleic, and have low concentrations of saturated fatty acids. In addition, the oil fraction presents significant amounts of tocopherols and phytosterols. The cake and shell, coproducts obtained from nuts, are rich in polyphenols. Besides the lipids and bioactive substances, nuts and their coproducts are also rich in other macronutrients (protein and fiber) and micronutrients (vitamins and minerals). The composition and concentration of bioactive compounds vary according to the type of nut and its coproducts [1].

Brazil has a great diversity of nut trees distributed in five of its six biomes Amazon, Cerrado, Atlantic Forest, Caatinga, Pampas, and Pantanal. **Figure 1** presents the Brazilian biomes where the nuts discussed in this chapter are found.

Brazil nuts and cashew are commercial nuts native to Brazil. The cashew tree (*Anacardium occidentale* L.), which is well adapted in tropical and subtropical regions, is present in Cerrado, a biome known for long periods of drought. The nut,

Figure 1.
Distribution of different nut trees in Brazilian biomes.

which is obtained from the cashew fruit, is composed of shell, skin, and almond. The largest producer of cashew nuts in 2018 was Vietnam (863,060 tons), followed by India (745,000 tons), Côte d'Ivoire (711,000 tons), Philippines (222,541 tons) and Brazil (141,418 tons) [2]. The Brazilian state of Ceará led the production with 61.9% of the planted area of cashew trees, followed by Rio Grande do Norte (15.4%), and Piauí (15.2%) [2].

Brazil nut tree (*Bertholletia excelsa*) is present in the Amazon, where it grows in well-drained soil along the Amazon River without the use of pesticides and herbicides. Brazil is the largest producer of Brazil nuts in the world (36,923 tons), followed by Bolivia, Ivory Coast, and Peru (25,749; 19,356; and 6042 tons, respectively) [2].

On the other hand, the pecan nut (*Carya illinoinensis*) is native to northern Mexico and southern United States. It was introduced in Brazil in 1866, and it is cultivated in the biomes Atlantic Forest, and mainly in the Pampas in the southeast and south regions. The tree is large-sized and can grow up to 20-40 m, and its fruit is technically classified as a drupe, characterized by a single pit surrounded by husk [3, 4].

These commercial or named conventional nuts in Brazil are usually consumed in a variety of ways such as raw, roasted and salted, caramelized, and coated. In addition, they can be incorporated into other food products, such as cereal bars, chocolates, bakery goods and spreads, among other foods. They can also be used for the extraction of specialty oils and for obtaining plant-based milk [1, 5–7].

However, the availability of nuts in Brazil is not limited to cashew, pecan, and Brazil nuts. A large variety of nut trees is available, especially in the Amazon region. These relatively unknown indigenous nuts represent a vast potential for the introduction into the diet of Brazilians and other consumers around the world. Sapucaia, chichá, monguba, and pracaxi are some of the nuts, which are not commercialized on a large scale and are usually consumed in their natural form by the local population. Therefore, data on their production is not found in the literature. A few reports showed that such nuts are also rich in macro and micronutrients, including phenolic compounds [8–10].

Sapucaia (*Lecythis Pisonis*) grows in the Atlantic Forest biome [9], while monguba (*Prachira aquatica*) and pracaxi (*Pentaclethra macroloba* (Wild.) Kuntze) can be found in the Amazon. The monguba tree is cultivated in different regions of the Amazon biome as an ornamental plant. Its fruits are oval-shaped, surrounded by a brown wooden peel where large-sized seeds are contained. The seeds are edible and can be consumed in a variety of ways, such as roasted, boiled, or fried [10]. The pracaxi tree produces a pod-shaped fruit containing edible seeds, from which phenolic-rich oils can be obtained [11]. Chichá (*Sterculia striata*), a nut tree originated from India and Malaysia, has thrived in the semi-arid conditions of the Cerrado, yielding nuts rich in phenolic compounds and lower lipid content [8].

This chapter presents the nutritional composition, phytochemical properties, and bioactive compounds of commercial (cashew, Brazil nuts, and pecan) and non-commercial (sapucaia, chichá, monguba, and pracaxi) nuts found in Brazil. The health benefits associated with their consumption, as well as novel products based on these nuts and their coproducts will be discussed with emphasis on their functional properties and nutritional profile.

2. Nutritional composition of native Brazilian nuts

Table 1 shows the macronutrient composition of conventional and non-conventional nuts native to Brazil. Brazil nuts present the highest lipid content and energy value. On the other hand, cashew nuts have higher protein content followed by chichá, which has the highest content of carbohydrates and the lower content of lipids. The sapucaia nut presents the highest fiber content among all nuts showed in **Table 1**.

Table 2 shows that the conventional and non-conventional nuts are rich in lipids considered beneficial to health, such as monounsaturated fatty acids (MUFAS) and polyunsaturated fatty acids (PUFAS). Pecan nut presents the highest content of the MUFA oleic acid (C18:1, ω9) followed by the cashew and pracaxi. Besides the nutritional benefits, another advantage of oleic acid is related to its higher oxidative stability compared to PUFAS. On the other hand, Brazil nuts show a high PUFA

Component (g/100 g)	Conventional			Non-conventional			
	Brazil nuts [12, 13]	Cashew [14]	Pecan [15]	Chichá [8, 16]	Monguba [10, 17]	Pracaxi [11]	Sapucaia [18, 19]
Ashes	3.3	2.8-4.1	1.8	3.0-3.2	2.3-2.7	1.9	2.9-3.5
Moisture	3.1-3.2	2.7-8.4	3.3	8.2-11.4	5.3-8.3	4.0	4.1-4.2
Protein	14.4-16.2	19.7-24.5	8.6	18.5-22.5	13.3-15.4	15.5	15.5-20.5
Lipids	64.9-67.3	39.8-47.1	62.2	24.5-28.6	41.9-45.6	53.4	58.7-60.8
Total fiber	7.5-8.0	2.5-4.2	10.9	4.6-5.8	4.7-6.1	—	16.5
Carbohydrate	10.9-15.9	27.1-34.9	13.4	40.5-45.8	34.3-36.2	25.2	4.9-13.8
Energy value (kcal/100 g)	659-715	499-707	633.9	456-530	557-677	644	616-665

—: not presented.

Table 1.
Composition of macronutrients of conventional and non-conventional nuts native from Brazil.

Fatty acids (%)	**Conventional**			**Non-conventional**			
	Brazil nuts [12, 13, 20]	**Cashew [21, 22]**	**Pecan [4, 15]**	**Chichá [8]**	**Monguba [10]**	**Pracaxi [11, 23]**	**Sapucaia [18]**
Miristic (C14:0)	0-0.1	—	—	—	—	—	0.1
Palmitic (C16:0)	14.9-16.7	10.3	5.4	26.5	60.9	1.4-1.5	12.9-15.2
Palmitoleic (C16:1 cis 9)	0-0.4	0.3	—	2.4	—	—	0.2-0.3
Margaric (C17:0)	—	0.1	—	—	—	—	0-0.1
Stearic (C18:0)	9.9-11.9	9.8	1.4	4.0	1.8	2.5-2.7	7.7-8.4
Oleic (C18:1 cis 9)	28.5-36.3	60.6	71.8	37.8	7.7	53.2-53.5	39.7-44.4
Linoleic (C18:2 cis 9,12)	36-37.5	17.0	20.2	11.2	6.6	12.1-12.2	32.2-40.0
Linolenic (C18:3cis 9,12,15)	0.1-0.2	0.2	0.8	0.3	—	0.1	0.3-0.4
Arachidic (C20:0)	0-0.2	0.7	—	0.7	—	—	—
Gondoic (C20:1 cis 11)	0-0.1	0.2	—	—	—	—	0.1
Behenic (C22:0)	—	0.1	—	0.3	—	16.4-16.5	—
Lignoceric (C24:0)	—	—	—	—	—	11.1-11.6	—
∑ Saturated	24.9-28.9	21.3	7.0	31.5	62.7	33.3-33.6	21.5-23.3
∑Monounsaturated	28.5-36.8	61.1	71.8	40.4	14.2	54.1-54.3	40.1-45.7
∑Polyunsaturated	36.1-37.7	17.2	21.0	12.2	—	12.2-12.3	32.7-40.4
Tocopherol (μg/g)							
α-Tocopherol	72.5	78.4	1.7	16.6	ND	ND	11.2
β-Tocopherol	—	1329.8	—	1.1	ND	ND	ND
γ-Tocopherol	74.4	300.3	26.8	88.5	5.1	416.1	285.0
δ-Tocopherol	5.9	6.3	—	21.0	ND	7.8	2.8
Total	152.8	1714.8	28.7	127.0	5.1	423.9	299.0
Phytosterols (mg/100 g)							
Brassicasterol	1.5	101.4	—	ND	—	—	ND
β-Sitosterol	39.5	7.8	—	184.9	—	—	93.7
Campesterol	4.0	—	—	18.6	—	—	8.6
Stigmasterol	11.3	—	—	54.2	—	—	11.2
Sitostanol	39.5	8.6	—	—	—	—	—
δ5-Avenasterol + δ7-stigmasterol	6.7	—	—	—	—	—	—
Total	47-148	117.8	0.2-0.3	257.7	—	—	113.5

—: not presented.

Table 2.
Lipid composition of conventional and non-conventional nuts.

linoleic acid (C18:2, ω6) content (36-37%), which is an essential fatty acid. It also presents a balanced fatty acid composition containing significant concentrations of MUFAs and saturated fatty acids (SFAs). The main fatty acids present in Brazil nuts is oleic (C18:1, ω9), linoleic (C18:2, ω6), and palmitic (C16:0) acids. In addition, most of the nuts have a low concentration of saturated fatty acids. The exception is monguba, carrying 63% saturated fatty acid, with palmitic acid being the main fatty acid.

Between the nuts shown in **Table 2**, cashew has the highest tocopherol content (1714.80 μg/g) followed by the pracaxi nuts (423 μg/g). The main tocopherols identified in the cashew nut were β-tocopherol > γ-tocopherol > α-tocopherol > δ-tocopherol. The chichá nut presents the highest content of phytosterols (257.7 mg/100 g), followed by the cashew nut (117 mg/100 g). β-sitosterol and brassicasterol were the main phytosterols found in these nuts. The bioactive composition of Brazilian nuts is presented in item 3 of this chapter.

Table 3 shows that Brazil and sapucaia nuts are a source of selenium. Considering the level of Se detected in Brazil (36.1 μg/g), and sapucaia (46.9 μg/g) nuts, a daily intake of more than one or two nuts can exceed the recommended daily dose of selenium for adults (55 μg/day) [18]. A selenium intake higher than 400 μg/day has been associated with toxic effects, including selenosis, which symptoms are hair loss, skin damage, and nervous system disorders [12]. Sapucaia nuts are also rich in magnesium (1572 μg/g) and calcium (1168 μg/g). Chichá is rich in potassium (8718 μg/g) and presents the highest concentration of zinc (24.2 μg/100 g). On the other hand, cashew is rich in potassium (K) and pecan in phosphorus (P).

Minerals play fundamental roles in several functions in the human body, acting as cofactors in enzymatic processes, structural elements, and participating in the regulation of acid–base balance, nerve impulse, and muscle activity [24]. The consumption of nuts as part of a balanced diet can contribute to a proper consumption of minerals, which play an important role in maintaining good health.

Mineral (μg/g)	**Conventional**			**Non-conventional**			
	Brazil nuts [25]	**Cashew [26]**	**Pecan [27]**	**Chichá [8, 16]**	**Monguba [10, 17]**	**Pracaxi [28]**	**Sapucaia [18, 19]**
Al	—	—	—	6.2	—	—	1.7
Ca	7432.8 ± 10.2	35.8 ± 0.7	70.0	149.1	558.9	8.9 ± 0.2	1168.0
Cd	—	—	—	0.4	—	—	0.3
Cr	1.3 ± 0.2	—	—	0.4	—	—	0.5
Cu	59.4 ± 0.5	9.5 ± 0.2	—	7.5	7.5	—	16.9
Fe	74.3 ± 0.5	26.1 ± 0.5	2.5	218.2	4.4	1.7 ± 0.1	—
K	—	1443.3 ± 26.2	4.1	8718.0	—	—	—
Mg	9678.5 ± 68.5	612.7 ± 13.2	121.0	1327.0	875.3	5.6 ± 0.2	1572.0
Mn	3.4 ± 0.4	7.6 ± 0.2	4.5	32.1	2.0	—	41.4
Na	—	148.5 ± 2.6	—	—	11.4	—	0.5
Ni	—	0.6 ± 0.5	—	0.4	—	—	—
P	—	—	277.0	—	—	18.1	—
Pb	—	—	—	0.5	—	—	0.5
Se	36.1 ± 0.4	0.2 ± 0.7	—	—	—	—	46.9
Sn	—	—	—	16.6	—	—	11.1
Zn	110.3 ± 1.3	28.9 ± 1.2	4.5	24.2	9.9	—	20.9

—: not presented.

Table 3.
Mineral composition of conventional and non-conventional nuts.

On the other hand, an excessive consumption of Brazil and sapucaia nuts is not recommended due to the high levels of selenium.

3. Bioactive composition of Brazilian nuts

Tocopherols and tocotrienols are a group of eight compounds widely spread in nature. They are monophenols chemically characterized by a chromanol ring, in which a hydroxyl group is attached. The configuration of the side hydrocarbon chain determines whether the compound is either tocopherol or tocotrienol (saturated side chain or three double bonds, respectively). Both tocopherols and tocotrienols have four homologs each (α, β, γ, and δ), which are defined by the methylation pattern on the chromanol ring. These compounds can act as antioxidants by the donation of a hydrogen atom from the hydroxyl group to free radicals, stabilizing them and reducing oxidative stress. α-Tocopherol presents the highest *in vivo* antioxidant capacity and 100% of vitamin E activity. The activity of α-tocopherol is related to the transfer protein (α-TTP) in the liver, involved in the absorption of tocopherols [29].

Nuts, seeds, and vegetable oils are good sources for tocopherols, and α-tocopherol is the most abundant one in photosynthetic tissues. On the other hand, seeds accumulate about 10–20 times more γ-tocopherol. Tocopherols can be found in considerable amounts in Brazil nuts, where they are concentrated in the oil fraction due to their lipophilic character, as reported by Costa et al. [20]. The authors reported a total tocopherol content of 152.80 μg/g for Brazil nuts, composed of α-tocopherol (72.55 μg/g), γ-tocopherol (74.35 μg/g), and δ-tocopherol (5.90 μg/g). This profile may change according to the region where the nuts are grown. Funasaki et al. [30] reported the tocopherol composition of Brazil nuts from seven different Amazon rainforest areas. α-Tocopherol content ranged from 37.92 μg/g (Manicoré 2-AM) to 74.48 μg/g (Manicoré 1-AM) and γ-tocopherol levels varied between 106.88 μg/g (Manicoré 2-AM) and 171.80 μg/g (Xapuri – AC). The differences in the tocopherol content can be related to climate variations and post-harvest handling.

As natural antioxidants, tocopherols play a role in the protection of nuts against oxidation. Zajdenwerg et al. [31] showed a correlation between the decrease in tocopherol concentration and the appearance of secondary oxidation products in Brazil nuts after a period of storage of 16 days at 80°C γ-tocopherol was depleted by 50%, and α-tocopherol was completely consumed, while aldehydes started to build up from hydroperoxide breakdown. The authors suggested that α-tocopherol acted as a primary antioxidant, which would explain why this homolog was depleted first. α-Tocopherol is less polar than γ-tocopherol due to the presence of three methyl groups on its chromanol ring, which gives it higher antioxidant efficiency in non-polar systems [31].

Cashew nuts, on the other hand, present a lower concentration of tocopherols when compared with Brazil nuts. Ryan et al. [32] reported α-tocopherol and γ-tocopherol contents of 3.6 mg/g and 57.2 mg/g, respectively, for cashew oil. Meanwhile, Brazil nut oil was composed of 82.9 mg/g of α-tocopherol and 116.2 mg/g of γ-tocopherol. However, according to the same study, cashew oil is richer in other bioactive compounds than Brazil nut oil, namely some types of phytosterols, such as β-sitosterol (1768 mg/g against 1325 mg/g from Brazil nut) and campesterol (105.3 mg/g against 26.9 mg/g for Brazil nut). Nevertheless, stigmasterol is more concentrated in Brazil nuts oil (577.5 mg/g) than cashew oil (116.7 mg/g). The consumption of phytosterols has been associated with a decrease in LDL-cholesterol levels. The reason behind this bioactivity relies on the higher hydrophobicity of phytosterols compared with cholesterol, which would give them an advantage in being transported by micelles and later exerted once humans do not absorb them. This would prevent cholesterol from accumulating in the enterocytes and further reaching the bloodstream [32]. Pecan nut is rich in γ-tocopherol, with 23.8-38.1 mg/100 g [4].

The bioactive composition of cashew, Brazil and pecan nuts goes beyond the presence of tocopherols and phytosterols in their oil fraction. Both are also rich sources of polyphenols, an extensive class of secondary plant metabolites with antioxidant properties. Polyphenols primarily act by donating a hydrogen atom to free radicals in order to interrupt oxidation reaction chains. As a minor antioxidant mechanism, phenolics also can chelate transition metals, preventing these prooxidant agents from initiating the reaction chain that originates from the oxidative process. Based on structural differences, this large group can be further divided into subgroups, with the main ones being flavonoids, hydroxycinnamic and hydroxybenzoic acids, hydrolysable tannins, and proanthocyanidins. In nuts, phenolics can be present as free soluble compounds, esterified to fatty acids (soluble esters), or insoluble-bound to macromolecules (e.g., cellulose, structural protein, pectin) [1].

John et al. [1] reported the phenolic composition and antioxidant activity of Brazil nut. Soluble phenolics (free plus esterified) were found to be the predominant state present in the whole nut (519.11 mg/100 g), as well as the kernel (406.83 mg/100 g) and the brown skin (1236.07 mg/100 g). However, a significant amount (352.48 mg/100 g) of insoluble-bound phenolics were detected in the brown skin of Brazil nuts. Gallocatechin, protocatechuic acid, catechin, and vanillic acid were the main phenolics identified in the bound fraction. Insoluble-bound polyphenols are related to beneficial effects on gut health, once they are associated with a decrease in the colonic pH, preventing the growth of harmful microorganisms [33]. The study of John et al. [1] also demonstrated that the brown skin extract of Brazil nuts showed the highest *in vitro* antioxidant activity, measured by DPPH, hydroxyl radical scavenging activity, reducing power and ORAC assay. The brown skin showed the highest amount of soluble and insoluble-bound polyphenols and this coproduct of Brazil nuts can be considered an economical source of natural antioxidants.

Yang et al. [34] compared the phenolic content and antiproliferative activities of Brazil nuts and cashew. Cashew showed a higher concentration of both soluble (86.7 mg/100 g) and insoluble-bound (229.7 mg/100 g) phenolics when compared to Brazil nuts (46.2 mg/100 g of soluble and 123.1 mg/100 g of insoluble-bound). Brazil nuts did not display antiproliferative activity against HepG2 (human liver cancer cells) at the doses tested in the study (from 1 to 200 mg/mL). On the other hand, cashew displayed the effect at high doses (100-200 mg/mL). In addition, neither of them showed inhibition for human colon cancer cells (Caco-2) proliferation.

The roasting, which is a common process in nut preparation, has shown evidence to impact the phenolic composition of cashew and Brazil nuts in different ways. Özcan et al. [35] reported that the total phenolic content of Brazil nuts significantly decreased from 68.97 mg/100 (raw nut) to 66.47 mg/100 g when oven-roasted (130°C for 20 min). The microwave-roasted (720 W for 5 min) showed the highest impact on the phenolic compounds, decreasing from 68.97 mg/100 to 25.88 mg/100 g. The DPPH radical scavenging capacity of the phenolic extracts also declined from 81.77% in the raw nut to 40.66% in the oven-roasted nut and 34.60% in the microwave-roasted nut. On the other hand, Chandrasekara and Shahidi [21] demonstrated that oven-roasting cashew at 130°C for 33 min increased the total phenolic content (both soluble and insoluble-bound) of the whole nut, as well as the kernel and the testa. At the same time, the levels of proanthocyanidins decreased, except for soluble phenolic extract from the kernel. In particular, the concentrations of the flavonoids (+)-catechin, (−)-epicatechin, and epigallocatechin were enhanced after the roasting process, which also positively affected the antioxidant capacity of such extracts. According to the authors, these findings could be owed to the release of phenolics because of temperature, making them readily soluble in the extraction solvent. Although relatively high, the temperature was applied for a short period, which prevented an extensive degradation of polyphenols.

Pecan nutshell, a coproduct, have demonstrated great potential due to its rich phenolic composition and high antioxidant capacity. According to Hilbig et al. [36], extracts from this coproduct obtained under optimum conditions yielded total phenolic contents of 426.15-581.9 mg GAE/g, resulting in antioxidant activities of 2574.32-2573 μmol TEAC/g and 1268.03-1287.08 μmol TEAC/g measured by ABTS and DPPH assays, respectively. The extracts showed a variety of 29 different phenolic compounds, with gallic acid as the predominant one.

Demoliner et al. [9, 18] investigated the nutritional and phytochemical composition of sapucaia nut. The oil extracted from the nuts presented a considerable amount of total tocopherols (21.8-29.9 mg/100 g), with γ-tocopherol identified as the primary homolog (19.2-28.5 mg/100 g). α- and δ-Tocopherols were also identified in smaller quantities. Mounting evidence has shown that γ-Tocopherol excels in scavenging reactive oxygen species, which play an essential role in the development of chronic diseases [37–39]. Sapucaia nut oil also showed a significant concentration of phytosterols, namely β-sitosterol (92.8-193.9 mg/100 g), stigmasterol (9.92-13.2 mg/100 g), and campesterol (8.42-9.63 mg/100 g) [9].

Demoliner et al. [9] extracted the phenolic compounds from sapucaia nut and shell and used LC-ESI-MS/MS to identify and quantify the individual polyphenols present. The nut extracts were composed of 14 compounds, mainly phenolic acids, and flavonoids, with myricetin, vanillic, ferulic, and ellagic acid showing the highest amounts. Interestingly, shell extracts demonstrated to carry a greater variety of phenolics, with 22 compounds identified, with high levels of phenolic acids (gallic, protocatechuic, vanillic, ferulic, and ellagic) and flavonoids (epigallocatechin, catechin, epicatechin, taxifolin, myricetin, and vanillin). Shell extracts were also superior to the nut extracts in terms of *in vitro* antioxidant activity, measured by DPPH, FRAP, and ABTS. These outcomes highlight the potential for using nuts coproducts as sources of natural antioxidants.

Similar to sapucaia, the coproducts of chichá, namely the pellicle (26.26 mg GAE/g) and the shell (21.42 mg GAE/g), have been reported to be richer in total phenolic compounds than the nut (16.85 mg GAE/g). The extracts presented a wide variety of phenolic acids, such as ellagic, ferulic, salicylic, protocatechuic, and rosmarinic acid [8].

It has been reported that monguba oil is source of γ-tocopherol (513.5 mg/kg). In addition, ten phenolic compounds have been identified, mainly phenolic acids and flavonoids. The majority of phenolics (74.58%) were in the esterified form, followed by the glycosylated (13.02%), free (8.22%), and insoluble bound (4.18%) forms. Caffeic acid Monguba was the main phenolic compound found in this raw material (57.5%) [10].

Teixeira et al. [11] reported total phenolic contents ranging from 31.92 to 54.05 mg GAE/kg for pracaxi oil extracted by supercritical CO_2 extraction. The highest phenolic content and in vitro antioxidant activity of the extracts was obtained using 200 bar at 40-60°C. The oils obtained under low pressure demonstrated high antioxidant capacity measure by the CUPRAC method, indicating a significant content of both hydrophilic (phenolics) and lipophilic (carotenoids and tocopherols) antioxidants. The presence of natural antioxidants may also have positively affected the stability oxidative index (OSI of 11.38 h at 110°C and 10.83 at 120°C), which suggests a prolonged shelf life for the oil [11].

4. Bioavailability and health-promoting benefits

The nutritional profile and bioactive compounds present in Brazilian nuts are certainly a positive characteristic of these raw materials. However, the presence of these nutrients and minor compounds do not guarantee their conversion into health-promoting benefits. They need to be efficiently released from the food matrix and be absorbed in sufficient amounts to be converted into biologically active metabolites for having and positive impact on human health. In addition, the potentially toxic compounds should be assessed to ensure its safety. That is especially true when considering expanding the commercialization of relatively unknown nuts, like the ones we have been presenting throughout this chapter.

Moreda-Piñeiro et al. [40] assessed the *in vitro* bioavailability of essential and toxic metals in several nuts and seeds, including Brazil nuts and cashew. Essential metals (Ba, Ca, Cd, Co, Cu, K, Li, Mg, Mn, Mo, P, Pb, Se, Sr., Tl, and Zn) consistently presented moderate bioavailability (dialyzability of 2.1-40.7%) among the samples. The exception was iron, which presented a low dialyzability (0.70-3.7%). On the other hand, toxic metals (Al, Ba, Cd, and Hg) demonstrated poor bioavailability (dialyzability ratios between 0.35-16.8%). The results suggest that the consumption of these nuts can be considered as safe and beneficial because of the high bioavailability of the essential metals. The authors also found a positive correlation between carbohydrate content and dialyzability ratio, meaning that the higher the carbohydrate concentration, the greater the bioavailability. On the other hand, nuts with high-fat content were found to present lower mineral bioavailability.

Nascimento et al. [41] reported the *in vitro* bioavailability of Cu and Fe in cashew nuts using simulated gastric and intestinal fluids as well as dialysis procedures. The results showed that 83% of Cu and 78% of Fe were recovered during the experiment, indicating a high level of bioavailability for these minerals in cashew nuts. These two minerals are essential for a variety of physiological and metabolic processes. Cu deficiency can lead to high blood pressure and infertility, while Fe deficiency can cause anemia.

Using an *in vitro* dialyzability approach, Herbello-Hermelo et al. [42] evaluated the bioavailability of the polyphenol fraction of selected nuts and seeds. Brazil nuts and cashew were among the samples with the highest recovery of phenolics in the digested extracts (81 and 89%, respectively). The authors also showed that polyphenol bioavailability was dependent on Cu content. This can be explained by the strong binding ability between Cu^{+} and phenolics, with the latter being reported to reduce Cu^{2+} to Cu^{+}, even though the mechanism by which this happens is not entirely understood yet.

Brazil nuts are one of the main sources of Se in nature, which are constituents of selenoproteins (e.g., glutathione peroxidases – GSH-Px), enzymes that are part of the endogenous antioxidant defenses system. A high blood concentration of such enzymes is associated with a lower risk of cardiovascular diseases. Stockler-Pinto et al. [43] conducted a human trial on the supplementation of Brazil nuts to patients on hemodialysis, which produces a large amount of reactive oxygen species. The administration of one nut per day for three months increased the subjects' GSH-Px activity. Before the supplementation, 11% of the patients presented GSH-Px levels below the normal range. After the supplementation, all subjects showed results within the normal range.

Inflammation processes are essential biological responses when the organism needs to fight intrusive agents. However, inflammatory disorders are extremely damaging and can lead to conditions such as cancer, type 1 diabetes, and rheumatoid arthritis, among others [44]. Colpo et al. [45] monitored the activity of inflammation markers upon the intake of different portions of Brazil nuts (from 0 to 50 g) by healthy individuals. The trial revealed that the consumption of 20 or 50 g of Brazil nuts was responsible for a decrease in serum levels of inflammation markers (IL-1, IL-6, TNF-α, and INF-γ).

Nanoemulsions prepared from cashew nut shell liquid showed cytotoxicity against human breast cancer cell line MCF-7. The cells treated with the nanoemulsion presented reduced viability, primarily through apoptosis or necrosis [46]. Cashew nut has also demonstrated a positive impact on cholesterol levels and systolic blood pressure. In a trial conducted with 300 type 2 diabetic adult patients, the administration of 30 g of cashew nuts daily for 12 weeks could reduce the subjects' blood pressure while the concentration of plasma HDL cholesterol increased.

Müller et al. [47] reported a potential biological effect of pecan nutshell extracts *in vivo* and *in vitro* studies. The phenolic-rich extracts from pecan nutshell demonstrated protection against liver damage induced by ethanol in rats by increasing the levels of endogenous antioxidant defenses, such as glutathione (by 33%), superoxide dismutase (by 47%) and catalase (by 47-73%). Hilbig et al. [48] showed that aqueous extracts from pecan nutshell were able to reduce the viability of MCF-7 breast cancer cells. The effect was attributed to the high abundance of phenolic compounds in the extract, such as gallic, ellagic, and chlorogenic acid, as well as catechin, epicatechin, epigallocatechin, among others.

5. Nut-based products

Nuts are usually consumed as a snack, in their natural form, as well as toasted, salted, or caramelized. The Brazil nuts, cashew and pecan and are used as ingredients in several industrialized food products such as bakery goods, sweets, chocolates and ice creams, among others. The underexplored nuts, such as chichá, monguba, pracaxi, and sapucaia, although not reaching the same level of commercialization, have shown to be potentially suitable for these types of application.

5.1 Nut oil

Nut oils are usually obtained by pressing, and since they are not refined, they are classified as extra virgin. The main steps involved in the processing of an extra virgin nut oil are harvesting, pre-drying, peeling, drying, oil extraction, and centrifugation. Because of the appreciable sensorial attributes and elevated price compared with other vegetable oils, the nut oils are considered gourmet oils. The price will differ according to the nut type, its availability, and the processing used.

Brazil nut, pecan and cashew nut oils are usually found in specialized stores of natural products, in pharmacies (for use as cosmetics) and are available for online purchase, where their price ranges from 8 US $ to 17 US $. The pracaxi oil can be found for commercialization, mainly as a cosmetic, due to its emollient properties. On the other hand, chichá, sapucaia and monguba oils have not been commercially produced yet. However, studies show that these oils are rich in monounsaturated fatty acids and possess high oxidative stability, thus showing great potential for commercialization [8, 18]. Monguba oil, which is rich in palmitic acid, has a great potential for food applications as a substitute for cocoa butter [10].

5.2 Cereal bars

Cereal bars are consumed mainly as a diet substitute for sugar-dense snacks, as well as an energy and protein source for athletes. In the formulation of cereal bars, it is important to take into account the cereal choice (oats, wheat, rice, barley, maize). In addition, the selection of the appropriate carbohydrate to maintain a balance between taste and shelf life, the nutrient profile, the dietary fiber, and the processing stability [49] are also important. Cereal bars made with nuts are widely accepted by consumers. Besides their nutritional and sensory quality, they remain stable during product storage. Currently, we can easily find cereal bars containing Brazil nuts and cashew as ingredients.

5.3 Nut flour

Nuts can be ground into flour for use in bakery goods. Cake, the residue left after nut oil extraction, can be used as raw material to produce defatted nut flour. This is considered a sustainable process since it adds value to a coproduct, reducing waste generation [50]. Nut flour is rich in dietary fiber, protein, vitamins, minerals, and bioactive compounds. The application of partially defatted nut flour has been reported to improved consumer acceptance of gluten-free bakery products [50]. The combination of rich nutritional composition with appreciable physical properties makes these flours suitable for bakery products, such as cookies, bread, cakes, sweets, among others.

Physical properties, such as water or oil retention capacity, foam properties, emulsifying activity, and emulsion stability, are very important for the incorporation of flour into bakery products. Sanchez et al. [51] reported that pistachio and cashew nut flours with thermal trapping in autoclave showed interesting emulsifying and water rectifying properties for application in bakery products. In another study, Teixeira et al. [19] reported functional properties of defatted sapucaia flour. The parameters of emulsion formation and stability indicated that it could be applied as an ingredient in emulsified products such as cakes, creams, sauces and sausages, among others.

5.4 Nut-based milk alternative

Plant-based milk alternatives are beverages obtained from crushing a specific feedstock homogenized in water. The resulting particle size distribution should be between 5 and 20 μm to mimic cow's milk in appearance and consistency. Sethi et al. [5] reported that the vegetable-based milks can be divided in five categories according to the raw material used: cereals (oats, rice, corn, spelled), vegetables (soy, peanuts, lupines, cowpea), nuts (almonds, coconut, hazelnuts, pistachios, walnuts), oilseeds (sesame, flax, hemp, sunflower), and pseudo-cereals (quinoa, teff, amaranth).

The improved diagnosis of conditions related to cow's milk consumption, such as milk protein allergy and lactose intolerance, increased the demand for non-dairy milk alternatives. Besides, vegans also prefer these beverages, and it is being considered a continuously growing niche [5]. Nuts, such as walnuts, chestnuts, and almonds, are used to produce plant-based milk alternatives, mainly due to their functionality and better sensory characteristics [5]. The allergenic potential and high cost are some of the limiting factors presented by the nut-based milks.

The addition of Brazil nuts extract in a soybean-based drink positively influenced the sensory characteristics of the product [52]. On the other hand, prebiotic drinks based on cashew nuts and fruit juice have proved to be a viable alternative for the development of functional products [53]. Bruno et al. [6] reported that cashew nut-based milk alternative was a good matrix for the development of probiotics. They showed that *Bifidobacterium animalis*, *Lactobacillus acidophilus* and *L. plantarum* remained viable during 30 days at 4°C. The probiotic drink based on cashew nuts also achieved a good sensory acceptance, without significant changes in whiteness and microbiological quality.

Plant-based milk of sapucaia nut cake using block freeze concentration, which was done in five consecutive stages, allowed the concentration of the phenolic compounds (gallic, vanillic, ferulic, sinapic and salicylic acids, catechin, taxifolin and sinapaldehyde) and minerals in all the five fractions [54]. Studies with different raw materials and formulations, especially non-conventional nuts, are still scarce. The knowledge about such products should contribute to encouraging their consumption.

6. Nuts shell and cake

Industrial nut processing results in a large amount of shells, a coproduct that can represent approximately 40-50% of the original total mass. The reuse of these coproducts, which are usually discarded, should be encouraged for reducing waste disposal, preserving the environment and adding value to the raw materials.

Some nutshells, such as pecan, are sold in pieces to make tea [4]. The ethnopharmacological use of nutshell tea includes the prevention and treatment of various diseases, such as type 2 diabetes, obesity, hypertension, hypercholesterolemia, cancer and, inflammatory diseases [24]. The therapeutic effects of the tea made with pecan nutshell have been associated with the presence of several phenolic compounds, such as phenolic acids, flavonoids, and proanthocyanidins. The antioxidant, antimicrobial, and antitumor activity of pecan peel extract have been reported [4, 50]. Other nutshells may have health benefits due to their bioactive composition.

Cashew nut is covered with a thin antioxidant-rich layer of reddish-brown color, known as testa. This fraction is an excellent source of hydrolysable tannins and polymeric proanthocyanidins. It comprises of phenolic acids like syringic acid, gallic acid, and *p*-coumaric acid as the major components. The concentration of catechin and epicatechin were found as 5.70 and 4.46 g/kg of dry matter, respectively [55].

The process of mechanical oil extraction generates a partially defatted by-product known as cake. The agro-industrial use of the defatted nut cake has great nutritional value, based on its high lipid and protein contents and the functional aspect present in this material, adding value to food products. The nuts cake can be used in bakery (as presented in item 5.3), but also in sweets. In the study by Lima et al. [56], the cashew nut cake was used to replace peanuts in the production of a sweet known as *paçoca,* made by mixing ground peanuts with other ingredients, such as corn flour, sugar, honey, and oil. The product showed physicochemical and microbiological stability, as well as good sensory acceptance. In addition, the nut cake presents bioactive compounds. For instance, Maciel et al. [57] demonstrated that pecan nut cake is rich in phenolic compounds with antioxidant activity.

7. Future perspectives

Commercial and non-commercial Brazilian nuts are nutritionally rich in macronutrients and bioactive compounds, with considerable amounts of natural antioxidants. Such substances are related to a myriad of health benefits since they are able to reduce oxidative stress. A significant number of studies have investigated the bioavailability of these bioactive compounds of Brazil nuts and cashew with promising results. The effect of the nuts on health using human trials had positive outcomes.

However, relatively unknown Brazilian nuts, such as sapucaia, chichá, pracaxi, and monguba remain an underexplored topic. Information about their bioavailability aspects and their impact on human health is still necessary. Therefore, more studies should be done in order to stimulate large-scale commercialization.

Finally, nuts are incredibly versatile with great economic potential. The incorporation of the coproducts, which are rich in nutrients and bioactive compounds, is an opportunity to enrich food formulation with these cost-effective ingredients, diminishing the waste generated by nut processing.

Acknowledgements

The authors are grateful to National Council for Scientific and Technological Development (CNPq) for the scientific productivity fellowship of Jane Mara Block (process 311070/2018-3), and to Coordination for the Improvement of Higher Education Personnel (CAPES) for the doctoral scholarship of Gabriela Polmann (process 88882.344944/2019-2101).

Author details

Renan Danielski[1], Gabriela Polmann[2] and Jane Mara Block[2*]

1 Department of Biochemistry, Memorial University of Newfoundland, St. John's, NL, Canada

2 Department of Food Science and Technology, Federal University of Santa Catarina, Florianópolis, SC, Brazil

*Address all correspondence to: janeblock@gmail.com; jane.block@ufsc.br

References

[1] John JA, Shahidi F. Phenolic compounds and antioxidant activity of Brazil nut (*Bertholletia excelsa*). Journal of Functional Foods. 2010; 2: 196-209. https://doi.org/10.1016/j.jff.2010.04.008.

[2] FAOSTAT. Nuts production 2018. Available from: < http://www.fao.org/faostat/en/#data/QC>

[3] Chang SK, Alasalvar C, Bolling BW, Shahidi F. Nuts and their co-products: The impact of processing (roasting) on phenolics, bioavailability, and health benefits – A comprehensive review. Journal of Functional Foods. 2016; 26: 88-122. https://doi.org/10.1016/j.jff.2016.06.029.

[4] Prado ACP, Manion BA, Seetharaman K, Deschamps FC, Barrera Arellano D, Block JM. Relationship between antioxidant properties and chemical composition of the oil and the shell of pecan nuts [*Carya illinoinensis* (Wangenh) C. Koch]. Industrial Crops and Products. 2013; 45: 64-73. https://doi.org/10.1016/j.indcrop.2012.11.042

[5] Sethi S, Tyagi SK, Anurag RK. Plant-based milk alternatives an emerging segment of functional beverages: a review. Journal of Food Science and Technology. 2016; 53: 3408-3423. https://doi.org/10.1007/s13197-016-2328-3

[6] Bruno LM, Lima JR, Wurlitzer NJ, Rodrigues TC. Non-dairy cashew nut milk as a matrix to deliver probiotic bacteria. Food Science and Technology. 2019.

[7] Lima JR, Garruti DS, Machado TF, Araújo IMS. Vegetal burgers of cashew fiber and cowpea: formulation, characterization and stability during frozen storage. Revista Ciencia Agronomica. 2018; 49: 708-714.

[8] Policarpi PB, Turcatto L, Demoliner F, Ferrari RA, Bascuñan VLAF, Ramos JC, Block JM. Nutritional potential, chemical profile and antioxidant activity of Chichá (*Sterculia striata*) nuts and its by-products. Food Research International. 2018; 106: 736-744. https://doi.org/10.1016/j.foodres.2017.12.069

[9] Demoliner F, Policarpi P, Vasconcelos LFL, Vitali L, Micke GA, Block JM. Sapucaia nut (*Lecythis pisonis* Cambess) and its by-products: A promising and underutilized source of bioactive compounds. Part II: Phenolic compounds profile. Food Research International. 2018; 112: 434-442. https://doi.org/10.1016/j.foodres.2018.06.050.

[10] Rodrigues AP, Pereira GA, Tomé PHF, Arruda HS, Eberlin MN, Pastore GM. Chemical Composition and Antioxidant Activity of Monguba (*Pachira aquatica*) Seeds. Food Research International. 2019; 121: 880-887. https://doi.org/10.1016/j.foodres.2019.01.014.

[11] Teixeira GL, Maciel LG, Mazzutti S, Gonçalves CB, Ferreira SRS, Block JM Composition, thermal behavior and antioxidant activity of pracaxi (*Pentaclethra macroloba*) seed oil obtained by supercritical CO_2. Biocatalysis and Agricultural Biotechnology. 2020; 24. https://doi.org/10.1016/j.bcab.2020.101521.

[12] Cardoso BR, Duarte GBS, Reis BZ, Cozzolino SMF. Brazil nuts: Nutritional composition, health benefits and safety aspects. Food Research International. 2017; 100: 9-18. https://doi.org/10.1016/j.foodres.2017.08.036

[13] Santos OV, Corrêa NCF, Carvalho RN, Costa CEF, França LFF, Lannes SCS. Comparative parameters of the nutritional contribution and functional claims of Brazil nut kernels,

oil and defatted cake. Food Research International. 2013; 51: 841-847. https://doi.org/10.1016/j.foodres.2013.01.054

[14] Janice R L, Ana CON, Hilton CRM, Raimundo NMS. Chemical composition and fatty acid profile of kernels from different Brazilian cashew tree genotypes. African Journal of Food Science. 2015; 9: 390-394. https://doi.org/10.5897/ajfs2015.1316

[15] Polmann G, Badia V, Frena M, Teixeira GL, Rigo E, Block JM, Feltes MMC. Enzyme-assisted aqueous extraction combined with experimental designs allow the obtaining of a high-quality and yield pecan nut oil. LWT. 2019; 113: 108283. https://doi.org/10.1016/j.lwt.2019.108283

[16] Oliveira JTA, Vasconcelos IM, Bezerra LCNM, Silveira SB, Monteiro ACO, Moreira RA. Composition and nutritional properties of seeds from *Pachira aquatica* Aubl, *Sterculia striata* St Hil et Naud and *Terminalia catappa* Linn. Food Chemistry. 2000; 70: 185-191. https://doi.org/10.1016/S0308-8146(00)00076-5

[17] Becker MM, Mandaji CM, Catanante G, Marty JL, Nunes GS. Mineral and bromatological assessment and determination of the antioxidant capacity and bioactive compounds in native Amazon fruits. Brazilian Journal of Food Technology. 2018; 21. https://doi.org/10.1590/1981-6723.02218

[18] Demoliner F, Policarpi P, Ramos JC, Bascuñan VLAF, Ferrari RA, Jachmanián I, Block JM. Sapucaia nut (*Lecythis pisonis* Cambess) and its by-products: A promising and underutilized source of bioactive compounds. Part I: Nutritional composition and lipid profile. Food Research International. 2018; 108: 27-34. https://doi.org/10.1016/j.foodres.2018.03.028

[19] Teixeira GL, Ávila S, Hornung PS, Barbi RCT, Ribani RH. Sapucaia nut (*Lecythis pisonis* Cambess.) flour as a new industrial ingredient: Physicochemical, thermal, and functional properties. Food Research International. 2018; 109: 572-582. https://doi.org/10.1016/j.foodres.2018.04.071

[20] Costa PA, Ballus CA, Teixeira-Filho J, Godoy HT. Phytosterols and tocopherols content of pulps and nuts of Brazilian fruits. Food Research International. 2010; 43: 1603-1606. https://doi.org/10.1016/j.foodres.2010.04.025

[21] Chandrasekara N, Shahidi F. Effect of roasting on phenolic content and antioxidant activities of whole cashew nuts, kernels, and testa. Journal of Agricultural and Food Chemistry. 2011; 59: 5006-5014. https://doi.org/10.1021/jf2000772.

[22] Islam MA, Jeong BG, Jung J, Shin EC, Choi SG, Chun J. Phytosterol Determination and Method Validation for Selected Nuts and Seeds. Food Analytical Methods. 2017; 10: 3225-3234. https://doi.org/10.1007/s12161-017-0877-3

[23] Serra JL, Rodrigues AMC, Freitas RA, Meirelles AJA, Darnet SH, Silva LHM. Alternative sources of oils and fats from Amazonian plants: Fatty acids, methyl tocols, total carotenoids and chemical composition. Food Research International. 2019; 116: 12-19. https://doi.org/10.1016/j.foodres.2018.12.028

[24] Damasceno NRT, Sala-Vila A, Cofán M, Pérez-Heras AM, Fitó M, Ruiz-Gutiérrez V, Ros E. Mediterranean diet supplemented with nuts reduces waist circumference and shifts lipoprotein subfractions to a less atherogenic pattern in subjects at high cardiovascular risk. Atherosclerosis. 2013; 230:

347-353. https://doi.org/10.1016/j.atherosclerosis.2013.08.014

[25] Moodley R, Kindness A, Jonnalagadda SB. Elemental composition and chemical characteristics of five edible nuts (almond, Brazil, pecan, macadamia and walnut) consumed in Southern Africa. Journal of Environmental Science and Health - Part B Pesticides, Food Contaminants, and Agricultural Wastes. 2007; 42: 585-591. https://doi.org/10.1080/03601230701391591

[26] Soares DJ, De Sousa Sabino LB, De Sousa MSML, De Carvalho Magalhães CE, Almeida MMB, De Sousa PHM, De Figueiredo RW. Mineral content, based in the Recommended Daily Intake, in cashew nut obtained from conventional and organic cultivation in different stages of processing. Semina: Ciencias Agrarias. 2012; 33: 1869-1876. https://doi.org/10.5433/1679-0359.2012v33n5p1869

[27] USDA. National Nutrient Database for Standard Reference, Agricultural Research Service, U.S. Department of Agriculture. 2020. Available from: https://fdc.nal.usda.gov/fdc-app.html#/food-details/170182/nutrients

[28] Pereira E, Ferreira MC, Sampaio KA, Grimaldi R, Meirelles AJA, Maximo GJ. Physical properties of Amazonian fats and oils and their blends. Food Chemistry. 2019; 278: 208-215. https://doi.org/10.1016/j.foodchem.2018.11.016

[29] Shahidi F, de Camargo A. Tocopherols and Tocotrienols in Common and Emerging Dietary Sources: Occurrence, Applications, and Health Benefits. International Journal of Molecular Sciences. 2016; 17. https://doi.org/10.3390/ijms17101745.

[30] Funasaki M, Menezes IS, Barroso HDS, Zanotto SP, Carioca CRF. Tocopherol profile of Brazil nut oil from different geographic areas of the Amazon region. Acta Amazonica. 2013; 43: 505-509. https://doi.org/10.1590/S0044-59672013000400012.

[31] Zajdenwerg C, Branco G, Alamed J, Decker E, Castro I. Correlation between sensory and chemical markers in the evaluation of Brazil nut oxidative shelf-life. European Food Research and Technology. 2011; 233: 109-116. https://doi.org/10.1007/s00217-011-1493-x.

[32] Ryan E, Galvin K, O'Connor TP, Maguire AR, O'Brien NM. Fatty acid profile, tocopherol, squalene and phytosterol content of brazil, pecan, pine, pistachio and cashew nuts. International Journal of Food Sciences and Nutrition. 2006; 57: 219-228. https://doi.org/10.1080/09637480600768077.

[33] Shahidi F, Yeo JD. Insoluble-Bound Phenolics in Food. Molecules (Basel, Switzerland). 2016; 21(9). https://doi.org/10.3390/molecules21091216.

[34] Yang J, Liu RH, Halim L. Antioxidant and antiproliferative activities of common edible nut seeds. LWT - Food Science and Technology. 2009; 42: 1-8. https://doi.org/10.1016/j.lwt.2008.07.007.

[35] Özcan M, Juhaimi F, Uslu N. The effect of heat treatment on phenolic compounds and fatty acid composition of Brazilian nut and hazelnut. Journal of Food Science and Technology. 2018; 55: 376-380. https://doi.org/10.1007/s13197-017-2947-3.

[36] Hilbig J, Alves VR, Müller CMO, Micke GA, Vitali L, Pedrosa RC, Block JM. Ultrasonic-assisted extraction combined with sample preparation and analysis using LC-ESI-MS/MS allowed the identification of 24 new phenolic compounds in pecan nut shell [*Carya illinoinensis* (Wangenh) C. Koch] extracts. Food Research International.

2018; 106: 549-557. https://doi.org/10.1016/j.foodres.2018.01.010.

[37] Alberdi-Cedeño J, Ibargoitia ML, Guillén MD. Effect of the Enrichment of Corn Oil With alpha - or gamma -Tocopherol on Its In Vitro Digestion Studied by 1 H NMR and SPME-GC/MS; Formation of Hydroperoxy-, Hydroxy-, Keto-Dienes and Keto- E -epoxy- E -Monoenes in the More alpha -Tocopherol Enriched Samples. Antioxidants. 2020; 9(3). https://doi.org/10.3390/antiox9030246.

[38] Figueroa Gonzalez D, Young F. Gamma Tocopherol Reduced Chemotherapeutic-Induced ROS in an Ovarian Granulosa Cell Line, But Not in Breast Cancer Cell Lines *In Vitro*. Antioxidants (Basel, Switzerland). 2020; 9. https://doi.org/10.3390/antiox9010051.

[39] Pahrudin Arrozi A, Shukri SNS, Wan Ngah WZ, Mohd Yusof YA, Ahmad Damanhuri MH, Jaafar F, Makpol S. Comparative Effects of Alpha- and Gamma-Tocopherol on Mitochondrial Functions in Alzheimer's Disease *In Vitro* Model. Scientific reports. 2020; 10: 8962-8962. https://doi.org/10.1038/s41598-020-65570-4.

[40] Moreda-PiñeiroJ,Herbello-HermeloP, Domínguez-GonzálezR,Bermejo-BarreraP, Moreda-Piñeiro A. Bioavailability assessment of essential and toxic metals in edible nuts and seeds. Food Chemistry. 2016; 205: 146-154. https://doi.org/10.1016/j.foodchem.2016.03.006.

[41] Nascimento AN, Naozuka J, Oliveira PV. *In vitro* evaluation of Cu and Fe bioavailability in cashew nuts by off-line coupled SEC–UV and SIMAAS. Microchemical Journal. 2010; 96: 58-63. https://doi.org/10.1016/j.microc.2010.01.016.

[42] Herbello-Hermelo P, Lamas JP, Lores M, Domínguez-González R, Bermejo-Barrera P, Moreda-Piñeiro A. Polyphenol bioavailability in nuts and seeds by an *in vitro* dialyzability approach. Food Chemistry. 2018; 254: 20-25. https://doi.org/10.1016/j.foodchem.2018.01.183.

[43] Stockler-Pinto MB, Mafra D, Farage NE, Boaventura GT, Cozzolino SMF. Effect of Brazil nut supplementation on the blood levels of selenium and glutathione peroxidase in hemodialysis patients. Nutrition. 2010; 26: 1065-1069. https://doi.org/10.1016/j.nut.2009.08.006.

[44] Del Rio D, Rodriguez-Mateos A, Spencer JPE, Tognolini M, Borges G, Crozier A. Dietary (poly)phenolics in human health: Structures, bioavailability, and evidence of protective effects against chronic diseases. Antioxidants and Redox Signaling. 2013; 18: 1818-1892. https://doi.org/10.1089/ars.2012.4581

[45] Colpo E, Dalton DA, Vilanova C, Reetz LGB, Duarte MMMF, Farias ILG, Meinerz DF, Da Rocha JBT. Brazilian nut consumption by healthy volunteers improves inflammatory parameters. Nutrition. 2014; 30: 459-465. https://doi.org/10.1016/j.nut.2013.10.005.

[46] Al-Hazzani A, Periyasamy V, Subash-Babu P, Alshatwi A. Formulation of cashew nut shell liquid (CSNL) nanoemulsion, a potent inhibitor of human MCF-7 breast cancer cell proliferation. Medicinal Chemistry Research. 2012; 21: 1384-1388. https://doi.org/10.1007/s00044-011-9657-6.

[47] Müller LG, Pase CS, Reckziegel P, Barcelos RCS, Boufleur N, Prado ACP, Burger ME. Hepatoprotective effects of pecan nut shells on ethanol-induced liver damage. Experimental and Toxicologic Pathology. 2013; 65: 165-171. https://doi.org/10.1016/j.etp.2011.08.002.

[48] Hilbig J, Policarpi PB, GriniciusVMAS,MotaNSRS,ToaldoIM, Luiz MTB, Block JM. Aqueous extract from pecan nut [*Carya illinoinensis* (Wangenh) C. Koch] shell show activity against breast cancer cell line MCF-7 and Ehrlich ascites tumor in Balb-C mice. Journal of Ethnopharmacology. 2018; 211:256-266. https://doi.org/10.1016/j.jep.2017.08.012.

[49] Gutkoski LC, Bonamigo JMDA, Pedó I, Teixeira DMDF. Development of oat based cereal bars with high dietary fiber content. Ciencia e Tecnologia de Alimentos. 2007; 27: 355-363. https://doi.org/10.1590/S0101-20612007000200025

[50] Barreira JCM, Nunes MA, da Silva BV, Pimentel FB, Costa A S G, Alvarez-Ortí M, Oliveira MBPP. Almond cold-pressed oil by-product as ingredient for cookies with potential health benefits: Chemical and sensory evaluation. Food Science and Human Wellness. 2019; 8: 292-298. https://doi.org/10.1016/j.fshw.2019.07.002

[51] Sanchiz A, Pedrosa MM, Guillamón E, Arribas C, Cabellos B, Linacero R, Cuadrado C. Influence of boiling and autoclave processing on the phenolic content, antioxidant activity and functional properties of pistachio, cashew and chestnut flours. Lwt. 2019; 105: 250-256. https://doi.org/10.1016/j.lwt.2019.02.035

[52] Felberg I, Antoniassi R, Deliza R, Freitas SC, Modesta RC. Della. Bebida de soja e castanha do Brasil: Processamento, composição, avaliação sensorial e de cor. Ciencia e Tecnologia de Alimentos. 2009; 29: 609-617. https://doi.org/10.1590/S0101-20612009000300024

[53] Rebouças MC, Rodrigues MCP, Laurentino LS, Aguiar EC, Penha MFA. Efeito de atributos externos na aceitação de bebida de castanha de caju. Arquivos Brasileiros de Alimentação. 2018; 1(1), 26-41.

[54] Demoliner F, Carvalho LT, Liz GR, Prudêncio E.S, Ramos JC, Bascuñan VLAF, Block JM. Improving the nutritional and phytochemical compounds of a plant-based milk of sapucaia nut cake using block freeze concentration. International Journal of Food Science and Technology. 2020; 1-12. https://doi.org/10.1111/ijfs.14568

[55] Sharma P, Gaur VK, Sirohi R, Larroche C, Kim SH, Pandey A. Valorization of cashew nut processing residues for industrial applications. Industrial Crops and Products. 2020; 152: 112550. https://doi.org/10.1016/j.indcrop.2020.112550

[56] Lima JR, Garruti DDS, Bruno LM, Araújo ÍMS, Nobre ACO, Garcia LGS. Replacement of Peanut by Residue from the Cashew Nut Kernel Oil Extraction to Produce a Type Paçoca Candy. Journal of Food Processing and Preservation. 2017; 41: 1-5. https://doi.org/10.1111/jfpp.12775

[57] Maciel LG, Ribeiro FL, Teixeira GL, Molognoni L, Nascimento dos Santos J, Larroza Nunes I, Block JM. The potential of the pecan nut cake as an ingredient for the food industry. Food Research International. 2020; 127: 108718. https://doi.org/10.1016/j.foodres.2019.108718

Chapter 9

Bioactive Peptides from Agriculture and Food Industry Co-Products: Peptide Structure and Health Benefits

Jirawat Yongsawatdigul and Ali Hamzeh

Abstract

Co-products from food processing are typically disposed or turned into low value animal feed. Proteinaceous co-products can be converted to bioactive peptides exerting health benefits, which can lead to development of nutraceuticals and functional foods. This is an effective means for valorization of these co-products. The release of encrypted peptides exhibits various bioactivities, including anti-hypertension, antioxidant, immunomodulatory activities among others, *in vitro*, and some activities have been demonstrated *in vivo*. Structure modification of bioactive peptides occurring under gastrointestinal digestion and cellular transport remains the important factor determining the health benefits of bioactive peptides. Understanding peptide transformation in gastrointestinal tract and in blood circulation before reaching the target organs would shed some lights on its bioavailability and subsequently ability to exert physiological impact. In this chapter, the potential health promoting properties of peptides encrypted in various sources of co-products will be reviewed based on evidence on *in vitro*, *in vivo* and clinical trial studies. Structural changes of bioactive peptides under physiological condition will also be discussed in relation to its bioactivities.

Keywords: bioactive peptide, transepithelial transport, bioavailability, peptide sequence

1. Introduction

A large number of food co-products is annually generated from various plant- and animal-based food processing plants. The Food and Agriculture Organization (FAO) recently reported that 14% of global food production is lost prior to getting the retails [1]. There is also an estimation of generating co-products and waste of 38 percent in food processing industries [2]. These co-products are usually disposed, landfilled, incinerated and/or processed into animal feed and other low value products such as compost and fertilizer [3], which would cause economic losses and environmental concerns.

Bioactive peptide production is a promising approach to fully utilize proteinaceous co-products. These peptides could also exert physiological functions, leading to health benefits, such as antioxidant, antihypertension, antidiabetic, immunomodulatory activities among others. But these peptides are encrypted in the protein chain and needed to be released to exert bioactivities. Fermentation, enzymatic and chemical (by acid or alkali) hydrolysis could be applied to produce bioactive peptides as well as using solvents which is normally used to extract natural peptides. Among them, enzymatic hydrolysis known to be more effective as it is a mild, ecofriendly, and controllable process. Bioactivities of the released peptides are affected by their size, hydrophobicity, charge, amino acid composition and their sequence which are different based on various enzyme and substrate as well as the hydrolysis conditions. After production of peptides with the certain bioactivity, their structural modification upon gastrointestinal digestion and epithelial transportation and absorption must be taken into consideration to determine their potential bioavailability. Several bioactivities and physiological functions of peptides derived from both animal and plant-based agricultural co-products have been reported, which are mainly focused on their antioxidant, antihypertensive, antidiabetic and antibacterial properties. In this chapter, their main biological activities along with the associated structure will be considered. Besides, the structure-bioavailability relation of the peptides will be demonstrated and procedures to keep them intact upon gastrointestinal digestion and transepithelial transportation will also be proposed.

2. Agricultural co-products as a source of bioactive peptides

Agricultural co-products involve a vast variety of materials with great potential as a substrate for bioactive peptide production due to the low value and high protein content. It is estimated that about 9000 tons in dairy, 3 and 9 million tons in seafood and livestock industries, respectively, are annually discarded [4]. Skins, bones and heads are rich in collagen. Collagen is known as a promising source of peptides with several bioactivities which could exert physiological functions. Collagen comprises of hydrophobic amino acids, particularly proline and hydroxyproline, offering a higher chance of being absorbed through epithelial membrane. These amino acids also make them stable against proteases in gastrointestinal digestion tract and brush border, so that the intact peptides could be absorbed and reach the target organ. Blood is also high-protein co-product obtained in a large volume in a slaughterhouse. It has been reported that about 2.5 billion tons of blood are annually generated only in Europe [5]. Hemoglobin in red corpuscles and albumin, globulins, and fibrinogen in plasma are the major proteins [6]. Whey is a predominant co-product of dairy industry with an annual production of approximately 180–190 million tons [7]. The whey-derived peptides have also lower allergenicity beyond their bioactivities, because β-lactoglobulin (β-lg) as a main whey protein contributing to allergenicity would degrade within hydrolysis process [8].

Production of bioactive peptide from some plan-based materials are restricted due to difficulties in protein recovery from their unique rigid structure of polysaccharides. To meet the challenge, pretreatment of these substrates would improve protein recovery and provide other benefits such as reduction of time and energy consumption, leading to a higher efficacy in bioactive peptide production [9]. Corn gluten meal was immersed in alkali solution, treated with α-amylase and cooked prior to proteolysis [10]. Ultrasound exposure of watermelon seed caused structural changes, leading to production of peptides with higher antioxidant activity than those without any treatment. Ultrasound could degrade interactions in matrix and unfold proteins, so that more hydrophobic residues and reactive sites are exposed, resulting in a higher efficacy in protein hydrolysis [9]. Oat bran polysaccharide was digested by cellulase and viscozyme to ease protein recovery prior to proteolysis and production of antioxidant peptides [11].

Bran from rice and wheat, is a main protein-rich co-product in cereal industry with production of about 50–60 and 90 million tons every year, respectively. Albumin, globulin, prolamin and glutelin are the major proteins in bran with a total protein content of 12–20% [12]. Seeds are also another source of bioactive peptides, among them those harvested for oil production generate a large mass of co-product. Soybean, rapeseed and canola meals with the crude protein contents of about 48, 36 and 38%, respectively, are important cases of protein-rich co-products obtained from oil industries [13]. Corn gluten meal is also a major proteinaceous co-product generated in a large quantity from corn wet milling process in which, zein and glutelin are the main proteins (68 and 28%, respectively). Zein composes of mainly hydrophobic amino acids by which, more hydrophobic peptides are likely produced [14]. Hydrophobicity is an important feature of peptides to exert some bioactivities such as antioxidant and antihypertensive properties.

Thus far, peptides with various bioactivities *in vitro* and *in vivo* levels have been recognized from these low value co-products. Conversion of these co-products to more value-added hydrolysates and/or peptides would ultimately lead to development of health promoting food products.

3. Biological activities

Several bioactivities and physiological functions of peptides derived from both animal- and plant- based agricultural co-products have been reported. These activities include antioxidant, antihypertensive, antidiabetic and antibacterial properties.

3.1 Antioxidant activity

Reactive oxygen species (ROS) are formed during normal cellular metabolism of providing energy, respiration and also when cells are exposed to exogenous oxidative stress [15, 16]. Normally, these products are neutralized by endogenous antioxidant defense systems, such as antioxidant enzymes (superoxide dismutase: SOD, glutathione peroxidase: GPx and catalase), glutathione (GSH) and others. But, the excessive level of ROS could result in many health disorders, such as cancer, cardiovascular, respiratory, neurodegenerative and other diseases [16, 17]. Therefore, antioxidant-containing diet could help to overcome ROS and subsequently their corresponding disorders. Nowadays, attempts have been made to find new antioxidant compounds from natural resources due to their benefits over synthetic compounds. To that direction, antioxidant peptides from agricultural co-products are gaining more attractions, because of their nontoxicity and safety besides their nutritional properties. Antioxidant hydrolysates/peptides production from agricultural co-products including skins, bones, viscera, whey as animal-based and brans, seeds, leaves, gluten as plant-based co-products have been extensively reviewed [17–21].

Lower molecular weight (MW) peptides with hydrophobic and aromatic amino acids (HAAs and AAA, respectively) have been generally reported to exert good antioxidant activities [22]. The HAAs could improve peptides accessibility toward ROS through binding with lipid and reaching to free radicals, so that the peptides could quench them effectively. The AAAs including Trp, Phe and Tyr are also correlated with the strong antioxidant activity via their high electron transferring capacities of their aromatic rings. Besides HAAs and AAAs, some of hydrophilic amino acids, such as His could improve the activity through its imidazole ring which has been indicated as strong electron donator [23]. Presence of charged amino acids in peptide structure could also improve the activity. A higher negatively charged amino acids (NCAA) have been observed in plasma hydrolysates prepared from chicken blood that showed higher antioxidant activity than those prepared from blood corpuscles with lower NCAAs [6]. Presence of NCAAs are reported to correlate with the strong antioxidant activity, because they can neutralize free radicals by giving their excess electrons.

Although antioxidant activity of agricultural co-product has been widely evaluated, few studies were conducted to assess the activity *in vivo* and there is still a gap for clinical trial. The chemical *in vitro* studies are not able to reflect the activity in biological systems due to their complicated physiological conditions. In addition, cellular evaluation might be able to provide a comparable environment to biological systems. Antioxidant properties of peptides from fish sauce increased with the increasing of peptides concentration based on chemical assays, while these peptides could act as pro-oxidants in higher concentration (>50 μg L-leucine equivalent/ml) in cellular experiments [24]. Hydrolysates prepared from corn gluten meal and ham seed meal as well as ACFL, a peptide from horse mackerel viscera, increased the level of antioxidant enzymes based on *in vivo* models upon exposure to oxidative stress [10, 25, 26]. In a human trial study, a reduction in plasma malondialdehyde and an increase in SOD level was observed after daily ingestion of 4.5 g black soy derived peptides for 8 weeks [27].

3.2 Antihypertensive activity

Hypertension is a major risk for many disorders including coronary heart disease, stroke, heart failure, vision loss, chronic kidney disease, and dementia [28, 29]. Renin-angiotensin system (RAS) is responsible in blood pressure regulation. The liver-released angiotensin is converted to angiotensin-I (Ang-I) by renin, then the Ang-I is easily degraded to Ang-II, a potent vasoconstrictor, by angiotensin converting enzyme (ACE), leading to vasoconstriction and hypertension. Thus, hypertension can be controlled using renin and ACE inhibitors, in which the latter is more important due to its dual functions, catalyzing Ang-II (a potent vasoconstrictor) and inactivation of bradykinin (a vasodilator) [30]. However, controlling blood pressure by peptides from agricultural co-products also reported to be achieved through other mechanisms, such as nitric oxide production and blocking calcium channel and Ang-II receptor. The receptor blockers are able to hinder vasoconstriction and other functions mediated by Ang-II. Calcium channel blocker can avoid calcium availability in blood vessel cell wall and heart which make them to have a lower extent contraction, leading to relaxation and lower blood pressure [27, 31]. Arg-containing peptides have also been reported to suppress hypertension as Arg is a precursor in nitric oxide production. Nitric oxide (NO) is synthesized by the reaction of Arg and oxygen in the presence of nitric oxide synthase as a catalyst. NO is a vasodilator which act against Ang-II and a balanced level of these two compounds could lead to normal blood pressure [32]. Peptides namely, LIWKL, RPYL, RRWQWR, blocking the Ang-II receptor and HRW, a calcium channel blocker, have been reported to exert antihypertensive effects [33, 34].

The antihypertensive activity of peptides depends on several factors including amino acid composition and their position, molecular weight and charge. Normally, peptides with lower molecular weight exert higher activity due to their higher affinity to bind with ACE, in which the most potent reported peptides are di- and tri-peptide. Moreover, small peptides can stay intact through gastrointestinal digestion and epithelial transportation, reaching to blood circulation system and the target organ [31]. The structure–activity relationship of 168 dipeptides and 140 tripeptides with ACE inhibitory activity was assessed [35]. The authors reported that presence of bulky side chain and hydrophobic amino acids in dipeptides could result in higher activity, while presence of aromatic amino acids at C-terminus and positively charged amino acids in the middle and hydrophobic residues at N-terminus brought them higher ACE inhibitory activity. The interaction of peptides with zinc ion in active site of ACE could effectively deactivate the enzymes. It has been reported that Leu could bind with Zn^{2+} by its carboxyl group and inhibit the enzyme activity [36].

Antihypertensive activity of hydrolysates/peptides from agricultural co-products was extensively explored by *in vitro* and *in vivo* studied, mostly based on their inhibition capacity against ACE and rat with hypertension (either SHR or hypertensive-induced rat), respectively (**Table 1**). However, clinical studies are still needed to confirm their health benefits. Clinical studies on the antihypertensive effect of IPP and VPP showed a controversial result as it reduced systolic and diastolic blood pressure in Asian case studies to a higher extent than Caucasians, while no effect was observed in Dutch and Danish cases [46, 47]. They reported that the difference could be associated to variations in genetics and dietary habit which should be taken into consideration. Kwak et al. [27] reported antihypertensive effect of black soy peptide in human trial, in which systolic blood pressure decreased in hypertensive subjects likely through reduction in ACE activity and increase in nitric oxide production. The authors concluded that the activity of black soy peptides might also be associated with higher arginine content which is a substrate for nitric oxide formation, known as a strong vasodilator.

Source	**Enzyme**	**Hydrolysate/ Peptide**	**Activity**		**Ref.**
			In vivo	***In vitro***	
Animal based co-products					
Bovine whey lactoferrin	Pepsin	Partially purified hydrolysate (<3 kDa) LIWKI, RPYL and RRWQWR		ACE inhibitory activity, Blocking Ang-II receptor	[33]
Fish (Cobia) skin	Protamex	Hydrolysate WAA, AWW, IWW, WL	Systolic and diastolic blood pressure reduction at 600 mg/kg: −21.9 and – 15.5 mmHg, respectively, after 4 h in SHR	ACE inhibitory activity ACE inhibitory activity IC_{50}[2]: 41, 4.3, 0.2 and 8.5, respectively	[37]
Chicken bone	Pepsin	YYRA	Systolic blood pressure reduction at 10 mg/ kg in SHR: −2 mmHg after 6 h	ACE inhibitory activity IC_{50}: 33.9	[38]
Bovine bone gelatin	Alcalase	RGL-(Hyp)-GL and RGM-(Hyp)-GF	Systolic blood pressure reduction at 30 mg/kg: −31.3 and −38.6 mmHg for RGL and RGM after 4 and 6 h, respectively, in SHR	ACE inhibitory activity IC_{50}: 0.9 and 6.9, respectively	[39]
Bovine blood plasma	Flavourzyme	HPY		ACE inhibitory activity IC_{50}: 0.7	[40]

Source	Enzyme	Hydrolysate/ Peptide	Activity		Ref.
			In vivo	*In vitro*	
Poultry viscera	Autolysis	ARIYH, LRKGNLE and RVWCP		ACE inhibitory activity IC_{50}: 8.9, 8.9 and 4.9, respectively	[29]
Whey		IW	ACE activity reduction in human plasma by 32% at 50 mg administration		[41]
Plant based co-products					
Wheat bran	Alcalase	Hydrolysate, peptides <1 kDa including NL, QL, FL, HAL, AAVL, AKTVF, TPLTR	Systolic blood pressure reduction at 100 mg/kg in SHR[1]: −20 and − 35 mmHg after 6 h for hydrolysates and the peptides <1 kDa, respectively	Renin and ACE inhibition activity	[42]
Corn gluten meal	Trypsin	AY	Systolic blood pressure reduction at 50 mg/kg in SHR: −9.5 mmHg after 2 h	ACE inhibitory activity IC_{50}: 3.6	[28]
Cottonseed	Papain	FPAIGMK		ACE inhibitory activity IC_{50}: 46.7	[43]
Flaxseed protein isolate	Thermoase	Partially purified hydrolysate (3–5 kDa)	Systolic blood pressure reduction at 200 mg/kg in SHR: −37 mmHg after 8 h	Renin and ACE inhibitory activity	[44]
Red seaweed (*Porphyra columbina*) by-product	Trypsin and Alcalase	Hydrolysate		ACE inhibitory activity	[45]

[1]*Spontaneously hypertensive rat.*
[2]*IC_{50} based on μg/ml.*

Table 1.
Peptides/hydrolysates derived from agricultural co-products involved in antihypertension activity.

3.3 Antidiabetic activity

The diabetes is a chronic health problem which involved 463 million adults in 2019 and it is estimated to reach 700 million by 2045. In the health disorder, the elevated blood glucose cannot be treated properly due to either pancreas failure in insulin production (type-I) or insulin resistance in the body (type-II), in which the latter comprised the majority of about 90–95% [48].

The diabetes type-I treatment is associated to insulin injection, while the diabetes type-II can be prevented by controlling the pathways, by which the blood glucose elevates. The enzymes, α-amylase and α-glucosidase, play roles in carbohydrate digestion through breaking them down to oligosaccharides and subsequently to glucose, which is easily absorbed from intestine to blood, leading to hyperglycemia. In addition, dipeptidyl peptidase IV (DPP-IV), a protease located on endothelial, epithelial and some other cells, could easily degrade hormones stimulating insulin secretion during food ingestion, including glucagon-like peptide-1 (GLP-1) and glucose-dependent insulinotropic peptide (GIP), and cause dysregulation of blood glucose. Therefore, many studies have attempted to find bioactive peptides to inhibit these enzymes, so that the diabetes type-II would be cured or prevented. Bioactive peptide with *in vitro* enzyme inhibitory and *in vivo* antidiabetic activities derived from agricultural co-products are summarized in **Table 2**.

Source	**Enzyme**	**Hydrolysate/ Peptide**	**Activity**		**Ref.**
			In vivo	***In vitro***	
Animal based co-products					
Chicken feet	Neutrase	Hydrolysate	Reduction of glycemia in glucose-intolerant rats at 300 mg/kg	DPP-IV inhibitory activity	[49]
Collagen from pig and cattle skin, fish scale and chicken feet	Collagenase	GA-Hyp GPA GP-Hyp		DPP-IV inhibitory activity IC_{50}: >20, 5.03 and 2.51 mM, respectively	[50]
Mare whey	Papain	NLEIILR TQMVDEEIM-EKFR		DPP-IV inhibitory activity IC_{50}: 86.3 and 69.84 μM, respectively	[51]
β-Lactoglobulin from bovine whey	Trypsin	VAGTWY	Reduction of glucose level in mice at 300 mg/kg	DPP-IV inhibitory activity IC_{50}: 210 μM	[52]
	Trypsin	IPAVF		DPP-IV inhibitory activity IC_{50}: 44.7 μM	[53]
Atlantic salmon skin	Flavourzyme	GPAE, GPGA		DPP-IV inhibitory activity IC_{50}: 49.6 and 41.9 μM, respectively	[54]
Cuttlefish viscera	Crude protease extracts from smooth hound and cuttlefish hepatopancreas	Hydrolysate		Stimulation cholecystokinin (CCK) and GLP-1 release in enteroendocrine STC-1 cells, DPP-IV inhibitory activity	[55]
Bovine haemoglobin	*in vitro* GI digestion by pepsin and pancreatin	VAAA KAAVT, YGAE, ANVST and TKAVEH		DPP-IV inhibitory activity, IC_{50}: 141 μM Stimulation of GLP-1 secretion in STC-1	[56]

Source	Enzyme	Hydrolysate/ Peptide	Activity		Ref.
			In vivo	*In vitro*	
Plant based co-products					
Wheat gluten	Debitrase HYW20	Pro-containing peptides		DPP-IV inhibitory activity	[57]
	Protease from ginger	QPQ, QPG, QPF, LPQ, SPQ		DPP-IV inhibitory activity IC_{50}: 79.8, 70.9, 71.7, 56.7, and 78.9 μM, respectively	[58]
Oat globulin	Trypsin	LQAFEPLR		DPP-IV inhibitory activity IC_{50}: 103.5 μM	[59]
Luffa cylindrica seed	Pepsin Trypsin Alcalsae	Hydrolysate		α-amylase and α-glucosidase inhibitory activity	[60]
Rice bran	UmamizymeG Alcalase	IP 13 peptides containing 6–32 amino acids		DPP-IV inhibitory activity IC_{50}: 0.41 mM α-amylase and β-glucosidase inhibitory activity	[61] [62]
Hemp seed meal	Alcalase	LR, PLMLP		α-glucosidase inhibitory activity	[63]

Table 2.
Examples of agricultural by-product peptides and hydrolysates exhibiting antidiabetic activity.

In a clinical study, Goudarzi and Madadlou [64] indicated that hydrolysate prepared from whey proteins stimulated insulin production, so that plasma glucose got back to normal level in postprandial hyperglycaemia cases, while the hydrolysate had no effect in prehypertensive cases. Although studies indicated that hydrolysates/peptides might stimulate secretions of hormones involving in insulin production [55, 56, 64], most studies have focused on major enzymes involving in carbohydrate digestion and DPP-IV. The structure–activity relation of peptides possessing the diabetes-involving-enzyme inhibition has not been completely understood yet. Nongonierma et al. [57] identified di- and tri-peptides inhibiting DPP-IV in wheat gluten hydrolysate. These peptides had some main characteristics including the presence of Pro at carboxyl terminus or penultimate position and Phe or Leu at amino terminus. Li-Chan et al. [54] described that peptides with DPP-IV inhibitory activity required presence of hydrophobic amino acids, particularly Pro, as Pro placed at 1–4 (preferably at second) positions from N-terminal end and bounded with Leu, Val, Phe, Ala and Gly. Dipeptides as X-Pro with X as a small size hydrophobic amino acid would likely be an effective inhibitor. Presence of hydrophobic and aromatic amino acids at N-terminal end of peptides with DPP-IV inhibitory activity was also reported by Lima et al. [65]. Ren et al. [63] evaluated the α-glucosidase inhibitory capacity of peptides from hemp seed and indicated that hydrophobicity of peptides was a prime factor affecting inhibitory activity and molecular weight as a second priority. The authors have also reported that larger molecular weight peptides could also enhance α-glucosidase activity. α-Amylase is another enzyme involving in carbohydrate digestion and it has been reported that presence of branched and aromatic amino acids such as Lys, Phe, Tyr and Trp and positively charged amino acid could help to inhibit the enzyme [60].

3.4 Antibacterial activity

The antibacterial activity of hydrolysates/peptides has been studied to a lesser extent when compared to other aforementioned properties. Hydrolysates/peptides with antibacterial activity have been obtained from co-products of milk, seafood, meat and others which are summarized in **Table 3**. Conventional antibiotics and preservatives are extensively applied to control pathogens, which lead to antibiotic-resistant strains. Therefore, an alternative antimicrobial agent has been sought. Peptides with antibacterial properties could be one of alternative agents as they are non-toxic and could act against both Gram-negative and Gram-positive as well as antibiotic-resistant bacteria [76]. Typically, chemical antibiotics have specific targets and bacteria can develop various defense strategies towards antibiotics. In contrast, antimicrobial peptides target cell membrane and can cause serious damage which make it difficult to develop resistance [77].

The antibacterial activity of these peptides is associated to their molecular weight, charge and hydrophobicity [67]. Peptides are attached by negatively-charged residues of cell membrane, like lipopolysaccharides and lipoteichonic acid on Gram-negative and Gram-positive bacteria, respectively, through electrostatic

Source	Peptide/Hydrolysate	Test bacteria	Activity		Ref.
Animal based by-products					
Rainbow trout Viscera	Hydrolysate prepared by pepsin	*Flavobacterium psychrophilum*, *Renibacterium salmoninarum*	MIC[1] (mg/ml)	2 5	[66]
Yellowfin tuna viscera	Partially purified (<3 kDa) hydrolysate prepared by Protamex	*Listeria. monocytogenes*, *Staphylococcus aureus*, *Escherichia coli*, *Pseudomonas aeruginosa*	MIC (mg/ml)	0.5 0.5 0.5 0.5	[67]
Bovine hemoglobin	TSKYR obtained by pepsin-hydrolyzed hemoglobin	Total viable colonies Coliform bacteria Yeasts Molds	Reduce microbial counts in ground beef stored in refrigerator		[68]
Porcine blood proteins: albumin globulin	Hydrolysate prepared by Alcalase, Flavourzyme, Protamex, trypsin and papain	*Bacillus cereus*	IZ[2] (mm): 2.30–2.55		[69]
Camel whey	Crude hydrolysate prepared by trypsin/ partially purified (<3 kDa)	*E. coli* *S. aureus* *Salmonella typhimurium* *Streptococcus mutans*	MIC (mg/ml)	130/65 260/130 NI[3]/260 NI/260	[70]
Cow whey	Crude hydrolysate prepared by trypsin/ partially purified (<3 kDa)	*E. coli* *S. aureus* *S. typhimurium* *S. mutans*	MIC (mg/ml)	260/130 260/130 NI/260 NI/260	
Snow crab co-products	Hydrolysate prepared by Protamex	*E. coli* *L. innocua*	IZ (mm)	9 7.5	[71]

Source	Peptide/Hydrolysate	Test bacteria	Activity		Ref.
Plant-based co-products					
Palm kernel cake	Hydrolysate prepared by Alcalase	*Lisinibacillus sphaericus* *Bacillus thuringiensis* *B. cereus* *Clostridium perfringens* *B. subtilis*	MIC (μg/ml)	150 150 200 250 450	[72]
Jatropha curcas meal	CAILTHKR obtained by Protamex-hydrolyzed meal	*E. coli* *Shigella dysenteriae* *P. aeruginosa* *S. aureus* *B. subtilis* *S. pneumoniae*	MIC (μg/ml)	29 46 58 45 34 68	[73]
Rice bran	KVDHFPL obtained by bromelain-hydrolyzed bran	*Listeria monocytogenes* *L. monocytogenes* biofilm	MIC (μg/ml)	0.25 8	[74]
	LRRHASEGGHGPHW EKLLGKQDKGVIIRA SSFSKGVQRAAF obtained by pepsin-hydrolyzed bran	*Porphyromonas gingivalis* (PG), *Candida albicans* (CA)	IC_{50}: (PG, CA; μM)	289, ND[4] ND, 75.6 ND, 78.5	[75]

[1]*Minimum inhibitory concentration.*
[2]*Inhibition zone.*
[3]*No inhibition.*
[4]*Not detected.*

Table 3.
Antibacterial peptide/hydrolysate prepared from agricultural co-products.

interactions, by which the structure of cell surface was disrupted. Subsequently, peptides could permeate to the cell and reach to cytoplasmic membrane, causing leakage of cytoplasmic fluid [78].

Antibacterial activity of hydrolysates and/or peptides has mostly evaluated by their directly exposure to pathogens and less studies have conducted to assess their application in food. A peptide, TSKYR, obtained from bovine hemoglobin [68] and hydrolysate prepared from yellowfin tuna waste [67] were added to ground beef and minced fish, respectively. The peptide, TSKYR, could reduce total viable colonies, yeasts, molds and particularly coliform bacteria within 14 days storage in a refrigerator. Moreover, the peptide (0.5% w/w) in ground beef was able to diminish lipid oxidation by 60% which was reported to be comparable to BHT. Pezeshk et al. [67] reported that hydrolysates prepared from yellowfin tuna were able to increase fish (silver carp) mince shelf-life in a refrigerator through inhibition of psychrophilic and total count bacteria as well as prevention of oxidative degradation.

4. Bioavailability

Studies revealed that peptides prepared from agricultural co-products have great potential of health promoting properties. However, bioavailability of these peptides is a challenge, in which they need to stay intact within gastrointestinal tract (GIT) and epithelial transportation to reach their target organs and exert physiological functions. Proteases and peptidases in GIT, brush border and cytoplasm are able to break down the peptide bond to a higher extent, leading to changes in structure and subsequently the activity. However, some peptides have been reported to be stable within the digestion and transportation. There are several factors affecting the peptides stability, which are associated to proteases specificities in GIT. Lower molecular weight, negatively charged, hydrophilic and acidic amino acid containing peptides are reported to be more stable against GI digestion. Negatively charged peptides from milk are more stable against GI digestion followed by positively charged and neutral peptides [79, 80]. Hydrophobic peptides reported to have less stability, which might be due to pepsin specificity towards hydrophobic amino acids [81]. Peptides containing more acidic amino acids, and also those with lower molecular weight showed more stability against GI digestion [6, 79]. Peptides with the molecular weight of larger than 3 kDa were easily digested by GI proteases, while peptides with <1 kDa mostly survived and no change in their antioxidant activity upon GI digestion [82]. Savoie et al. [83] reported that peptides from animal- (casein and cod fish) and plant- (soy and gluten) based substrates with Pro and Glu showed higher stability. Pro has a rigid ring structure bonded to β-carbon which makes it resistant against proteolytic degradation [80]. Thus, Pro containing peptides, IAGRP and PTPVP, have been reported to stay intact after in vitro GI digestion [84].

Epithelial permeation of bioactive peptides into blood circulation system is another challenge that affects physiological activities. Peptides may undergo some structural modification induced by brush border proteases (**Table 4**). For instance, a peptide with ACE inhibitory activity, KPLL, can be degraded to KP and LL within epithelial permeation, resulting in lower activity than the intact form [96]. The permeation could occur through four pathways, including peptide transporter 1 (PepT1), passive paracellular transportation through tight junctions, transcytosis and simple passive transcellular diffusion. Peptide properties such as size, hydrophobicity, charge and amino acid sequence are important factors affecting their absorption. Briefly, small (di- and tri-) peptides can be transported via PepT1 route, however peptide properties have effects on its efficacy. Non charged and hydrophobic peptides have higher affinity towards PepT1. Hydrophilic and negatively charged low molecular weight peptides can pass through energy-independence paracellular route. Transcytosis is an energy-dependent route, by which long chain peptides, particularly hydrophobic, can be transported. A highly hydrophobic peptide is likely transported through simple passive transcellular diffusion. To evaluate the effect of molecular weight on the permeation, Wang and Li [97] reported that hydrolysates with the molecular weight lower than 500 Da (mostly di- and tri-peptides) showed higher bioavailability and were able to pass through Caco2 cell via PepT1 route, while those with the molecular weight ranging 500–1000 and 1300–1600 Da permeated through paracellular route. Besides molecular weight, peptide sequence also affects its bioavailability. A Pro-containing peptide has more stability towards brush border proteases and peptides with Leu at N-terminus have been reported to be highly susceptible to hydrolysis [86, 90, 91].

Peptides structural changes usually occur in GI tract and transepithelial transportation that would likely have effects on their physiological functions. To meet the challenge, some approaches have been applied to improve the stability of these peptides such as using permeation enhancer, enzyme inhibitor and encapsulation. Sodium caprate has been used to improve the permeability of two antihypertensive peptides, IPP and LKP, through paracellular route in Caco2 cell [98]. The authors reported that sodium caprate could intensify the peptides absorption via paracellular mechanism and inhibited PepT1 route, leading to antihypertensive effect in SHRs model. An antihypertensive peptide, RLSFNP, would degrade to RLSF, SFNP, FNP and F during the epithelial transportation. Permeation enhancers including

Peptide	Bioactivity	Permeated peptides	Route of transport	Ref.
VLPVPQK	ACE inhibitor Antioxidant	VLPVPQK, VLPVPQ	PepT1/ SOPT2	[85]
LHLPLP	ACE inhibitor Antihypertensive	LHLPLP, HLPLP	Paracellular	[86]
IPP, LKP	ACE inhibitor	IPP, LKP	PepT1 and paracellular	[87]
IWHHT	ACE inhibitor Antihypertensive Antioxidant Anti-inflammatory	IWHHT, IWH, IW	Paracellular, PepT1 and paracellular, PepT1	[88]
YQEPVLGPVRGPFPIIV	Immunomodulatory	YQEPVLGPVRGPFPIIV, QEPVLGPVRGPFPIIV, YQEPVLGPVR-GPFPII	Transcytosis	[89]
LSW	ACE inhibitor Anti-inflammatory	LSW, SW	PepT1 and paracellular	[90]
RRWQWR	ACE inhibitory Antihypertensive	RWQ, WQ	Paracellular	[91]
WDHHAPQLR	Antioxidant	WDHHAPQLR, DHHAPQLR, WDHHAP, QLR	Paracellular PepT1	[92]
YWDHNNPQIR	Antioxidant	YWDHNNPQIR, WDHNNPQIR, DHNNPQIR and YWDHNNPQ	Transcytosis	[93]
YFCLT, GLLLPH	Antioxidant	YFCLT, GLLLPH	Paracellular and transcytosis	[94]
WGAPSL	Cholesterol- Lower	WGAPSL, WGAPS, WGAP, GAPSL, GAP, SL	Paracellular	[95]

Table 4.
Peptide modification within epithelial permeation and their transportation route across Caco-2 cell.

sodium glycocholate hydrate, sodium deoxycholate and Na_2EDTA as well as enzyme inhibitors, bacitracin and leupeptin, have been applied to improve the intact peptide bioavailability [99]. Na_2EDTA was the most effective to enhance RLSFNP absorption through enlarging intracellular junctions. They also reported that bacitracin could exert permeation enhancer activity beyond its protease inhibitory effect. Permeation enhancer is believed to cause damages in cell membrane in case of long-term usage, leading to inflammation. However, major destructive effects were not observed by using bacitracin in rat intestine [100]. Besides, encapsulation of RLSFNP by liposome could also facilitate the intact peptide transportation through transcytosis in Caco2 cell [101]. In addition, Li et al. [102] used nano-encapsulation of antidiabetic peptides made by chitosan coated liposome to maintain the stability of peptides.

5. Conclusions

Co-products are inevitably generated in food production, distribution, processing and consumption. These protein rich and low value materials could provide a great source of bioactive peptides production. Many peptides with various activities have been purified and identified, however, their activities need to be confirmed via *in vivo* studies and human trials, so that they could be further develop to functional food products. The main obstacle of developing these peptides in functional foods is their structural modification after ingestion via proteolytic degradation in gastro-intestinal tract and epithelial absorption, which would likely lead to a reduction in bioactivity. Although some peptides, particularly those containing Pro, could stay intact within digestion and absorption, structural modification usually happen in the route. Encapsulation of susceptible peptides or applying protease inhibitor as well as permeation enhancer in epithelial cells could facilitate the intact peptides absorption. Although these strategies might allow peptides to reach the target organ and exert certain physiological effect, their safety, particularly the use of protease inhibitors, needs further investigation regarding their side effects under physiological condition.

Acknowledgements

This work was financially supported by the National Research Council of Thailand under the project Food Innovation for Safety and Value Creation of Nakhonchaiburin (SUT3-305-61-12-06). Postdoctoral fellowship to AH supported by Suranaree University of Technology is also greatly appreciated.

Conflict of interest

The authors declare no conflict of interest.

Author details

Jirawat Yongsawatdigul* and Ali Hamzeh
School of Food Technology, Institute of Agricultural Technology,
Suranaree University of Technology, Nakhon Ratchasima, Thailand

*Address all correspondence to: jirawat@sut.ac.th

References

[1] The state of food and agriculture. 2019. The Food and Agriculture Organization: www.FAO.org

[2] Bharat Helkar P, Sahoo A. Review: Food Industry By-Products used as a Functional Food Ingredients. International Journal of Waste Resources. 2016; 6:3. DOI:10.4172/2252-5211.1000248

[3] Lin CSK, Pfaltzgraff LA, Herrero-Davila L, Mubofu EB, Abderrahim S, Clark JH, et al. Food waste as a valuable resource for the production of chemicals, materials and fuels. Current situation and global perspective. Energy and Environmental Science. 2013; 6: 426-64. DOI: 10.1039/C2EE23440H

[4] Moreno-Hernández JM, Benítez-García I, Mazorra-Manzano MA, Ramírez-Suárez JC, Sánchez E. Strategies for production, characterization and application of protein-based biostimulants in agriculture: A review. Chilean Journal of Agricultural Research. 2020; 80: 274-89. DOI: 10.4067/S0718-58392020000200274

[5] Toldrà M, Lynch SA, Couture R, Álvarez C. Blood proteins as functional ingredients. In: Galanakis CM, editor. Sustainable Meat Production and Processing. Academic Press; 2019. p. 85-101. DOI: 10.1016/C2017-0-02230-9

[6] Hamzeh A, Wongngam W, Kiatsongchai R, Yongsawatdigul J. Cellular and chemical antioxidant activities of chicken blood hydrolysates as affected by in vitro gastrointestinal digestion. Poultry Science. 2019;98(11): 6138-6148. DOI: 10.3382/ps/pez283

[7] Oliveira D, Fox P, O'Mahony J. Byproducts from dairy processing. In: Simpson BK, Aryee ANA, Toldrá F, editors. Byproducts from Agriculture and Fisheries: Adding Value for Food, Feed, Pharma and Fuels. Wiley; 2020 p. 57-106. DOI: 10.1002/9781119383956

[8] Brandelli A, Daroit DJ, Corrêa APF. Whey as a source of peptides with remarkable biological activities. Food Research International. 2015; 73: 149-61. DOI: 10.1016/j.foodres.2015.01.016

[9] Wen C, Zhang J, Zhang H, Duan Y, Ma H. Effects of divergent ultrasound pretreatment on the structure of watermelon seed protein and the antioxidant activity of its hydrolysates. Food Chemistry. 2019; 299: 125165. DOI: 10.1016/j.foodchem.2019.125165

[10] Liu X, Zheng X, Song Z, Liu X, Kopparapu NK, Wang X, Zheng Y. Preparation of enzymatic pretreated corn gluten meal hydrolysate and in vivo evaluation of its antioxidant activity. Journal of Functional Foods. 2015; 18: 1147-57. DOI: 10.1016/j.jff.2014.10.013

[11] Esfandi R, Willmore WG, Tsopmo A. Antioxidant and anti-apoptotic properties of oat bran protein hydrolysates in stressed hepatic cells. Food 2019; 8: 1-12. DOI: 10.3390/foods8050160

[12] Akanbi TO, Dare KO, Aryee ANA. High-Value Products from cereal, nuts, fruits, and vegetables wastes. In: Simpson BK, Aryee ANA, Toldrá F, editors. Byproducts from Agriculture and Fisheries: Adding Value for Food, Feed, Pharma and Fuels. Wiley; 2020 p. 349-368. DOI: 10.1002/9781119383956

[13] Yun HM, Lei XJ, Lee SI, Kim IH. Rapeseed meal and canola meal can partially replace soybean meal as a protein source in finishing pigs. Journal of Applied Animal Research. 2018; 2119: 195-199. DOI: 10.1080/09712119.2017.1284076

[14] Yang Y, Guanjun TAO, Ping LIU, Liu JIA. Peptide with angiotensin

I-converting enzyme inhibitory activity from hydrolyzed corn gluten meal. Journal of Agricultural and Food Chemistry 2007; 55: 7891-95. DOI: 10.1021/jf0705670

[15] Poljsak B, Šuput D, Milisav I. Achieving the balance between ROS and antioxidants: When to use the synthetic antioxidants. Oxidative Medicine and Cellular Longevity. 2013; 956792. DOI: 10.1155/2013/956792

[16] Liu Z, Ren Z, Zhang J, Chuang CC, Kandaswamy E, Zhou T, et al. Role of ROS and nutritional antioxidants in human diseases. Frontiers in Physiology 2018; 9: 477. DOI: 10.3389/fphys.2018.00477

[17] Liu R, Xing L, Fu Q, Zhou G, Zhang W. A Review of Antioxidant Peptides Derived from Meat Muscle and By-Products. Antioxidants. 2016; 5: 32. DOI: 10.3390/antiox5030032

[18] Sila A, Bougatef A. Antioxidant peptides from marine by-products: Isolation, identification and application in food systems. A review. Journal of Functional Foods. 2016; 21: 10-26. DOI: 10.1016/j.jff.2015.11.007

[19] Corrochano AR, Buckin V, Kelly PM, Giblin L. Invited review: Whey proteins as antioxidants and promoters of cellular antioxidant pathways. Journal of Dairy Science. 2018; 101: 4747-61. DOI: 10.3168/jds.2017-13618

[20] Görgüç A, Gençdağ E, Yılmaz FM. Bioactive peptides derived from plant origin by-products: Biological activities and techno-functional utilizations in food developments – A review. Food Research International. 2020; 136: 109504. DOI: 10.1016/j.foodres.2020.109504

[21] Esfandi R, Walters ME, Tsopmo A. Antioxidant properties and potential mechanisms of hydrolyzed proteins and peptides from cereals. Heliyon. 2019; 5: e01538. DOI: 10.1016/j.heliyon.2019.e01538

[22] Silveira Coelho M, de Araujo Aquino S, Machado Latorres J, de las Mercedes Salas-Mellado M. *In vitro* and *in vivo* antioxidant capacity of chia protein hydrolysates and peptides. Food Hydrocolloids 2019; 91: 19-25. DOI: 10.1016/j.foodhyd.2019.01.018

[23] Nwachukwu ID, Aluko RE. Structural and functional properties of food protein-derived antioxidant peptides. Journal of Food Biochemistry. 2019: 1-13. DOI: 10.1111/jfbc.12761

[24] Hamzeh A, Noisa P, Yongsawatdigul J. Characterization of the antioxidant and ACE-inhibitory activities of Thai fish sauce at different stages of fermentation. Journal of Functional Foods. 2020; 64: 103699. DOI: 10.1016/j.jff.2019.103699

[25] Sampath Kumar NS, Nazeer RA, Jaiganesh R. *In vivo* antioxidant activity of peptide purified from viscera protein hydrolysate of horse mackerel (*Magalaspis cordyla*). International Journal of Food Science and Technology. 2012; 47: 1558-62. DOI: 10.1111/j.1365-2621.2012.03002.x

[26] Girgih AT, Alashi AM, He R, Malomo SA, Raj P, Netticadan T, et al. A novel hemp seed meal protein hydrolysate reduces oxidative stress factors in spontaneously hypertensive rats. Nutrients 2014; 6: 5652-66. DOI: 10.3390/nu6125652

[27] Kwak JH, Kim M, Lee E, Lee SH, Ahn CW, Lee JH. Effects of black soy peptide supplementation on blood pressure and oxidative stress: A randomized controlled trial. Hypertension Research. 2013; 36: 1060-66. DOI: 10.1038/hr.2013.79

[28] Yang Y, Guanjun TAO, Ping LIU, Liu JIA. Peptide with angiotensin

I-converting enzyme inhibitory activity from hydrolyzed corn gluten meal. Journal of Agricultural and Food Chemistry 2007; 55: 7891-95. DOI: 10.1021/jf0705670

[29] Mane S, Jamdar SN. Purification and identification of Ace-inhibitory peptides from poultry viscera protein hydrolysate. Journal of Food Biochemistry. 2017; 41: 1-10. DOI: 10.1111/jfbc.12275

[30] Shih YH, Chen FA, Wang LF, Hsu JL. Discovery and study of novel antihypertensive peptides derived from *Cassia obtusifolia* seeds. Journal of Agricultural and Food Chemistry. 2019; 67: 7810-20. DOI: 10.1021/acs.jafc.9b01922

[31] Abdelhedi O, Nasri M. Basic and recent advances in marine antihypertensive peptides: Production, structure-activity relationship and bioavailability. Trends in Food Science and Technology. 2019; 88: 543-57. DOI: 10.1016/j.tifs.2019.04.002

[32] Udenigwe CC, Adebiyi AP, Doyen A, Li H, Bazinet L, Aluko RE. Low molecular weight flaxseed protein-derived arginine-containing peptides reduced blood pressure of spontaneously hypertensive rats faster than amino acid form of arginine and native flaxseed protein. Food Chemistry. 2012; 132: 468-75. DOI: 10.1016/j.foodchem.2011.11.024

[33] Fernández-Musoles R, Castelló-Ruiz M, Arce C, Manzanares P, Ivorra MD, Salom JB. Antihypertensive mechanism of lactoferrin-derived peptides: Angiotensin receptor blocking effect. Journal of Agricultural and Food Chemistry. 2014; 62: 173-81. DOI: 10.1021/jf404616f

[34] Tanaka M, Watanabe S, Wang Z, Matsumoto K, Matsui T. His-Arg-Trp potently attenuates contracted tension of thoracic aorta of Sprague-Dawley rats through the suppression of extracellular Ca^{2+} influx. Peptides 2009; 30: 1502-7. DOI: 10.1016/j.peptides.2009.05.012

[35] Wu J, Aluko RE, Nakai S. Structural requirements of angiotensin I-converting enzyme inhibitory peptides: Quantitative structure-activity relationship study of Di- and tripeptides. Journal of Agricultural and Food Chemistry. 2006; 54: 732-38. DOI: 10.1021/jf051263l

[36] Pan D, Cao J, Guo H, Zhao B. Studies on purification and the molecular mechanism of a novel ACE inhibitory peptide from whey protein hydrolysate. Food Chemistry. 2012; 130: 121-26. DOI: 10.1016/j.foodchem.2011.07.011

[37] Lin YH, Chen CA, Tsai JS, Chen GW. Preparation and identification of novel antihypertensive peptides from the in vitro gastrointestinal digestion of marine cobia skin hydrolysates. Nutrients 2019; 11. DOI: 10.3390/nu11061351

[38] Nakade K, Kamishima R, Inoue Y, Ahhmed A, Kawahara S, Nakayama T, et al. Identification of an antihypertensive peptide derived from chicken bone extract. Animal Science Journal. 2008; 79: 710-15. DOI: 10.1111/j.1740-0929.2008.00584.x

[39] Cao S, Wang Y, Hao Y, Zhang W, Zhou G. Antihypertensive effects in vitro and in vivo of novel angiotensin-converting enzyme inhibitory peptides from bovine bone gelatin hydrolysate. Journal of Agricultural and Food Chemistry. 2020; 68: 759-68. DOI: 10.1021/acs.jafc.9b05618

[40] Lee SH, Song KB. Isolation of an angiotensin converting enzyme inhibitory peptide from irradiated bovine blood plasma protein hydrolysates. Journal of Food Science. 2003; 68: 2469-72. DOI: 10.1111/j.1365-2621.2003.tb07047.x

[41] Kaiser S, Martin M, Lunow D, Rudolph S, Mertten S, Möckel U, et al. Tryptophan-containing dipeptides are bioavailable and inhibit plasma human angiotensin-converting enzyme *in vivo*. International Dairy Journal. 2016; 52: 107-14. DOI: 10.1016/j.idairyj.2015.09.004

[42] Zou Z, Wang M, Wang Z, Aluko RE, He R. Antihypertensive and antioxidant activities of enzymatic wheat bran protein hydrolysates. Journal of Food Biochemistry. 2020; 44: 1-13. DOI: 10.1111/jfbc.13090

[43] Gao D, Zhang F, Ma Z, Chen S, Ding G, Tian X, et al. Isolation and identification of the angiotensin-I converting enzyme (ACE) inhibitory peptides derived from cottonseed protein: optimization of hydrolysis conditions. International Journal of Food Properties. 2019; 22: 1296-1309. DOI: 10.1080/10942912.2019.1640735

[44] Nwachukwu ID, Girgih AT, Malomo SA, Onuh JO, Aluko RE. Thermoase-derived flaxseed protein hydrolysates and membrane ultrafiltration peptide fractions have systolic blood pressure-lowering effects in spontaneously hypertensive rats. International Journal of Molecular Sciences. 2014; 15: 18131-47. DOI: 10.3390/ijms151018131

[45] Cian RE, Martínez-Augustin O, Drago SR. Bioactive properties of peptides obtained by enzymatic hydrolysis from protein byproducts of *Porphyra columbina*. Food Research International. 2012; 49: 364-72. DOI: 10.1016/j.foodres.2012.07.003

[46] Engberink MF, Schouten EG, Kok FJ, Van Mierlo LAJ, Brouwer IA, Geleijnse JM. Lactotripeptides show no effect on human blood pressure: Results from a double-blind randomized controlled trial. Hypertension 2008; 51: 399-405. DOI: 10.1161/HYPERTENSIONAHA.107.098988

[47] Martínez-Maqueda D, Miralles B, Recio I, Hernández-Ledesma B. Antihypertensive peptides from food proteins: A review. Food and Function. 2012; 3: 350-61. 10.1039/c2fo10192k

[48] Diabetes facts and figures. 2020. International Diabetes Federation: www.idf.org

[49] Casanova-Martí À, Bravo FI, Serrano J, Ardévol A, Pinent M, Muguerza B. Antihyperglycemic effect of a chicken feet hydrolysate: Via the incretin system: DPP-IV-inhibitory activity and GLP-1 release stimulation. Food and Function. 2019; 10: 4062-70. DOI: 10.1039/c9fo00695h

[50] Hatanaka T, Kawakami K, Uraji M. Inhibitory effect of collagen-derived tripeptides on dipeptidylpeptidase-IV activity. Journal of Enzyme Inhibition and Medicinal Chemistry. 2014; 29: 823-28. DOI: 10.3109/14756366.2013.858143

[51] Song JJ, Wang Q, Du M, Ji XM, Mao XY. Identification of dipeptidyl peptidase-IV inhibitory peptides from mare whey protein hydrolysates. Journal of Dairy Science. 2017; 100: 6885-94. DOI: 10.3168/jds.2016-11828

[52] Uchida M, Ohshiba Y, Mogami O. Novel dipeptidyl peptidase-4-inhibiting peptide derived from β-lactoglobulin. Journal of Pharmacological Sciences 2011; 117: 63-66. DOI: 10.1254/jphs.11089SC

[53] Silveira ST, Martínez-Maqueda D, Recio I, Hernández-Ledesma B. Dipeptidyl peptidase-IV inhibitory peptides generated by tryptic hydrolysis of a whey protein concentrate rich in β-lactoglobulin. Food Chemistry. 2013; 141: 1072-77. DOI: 10.1016/j.foodchem.2013.03.056

[54] Li-Chan ECY, Hunag SL, Jao CL, Ho KP, Hsu KC. Peptides derived from Atlantic salmon skin gelatin as dipeptidyl-peptidase IV inhibitors.

Journal of Agricultural and Food Chemistry. 2012; 60: 973-78. DOI: 10.1021/jf204720q

[55] Cudennec B, Balti R, Ravallec R, Caron J, Bougatef A, Dhulster P, et al. In vitro evidence for gut hormone stimulation release and dipeptidyl-peptidase IV inhibitory activity of protein hydrolysate obtained from cuttlefish (*Sepia officinalis*) viscera. Food Research International. 2015; 78: 238-45. DOI: 10.1016/j.foodres.2015.10.003

[56] Caron J, Cudennec B, Domenger D, Belguesmia Y, Flahaut C, Kouach M, et al. Simulated GI digestion of dietary protein: Release of new bioactive peptides involved in gut hormone secretion. Food Research International. 2016; 89: 382-90. DOI: 10.1016/j.foodres.2016.08.033

[57] Nongonierma AB, Hennemann M, Paolella S, Fitzgerald RJ. Generation of wheat gluten hydrolysates with dipeptidyl peptidase IV (DPP-IV) inhibitory properties. Food and Function. 2017; 8: 2249-57. DOI: 10.1039/c7fo00165g

[58] Taga Y, Hayashida O, Kusubata M, Ogawa-Goto K, Hattori S. Production of a novel wheat gluten hydrolysate containing dipeptidyl peptidase-IV inhibitory tripeptides using ginger protease. Bioscience, Biotechnology and Biochemistry. 2017; 81: 1823-28. DOI: 10.1080/09168451.2017.1345615

[59] Wang F, Yu G, Zhang Y, Zhang B, Fan J. Dipeptidyl Peptidase IV Inhibitory peptides derived from oat (*Avena sativa* L.), buckwheat (*Fagopyrum esculentum*), and highland barley (*Hordeum vulgare trifurcatum* (L.) Trofim) proteins. Journal of Agricultural and Food Chemistry. 2015; 63: 9543-49. DOI: 10.1021/acs.jafc.5b04016

[60] Arise RO, Idi JJ, Mic-Braimoh IM, Korode E, Ahmed RN, Osemwegie O. In vitro Angiotesin-1-converting enzyme, α-amylase and α-glucosidase inhibitory and antioxidant activities of *Luffa cylindrical* (L.) M. Roem seed protein hydrolysate. Heliyon. 2019; 5: e01634. DOI: 10.1016/j.heliyon.2019.e01634

[61] Hatanaka T, Inoue Y, Arima J, Kumagai Y, Usuki H, Kawakami K, et al. Production of dipeptidyl peptidase IV inhibitory peptides from defatted rice bran. Food Chemistry. 2012; 134: 797-802. DOI: 10.1016/j.foodchem.2012.02.183

[62] Uraipong C, Zhao J. Rice bran protein hydrolysates exhibit strong in vitro α-amylase, β-glucosidase and ACE-inhibition activities. Journal of the Science of Food and Agriculture. 2016; 96: 1101-10. DOI: 10.1002/jsfa.7182

[63] Ren Y, Liang K, Jin Y, Zhang M, Chen Y, Wu H, et al. Identification and characterization of two novel α-glucosidase inhibitory oligopeptides from hemp (*Cannabis sativa* L.) seed protein. Journal of Functional Foods. 2016; 26: 439-50. DOI: 10.1016/j.jff.2016.07.024

[64] Goudarzi M, Madadlou A. Influence of whey protein and its hydrolysate on prehypertension and postprandial hyperglycaemia in adult men. International Dairy Journal. 2013; 33: 62-66. DOI: 10.1016/j.idairyj.2013.06.006

[65] Lima RDCL, Berg RS, Rønning SB, Afseth NK, Knutsen SH, Staerk D, et al. Peptides from chicken processing by-product inhibit DPP-IV and promote cellular glucose uptake: Potential ingredients for T2D management. Food and Function. 2019; 10: 1619-28. DOI: 10.1039/c8fo02450b

[66] Wald M, Schwarz K, Rehbein H, Bußmann B, Beermann C. Detection of antibacterial activity of an enzymatic hydrolysate generated by processing rainbow trout by-products with trout pepsin. Food Chemistry.

2016; 205: 221-28. DOI: 10.1016/j.foodchem.2016.03.002

[67] Pezeshk S, Ojagh SM, Rezaei M, Shabanpour B. Fractionation of protein hydrolysates of fish waste using membrane ultrafiltration: investigation of antibacterial and antioxidant activities. Probiotics and Antimicrobial Proteins. 2019; 11: 1015-22. DOI: 10.1007/s12602-018-9483-y

[68] Przybylski R, Firdaous L, Châtaigné G, Dhulster P, Nedjar N. Production of an antimicrobial peptide derived from slaughterhouse by-product and its potential application on meat as preservative. Food Chemistry. 2016; 211: 306-13. DOI: 10.1016/j.foodchem.2016.05.074

[69] Jin SK, Choi JS, Yim DG. Hydrolysis conditions of porcine blood proteins and antimicrobial effects of their hydrolysates. Food Science of Animal Resources. 2020; 40: 172-82. DOI: 10.5851/kosfa.2020.e2

[70] Wang R, Han Z, Ji R, Xiao Y, Si R, Guo F, He J, Hai L, Ming L, Yi L. Antibacterial activity of trypsin-hydrolyzed camel and cow whey and their fractions. Animals 2020; 10(2):327. DOI: 10.3390/ani10020337

[71] Doyen A, Saucier L, Beaulieu L, Pouliot Y, Bazinet L. Electroseparation of an antibacterial peptide fraction from snow crab by-products hydrolysate by electrodialysis with ultrafiltration membranes. Food Chemistry. 2012; 132: 1177-84. DOI: 10.1016/j.foodchem.2011.11.059

[72] Tan YN, Ayob MK, Wan Yaacob WA. Purification and characterisation of antibacterial peptide-containing compound derived from palm kernel cake. Food Chemistry. 2013; 136: 279-84. DOI: 10.1016/j.foodchem.2012.08.012

[73] Xiao J, Zhang H. An *escherichia coli* cell membrane chromatography-offline LC-TOF-MS method for screening and identifying antimicrobial peptides from Jatropha curcas meal protein isolate hydrolysates. Journal of Biomolecular Screening. 2012; 17: 752-60. DOI: 10.1177/1087057112442744

[74] Pu C, Tang W. The antibacterial and antibiofilm efficacies of a liposomal peptide originating from rice bran protein against: *Listeria monocytogenes*. Food and Function. 2017; 8: 4159-69. DOI: 10.1039/c7fo00994a

[75] Taniguchi M, Saito K, Nomoto T, Namae T, Ochiai A, Saitoh E, et al. Identification and characterization of multifunctional cationic and amphipathic peptides from soybean proteins. Biopolymers 2017; 108: 287-96. DOI: 10.1002/bip.23023

[76] Hu F, Wu Q, Song S, She R, Zhao Y, Yang Y, et al. Antimicrobial activity and safety evaluation of peptides isolated from the hemoglobin of chickens. BMC Microbiology. 2016: 1-10. DOI: 10.1186/s12866-016-0904-3

[77] Bahar AA, Ren D. Antimicrobial peptides. Pharmaceuticals. 2013; 6: 1543-75. DOI: 10.3390/ph6121543

[78] Perez Espitia PJ, de Fátima Ferreira Soares N, dos Reis Coimbra JS, de Andrade NJ, Souza Cruz R, Alves Medeiros EA. Bioactive peptides: synthesis, properties, and applications in the packaging and preservation of food. Comprehensive Reviews in Food Science and Food Safety. 2012; 11: 187-204. DOI: 10.1111/j.1541-4337.2011.00179.x

[79] Picariello G, Ferranti P, Fierro O, Mamone G, Caira S, Di Luccia A, et al. Peptides surviving the simulated gastrointestinal digestion of milk proteins: Biological and toxicological implications. Journal of Chromatography B: Analytical Technologies in the Biomedical and

Life Sciences. 2010; 878: 295-308. DOI: 10.1016/j.jchromb.2009.11.033

[80] Mirzaei M, Mirdamadi S, Safavi M, Soleymanzadeh N. The stability of antioxidant and ACE-inhibitory peptides as influenced by peptide sequences. Lwt. 2020; 130: 109710. DOI: 10.1016/j.lwt.2020.109710

[81] Xie N, Wang B, Jiang L, Liu C, Li B. Hydrophobicity exerts different effects on bioavailability and stability of antioxidant peptide fractions from casein during simulated gastrointestinal digestion and Caco-2 cell absorption. Food Research International. 2015; 76: 518-26. DOI: 10.1016/j.foodres.2015.06.025

[82] Chen M, Li B. The effect of molecular weights on the survivability of casein-derived antioxidant peptides after the simulated gastrointestinal digestion. Innovative Food Science and Emerging Technologies. 2012; 16: 341-48. DOI: 10.1016/j.ifset.2012.07.009

[83] Savoie L, Agudelo RA, Gauthier SF, Marin J, Pouliot Y. Free amino acids during the digestion of food proteins. Journal of AOAC International. 2005; 88: 935-48. DOI: 10.1093/jaoac/88.3.935

[84] Escudero E, Mora L, Toldrá F. Stability of ACE inhibitory ham peptides against heat treatment and in vitro digestion. Food Chemistry. 2014; 161: 305-11. DOI: 10.1016/j.foodchem.2014.03.117

[85] Vij R, Reddi S, Kapila S, Kapila R. Transepithelial transport of milk derived bioactive peptide VLPVPQK. Food Chemistry. 2016; 190: 681-88. DOI: 10.1016/j.foodchem.2015.05.121

[86] Quirós A, Dávalos A, Lasunción MA, Ramos M, Recio I. Bioavailability of the antihypertensive peptide LHLPLP: Transepithelial flux of HLPLP. International Dairy Journal. 2008; 18: 279-86. DOI: 10.1016/j.idairyj.2007.09.006

[87] Gleeson JP, Brayden DJ, Ryan SM. Evaluation of PepT1 transport of food-derived antihypertensive peptides, Ile-Pro-Pro and Leu-Lys-Pro using *in vitro*, *ex vivo* and *in vivo* transport models. European Journal of Pharmaceutics and Biopharmaceutics. 2017; 115: 276-284. DOI: 10.1016/j.ejpb.2017.03.007

[88] Fan H, Xu Q, Hong H, Wu J. Stability and transport of spent hen-derived ACE-inhibitory peptides IWHHT, IWH, and IW in human intestinal Caco-2 cell monolayers. Journal of Agricultural and Food Chemistry. 2018; 66: 11347-11354. DOI: 10.1021/acs.jafc.8b03956

[89] Regazzo D, Mollé D, Gabai G, Tomé D, Dupont D, Leonil J, et al. The (193-209) 17-residues peptide of bovine β-casein is transported through caco-2 monolayer. Molecular Nutrition and Food Research. 2010; 54: 1428-35. DOI: 10.1002/mnfr.200900443

[90] Lin Q, Xu Q, Bai J, Wu W, Hong H, Wu J. Transport of soybean protein-derived antihypertensive peptide LSW across Caco-2 monolayers. Journal of Functional Foods. 2017; 39: 96-102. DOI: 10.1016/j.jff.2017.10.011

[91] Fernández-Musoles R, Salom JB, Castelló-Ruiz M, Contreras M del M, Recio I, Manzanares P. Bioavailability of antihypertensive lactoferricin B-derived peptides: Transepithelial transport and resistance to intestinal and plasma peptidases. International Dairy Journal. 2013; 32: 169-74. DOI: 10.1016/j.idairyj.2013.05.009

[92] Xu F, Zhang J, Wang Z, Yao Y, Atungulu GG, Ju X, et al. Absorption and metabolism of peptide wdhhapqlr derived from rapeseed protein and inhibition of HUVEC apoptosis under oxidative stress. Journal of Agricultural and Food Chemistry. 2018; 66: 5178-89. DOI: 10.1021/acs.jafc.8b01620

[93] Xu F, Wang L, Ju X, Zhang J, Yin S, Shi J, et al. Transepithelial transport of

YWDHNNPQIR and its metabolic fate with cytoprotection against oxidative stress in human intestinal Caco-2 cells. Journal of Agricultural and Food Chemistry. 2017; 65: 2056-65. DOI: 10.1021/acs.jafc.6b04731

[94] Ding L, Wang L, Zhang T, Yu Z, Liu J. Hydrolysis and transepithelial transport of two corn gluten derived bioactive peptides in human Caco-2 cell monolayers. Food Research International. 2018; 106: 475-80. DOI: 10.1016/j.foodres.2017.12.080

[95] Zhang H, Duan Y, Feng Y, Wang J. Transepithelial transport characteristics of the cholesterol- Lowing soybean peptide, WGAPSL, in Caco-2 cell monolayers. Molecules 2019; 24. DOI: 10.3390/molecules24152843

[96] Sangsawad P, Roytrakul S, Choowongkomon K, Kitts DD, Chen XM, Meng G, et al. Transepithelial transport across Caco-2 cell monolayers of angiotensin converting enzyme (ACE) inhibitory peptides derived from simulated in vitro gastrointestinal digestion of cooked chicken muscles. Food Chemistry. 2018; 251: 77-85. DOI: 10.1016/j.foodchem.2018.01.047

[97] Wang B, Li B. Effect of molecular weight on the transepithelial transport and peptidase degradation of casein-derived peptides by using Caco-2 cell model. Food Chemistry. 2017; 218: 1-8. DOI: 10.1016/j.foodchem.2016.08.106

[98] Gleeson JP, Frías JM, Ryan SM, Brayden DJ. Sodium caprate enables the blood pressure-lowering effect of Ile-Pro-Pro and Leu-Lys-Pro in spontaneously hypertensive rats by indirectly overcoming PepT1 inhibition. European Journal of Pharmaceutics and Biopharmaceutics. 2018; 128: 179-87. DOI: 10.1016/j.ejpb.2018.04.021

[99] Guo Y, Gan J, Zhu Q, Zeng X, Sun Y, Wu Z, et al. Transepithelial transport of milk-derived angiotensin I-converting enzyme inhibitory peptide with the RLSFNP sequence. Journal of the Science of Food and Agriculture. 2018; 98: 976-83. DOI: 10.1002/jsfa.8545

[100] Muheem A, Shakeel F, Jahangir MA, Anwar M, Mallick N, Jain GK, et al. A review on the strategies for oral delivery of proteins and peptides and their clinical perspectives. Saudi Pharmaceutical Journal. 2016; 24: 413-28. DOI: 10.1016/j.jsps.2014.06.004

[101] Zhang T, Su M, Jiang X, Xue Y, Zhang J, Zeng X, et al. Transepithelial transport route and liposome encapsulation of milk-derived ace-inhibitory peptide Arg-Leu-Ser-Phe-Asn-Pro. Journal of Agricultural and Food Chemistry. 2019; 67: 5544-51. DOI: 10.1021/acs.jafc.9b00397

[102] Li Z, Paulson AT, Gill TA. Encapsulation of bioactive salmon protein hydrolysates with chitosan-coated liposomes. Journal of Functional Foods. 2015; 19: 733-43. DOI: 10.1016/j.jff.2015.09.058

Chapter 10

The Potential of Lutein Extract of *Tagetes erecta* L. Flower as an Antioxidant and Enhancing Phagocytic Activity of Macrophage Cells

Kusmiati Kusmiati, Fifi Afiati, Chrisna Widhiani, Alpinna Aditia, Destia D. Elviani and Atit Kanti

Abstract

Marigold flower (*Tagetes erecta* L.) produces lutein compounds which present biological activities such as antioxidant, antiinflammatory, antimutagenicity, and immunomodulatory effects. The study was to investigate the antioxidant activity of the lutein of *T. erecta* L. and the effect of lutein on the activity and phagocytic capacity of macrophage cells. The antioxidant screening was carried out using diphenyl picrylhydrazyl (DPPH), 2,2′-and-bis(3-ethylbenzothiazoline-6-sulphonic acid (ABTS) radical scavenging assay with serial concentrations and ferric-reducing antioxidant power (FRAP) method. For the observation of activity and phagocytic capacity of peritoneal macrophages, twenty-eight mice were used and divided into seven groups each comprising four replicates, i.e., Group (I) normal controls, mice were untreated (II) a negative control, mice were induced by *Staphylococcus aureus* (III) positive control, mice were induced by *S. aureus* and treatment of meniran extract (*Phyllanthus niruri*). The treatment group (IV-VII) mice were induced by *S. aureus* and treated crude lutein, respectively: 0.15 mg, 0.30 mg, 0.60 mg, and 0.90 mg. 20 g^{-1} of body weight. The lutein extracted from *T. erecta* shows an antioxidant activity against DPPH radical with an IC50 value of 53.58 $\mu g.ml^{-1}$, while the antioxidant activity against ABTS has an IC50 value of 72.91 $\mu g.ml^{-1}$. The antioxidant activity test results by the FRAP method at each lutein concentrations of 10, 25, 50, and 75 ppm were obtained respectively of 33, 88, 185.5, and 288.5 μmol Fe^{2+}/g extract. The data were analyzed using one-way ANOVA and Duncan's multiple range test (DMRT) after. The phagocytic activity was 45.5%; 54.75%; 57.50% and 67.0%, respectively, while the phagocytic capacity values were 355; 519; 611 and 767 *S. aureus* bacterial cells per 50 macrophage cells. The lutein from marigolds (*T. erecta* L.) is capable of scavenging free radicals and reducing oxidants. Lutein can increase the activity and capacity of phagocytic of peritoneum macrophage cells in mice.

Keywords: *Tagetes erecta* L., lutein, antioxidant, peritoneal macrophages, phagocytic activity

1. Introduction

Marigold flower (*Tagetes erecta* L.) is an annual herbaceous plant commercialized worldwide as an ornamental plant and a natural source of pigment from its yellow/orange flowers. The flower is rich in carotenoids, the extract of which is used as a colorant in a wide variety of food products, including cake mixture, drinks, and ordinary drinks, cereals, chewing gum, dairy analogs, egg products, fats and oils, dairy products, processed fruit and fruit juices, and soups [1, 2]. Marigold flowers are classified as medicinal plants from the Compositae tribe. People empirically use it to treat asthma, bronchitis, fever, ulcers, burns, and swelling [3]. The pigment content in marigold flowers are classified as carotenoids, namely lutein pigments that are yellow to orange [4]. *T. erecta* L. is a major source of lutein for commercial use. In 2010, lutein occupied a $ 233 million share of the world carotenoid market [1, 2, 5]. Lutein is a primary pigment because this pigment is not produced synthetically. It is due to its production, which requires a long process. Lutein from *T. erecta* L. is a pure extract obtained from marigold oleoresin, extracted from the petals of marigold flowers with an organic solvent. After the saponification process, the final product contains the main component lutein and a fraction of zeaxanthin. Lutein (3R, 3'R, 6'R-βε-carotene-3,3′-diol) is a member of the pigment group known as xanthophylls and lacks provitamin A activity. Lutein generally coexists in nature with its stereoisomer zeaxanthin and the double bonds of the isoprene backbone can exist in the all-trans (**Figure 1**). Pure lutein typically appears as a yellow-orange crystalline, lipophilic, solid with the chemical name β,ε-carotene-3,3′-diol (C40H56O2) [3].

Lutein pigments are found in egg yolks, fruits, and vegetables, including tomatoes, carrots, pumpkin, corn, and various green plants [6]. Some research results reported that lutein is efficacious to protect the eyes from macular degeneration and epithelial cancer, and has antioxidant properties [7], and can increase the immune system in the body [8]. Unfortunately, lutein cannot be synthesized in the body; so, it depends on plants [9]. In plants, lutein can be in the form of free lutein such as in spinach, cabbage, and broccoli, or in the form of esters with the fatty acids in the following fruits and vegetables: mango, orange, papaya, red or green peppers, yellow corn etc. [10–12]. Lutein content in natural sources depends on the type, variety, level of maturity, part of the fruit, and also on processing processes such as heating, preservation, or storage [12]. Dried marigold flowers contain 0.1%–0.2% carotenoids with a composition of 80% as lutein ester [4].

Lutein is an antioxidant because it can reduce free radicals and reactive molecules that can inhibit cell damage [13]. Free radicals can be produced from the body's metabolism and also from outside the body, such as the smoke of a cigarette, environmental pollution, radiation, drugs, pesticides, and ultraviolet rays [14]. Free radicals can cause damage because they can react with cell components that are important for maintaining cell life, both structural components (for example, the molecules making up membranes) or functional components (for example, DNA enzymes) [15]. Antioxidants are chemical compounds that can neutralize free

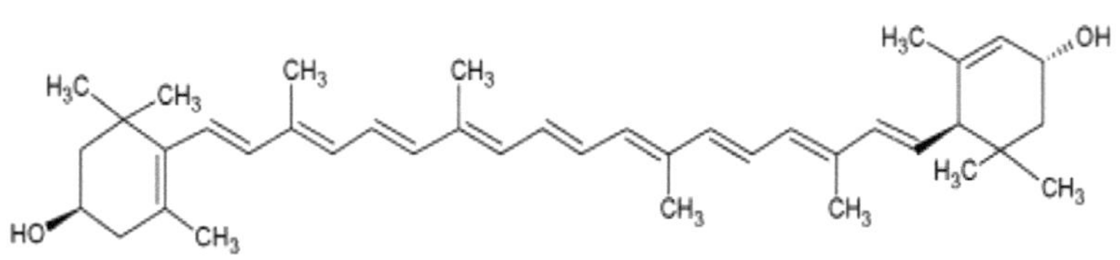

Figure 1.
Structural of all-trans lutein.

radicals by donating one or more electrons to free radicals; so, these free radicals can be suppressed [16]. Lutein is very effective as an antioxidant to protect the eyes because it is able to neutralize free radicals formed by the action of ultraviolet radiation on the retina of the eye and reduce the risk of cataracts due to aging [17]. Humans are unable to synthesize lutein, so they can obtain it by consuming fruits, vegetables, and dietary supplements [12]. Testing the antioxidant potential of lutein compounds from marigold flowers (*T. erecta* L.) can be carried out using DPPH, ABTS, and FRAP methods.

The lutein pigment in marigold flowers (*T. erecta* L.) can act as an immunomodulator, especially to protect the eyes from pathogenic elements such as viruses, bacteria, and infectious diseases intra-ocular inflammation [18]. Humans have resistance to a disease/infection from microorganisms or foreign substances, which is called the immune system. The body's immune system's decreased function can be caused by stress, an unhealthy lifestyle, aging, and chronic disease. The human body's immune system consists of specific and non-specific immune systems. One of the defenses made by the body's non-specific immune system in preventing the entry of antigens is by doing phagocytosis. Phagocytosis is the process of ingesting, digesting, and destroying antigens/microbes. The cells that play their role in carrying out phagocytosis are macrophage cells [19].

Previous research conducted *in vivo* showed that lutein derived from plants with a yellow pigment had an immunomodulatory activity increasing the regeneration of the immune system [8]. Immunomodulators are substances that can restore and repair the immune system, which function is disturbed or to suppress its excessive function [19]. In previous studies, researcher found that lutein activity was able to stimulate an immune response. In addition, it was found that 10 mg of lutein consumed daily by cats for 12 weeks led to an increase in the percentage of CD4+ and CD21+ lymphocytes, plasma IgG concentrations, and NK cell activation. It showed that lutein could stimulate mediator cells and humoral immunity in cats, which was done *in vivo*; so, stimulating the activity and capacity of macrophage cells in the body as an immune response [8]. Testing of immunomodulatory activity was carried out using observing the activity and capacity of macrophage cells from test animals (mice) induced by *S. aureus* bacteria [20].

This research has two objectives. First, this study examined the potential of lutein compounds from marigold flowers (*T. erecta* L.) as antioxidants using the DPPH, ABTS, and FRAP methods. Second, this study examined the possibility of lutein extract as an immunomodulator through a non-specific immune system, namely the activity and capacity of phagocytosis of mice's peritoneal macrophage cells (*in vivo*) induced by *S. aureus*.

2. Experiment

2.1 Material collection

Marigold flower (*T. erecta* L.) belongs to the Asteraceae family obtained from Taman Bunga Nusantara, Cipanas - West Java. Plant determination was carried out at Herbarium Bogorienses, Research Center for Biology, Indonesian Institute of Sciences (LIPI).

2.2 Lutein extract from marigold flowers (*T. erecta* L.)

Dry powder of Marigold flower crown was weighed as much as 20 g into Erlenmeyer to extract of lutein [21, 22]. The weighing was carried out six times.

Figure 2.
Extraction of lutein from the crown of marigold flowers (T. erecta L.).

Each Erlenmeyer was added with 300 ml of n-hexane (Merck) and was shaken out for 24 h. The extracted liquid was filtered with filter paper to separate the filtrate and residue. The residue was remastered by adding 200 ml of n-hexane and was shaken out for 24 h, then was filtered, and the filtrate was combined.

The filtrate was concentrated with a rotary evaporator. The extract was obtained, and it was dried at 40°C until hexane extract was obtained. The n-hexane extract was digested by adding isopropanol solution, then was stirred for one hour at 50°C. At the same temperature, it was saponified with 50% NaOH solution, stirred homogeneously for one hour until two layers were formed, i.e., semisolid and liquid. Next semisolid and liquid were cooled, the semisolid part is separated from the liquid. The semisolid solution was digested with distilled water repeatedly until the other carotenoids were completely separated. The solution was centrifuged for ten minutes at a speed of 3000 rpm; yellow deposits were formed, then the filtrate was removed. The precipitate was dried in a water bath at 40°C; so, the lutein extract was obtained (**Figure 2**).

2.3 Antioxidant activity of lutein extract from *T. erecta* L.

2.3.1 DPPH free radical-scavenging activity

The ability of Lutein extract of *T. erecta* L. to scavenge the DPPH radical was estimated using the method described by Kushwaha and Verma [23]. An aliquot of 50 μL of various sample concentrations was added to a volume of 2 mL from the DPPH methanolic solution (60 μM). The reaction mixture was well shaken and incubated for 20 min at room temperature in the dark and the absorbance was recorded at 517 nm. The blank was constituted by methanol instead of the extract. The percentage inhibition of the DPPH radical by the samples was calculated using the following equation:

$$\%\,inhibition = \frac{Ao - As}{Ao} x100. \tag{1}$$

where A_0 is the absorbance of blank and As is the absorbance of the lutein sample. The sample concentration providing 50% of inhibition (IC50) was determined from the plotted curve of inhibition using several concentrations. Vitamin E were used as a comparative antioxidant compound.

2.3.2 ABTS [2,2′-azinobis (3-ethylbenzothiazoline-6-sulphonic acid)] free radical scavenging activity assay

The ABTS● + stock solution was prepared by reacting ABTS (Sigma Aldrich aqueous solution 7 mM) with 2.45 mM aqueous solution of potassium persulfate (Merck) in equal quantities; the mixture was allowed to stand in the dark at room temperature for 12–16 hours before use [24–25]. The working solution of

ABTS● + was obtained by diluting the stock solution in methanol to give an absorbance of 0.70 ± 0.02 at 734 nm. Then, 1.0 mL of ABTS● + solution was mixed with 0.5 mL of the Lutein extracts at different concentrations (25–100 ppm). The mixture was then incubated at room temperature for exactly 10 min in the dark. The control was prepared by mixing 1.0 mL of ABTS● + solution with 0.5 mL of distilled water. The absorbance was measured against a blank at 734 nm using spectrophotometer (Shimadzu UV-160). The percentage of scavenging activity of each extract on ABTS● + was calculated as % inhibition (I%) using the following equation:

$$\% \, Inhibition = \left[\frac{(Ao - As)}{Ao} \right] x\, 100 \qquad (2)$$

Where Ao is the absorption of blank and As is the absorption of the lutein extract solution.

2.3.3 Ferric ion reducing antioxidant power (FRAP) assay

The ferric reducing capacity of extracts was investigated by using the potassium ferricyanide-ferric chloride method [26]. Briefly, Reagents for FRAP assay a) Acetate buffer 300 mM pH 3.6: b) TPTZ (2, 4, 6-tripyridyl-s- triazine): (M.W. 312.34), 10 mM in 40 mM HCl (M.W. 36.46). c) FeCl3. 6 H2O: (M.W. 270.30), 20 mM. The working FRAP reagent was prepared by mixing a, b and c in the ratio of 10:1:1 just before testing. All the reagents were prepared from Merck (Germany) company. FRAP solution (3.6 mL) add to distilled water (0.4 mL) and incubated at 37̊C for 5 min. Then this solution was mixed with certain concentration of the plant extract (80 mL) and incubated at 37̊C for 10 min. The absorbance of the reaction mixture was measured at 593 nm. For construction of the calibration curve, five concentrations of FeSO4, 7H2O (25, 50, 75, 100, 150, 200, 300 μmol) were used and the absorbance values were measured as for sample solutions. The sample concentration providing 0.5 of absorbance (IC50) was calculated by plotting absorbance against the corresponding sample concentration [27, 28]. Vitamin E were used as a comparative antioxidant compound.

2.4 Measurement of *in vivo* phagocytosis activities and capacity of macrophage

Activity and capacity were observed under a 10x100 magnificent light microscope [29]. Animal experimental was performed in the Animal laboratory of the Faculty of Veterinary Medicine, IPB University- Dramaga Bogor. The experimental animal used was male mice (*Mus musculus*) strain DDY. A total of 28 mice were used in 6–8 weeks old, with 18–21 g of body weight. The mice were acclimated for seven days ahead. The mice were randomly divided into seven experimental groups of four mice in each group: (I) Normal control (mice were given vegetable oil), (II) Negative control (mice were given vegetable oil + *S aureus*, (III) Positive control (mice were given immunomodulator stimulator 0.078 mg + *S. aureus*), (IV) Lutein extract dose 0.15 mg/20 g BW + *S. aureus*, (V) Lutein extract dose of 0.30 mg/20 g BW + *S. aureus*, (VI) Lutein extract dose of 0.60 mg/20 g BW + *S. aureus*, (VII) Lutein extract dose 0.90 mg/20 g BW + *S. aureus*. Lutein extract was administered orally for 14 days. On the 15th day, all experimental animals were injected intraperitoneally with 0.5 ml of *S. aureus* bacterial suspension (1 x 10^6 CFU/ml), except the mice in normal control group. Those experimental animals were euthanatized one hour after infection, and then their peritoneal liquid was collected. Swab preparation was made for all samples, and then fixated using methanol for 5 minutes, stained with Giemsa and left for 20 minutes, and then washed using distilled water.

2.4.1 Phagocytosis activity

Phagocytosis activity value is the number of macrophage cells that actively phagocyte *S. aureus* in 100 macrophage cells. Phagocytosis activity was presented in percent, with the formulation [30] as the following:

$$\% \, Activity = \frac{The\ number\ of\ active\ macrophage}{The\ number\ of\ whole\ macrophage} x\,100 \tag{3}$$

2.4.2 Phagocytosis capacity

Phagocytosis capacity value is the number of bacteria ingested by 50 active macrophages [30]. This parameter was observed by staining with Giemsa and then the bacteria number was counted under microscope. Phagocytosis is a devouring process on bacteria or strange objects by enfolding those things using macrophage cytoplasm.

2.5 Data analysis

Determining of immunomodulatory activity of crude lutein from *T. erecta* L. was through macrophage activity and capacity induced by *Staphylococcus aureus* bacteria by *in vivo*. Data were analyzed by one way ANOVA with 4 replications using SPSS ver 22.0 with P = 0.05. This analysis was then followed by Duncan Multiple Range Test.

3. Results

3.1 Antioxidant activity of lutein extract from *T. erecta* L.

Antioxidant activity is a complex procedure usually happening through several mechanisms and is influenced by many factors, which cannot be fully described with one method. Therefore, it is essential to perform more than one type of antioxidant capacity measurement to consider the various mechanisms.

3.1.1 DPPH free radical-scavenging activity and ABTS radical cation scavenging activity assay

The potential test of lutein extract as an antioxidant using the DPPH method has some advantages, such as it is easy to conduct, and the process is relatively fast. The test method using the DPPH was based on a decrease in absorbance caused by a change in the DPPH solution's color. DPPH will react with hydrogen atoms from free radical absorbing compounds to form DPPH Hydrazine, which is more stable. DPPH reagent that reacts with antioxidants will change color from purple to yellow; the intensity of the color formed depends on the antioxidants' ability [31]. The data processing technique was carried out by comparing the concentration with the percentage value of antioxidant activity for each sample in a regression graph (**Figure 3**).

The results of the marigold flower lutein extract antioxidant test using the DPPH and ABTS methods can be seen in **Table 1**.

IC50 is a number that shows a concentration (μg/ml) which can inhibit the oxidation process by 50%. The smaller the IC50 value, the higher the antioxidant activity is. Based on the IC50 value, if it is less than 50 ppm, the antioxidant is very strong, 50 ppm–100 ppm is strong, 100 ppm–150 ppm is moderate, and 150 ppm–200 ppm is weak [31].

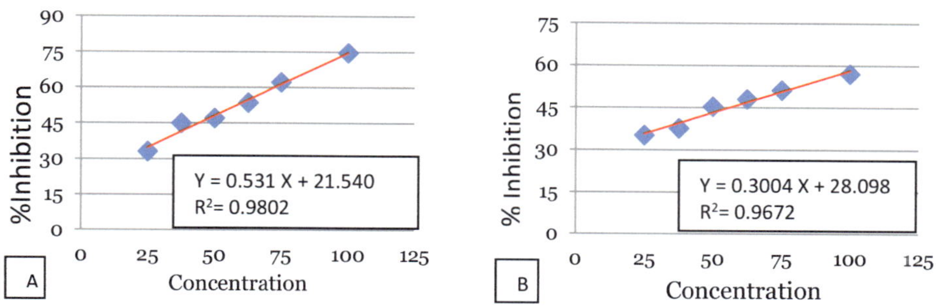

Figure 3.
Relationship between the concentration of lutein extract from Marigold flower (T. erecta L.) (ppm) with % inhibition of (A) DPPH, and (B) ABTS.

Lutein Conc. (ppm)	DPPH		ABTS	
	% Inhibition	IC 50(μg/ml)	% Inhibition	IC 50(μg/ml)
25	33.15	53.58	35.07	72.91
37.5	44.88		37.50	
50	46.96		45.28	
62.5	53.44		47.88	
75	62.07		51.10	
100	74.64		56.92	

Table 1.
Antioxidant activity values of marigold flower (T. erecta L.) using different assays.

Vit E Conc. (ppm)	DPPH		ABTS	
	% Inhibition	IC 50(μg/ml)	% Inhibition	IC 50(μg/ml)
4	34.97	9.60	34.43	12.36
6	40.47		37.81	
8	45.06		42.30	
10	51.14		44.73	
12	56.36		49.76	

Table 2.
Antioxidant activity values of vitamin E using different assays.

The results of testing for antioxidant potential with the DPPH suppression method showed that the lutein extract of flowers (*T. erecta* L.) had strong antioxidant activity with an IC50 value of 53.58 μg/ml (**Table 1**). The IC50 value of Vitamin E, i.e. 9.60, is categorized as a very strong antioxidant (**Table 2**). The antioxidant activity was measured based on the reduction in the purple DPPH intensity which was proportional to the reduction in the concentration of DPPH solution. The color change from purple to reddish purple then to yellow occurred after 30 minutes of giving lutein extract into the DPPH solution. The color change was caused by the reaction between diphenyl picryl hydrazine and the hydrogen atom released by one molecule of the lutein extract component; then diphenyl picryl hydrazine compound was [32].

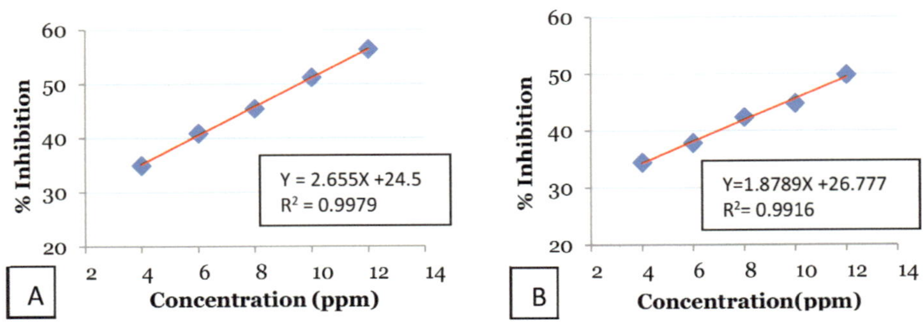

Figure 4.
Relationship between the concentration of vitamin E (ppm) with % inhibition of (A) DPPH and (B) ABTS.

Antioxidant activity testing could be done using the compound 2.2-Azinobic Acid (3-ethylbenzatiazolin-6 sulfonate). ABTS was a water-soluble and chemically-stable compound. This method was used for research on antioxidants that are water soluble, fat soluble, and dissolve in pure compounds. ABTS compounds were converted into radical cations (ABTS ●+) with the addition of potassium per sulfate. Free radical cations were blue green which could absorb light at a wavelength of 734 nm [33].

Maximum absorbance could also occur at other wavelengths. Radical cations were reactive to most antioxidants. ABTS which was blue-green during the reaction turned colorless which was analyzed using a spectrophotometer. From measurements using a UV–VIS spectrophotometer, the absorption value was obtained, and the % resistance was calculated. Furthermore, the relationship between concentration and % inhibition of marigold flower lutein extract (*T. erecta* L.) or vitamin E was plotted and the linear regression was calculated. The results can be seen in **Figures 3** and **4**. Furthermore, from the Eq. 50% activity was plotted, so that the concentration value (IC50) of the marigold flower lutein extract (*T. erecta* L.) was obtained, i.e. 72.91 μg/ml which was included in the strong antioxidant category because it had an IC50 value of around 50–100 μg/ml. The IC50 vitamin E yield of 12.36 μg/ml was in the very strong category. The use of the two DPPH and ABTS methods for testing the antioxidants of lutein and vitamin E extracts showed that there were differences in the IC50 obtained. Thus, the use of the DPPH method was more suitable and showed more sensitive results.

3.1.2 Ferric ion reducing antioxidant power (FRAP) of lutein extracts and vitamin E assay

The FRAP assay is based on the measurement of the ability of the substance to reduce Fe3+ to Fe2+ resulting in the change of color from yellow to blue colored solution of Fe2 + – TPTZ complex (Fe2+ tripyridyltriazine) which has a high absorbance at 593 nm. The FRAP assay provides a reliable method.

FRAP value was obtained by comparing the changes of absorbance values in the sample mixture with those obtained from the increased concentrations of Fe3+. FRAP values were expressed as μmol of Fe2+ equivalents per g sample. **Figure 5** shows the relationship between Fe (II) sulfate concentration and absorption at the wavelength 593 nm. **Table 3** FRAP values of lutein extract of marigold flower and Vitamin E in several concentration. The FRAP value is directly proportional to the concentration of sample. The higher the sample concentration exhibited the higher the FRAP value.

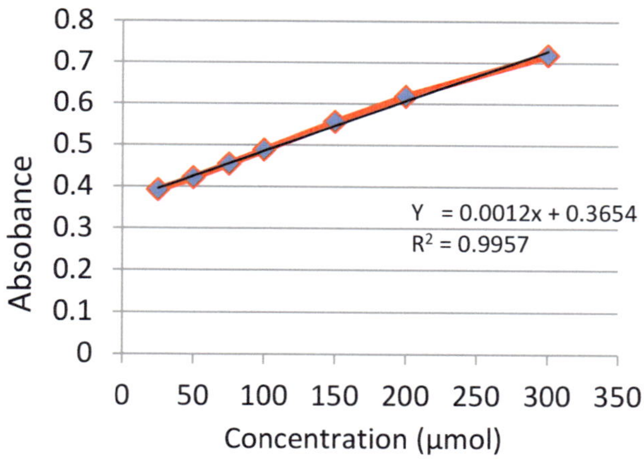

Figure 5.
Standard curve of FeSO4.7H2O.

No	Sample	Concentration of Sample (ppm)	Conc. of Fe(II) (µmol)
1	Lutein Extract of *T. erecta* L.	10	33.0
		25	88.0
		50	185.5
		75	288.5
2	Vitamin E	4	12.16
		6	18.83
		8	25.50
		10	33.83

Table 3.
Antioxidant activity of lutein extracts from marigold flower (T. erecta L.), using FRAP assay.

Generally, the reducing properties are associated with the presence of compounds, which exert their action by breaking the free radical chain through donating a hydrogen atom [34, 35].

3.2 Activity and capacity of phagocytosis of macrophage cells in mice peritoneum liquids

The results of the immunomodulatory activity test upon lutein extract of kenikir flower (*T. erecta* L.) based on the value of phagocytic activity and the phagocytic capacity of mice peritoneal fluid macrophage cells can be seen in **Figure 6** and **Table 4**.

Remarks:

I. Group normal control, mice were given vegetable oil.

II. Group negative control, mice were given vegetable oil+0.5 ml *S. aureus.*

III. Group positive control, mice were given stimuno®0.078 mg/20 g bw + 0.5 ml *S. aureus*.

IV. Group mice were given the extract lutein 0.15 mg/20 g bw/day +0.5 mL *S. aureus.*

V. Group mice were given the extract lutein 0.30 mg/20 g bw/day +0.5 mL *S. aureus.*

VI. Group mice were given the extract lutein 0.60 mg/20 g bw/day +0.5 mL *S. aureus.*

VII. Group mice were given the extract lutein 0.90 mg/20 g bw/day+0.5 mL *S. aureus.*

The results of the Duncan Multiple Range Test on the value of phagocytosis activity for each test group showed that there was no significant difference between the negative control group and the 0.15 mg/20gBW lutein extract treatment group. Treatment of lutein extract 0.30 mg/20gBW, was not significantly different from treatment of lutein extract 0.60 mg/20gBW, while lutein treatment 0.90 mg/20gBW showed a significant difference. The normal control group and the positive control group showed significantly-different phagocytic activity. The lowest phagocytosis activity was found in the normal control group because in the normal control group mice were not given intra-peritoneal *S. aureus* suspension; so, macrophages were not triggered to perform phagocytosis. Macrophages are activated by various stimuli originating from exogenous antigens and microorganisms.

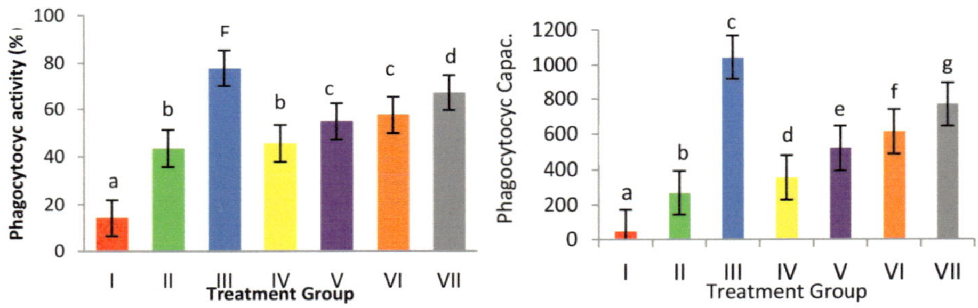

Figure 6.
Histogram of activity and capacity of phagocytosis macrophage cells in mice peritoneum liquids on each test group.

No.	Test Group	Phagocytosis Activity (%)	Phagocytosis Capacity *S. aureus* cells/ 50 macrophage cells
I	Normal Control	$14.25^{a} \pm 2.22$	$46.50^{a} \pm 6.61$
II	Negative Control	$43.50^{b} \pm 2.65$	$268.00^{b} \pm 5.09$
III	Positive Control	$77.75^{e} \pm 4.11$	$1041.50^{c} \pm 15.06$
IV	Lutein Extract 0.15 mg/20 g bw	$45.50^{b} \pm 1.29$	$354.75^{d} \pm 3.59$
V	Lutein Extract 0.30 mg/20 g bw	$54.75^{c} \pm 4.35$	$519.25^{e} \pm 5.12$
VI	Lutein Extract 0.60 mg/20 g bw	$57.50^{c} \pm 2.65$	$611.25^{f} \pm 4.57$
VII	Lutein Extract 0.90 mg/20 g bw	$67.00^{d} \pm 2.58$	$766.75^{g} \pm 6.50$

Note: The numbers followed by the same letter are not significantly different ($P < 0.05$).

Table 4.
Phagocytosis activity of macrophage cell (per 100 cells) and phagocytosis capacity (the number of bacteria devoured per 50 active macrophage cells).

Both will be recognized, then captured, eaten, and digested (phagocytes), and finally degraded and disappeared from the body. Peritoneal macrophages are freely available in the peritoneal fluid, so that pathogens entering the body [19] are possible to be captured. The highest value of phagocytosis activity was found in the immunosuppressive positive control group and in the treatment group lutein extract 0.90 mg/20gBW. Oral administration of lutein with different doses, namely 0.15 mg, 0.30, 0.60, and 0.90 mg per 20 g BW, resulted an increase in phagocytic activity, respectively 45.50%, 54.75%, 57.50%, and 67.0%. The increase in phagocytic activity of peritoneal fluid macrophage cells of mice occurred simultaneously with the increase in lutein dose.

The results of Duncan's test on the value of phagocytosis capacity in each of the 50 active macrophage cells showed that there were significant differences between groups. The lowest macrophage cell phagocytosis capacity was found in the normal control group. It was due to the fact that in a healthy body condition, macrophages became inactive; so, they did not show any activity and capacity to carry out phagocytosis. The highest value of macrophage phagocytosis capacity was found in the normal control group. The lutein extract treatment group with the highest phagocytosis capacity value was found in the lutein treatment group 0.90 mg/20gBW. The administration of lutein extract for 14 days to experimental animals, at a dose of 0.15 mg/20gBW, 0.30 mg/20gBW, 0.60 mg/20gBW and 0.90 mg/20gBW, was able to increase the phagocytosis capacity of every 50 peritoneum macrophage cells, respectively 355; 519; 611 and 767 of *S. aureus* cells.

Phagocytosis is an important process for nutrition in unicellular organisms, while in multicellular organisms it is found in specialized cells called phagocytes. Phagocytes can ingest microbial pathogens, but importantly also apoptotic cells. Phagocytosis becomes essential not only for microbial elimination, but also for tissue homeostasis [36]. Macrophages are one of the cells that play an important role in the phagocytosis process or as antigen presenting cells (APC) [19]. The phagocytosis ability of macrophage cells is a form of the body's immune system response to overcome the incoming antigen. Bacteria *S. aureus* will stimulate macrophages to produce IL-6 which will activate NK cells and then will secrete IFN-β which will activate macrophages. Macrophages will also be stimulated to produce TNF-α which will activate other macrophages. Experiments of administering lutein extract for 14 days resulted the increased activity and capacity of mice peritoneal fluid macrophage macrophages. It showed that lutein could boost the body's immune system or could act as an immunomodulator. The lutein mechanism could increase macrophage phagocytosis activity, namely by inducing ROS (Reactive Oxygen Species), intracellular and iNOS (inducible nitric oxide synthase), activating MAPK (mitogen activated protein kinase), and RAR β (retinoic acid receptor beta) in macrophage cells that could increase the phagocytosis ability of macrophages [37].

4. Conclusions

The work confirms that lutein extract from marigold flower (*T. erecta* L.) had strong an antioxidant potential. The three protocols used i.e. DPPH, ABTS and FRAP provide a good selection method to use for antioxidant measurements. Determination of antioxidant lutein extract will use to developed new drugs in pharmaceutical fields. The results of *in vivo* experiments showed that Lutein extract was able to increase the activity and phagocytic capacity of *S. aureus*-induced rat peritoneal macrophages. Phagocytic processes in immune cells contribute in particular to homeostasis and defense against disease.

Acknowledgements

The authors are grateful to the RC Biotechnology and InaCC-RC Biology-Indonesian Institute of Sciences (LIPI) for their valuable support to complete this study. We express our gratitude to Dra. Herdini, M.Si., Apt. and Ika M. Kusuma SP. M.Si., Lecturer at Fac. of Pharmacy-ISTN, Jakarta for suggestions during the research. We are also much thankful to the Faculty of Veterinary Medicine- IPB University, for providing us facilities and collaboration.

Conflict of interest

The authors declare no conflict of interest

Animal rights

The institutional and international ethical guidelines for the Care and Use of Laboratory Animals were followed. See the experimental parts for details.

Author details

Kusmiati Kusmiati[1]*, Fifi Afiati[2], Chrisna Widhiani[3], Alpinna Aditia[3], Destia D. Elviani[3] and Atit Kanti[1]

1 Indonesian Culture Collection (InaCC)-Research Center for Biology, Indonesian Institute of Sciences (LIPI), Indonesia

2 Research Center for Biotechnology, Indonesian Institute of Sciences (LIPI), Indonesia

3 Faculty of Pharmacy, Institute Science and Technology National, Jakarta, Indonesia

*Address all correspondence to: kusmiati02@yahoo.com

References

[1] Cantrill R., Lutein from *Tagetes erecta*. Chemical and Technical Assessment (CTA). Expert Committee on Food Additives, FAO. 2004. http://www.fao.org/fileadmin/templates/agns/pdf/jecfa/cta/63/Lutein.pdf

[2] Heuzé V, Tran G, Hassoun P, Lebas F, 2017. Mexican marigold (*Tagetes erecta*). In: Feedipedia, INRA, CIRAD, AFZ and FAO.https://www.feedipedia.org/node/90

[3] Miroslav Šivel, Stanislav Kráčmar, Miroslav Fišera. Bořivoj Klejdus and Vlastimil Kubáň. Lutein Content in Marigold Flower (*Tagetes erecta* L.) Concentrates used for Production of Food Supplements. Czech J. Food Sci. Vol. 32, 2014, No. 6: 521-525

[4] Kusmiati, Swasono R. Tamat, Tria Ajeng Ilmiarti. 2015. Isolation of Lutein from Marigold Flowers (*Tagetes erecta* L.) and Identification by Fourier Transformed Infrared and Liquid Chromatography Mass Spectrometry. Indonesian Journal of Pharmaceutical Sciences, September 2015, Vol. 13, No. 2 p123-130. ISSN 1693-1831.

[5] Demers KJ, 2015. Screening of herbicides for selective weed control in African marigold (*Tagetes erecta*). Graduate Dissertation. Ames, Iowa, USA: Iowa State University. http://lib.dr.iastate.edu/etd/14321

[6] El-Sayed M. Abdel-Aal, Humayoun Akhtar , Khalid Zaheer and Rashida Ali. Dietary Sources of Lutein and Zeaxanthin Carotenoids and Their Role in Eye Health. Nutrients 2013, 5, 1169-1185; doi:10.3390/nu5041169

[7] Ćetković, G. S., Djilas, S. M., Čanadanović-Brunet, J. M., & Tumbas, V. T. (2004). Antioxidant properties of marigold extracts. Food Research Inte rnational,37(7),643650.http://dx.doi.org/10.1016/j.foodres.2004.01.010.

[8] Otang WM, Grierson DS, Ndip RN. Phytochemical studies and antioxidant activity of two South African medicinal plants traditionally used for the management of opportunistic fungal infections in HIV/AIDS patients. BMC Complementary Alternative Medicine. 2012;12(43):1-7. DOI: 10.1186/1472-6882-12-43

[9] Trumbo PR, Ellwood KC. Lutein and zeaxanthin intakes and risk of age-related macular degeneration and cataracts: An evaluation using the food and drug administration's evidence-based review system for health claims. The American Journal of Clinical Nutrition. 2006;84(5):971-974. DOI: 10.1093/ajcn/84.5.971

[10] Khachik F., Beecher G.R., Whittaker N.F. (1986): Separation, identification, and quantification of the major carotenoid and chlorophyll constituents in concentrates of several green vegetables by liquid chromatography. Journal of Agricultural and Food Chemistry, 34: 603-616

[11] Breithaupt D.E., Bamedi A. (2001): Carotenoid esters in vegetables and fruits: a screening with emphasis on β-cryptoxanthin esters. Journal of Agricultural and Food Chemistry, 49: 2064-2070.

[12] Calvo M.M. (2005): Lutein: a valuable ingredient of fruit and vegetables. Critical Reviews in Food Science and Nutrition, 45: 671-696

[13] Edakkadath R. Sindhu, Korengath C. Preethi & Ramadasan Kutta. Antioxidant activity of carotenoid lutein in vitro and in vivo. Indian Journal of Experimental Biology Vol 48, August 2010, pp. 843-848

[14] Gong, Y., Liu, X., He, W. H., Xu, H. G., Yuan, F., & Gao, Y. X. Investigation into the antioxidant activity and

chemical composition of alcoholic extracts from defatted marigold (*Tagetes erecta* L.) residue. Fitoterapia, 2012; 83(3), 481-489. http://dx.doi.org/10.1016/j.fitote.2011.12.013. PMid:22223143. [Links]

[15] Kumar S, Pandey AK. Chemistry and Biological Activities of Flavonoids: An Overview. The Scientific World Journal. 2013;(2013):1-16. DOI: 10.1155/2013/162750

[16] Birasuren B, Kim NY, Jeon HL, Kim MR. Evaluation of the antioxidant capacity and phenolic content of *Agriophyllum pungens* seed extracts from Mongolia. Preventive Nutrition Food Science. 2013;18(3):188-195

[17] Fernández-Sevilla, J. M., Acién Fernández, F. G., & Molina Grima, E. Biotechnological production of lutein and its applications. Applied Microbiology and Biotechnology, 2010;86(1), 27-40. http:// dx.doi.org/10.1007/s00253-009-2420-y. PMid:20091305. [Links]

[18] Harsha Hirdyani And Mini Sheth. Lutein – The Less Explored Carotenoid. World Journal of Pharmaceutical Research. Vol 6, Issue 6, 528-553. 2017. ISSN 2277-7105. Doi: 10.20959/Wjpr20176-8671

[19] Ferreira C.K.O., J. Prestes, F.F. Donato. R. Verlengia, J.W. Navalta, C. R. Cavaglieri. Phagocytic responses of peritoneal macrophages and neutrophils are different in rats following prolonged exercise. CLINICS 2010;65(11):1167-1173. DOI:10.1590/S1807-59322010001100020

[20] Robert E. Baughn And Peter F. Bonventre. Phagocytosis and intracellular Killing of *Staphylococcus aureus* by Normal Mouse Peritoneal Macrophages. Infection and Immunity, Aug. 1975; Vol. 12 (2): 346-352.

[21] Khachik F. Process for isolation, purification, and recrystallization of lutein from saponified marigold oleoresin and uses thereof. United States Patent.1995; 382:14.

[22] Madhavi. D.L., dan Kagan D.I., Process for The Isolation of Mixed Carotenoids from Plants. United States Pantent Documents, United States. 6,380,442. 2002.

[23] Kushwaha D., and Y. Verma. Evaluation of Antioxidant and Free Radical Scavenging Activity of *Tagetes patula*. Annual Research & Review in Biology 2017. 13(6): 1-8, 2017; Article no. ARRB.34349 ISSN: 2347-565X, NLM ID: 101632869

[24] Re R, Pellegrini N, Proteggente A, Pannala A, Yang M & Rice-Evans C. Antioxidant activity applying an improved ABTS radical cation decolorization assay. Free Rad. Biol. Med. 1999: 26, 1231-1237.

[25] Pulido R, Bravo L & Saura-Calixto F. Antioxidant activity of dietary polyphenols as determined by a modified ferric reducing/antioxidant power assay. J. Agric. Food Chem. 2000: 48, 3396-3402

[26] Al-Duais, M., Müller, L., Böhm, V., & Jetschke, G. Antioxidant capacity and total phenolics of *Cyphostemma digitatum* before and after processing: use of different assays European Food Research and Technology, 2009;228(5), 813-821. http://dx.doi.org/10.1007/s00217-008-0994-8. [Links]

[27] Benzie, I. F. F., & Strain, J. J. The ferric reducing ability of plasma (FRAP) as a measure of antioxidant power: the FRAP assay. Analytical Biochemistry, 1996;239(1),70-76. http://dx.doi.org/10.1006/abio.1996.0292. PMid:8660627. [Links]

[28] Thaipong K, Booprakob U, Crosby K, Cisneros-Zevallos L, Byrne DH. Comparison of ABTS, DPPH, FRAP, and ORAC assays for estimating

antioxidant activity from guava fruit extracts. J Food Compos Anal, 2006; 19:669-75

[29] Ramadan G., A. Al Neyadi, A. Mashli, R. Iratni, and S. AbuQamar. Isolation, Culture Procedures and Microscopy of the Mouse Peritoneal Cells for Assaying Phagocytosis: An Undergraduate Laboratory Experiment. International Research Journal of Education and Innovation (IRJEI) Vol 1 No 8 August 2015

[30] Ranjith, M.S., A.J.A. Ranjitsingh, S.G. Shankar, G.S. Vijayalaksmi, K. Deepa and H.S. Sidhu, 2008. Enhanced phagocytosis and antibody production by Tinospora cordifolia: A new dimension in immunomodulation. Afr. J. Biotechnol., 7: 81-85

[31] Molyneux, P. The use of the stable free radical diphenylpicrylhydrazyl (DPPH) for estimating antioxidant activity. Journal of Science Technology.2004; 26(2), 211-219.

[32] Brand-williams, W., Cuvelier, M. E. & Berset, C. Use of a free radical method to evaluate antioxidant activity. LWT-Food Sci Technol 28(1), 25-30 (1995).

[33] Kusmiati and Ni Wayan Sri Agustini. Potential Lutein Extract of Broccoli (*Brassica oleracea* L. var. italica) as Antiradical ABTS (2,2-Azinobis Acid, 3-Ethyl Benzothiazoline-6-Sulfonic Acid). ICBS Conference Proceedings, International Conference on Biological Science. 2015; KnE Life. Sciences, pages 125-135. DOI 10.18502/kls.v3i4.696

[34] Ingkasupart P., B. Manochai., W. Tae Song, J. H. Hong. Antioxidant activities and lutein content of 11 marigold cultivars (*Tagetes* spp.) grown in Thailand. Food Sci. Technol (Campinas) vol.35 no.2 Campinas April/June 2015. http://dx.doi.org/10.1590/1678-457X.6663

[35] Ngoitaku, C., Kwannate, P. and Riangwong, K. Total phenolic content and antioxidant activities of edible flower tea products from Thailand. International Food Research Journal (2016;23(5): 2286-2290

[36] Carlos Rosales and Eileen Uribe-Querol. Phagocytosis: A Fundamental Process in Immunity. Hindawi. BioMed Research International. Volume 2017, Article ID 9042851, 18 pages https://doi.org/10.1155/2017/9042851

[37] Lo H.M. The Carotenoid Lutein Enhances Matrix Mettaloproteinase-9 Production and Phagocytosis Through Intracellular ROS Generation and ERK1/2, p38 MAPK and RAR β Avtivation In Murine Macrophages. May 2013. 93 (5). 723-735 https://doi.org/10.1189/jlb.0512238

Chapter 11

Utilization of Starch in Food and Allied Industries in Africa: Challenges and Prospects

Akeem Olayemi Raji

Abstract

The shortage of food supply has affected the food situation in most developing tropical countries, resulting into a high incidence of hunger and malnutrition. This has also affected the attainment of self sufficiency in starch production for food, pharmaceutical and industrial usage. The review critically appraised the challenges that food and allied industries are facing on the utilization of starch as their major raw material. Information on various conventional and non conventional starch sources were provided, starch forms, properties and recent advances in starch modification methods were discussed. Starch applications in food and allied industries were mentioned and various challenges facing common starch sources were stated. Possibly, utilization of unconventional lesser known crops as starch sources might broadening the present narrow commonly cultivated starch sources, while value addition and good agricultural practices might improve the productivity of conventional starch sources.

Keywords: starch, sources, utilization, challenges, prospects

1. Introduction

Starch or amylum can be defined as a polysaccharide carbohydrate consisting of a large number of glucose units joined together by glycosidic bonds [1]. It comprises of two main components which are mainly linear amylose and highly branched amylopectin. It exists as a stored discrete semi crystalline granules in higher plants [2]. Starch is an important energy source for humans, produced by all green plant as an energy store [2]. Its production in the chloroplast occurs in the daylight and it was rapidly produced by plants. However, glucose chain produced biochemically by photosynthesis in plant cells is responsible for the synthesis of starch [3].

Amylose contains 1–4 D-glucopyranosyl units and it constitutes about 15–30% of common starch, while amylopectin possesses a large number of short chains linked together at their reducing end side by 1–6 glucosidic linkage [4, 5]. However, the formation of crystalline lamella in starch is linked to amylopectin and their branching points are part of the amorphous nature [6]. Starch is regarded as a semi crystalline entity because of the presence of amorphous and crystalline regions in starch granules [7]. Starch also contains some minor components such as lipids and proteins aside amylase and amylopectin. The sizes of starch granules range from 1 to 100 mm, while its shape and composition depends on their botanical source [8].

Starch is derived from a range of raw materials such as corn, wheat, pea, potato, and cassava roots [9] and it has a wide range of applications beyond the food industry. It is also used in the paper/board sector for wet-end addition, size press, surface coating and in the production of recycled paper. It is also used as a binding agent in the pharmaceuticals sector, as an adhesive in industrial binding sector and as a stiffener in the textile sector [10]. Other non food applications of starch include utilization as alcohol-based fuel, low-calorie substitutes, biodegradable packaging materials, thin films and thermoplastic materials with improved thermal and mechanical properties [5].

Starch is the basis of our food and industrial economy, but the food situation in most developing tropical countries is alarmingly worsening owing to increasing population and shortage of fertile land [11]. The shortage of food supply has resulted into a high incidence of hunger and malnutrition [12, 13]. It also affected the demand for starch as food, pharmaceutical and industrial uses coupled with the need to attain self-sufficiency in starch production. However, the focus of this review is to critically appraise the challenges that food and allied industries are facing on the utilization of starch as their major raw materials and to suggest possible way outs.

2. Sources of starch

Starch is a source of carbon and energy, it is predominantly found in different part of a plant such as stem, roots, flowers, seeds, and fruits [3]. Starch can be derived from tubers and cereals, such as cassava, potatoes, maize, yam, rice, etc. [14]. **Table 1** shows the starch compositions of some crops. The starch contents of cereals and tubers presented in **Table 1** ranged from 20.48–77.90%, with soft wheat having the highest value and bitter yam having the lowest value.

2.1 Cassava

Cassava (*Manihot esculenta* Crantz) is a tropical valuable root crop, used both as food and industrial raw material, due to the high starch content in tubers [22, 23]. It is one of the world's most important food crops, with annual global production at approximately 276 million metric tons in 2013 [24]. Its price is often decided by the industrialists based on the percentage of starch in the tubers. In 2013, the top producing countries globally were: Nigeria (accounting for 19% of the total), Thailand (11%), Indonesia (9%), Brazil (8%) and Democratic Republic of Congo (6%) [24]. It is a source of livelihood for at least 300 million people. Nearly 90% of cassava produced in Africa is used as a staple food for human consumption, which provides calories for 500 million people and it constitutes 37% of the population's dietary energy requirements [24]. Cassava is perceived in most African countries both as a food security crop and also as a raw material for various types of industries. It can be transformed into several types of products ranging from traditional foods and feeds to novel food products [24, 25].

2.2 Potatoes

Potato is one of the top four staple food crops in the world. Majority of its production is meant for human consumption (50–60%), while the rest are used for production of animal feeds, industrial products and as seed tubers [26, 27]. The tuber is a carbohydrate reserve and it also contains high quality protein, substantial amounts of vitamins, minerals and trace elements. Potatoes grow best in moderate

Food Plants	Starch Contents (%)	Sources
Cereals		
Waxy rice	74.76	Moongngarm [15]
Hard wheat	77.40	Ragaee *et al.* [16]
Soft wheat	77.90	Ragaee *et al.* [16]
Barley	53.60	Ragaee *et al.* [16]
Millet	67.40	Ragaee *et al.* [16]
Rye	58.00	Ragaee *et al.* [16]
Sorghum	67.70	Ragaee *et al.* [16]
Fonio	68.00	Cruz *et al.* [17]
Sweet Corn	36.23	Moongngarm [15]
Roots/ Tubers		
Sweet potato	52.54	Moongngarm [15]
Lesser yam	54.70	Moongngarm [15]
White yam	58.02	Moongngarm [15]
Water Yam	31.90	Brunnschweiler *et al.* [18]
Bitter yam	20.48	Ezeocha and Okafor [19]
Potato yam	38.10	Shajeela *et al.* [20]
Yellow yam	41.72	Oladebeye *et al.* [21]
Taro	63.74	Moongngarm [15]
Cassava	65.71	Moongngarm [15]

Table 1.
Starch compositions of cereals and tubers.

climate, where continuous cultivation throughout the year is impossible [26]. Among the potato varieties, sweet potato (*Ipomoea batatas* L (Lam)) is regarded as one of the most economically important species, as it can grow in great abundance on marginal soils [27]. It is rich in starch (58–76% on a dry basis) and the starch is widely used in starch noodles, bakery foods, snack foods and confectionary products [27].

2.3 Yams

Yams belong to the family *Dioscoreaceae* and they are annual or perennial tuber-bearing and climbing plants. It is an important major food crop in Nigeria [28] and ranked as the fourth major root crop in the world after cassava, potatoes and sweet potatoes, having an annual production of above 28 million metric tonnes [29]. However, Nigeria was rated as the world's largest producer of yams, with *Dioscorea rotundata* and *Dioscorea alata* as the two most cultivated yam species [30]. Yam can be explored for commercial starch production because of their high starch content of about 70 to 80% dry weight [31], and it plays a prominent role in ensuring food and livelihood security of at least 60 million people in West Africa. Globally, roughly about 57 million tons of yams (representing 93% of annual global production) are produced on 4.7 million hectares annually in this sub-region which comprise of Benin, Côte d'Ivoire, Ghana, Nigeria, and Togo [24, 32]. The yam species that are majorly grown globally include: *Dioscorea alata* (water yam), *Dioscorea bulbifera* (potato yam), *Dioscorea cayenensis* (yellow yam), *Dioscorea dumetorum* (bitter yam), *Dioscorea esculenta* (lesser yam), *Dioscorea opposita* (Chinese yam), *Dioscorea rotundata* (white yam), and *Dioscorea trifida* (cush-cush yam) [33]. Among the species mentioned above, *Dioscorea rotundata* and *Dioscorea cayenensis* are the most commonly grown for consumption and commercial production. Yam contains mainly carbohydrates with little amount of proteins, lipids and vitamins, and it can provide around 110 calories per 100 grams of products [34]. Yam possesses high in moisture, dry matter, starch, dietary fiber, vitamin B_6, but low in saturated fat, sodium and vitamin A contents. Yams contain about 5–10 mg/100 g of vitamin C, and the limiting amino acids are isoleucine and sulfur containing amino acids. They also contain a compound called "diosgenin", which can be extracted and used as base for drugs such as cortisone and hormonal drugs. Some species contain alkaloids (e.g. dioscorine $C_{13}H_{19}O_2N$) and steroid derivatives [34]. However, the nutrient content of yam is compared with other crops in **Table 2**.

2.4 Yam varieties

2.4.1 *Dioscorea alata*

It is widely called "water yam", "winged yam" and "purple yam". It is the most widely cultivated specie globally and it is second to white yam in popularity [36, 37]. Water yam is economically important yam specie which serve as a staple food for millions of people in tropical and subtropical countries, and with a great potential for increase in consumers demand due to its low sugar content necessary for diabetic patients [38]. Aside from being source of carbohydrate, it also possesses higher content of protein and low lipids than *D. cayenensis*, *D. escunlenta*, *D. rotundata* and *D. trifida* [39].

2.4.2 *Dioscorea bulbifera*

It is usually cultivated in Africa and Asia, with slight differences between those found in each place. It is a large vine of about 6 meters (20 ft.) or more in length and produces tubers (bulbils) which grow at the base of its leaves. It is an important food product and it is about the size of potatoes (hence the name "air potato"), weighing from 0.5 to 2 kg (1 to 5 lbs.) [40]. Some known varieties of air potato can be eaten raw while some need to be detoxified by either soaking or boiling before eating. Its growth for commercial purpose is hampered by its moderately unpleasant

	Crops						
Staple	**Maize/Corn**	**Rice**	**Wheat**	**Potato**	**Cassava**	**Soybean (Green)**	**Yam**
COMPONENT (PER 100G PORTION)							
Water (g)	10	12	13	79	60	68	70
Protein (g)	9.4	7.1	12.6	2	1.4	13	1.5
Fat (g)	4.74	0.66	1.54	0.09	0.28	6.8	0.17
Carbohydrate (g)	74	80	71	17	38	11	28
Fiber (g)	7.3	1.3	12.2	2.2	1.8	4.2	4.1
Sugar (g)	0.64	0.12	0.41	0.78	1.7	0	0.5

Source: Kumar et al. [35].

Table 2.
Nutrient content of white yam in comparison with other crops.

flavor, making other yam to be mostly preferred by consumers [41, 42]. The aerial or air yam is popular in home vegetable gardens, because it produces its first bubils only after four months of growth and thereafter throughout the life of the vine, as long as two years [41, 43].

2.4.3 Dioscorea cayenesis

This yam specie got its common name from its yellow flesh, due to the presence of carotenoids. It is a West African native; it has a longer period of vegetation and a shorter dormancy than white yam [44]. Yellow and white yam were considered in past as two separate species, but now they are been considered as same specie by most taxonomists with about over 200 cultivated varieties between them [44]. They are large and their vines are as long as 10 to 12 meters (35 to 40 feet). The tubers can weigh up to about 2.5 to 5 kg (6 to 12 lbs.) each for an average size, but extra large tubers can weigh as much as 25 kg (60 lbs.) [40]. The maturation stage after planting is 7 to 12 months [45]. It is commonly used in Africa for the traditional popular dish known as "iyan" (pounded yam) [46].

2.4.4 Dioscorea dumetorum

The tuber is commonly referred to as cluster yam, bitter yam or trifoliate yam. It is called *"ona"* by the Igbos of Southeast Nigeria and 'esuru' by the Yorubas in Southern Nigeria [42, 47]. The tuber contains fleshy edible part having a yellowish or whitish color which can be boiled and eaten as snacks. Extract from the tubers can be used for the treatment of *diabetes mellitus* in traditional medicine [48, 49]. It has been reported that the tuber is rich in fiber and contains an alkaloid, dioscorentine, which possesses hypoglycemic activity [50, 51]. It is fairly high in protein and possesses a well balanced amino acid profile, making it the most nutritious of the commonly consumed yam species. It has been established that trifoliate yam contain 15–38% starch on wet weight basis and 70–80% dry weight basis [52].

2.4.5 Dioscorea esculenta

This type of tuber is popularly referred as 'Chinese yam'; it is smaller than other known yam species grown in Africa [24]. It possesses ability to tolerate cold temperature and can be grown in much cooler conditions than other yams. It is commonly grown in China, Korea, and Japan and it was introduced to Europe in the 1800s when the potato crop was affected by diseases [53]. In France, it is still grown for Asian food market. The tubers mature in 6 months and can be eaten immediately after harvest, while others are used as ingredients for other dishes, including noodles and traditional medicines [24, 54].

2.4.6 Dioscorea rotundata Poir (white yam)

This yam specie is native to Africa. It produces edible tubers that possess economic importance [55]. It is among the most important cultivated yam species and it is mostly consumed in West Africa and Cameroon [56]. It has a long shelf-life which makes it available all year round. White yam are subjected to different cooking methods in Western and Central Africa, however, the most common ones are boiling, frying and roasting [29]. It is mostly used in Africa for the traditional popular dish known as "iyan" (pounded yam). Sun-drying of parboiled yam pieces and then milled into a light-brownish powder (*elubo*) is another method of processing in Nigeria. A thick brown starchy paste known as" *amala*" which is consumed with local soups and sauces can be prepared from the light brown powder [55].

2.4.7 Dioscorea trifida (cush-cush yam)

This specie is a native of Guyana region of South America and it is the mostly referred to as "New World" yam, their growth cycle is less related to seasonal changes than other yam, since they originated from the tropical rain forest. They can be easily cultivated when compared to other yams and also they have good flavor, which may serve as a great potential for increase production [54, 57]. Its starch contains amylose content in the 34.7–43.3% range for white and purple varieties [57].

2.4.8 Dioscorea villosa (wild yam)

It is a perennial vine that grows in moist thickets and hedges [57]. It possesses a reddish-brown stem, having heart-shaped leaves with prominent veins and inconspicuous greenish yellow flowers that flourish from September to October. It is usually cultivated as food source and its roots are harvested in the fall. Its taste is usually bland and then acrid, but under special preparation herbalist use its fresh and dried roots for medicine [58]. Also, it is a common herbal remedy for pains associated with rheumatism and arthritis, colic and intestinal cramps, as well as a reliable antispasmodic and anti-inflammatory [59]. It is also applicable in contraceptive manufacture. High content of saponin in species of *Dioscorea villosa* made them useful for the preparation of steroids in the pharmaceutical industry [60, 61]. Studies also revealed that *Dioscorea villosa* had antioxidant activities [62] and the anti-inflammatory activity could be linked to the antiphlogistic effect of the steroidal saponins.

2.5 Cereals

Cereals are edible grains or seeds that belong to the grass family *Gramineae* (**Figure 1**). Grains are developed from flowers or florets and their structures vary from one to another with some typical features [64, 65]. They possess *embryo* (or germ) which is a thin-walled structure, containing the new plant. The embryo is separated by the *scutellum* (which is involved in mobilization of food reserves of the grain during germination) from the main part of the grain. The endosperm surrounded by thin-walled cells (*aleurone*), packed with starch grains [66, 67]. The *aleurone* layer present in grains consist of one or three cell layers (wheat, rye, oats,

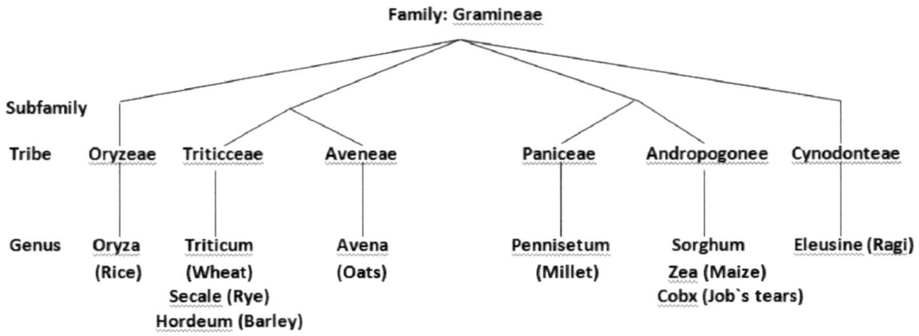

Figure 1.
Taxonomy of the Graminae family. Source: Shewry et al. *[63].*

maize and sorghum have one; rice and barley have three). The pericap is the outer layers of the grain (derived from the ovary of the flower) that surround the *seed coat* (the testa), while the bran is formed from the outer thick-walled structures. They are staple foods both for direct human consumption and indirect through livestock feed [68]. Cereal based foods are good sources of energy, protein, B vitamins and minerals for the world population. Cereals are inexpensive to produce, they are easy to store and transported, and they do not deteriorate readily if kept dry.

However, in cereal products, a proportion of this starch is not digested and absorbed in the small intestine. This is referred to as resistant starch and it appears to act in a similar way to dietary fiber. Four categories of resistance have been defined [65, 69].

- RS1 refers to starch that is physically inaccessible for digestion as it is 'trapped' (e.g. intact whole grains and partially milled grains).
- RS2 refers to native resistance starch granules (e.g. found in high amylose maize starch).
- RS3 refers to retrograded starch (e.g. found in cooked and cooled potatoes, bread and some types of corn-flakes).
- RS4 refers to chemically modified starch (e.g. commercially manufactured starches).

2.5.1 Wheat

Wheat is ranked as a major cereal crop in many parts of the world and belongs to the *Triticum* family of which there are many thousands of species [65, 67]. Among the known species, *T. aestivum* subspecies *vulgare* and the hard wheat *T. durum* are the most commercially viable [65, 70]. Wheat is grown in both winter and spring but their cultivation in winter or spring is dependent on the species, varieties and their adaptability. However, wheat is grown in many countries around the world, but the most prominent great wheat-producing countries are USA, China and Russia. Wheat grains are also present extensively in India, Pakistan, the European Union (EU), Canada, Argentina and Australia. Estimation showed that 556.4 million tons of wheat might have been produced in 2003; resulting to 30% of the world's cereal production [71]. The wheat grain is sandwiched between the lemma and the palea of each spikelet, possessing an elliptical, oval or ovate shape, and has short or long brush hairs.

2.5.2 Rice

Rice is an important staple food crop for many of the world's population, especially those living in Asia. Rice is cultivated mainly for human consumption and this include utilization as breakfast cereals, and its use in Japan as brew saké [65, 67]. There are a lot of rice varieties but only a few are grown widely (*e.g.* varieties of the improved semi-dwarf plant type with erect leaves). The rice grain possesses an outer protective coating which is referred to as the hull or husk and the edible rice caryopsis. Also, the brown rice contains an outer layer called pericarp (which contains pigment), seed coat, the embryo and the endosperm. The endosperm comprises of the aleurone layer which encloses the embryo, sub aleurone layer and the starchy or inner endosperm. However, wild rice is less common and it is the grain of a North American plant, *Zizania aquatic*. It is difficult to harvest and it is more expensive than other grains. It possesses higher protein content than rice [64].

2.5.3 Maize

Zea mays L., alternatively referred to as corn is a native of the Western Hemisphere and its production in the USA exceeds that of others countries [72]. The kernel which is the reproductive seed of the plant has four main parts, and these are the germ, the endosperm, the pericarp and the tip cap. It is an inexpensive source of starch and it is a major source of energy for animal feed [65, 70]. Among the hundreds of different varieties of wheat grains, only four varieties listed below are of commercial importance:

1. Dent maize (identified by the dent in the crown of the kernel);

2. Flint maize (hard, round kernels);

3. Sweet corn (a dent-type maize);

4. Popcorn (flint-type maize which expands when heated).

2.5.4 Barley

Barley is a resilient plant that can tolerate a wide range of conditions, which might have been cultivated since 15,000 BC [72]. The head or spike of barley is made up of spikelets that are attached to the rachis in an alternating pattern. The barley kernel's outer layers consist of the husk, the pericarp (to which the husk is tightly joined in most species), the testa or seed coat and the aleurone which completely covering the grain. Barley (*Hordeum vulgare*) is mainly cultivated for malting and brewing in the manufacture of beer and for distilling into whisky manufacture [65, 70]. The spent grains from the brew are used as animal feed, especially for pigs. However, a small amount of barley is used for food. Pearled barley is used in cooking soups and stews in the UK and in the far Middle East. Barley flour is also used in bread production and can be cooked as porridge in some countries [67].

2.5.5 Oats

Among the several different species of oats, spring or white oat (*Avena sativa* L.) is the most important cultivated form. *Avena byzantine* is a red-oat type or alternatively known as a winter oat is cultivated in warmer climates [65, 70]. A spikelet of oat consists of oat kernels. A hall which is made up of two layers called the lemma and palea, enclosed each kernel, the hall is only loosely attached to the groat. However, the groat is made up of 65–85% of the oat kernel and it is enveloped by bran layers (pericarp, seed coat and aleurone cells) [65]. It grows well on poor soil and in cool, moist climates and has mainly been grown for animal feed. Its cultivation for human consumption is at minimal level, products made from oats are oatmeal for porridge and oatcakes, rolled oats for porridge and oat flour for baby foods and for ready-to-eat (RTE) breakfast cereals [67]. Non-food uses of oats include their utilization in the production of cosmetics and adhesives [70].

2.5.6 Rye

Rye is generally a hardy plant and it is wildly cultivated in cool temperature zones, where other cereals can merely survive. Rye can also be grown in semi-arid areas and at high altitudes. It is cultivated as a winter crop, which is sown in early autumn and harvested in early summer [65]. The plant varies in height from 30 cm to more than 2 m. It is a major crop grown in Russia, Poland, Germany and the Scandinavian countries for crispy bread, alcohol animal feed production [67]. The grain is protected with a bearded *glume* husk, arranged in an alternating pattern along the rachis. Rye is thinner and more elongated than wheat; it possesses a grayish yellow color and varies from 1.5 mm to 3.5 mm in size. The grain comprises of the starchy endosperm, the pericarp and the testa. Other constituents include the bran, accounting for 10% of the grain with the remainder consisting of the germ (the embryo and scutellum).

2.5.7 Millet

Millet is an annually cultivated cereal which possesses different species of small grains [64, 70]. Among the various species of millet, pearl millet is the only one that is mostly of economic concern. Other known species of millet include, finger (or ragi), proso and foxtail, but the minor millets account for less than 1% of the grains produced for human consumption, they are less important in terms of world food production [65, 70]. Cultivation of millet is of utmost importance in certain locations in Africa and Asia where major cereals cannot be solely depended on to provide sustainable yields [73]. The species type influence climatic and soil requirements, length of growing period, grain consistency, size and taste of millet [64].

2.5.8 Sorghum

Sorghum (*Sorghum bicolor* L. Moench) is wild regarded as a warm season crop, intolerant of low temperatures but fairly resistant to serious pests and diseases. The known varieties are great millet and guinea corn in West Africa, kafir corn in South Africa, jowar in India and kaoliang in China. It is one of the staple foods in many parts of Africa, Asia, and parts of the Middle East [65, 70]. However, most of the sorghum grains produced in North and Central America, South America and Oceania are used for animal feed [73]. The grain comprises of a naked caryopsis, which is made up of a pericarp, endosperm and germ. Based on huge range of physical diversities, such as the color of the pericarp (white, yellow or red) and presence/absence of pigmented testa (with/without tannins), sorghums are classed into one of four groups and these are (1) grain sorghum, (2) forage sorghum, (3) grass sorghum or (4) Sudan sorghums and broom corn [70].

2.5.9 Acha

Acha, which is also referred to as "fonio", "findi", "fundi", "pom, and kabug" "hungry rice" and "petit mil", is a small-grained cereal that is native to West Africa, which is generally classified among the millet [74, 75]. It is cultivated in various parts of Nigeria, Sierra Leone, Ghana, Guinea Bissau and Benin Republic on less fertile sandy soils that could not support the growth of other more demanding cereals. There are two known varieties of acha which are *Digitaria exilis Kippis stapf.* and *Digitaria iburua Kippis Stapf.* It is regarded as one of the lost crops in the West Africa sub region. Its production is important to West African Farmers, though hindered by several factors among which are poor agronomic performance because of unimproved seeds and husbandary practices. Its West Africa annual production is about 250,000 tonnes [17]. The annual yields of 3098 metric tonnes, 112,000 metric tonnes and 126,000 metric tonnes of fonio were reported in Nigeria [76]. The economic returns of acha when compared with other crops like rice, sorghum, and cowpea showed that it was profitable to grow the crop on a commercial scale [77]. Fonio are the most nutritious and testiest of all grains [77] and it contains 7%

crude protein that is high in leucine (19.8%), methionie and cystine (7%) and valine (5.8%) [78]. Fonio grains are mostly consumed wholly, they are also milled into flour and constitute a versatile raw material for preparation varieties of food such as gruels, porridges, couscous, bread, beer, and beverages [79]. Starch extracted from fonio possesses good disintegrant and binding properties [80] and it also has good glidant properties [81].

2.6 Food starch properties

The starch granules sizes obtained from different crops vary in properties, because of their sources, extraction methods and cultivars [82]. The purity as well as the granules size can be determined using scanning electron microscopy [83]. The amylose content of starch is one of the most important factors influencing the cooking and textural qualities of whole storage root, and quality of starch-based foods [84]. *Dioscorea* starch granules possess varying shapes, which are spherical, oval and polygonal, depending on the species with granule size varying from 2 to 50 mm [85]. Also the X-ray diffraction pattern of the *Dioscorea* starch granules range from the B to C-type, depending on the *Dioscorea* specie [18]. *Dioscorea* starches contain18 to 30% amylose contents and their gelatinization temperature vary from 70 to 92°C [84].

Starch particle sizes obtained from white and yellow yam varieties showed similar patterns with a single symmetrical distribution centered at approximately 32 and 35 mm respectively. Sources, varietal differences and growing conditions significantly influence size and shape of starch granules. However, due to better mouth feel, small sized starch granules has been suggested as possible lipid substitute in food systems [86]. They are also used as laundry-stiffening agents because they possess good fabric penetration ability, better glossiness and stiffness, which are required in textile industries [87].

Sweet potato starch granules possess spherical, oval, and polygonal shapes, and they are about 2–46 mm in size. The X-ray diffraction patterns of the sweet potato starch granules are of type A and they also exhibit 38% crystallinity [88]. The starch in sweet potato is made of 16.1–24.4% amylose, having a swelling power of 80% at 90°C and gelatinization temperature of 64.6–84.6°C [84]. Cassava starch possesses small spherical granules, having an average granule size of 14.7 mm. Its amylose content ranges from 13.6 to 23.8% [84, 89] and its crystallinity has been reported to be 38% [84], while the gelatinization temperature varies from 59.6 to 87.2°C [84, 89]. The X-ray diffraction patterns of the cassava starch granules depict type A [90]. The cocoyam granules (Taro) are small rounded and ellipsoidal-truncated, with their sizes varying from 0.5 to 5 mm in diameter, making them to be more easily digestible [84]. Taro starch has been used in the preparation of some baby foods and diets of people who are allergic to cereals [91]. The X-ray diffraction patterns of the taro starch granules exhibit the typical A-type pattern, while the starch contains 14.0–19% amylose with the pasting temperatures varying from 81 to 85°C [92]. Interestingly, starch from Tannia (*Xanthosoma sagittifolium*) comprises of small rounded and large truncated ellipsoidal-shaped granules, which possesses granular diameters which range from 2 to 50 mm [91]. The amylose contents of different cultivars range from 21.3 to 25.4% [93]. Tannia starch possesses a type A X-ray diffraction pattern, higher pasting temperatures and lower paste viscosity than those of other starches, such as potato starch [89]. Also, it has higher swelling power and solubility at relatively high temperatures than sweet potato starch granule [84]. The maize starch granules exhibit polyhedral granule shapes and differences in their mean granule size range from 2.3 to 19.5 μm. The starch samples show A-type diffraction pattern with strong reflection at 15.25, 18.11, and 23.33° [94]. The gelatinization temperatures of maize starch range from 69.16 to 76.98°C [95] and the amylose content varies from 24.74 to 30.32% [94]. Fonio starch granules are polygonal in shape with diameter ranging from 2.0–14.3 mm. Their X-ray diffraction patterns are of the type-A crystalline form [96], with the amylose contents ranging from 18.7% to 19.6%. The millet starch granules are small, spherical to polygonal in shape, but may vary from specie to another. The granular sizes range from 0.8 to 10 mm. The gelatinization temperatures of millet starch ranged from 75.8 to 84.9°C and the amylose content vary from 16.0 to 27.1% [84]. The starch granules obtained from sorghum are typically 3–27 mm and the gelatinization temperature range 61.1–81.2°C [97]. The starch showed the A-type crystalline diffraction pattern and its amylase content varies from 11.2–28.5%, depending on both genetic and environmental factors [98].

2.7 Forms of starch

2.7.1 Native starch

Native starches are crude starches that are extracted directly from their sources and they are mainly used as food, but irrespective of their sources they are undesirable for many applications because of their inability to withstand processing conditions. Each starch has unique functional properties, and much of the starch used industrially is modified before use, giving a wider range of useful products [99]. Starch can be readily converted chemically and biologically into many useful and diverse products such as paper, textiles, adhesives, beverages, confectionery, pharmaceuticals, and plastics, so as to improve desirable functional properties in order to meet the requirements of specific industrial processes [13].

2.7.2 Extraction techniques

Starch granules' settling is often prevented by presence of various components like mucilage and latex, which may lead not only to loss of starch, but also reducing the quality of extracted starch. However, microbial growth can also be promoted if the extraction residence time is prolonged than necessary, which may also result into breakdown of starch and resultant loss of starch quality. It also affects the color of the starch limiting its utilization in food and industrial applications. Therefore, optimum recovery of starch having physicochemical and functional qualities coupled with economical extraction of starches from cereals and tubers is important. Extraction of starch with water is the most common form of starch extraction, but this has been improved upon over time. The Central Tuber Crops Research Institute, Trivandrum, India research on various chemicals that could improve the yield of starch from various tubers [100, 101]. It was discovered that ammoniacal solutions gave the best results. However, when aqueous ammonia (0.03 M) was used for starch extraction, the yield, paste viscosity and swelling capacity of the extracted starch improved. Ammonia formed a complex with the mucilagenous material in the slurry, thereby releasing the starch granules and promoting faster settling of starch in less viscous slurry, which prevents microbiological damage of the starch due to short residence time. Moorthy [101] observed that lactic and citric acids improved the yield and color of starch from sweet potato tubers. Interestingly, an enzymatic method for enhancing the recovery (26% increase) of starch from cassava tubers using pectinase and cellulase enzymes [101]. The aforementioned enzymes work by altering the integrity of the pectin-cellulosic matrix of cell membranes and thereby promoting the release of the starch granules. The same technique was used to promote starch recovery from sweet potatoes by 20% without affecting starch properties [101].

2.7.3 Modified starch

Modified starches are native starches whose physical and chemical characteristics have been altered in order to improve their functional characteristics [102]. However, modification of starch is done to tailor starch to a specific food application, to stabilize starch granules during processing and make it suitable for many foods and industrial applications. During modification, starch properties are altered, including solution viscosity, gelatinization properties, pasting properties, retrogradation behavior, association behavior, and shelf-life stability in final products [102, 103]. Modification can be achieved through etherification, esterification, cross-linking/grafting of starch, decomposition (acid or enzymatic hydrolysis and oxidization of starch) and physical treatment of starch using heat or moisture [104].

Modified starches produced from different methods and sources are usually used as thickening agents to provide desired structures in food products [105]. Acid-thinned starch is starch that has been treated in acid slurry [1]. This starch possesses faster gelatilization, low viscosity, and could produce a weak gel (Abbas *et al*., 2010). Oxidized starch is a starch which has undergone oxidation and it has a low hot paste viscosity. Both of acid-thinned and oxidized starches are suitable for confections because they allow rapid and efficient cooking of starch solution in the presence of concentrated sugar syrups [106]. Stabilized starch is a type of modified starch that possesses a resistant property against acid degradation under dry acidic storage. Stabilized starch can be produced by adding buffer (pH 6–9) to starch slurry and drying the starch-buffer slurry in a conventional oven [107]. Stabilized starch has a reduced starch gelatinization temperature; it's easier to cook and allows for the formation of stronger gel with increased clarity and longer self-life. Converted starch, such as dextrin, which possesses a good film forming capacity, can be used with high-sugar solutions to produce stable and flexible coatings [108]. Thin boiling starch is a modified starch that can be produced by treatint the starch with amylase enzyme to hydrolize the α-1, 4 glycosidic bonds. This starch possesses gelling properties and low hot paste viscosity, which make it suitable for gum drops since this type of starch allows better evaporation and pouring [104].

Octenyl succinic anhydride (OSA) starch is a recent type of chemically modified starch, possessing surface active properties [109]. Octenyl succinic anhydride (OSA) starch is produced by esterification of different sources of starch with anhydrous octenyl succinic acid under alkaline conditions [110]. The adsorption of the OSA starch molecules at the oil/water (O/W) interface might be as a result of hydrophobic short octenyl succinate side chains in octenyl succinic anhydride (OSA) starch molecules and the long amylopectin backbone which protect the droplets against flocculation by the mechanism of steric stabilization [111]. OSA behaves like a typical surfactant and forms a strong film at the O/W interface which provides a good resistance against re-agglomeration [111]. The increase in the viscosity of the continuous phase in conjunction with its ability of adsorption at interfaces enables OSA starch to behave as a stabilizer and also as an emulsifier in O/W emulsion systems. OSA starch type of modified starch has been approved as a food additive by the Food and Drug Administration (FDA) and European Union [112].

2.8 Utilization of starch in food industries

Starch plays a crucial role in food systems by stabilizing and creating the food structure. Starch also co-exists with other components to deliver or maintain nutrient and flavor [113]. However, starch importance in some food applications are elaborated below.

2.8.1 Snack foods

Starch is generally used in the production snack foods to achieve desired textural and sensory attributes by improving crispiness, oil binding properties, expansions, and overall eating quality [113]. The properties of the amylopectin and amylose of starch are important for the texture creating of this kind of foods. The highly branched amylopectin might increase dough expansion and viscosity, which could result to production of light, crispy, and expanded products. The amylase on the other part strengthens the dough and improves its forming and cutting properties. Consequentially, a more crunchy final texture could be obtained [114]. Furthermore, a high quality fiber-fortified snack could be developed by incorporating modified starch [108].

2.8.2 Baked products

Starch contribution to baked products quality is through its important properties, such as gelatinization, water absorption, and retrogradation [115]. The gelatinization property of starch is very important in building the structure and texture of baked products. However, starch ability to bind water could reduce the stickiness of dough, improve handling, and increase cake volume. It could also improve the moistness and softening the texture of baked products [108].

2.8.3 Confectionery

Starch behaves as a structure builder in coatings and also acts as a medium for molding to support desired shapes in confectionaries. Starch is used in the production of pastes, gums, and molds; it is also in the manufacture of dusting sweets to prevent them from sticking together. Starch is selected because of its ease of cooking in high-sugar environments and also based on its ease of handling during processing [108]. In most cases, starch is modified to possess specific properties that suit certain applications.

2.8.4 Gravies, soups, and sauces

Starch application in gravies, soup, and sauces depends on the production process, which is usually influenced by pH of products and heat during processing [114]. Compared to neutral products, high acid products (pH < 4.5) require a higher-degree of cross-linking starch. The production of cross-linked starch is achieved by the reaction in which a small number of hydroxyl groups on the glucose units of amylose and amylopectin, mostly in the amorphous are modified without destroying the granular nature of starch [116]. Since sterilization of acidic products require shorter processing time and lower temperature, other factors such as shelf-life requirements, fill-viscosity, and heat penetration also influence the suitability of starch used in these types of food products. Hydroxypropylated starch which possesses high freeze thaw stability is suitable for chilled and frozen foods [108]. Hydroxypropylated starch is produced by reaction of starch with propylene oxide. Aside its better freeze thaw stability, modification improves hydroxypropylated starch shelf-life, cold-storage stability, past clarity and texture properties of its paste. Generally, gravies, soups, and sauces require starch with opaque paste [108].

2.8.5 Mayonnaise and salad dressing

The function of starch in these products is mainly to thicken and stabilize the dispersed phase [108]. Basically, these food products are produced under acidic condition and the process involves application heat and shear thinning. Therefore, starch which has ability to tolerate acidity, heat, and shear is suitable for these products. Also, lipophilic starch which has potential to stabilize emulsions, other modified starches such as cross-linked starch and stabilized starch are the most commonly used for these products [117]. Lipophilic starch is produced by esterification with n-octenyl succinic anhydrate, which led to a starch structure comprising both hydrophilic and lipophilic properties. This starch can be used for mayonnaise and salad dressing, lipophilic starch might be used to replace animal-derived sodium caseinate and gum Arabic [102].

2.8.6 Emulsion stabilizer

A scientific study suggested that starch could be used as a stabilizer in oil-in-water emulsions [117]. The study further revealed that emulsion prepared with equivalent volume ratio of water and paraffin could be stabilized by adding an aliquot of starch dispersion [117]. Kim *et al.* [117] established that the introduction of starch nano particles of more than 0.02% wt kept the emulsion longer than 2 month of storage without coalescence of oil droplets. Starch might be used in various emulsions not only for foods but also for cosmetics and pharmaceuticals [117].

2.9 Utilization of starch in allied industries

Some non food uses of starch in allied industries are described below:

2.9.1 Packaging component

Nanoparticle from starch can be used as a material in barrier films for food packaging. The primary focus of the barrier properties of the packaging films is on water vapor transmission and oxygen permeability [117]. However, a decrease in water vapors permeability was observed when the maize starch was used. Kristo and Biliaderis [118] revealed addition of 30–40% waxy maize starch significantly decreased the water vapor permeability in the component. Similarly, García *et al.* [119] observed a 40% decrease in the permeability of cassava starch film, reinforced with 2.5% starch nanoparticles.

2.9.2 Drug carrier and tablet binder

Starch has been described as a nontoxic, biodegradable, and biocompatible polymer [117]. Its nanoparticles possess the ability to deliver an ample range of molecules to any location in human body for sustained periods of time. Starch has been useful in tablet formulation and binding due to its relative inertness, abundance, low cost, and suitable physicochemical properties [31].

2.9.3 Binders in paper making and paper coating

Starch possesses a binding property and can be used as a binder in papermaking and paper coating. Cooked starch is widely used as paper-making additive and the retention of the cooked starch on the paper matrix is based on starch absorption [108]. However, the amount of starch absorbed is limited by the cellulose substrates absorption saturation. Another problem which might cause operational problems is the high viscosity of the starch paste after cooking. Bloembergen *et al.* [120] revealed that the modified starch possessed better paper binding capacity than cooked starches.

2.9.4 Starch as energy source

Starch bioconversion into ethanol is a two-step process. Saccharification is the first step, which involves conversion of starch into sugar using an amylolytic microorganism or enzymes such as glucoamylase and α-amylase. Fermentation is the second step, which involves conversion of sugar into ethanol using *Saccharomyces cerevisiae* [121]. An alternative to the conventional multistage process of starch fermentation which offers poor economic feasibility is the use of amylolytic yeasts for the direct fermentation of starch. Despite the fact that there are over 150 amylolytic yeast species, they possess limited industrial use because of their low ethanol tolerance [121]. Therefore, most research effort has been geared towards the development of genetically engineered amylolytic strains of *S. cerevisiae*, and in these strains, heterologous genes encoding α-amylase and glucoamylase from various organisms have been produced including their products and their products [121].

3. Challenges facing common starch sources and way outs

Starch is the basis of our food and industrial economy, but the food situation in most developing tropical countries is alarmingly worsening owing to increasing population, fragmented farms with rudimentary technologies, poor pre-harvest and postharvest farm practices and shortage of fertile land [11, 24, 108]. The shortage of food supply has resulted into a high incidence of hunger and malnutrition [12, 13]. It also affected the demand for starch as food, pharmaceutical and industrial uses coupled with the need to attain self-sufficiency in starch production.

However, self-sufficiency in starch production could be attained if the following suggestions described below are given utmost attention:

i. Some lesser known and unconventional crops could be good sources of nutrients and starch, and even have the potential of broadening the present narrow food base of the human species (Viano *et al*., 1995). *Dioscorea villosa* L (common name wild yam or atlantic yam) is one of the underutilized starch sources, and it is less explored due the presence of anti nutritional factors. Processing techniques can reduce the anti-nutritional factors in food and make them readily available for consumption.

ii. Efficient and effective development of the various starch based crop value chains such as cassava in order to increase their productivity [24].

iii. Encouraging small and medium scale investments in order to enhance production, processing and delivery of high quality and quantity of cassava products to the larger industries [24].

iv. Promotion of effective pest/pathogens integrated management programs and reduction of the occurrence of mycotoxin in starch based crops, to prevent emerging and endemic pests and diseases as well as reduction of mycotoxin contamination in starch [108].

v. Access to high yielding, climate-resilient and affordable quality seeds, advances in breeding technology using genomic approaches that is cost-effective and field-relevant high throughput phenotyping approaches will accelerate and contribute significantly to mass production starch-based crops [24].

4. Conclusion

Conclusively, suggested way outs should be given utmost attention, most especially research efforts should be geared towards utilization of lesser known non conventional crops as sources starch such as wild yams. Sufficient to say, the current level of research on wild yams starches, such *Dioscorea villosa*, *Dioscorea bulbifera* and other lesser known starchy crops is limited. Therefore, the properties and potential utilization of starch obtained from *D. villosa*, *D. bulbifera* and other lesser known starchy crops should be established in order to broadening the present narrow commonly cultivated starch sources. Also, value addition and conceptual efforts towards good agricultural practices might enhance the productivity of conventional starch sources such as cassava and maize.

Conflict of interest

The author declared that there is no conflict of interest.

Author details

Akeem Olayemi Raji
Department of Food Science and Technology, College of Agriculture and Veterinary Medicine, Kwara State University, Malete, Ilorin, Nigeria

*Address all correspondence to: akeem.raji@kwasu.edu.ng

References

[1] Abbas KA, Khalil SK, Hussin AS. Modified starches and their usages in selected food products: A review study. The Journal of Agricultural Science. 2010;**2**(2):90

[2] Korma SA, Niazi S, Ammar AF, Zaaboul F, Zhang T. Chemically modified starch and utilization in food stuffs. International Journal of Nutrition and Food Sciences. 2016;**5**(4):264

[3] Castro C, Corraze G, Pérez-Jiménez A, Larroquet L, Cluzeaud M, Panserat S, et al. Dietary carbohydrate and lipid source affect cholesterol metabolism of European sea bass (Dicentrarchus labrax) juveniles. The British Journal of Nutrition. 2015;**114**(8):1143-1156

[4] Biliaderis CG. Structures and phase transitions of starch polymers. ChemInform. 1998;**24**:29(47)

[5] Kaur B, Ariffin F, Bhat R, Karim AA. Progress in starch modification in the last decade. Food Hydrocolloids. 2012;**26**(2):398-404

[6] Vilar M. Starch-Based Materials in Food Packaging: Processing, Characterization and Applications. Academic Press; 2017

[7] Pérez S, Bertoft E. The molecular structures of starch components and their contribution to the architecture of starch granules: A comprehensive review. Starch-Stärke. 2010;**62**(8):389-420

[8] Halley PJ, Truss RW, Markotsis MG, Chaleat C, Russo M, Sargent AL, et al. A review of biodegradable thermoplastic starch polymers. In:

[9] Gozzo A, Glittenberg D. Starch: A versatile product from renewable resources for industrial applications. Sustainable Solutions for Modern Economies. 2009;**20**(4):238

[10] Yazid NS, Abdullah N, Muhammad N, Matias-Peralta HM. Application of starch and starch-based products in food industry. Journal of Science and Technology. 2018;**10**(2)

[11] Sadik N. Population growth and the food crisis. Food, Nutrition and Agriculture. 1991;**1**(1):3-6

[12] Weaver LT. Feeding the weanling in the developing world: Problems and solutions. International Journal of Food Sciences and Nutrition. 1994 Jan 1;**45**(2):127-134

[13] Afolayan MO, Omojola MO, Onwualu AP, Thomas SA. Further physicochemical characterization of anchomanes difformis starch. Agriculture and Biology Journal of North America. 2012;**3**(1):31-38

[14] Riley CK, Wheatley AO, Asemota HN. Isolation and characterization of starches from eight *Dioscorea alata* cultivars grown in Jamaica. African Journal of Biotechnology. 2006;**5**(17)

[15] Moongngarm A. Chemical compositions and resistant starch content in starchy foods. American Journal of Agricultural and Biological Sciences. 2013 Apr 1;**8**(2):107

[16] Ragaee S, Abdel-Aal ES, Noaman M. Antioxidant activity and nutrient composition of selected cereals for food use. Food Chemistry. 2006 Jan 1;**98**(1):32-38

[17] Cruz JF. Fonio: A small grain with potential. LEISA-LEUSDEN. 2004 Mar;**20**:16-17

[18] Brunnschweiler J, Luethi D, Handschin S, Farah Z, Escher F, Conde-Petit B. Isolation, physicochemical characterization and application of yam (Dioscorea

spp.) starch as thickening and gelling agent. Starch-Stärke. 2005 Apr;**57**(3-4):107-117

[19] Ezeocha CV, Okafor JC. Evaluation of the chemical, functional and pasting properties of starch from trifoliate yam (Dioscorea dumetorum) landraces. Eur. J. Adv. Res. Biol. Life Sci. 2016;**4**

[20] Shajeela PS, Mohan VR, Louis Jesudas L, Tresina SP. Evaluación del valor nutricional y factores antinutricionales de Dioscorea spp. silvestre. Tropical and Subtropical Agroecosystems. 2011 Aug;**14**(2):723-730

[21] Oladebeye AO, Oshodi AA, Amoo IA, Abd KA. Selective comparison of native and nanocrystals of White and yellow yam starches. Walailak Journal of Science and Technology (WJST). 2015 Jan 6;**12**(12):1137-1145

[22] Moorthy SN, Padmaja G. Starch content of cassava tubers. Journal of Root Crops. 2002;**28**(1):30-37

[23] Teixeira EDM, Pasquini D, Curvelo AA, Corradini E, Belgacem MN, Dufresne A. Cassava bagasse cellulose nanofibrils reinforced thermoplastic cassava starch. Carbohydrate Polymers. 2009;*78*(3):422-431

[24] Sanginga N, Mbabu A. Root and tuber crops (cassava, yam, potato and sweet potato). Inproceedings of the an action plan for African Agricultural Transformation Conference, Dakar, Senegal, 2015 Oct 21 (pp. 21-23).

[25] Nyirenda DB, Chiwona-Karltun L, Chitundu M, Haggblade S, Brimer L. Chemical safety of cassava products in regions adopting cassava production and processing–experience from southern Africa. Food and Chemical Toxicology. 2011 Mar 1;**49**(3):607-612

[26] Sonnewald U. Control of potato tuber sprouting. Trends in Plant Science. 2001 Aug 1;**6**(8):333-335

[27] Iheagwara MC. Isolation, modification and characterization of sweet potato (Ipomoea batatas L (lam)) starch. Journal of Food Processing & Technology. 2013;**4**(1):1-6

[28] Okigbo RN, Ogbonnaya UO. Antifungal effects of two tropical plant leaf extracts (*Ocimum gratissimum* and *Aframomum melegueta*) on postharvest yam (Dioscorea spp.) rot. African Journal of biotechnology. 2006;5(9).

[29] Akinoso R, Olatoye KK. Energy utilization and conservation in instant-pounded yam flour production. International Food Research Journal. 2013 Mar 1;**20**(2):575

[30] Polycarp D, Afoakwa EO, Budu AS, Otoo E. Characterization of chemical composition and anti-nutritional factors in seven species within the Ghanaian yam (Dioscorea) germplasm. International Food Research Journal. 2012 Jul 1;**19**(3):985

[31] Bharath S, Reddy PM, Deveswaran R, Basavaraj BV, Madhavan V. Extraction of polysaccharide polymer from Dioscorea trifida and evaluation as a tablet binder. International Journal of Pharmacy and Pharmaceutical Sciences. 2012;**4**(3):347-352

[32] Nwakor FN, Olaniyi AI, Ezebuiro NC, Kadurumba C. Participation and performance of root crops scientists on cassava research and development in south eastern Nigeria. Journal of Agriculture and Social Research (JASR). 2016;**16**(2):1-8

[33] Food F. Agriculture Organization (1998). Strategies for Crop Insurance:1997

[34] Babalola M, Oyenuga VA. Value of yam (Dioscorea spp.) in Nigeria food and feeding stuff, Ibadan.

[35] Kumar S, Das G, Shin HS, Patra JK. Dioscorea spp.(a wild edible tuber): a study on its ethnopharmacological potential and traditional use by the local people of Similipal Biosphere Reserve, India. Frontiers in pharmacology. 2017 Feb 14;**8**:52

[36] Mignouna HD, Abang MM, Asiedu R. Harnessing modern biotechnology for tropical tuber crop improvement: Yam (Dioscorea spp.) molecular breeding. African Journal of Biotechnology. 2003;**2**(12):478-485

[37] Ezeocha VC, Ojimelukwe PC. The impact of cooking on the proximate composition and anti-nutritional factors of water yam (Dioscorea alata). Journal of Stored Products and Postharvest Research. 2012 Dec 31;**3**(13):172-176

[38] Udensi EA, Oselebe HO, Onuoha AU. Antinutritional assessment of D. alata varieties. Pakistan Journal of Nutrition. 2010;**9**(2):179-181

[39] Muzac-Tucker I, Asemota HN, Ahmad MH. Biochemical composition and storage of Jamaican yams (Dioscorea sp). Journal of the Science of Food and Agriculture. 1993;**62**(3):219-224

[40] Ofosu SB. *Assessment of three White Yam (Dioscorea rotundata) varieties for possible development into Flour* (Doctoral dissertation).

[41] Schultz GE. Element stewardship abstract for *Dioscorea bulbifera*, air potato. *Control methods—plants*. In: *Global Invasive Species Team (GIST). Arlington, VA: The Nature Conservancy*; 1993 *Available: http://www.invasive.org/gist/esadocs/documnts/diosbul.pdf [accessed 2009, November 2]*

[42] Ogbuagu MN. Nutritive and antinutritive composition of the wild (in-edible) species of Dioscorea bulbifera (potato yam) and Dioscorea dumentorum (bitter yam). Journal of Food Technology. 2008;**6**(5):224-226

[43] Rayamajhi MB, Pratt PD, Tipping PW, Lake E, Smith M, Rohrig E, et al. Seasonal growth, biomass allocation, and invasive attributes manifested by Dioscorea bulbifera L.(air-potato) plants generated from bulbils in Florida. Invasive Plant Science and Management. 2016 Jul;**9**(3):195-204

[44] Djeri B, Tchobo PF, Adjrah Y, Karou DS, Ameyapoh Y, Soumanou MM, et al. Nutritional potential of yam chips (Dioscorea cayenensis and Dioscorea rotundata Poir) obtained using two methods of production in Togo. African Journal of Food Science. 2015 May 27;**9**(5):278-284

[45] AdeOluwa OO. Yam (Dioscorea spp.). In: Encyclopedia of Applied Plant Sciences. Second ed. 2017. pp. 435-441

[46] Otegbayo B, Aina J, Asiedu R, Bokanga M. Pasting characteristics of fresh yams (Dioscorea spp.) as indicators of textural quality in a major food product–'pounded yam'. Food Chemistry. 2006 Jan 1;**99**(4):663-669

[47] Coursey DG. Potential utilization of major root crops, with special emphasis on human, animal, and industrial uses. In Tropical root crops: production and uses in Africa: proceedings of the Second Triennial Symposium of the International Society for Tropical Root Crops-Africa Branch held in Douala, Cameroon, 14-19 Aug. 1983 1984. IDRC, Ottawa, ON, Canada.

[48] Adeigbe OO, Ilori CO, Adewale BD. Phenotypic diversity and ploidy level of some Dioscorea dumetorum genotypes. IOSR JAVS. 2015;**8**:47-52

[49] Ojo JO, Ojo OI. A comparative study of effects of storage on basic nutritional composition of two major edible Dioscorea dumentorum varieties.

Global Journal of Pure and Applied Sciences. 2009;**15**(3-4)

[50] Maurice MD. The hypoglycemic principle of dioscorea dumetorum. Planta Medica. 1990;**56**(1):119

[51] Adaramola TF, Sonibare MA, Sartie A, Lopez-Montes A, Franco J, Albach DC. Integration of ploidy level, secondary metabolite profile and morphological traits analyses to define a breeding strategy for trifoliate yam (Dioscorea dumetorum (Kunth) Pax). Plant Genetic Resources: Characterization and Utilization. 2014;**1**:10

[52] Agbor-Egbe T, Trèche S. Evaluation of the chemical composition of Cameroonian yam germplasm. Journal of Food Composition and Analysis. 1995 Sep 1;**8**(3):274-283

[53] Coursey DG, The cultivation and use of yams in West Africa. An Economic History of Tropical Africa: Volume One. The Pre-Colonial Period. Vol. 112013 Jan. p. 31

[54] Kay, D.E. Root crops (Revised).*Crop and Product Digest* No. 2. (Second Edition). Tropical Development and Research Institute. London. 1987:380

[55] Omonigho SE, Ikenebomeh MJ. Effect of temperature treatment on the chemical composition of pounded white yam during storage. Food Chemistry. 2000 Nov 1;**71**(2):215-220

[56] Lape IM, Treche S. Nutritional quality of yam (Dioscorea dumetorum and D rotundata) flours for growing rats. Journal of the Science of Food and Agriculture. 1994 Dec;**66**(4):447-455

[57] Foster S, Duke JA. A field guide to medicinal plants and herbs of eastern and central North America. Houghton Mifflin Harcourt; 2000

[58] Hoffmann D. Medical Herbalism: The Science and Practice of Herbal Medicine. Simon and Schuster; 2003 Oct 24

[59] Howell P. *Medicinal Plants of the Southern Appalachians*. Mountain City, GA: BotanoLogos Books; 2006

[60] Wren RC. Potter's New Cyclopaedia of Botanical Drugs and Preparations. Saffron Walden, Essex, England: CW Daniel Co; 1994

[61] Okukpe KM, Adeloye AA, Belewu MA, Alli OI, Adeyina OA, Annongu AA. Investigation of phytohormonal potential of some selected tropical plants. Research Journal of Medicinal Plant. 2012;**6**(6):425-432

[62] Araghiniknam M, Chung S, Nelson-White T, Eskelson C, Watson RR. Antioxidant activity of dioscorea and dehydroepiandrosterone (DHEA) in older humans. Life Sciences 1996 Aug 9;59(11):PL147-57.

[63] Shewry PR, Halford NG, Tatham AS. High molecular weight subunits of wheat glutenin. Journal of Cereal Science. 1992 Mar 1;**15**(2):105-120

[64] Bender DA, Bender AE. Benders' Dictionary of Nutrition and Food Technology. 7th ed. Abington: Woodhead Publishing; 1999

[65] McKevith B. Nutritional aspects of cereals. Nutrition Bulletin. 2004 Jun;**29**(2):111-142

[66] FAO (Food and Agriculture Organisation). Cereal and Grain-legume Seed Processing.Technical guide. FAO Plant Production and Protection Series No 21. Rome: FAO; 1991

[67] Kent NL. Kent's Technology of Cereals: An Introduction for Students of

Food Science and Agriculture. Elsevier; 1994

[68] FAO (Food and Agriculture Organisation). World Agriculture: Towards 2015/2030. Summary report. Rome: FAO; 2002

[69] Baghurst PA, Baghurst KI, Record SJ. Dietary fibre, non-starch polysaccharides and resistant starch: A review. Food Australia. 1996;**48**(3):S3-S5

[70] Macrae R, Robinson RK, Sadler MJ. Encyclopaedia of Food Science, Food Technology, and Nutrition. Academic Press; 1993

[71] FAO (Food and Agriculture Organisation). Food Outlook. No. 4.FAO Global Information and Early Warning System on Food and Agriculture. Rome: FAO; 2003

[72] Fast RB, Caldwell EF. Breakfast Cereals and How they Are Made. American Association of Cereal Chemists; 2000

[73] FAO (Food and Agriculture Organisation). World Agriculture: Towards 2010. An FAO study. Rome; 1995

[74] Carcea M, Acquistucci R. Isolation and physicochemical characterization of Fonio (Digitaria exilis Stapf) starch. Starch-Stärke. 1997;**49**(4):131-135

[75] Jideani IA. Digitaria exilis (acha/fonio), Digitaria iburua (iburu/fonio) and Eluesine coracana (Tamba/finger millet) non-conventional cereal grains with potentials. Scientific Research and Essays. 2012 Nov 19;**7**(45):3834-3843

[76] Dauda A, Luka D. status of Acha (*Digitaria exilis*) production in Bauchi State, Nigeria. In: Kwon-Ndung, EH, Bright, EO & Vodouhe, R., eds., Proceedings of the first National Acha stakeholders workshop at PADP, Jos (9-11th March 2003), 2003.

[77] Dachi SN, Gana AS. Adaptability and yield evaluation of some Acha (Digitaria exilis and Digitaria iburua kippis Stapf) accessions at Kusogi–Bida, Niger State, Nigeria. African Journal of General Agriculture. 2008;**4**(2):73-77

[78] Temple VJ, Bassa JD. Proximate chemical composition of Acha (Digitaria exilis) grain. Journal of the Science of Food and Agriculture. 1991;**56**(4):561-563

[79] Jideani AI, Akingbala JO. Some physicochemical properties of acha (Digitaria exilis Stapf) and iburu (Digitaria iburua Stapf) grains. Journal of the Science of Food and Agriculture. 1993;**63**(3):369-374

[80] Musa H, Muazu J, Bhatia PG, Mshelbwala K. Investigation into the use of fonio (*Digitaria exilis*) starch as a tablet disintegrant. Nigerian Journal of Pharmaceutical Sciences. 2008 Mar;**7**(1)

[81] Muazu J, Musa H, Bhatia PG. Evaluation of the glidant property of fonio starch. Research Journal of Applied Sciences, Engineering and Technology. 2010 Mar 10;**2**(2):149-152

[82] Adedokun MO, Itiola OA. Material properties and compaction characteristics of natural and pregelatinized forms of four starches. Carbohydrate Polymers. 2010 Mar 17;**79**(4):818-824

[83] Piecyk M, Drużyńska B, Worobiej E, Wołosiak R, Ostrowska-Ligęza E. Effect of hydrothermal treatment of runner bean (Phaseolus coccineus) seeds and starch isolation on starch digestibility. Foodservice Research International. 2013 Jan 1;**50**(1):428-437

[84] Odeku OA. Potentials of tropical starches as pharmaceutical

excipients: A review. Starch-Stärke. 2013;65(1-2):89-106

[85] Jiang Q, Gao W, Li X, Xia Y, Wang H, Wu S, et al. Characterizations of starches isolated from five different Dioscorea L. species. Food Hydrocolloids. 2012 Oct 1;**29**(1):35-41

[86] Oyeyinka SA, Adeleke OF, Dauda AO, Abiodun OA, Kayode RM, Adejuyitan JA. Flour composition and physicochemical properties of white and yellow bitter yam (Dioscorea dumetorum) starches. Industrial Crops and Products. 2018 Sep 15;**120**:135-139

[87] Otegbayo B, Oguniyan D, Akinwumi O. Physicochemical and functional characterization of yam starch for potential industrial applications. Starch-Stärke. 2014 Mar;**66**(3-4):235-250

[88] Lu G, George MS, Zhou W. Genotypic variation of sweetpotatoes grown under low potassium stress. Journal of Plant Nutrition. 2003 Mar 1;**26**(4):745-756

[89] Pérez EE, Breene WM, Bahnassey YA. Ariations in the gelatinization profiles of cassava, Sagu and arrowroot native starches as measured with different thermal and mechanical methods. Starch-Stärke. 1998 Mar;**50**(2-3):70-72

[90] Atichokudomchai N, Varavinit S. Characterization and utilization of acid-modified cross-linked tapioca starch in pharmaceutical tablets. Carbohydrate Polymers. 2003 Aug 15;**53**(3):263-270

[91] Pérez E, Schultz FS, de Delahaye EP. Characterization of some properties of starches isolated from Xanthosoma sagittifolium (tannia) and Colocassia esculenta (taro). Carbohydrate Polymers. 2005 May 6;**60**(2):139-145

[92] Moorthy SN, Pillai PT, Unnikrishnan M. Variability in starch extracted from taro. Carbohydrate Polymers. 1993 Jan 1;**20**(3):169-173

[93] Lawal OS. Composition, physicochemical properties and retrogradation characteristics of native, oxidised, acetylated and acid-thinned new cocoyam (Xanthosoma sagittifolium) starch. Food Chemistry. 2004 Sep 1;**87**(2):205-218

[94] Mir SA, Bosco SJ, Bashir M, Shah MA, Mir MM. Physicochemical and structural properties of starches isolated from corn cultivars grown in Indian temperate climate. International Journal of Food Properties. 2017 Apr 3;**20**(4):821-832

[95] Paraginski RT, Vanier NL, Moomand K, de Oliveira M, da Rosa Zavareze E, e Silva RM, et al. Characteristics of starch isolated from maize as a function of grain storage temperature. Carbohydrate Polymers. 2014 Feb 15;**102**:88-94

[96] Musa H, Gambo A, Bhatia PG. Studies on some physicochemical properties of native and modified starches from Digitaria iburua and Zea mays. International Journal of Pharmacy and Pharmaceutical Sciences. 2011;**3**(1):28-23

[97] Chanapamokkhot H, Thongngam M. The chemical and physico-chemical properties of sorghum starch and flour. Agriculture and Natural Resources. 2007 Dec 31;**41**(5):343-349

[98] Singh H, Sodhi NS, Singh N. Characterisation of starches separated from sorghum cultivars grown in India. Food Chemistry. 2010 Mar 1;**119**(1):95-100

[99] Lawal OS, Ogundiran OO, Awokoya K, Ogunkunle AO. The low-substituted propylene oxide etherified plantain (Musa paradisiaca normalis) starch: Characterization and functional

parameters. Carbohydrate Polymers. 2008 Nov 4;**74**(3):717-724

[100] Rani VS, John JK, Moorthy SN, Raja KC. Effect of pretreatment of fresh Amorphophallus paeoniifolius on physicochemical properties of starch. Starch-Stärke. 1998 Mar;**50**(2-3):72-77

[101] Moorthy SN. Physicochemical and functional properties of tropical tuber starches: A review. Starch-Stärke. 2002 Dec;**54**(12):559-592

[102] Özden Ö, Sönmez S. Starch usage in paper industry. Research and Reviews in Engineering. 2019;**2019**:207

[103] Raji AO, Ismael SD, Sani M, Raji MO, Adebayo OF. Value addition and influence of succinylation levels on the quality of wild yam (Discorea villosa) starch. Food & Function. 2020;**11**(9):8014-8027

[104] Singh J, Kaur L, McCarthy OJ. Factors influencing the physico-chemical, morphological, thermal and rheological properties of some chemically modified starches for food applications—A review. Food Hydrocolloids. 2007 Jan 1;**21**(1):1-22

[105] Silva EK, Zabot GL, Cazarin CB, Maróstica MR Jr, Meireles MA. Biopolymer-prebiotic carbohydrate blends and their effects on the retention of bioactive compounds and maintenance of antioxidant activity. Carbohydrate Polymers. 2016 Jun 25;**144**:149-158

[106] Taggart P, Mitchell JR. Starch. InHandbook of hydrocolloids 2009 Jan 1 (pp. 108-141). Woodhead Publishing.

[107] Tharanathan RN. Starch—Value addition by modification. Critical Reviews in Food Science and Nutrition. 2005 Jul 1;**45**(5):371-384

[108] Aprianita A, Vasiljevic T, Bannikova A, Kasapis S. Physicochemical properties of flours and starches derived from traditional Indonesian tubers and roots. Journal of Food Science and Technology. 2014 Dec 1;**51**(12):3669-3679

[109] Zhu J, Li L, Chen L, Li X. Nano-structure of octenyl succinic anhydride modified starch micelle. Food Hydrocolloids. 2013 Jul 1;**32**(1):1-8

[110] Dokić P, Dokić L, Dapčević T, Krstonošić V. Colloid characteristics and emulsifying properties of OSA starches. In: Colloids for nano-and biotechnology. Berlin, Heidelberg: Springer; 2008. pp. 48-56

[111] Timgren A, Rayner M, Dejmek P, Marku D, Sjöö M. Emulsion stabilizing capacity of intact starch granules modified by heat treatment or octenyl succinic anhydride. Food Science & Nutrition. 2013 Mar;**1**(2):157-171

[112] Tesch S, Gerhards C, Schubert H. Stabilization of emulsions by OSA starches. Journal of Food Engineering. 2002 Sep 1;**54**(2):167-174

[113] Cui SW, editor. Food carbohydrates: chemistry, physical properties, and applications. CRC Press; 2005 May 23

[114] Taggart P. Starch as an ingredient: Manufacture and applications. InStarch in food. Woodhead Publishing; 2004. pp. 363-392

[115] Park JH, Kim DC, Lee SE, Kim OW, Kim H, Lim ST, et al. Effects of rice flour size fractions on gluten free rice bread. Food Science and Biotechnology. 2014 Dec 1;**23**(6):1875-1883

[116] Neelam K, Vijay S, Lalit S. Various techniques for the modification of starch and the applications of its derivatives. International Research Journal of Pharmacy. 2012;**3**(5):25-31

[117] Kim HY, Park SS, Lim ST. Preparation, characterization and

utilization of starch nanoparticles. Colloids and Surfaces. B, Biointerfaces. 2015 Feb 1;**126**:607-620

[118] Kristo E, Biliaderis CG. Physical properties of starch nanocrystal-reinforced pullulan films. Carbohydrate Polymers. 2007 Mar 1;**68**(1):146-158

[119] García NL, Ribba L, Dufresne A, Aranguren M, Goyanes S. Effect of glycerol on the morphology of nanocomposites made from thermoplastic starch and starch nanocrystals. Carbohydrate Polymers. 2011 Feb 11;**84**(1):203-210

[120] Bloembergen S, Lee DI, McLennan IJ, Wildi RH, Van Egdom E, inventors; Ecosynthetix Inc, assignee. Process for producing biopolymer nanoparticle biolatex compositions having enhanced performance and compositions based thereon. United States patent application US 12/630,526. 2010 Jun 10.

[121] Öner ET, Oliver SG, Kırdar B. Production of ethanol from starch by respiration-deficient recombinant Saccharomyces cerevisiae. Applied and Environmental Microbiology. 2005 Oct 1;**71**(10):6443-6445

Chapter 12

Genes Involved in Lipid Metabolism in Coconut

Wei Xia

Abstract

Coconut palm (*Cocos nucifera* L) is an economically important monocot plant grown in tropical and subtropical regions. Coconut oil is stored in a solid endosperm and has 47.48–50.5% fatty acid component as lauric acid (C12:0). Present research showed that acyl-acyl carrier protein thioesterases (FatA/B) and lysophosphatidic acid acyltransferase (LAAPT) are key enzymes determining medium-chain fatty acid accumulation in coconut oil. Among five *CnFatB* genes, *CnFatB3* expressed specifically in endosperm and *in vitro* experiment showed that this gene made mainly lauric acid (C12:0) and tetradecenoic acid (C14:1). Overexpression of *CnFatB3* in *Arabidopsis* increased the amounts of C12:0 and C14:0 in transgenic plant. *CnLPAAT* gene that is expressed specifically in coconut endosperm showed a preference for using acyl-CoAs containing C10:0, C12:0, and C14:0 acyl groups as acyl-donor substrates. Coconut and oil palm are closely related species with approximately 50% lauric acid (C12:0) in their endosperm. The two species have a close evolutionary relationship between predominant gene isoforms and high conservation of gene expression bias in the lipid metabolism pathways. Moreover, since no stable transformation system has been constructed in coconut palm, gene function validations have been done in vitro, or genes transformed into a heterologous system.

Keywords: medium-chain fatty acid, lipid metabolism, coconut endosperm, gene evolution, de novo fatty acid synthesis, TAG biosynthesis

1. Introduction

Coconut palm (*Cocos nucifera* L), belonging to the Arecaceae family, is an economically important monocot plant grown in tropical and subtropical regions. Coconut kernels have approximately 63.1% oil content in a solid endosperm (copra) [1]. A noticeable feature of coconut oil is that 47.48–50.5% of its fatty acid component is lauric acid (C12:0), which is a type of medium-chain fatty acid (MCFA) [2]. A closely related species of coconut, the African oil palm, also contains 50% lauric acid in its kernel oil [3]. The lauric acid content of coconut oil makes it useful for a range of edible and nonedible purposes. A number of genes differentially expressed in coconut endosperm have been identified by suppression subtractive hybridization [4]. *Arabidopsis* has more than 600 genes involved acyl-lipid metabolism, and Xiao et al. [5] identified 806 orthologous genes of these *Arabidopsis* genes in coconut palm based on the first version of coconut genome sequences [1, 2]. A better understanding of lipid biosynthesis and tissue-specific transcription could help breeding efforts to improve the content and composition of coconut oil used

IntechOpen

Coconut gene ID	Protein	Annotation	Gene validated in reference
CCG009120.1	PDH-E1α	E1-alpha component of pyruvate dehydrogenase complex	
CCG003104.1	PDH-E1β	E1-beta component of pyruvate dehydrogenase complex	
CCG004328.1,CCG016344.1,CCG020484.1,CCG022353.1	LTA1	Dihydrolipoamide acetyltransferase, E2 component of pyruvate dehydrogenase complex	
CCG022878.1	LTA2	Dihydrolipoamide acetyltransferase, E2 component of pyruvate dehydrogenase complex	
CCG016999.1	LPD2	Dihydrolipoamide dehydrogenase, E3 component of pyruvate dehydrogenase complex	
CCG004885.1	CT-α	Carboxyltransferase-alpha; subunit of heteromeric ACCase	
CCG016556.1	BCCP	Biotin carboxyl carrier protein of heteromeric ACCase	
CCG014874.2,CCG016422.1	BC	Biotin carboxylase of heteromeric ACCase	
CCG000475.1,CCG019561.1	MCMT	Malonyl-CoA: ACP malonyltransferase	
CCG001191.1,CCG001193.1,CCG001194.1,CCG024029.1,CCG026381.1	KASI	Ketoacyl-ACP Synthase I	[6]
CCG000907.1,CCG015780.2,CCG023608.3	KASII	Ketoacyl-ACP Synthase II	
CCG003289.1,CCG025932.1	KASIII	Ketoacyl-ACP Synthase III	
CCG006105.2,CCG025988.1,CCG014527.1,CCG024266.1	KAR	Ketoacyl-ACP Reductase	
CCG007292.1,CCG001741.1	HAD	Hydroxyacyl-ACP Dehydratase	
CCG019022.2,CCG019145.2	ENR1	Enoyl-ACP Reductase	
CCG001923.1	ACP1	Acyl carrier protein	
CCG000806.1,CCG000980.1,CCG017093.1,CCG026999.1,CCG027016.1 ,CCG027238.1	ACP4		

DOI: http://dx.doi.org/10.5772/intechopen.90998

Coconut gene ID	Protein	Annotation	Gene validated in reference
CCG025689.1	DES6	Stearoyl-ACP desaturase	[7]
CCG005175.1,CCG011462.1,CCG019622.1	FAB2		
CCG017191.1,CCG017192.1,CCG017193.1,CCG021345.1	DES5		
CCG005178.1,CCG012754.1	FatA	Acyl-ACP thioesterase A	
CCG006479.1,CCG007799.1,CCG011598.1,CCG015192.1,CCG019705.1	FATB	Acyl-ACP thioesterase B	[8, 9]
CCG005500.1	HACPS	Holo-ACP synthase	
CCG001744.2,CCG007290.1,CCG007291.1	LACS9	Long-chain Acyl-CoA synthetase	

Table 1.
Coconut genes belong to de novo fatty acid biosynthesis.

Coconut gene ID	Protein	Annotation	Gene validated in reference
CCG004869.3,CCG020141.3,CCG023968.1,CCG027042.1	GPDH	NAD-dependent glycerol-3-phosphate dehydrogenase	
CCG019614.2	GPAT9	Glycerol-3-Phosphate acyltransferase (mammalian homolog)	
CCG006531.1,CCG015599.1,CCG016821.1	LPAAT2	1-Acylglycerol-3-phosphate acyltransferase	[6, 10–13]
CCG022695.1	PAH1	Phosphatidate phosphatase	
CCG009829.1,CCG016247.2	PAH2	Phosphatidate phosphatase	
CCG007725.1,CCG026806.1	LPP-β	Phosphatidate phosphatase	
CCG003641.1,CCG010800.1	LPP-δ	Long chain base 1-phosphate phosphatase	
CCG015429.1,CCG019248.1	DGAT1	Acyl-CoA:diacylglycerol acyltransferase	
CCG004186.1,CCG026159.1	DGAT2	Acyl-CoA:diacylglycerol acyltransferase	[14]
CCG015380.1	DAcT	Wax synthase-like	
CCG005217.1	PDAT1	Phospholipid:diacylglycerol acyltransferase	
CCG019998.1,CCG019999.1,CCG020055.1	PDAT-related?	Phospholipid:acyl acceptor acyltransferase	
CCG011285.1,CCG021291.1	LPEAT1	1-Acylglycerol-3-phosphoethanolamine acyltransferase	
CCG000909.1,CCG000910.1	LPEAT2		
CCG002335.2,CCG015142.3	LPCAT	1-Acylglycerol-3-phosphocholine acyltransferase	
CCG017936.1	PDCT/ROD1	Phosphatidylcholine:diacylglycerol cholinephosphotransferase	
CCG019021.1,CCG019148.1	FAD2	Oleate desaturase	
CCG003640.4,CCG010801.1	CDP-DAGS	CDP-DAG synthase	
CCG021791.1	DAG-CPT	Diacylglycerol cholinephosphotransferase	
CCG009590.1,CCG024101.3,CCG025115.1	CK	Choline kinase	
CCG007754.1,CCG019356.1,CCG026050.3	CCT2	Choline-phosphate cytidylyltransferase	

DOI: http://dx.doi.org/10.5772/intechopen.90998

Coconut gene ID	Protein	Annotation	Gene validated in reference
CCG021844.1	ACBP2	Acyl CoA binding protein	
CCG005041.1,CCG008659.2,CCG018700.1	ACBP3		
CCG009417.1,CCG020854.1,CCG026758.2	ACBP4		
CCG000884.2,CCG026958.1	ACBP6		
CCG009767.1,CCG016753.1,CCG016754.2	LACS4	Long-chain Acyl-CoA synthetase	
CCG027986.1,CCG027990.1	NMT1	Phosphoethanolamine N-methyltransferase	
CCG009861.1	PIS2	Phosphatidylinositol synthase	
CCG026466.1	PSD1	Phosphatidylserine decarboxylase	
CCG023785.2	PSD3		
CCG005386.2,CCG012449.1,CCG015191.4	PSS	Base-exchange-type phosphatidylserine synthase	
CCG001187.2,CCG026384.1	EK	Ethanolamine kinase	
CCG000220.1,CCG001400.4,CCG005823.1,CCG026528.2	PECT1	CDP-ethanolamine synthase	

Table 2.
Coconut genes involved in TAG biosynthesis.

for food and other applications. The most noticeable feature of coconut oil is that the major components of fatty acids are medium-chain fatty acid. This feature has attracted the attention of researchers and become the focus of coconut oil research. What genes related with the accumulation of medium-chain fatty acid in endosperm? How these genes were evolved and related to a closely related species—oil palm (*Elaeis guineensis*), which also has MCFA as its main fatty acid component in endosperm?

We had reviewed three parts of research related to coconut lipid metabolism in this chapter. Firstly, we summarized key genes related to MCFA accumulation in coconut endosperm. Secondly, we summarized the evolutionary relationship between coconut palm and oil palm for MCFA accumulation. Thirdly, we include descriptions of in vivo and in vitro gene validation experiments. Two tables provide coconut genes related to de novo fatty acid biosynthesis (**Table 1**) and triacylglycerols (TAG) biosynthesis (**Table 2**).

2. Genes related to lipid metabolism in coconut palm

Coconut palm stores oil in endosperm tissues, and its fatty acid composition changes in different developing stages of endosperm [4, 5]. The proportion of lauric acid increases with the maturing process of coconut fruit and reaches the peak when the fruit matures. The comparison of gene expression for different developing stages of endosperm indicated that the expression levels of stearoyl-acyl carrier protein desaturase, acyl-ACP thioesterase B (FatB), and lysophosphatidic acid acyltransferase (LPAAT) arose along with the endosperm development [4]. Xiao et al. [5] identified 71 genes belonging to plastidial fatty acid synthesis pathway in coconut, and 62 enzymes catalyze the conversion of pyruvate to fatty acid (**Table 1**). Moreover, the 17 plastidial proteins involved in the conversion of pyruvate to fatty acids were five- to sixfold higher in the endosperm than in the leaf or embryo tissue, such as acyl carrier protein (ACP), ketoacyl-ACP reductase (KAR), hydroxyacyl-ACP dehydratase (HAD), and pyruvate dehydrogenase complex (PDHC). TAG is a compact molecule for energy and carbon storage in organisms. Thus, another key pathway for oil storage—triglycerides (TAG) synthesis is analyzed for coconut palm and 69 genes were identified (**Table 2**). Key genes in the two pathways were deeply analyzed through in vivo and in vitro assays, including FatB, LPAAT, and orthologs of *Arabidopsis WRINKLED 1* (*WRI 1*) [9, 11, 14, 15].

2.1 Genes related to MCFA accumulation in coconut endosperm

2.1.1 Acyl-acyl carrier protein thioesterases

Acyl-acyl carrier protein thioesterases (acyl-ACP TEs) terminate acyl chain elongation during de novo fatty acid biosynthesis. This reaction is the biochemical determinant of the fatty acid compositions of storage lipids. There are two classes of acyl-ACP TEs—FatA and FatB. Since 1996, researchers have cloned acyl-ACP TEs from California bay laurel (*Umbellularia californica*) and validated its role in accumulating MCFA by transforming it into rapeseed (*Brassica napus*). Further research has classified FatB genes into three classes based on their specificities: class I acyl-ACP TEs act primarily on 14- and 16-carbon acyl-ACP substrates; class II acyl-ACP TEs have broad substrate specificities, with major activities toward 8- and 14-carbon acyl-ACP substrates; and class III acyl-ACP TEs act predominantly on 8-carbon acyl-ACPs.

Coconut palm has two acyl-ACP thioesterase A (FatA) genes in coconut palm and five FatB genes, which were CnFatB1 (CCG011598.1), CnFatB2–1 (CCG006479.1), CnFatB2–2 (CCG007799.1), CnFatB3 (CCG019705.1), and CnFatB4 (CCG015192.1). Three FatB genes were highly expressed in more than one analyzed tissue: CnFatB2–1 (leaf and embryo), CnFatB2–2 (leaf, embryo, and endosperm), and CnFatB3 (embryo and endosperm). Three acyl-ACP TEs of coconut (CnFatB1, CnFatB2, and CnFatB3) indicated divergent specificity: CnFatB1 (JF338903) and CnFatB2 (JF338904) produced major fatty acids as myristic acid (C14:0) and palmitoleic acid (C16:1); CnFatB3 (JF338905) made mainly lauric acid (C12:0) and tetradecenoic acid (C14:1) [14]. Yuan et al. transformed and overexpressed CnFatB3 in *Arabidopsis*, and the transgenic plants increased the amounts of 12:0 (lauric acid), 14:0 (myristic acid), 16:0 (palmitic acid), and 18:0 (stearic acid) by 30-, 80-, 4-, and 2-fold, respectively [6].

2.1.2 Lysophosphatidic acid acyltransferase

Coconut oil has 92% saturates and most of its TAGs are trisaturated. Moreover, laurate is found enriched at sn-2 position, which is catalyzed by membrane-bound lysophosphatidic acid acyltransferase (LPAAT) enzyme. Davies et al. detected an enzyme from coconut endosperm, which is a laurate-CoA-preferring LPAAT and active during endosperm maturation [9]. The LPAAT enzyme prefers acyl-CoAs containing C10:0, C12:0, and C14:0 acyl groups as acyl-donor substrates [9]. Knutzon et al. [11] performed the LPAAT protein purification and cloned the corresponding cDNA of this gene from coconut. The gene was then transformed and expressed in *Escherichia coli*, and substrate activity profile of this gene matched that of the coconut enzyme. This copy of LPAAT is the gene named as CCG001603.1 in the first version of coconut genome sequence [5]. Knutzon et al. transformed this gene into a rapeseed transgenic gene line which is expressed of a California bay laurel (*Umbellularia californica*) 12:0-acyl carrier protein thioesterase (BET) and contained up to 50% laurate in its developing seeds [11]. In this transgenic rapeseed with BTE, laurate is found almost exclusively at the *sn-1* and *sn-3* positions of the triacylglycerols. Coexpression of the coconut *LPAAT* gene in the transgenic rapeseeds facilitates efficient laurate deposition at the *sn-2* position and caused the accumulation of trilaurin [11].

Xu et al. cloned the promoter sequence of the LPAAT gene and characterized the promoter by constructing a series of plasmids with promoter sequences with varied length of deletions to promote a β-glucuronidase (GUS) gene. The plasmids were transformed into rice, and the transgenic plants showed that reporter genes with these promoter fragments tend to express specifically in rice endosperm [12]. Yuan et al. transformed *CnLPAAT* into yeast, and tested fatty acid composition indicated that the gene increased the levels of C12:0 and C14:0 in a CnLPAAT-pYES2 transformant [16]. However, heterologous overexpression of CnLPAAT in tobacco (*Nicotiana tabacum* L.) decreased the contents of C12:0 and C14:0 in transgenic tobacco seeds, which could result from low contents of short- and medium-chain FAs (0.22%), which are available in tobacco seeds of the total FAs.

2.1.3 Diacylglycerol acyltransferase

Besides genes important for MCFA accumulation, there are key genes in TAG biosynthesis pathway that influence oil contents and FA composition. Diacylglycerol acyltransferases (DGAT) and phospholipid:diacylglycerol acyltransferases (PDAT) catalyze diacylglycerol (DAG) to form TAG as the final step in TAG synthesis, using either acyl-CoAs or phospholipids. DAG is an important branch point between storage and membrane lipid synthesis. Coconut palm has three orthologs of AT2G19450 (*AtDGAT1*) and two orthologs of AT3G51520 (*AtDGAT2*). Coconut *DGATs* genes had higher expression level in coconut endosperm than in the leaf and embryo, especially for *DGAT1* isoform CCG007098.3 and *DGAT2* isoform CCG026159.1 [5].

Zheng et al. cloned a DGAT2 gene from coconut pulp and transferred the gene into the deficient yeast H1246 and *Arabidopsis* [13]. The DGAT2 gene that is expressed in the deficient yeast had DGAT catalysis activity and restored TAG synthesis in the yeast. Further lipid composition analysis showed that *CnDGAT2* has a substrate preference for two UFAs (C16:1 and C18:1) in yeast and linoleic acid (C18:2) in transgenic plants. These results provide knowledge on CnDGAT2 and offer new insights into TAG assembly in coconut.

2.2 Transcription factors regulating fatty acid biosynthesis

WRINKLED1 (WRI1, AT3G54320) directly controls the transcriptional activation of the fatty acid biosynthetic pathway in *Arabidopsis* and belongs to the APETALA2-ethylene-responsive element-binding protein (AP2-EREBP) family [17]. The ortholog of *AtWRI1* in oil palm was validated as a key transcription factor associated with lipid synthesis [3]. In coconut, three AT3G54320 orthologs were found—CCG005292.1, CCG012597.1, and CCG019741.1. CCG005292.1 and CCG012597.1 were expressed in the endosperm but had low expression in leaf and endosperm, while CCG019741.1 has no expression in leaf, embryo, or endosperm [5]. The *CnWRI1* gene copy (CCG012597.1) validated its interaction with the promoter sequence of acetyl-CoA carboxylase by yeast one-hybrid system [15]. Overexpression of *CnWRI1* (CCG012597.1) specifically in *Arabidopsis* seed showed an increase of palmitic acid (C16:0) and linolenic acid (C18:3) but a decrease in oleic acid content [15].

3. Evolutionary relationship between coconut palm and oil palm

Coconut and oil palm are important oil trees grown in tropical region and closely related species with approximately 50% lauric acid (C12:0) in their endosperm. There are 806 and 840 lipid-related genes annotated for coconut and oil palm, respectively [13]. The majority of lipid-related genes between coconut and oil palm were homologous genes, while 72.8% (438/601) of genes in coconut palm were located in homologous segments with oil palm. The two species have a close evolutionary relationship between predominant gene isoforms and high conservation of gene expression bias in the lipid metabolism pathways.

Since coconut and oil palm have high lauric acid (C12:0) in their endosperm, key genes responsible for MCFA also shared high homology in gene copy and expression pattern. Both coconut and oil palm have five FATB genes, but only three *EgFatB* genes highly expressed in oil palm mesocarp or endosperm and four *CnFatB* genes were highly expressed in endosperm or embryo. Homologous gene pair—*CnFatB3* and *EgFatB3*—were both highly expressed in their endosperms, which were validated as key genes for MCFA biosynthesis [3, 6]. Another key enzyme—LPAAT, three AtLPAAT1, or AtLPAAT2 orthologs were found in each of coconut and oil palm [5]. The LPAAT1 genes were clustered into class I and class II, and the class I genes of both species had higher expression levels in endosperm tissue. Moreover, the LPAAT2 genes were also clustered into two classes, and genes in class II had low or no expression.

For the key transcription factor associated with lipid synthesis—*WRI1*, Xiao et al. [5] identified three *WRI1* genes in coconut and six in oil palm and classified the genes into three groups based on conserved amino acid sequences. The coconut and oil palm *WRI1* genes in the same group indicated the same expression pattern: group I was highly expressed both in the coconut endosperm and the oil palm endosperm/mesocarp; group II or III has very low or no expression.

4. Methods used in validation gene function in coconut palm

Coconut palm has a long life cycle and takes 5–10 years to start reproductive stage. Since that, using gene overexpression or knockout to analyze gene function in its own plant system will take years to observe the traits related to fruits. At present, no stable transformation system has been constructed in coconut palm. The convenient ways to validate gene function in coconut are testing biochemical feature of proteins *in vitro* or transforming gene into a heterologous system, such as *Arabidopsis*, rice, yeast or *Escherichia coli (E. coli)*.

4.1 Testing enzyme activity in vitro

Lipid metabolism is composed of more than 120 enzymatic reactions. Validation of gene function related to lipid metabolism could be done by testing enzyme activity in vitro. Davies et al. have isolated CnLPAAT protein from immature coconut seeds and tested the LPAAT activity by adding Acyl-CoA and LPA as substrates [10].

Laurate is found enriched in sn-2, which indicates that a laurate-CoA-preferring LPAAT is active during endosperm maturation. Davies et al. were able to detect such an enzyme from this tissue, which allowed Knutzon et al. [11] to perform protein purification and cloning of a cDNA encoding the 299-amino acid CLP protein from coconut. When expressed in *E. coli*, and using 12:0-LPA as an acceptor, this enzyme preferred medium-chain CoAs over 18:1-CoA as acyl donors. This is a direct evidence that in coconut endosperm, not only had the common fatty acid biosynthesis pathway been modified to produce almost entirely saturated medium chains but at least one enzyme of lipid biosynthesis (LPAAT) had been modified.

4.2 Testing enzyme activity in vivo

Gene function validation has been conducted through gene overexpression in heterologous plant systems which have stable gene transformation system, such as *Arabidopsis*, rice, and tobacco. Functional characterization of *CnWRI1* was done by gene overexpression in *Arabidopsis* and rice [15]. Overexpression of CnWRI1 in *Arabidopsis* seeds caused fatty acid composition changes but not for oil content, while overexpression of the gene in rice endosperm increased the starch content and decreased the protein contents [15]. For gene function validation of CnLPAAT (CCG001603.1), this gene was overexpressed in a transgenic oilseed (*Brassica napus*) plant, which expressed a 12:0-ACP thioesterase from California bay laurel (*Umbellularia californica*). The transgenic lines that coexpressed a 12:0-ACP thioesterase and CnLPAAT had increase laurate content from 50 mol% to total laurate levels, which suggested that CnLPAAT facilitates efficient laurate deposition at the sn-2 position [11].

Transient transgenic expression system of tobacco is also widely used for gene function analysis. Genes belonging to lipid metabolism were also validated by this system, investigating the possibility of oil production in non-sees biomass [18].

Escherichia coli (*E. coli*) strains are commonly used in molecular biology, because the introduction of DNA into *E. coli* is convenient. Since lipid metabolism is basic in all living cells, specific *E. coli* strain with gene mutation could be used for analyzing enzyme functions. Knutzon et al. [11] cloned the CnLAAPT gene copy (CCG001603.1) from coconut endosperm and tested enzyme activity by introducing the gene into *E. coli* strain K27 that has a mutation in the *fadD* gene as well as β-oxidation of fatty acids. Overexpression of this *CnLAAPT* gene copy caused the accumulation of free fatty acids in the growth medium. Enzymic specificity of three acyl-ACP TEs of coconut (CnFatB1, CnFatB2, and CnFatB3) have been tested by transforming and expressing in *E. coli* K27 and analyzing free fatty acids accumulated in the medium [14].

5. Conclusions

Coconut palm (*Cocos nucifera* L) is an economically important monocot plant grown in tropical and subtropical regions. Coconut oil is stored in a solid endosperm and has 47.48–50.5% fatty acid component, which is a medium-chain fatty acid (MCFA) such as lauric acid (C12:0). Present research showed that acyl-acyl carrier protein thioesterases and lysophosphatidic acid acyltransferase are key enzymes determining MCFA accumulation in coconut oil (**Figure 1**). In this chapter, we reviewed three aspects of research related to coconut lipid metabolism. Firstly, we summarized key genes related to MCFA accumulation in coconut endosperm. Secondly, we summarized evolutionary relationship between coconut palm and oil palm for MCFA accumulation. Thirdly, we described studies using in vivo and in vitro gene validation experiments in coconut palm.

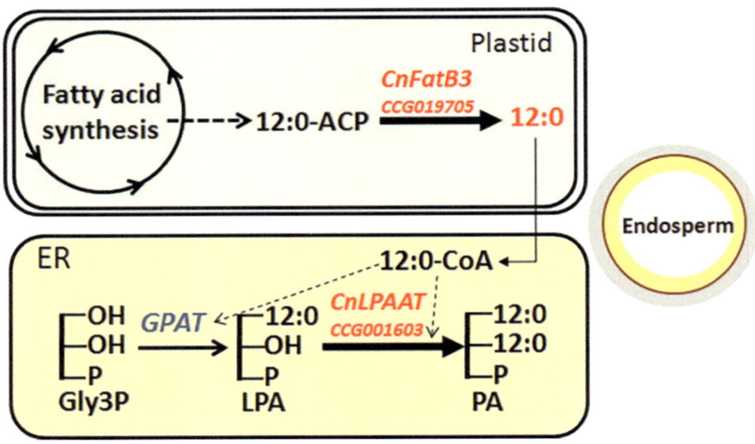

Figure 1.
The diagram of key genes involved in medium-chain fatty acid accumulation in coconut endosperm.

Conflict of interest

The authors declare no conflict of interest.

Appendices and nomenclature

ACP	acyl carrier protein
DGAT	diacylglycerol acyltransferases

FA	fatty acid
FATA	acyl-ACP thioesterase A
FATB	acyl-ACP thioesterase B
HAD	hydroxyacyl-ACP dehydratase
LPAAT	lysophosphatidic acid acyltransferase
PDHC	pyruvate dehydrogenase complex
TAG	triacylglycerols
WRI1	WRINKLED1

Author details

Wei Xia
Hainan Key Laboratory for Sustainable Utilization of Tropical Bioresources, College of Tropical Crops, Hainan University, Haikou, P.R. China

*Address all correspondence to: saizjxiawei@hainanu.edu.cn

References

[1] Lee ST, Radu S, Ariffin A, Ghazali HM. Physico-chemical characterization of oils extracted from noni, spinach, lady's finger, bitter gourd and mustard seeds, and copra. International Journal of Food Properties. 2015;**18**(11):2508-2527

[2] Laureles LR, Rodriguez FM, Reano CE, Santos GA, Laurena AC, Mendoza EM. Variability in fatty acid and triacylglycerol composition of the oil of coconut (*Cocos nucifera* L.) hybrids and their parentals. Journal of Agricultural and Food Chemistry. 2002;**50**(6):1581-1586

[3] Dussert S, Guerin C, Andersson M, Joët T, Tranbarger TJ, Pizot M, et al. Comparative transcriptome analysis of three oil palm fruit and seed tissues that differ in oil content and fatty acid composition. Plant Physiology. 2013;**162**:1337-1358

[4] Liang Y, Yuan Y, Liu T, Mao W, Zheng Y, Li D. Identification and computational annotation of genes differeentially expressed in pulp development of *Cocos nucifera* L. by suppression subtractive hybridization. BMC Plant Biology. 2014;**14**:205

[5] Xiao Y, Xia W, Mason AS, Cao Z, Fan H, Zhang B, et al. Genetic control of fatty acid composition in coconut (*Cocos nucifera*), African oil palm (*Elaeis guineensis*), and date palm (*Phoenix dactylifera*). Planta. 2019;**249**(2):333-350

[6] Yuan Y, Liang Y, Li B, Zheng Y, Luo X, Dongdong L. Cloning and function characterization of a β-ketoacyl-acyl-ACP synthase I from coconut (*Cocos nucifera* L.) endosperm. Plant Molecular Biology Reporter. 2015;**33**(4):1131-1140

[7] Gao L, Sun R, Liang Y, Zhang M, Zheng Y, Li D. Cloning and functional expression of a cDNA encoding stearoyl-ACP Δ9-desaturase from the endosperm of coconut (*Cocos nucifera* L.). Gene. 2014;**549**(1):70-76

[8] Yuan Y, Liang Y, Gao L, Sun R, Zheng Y, Li D. Functional heterologous expression of a lysophosphatidic acid acyltransferase from coconut (*Cocos nucifera* L.) endosperm in *Saccharomyces cerevisiae* and *Nicotiana tabacum*. Scientia Horticulturae. 2015;**192**:224-230

[9] Yuan Y, Chen Y, Yan S, Liang Y, Zheng Y, Dongdong L. Molecular cloning and characterisation of an acyl carrier protein thioesterase gene (CocoFatB1) expressed in the endosperm of coconut (*Cocos nucifera*) and its heterologous expression in *Nicotiana tabacum* to engineer the accumulation of different fatty acids. Functional Plant Biology. 2014;**41**(1):80-86

[10] Davies HM, Hawkins DJ, Nelsen JS. Lysophosphatidic acid acyltransferase from immature coconut endosperm having medium chain length substrate specificity. Phytochemistry. 1995;**39**(5):989-996

[11] Knutzon DS, Hayes TR, Wyrick A, Xiong H, Davies HM, Voelker TA. Lysophosphatidic acid acyltransferase from coconut endosperm mediates the insertion of laurate at the sn-2 position of triacylglycerols in lauric rapeseed oil and can increase total laurate levels. Plant Physiology. 1999;**120**(3):739-746

[12] Xu L, Ye R, Zheng Y, Wang Z, Zhou P, Lin Y, et al. Isolation of the endosperm-specific LPAAT gene promoter from coconut (*Cocos nucifera* L.) and its functional analysis in transgenic rice plants. Plant Cell Reports. 2010;**29**(9):1061-1068

[13] Zheng Y, Jin Y, Yuan Y, Feng D, Chen L, Li D, et al. Identification and function analysis of a type 2

diacylglycerol acyltransferase (DGAT2) from the endosperm of coconut (*Cocos nucifera* L.). Gene. 2019;**702**:75-82

[14] Jing F, Cantu DC, Tvaruzkova J, Chipman JP, Nikolau BJ, Yandeau-Nelson MD, et al. Phylogenetic and experimental characterization of an acyl-ACP thioesterase family reveals significant diversity in enzymatic specificity and activity. BMC Biochemistry. 2011;**12**:44

[15] Reynolds KB, Taylor MC, Zhou XR, Vanhercke T, Wood CC, Blanchard CL, et al. Metabolic engineering of medium-chain fatty acid biosynthesis in *Nicotiana benthamiana* plant leaf lipids. Frontiers in Plant Science. 2015;**6**:164

[16] Yuan Y, Gao L, Sun R, Yu T, Liang Y, Li D, et al. Seed-specific expression of an acyl-acyl carrier protein thioesterase CnFatB3 from coconut (*Cocos nucifera* L.) increases the accumulation of medium-chain fatty acids in transgenic Arabidopsis seeds. Scientia Horticulturae. 2017;**223**:5-9

[17] Cernac A, Benning C. *WRINKLED1* encodes an AP2/EREB domain protein involved in the control of storage compound biosynthesis in *Arabidopsis*. The Plant Journal. 2004;**40**(4):575-585

[18] Sun R, Ye R, Gao L, Zhang L, Wang R, Mao T, et al. Characterization and ectopic expression of *CoWRI1*, an AP2/EREBP domain-containing transcription factor from coconut (*Cocos nucifera* L.) endosperm, changes the seeds oil content in transgenic *Arabidopsis thaliana* and Rice (*Oryza sativa* L.). Frontiers in Plant Science. 2017:8

Chapter 13

Gums—Characteristics and Applications in the Food Industry

Diego Aires da Silva, Giselle Cristine Melo Aires and Rosinelson da Silva Pena

Abstract

Gums, or polysaccharides, are complex carbohydrates, soluble in water, which can form gels and mucilages. They have high molar mass and can be formed by galactose, arabinose, rhamnose, xylose, galacturonic acid, among others. They have gelling characteristics, thickening, moisture retention, emulsification and stabilization. Polysaccharides are widely used in the formulation of food products, due to their wide versatility. Its diversity of applications is closely linked to its chemical structures. The characterization of structural molecules allows the knowledge of the properties of polysaccharides or glycoconjugates. In this sense, this chapter addresses knowledge about chemical, molecular, rheological, thermodynamic characteristics that are extremely important to identify the use and applications of polysaccharides in the context of elaboration and innovation in the food industry.

Keywords: gum, hydrocolloids, carbohydrate

1. Introduction

The food and beverage industries face increasingly challenging scenarios, as they need to meet consumers' desires, and use ingredients that are natural, and that fulfill their technological roles in processed foods. Among these ingredients, gums and hydrocolloids are the compounds most widely used as agents of innovation in the food industry.

Gums, also known as hydrocolloids or polysaccharides, are very versatile biopolymers, extensively used in the food sector as ingredient or additive, which fulfill several technological and, sometimes, nutritional functions. This versatility is intrinsically related to their molecular composition, which gives these polysaccharides certain properties such as gelling, thickening, moisture retention, emulsification, and stabilization. In the food industry, they are widely used in confectionery, as ice cream stabilizers, food emulsions, in the microencapsulation of flavors and dyes, clarifiers, and beverage stabilizers.

Therefore, information on the molecular structure, thermal stability, interaction with water, and rheological behavior are essential knowledge for prospecting and developing applications for each type of polysaccharide, whether isolated or in mixtures.

Another important fact, in this sense, is the constant search for new sources of polysaccharides that might have similar and/or better effects than those already known. This is important because it also shows regional valorization, source of income, and new business opportunities.

IntechOpen

Thus, this chapter aims to discuss the physical, chemical, and molecular knowledge of polysaccharides, in addition to their versatility of applications in the food industry.

2. Gums: origin and definition

The term gum is generally used to define hydrophilic or hydrophobic molecules of high molar mass, which have colloidal properties [1]. Classified according to origin, behavior, and chemical structure, gums can be derived from plant seed endosperm (guar gum) [2], plant exudates (tragacanth), shrubs or trees (gum arabic, karaya gum, cashew gum) [2–5], algae extracts (agar) [6], bacteria (xanthan gum), animal source (chitin), and others [7–10].

Vegetable exudates are fluids that flow spontaneously from trees, due to adaptations to climatic conditions (physiological gummosis) or in response to any injury suffered, whether mechanical, such as cutting, or by the action of microorganisms, which dry out when exposed to air [11].

Hillis [12] describes in detail the differences between exudates from tree trunks, specifically the differences between resins and gums, and their formation. The author defines resins as materials composed largely by terpenoids, and that may contain phenolic compounds (coumaric, caffeic, and ferulic acids), with few fatty acids and glycerides. They may be formed within plastids present in epithelial cells of plants [13] or even synthesized in spherosomes, both in resin duct cells and in parenchymal cells [14].

Hillis [12] also defines gums as products composed mainly of complex carbohydrates, soluble in water, which can form gels and mucilages. They have high molar mass and can be formed by galactose, arabinose, rhamnose, xylose, galacturonic acid, and other compounds. In some species, they are secreted by organelles present in the bark or between barks, whose main function is protecting the plant from injuries caused by cuts or microbial attack [15–17].

The interest in gums exuded from plants is due to their structural properties and respective functions in food, pharmaceutical, cosmetic, textile, and biomedical products [18]. Water-soluble gums, also known as hydrocolloids, can have various applications such as: dietary fibers, texture modifiers, gelling agents, thickeners, stabilizers, emulsifiers, coatings, films, and as encapsulants [19, 20]. There has been a strong trend towards replacing synthetic materials by natural gums due to their non-toxicity, low cost, safety, and availability [21].

3. Structural aspects of gums

All the properties and applications of gums are closely linked to their chemical structures. Gums can be formed by numerous sugars, in their main chains and/or side chains, and can be more or less branched, which determines, in general, their complexity [15].

Among the most well-known and commercialized gums [22], the gum arabic, produced by the species *Acacia senegal*, presents in its structure a main chain formed by β-D-galactopyranose joined by bonds (1➔3), alternated by highly branched bonds (1➔6), and shows lateral chains constituted by 4-O-methyl-glucuronic acid (1.5%), glucuronic acid (17.5%), galactose (39%), arabinose (28%), and rhamnose (14%) [23]. Anderson; Hirst; Stoddart [24] proposed the structure presented in **Figure 1** for acacia gum. The authors indicated, as possible replacement units, those represented by the radical "R": (L-Araf); (L-Araf 1➔3 L-Araf); (β-L-Arap 1➔3 L-Araf);

(L-Araf 1→3 L-Araf 1→3 L-Araf); (β-L-Arap 1→3 L-Araf 1→3 L-Araf); (β-D-Galp 1→3 L-Araf). Arabinofuranoside is Araf, arabinopyranoside is Arap, and galactopyranoside is Galp. The radicals "R" are not shown in **Figure 1B**.

Gum ghatti is also important among exudate gums because of its high emulsifying capacity [25]. It is extracted from the trunk of *Anogeissus latifolia*, an abundant tree in India [26]. Its molecular structure is formed by a main chain of (1→6)-β-Galactose bonds, whose branches at positions O-3 and O-4 are replaced, consisting of →2)-Araf-(1→4)-GlcpA-(1→6)-Galp-(1→6)-Galp-(1→. The terminal

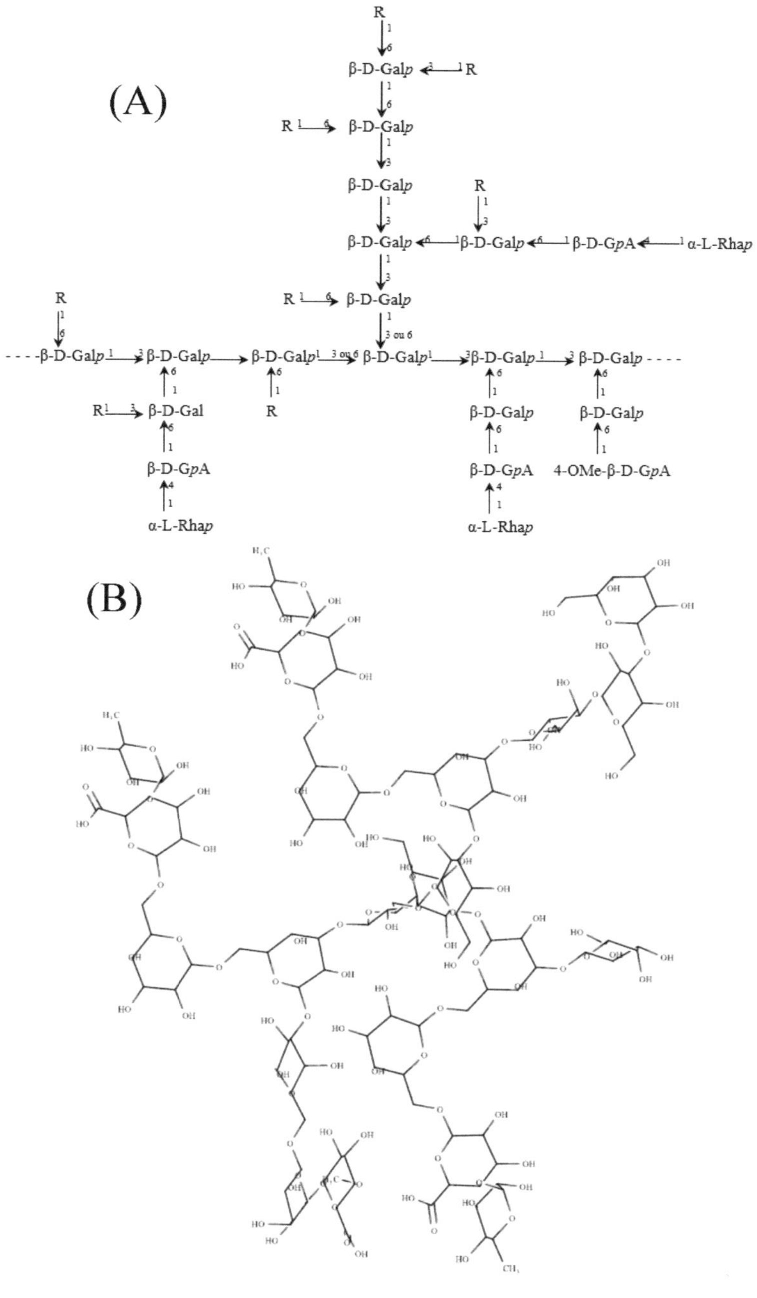

Figure 1.
*Structural fragment of gum arabic (*Acacia senegal*). (A) Scheme and (B) three-dimensional structure referring to the fragment shown.*

lateral chains are formed by residues of arabinofuranoside (Araf) and occasionally by rhamnopyranoside (Rhap), arabinopyranoside (Arap), galactopyranoside (Galp) or glucuronopyranoside (GlcpA) [27, 28]. The structure of gum ghatti is shown in **Figure 2**.

Karaya gum is also on the list of exudates from commercially interesting plants, and is extracted from *Sterculia urens* tree. Structurally, it is a complex, partially acetylated polysaccharide, composed of 55–60% of rhamnose and galactose, 8% of acetyl groups, and 37–40% of uric acid residues (galacturonic and glucuronic acids) [29]. Its structure can be seen in **Figure 3**.

3.1 Gum structure of exudates from arecaceae family species

The Arecaceae (Palmae) family consists of a large variety of monocot plants found predominantly in tropical and subtropical environments, mostly in South America, and contains 457 palm species distributed in 50 genera [30, 31].

Nussinovitch [26] described, in general, three types of gum from plants of the Arecaceae family, with sensory information about them. According to the author, *Borassus flabellifer* palm gum is a black glassy exudate, which swells and is insoluble in water; *Cocos nucifera* L. gum has coloration ranging from light brown to red, and in water, it presents certain insolubility, forms gel, and has low adhesiveness; *Corypha utan Lam*. gum has sweet odor and brown coloration, being used in medicine.

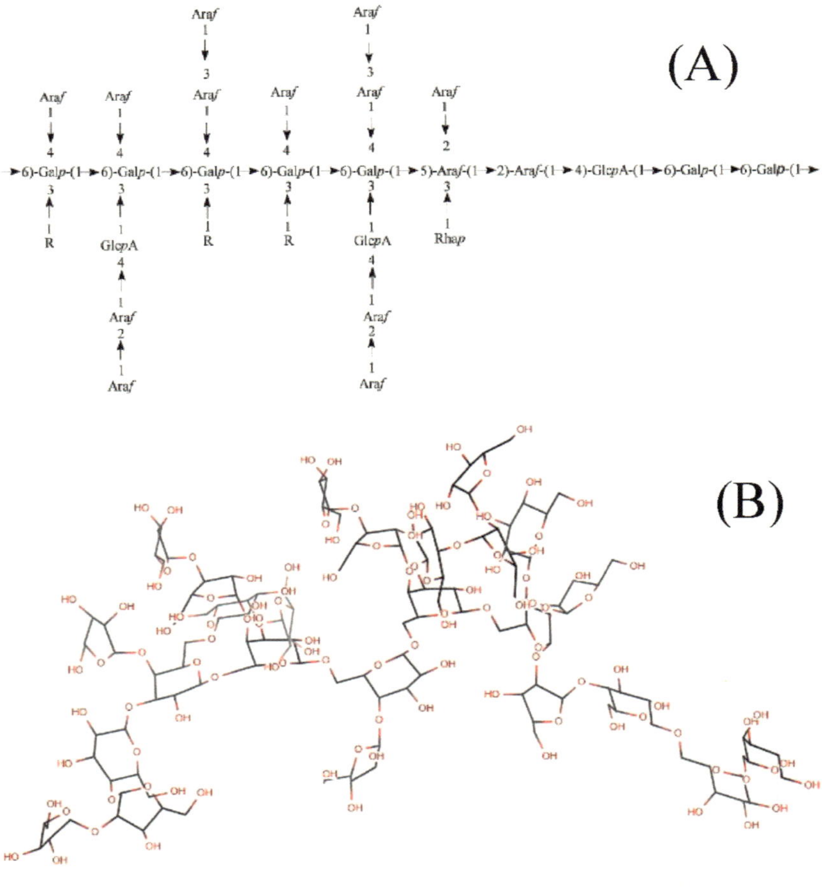

Figure 2.
Structural fragment of gum ghatti. (A) Scheme and (B) three-dimensional structure referring to the fragment shown.

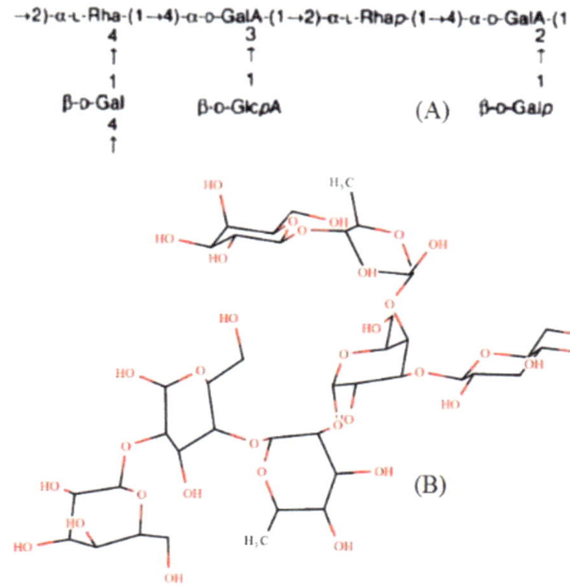

Figure 3.
Structural fragment of karaya gum. (A) Scheme and (B) three-dimensional structure referring to the fragment shown.

Gums from exudates of Chinese fan palm trunk (*Livistona chinensis*) [32] and jerivá (*Syagrus romanzoffiana*) [33] were presented as heteroxylans, whose main chain is joined by β-(1➔4) bonds, highly substituted at O-2 and O-3 positions by units of arabinose, xylose, and terminal fucose, as shown in **Figure 4**.

The exudate from Uricuri palm (*Scheelea phalerata*) was also identified by Fernanda F. Simas et al., [34]. The authors found a water-insoluble polysaccharide with a branched structure. Units of Xylp (~8%) were replaced at O-2, whereas Araf units (12%) were replaced at O-3. They also found non-reducing units of Araf (15%), Fucp (fucopyranose - 10%), Xylp (4%), and Arap (6%) as side chains attached to the main chain composed of Xylp units joined by β-(1➔4) bonds, which were replaced at 3-O-(9%), 2-O-(13%), and 2,3-di-O-(13%) positions.

The structure of the gum obtained from coconut tree trunk exudate (*Cocos nucifera*) was elucidated by Simas-Tosin et al., [35]. This gum is a glucurono-arabinoxylan composed of Fuc, Ara, Xyl, and GlcpA at molar ratio of 7:28:62:3.

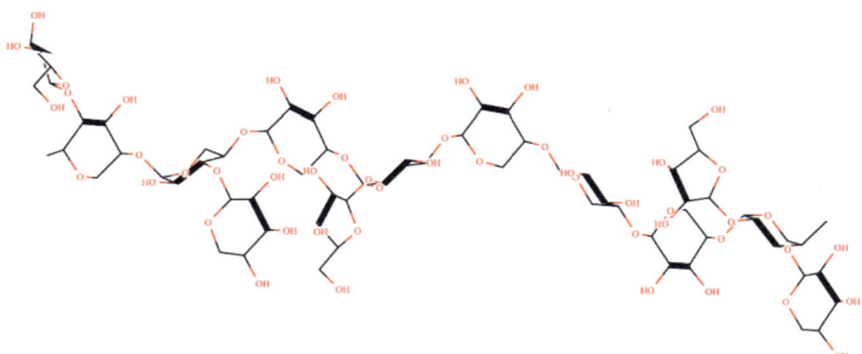

Figure 4.
Three-dimensional representation of the heteroxylan present in Scheelea phalerata *(Uricuri) palm gum, with β-(1➔4) bonds. Main chain branches are substituted at O-2 or O-3 positions by arabinose and xylose units.*

Non-reducing units substituted at 3-O (Araf - 8%); 3,4-di-O-(15%); 2,4-di-O (5%); and 2.3.4-tri-O (Xylp 17%) positions were also found, attached to a main chain composed of Xylp joined by β-(1→4) bonds.

4. Gum characterization

4.1 Spectroscopic methods for gum characterization

"Structure is the key to everything in chemistry. The properties of a substance depend on the atoms it contains and how these atoms are bound. Less obvious, but very powerful, is the idea that someone with knowledge of chemistry can look at the structural formula of a substance and say several things about its properties" [36]. "Looking at the structural formula" inevitably refers to the use of techniques that assist in the chemical and structural knowledge of organic molecules, and in this context, spectroscopic techniques can be a very important tool to fulfill such function [37].

In order to know the properties of polysaccharides or glycoconjugates, it is essential to elucidate and characterize the structural and dynamic aspects of their molecules [38]. Carbohydrate chemistry can rely on one of the most efficient spectroscopic techniques for investigating organic compounds in solution: Nuclear Magnetic Resonance (NMR), which has advanced methods, and becomes essential in the characterization of polysaccharides with complex structures [39, 40].

The commonly used NMR techniques are hydrogen (^{1}H), carbon-13 (^{13}C), homonuclear correlations (^{1}H-^{1}H), COSY (homonuclear Correlation Spectroscopy), and ^{13}C-^{1}H HMQC (Heteronuclear Multiple Quantum Coherence) [41].

The elements that are most common in organic molecules (carbon and hydrogen) have isotopes (^{1}H and ^{13}C) capable of providing NMR spectra rich in structural information. A proton nuclear magnetic resonance spectrum (1H NMR) provides information about the environments of the various hydrogens present in a molecule. A carbon-13 nuclear magnetic resonance spectrum (^{13}C NMR) does the same for carbon atoms [36, 38].

NMR spectrum of coconut trunk gum (*Cocos nucifera*), obtained by alkaline extraction, presented approximately 10 signs in the anomeric region, which reveals a complex structure. The signals made reference to the presence of L-Araf (δ 108.6–107.0); α-Arap (δ 103.1); β-Xylp (δ 101.6), and also α-Fucp and α-Glcp units (δ 100.5–99.2), bonded to C-4. Reducing terminals were bonded to C-5 [35].

Peach gum (*Prunus persica*) was also considered as a complex molecule, as it shows 8 signs in the anomeric region (δ 110–90). The main sign in δ 103.2 refers to β-D-Galp units in the main chain, and the sign in δ 102.8 suggests the presence of β-D-GlcAp. In the substituted carbon region, the signs in δ 84.1 and δ 82.0–82.5 refer to C-3 of the replaced units α-L-Araf and β-D-Galp 3-O-, respectively [42]. These are examples that demonstrate that the NMR technique is an indispensable tool for the knowledge of polysaccharides and their properties.

Another technique widely used for the structural identification of polysaccharides, even before the advent of NMR, is the Fourier-Transform Infrared Spectroscopy (FTIR) [36]. Although NMR gives more information about the structure of an unknown compound, infrared is important because it can identify certain functional groups. Structural units, including functional groups, vibrate in characteristic ways, and this sensitivity to group vibrations forms the basis of infrared spectroscopy [43].

Molecular movements are described by two types of vibrations: deformation and stretching (**Figure 5**). The deformation causes a bond angle change that can occur in or out of the molecular plane of symmetry; and the stretching is a linear

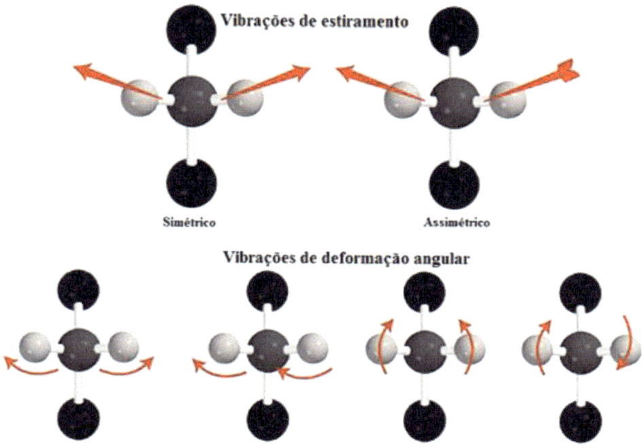

Figure 5.
Aspects of the molecule vibrations observed in infrared spectroscopy.

intermittent movement so that the interatomic distance changes constantly. It can be symmetrical or asymmetrical [44].

When irradiated by infrared light, the atoms of the molecular structure of a given sample absorb it. The vibration or rotation will depend on the type of chemical bond formed by these atoms B [45, 46]. **Table 1** shows some bands of infrared

Bands	Associated vibrations	Possible assignments to bands	References
≈1650 and 1550 cm^{-1}	v(C=O) γ(CN) δ(NH) $(CCN)_{deform.}$ v(C=C) v(COO^-)	Amide I and II of proteins, respectively	[47–50]
1640–1600 and 1420 cm^{-1}		Carboxylic acids deprotonated in uronic acid	[48, 50]
1444, 1371, 975–978, and 923 cm^{-1}	δ(CH_3) $(CH)_{deform.}$ δ(CO) δ(NH) δ(C–O) $δ(OH)_{COOH}$ v(C–O–C) v(CN)	Methyl ester groups (CH_3) in pectins	[51]
1280 and 1220 cm^{-1}		Methyl ester groups (CH3) in pectates	[51]
1280–1260 cm^{-1}		Phenolic esters bonded to cell walls groups	[52]
≈1230 cm^{-1}		Amide III of protein secondary structures	[49, 52]
Fingerprint region in polysaccharides			[53]
1155–1038 cm^{-1}	v(C–O–C) v(C–OH) v(C–O) v(C–C) v(O–CH_3) (CH_3) (C_1–H) δ(OH) δ(CCH) δ(COH)	Galactan attached to main chain β 1➔6 Galp	[53]
1141–1039 cm^{-1}		Arabinans connected to the main and side chains of Araf	[53]
1139–985 cm^{-1}		Arabinogalactans linked to the main chain of β 1➔3 Galp, and side chain of α 1➔3 Araf (8%) and β 1➔6 Galp (92%)	[53]
1140–975 cm^{-1}		Arabinogalactan-rhamnoglycan attached to the main chain β 1➔6 Galp (24%) and α 1➔4 Rhap (42%), and side chain of α-Araf and α 1➔5 Arap (34%)	[53]
900–870 cm^{-1}		B-type bonds between monosaccharides	[54, 55]

Table 1.
Infrared Fourier transform bands in plane (δ); out of plane (γ) and stretching (v), and assignments related to functional groups.

spectroscopy and their respective functional groups present in polysaccharides. It is also possible to see that FTIR can provide information on important functional groups in polysaccharides in the fingerprint region [44, 46].

In polysaccharides, the infrared spectroscopy can be used to qualitatively observe possible structural changes. Quelemes et al., [56] demonstrated the structural change in cashew gum when submitted to quaternary ammonium reagent, which also improved some properties such as biocompatibility and antimicrobial action. FTIR was also efficient to demonstrate that the interaction of gum arabic and chitosan was formed by electrostatic complexes, a result of the interaction between functional groups ($NH3^+$ and –COO-) of both macromolecules. Also, it improved viscoelastic characteristics at different pH's, demonstrating its complex versatility for use as food additives [57].

4.2 Thermal analysis of gums

Most polymers, synthetic or natural, suffer degradation when subjected to thermal stress [58]. This is attributed to chain depolymerization, point splits, or even the elimination of low molecular weight fragments, which cause mass loss due to the increase in temperature [59]. They cause thermal effects related to physical or chemical changes, and are associated with thermodynamic events [58]. These changes in energy and mass can be measured by thermogravimetry (TG), derivative thermogravimetry (DTG), differential thermal analysis (DTA) and differential scanning calorimetry (DSC), which make it possible to obtain information such as changes in the crystalline structure, reaction kinetics, melting and boiling point, glass transition, and others [60]. Changes in mass as a function of temperature and/or time [61] and continuous registration of mass subjected to heating or cooling [62] are definitions attributed to thermogravimetry.

Being the combination of an electronic microbalance and an oven, associated with a linear temperature programmer, thermogravimetric analysis consists of submitting a known mass of sample inside a crucible, suspended by a platinum wire, to a programmed temperature gradient, for a predefined time, which is automatically registered, simultaneously with the sample mass [63].

In DTG, the mass variation derivative (dm/dt) is registered as a function of temperature or time. In this method, the levels observed in TG are replaced by peaks that delimit areas which are proportional to the changes in mass suffered by the sample and can indicate the exact initial temperatures and maximum speed of reactions. DTG allows a clear distinction of successive reactions (not detected by TG), by quantitative determinations of loss or gain of mass which are associated with the peak areas [60].

DSC and DTA are analyses that measure energy gradients between the sample and a reference material subjected to controlled temperature. DSC is a calorimetric method in which energy differences are measured, whereas in DTA, temperature differences between the sample and the reference material are registered [59]. DTA provides a qualitative analysis of the thermal events experienced by the sample, whereas DSC is able to quantify such events because it measures the heat flow through a temperature gradient [64].

Changes in composition, food processing temperatures or ingredients result in changes in phase transitions of the product [65]. Quantifying the variables involved in these phenomena, such as temperature or thermodynamic quantities, is important for understanding processes such as evaporation, dehydration, and freezing [66]. Being the responsible for plasticizing effects and important component of food, water and its state transitions (gaseous or crystalline) guide such processes, and can also be used to describe the effects of temperature on physical properties [59].

4.3 Gum rheology

Natural polymers are of particular interest in rheological studies [67]. Their thickening, emulsifying, gelling, and stabilizing properties, which enable them to be used in food, pharmaceutical, and cosmetic industries are supported by a series of inter and intramolecular association mechanisms inherent to each polymer. Such mechanisms lead them to particular applications in different processes and products [68].

Gum arabic *(Acacia senegal)* 3% (m/v), originating from African regions such as Sudan, Senegal, and Mali, has typical behavior of a liquid. Sanchez, Renard, Robert, Schmitt, & Lefebvre, [69] investigated G' and G" in gum arabic, where G' is the storage modulus and indicates the portion of energy (from the applied voltage) that is temporarily stored during the test, and it provides information on the elastic characteristic of the fluid. On the other hand, G" is the loss modulus, which indicates the portion of energy used to initiate flow. It is irreversibly transferred in the form of heat and provides information on the viscous characteristics of the fluid [70]. The authors state that gum arabic presented a viscous modulus (G') greater than its elastic modulus (G'), but after 5 hours of rest, gel characteristics were identified, consequently showing a more elastic structure [69].

Acacia tortuosa gum, originating from species located in South America (Venezuela) (15% m/v), presented elastic modulus (G') greater than its viscous modulus (G"), indicating the occurrence of a gel material that became progressively weaker with increasing temperature [71]. In both studies, gums showed transition from Newtonian to non-Newtonian behavior with increasing concentration. Also, the influence of inter and intramolecular structural interactions as agents responsible for rheological changes was observed [69, 71].

The emulsifying and rheological characters of chemically modified gum arabic (*Acacia senegal*) (esterified with octenyl succinic anhydride (OSA) at different concentrations) was measured by [72]. The study revealed that the gum presented an increase in its emulsifying capacity and a gradual increase in apparent viscosity with increasing OSA content, indicating satisfying emulsion stability and potential use as microencapsulant. The electrostatic interaction between gum arabic and soy protein β-conglycinin was the mechanism that improved the flocculating action of *Acacia senegal*, in addition to providing greater elasticity at the oil/water interface of the gum, consequently improving its emulsifying capacity [73]. The interaction of gum arabic with native tapioca starch also provided improved product elasticity and adhesiveness [74]. Chenlo, Moreira, & Silva, [75], studied the rheology of aqueous dispersions of tragacanth gum and guar gum (10 g/L) during storage for 47 days. In general, the apparent viscosity decreased significantly ($\alpha = 0.05$) for both systems at low values of $\dot{\gamma}$ ($< 10s^{-1}$) and remained constant above this value. The decrease in viscosity was lower for tragacanth gum and lasted until the 15th day, whereas for guar gum, the decrease occurred until the 20th day.

Mixtures of corn starch (5% m/m) and locust bean gum (0; 0.125; 0.25; 0.50; and 1% m/v) were rheologically evaluated by Hussain, Singh, Vatankhah, & Ramaswamy, [76], who found that the addition of locust bean gum at low concentrations (0.125%) made the mixture behave as a liquid at low oscillatory frequencies (0.1 to 10 rad/s). It also presented increased elasticity, with typically solid behavior at concentrations of 0.5 to 1%, at higher frequencies (15 to 100 rad/s). Thus, locust bean gum has potential to specifically modify the structure and texture of corn starch products.

The research results showed that there are many variables that influence the rheological characteristics of gums. Among them, the fine chemical structure of the polysaccharide, their interactions, and molecular conformations can be highlighted, which confirms the importance of characterizing the structure of new gums.

5. Thermodynamic relations between gums and water

The functions derived from the physical and chemical properties of gums are closely related to the interactions of polysaccharides with water. The relationship between the water content of a product and its relative humidity at equilibrium, at constant temperature, can be expressed by characteristic curves called moisture sorption isotherms [77, 78]. In fact, the thermodynamic properties of sorption, such as water-solute affinity and spontaneity of the sorption process provide a better understanding of the water-solute equilibrium that is present in the product [79]. In addition, they facilitate the definition of order and disorder existing in water-solute systems [80].

The differential enthalpy or isosteric heat of sorption defines the amount of heat released or absorbed in the sorption process at constant pressure, and is used as an indicator of the binding force between the water and solutes of the product [81]. When the free water latent heat of vaporization is added, the integral isosteric heat of sorption is obtained, which is the total energy necessary to transfer the water molecules in the vapor state to a solid surface, or vice versa [79, 82]. Also, the differential entropy of a material is proportional to the number of available sorption sites, corresponding to a specific energy level, and indicates the mobility state of the water molecules present in the product [81]. Entropy describes the degree of disorder and randomness in the movement of water molecules, and has been used to explain how water sorption in biological materials occurs [83].

Thermodynamic properties, such as enthalpy and entropy, are necessary to design a process and to qualitatively understand the water state at a certain food surface. Alterations in enthalpy provide the energy variation of the interaction between water molecules and the adsorbent. Entropy, in contrast, may be associated with the binding or repulsion of forces and, consequently, with the spatial arrangement of the water-adsorbent relationship. Thus, entropy characterizes the degree of order or disorder existing in the water-adsorbent system [84]. Gibbs free energy, in turn, is influenced by the thermodynamic properties enthalpy and entropy, and indicates the energetic spontaneity of the water-adsorbent interaction, providing the availability of process energy. If the value of this property is negative, the process is spontaneous, and if it is positive, the process is nonspontaneous. In systems with many constituents, such as food and polysaccharides, Gibbs-free energy depends not only on pressure and temperature, but also on the amount of each component [80].

6. Gum applications

The applications of gums from plant exudates are very diversified, and can be present in various areas of the food industry: confectionery (lollipops, chocolates, jelly beans, pastilles, and others), in which there is a high sugar content and low humidity; to prevent sugar crystallization; in salad dressings (thickeners and emulsion stabilizers) [85]; in frozen products (pasta, popsicles, ice cream) [1]; in dehydrated products, such as juices obtained by spray drying, protecting important compounds such as vitamin C, anthocyanins, and improving solubility, or also as microencapsulants for colors, flavors, and oils [86]; in wine clarification; flavor fixatives and emulsifiers; and in beverages and meat products [87, 88] (**Table 2**).

In adhesion functions, gums are used as fixatives of skin bioelectrodes, dentures, ostomy devices, and transdermal membrane systems, which perform controlled release of drugs through the skin [7, 121, 122]. They are used as adhesive materials in wood-based industry, and obviously, in adhesive industries in general [123]. Gums have applicability in the pharmaceutical area as emulsifiers and reducing agents for suspended particles, laxatives, in the preparation of antiseptics, binders for tablets and pills, and

Common name	Scientific name	Main chemical compounds	Application	Reference
Gums from fruits				
Date palm mucilage	*Phoenix dactylifera*	Fructose, sucrose, mannose, glucose, and maltose	Anti-cancer action	[89]
"Erva Baleeira" Mucilage	*Cordia obliqua*	Arabinose, galactose, and pyrralinose	Expectorant, tablet binder, emulsifier	[90]
Jackfruit	*Artocarpus heterophyllus*	Galactomannan, starch	Suspension stabilizer, emulsifier, binder, mucoadhesive	[91, 92]
Gums from seeds				
Tamarind gum	*Tamarindus indica*	Glucose:xylose:galactose (3:2:1)	Tablet formulation, biodegradable support for controlled drug release (colon), bioadhesive	[93, 94]
Fenugreek mucilage	*Trigonella foenum-graceum*	Galactomannan	Textural and sensory properties of soup powder/ anthocyanin encapsulation	[95, 96]
Locust bean gum	*Ceretonia Siliqua*	D-galacto-D-manoglycan, cellulose, galactomannan	Superdisintegrant in controlled drug delivery system	[97, 98]
Tara gum	*Caesalpinia spinosa*	Mannose:Galactose (3:1)	Smart food packaging	[99]
Gleditsia triacanthos gum	*Gleditsia triacanthos*	Galactomannan	Matrix formulation for tablets	[100]
Cassia tora Mucilage	*Cassia tora*	Arabinose and glucose	Suspension stabilizer, binder	[101]
Flamboyant gum	*Mimosa scabrella*	Mannose:Galactose (3.65:1)	Dietary fiber, probiotic viability in milk drink	[102]
Guar gum	*Ocimum americanum*	Xylose, arabinose, rhamnose, and galacturonic acids	Guar gum nanocomposite films	[103]
Gums obtained from tree trunks exudates				
Albizia stipulata Boiv. gum	*Albizia stipulata* Boiv.	Arabinose, galactose, and rhamnose	Antioxidant properties	[104]
Almond gum	*Prunus amygdalus*	Aldobionic acid, L-arabinose, L-galactose, and D-mannose	Emulsifier, suspension stabilizer, binder, thickener	[105]
Cashew gum and cashew nut gum	*Anacardium occidentale*	Galactose, arabinose, rhamnose, glucose, glucuronic acid	Encapsulation of a lipid shrimp waste extract, anti-inflammatory effect	[86, 106]

Common name	Scientific name	Main chemical compounds	Application	Reference
Cherry gum	*Prunus avium*	Arabinogalactan	Coating film	[107]
Raphia hookeri gum	*Raphia hookeri*	Mannose and galactose	Aluminum anti-corrosion agent in acid medium	[108]
Tragacanth gum	*Astragalus gummifer*	D-galacturonic acid, D-galactose, L-fucose (6-deoxy-L-galactose), D-xylose, L-arabinose, and L-rhamnose	Catalyst in the production of nanoparticles	[109]
Gum kondagogu	*Cochlospermum gossypium*	Rhamnogalacturonan	Production of biocompatible and antimicrobial scaffold for bandages	[110]
Gums obtained from leaves				
Cocculus hirsutus mucilage	*Cocculus hirsutus*	Polysaccharides and gelatinous materials	Binding agent, gelling agent (drugs)	[111]
Hibiscus mucilage	*Hibiscus rosa-sinensis*	L-rhamnose, D-galactose, Dgalactouronic acid, and D- glucuronic acid	Controlled drug release	[112, 113]
Gums obtained from microorganisms				
Curdlan gum	*Agrobacterium* spp.	Glucose	Food additive, thickener, gelling agent	[114]
Gellan gum	*Sphingomonas* spp.	Glucose, rhamnose, and glucuronate	Emulsion stabilizer, ophthalmic hydrogel	[115, 116]
Cholic acid	*Escherichia coli*	Fucose, glucose, glucuronate, and galactose	Viscosity enhancer	[114]
Xanthan gum	*Xanthomonas spp.*	D-glucose, D-mannose, and glucuronic acid	Carotenoid encapsulation for use in yogurts	[117]
K30 antigen	*Escherichia coli*	Mannose, galactose, and glucuronate	Viscosity enhancer/ controlled drug release	[114]
Gums obtained from tubers				
Konjac glucomannan	*Amorphophallus konjac*	D-Glucose and D-mannose	Gelling agent, controlled drug release	[118, 119]
Taro	*Colocasia Esculenta*	Galactose and arabinose	Gelling agent, mucoadhesives	[120]

Table 2.
Applications of gums from various origins.

in the cosmetics area (perfume fixers, skin cleansers, and repellents) [124–127]. Also, in the medical field, gums are used to control osmotic pressure, in addition to having activity against *Leishmania amazonensis* and antifungal properties [128].

The most recent studies have shown that the versatility of gum use has increased. The beverage industry, for instance, is always seeking products with greater stability. Some polysaccharides are excellent stabilizers, such as tara gum, which is often used to stabilize casein aggregation in dairy drinks, improving phase separation. This occurs because tara gum makes it difficult to approach casein molecules, providing greater stability and improving the sensory acceptance of the product [125].

Carrageenan gum, xanthan gum, guar gum, sodium alginate, carboxymethyl cellulose, gum arabic, and pectin were tested to prevent the formation of turbidity, caused by protein-polyphenol complexation, in packaged beverages. Among them, pectin, xanthan gum, and guar gum showed the best results [126]. These polysaccharides, when present in low concentrations: 0.5, 0.05, and 0.01 mg/mL, compete with proteins to bind polyphenols, which decrease protein-polyphenol aggregation; or they can form a ternary complex (protein-tannin-polysaccharide) to increase the solubility of protein- polyphenol systems. This mechanism promotes the reduction of unwanted turbidity in such products [127].

The use of gums and polysaccharides in film production is also an area of great concentration of studies. Active, functional, and biodegradable packagings are examples which may have antibacterial activity.

Tragacanth gum, for instance, showed excellent results in the production of nanocomposite biofilms, and can be applied in the prevention of lipid oxidation in high-fat foods, with antimicrobial action and excellent responses to biodegradability tests [128, 129]. In addition, chemically modifying the gums to improve their hydration control, gel formation, and swelling can also be an interesting way to use these polysaccharides to produce biodegradable films, which have a good response in prolonging food quality [130].

Gums can offer great innovation opportunities for the food sector. Its use is reported in wastewater treatment and in the production of nanoemulsions, and micro and nano encapsulation of dyes, essential oils, and probiotics [131–136].

Therefore, it is important to encourage the search for new sources of gums and polysaccharides from biodiversity, as their applicability and benefits can and, obviously, should be explored.

7. Conclusion

Gums have incredible versatility and are a rich source of innovation in food formulations and elaborations in the industry. They can be used both in isolation and in mixtures and can be modulated to deliver not only taste and nutrition, but also a new consumption experience, whether due to texture or applied technology. It is important that new sources of these carbohydrates are increasingly known, as there is still much to explore in this area.

Acknowledgements

The authors acknowledge *Conselho Nacional de Desenvolvimento Científico e Tecnológico* (CNPq, Brazil) for the financial support and *Coordenação de Pessoal de Nível Superior* (CAPES, Brazil), University of Pará State (UEPA) and Federal University of Pará (UFPA).

Author details

Diego Aires da Silva[1*], Giselle Cristine Melo Aires[1] and Rosinelson da Silva Pena[2]

1 Food Technology Department, Natural Science and Technology Center, University of Pará State (UEPA), Belém, PA, Brazil

2 Graduate Program in Food Science and Technology, Technology Institute, Federal University of Pará (UFPA), Belém, PA, Brazil

*Address all correspondence to: diegoaires@uepa.br

References

[1] Kurt A, Cengiz A, Kahyaoglu T. The effect of gum tragacanth on the rheological properties of salep based ice cream mix. Carbohydr Polym [Internet]. 2016;143:116-23. Available from: http://dx.doi.org/10.1016/j.carbpol.2016.02.018

[2] Gooch J. Dicionário enciclopédico de polímeros. 2010 [cited 2020 Aug 20]; Available from: https://books.google.com.br/books?hl=pt-BR&lr=&id=HRgy8iHQtdwC&oi=fnd&pg=PA967&dq=Encyclopedic+Dictionary+of+Polymers&ots=fygVFnDbJL&sig=hJi9CymJldRQq883ZRCiXjeK5c4

[3] Anderson DMW, Bell PC, Millar JRA. Composition of gum exudates from Anacardium occidentale. Phytochemistry. 1974;13(10):2189-93.

[4] Williams PA. Polysaccharide Ingredients: Gum Arabic. In: Reference Module in Food Science. Elsevier; 2016.

[5] Luo Y, Hu Q. Food-derived biopolymers for nutrient delivery. In: Nutrient Delivery. Elsevier; 2017. p. 251-91.

[6] Nazir A, Asghar A, Aslam Maan A. Food Gels: Gelling Process and New Applications. In: Advances in Food Rheology and Its Applications. Elsevier Inc.; 2017. p. 335-53.

[7] Ben-Zion O, Nussinovitch A. Physical properties of hydrocolloid wet glues. Food Hydrocoll. 1997;11(4):429-42.

[8] Laaman TR. Hydrocolloids in Food Processing. Hydrocolloids in Food Processing. Wiley-Blackwell; 2011.

[9] Di Palma TM, Migliardini F, Caputo D, Corbo P. Xanthan and κ-carrageenan based alkaline hydrogels as electrolytes for Al/air batteries. Carbohydr Polym. 2017 Feb 10;157:122-7.

[10] Xu J, He Z, Zeng M, Li B, Qin F, Wang L, et al. Effect of xanthan gum on the release of strawberry flavor in formulated soy beverage. Food Chem. 2017 Aug 1;228:595-601.

[11] Nussinovitch A. Plant Gum Exudates of the World: Sources, Distribution, Properties, and Applications. CRC PRESS, editor. 2010.

[12] Hillis W. Heartwood and tree exudates. 2012 [cited 2020 Aug 20]; Available from: https://books.google.com.br/books?hl=pt-BR&lr=&id=w2DwCAAAQBAJ&oi=fnd&pg=PA1&dq=Heartwood+and+tree+exudates+ISBN+3642725341&ots=HdyZkY8xzw&sig=VHYV9yFUJK7QhmKTaUnp92UyyaM

[13] Dell B, McComb AJ. Plant Resins—Their Formation, Secretion and Possible Functions. Adv Bot Res. 1979 Jan 1;6(C):277-316.

[14] Werker E, Fahn A. Site of resin synthesis in cells of *Pinus halepensis* mill. (34). Vol. 218, Nature. 1968. p. 388-9.

[15] Nussinovitch A, Hirashima M. Cooking innovations: Using hydrocolloids for thickening, gelling, and emulsification. 2013 [cited 2020 Aug 20]; Available from: https://books.google.com.br/books?hl=pt-BR&lr=&id=YUfGAAAAQBAJ&oi=fnd&pg=PP1&dq=Cooking+innovations:+using+hydrocolloids+for+thickening,+gelling,+and+emulsification&ots=PkqioGQyAu&sig=F-WAP4i_6E-mSoieKWfH0MELHQ8

[16] Martínez M, Beltrán O, Rincón F, León De Pinto G, Igartuburu JM. New structural features of *Acacia tortuosa* gum exudate. FOOD Chem [Internet]. 2015 [cited 2020 Aug 20];182:105-10. Available from: http://dx.doi.org/10.1016/j.foodchem.2015.02.124

[17] Shah JJ, Nair GM, Kothari IL. Ultrastructural changes in the gum-resin ducts of the bark of Commiphora wightii (Arnott) Bhandari* induced by mechanical injury. IAWA J. 1982;3(3-4):185-92.

[18] Li JM, Nie SP. The functional and nutritional aspects of hydrocolloids in foods. Food Hydrocoll. 2016 Feb 1;53:46-61.

[19] Martinetti L, Mannion AM, Voje WE, Xie R, Ewoldt RH, Morgret LD, et al. A critical gel fluid with high extensibility: The rheology of chewing gum. J Rheol [Internet]. 2014 Jul [cited 2020 Aug 20];58(4):821-38. Available from: http://dx.doi.org/10.1122/1.4874322]

[20] Porto BC, Cristianini M. Evaluation of cashew tree gum (Anacardium occidentale L.) emulsifying properties. LWT - Food Sci Technol. 2014;59(2P2):1325-31.

[21] Mirhosseini H, Amid BT. A review study on chemical composition and molecular structure of newly plant gum exudates and seed gums. Food Res Int [Internet]. 2012;46(1):387-98. Available from: http://dx.doi.org/10.1016/j.foodres.2011.11.017

[22] Burger K. Private versus communal tenure systems in gum Arabic collection. In: Dryland Forests. Springer International Publishing; 2016. p. 53-69.

[23] Defaye J, Wong E. Structural studies of gum arabic, the exudate polysaccharide from *Acacia senegal*. Carbohydr Res. 1986 Aug 1;150(1):221-31.

[24] Anderson DMW, Hirst E, Stoddart JF. Studies on uronic acid materials. Part XVII. Some structural features of *Acacia senegal* gum (gum arabic). J Chem Soc C Org. 1966;1959-66.

[25] Kang J, Guo Q, Phillips GO, Cui SW. Understanding the structure-emulsification relationship of gum ghatti-a review of recent advances. Elsevier [Internet]. 2014 [cited 2020 Aug 20]; Available from: http://krex.ksu.edu

[26] Nussinovitch A. Plant Gum Exudates of the World. Plant Gum Exudates of the World. CRC Press; 2009.

[27] Kang J, Cui SW, Phillips GO, Chen J, Guo Q, Wang Q. New studies on gum ghatti (*Anogeissus latifolia*) Part III: Structure characterization of a globular polysaccharide fraction by 1D, 2D NMR spectroscopy and methylation analysis. Food Hydrocoll. 2011 Dec;25(8):1999-2007.

[28] Kang J, Cui SW, Phillips GO, Chen J, Guo Q, Wang Q. New studies on gum ghatti (*Anogeissus latifolia*) part II. Structure characterization of an arabinogalactan from the gum by 1D, 2D NMR spectroscopy and methylation analysis. Food Hydrocoll. 2011 Dec;25(8):1991-8.

[29] Mirhosseini H, International BA-FR, 2012 undefined. A review study on chemical composition and molecular structure of newly plant gum exudates and seed gums. Elsevier [Internet]. [cited 2020 Aug 20]; Available from: https://www.sciencedirect.com/science/article/pii/S0963996911006417

[30] Smith *N. palms* and People in the Amazon [Internet]. Books. 2015 [cited 2020 Aug 20]. 484 p. Available from: http://link.springer.com/10.1007/978-3-319-05509-1

[31] Brokamp G, Weigend M, Hilger HH. Relevance and sustainability of wild plant collection in NW South America: Insights from the plant families Arecaceae and Krameriaceae. Relevance and Sustainability of Wild Plant Collection in NW South America: Insights from the Plant Families

Arecaceae and Krameriaceae. Springer Science+Business Media; 2015. 1-217 p.

[32] Maurer-Menestrina J, Sassaki GL, Simas FF, Gorin PAJ, Iacomini M. Structure of a highly substituted β-xylan of the gum exudate of the palm *Livistona chinensis* (Chinese fan). Carbohydr Res. 2003 Sep 1;338(18):1843-50.

[33] Simas FF, Maurer-Menestrina J, Reis RA, Sassaki GL, Iacomini M, Gorin PAJ. Structure of the fucose-containing acidic heteroxylan from the gum exudate of *Syagrus romanzoffiana* (Queen palm). Carbohydr Polym. 2006 Jan 18;63(1):30-9.

[34] Simas FF, Gorin PAJ, Guerrini M, Naggi A, Sassaki GL, Delgobo CL, et al. Structure of a heteroxylan of gum exudate of the palm Scheelea phalerata (uricuri). Phytochemistry. 2004 Aug;65(16):2347-55.

[35] Simas-Tosin FF, Barraza RR, Maria-Ferreira D, Werner MFDP, Baggio CH, Wagner R, et al. Glucuronoarabinoxylan from coconut palm gum exudate: Chemical structure and gastroprotective effect. Carbohydr Polym [Internet]. 2014;107(1):65-71. Available from: http://dx.doi.org/10.1016/j.carbpol.2014.02.030

[36] Carey F. Química Orgânica-Vol. 1. 2009 [cited 2020 Aug 20]; Available from: https://books.google.com.br/books?hl=pt-BR&lr=&id=tNXT5dPulogC&oi=fnd&pg=PR8&dq=Química+Orgânica-Vol.+1&ots=OqriJqS1RB&sig=3U7aSanAhrnPGASMESwrsLaHduQ

[37] Van De Velde F, Knutsen SH, Usov AI, Rollema HS, Cerezo AS. 1H and 13C high resolution NMR spectroscopy of carrageenans: Application in research and industry. Trends Food Sci Technol. 2002 Mar 1;13(3):73-92.

[38] Bubb WA. NMR spectroscopy in the study of carbohydrates: Characterizing the structural complexity. Concepts Magn Reson Part A Bridg Educ Res. 2003 Sep;19(1):1-19.

[39] Homans SW. Oligosaccharide conformations: Application of NMR and energy calculations. Vol. 22, Progress in Nuclear Magnetic Resonance Spectroscopy. 1990. p. 55-81.

[40] Wormald MR, Petrescu AJ, Pao Y-L, Glithero A, Elliott T, Dwek RA. Conformational Studies of Oligosaccharides and Glycopeptides: Complementarity of NMR, X-ray Crystallography, and Molecular Modelling. ACS Publ [Internet]. 2002 [cited 2020 Aug 20];102(2):371-86. Available from: https://pubs.acs.org/doi/full/10.1021/cr990368i

[41] Duus J, Gotfredsen CH, Bock K. Carbohydrate structural determination by NMR spectroscopy: modern methods and limitations. Chem Rev. 2000 Dec;100(12):4589-614.

[42] Simas FF, Gorin PAJ, Wagner R, Sassaki GL, Bonkerner A, Iacomini M. Comparison of structure of gum exudate polysaccharides from the trunk and fruit of the peach tree (*Prunus persica*). Carbohydr Polym. 2008 Jan 24;71(2):218-28.

[43] Koenig J. Infrared and Raman spectroscopy of polymers. 2001 [cited 2020 Aug 20]; Available from: https://books.google.com.br/books?hl=pt-BR&lr=&id=6J6Vv3l6GvQC&oi=fnd&pg=PP6&dq=Infrared+and+Raman+spectroscopy+of+polymers&ots=Fhnf4uS9mY&sig=nzp316w9QRST3oWNatWTjaUv5w0

[44] Yates JT. Infrared Spectroscopy. In: Experimental Innovations in Surface Science [Internet]. Cham: Springer International Publishing; 2015 [cited 2020 Aug 20]. p. 321-32. Available from: http://link.springer.com/10.1007/978-3-319-17668-0_30

[45] technology BS-K encyclopedia of chemical, 2000 undefined. Infrared spectroscopy. Wiley Online Libr [Internet]. [cited 2020 Aug 20]; Available from: https://onlinelibrary.wiley.com/doi/abs/10.1002/0471238961.0914061810151405.a01.pub2

[46] Colapret A. Infrared Spectroscopy, University of Texas Organic Lecture Series (ch 12) [Internet]. University Lecture Notes Handout. John Wiley & Sons, Inc.; 2016 [cited 2020 Aug 20]. p. 19. Available from: http://colapret.cm.utexas.edu/courses/Chapter 12.pdf

[47] Barth A. Infrared spectroscopy of proteins. Vol. 1767, Biochimica et Biophysica Acta - Bioenergetics. 2007. p. 1073-101.

[48] Boulet JC, Williams P, Doco T. A Fourier transform infrared spectroscopy study of wine polysaccharides. Carbohydr Polym. 2007 May 1;69(1):79-85.

[49] Barth A, Zscherp C. What vibrations tell us about proteins. Vol. 35, Quarterly Reviews of Biophysics. 2002. p. 369-430.

[50] Manrique GD, Lajolo FM. FT-IR spectroscopy as a tool for measuring degree of methyl esterification in pectins isolated from ripening papaya fruit. Postharvest Biol Technol. 2002 May 1;25(1):99-107.

[51] Synytsya A, Čopíková J, Matějka P, Machovič V. Fourier transform Raman and infrared spectroscopy of pectins. Carbohydr Polym. 2003 Oct 1;54(1):97-106.

[52] Séné CFB, Mccann MC, Wilson RH, Crinter R. Fourier-Transform Raman and Fourier-Transform lnfrared Spectroscopy ' An lnvestigation of Five Higher Plant Cell Walls and Their Components [Internet]. Vol. 106, Plant Physiol. 1994 [cited 2020 Aug 20]. Available from: www.plantphysiol.org

[53] Kacuráková M, Capek P, Sasinková V, Wellner N, Ebringerová A. FT-IR study of plant cell wall model compounds: Pectic polysaccharides and hemicelluloses. Carbohydr Polym. 2000 Oct 1;43(2):195-203.

[54] Matsuhiro B. Vibrational spectroscopy of seaweed galactans. In: Fifteenth International Seaweed Symposium. Springer Netherlands; 1996. p. 481-9.

[55] Yuen SN, Choi SM, Phillips DL, Ma CY. Raman and FTIR spectroscopic study of carboxymethylated non-starch polysaccharides. Food Chem. 2009 Jun 1;114(3):1091-8.

[56] Quelemes P V, De Araújo AR, Plácido A, Delerue-Matos C, Maciel JS, Bessa LJ, et al. Quaternized cashew gum: An anti-staphylococcal and biocompatible cationic polymer for biotechnological applications [Internet]. Elsevier. 2017 [cited 2020 Aug 20]. Available from: https://www.sciencedirect.com/science/article/pii/S0144861716311869

[57] Espinosa-Andrews H, Sandoval-Castilla O, Vázquez-Torres H, Vernon-Carter EJ, Lobato-Calleros C. Determination of the gum Arabic-chitosan interactions by Fourier Transform Infrared Spectroscopy and characterization of the microstructure and rheological features of their coacervates. Carbohydr Polym. 2010 Feb 11;79(3):541-6.

[58] Tiwari A, Raj B. Reactions and mechanisms in thermal analysis of advanced materials. 2015 [cited 2020 Aug 20]; Available from: https://books.google.com.br/books?hl=pt-BR&lr=&id=bD5UCgAAQBAJ&oi=fnd&pg=PP1&dq=Reactions+and+mechanisms+in+thermal+analysis+of+advanced+materials&ots=Tv1OyWnnQJ&sig=vEAslXj7ME_cfTfVWV8ZeTAiZms

[59] Crompton T. Thermal methods of polymer analysis. 2013 [cited 2020

Aug 20]; Available from: https://books.google.com.br/books?hl=pt-BR&lr=&id=bQImDwAAQBAJ&oi=fnd&pg=PR1&dq=Thermal+methods+of+polymer+analysis+Crompton+T.&ots=W6D2Sh2ha8&sig=RjBmzzqzYT8M47Ko0OPJm9vG2o8

[60] Brown ME, Gallagher PK. Introduction to recent advances, techniques and applications of thermal analysis and calorimetry. In: Handbook of Thermal Analysis and Calorimetry. 2008. p. 1-12.

[61] Wendlandt W. Thermal methods of analysis. 1974 [cited 2020 Aug 20]; Available from: https://www.bcin.ca/bcin/detail.app?id=56419

[62] Šesták J. Thermophysical properties of solids: their measurements and theoretical thermal analysis. 1984;

[63] Turi E. Thermal characterization of polymeric materials. 2012 [cited 2020 Aug 20]; Available from: https://books.google.com.br/books?hl=pt-BR&lr=&id=F8UxsgyHtTMC&oi=fnd&pg=PP1&dq=Thermal+characterization+of+polymeric+materials&ots=P1b_rbyt2W&sig=eAFp40_tukp_Z198CpGTXwXr7pw

[64] Müller AJ, Michell RM. Differential scanning calorimetry of polymers. In: Polymer Morphology: Principles, Characterization, and Processing. wiley; 2016. p. 72-99.

[65] Pereira CG. Phase Transition in Foods [Internet]. Thermodynamics of Phase Equilibria in Food Engineering. Elsevier Inc.; 2019. 421-442 p. Available from: http://dx.doi.org/10.1016/B978-0-12-811556-5.00010-7

[66] Haines P. Thermal methods of analysis: principles, applications and problems. 2012 [cited 2020 Aug 20]; Available from: https://books.google.com.br/books?hl=pt-BR&lr=&id=CRntCAAAQBAJ&oi=fnd&pg=PA1&dq=Thermal+methods+of+analysis&ots=KdHeKs5Fjm&sig=JQ0TkY9kkTk47kV8suEv54Hqi34

[67] Dealy JM, Larson RG, Dealy JM, Larson RG. Structure and Rheology of Molten Polymers. In: Structure and Rheology of Molten Polymers. Carl Hanser Verlag GmbH & Co. KG; 2006. p. I–XIV.

[68] Chandran N, Sarathchandran C, Thomas S. Introduction to rheology. In: Rheology of Polymer Blends and Nanocomposites: Theory, Modelling and Applications. Elsevier; 2019. p. 1-17.

[69] Sanchez C, Renard D, Robert P, Schmitt C, Lefebvre J. Structure and rheological properties of acacia gum dispersions. Food Hydrocoll. 2002 May 1;16(3):257-67.

[70] Petrie C. Dynamics of polymeric liquids. Volume 1. Fluid mechanics (2nd ed). J Nonnewton Fluid Mech. 1988 Oct;30(1):93-5.

[71] Muñoz J, Rincón F, Carmen Alfaro M, Zapata I, de la Fuente J, Beltrán O, et al. Rheological properties and surface tension of *Acacia tortuosa* gum exudate aqueous dispersions. Carbohydr Polym. 2007 Sep 20;70(2):198-205.

[72] Shi Y, Li C, Zhang L, Huang T, Ma D, Tu Z cai, et al. Characterization and emulsifying properties of octenyl succinate anhydride modified *Acacia seyal* gum (gum arabic). Food Hydrocoll. 2017 Apr 1;65:10-6.

[73] Li W, Wang Y, Zhao H, He Z, Zeng M, Qin F, et al. Effects of soluble soy polysaccharides and gum arabic on the interfacial shear rheology of soy β-conglycinin at the air/water and oil/water interfaces. Food Hydrocoll. 2018 Mar 1;76:123-30.

[74] Singh A, Geveke DJ, Yadav MP. Improvement of rheological, thermal

and functional properties of tapioca starch by using gum arabic. LWT - Food Sci Technol. 2017 Jul 1;80:155-62.

[75] Chenlo F, Moreira R, Silva C. Rheological behaviour of aqueous systems of tragacanth and guar gums with storage time. J Food Eng. 2010 Jan;96(1):107-13.

[76] Hussain R, Singh A, Vatankhah H, Ramaswamy HS. Effects of locust bean gum on the structural and rheological properties of resistant corn starch. J Food Sci Technol. 2017 Mar 1;54(3):650-8.

[77] Chisté RC, Cardoso JM, da Silva DA, Da Pena RS. Hygroscopic behaviour of cassava flour from dry and water groups | Comportamento higroscópico das farinhas de mandioca tipos seca e d'água. Cienc Rural. 2015;45(8).

[78] Yogendrarajah P, Samapundo S, Devlieghere F, De Saeger S, De Meulenaer B. Moisture sorption isotherms and thermodynamic properties of whole black peppercorns (*Piper nigrum* L.). LWT - Food Sci Technol. 2015;64(1):177-88.

[79] Fasina OO. Thermophysical Properties of SweetPotato Puree at Freezing and Refrigeration Temperatures. Int J Food Prop [Internet]. 2005 [cited 2020 Aug 20];8(1):151-60. Available from: https://www.tandfonline.com/action/journalInformation?journalCode=ljfp20

[80] Rizvi S. Thermodynamic Properties of Foods in Dehydration. In: Engineering Properties of Foods, Fourth Edition. CRC Press; 2014. p. 359-436.

[81] Moreira R, Chenlo F, Torres MD, Vallejo N. Thermodynamic analysis of experimental sorption isotherms of loquat and quince fruits. J Food Eng. 2008 Oct 1;88(4):514-21.

[82] Slavutsky AM, Bertuzzi MA. Thermodynamic study of water sorption and water barrier properties of nanocomposite films based on brea gum. Appl Clay Sci. 2015 May 1;108:144-8.

[83] Mazza G, LeMaguer M. Water Sorption Properties of Yellow Globe Onion (*Allium cepa* L.). Can Inst Food Sci Technol J. 1978 Oct 1;11(4):189-93.

[84] Xiao Q, Tong Q. Thermodynamic properties of moisture sorption in pullulan-sodium alginate based edible films. Food Res Int. 2013 Dec 1;54(2):1605-12.

[85] Flory P. Principles of polymer chemistry. 1953 [cited 2020 Aug 20]; Available from: https://books.google.com.br/books?hl=pt-BR&lr=&id=CQ0EbEkT5R0C&oi=fnd&pg=PA3&dq=Principles+of+polymer+chemistry+flory&ots=16A05Rxuez&sig=ASzgyAPT6UrKF6lrlmcCLVqbgp4

[86] Venkataraju M, Gowda D, Rajesh K, Shivakumar H. Xanthan and Locust Bean Gum (from *Ceratonia siliqua*) Matrix Tablets for Oral Controlled Delivery of Metoprolol Tartrate. Curr Drug ther [Internet]. 2008 Jan 1 [cited 2020 Aug 21];3(1):70-7. Available from: http://www.eurekaselect.com/openurl/content.php?genre=article&issn=1574-8855&volume=3&issue=1&spage=70

[87] Fernández-Ferreiro A, Silva-Rodríguez J, Otero-Espinar FJ, González-Barcia M, Lamas MJ, Ruibal A, et al. In vivo eye surface residence determination by high-resolution scintigraphy of a novel ion-sensitive hydrogel based on gellan gum and kappa-carrageenan. Eur J Pharm Biopharm. 2017 May 1;114:317-23.

[88] Vilela JAP, Da Cunha RL. High acyl gellan as an emulsion stabilizer. Carbohydr Polym. 2016 Mar 30;139:115-24.

[89] Alves NO, Da Silva GT, Weber DM, Luchese C, Wilhelm EA, Fajardo AR.

Chitosan/poly(vinyl alcohol)/bovine bone powder biocomposites: A potential biomaterial for the treatment of atopic dermatitis-like skin lesions. Carbohydr Polym. 2016 Sep 5;148:115-24.

[90] Vashisth P, Nikhil K, Roy P, Pruthi PA, Singh RP, Pruthi V. A novel gellan-PVA nanofibrous scaffold for skin tissue regeneration: Fabrication and characterization. Carbohydr Polym. 2016 Jan 20;136:851-9.

[91] Anders R, Merkle HP. Evaluation of laminated muco-adhesive patches for buccal drug delivery. Int J Pharm. 1989;49(3):231-40.

[92] Chakravorty A, Barman G, Mukherjee S, Sa B. Effect of carboxymethylation on rheological and drug release characteristics of locust bean gum matrix tablets. Carbohydr Polym. 2016 Jun 25;144:50-8.

[93] Hamza SM, Supervisor - Mohammed El Mubarak Osman. Preparation of Mosquito Repellent Cream Using Gum Arabic and *Ocimum basilicum* L. Oil (Rehan). 2016 Nov 24 [cited 2020 Aug 20]; Available from: http://repository.sustech.edu//handle/123456789/15524

[94] Ghori MU, Alba K, Smith AM, Conway BR, Kontogiorgos V. Okra extracts in pharmaceutical and food applications. Food Hydrocoll. 2014 Dec 5;42(P3):342-7.

[95] Romano I, Ayadi F, Rizzello L, Summa M, Bertorelli R, Pompa PP, et al. Controlled antiseptic/eosin release from chitosan-based hydrogel modified fibrous substrates. Carbohydr Polym. 2015 Jun 27;131:306-14.

[96] Nishi KK, Antony M, Mohanan P V., Anilkumar T V., Loiseau PM, Jayakrishnan A. Amphotericin B-gum arabic conjugates: Synthesis, toxicity, bioavailability, and activities against Leishmania and fungi. Pharm Res [Internet]. 2007 Mar 20 [cited 2020 Aug 21];24(5):971-80. Available from: https://link.springer.com/article/10.1007/s11095-006-9222-z

[97] Ishurd O, Kennedy JF. The anti-cancer activity of polysaccharide prepared from Libyan dates (Phoenix dactylifera L.). Carbohydr Polym. 2005 Mar 15;59(4):531-5.

[98] Ahuja M, Kumar S, Kumar A. Evaluation of mucoadhesive potential of gum cordia, an anionic polysaccharide. Int J Biol Macromol. 2013 Apr;55:109-12.

[99] Sabale V, Paranjape A, Patel V. Isolation and characterization of jackfruit mucilage and its comparative evaluation as a mucoadhesive and controlled release component in buccal tablets. Int J Pharm Investig. 2012;2(2):61.

[100] Zhang Y, Zuo H, Xu F, Zhu K, Tan L, Dong W, et al. The digestion mechanism of jackfruit seed starch using improved extrusion cooking technology. Food Hydrocoll. 2021 Jan 1;110:106154.

[101] Gidley MJ, Lillford PJ, Rowlands DW, Lang P, Dentini M, Crescenzi V, et al. Structure and solution properties of tamarind-seed polysaccharide. Carbohydr Res. 1991 Jul 30;214(2):299-314.

[102] Mishra VK, Ooraikul B, Temelli F. Physical characterization and water sorption of freeze dried dulse *Palmaria palmata* powder. J Food Process Preserv. 1996;20(1):25-39.

[103] Ravindran G, Matia-Merino L. Starch-fenugreek (*Trigonella foenum-graecum* L.) polysaccharide interactions in pure and soup systems. Food Hydrocoll. 2009 May;23(3):1047-53.

[104] Yousefi S, Emam-Djomeh Z, Mousavi M, Kobarfard F, Zbicinski I. Developing spray-dried

powders containing anthocyanins of black raspberry juice encapsulated based on fenugreek gum. Adv Powder Technol. 2015 Mar 1;26(2):462-9.

[105] Nayak AK, Hasnain MS. Locust bean gum based multiple units for oral drug delivery. In: SpringerBriefs in Applied Sciences and Technology. Springer Verlag; 2019. p. 61-6.

[106] Ma Q, Du L, Wang L. Tara gum/ polyvinyl alcohol-based colorimetric NH3 indicator films incorporating curcumin for intelligent packaging. Sensors Actuators, B Chem. 2017;244:759-66.

[107] Üner M, Altinkurt T. Evaluation of honey locust (Gleditsia triacanthos Linn.) gum as sustaining material in tablet dosage forms. Farmaco. 2004 Jul;59(7):567-73.

[108] Singh R, Barreca D. Analysis of gums and mucilages. In: Recent Advances in Natural Products Analysis. Elsevier; 2020. p. 663-76.

[109] Buriti FCA, Dos Santos KMO, Sombra VG, Maciel JS, Teixeira Sá DMA, Salles HO, et al. Characterisation of partially hydrolysed galactomannan from *Caesalpinia pulcherrima* seeds as a potential dietary fibre. Food Hydrocoll. 2014 Mar 1;35:512-21.

[110] Saurabh CK, Gupta S, Bahadur J, Mazumder S, Variyar PS, Sharma A. Mechanical and barrier properties of guar gum based nano-composite films. Carbohydr Polym. 2015 Jun 25;124:77-84.

[111] Thanzami K, Malsawmtluangi C, Lalhlenmawia H, Seelan TV, Palanisamy S, Kandasamy R, et al. Characterization and in vitro antioxidant activity of Albizia stipulata Boiv. gum exudates. Int J Biol Macromol. 2015 Jul 4;80:231-9.

[112] Mahfoudhi N, Sessa M, Chouaibi M, Ferrari G, Donsì F, Hamdi S. Assessment of emulsifying ability of almond gum in comparison with gum arabic using response surface methodology. Food Hydrocoll. 2014 Jun 1;37:49-59.

[113] Yamassaki FT, Lenzi RM, Campestrini LH, Bovo F, Seyfried M, Soldera-Silva A, et al. Effect of the native polysaccharide of cashew-nut tree gum exudate on murine peritoneal macrophage modulatory activities. Carbohydr Polym [Internet]. 2015;125:241-8. Available from: http://dx.doi.org/10.1016/j.carbpol.2015.02.041

[114] Ayoubi Najafabadi A, Askari G, Mousavi M, Emamdjomeh Z. Novel Biodegradable Cast Film from Cherry Tree Gum, Development, Modification and Characterization. J Polym Environ [Internet]. 2017 Jun 1 [cited 2020 Aug 21];25(2):241-9. Available from: https://link.springer.com/article/10.1007/s10924-016-0803-6

[115] Umoren SA, Ebenso EE. Studies of the anti-corrosive effect of Raphia hookeri exudate gum-halide mixtures for aluminium corrosion in acidic medium. Pigment Resin Technol. 2008;37(3):173-82.

[116] Fardood ST, Ramazani A, Moradi S. Green synthesis of Ni–Cu–Mg ferrite nanoparticles using tragacanth gum and their use as an efficient catalyst for the synthesis of polyhydroquinoline derivatives. J Sol-Gel Sci Technol [Internet]. 2017 May 1 [cited 2020 Aug 21];82(2):432-9. Available from: https://link.springer.com/article/10.1007/s10971-017-4310-6

[117] Rathore HS, Sarubala M, Ramanathan G, Singaravelu S, Raja MD, Gupta S, et al. Fabrication of biomimetic porous novel sponge from gum kondagogu for wound dressing. Mater Lett. 2016 Aug 15;177:108-11.

[118] Rao M. Rheology of fluid, semisolid, and solid foods: principles and applications. 2013 [cited 2020

Aug 20]; Available from: https://books.google.com.br/books?hl=pt-BR&lr=&id=9-23BAAAQBAJ&oi=fnd&pg=PR5&dq=Rheology+of+fluid,+semisolid,+and+solid+foods:+principles+and+applications&ots=e3D8MMV_yK&sig=YzWpbFAt9qXI9-0gYNG47m0VECc

[119] Tomoda M, Gonda R, Shimizu N, Nakanishi S, Hikino H. A mucilage from *Hibiscus moscheutos* leaves. Phytochemistry. 1987 Jan 1;26(8):2297-300.

[120] Kaleemullah M, Jiyauddin K, Thiban E, Rasha S, Al-Dhalli S, Budiasih S, et al. Development and evaluation of Ketoprofen sustained release matrix tablet using Hibiscus rosa-sinensis leaves mucilage. Saudi Pharm J. 2017 Jul 1;25(5):770-9.

[121] Rutz JK, Borges CD, Zambiazi RC, Crizel-Cardozo MM, KuckLS,NoreñaCPZ.Microencapsulation of palm oil by complex coacervation for application in food systems. Food Chem. 2017 Apr 1;220:59-66.

[122] Alvarez-Manceñido F, Landin M, Lacik I, Martínez-Pacheco R. Konjac glucomannan and konjac glucomannan/xanthan gum mixtures as excipients for controlled drug delivery systems. Diffusion of small drugs. Int J Pharm. 2008 Feb 12;349(1-2):11-8.

[123] Davé V, McCarthy SP. Review of konjac glucomannan. Vol. 5, Journal of Environmental Polymer Degradation. 1997. p. 237-41.

[124] Jiménez-castellanos MR, Zia H, Rhodes CT. Mucoadhesive drug delivery systems. Drug Dev Ind Pharm [Internet]. 1993 [cited 2020 Aug 21];19(1-2):143-94. Available from: https://www.tandfonline.com/doi/abs/10.3109/03639049309038765

[125] Nejatian M, Abbasi S, Azarikia F. Gum Tragacanth: Structure, characteristics and applications in foods. Vol. 160, International Journal of Biological Macromolecules. Elsevier B.V.; 2020. p. 846-60.

[126] Zeng D, Xiao G, Xu Y, Zou B, Wu J, Yu Y. Protein and polyphenols involved in sediment formation in cloudy litchi juice. Food Sci Biotechnol [Internet]. 2019 Aug 1 [cited 2020 Nov 13];28(4):945-53. Available from: https://doi.org/10.1007/s10068-019-00567-y

[127] Wang L, Zhao S, Liao T, Shu X, Guo D, Huang Y, et al. Polysaccharide selection and mechanism for prevention of protein–polyphenol haze formation in beverages. J Food Sci [Internet]. 2020 Nov 20 [cited 2020 Nov 13];85(11):3776-85. Available from: https://onlinelibrary.wiley.com/doi/10.1111/1750-3841.15493

[128] Janani N, Zare EN, Salimi F, Makvandi P. Antibacterial tragacanth gum-based nanocomposite films carrying ascorbic acid antioxidant for bioactive food packaging. Carbohydr Polym. 2020 Nov 1;247:116678.

[129] Attaran SA, Hassan A, Wahit MU. Materials for food packaging applications based on bio-based polymer nanocomposites. Vol. 30, Journal of Thermoplastic Composite Materials. SAGE Publications Ltd; 2017. p. 143-73.

[130] Raj V, Bajpai A. Synthesis of hydrophobically modified guar gum film for packaging materials. In: Materials Today: Proceedings. Elsevier Ltd; 2020. p. 1132-42.

[131] Gheonea (Dima) I, Aprodu I, Cîrciumaru A, Râpeanu G, Bahrim GE, Stănciuc N. Microencapsulation of lycopene from tomatoes peels by complex coacervation and freeze-drying: Evidences on phytochemical profile, stability and food applications. J Food Eng. 2020 Jan 1;288:110166.

[132] Ribeiro JS, Veloso CM. Microencapsulation of natural dyes with biopolymers for application in food: A review. Vol. 112, Food Hydrocolloids. Elsevier B.V.; 2021. p. 106374.

[133] Yuan Y, Zhang X, Pan Z, Xue Q, Wu Y, Li Y, et al. Improving the properties of chitosan films by incorporating shellac nanoparticles. Food Hydrocoll. 2021 Jan 1;110:106164.

[134] Karaaslan M, Şengün F, Cansu Ü, Başyiğit B, Sağlam H, Karaaslan A. Gum arabic/maltodextrin microencapsulation confers peroxidation stability and antimicrobial ability to pepper seed oil. Food Chem. 2021 Feb 1;337:127748.

[135] Ma J, Xu C, Yu H, Feng Z, Yu W, Gu L, et al. Electro-encapsulation of probiotics in gum Arabic-pullulan blend nanofibres using electrospinning technology. Food Hydrocoll. 2021 Feb 1;111:106381.

[136] Maji B, Maiti S. Chemical modification of xanthan gum through graft copolymerization: Tailored properties and potential applications in drug delivery and wastewater treatment. Vol. 251, Carbohydrate Polymers. Elsevier Ltd; 2021. p. 117095.

Chapter 14

A Review on Effects of Pseudo Cereals Flour on Quality Properties of Biscuit, Cookies and Cake

Abu Saeid and Maruf Ahmed

Abstract

Gluten free products are currently highly demandable by those with different gluten intolerances. Pseudo cereals are a category of non-grass seeds used to manufacture various gluten free products, including bread, biscuits, cakes, and cookies. Pseudo cereal seeds contain high-quality proteins, rich quantities and unique characteristics of starch, vast amounts of micronutrients such as minerals, vitamins along with diverse bioactive compounds. This chapter is focused on other resentful research work on the characteristics of pseudo cereals seeds and pseudo cereals flour. It also reveals different effects of pseudo cereal flour on physical-chemicals properties of biscuit, cake, and cookie. We think that this study will have a significant influence on product developers and customers on the use of pseudo cereal seeds and pseudo cereals flour.

Keywords: pseudo cereals seeds, pseudo cereals flour, biscuits, cake, cookies

1. Introduction

Pseudo cereal is a category of non-greases that can be ground into flour and then used as cereals. Most of the pseudo cereals are amaranth (*Amaranthu* spp.), quinoa (*Chenopodium quinoa*), and buckwheat (*Fagopyrum esculentum* and *Fagopyrum tartaricum*). Pseudo cereals have high-quality proteins, rich in starch, minerals, vitamins, and bioactive compounds. Pseudo cereals could be the alternative option for developing gluten free food products for people suffering from various gluten intolerances. That is why interest in pseudo cereals has increased enormously since the turn of the century, and research has intensified [1].

Buckwheat (*Fagopyrume sculentum*) is nutritionally enriched due to its high amount of vitamin B_1 and B_2, proteins with significant amount of essential amino-acid. It is a rich source of flavonoids, phytosterols, soluble carbohydrates, and other substances like D-chiro-inositol, fagopyritols, or thiamine-binding proteins [2]. Buckwheat also contains a higher amount of rutin (quercetin-3-rutinoside) than other crops with significant antioxidant, anti-inflammatory, and anticarcinogenic properties. Food products made from buckwheat which have many different biological effects, include promoting intestinal microbiota and growth support of colonies of lactic acid bacteria in the gastrointestinal tract, inhibiting proteases' scavenging ability of free radical, glucose- and cholesterol-lowering effects [3].

IntechOpen

Amaranth (*Amaranthus* L.) has a higher amount of protein content (14–19%) than that of other traditional cereal crops with almost an acceptable proportion of essential amino acids which are rich in lysine and methionine [4]. The quality of starch content is low and there is no amylose (approximately 10% of starch, while amylopectin is 90%). Amaranth contains a good source of flavonoids and tocotrienols. Besides, lipid content is essential in amaranth seed included 6-7% of squalene compounds can reduce cancer risk, lipid metabolism control, anti-aging effects on the skin, and positive implications on the human immune system. Amaranth is also a rich source of magnesium, potassium, phosphorous and zinc minerals [5].

Quinoa (*Chenopodium quinoa* Willd) is rich in macronutrients, especially proteins which are analogous to the quality of the casein. Quinoa contains protein that is gluten free because of the lack of prolamins. Quinoas possess useful levels of lipids, such as monounsaturated fat (as oleic acid) and small quantities of omega-3 fatty acids such as alpha-linolenic acid, which are safe for health. It also contains higher fiber, mineral and carbohydrates such as polysaccharides that have a low glycemic index. Quinoa also is a pioneer in phytochemicals, antioxidants such as tocopherols and flavonoids such as quercetin and kaempferol [5–7]. Quinoa contains saponins and other valuable micro- and macronutrients [8].

Pseudo cereals have been widely recognized for many years due to their nutritional value by food scientists and food producers [1]. Many studies have investigated the use of pseudo cereals in the production of gluten free products rich in nutrients such as bread, pasta and confectionary products [8]. Flour, soup, cereal breakfast as well as beer are made using quinoa. Quinoa flour is used with wheat flour or corn meal to make biscuits, bread and processed food such as spaghetti [9]. On the other hand, buckwheat is used as a food supplement that can have a positive health impact and avoid foods being oxidized during processing. Buckwheat is recognized and recorded as part of wheat bread [10, 11]. Amaranth grain has high-quality protein, and flour is used in non-gluten formulations to obtain decent quality bread and cookies [12]. Hozova et al. [13] also proposed the use of amaranth flour to manufacture high-protein/energy-value gluten-free crackers and biscuits.

2. Types and characteristics of pseudo cereals

Pseudo cereals are non-grass species, eaten as grains, and having nutritional value is competitive or in most cases even better [14]. Amaranth (*Amaranthus L.*), quinoa (*Chenopodium quinoa*) and buckwheat (*Fagopyrum spp.)* are the main cereal-referential species (**Figure 1** and **Table 1**). Pseudo grain is not actual grain; it is dicotyledonous and pseudo grain is equivalent to true grain composition (**Table 2**). Hull (whether glossy or dull, brown, black or gray), testa, aleurone and starchy endosperm that occupy most of the seed (**Figure 2**) are the principal components of the buckwheat kernels (where there is perisperm absent). Pseudo cereal grain constitutes a healthy protein source, amino acids, vitamins, minerals and fatty acids [24]. In amaranth and quinoa seeds higher protein and fat levels are observed compared to common cereals due to high concentrations of amaranth and quinoa bran [25].

3. Nutritional compositions and application of pseudo cereals

3.1 Buckwheat

Buckwheat (BW), which belongs to the Polygonaceans families, is a typical Central and Eastern Europe and Asia crop. BW is widely used as a pseudo cereal

Figure 1.
Different types of leaves and seeds pseudo cereals: (a) quinoa; (b) buckwheat; (c) amaranth.

as an essential functional food. Grains of BW provides various valuable vitamins (B_1, B_2, B_6, and E) and minerals (P, Fe, Zn, K, and Mg) [26]. The biological value of BW proteins is high, but their digestibility is relatively low. The BW protein amino acids are well-balanced and rich in lysine [27] (**Table 2**). Besides, the contents of rutin, catechins, and polyphenols and their potential antioxidant activity are also of

	Cereals							Pseudo cereals			
Name	**Wheat**	**Rye**	**Barley**	**Oat**	**Rice**	**Corn**	**Sorghum**	**Millet**	**Amaranth**	**Quinoa**	**Buckwheat**
Class				*Monocotyledoneae*						*Dicotyledoneae*	
Order					*Poales*					*Caryophyllales*	
Family					*Poaceae*			*Amaranthaceae*			*Polygonaceae*
Subfamily		*Pooideae*		*Bambusoideae*				*Panicoideae*	*Amaranthoideae*	*Chenopodioideae*	*Polygonoideae*
Tribe	*Triticeae*		*Poeae*		*Oryzeae Andropogoneae*			*Paniceae Amarantheae*		*Chenopodieae*	*Fagopyreae*
Genus	*Trtiticum*	*Secale*	*Hordeum*	*Avena*	*Oryza*	*Zea*	*Sorghum*	*Pennisetum Panicum Setaria Paspalum*	*Amaranthus*	*Chemopodium*	*Fagoryrum*
Species	*T. aetivum, T.aestivum* spp., *spelta, T. durum*	*S. cereals*	*H. vulgare*	*A. sativa*	*O. sativa*	*Z. mays*	*S. bicolor*	*P. glaucum (pearl millte) P. meliaceum (proso millet) S. italica (foxitail millet) P. sacrobiculatum (kodo millet)*	*A. caudatus, A. cruentus, A. hypochondriacus*	*Ch. quinoa Wild, Ch. pallidicaule Aellen (kanigua/ kanihua/ canihua) Ch. nuttalia Safford*	*F. tartarum F. esculentum Moench*

Table 1.
Botanical classification of cereals and pseudo cereals [15, 16].

Parameters	Wheat flour	Quinoa flour	Buckwheat flour	Amaranth flour
Moisture (%)	10.12	9.97	9.72	6.21
Ash (%)	0.48	2.28	1.92	2.53
Protein (%)	11.2	12.4	11.5	17.0
Cellulose (%)	0.60	2.6	1.3	5.54
Fat (%)	0.70	5.3	2.4	4.90
Phytic acid/Phytate (mg/100 g)	196.0	1574	1335	237.75
Phytate phosphorus (mg/100 g)	38.35	443.88	376.48	—
Minerals (mg/100 g)				
Ca	20.05	30.5	17.96	1533.0
Cu	0.31	0.81	0.78	0.52
Fe	1.38	4.10	2.87	1.40
K	181.4	681.1	479.8	200.0
Mg	29.9	174.3	190.6	68.5
Mn	0.37	1.6	1.23	1.02
P	208.7	504.3	444.8	257.74
Zn	1.23	4.04	3.01	0.21

Table 2.
Chemical composition of wheat and pseudo cereals flours [17–21].

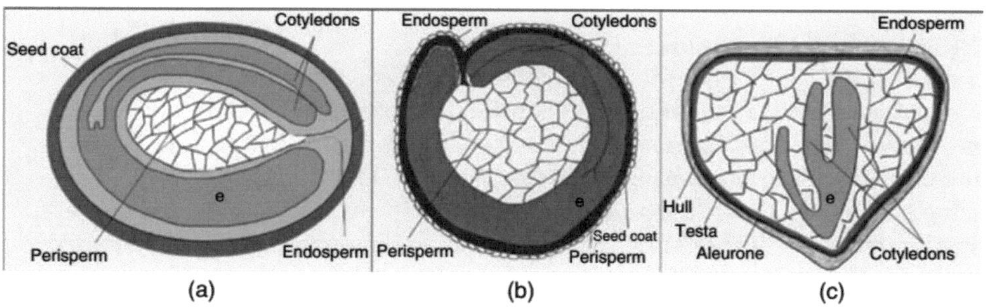

Figure 2.
Longitudinal sections of seed structures of the three major groups of pseudo cereals: (a) amaranth; (b) quinoa; (c) buckwheat [22, 23].

great significance [28]. BW does not contain gluten and can be used in celiac disease patients. Resistant starch content is a significant factor in BW 's preparation of low glycemic index food [28]. Some early studies also showed that BW flours and bran could prepare different bakery products, pasta-noodle, cookie, cake, crepe, breakfast cereal and soap formulations [11, 29, 30].

3.2 Amaranth

Amaranth (*Amaranthus* spp) is an indigenous pseudo grain domesticated in South America and has potential worldwide agronomic value [22]. Amaranth is a highly nutritious pseudo cereal that does not contain gluten. Due to its high protein content and its similar composition of essential amino acid [31], amaranth seed nutrient quality is superior to that of most cereal grains (**Table 2**). Additionally, lysine is abundant in amaranth grain, which is typically deficient in cereal grains. The overall content of minerals, especially calcium and magnesium, is generally more significant than observed in the grains [8]. Amaranth grain can be toasted, popped, extruded or milled into flour, thus eaten as various cereal products, including bread, cakes, muffins, cookies, dumplings, crepes, noodles and crackers. Some studies have found that amaranth grain could be used in gluten free goods such as crackers, maize tortillas, chips and bread [32, 33]. It is also used in foodstuffs to increase nutritional supplies that typically lack a celiac diet [34].

3.3 Quinoa

Quinoa belongs to the *chenopodiaceae* family, genus *chenopodium* is a pseudo cereal of the Andean regions of South America [35]. Quinoa seed can constitute a rich source for essential fatty acids, including linolenic (18:2n-6:52%) and linolenic (18:3n-6:40%) [36]. Quinoa is a good protein source, provided the nutritional profile of the material (12-18 g/100 g in dry weight), fiber, vitamins (such as C, E and B complex), calcium, magnesium, iron, copper and zinc have powerful content [35]. Several antinutrients, including saponins, phytic acid, tannins, protease inhibitors and others, have been found in quinoa [37]. The amino acid balance of quinoa is higher than the lysine based wheat and maize [38]. Quinoa flour can be added as a substitute for wheat flour as in bread (10–13% quinoa flour), noodles and pasta (30–40% quinoa flour), and sweet biscuits (60% quinoa flour) [39].

4. Impact of pseudo cereals flours on quality parameters of biscuits, cookies and snacks

4.1 Buckwheat biscuits and snacks

Increased market demand for composite flour based bakery products such as biscuits, snacks, or cereals has recently been noted. Buckwheat flour biscuits and corn snacks were produced by Wójtowicz et al. [40] and Baljeet et al. [41], using back wheat up to 20% and 30%, respectively. Buckwheat flour biscuits are nutritionally rich (**Table 3**) [41]. Filipčev et al. [43] have successfully integrated buckwheat flour (up to 50%) and made biscuits of ginger nuts and have found higher nutritional and bio-functional properties compared to control (**Table 4**). Biscuit thickness increased while the spread ratio and percent spread decreased due to decreased diameter with the inclusion of buckwheat flour (**Tables 3** and **4**) [41]. On the other hand, with the integration of buckwheat flour, biscuits' texture decreased in terms of (fracture strength) due to the decreasing gluten content in the buckwheat flours, the biscuits became soft with increasing BWF content [41]. The increase in the weight of biscuits was possibly due to buckwheat flour's ability to hold oil during baking [44]. Baljeet et al. [41] found that an improvement in the percentage of buckwheat flour in composite flour decreases the biscuit's sensory ranking.

4.2 Amaranth snacks and cookies

The protein and ash content of defatted amaranth snacks was higher, while the carbohydrate and lipid content were lower than maize snacks [45]. Compared to cookies made from wheat flour, the spread of cookies made with amaranth flour decreased significantly at 10-20%. Cookie thickness increased by up to 20% with the addition of amaranth flour, with marginal changes in thickness were observed.

Parameters	Control	Buckwheat		
		30%	40%	50%
Moisture content, g/100 g	9.47	8.94	8.56	8.70
Protein content, g/100 g	7.22	7.73	7.84	8.12
Fat content, g/100 g	8.12	8.43	8.45	8.59
Starch content, g/100 g	45.80	43.64	43.77	43.21
Total reducing sugar content	29.37	25.33	25.90	27.29
Total dietary fibers content, g/100 g	3.87	5.49	5.37	7.61
Zn mg/100 g	0.34	0.78	1.00	1.16
Cu, mg/100 g	0.07	0.15	0.16	0.22
Mn, mg/100 g	0.61	0.88	0.88	0.93
Fe, mg/100 g	0.57	1.39	1.66	1.72
Rutin content, mg/100 g	ND*	3.96	5.24	6.57
Quercetin content, mg/100 g	ND*	0.087	0.143	0.214
Total phenolics, mg GAE/100 g	157.06	196.35	202.58	238.92

ND-Not detected.*

Table 3.
Chemical composition of buckwheat supplemented ginger nut biscuits (dry basis) [42].

Parameters	Control	Buckwheat		
		30%	40%	50%
Antioxidative activity (AOA)	32.51	26.71	25.37	23.83
Reducing activity	29.36	28.46	28.00	26.2
DPPH scavenging activity	23.06	10.79	9.66	5.25
Chelating activity	11.24	11.84	11.35	11.21

Table 4.
Antioxidant potential profile (IC_{50}, mg/ml) of buckwheat supplemented ginger nut biscuits [42].

The breakage of the cookies decreased significantly with the addition of amaranth flour (**Tables 5** and **6**). Similar conclusions were observed in cookies from sorghum-wheat and oat-wheat mixtures. Hoseney and Rogers [47] reported that cookies' hardness is caused by protein and starch interactions with hydrogen bonding systems. It was noted that the diameter of composite cookies shows a rising trend along with the increasing degree of substitution of amaranth flour. This may be attributed to the lower viscosity of amaranth flour than wheat flour, as the viscosity decreases with the increase of the volume of amaranth flour and the spread rate. The results reveal that the spread ratio of the composite cookies displayed an increasing trend along with the increasing substitution level of amaranth flour. The decreased durability of amaranth flour replacement in cookies could be due to changes in gluten content. The delayed production of gluten matrices, which has contributed to an enormous decline in hardness, is also attributable to gluten reduction in cookie dough by the substitution of Amaranth flour [46]. Chauhan et al. [48] no major change was observed in color, aroma and texture of cookies made from mixtures with up to 100% amaranth flour. The sensory score for the taste decreased after addition of amaranth flour of 60%. This may be attributed to the bitter aftertaste of the amaranth flour. The overall acceptability score indicated that the cookies prepared up to 60% amaranth flour

Amaranth flour (%)	Weight (g)	Spread "W" (mm)	Thickness "T" (mm)	Spread ratio "W/T"	Breaking strength (kg)
0	19.46c ± 0.11	85.6a ± 0.66	10.9d ± 0.08	7.82a ± 0.03	4.935a ± 0.19
10	18.90d ± 0.21	81.30b ± 0.55	12.2a ± 0.08	6.69b ± 0.03	4.990a ± 0.28
20	19.72b ± 0.23	79.0bc ± 0.81	12.1ab ± 0.09	6.52c ± 0.14	4.887a ± 0.21
25	19.90a ± 0.13	77.6 cd ± 0.72	11.9b ± 0.11	6.52c ± 0.09	4.021bc ± 0.22
30	19.70b ± 0.18	75.9 dc ± 0.81	11.6bc ± 0.09	6.52c ± 0.10	3.948c ± 0.25
35	19.72b ± 0.15	75.0 dc ± 0.95	11.5c ± 0.13	6.52c ± 0.04	3.0293d ± 0.32
SEM (±)	0.10	0.15	0.13	0.08	0.12
Df	18	18	18	18	54

Note: Values for a particular column differ significantly when followed by different letters (p < 0.05); SEM, standard error of meant at 30 degrees of freedom.

Table 5.
Effect of replacement of wheat flour with amaranth flour on the sensory characteristic of cookies [46].

Amaranth flour (%)	Surface color (10)	Surface cracking (10)	Texture (10)	Mouthfeel (10)	Flavor (10)	Overall quality (50)
0	8.0c	7.8d	8.5b	8.5b	8.4c	40.7d
10	8.0c	8.0c	8.2c	8.5b	8.5b	41.2c
20	8.5b	8.2b	8.1 cd	8.4bc	8.5b	41.7c
25	8.8a	8.5a	8.6a	8.6a	8.7a	43.2a
30	8.9a	8.4a	8.5a	8.5b	8.6ab	42.9ab
35	8.7a	8.5a	8.5a	8.6ab	8.6b	42.4b
SEM (±)	0.10	0.12	0.11	0.14	0.14	0.23

Note: Values for a particular column differ significantly when followed by different letters (p < 0.05); SEM, standard error of meant at 30 degrees of freedom.

Table 6.
Effect of replacement of wheat flour with amaranth flour on the physical characteristic of cookies [46].

had most acceptable sensory attributes. This was against Sindhuja et al. [46], which showed that cookies with 25% amaranth flour were most acceptable for panelists. It has been reported that, no significant difference was demonstrated in color, smell, texture of biscuits made from 20, 30, 40 and 50% of amaranth flour but biscuit made from 40 and 50% of amaranth flour had significantly higher value than wheat flour (control). Overall acceptability score showed that biscuit made from maximum 40% amaranth flour best good sensory attributes [49].

4.3 Quinoa cookies and biscuits

The Demir and Kılınç study [50] that cookie samples have significantly increased ash, crude protein and crude fats ($p < 0.05$) with the addition of quinoa meal. In cookies made with different amounts of quinoa flour, the meaningful effect ($p < 0.05$) for the total content of K, Mg, Ca, Fe and Zn has been observed. Calcium, magnesium, iron and zinc are usually greater in quinoa than ordinary cereals, and their iron contents are very high [51, 52]. The use of quinoa flour was stated to lead to a slight increase in product thickness, but the cookie samples' spread ratio and diameter decreased [50]. When quinoa flour added, the hardness of cookies increased by up to 30% [53]. The sensory characteristics of cookie samples were influenced by quinoa flour. Added quinoa flour had statistically significant color, taste, crispness and total acceptability except odor ratings. Biscuits made of 100% quinoa flour ($p < 0.05$) vary considerably from the controls.

5. Conclusion

Pseudo cereal is a house of high-quality proteins with essential amino acids. It can be used to formulate gluten free food items as an alternative to wheat proteins in subjects suffering from celiac disease, due to the absence of gluten. In addition to the excellent nutrient profile, pseudo cereals are promising sources of phytochemical substances with significant health-promoting properties. Today pseudo cereals like buckwheat, quinoa and amaranth are incorporated successfully in bakery items such as biscuits, cookies, breads and snacks. Up to 50% buckwheat flours were used to produce nutritionally rich ginger-based biscuits with wheat flour. There was no loss of customer acceptance for the 60% amaranth flour used in the cookies production. Besides, the hardness of cookies increased by up to 30% of quinoa meal. This chapter also highlighted the actual color, taste, texture, and nutritional properties of pseudo cereal flour on biscuits, cookies, and the cake quality. Therefore, we assume that this study would considerably affect developers and customers and extensive understanding of pseudo cereal seeds and pseudo cereal flour.

Author details

Abu Saeid[1] and Maruf Ahmed[2]*

1 Department of Food Engineering, NPI University of Bangladesh, Manikganj, Bangladesh

2 Department of Food Processing and Preservation, Hajee Mohammad Danesh Science and Technology University, Dinajpur, Bangladesh

*Address all correspondence to: maruf@hstu.ac.bd; maruffpp@gmail.com

References

[1] Fabio AD, Origin PG. Production and utilization of pseudocereals. In: Haros CM, Schoenlechne R, editors. Pseudocereals Chemistry and Technology. Willey. Blackwell; 2017. pp. 1-27. DOI: 10.1002/9781118938256.ch1

[2] Przybylski R. Gruczynska E. a review of nutritional and nutraceutical components of Buckwheat. The European journal of plant science and. Biotechnology. 2009;**3**:10-22

[3] Giménez-Bastida JA, Zieliński H. Buckwheat as a functional food and its effects on health. Journal of Agricultural and Food Chemistry. 2015;**63**:7896-7913. DOI: 10.1021/acs.jafc.5b02498

[4] Narwade S, Pinto S. Amaranth - a functional food. Concepts of Dairy and Veterinary Sciences. 2018;**1**:1-6. DOI: 10.32474/CDVS.2018.01.000112

[5] Nascimento AC, Mota C, Coelho I, Gueifão S, Santos M, Matos AS, et al. Characterisation of nutrient profile of quinoa (Chenopodium quinoa), amaranth (Amaranthus caudatus), and purple corn (*Zea mays* L.) consumed in the north of Argentina: Proximates, minerals and trace elements. Food Chemistry. 2014;**148**:420-426. DOI: 10.1016/j.foodchem.2013.09.155

[6] Nowak V, Du J, Ruth CU. Assessment of the nutritional composition of quinoa (Chenopodium quinoa Willd.). Food Chemistry. 2015;**193**:47-54. DOI: 10.1016/j.foodchem.2015.02.111

[7] Wang S, Zhu F. Formulation and quality attributes of quinoa food products. Food and Bioprocess Technology. 2016;**9**:49-68. DOI: 10.1007/s11947-015-1584-y

[8] Alvarez-Jubete L, Auty M, Arendt EK, Gallagher E. Baking properties and microstructure of pseudo cereal flours in gluten free bread formulations. European Food Research and Technology. 2010;**230**:437-445. DOI: 10.1007/s00217-009-1184-z

[9] Bhargava A, Shukla S, Ohri D. Chenopodium quinoa–an Indian perspective. Industrial Crops and Products. 2006;**23**:73-87. DOI: 10.1016/j.indcrop.2005.04.002

[10] Fujarczuk M, Żmijewski M. Wheat bread quality depending on the addition of bran derived from various buckwheat varieties. Żywność Nauka Technologia Jakość. 2009;**6**:91-101

[11] Lin L, Liu H, Yu Y, Lin S, Mau J. Quality and antioxidant property of buckwheat enhanced wheat bread. Food Chemistry. 2009;**112**:987-991. DOI: 10.1016/j.foodchem.2008.07.022

[12] de la Barca AMC, Rojas-Martínez ME, Islas-Rubio AR, Cabrera-Chávez F. Gluten-free breads and cookies of raw and popped Amaranth flours with attractive technological and nutritional qualities. Plant Foods for Human Nutrition. 2010;**65**:241-246. DOI: 10.1007/s11130-010-0187-z

[13] Hozova´ B, Buchtova´ V, Dodok L, Zemanovic´ J. Microbiological, nutritional and sensory aspects of stored amaranth biscuits and amaranth crackers. Die Nahrung. 1997;**41**:151-158. DOI: 10.1002/food.19970410308

[14] Das S. Pseudo cereals: An efficient food supplement. In: Amaranthus: A Promising Crop of Future. Springer; 2016. pp. 5-11. DOI: 10.1007/978-981-10-1469-7_2

[15] Berghofer E, Schoenlechner R. Pseudocereals – An Overview. Department of Food Science and Technology: University of Natural Resources and Applied Life Sciences, Vienna; 2007

[16] Moreno ML, Comino I, Sousa C. Alternative grains as potential raw material for gluten–free food development in the diet of celiac and gluten–sensitive patients. Austin Journal of Nutrition and Food Sciences. 2014;**2**:1016.1-1016.9

[17] Bilgiçli N, İbanoğlu Ş. Effect of pseudo cereal flours on some physical, chemical and sensory properties of bread. Journal of Food Science and Technology. 2015;**52**:7525-7529. DOI: 10.1007/s13197-015-1770-y

[18] Shevkani K, Singh N, Kaur A, Rana JC. Physicochemical, pasting, and functional properties of Amaranth seed flours: Effects of lipids removal. Journal of Food Science. 2014;**79**:C1271-C1277. DOI: 10.1111/1750-3841.12493

[19] Vitali D, Amidžić Klarić D, Dragojević IV. Nutritional and functional properties of certain gluten-free raw materials. Czech Journal of Food Science. 2010;**28**:495-505. DOI: 10.17221/253/2009-CJFS

[20] Emire SA, Arega M. Value added product development and quality characterization of amaranth (*Amaranthus caudatus* L.) grown in East Africa. African. Journal of Food Science and Technology. 2012;**3**:129-141

[21] Escudero NL, Albarracín G, Fernández S, De Arellano LM, Mucciarelli S. Nutrient and antinutrient composition of Amaranthus muricatus. Plant Foods for Human Nutrition. 1999;**54**:327-336. DOI: 10.1023/a:1008149721435

[22] Valcárcel-yamani BV, da Silva Lannes SC. Applications of quinoa (Chenopodium Quinoa Willd.) and Amaranth (Amaranthus Spp.) and their influence in the nutritional value of cereal based foods. Food and. Public Health. 2012;**2**:265-275

[23] Prego I, Maldonado S, Otegui M. Seed structure and localization of reserves in Chenopodium quinoa. Annals of Botany. 1998;**82**:481-488. DOI: 10.1006/anbo.1998.0704

[24] Alemayehu FR, Bendevis MA, Jacobsen SE. The potential for utilizing the seed crop amaranth (Amaranthus spp.) in East Africa as an alternative crop to support food security and climate change mitigation. Journal of Agronomy and Crop Science. 2015;**201**:321-329. DOI: 10.1111/jac.12108

[25] Bressani R. Amaranth. In: Caballero B, editor. Encyclopedia of Food Sciences and Nutrition. 2nd ed. Oxford: Academic Press; 2003. pp. 166-173. DOI: 10.1016/B0-12-227055-X/00036-5

[26] Wijngaard HH, Buckwheat AEK. Cereal Chemistry. 2006;**83**:391-401. DOI: 10.1094/CC-83-0391

[27] Pomeranz Y, Robbıns GS. Amino acid composition of buckwheat. Journal of Agricultural and Food Chemistry. 1972;**20**:270-274. DOI: 10.1021/jf60180a029

[28] Fabjan N, Rode J, Kosir JI, Wang Z, Zhang Z, Kreft I. Tartary buckwheat (Fagopyrum tataricum Gaertn.) as a source of dietary rutin and quercitrin. Journal of Agricultural and Food Chemistry. 2003;**51**:6542-6455. DOI: 10.1021/jf034543e

[29] Bonafaccia G, Kreft I. Technological and qualitative characteristics of food products made with buckwheat. Fagopyrum. 1994;**14**:35-42

[30] Bilgicli N. Effect of buckwheat flour on chemical and functional properties of tarhana. Food Science and Technology. 2009;**42**:514-518. DOI: 10.1016/j.lwt.2008.09.006

[31] Oszvald M, Tamás C, Rakszegi M, Tömösközi S, Békés F, Tamás L. Effects

of incorporated amaranth albumins on the functional properties of wheat dough. Journal of the Science of Food and Agriculture. 2009;**89**:882-889. DOI: 10.1002/jsfa.3528

[32] Rai S, Kaur A, Chopra CS. Gluten-free products for celiac susceptible people. Frontiers in Nutrition. 2018;**5**:1-23. DOI: 10.3389/fnut.2018.00116

[33] Gebreil SY, Ali MIK, Mousa EAM. Utilization of Amaranth flour in preparation of high nutritional value bakery products. Food and Nutrition Sciences. 2020;**11**:336-354. DOI: 10.4236/fns.2020.115025

[34] Kupper C. Dietary guidelines and implementation for celiac disease. Gastroenterology. 2005;**128**:121-127. DOI: 10.1053/j.gastro.2005.02.024

[35] Valencia-Chamorro SA. Quinoa. In: Caballero B, editor. Encyclopedia of Food Science and Nutrition. Amsterdam: Academic Press; 2003. pp. 4895-4902

[36] Ogungbenle HN. Nutritional evaluation and functional properties of quinoa (Chenopodium quinoa) flour. International Journal of Food Science and Nutrition. 2003;**54**:153-158. DOI: 10.1080/0963748031000084106

[37] Vega-G'alvez A, Miranda M, Vergara J, Uribe E, Puenteb L, Mart'ınezc EA. Nutrition facts and functional potential of quinoa (Chenopodium quinoa willd.), an ancient Andean grain: A review. Journal of the Science of Food and Agriculture. 2010;**90**:2541-2547. DOI: 10.1002/jsfa.4158

[38] Rodriguez-Sandoval E, Sandoval G, Cortes-Rodríguez M. Effect of quinoa and potato flours on the Thermo mechanical and bread making properties of wheat flour. Brazilian Journal of Chemical Engineering. 2012;**29**:503-510. DOI: 10.1590/S0104-66322012000300007

[39] Sukeerthi SS, Singh DJK. Formulation and nutritional analysis of biscuits made from quinoa. flax seed and brown rice. International Journal of Food Science and Nutrition. 2017;**2**:27-30

[40] Wójtowicz A, Kolasa A, Mościcki L. Influence of buckwheat addition on physical properties, texture and sensory characteristics of extruded corn snacks. Polish Journal of Food and Nutrition Science. 2013;**63**:239-244. DOI: 10.2478/v10222-012-0076-2

[41] Baljeet SY, Ritika BY, Roshan LY. Studies on functional properties and incorporation of buckwheat flour for biscuit making. International Food Research Journal. 2010;**17**:1067-1076

[42] Bojana F, Olivera Š. Marijana Š, Ivana S, Pavle J, Mladenka P, Marija B. Feasibility of use of buckwheat flour as an ingredient in ginger nut biscuit formulation. Food Chemistry. 2011;**125**:164-170. DOI: 10.1016/j.foodchem.2010.08.055

[43] Filipčev B, Šimurina O, Sakač M, Sedej I, Jovanov P, Pestorić M, et al. Feasibility of use of buckwheat flour as an ingredient in ginger nut biscuit formulation. Food Chemistry. 2011;**125**:164-170. DOI: 10.1016/j.foodchem.2010.08.055

[44] Rufeng N, Enqi L, Chuangji C, Jiangping Z. A study of the production of healthy biscuit made with tartary buckwheat grown in North China. Current advances in buckwheat research. 1995:861-865

[45] Chávez-jáuregu RN, Santos RD, Macedo A, Chacra APM, Martinez TL, Arêas JAG. Effects of defatted amaranth (Amaranthus caudatusL.) snacks on lipid metabolism of patients with moderate hypercholesterolemia. Ciência e Tecnologia de Alimentos. 2008;**30**:1007-1010. DOI: 10.1590/S0101-20612010000400026

[46] Sindhuja A, Sudha ML, Rahim A. Effect of incorporation of amaranth flour on the quality of cookies. European Food Research Technology. 2005;**221**:597-601. DOI: 10.1007/s00217-005-0039-5

[47] Hoseney RC, Rogers DE. Mechanism of sugar functionality in cookies. In: Faridi H, editor. The Science of Cookie and Cracker Production. 1st Ed. American Association of Cereal Chemists. 1994. pp. 203-225

[48] Chauhan A, Saxena DC, Singh S. Physical, textural, and sensory characteristics of wheat and amaranth flour blend cookies. Cogent Food and Agriculture. 2016;**2**:1-8. DOI: 10.1080/23311932.2015.1125773

[49] Karki R, Mishra A, Ojha P, Subedi U. Comparative study on the sensory quality of prepared biscuit and cake from Amaranthus and sorghum. Journal of Food Science and Technology Nepal. 2016;**9**:79-84

[50] Demir MK, Kılınç M. Utilization of quinoa flour in cookie production. International Food Research Journal. 2017;**24**:2394-2401

[51] Jancurová M, Minarovicová L, Dandar A. Quinoa–a review. Czech Journal of Food Sciences. 2009;**27**:71-79

[52] Demir MK. Use of quinoa flour in the production of gluten-free tarhana. Food Science and Technology Research. 2014;**20**:1087-1092. DOI: 10.3136/fstr.20.1087

[53] Brito IL, de Souza EL, Felex SS, Madruga MS, Yamashita F, Magnani M. Nutritional and sensory characteristics of gluten-free quinoa (Chenopodium quinoa Willd)-based cookies development using an experimental mixture design. Journal of Food Science and Technology. 2015;**52**:5866-5873. DOI: 10.1007/s13197-014-1659-1

Chapter 15

Valorization Options of Strawberry Extrudate Agro-Waste. A Review

Juan Cubero-Cardoso, Antonio Serrano, Ángeles Trujillo-Reyes, Denys K. Villa-Gómez, Rafael Borja and Fernando G. Fermoso

Abstract

This review summarizes and critically analyzes the different types of potential valorization options for strawberry extrudate in order to have a broader overview of the potential management of this waste. Animal feed is commonly used as a management option for the strawberry extrudate; however, most of the strawberry extrudate is disposed in landfills. Strawberry extrudate contains different bioactive compounds that encourage the use of an alternative management approach than landfilled. The present review offers a complete comparative, including the advantages and drawbacks of each reviewed technique, to facilitate the selection of the most suitable technology for the different valorization scenarios. This review has been structured in three sections: 1. Composition of the strawberry extrudate and strawberry especially focused on their content in bioactive compounds. 2. The different techniques of extraction and purification of bioactive compounds. 3. The handling and management of the resulting biomass after the extraction process of bioactive compounds.

Keywords: strawberry extrudate, bioactive compounds, bioproducts, extraction techniques, purification

1. Introduction

In 2016, 8 million tons of strawberry were produced in the world with a value of agricultural gross production of 17,739 million US$ [1]. Besides its market as fresh product, strawberry is also used to produce many types of by-products, due to its peculiar flavor and aroma. Strawberry by-products are mainly formulated from a strawberry concentrate. The most common technology to obtain the strawberry concentrate is by extrusion. Strawberries are extruded by twin-screws up to several sieves with different mesh sizes. The sieves retain a residual fraction formed by the fibrous part and the achenes, named strawberry extrudate, which accounts about 7% of the manufactured strawberry [2].

Animal feed is commonly used as a management option for the strawberry extrudate, however, most of the strawberry extrudate is disposed in landfills, contributing to greenhouse emissions due to its high organic load [3]. Alternatives for strawberry extrudate management are required to avoid severe environmental impacts that cause landfills, such as negative effects on agricultural soil quality, polluting of aquatic ecosystems and atmospheric contamination [4].

Similar to strawberry, strawberry extrudate contains substances of high interest such as bioactive compounds. Some of these bioactive compounds have beneficial health effects on cardiovascular, neurological or cancerous disorders [5]. Due to their health benefits, bioactive compounds have an economic interest for different commercial sectors, such as the pharmaceutical, food and chemical industries [6]. Added to bioactive compounds, strawberry extrudate could be used to obtain other types of resources such as bioenergy [7, 8]. It is also well known the high phenolic composition in the achenes and in the pulp of the strawberry [9].

A general biorefinery scheme as a management option for the strawberry extrudate should look for synergies between unitary processes of extraction of bioactive compounds, purification and the management of the final biomass of the strawberry extrudate after extraction (**Figure 1**).

The extraction of bioactive compounds in agro-waste materials, such as the strawberry extrudate can be performed through various extraction techniques [10]. The main objective at this step consists of solubilizing the compounds of interest, with less possible impurities and making it an economically profitable technique. Extraction techniques in literature can be clustered into two groups: conventional extraction techniques, e.g. hydrothermal treatments, which are widely used at lab and full scale [11], and in recent years, more innovative techniques, e.g. enzyme assisted extraction. Additionally, combined extraction techniques between conventional and innovative techniques are being carried out to achieve high extraction yield [12]. All these extraction techniques will be revised and analyzed in the present chapter.

Any of these extraction techniques usually generate a liquid phase, with the bioactive compounds of interest, and a solid phase with a high amount of organic matter. After purification process of the liquid phase, a new liquid phase remains without the extracted bioactive compounds. Therefore, just the recovery of compounds of interest from the strawberry extrudate does not solve the problem of stabilization of the biomass of the strawberry extrudate and the use of a subsequent

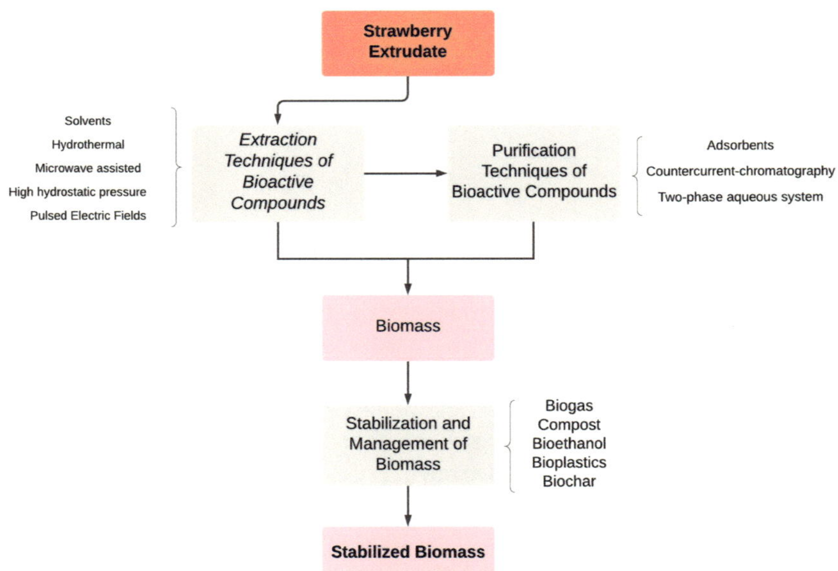

Figure 1.
General scheme of a biorefinery approach as a valorisation option for strawberry extrudate.

treatment is necessary for its stabilization [13]. The liquid phase after purification and the solid phase must undergo a new treatment for stabilization. In addition, extraction and purification processes consume energy which should be valued. The main options for assessing and stabilizing biomass after the extraction and purification process of the bioactive compounds should be focused on obtaining bioenergy and other bioproducts of interest [14].

The present chapter aims to summarize the bioactive compounds present in strawberries, to summarizes and critically analyzes the different extraction and purification techniques for the recovery of these bioactive compounds, as well as the different options for the management and stabilization of the strawberry extrudate after the extraction process.

2. Bioactive compounds in strawberry extrudate and strawberries

2.1 Nutrients

Strawberry extrudate presents similar nutrients composition than strawberry [15]. The strawberry has high concentration of dietary fibrous (2 g fibrous/100 g raw strawberry), such as lignin, hemicellulose, cellulose, and pectin, containing small amounts of protein (0.4–0.5 g protein/100 g raw strawberry) and fat (0.1 g fat/100 g raw strawberry) [16, 17].

The strawberry contains high concentrations of vitamin C, contributing to 24% to the antioxidant capacity of strawberries [16]. The recommended daily intake of vitamins (100–150 mg/day) can be satisfied with an average of 100 g of strawberries per day [18]. Furthermore, strawberry is a source of many other vitamins in smaller amounts, such as vitamin E, vitamin A, vitamin B6, vitamin K, thiamine, riboflavin, folate acid, and niacin (0.01–0.4 g vitamin/100 g raw strawberry) [5, 19]. Strawberry is also rich in manganese, potassium, and a good source of iodine, magnesium, copper, iron, and phosphorus [5, 16].

The sugar composition of strawberries varies with the degree of maturity of the fruit [20], being glucose, fructose, and sucrose the main sugars in strawberries [5]. Sugars in strawberries are involved in the taste of the fruit and are responsible for the caloric value of the strawberries. Organic fatty acids such as citric acid, malic acid, succinic acid, tartaric acid, oxalic acid, and fumaric acid are ones of the response of the taste, texture, pH, and color of the strawberry, and can alter the sensory quality of this fruit [6].

2.2 Phytochemicals

Figure 2 shows a general scheme for the classification of phytochemical compounds that can be founded in the strawberry extrudate. Phytochemicals are widely studied, mainly due to the extensive types of compounds that have potential biological benefits in humans. The main phytochemicals in strawberries are the flavonoids, followed by the hydrolysable tannins and the phenolic acids and, as minor constituents, the condensed tannins [5].

2.2.1 Flavonoids

Flavonoids are divided into two groups, i.e. anthocyanins, and anthoxanthines. Anthoxanthines, in turn, are grouped into five subclasses, i.e. flavones, flavonols, flavanones, flavanols, and isoflavones [21]. The three main classes of flavonoids in strawberries are anthocyanis, flavonols, and flavanols [22].

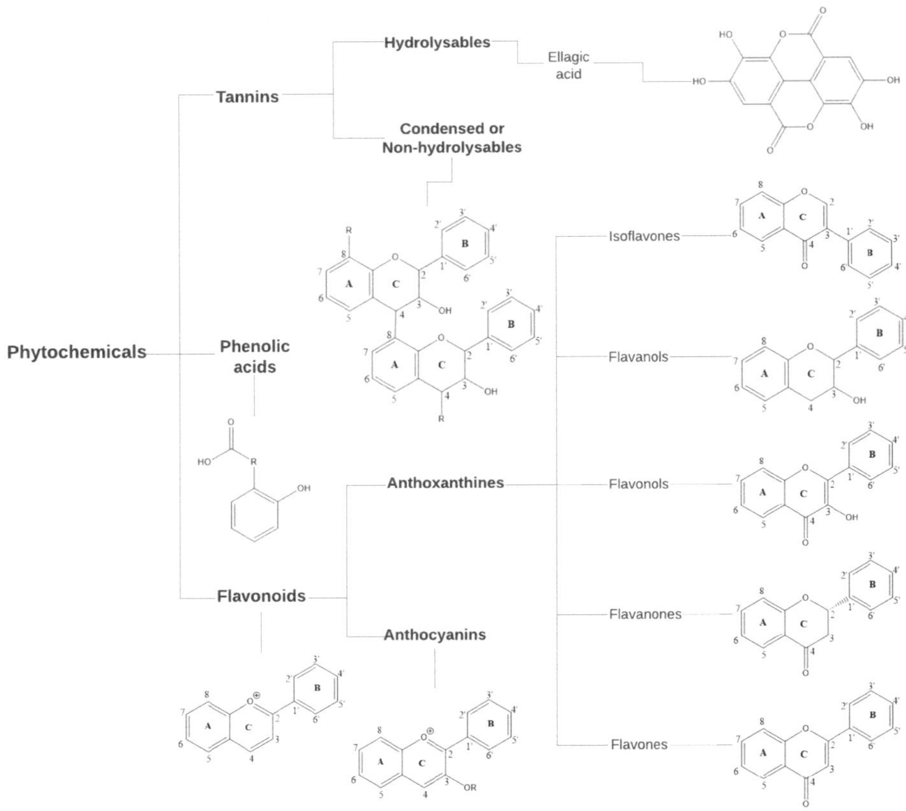

Figure 2.
General scheme of phytochemicals contained in strawberries.

The most relevant flavonoids present in strawberries are the anthocyanins due to their high concentration, approximately 20–47 mg/100 g raw strawberry [16]. More than 25 different pigments of anthocyanins have been described in the different varieties of strawberries [5]. Anthocyanins are responsible for the red color in strawberries [16]. The most important anthocyanins of the strawberry belong to the family of pelargonidin aglycones and cyanidin aglycones [23–25]. According to several studies, pelargonidin-3-glucoside is the dominant anthocyanin in strawberries [16, 24, 26–28]. The interest in anthocyanins have recently increased because of its pharmacological and therapeutic properties [5]. Anthocyanins have shown to positive effect toward reduction of coronary diseases, anticancer, antitumor, anti-inflammatory and anti-diabetic effects; as well as improving visual acuity and cognitive behavior [29]. These therapeutic effects of anthocyanins are connected to their high antioxidant activity [29]. In addition, anthocyanins can be used as a pigment in the food industry [29].

The second most important group of flavonoids in strawberries are flavonols, with approximately 1.5–3.4 mg/100 g raw strawberry [5, 16, 30, 31]. The most important flavonols of the strawberry belong to the family of quercetin and kaempferol, being the quercetin-derivatives the most abundant flavonols in strawberries [5, 17, 25–27]. Quercetin, in particular, is a potent antioxidant, cytoprotective, and anti-inflammatory [30].

Finally, the third group of flavonoids in strawberries are flavanols. Flavanols are the only class of flavonoids that do not naturally occur as glycosides. They are found in strawberries as monomeric compounds, such as catechins, and in polymeric form, which are called condensed or non-hydrolysable tannins [5, 25]. These compounds can be difficult to measure in the strawberry because they are usually presented as part of a complex mixture of phenolic substances. Because of this, the amount of catechins present is sometimes overestimated [31]. At low concentrations flavanols particularly the catechins, are used as sweetening and/or flavoring additives. These flavonols improve taste and sweetness but are not substitutes for sweeteners and flavorings as they do not have taste and are a little astringent. Some authors have pointed out that their role is to make the receptors in the mouth more sensitive to sweeteners, thus lowering the levels of the sweeteners and flavorings used [21, 32].

2.2.2 Tannins

Tannins are classified into two groups: non-hydrolysable or condensed tannins and hydrolysable tannins (**Figure 2**). The condensed tannins are also called proanthocyanins, and are bound to the flavanols [33]. The content of condensed tannins in strawberries is approximately 54–163 mg/100 g raw strawberry [16]. In strawberries, the most relevant condensed tannins are procyanidins from catechin and its polymers. Condensed tannins are commonly found in the pulp of strawberries and achenes [5]. Due to the variety of physiological activities, they have been reported to possess, directly and indirectly, antioxidant, antimicrobial, anti-allergic and antihypertensive properties, as well as to inhibit the activities of some enzymes and physiological receptors [34].

The most common hydrolysable tannins in strawberries are ellagitannins, specifically sanguiin H-6 and ellagic acid [5, 26–28]. The content of ellagitannins in strawberries is approximately 10–23 mg/100 g raw strawberry [16]. Ellagic acid is an ellagitannin present in the secondary metabolism of vegetables, its main characteristic is its antioxidant, antimicrobial, antimutagenic, anticarcinogenic and antiviral capacity [16]. The content of ellagic acid in strawberries is approximately 1–2 mg/100 g raw strawberry [16]. Due to the phenolic nature of ellagic acid, this compound tends to react by forming complexes with other molecules of proteins, alkaloids, and polysaccharides, so that it is usually found as ellagitannins esterified with glucose, because of this it is difficult to find it free [20]. The properties of ellagic acid are also exploited in the food industry, so it is used in the manufacture of nutraceutical drinks and food supplements. Likewise, the application of ellagic acid for food preservation is of great impact for the perishable food industry, using its antioxidant activity for microorganisms inhibition [35].

2.2.3 Phenolic acids

Strawberries contain a variety a of phenolic acids which are presented as derivatives of the hydroxycinnamic acid, such as caffeic acid, and hydroxybenzoic acids such as gallic acid [5]. The content of phenolic acids in strawberries is approximately 0.8–6.7 mg/100 g raw strawberry [16]. The major hydroxycinnamic acid in strawberries is p-coumaroylhexose, but ferulic acid and caffeic acid glycosides have also been identified in strawberries [26, 27]. Hydroxycinnamic acid derivatives are responsible for the bitter taste of the strawberry, and it is used in the manufacture of creams [33, 36]. The primary derivative of hydroxybenzoic acid is p-hydroxybenzoic glycoside [28]. The p-hydroxybenzoic glycoside is widely used in the synthesis of organic compounds and their esters, known as parabens, which are used as preservatives in cosmetics [37].

3. Extraction and purification of bioactive compounds in strawberry extrudate

3.1 Extraction techniques

3.1.1 Solvents extraction

The extraction with solvents is a technique to isolate a substance from a solid or liquid mixture. This technique is currently used in combination with other techniques such as microwaves and ultrasound since the solvent only extracts soluble compound. Due to the strawberry extrudate nature, the solid–liquid extraction can be carried out with a Soxhlet extractor, which is one of the most commonly used conventional extraction techniques [38]. The extraction efficiency depends mainly on the choice of solvents [39]. The polar character of the bioactive compounds allows their solubility in various solvents, such as water, alcohols, and acetone [40].

Recently, numerous studies have explored the extraction of bioactive compounds using deep eutectic solvents from various groups of natural sources [41]. The formation of eutectic solvents is the result of the complexation of a halide salt, which acts as a hydrogen bond receptor, and a hydrogen bond donor [41]. Some eutectic solvents have been developed from the combination of primary metabolites and bio-renewable starting materials, e.g., sugars, alcohols, amino acids and organic acids [41]. Eutectic solvents produce less adverse effects on the environment, allowing to replace conventional chemical methods [41].

There is a long variety of studies on solvent extraction focusing on the extraction of bioactive compounds. An evaluation of the effect of different solvents and acids in the extraction of anthocyanins from strawberry fruits concluded that acetone provided an efficient and reproducible extraction, avoiding problems with pectins and allowing the concentration of the sample at low temperature [42]. In another study, it was observed that the acetone/acetic acid mixture (99:1, v/v) reached good results for the qualitative and quantitative evaluation of polyphenols present in strawberries [43].

3.1.2 Hydrothermal extraction

Hydrothermal extraction is a process in which the matter is treated by adding hot water or water vapor [38]. Steam explosion is another kind of hydrothermal treatment where the matter is treated with saturated water vapor at high pressure followed by rapid depressurization [44]. The disadvantage of using hydrothermal treatments is that they affect thermosensitive compounds and might form undesirable compounds [45].

Hydrothermal extraction at low temperature, i.e. ranging between 50 and 90°C, mainly induces the de-flocculation of macromolecules [46]. Hydrothermal extraction at medium temperature, i.e. ranging between 150 and 180°C, solubilizes cellulosic and hemicellulose biomass [47]. The steam explosion treatment, with temperature ranging between 180 and 260°C and increase in pressure of 0.69–4.83 MPa, it is able to solubilize lignocellulose biomass [44].

Several studies confirm the successful extraction of bioactive compounds by these hydrothermal extractions [48, 49]. Extraction of bioactive compounds in strawberry extrudate has been studied by applying hydrothermal treatments in the range of 90–200°C, [15]. Thermal treatment at 150°C for 60 minutes was the most efficient process based on the solubilization of sugars and phenols as well as the antioxidant capacity of the liquid phase produced [15].

3.1.3 Microwave assisted extraction

Microwaves are electromagnetic fields in a frequency range of 300 MHz to 30 GHz, which are generally operated at a frequency of 2.45 GHz [10]. Microwaves can access biological matrices and interact with polar molecules, such as water, which vibrate or rotate by the effect of microwaves and generate heat and can enhance the processes of extraction of bioactive compounds [10, 50]. Microwave assisted extraction has been successfully applied in anthocyanin extraction processes in grape skins [51], the recovery of pectins from press residues of various berries, i.e. red and black currant, raspberry and elderberry [52] and to extract phenolic antioxidants from peanut skins [53].

3.1.4 High hydrostatic pressure extraction

High hydrostatic pressure extraction is a method that works at high pressures ranging from 100 to 1000 MPa. These high pressures cause cell deformation, cell membrane damage, protein denaturation, deprotonation of charged groups, and the breakdown of bonds, making bioactive compounds more accessible for extraction [54]. High hydrostatic pressure extraction is considered to be a faster and more efficient technique than other conventional extraction methods [55, 56]. In addition, high hydrostatic pressure extraction has the advantage of not increasing the temperature during the processing time, so it would be an ideal method to extract thermosensitive compounds.

Several high hydrostatic pressure extraction studies have been carried out with strawberries for the extraction of bioactive compounds. The impact of high hydrostatic pressure extraction on total strawberry puree phenols was observed by Patras et al., [56], which reported that the amount of total phenols increases as the pressure in high hydrostatic pressure extraction increases [56]. In another study, the change in kaempferol, and quercetin quantity in strawberries pulps were tested at different pressures and for different processing times [57]. According to this study, the change in the amount of kaempferol was not very significant and the amount of quercetin increased with increasing pressure [57]. Another study showed that the nutritional and sensory qualities of strawberry puree after high-pressure processing at 500 MPa and 50°C for 15 min were much better than after a heat treatment at 90°C for 15 min [58].

3.1.5 Pulsed electric fields extraction

Pulsed electric fields or high intensity pulsed electric fields consist of a short time electrical treatment, between nanoseconds to milliseconds, in which the material located between two electrodes is exposed to a strong electric pulse of intensity field of 100 to 300 V/cm Pulsed Electric Fields or 20 to 80 kV/cm high intensity pulsed electric field, the operation parameters being the duration and number of pulses [59, 60]. Pulsed electric fields can produce the electrical rupture of the cell membranes producing the formation of pores, what is known as electroporation [60, 61]. Pore formation improves cell permeability allowing the recovery of bioactive compounds [62]. Compared to other non-thermal treatments such as the high hydrostatic pressure extraction method, pulsed electric field extraction methods require a much shorter processing time, higher extraction efficiency and these techniques can be easily applied in continuous operation [59]. Therefore, pulsed electric fields is a promising technique for different applications in the food industry because they can improve extraction capacity and recovery of nutritionally valuable compounds as well as the bioavailability of micronutrients and compounds in a wide range of foods [59].

Several studies on the extraction of antioxidant compounds in agri-foods show enhanced yields with pulsed electric fields. For instance, a comparison study between a heat treatment at 90°C for 60 or 30 seconds and high intensity pulsed electric field in strawberries juice, showed that strawberry juice treated with high intensity pulsed electric field maintained greater amount of phenolic acids and total anthocyanins than thermally treated juices [63, 64]. Likewise, the recovery of phenols from the shell of the pomegranate by pulsed electric field has been assayed, resulting in a similar antioxidant extraction yields and an energy saving of 50% compared to an ultrasound extraction technique [65].

3.1.6 Extraction techniques comparison

After reviewing the different extraction techniques that have been applied to the strawberry and strawberry extrudate, a summary describing their most interesting aspects is shown in **Table 1**. The aspects that have been compared are: the specificity of the extraction techniques with the bioactive compounds, the possibility of combining with other extraction techniques, the ability to release bioactive compounds, the potential degradation of bioactive compounds, possibility of intracellular attack, bonds breakage and whether the technique has a high operational and investment cost. The choice of the best technique for the strawberry extrudate is a tailor-made solution for each situation that will depend of the investment capacity, target compounds to be recovered or the required extraction yield.

3.2 Purification techniques of bioactive compounds

3.2.1 Adsorbents

There are many studies that show the properties of adsorbents to separate, concentrate and purify various compounds [66, 67]. Functionality, porosity, irregularities, surface area, tightly bonded impurities, internal porous structure, particle size, ionic strength, pH, and temperature all influence physical adsorption [66]. The temperature influences the adsorption in two ways, increasing the transport

Extraction technique	Specificity	Possibility of combination	Ability to release compounds	Degradation of bioactive compounds	Intracellular attack	Breaks of bonds	High cost
Solvents extraction	x	x					
Hydrothermal extraction		x	x	x		x	
Microwave assisted extraction		x	x	x	x	x	
High hydrostatic pressure extraction		x	x	x	x	x	x
Pulsed electric fields extraction		x	x	x	x	x	

Table 1.
Summary table of characteristics for comparing extraction techniques.

speed through the outer boundary layer and inside the pores due to the decrease in the viscosity of the solution, and changing the capacity of the adsorbent. However, high temperatures can promote irreversible interactions [67]. Another important parameter for purification with adsorbents is pH. For example, at acid pH, the adsorption of phenolic compounds by different adsorbents increases because the phenols are not dissociated and dispersion interactions predominate [66]. At alkaline pH, the adsorption decreases due to the dissociation of hydroxyl groups and carboxyl groups [66]. There are many types of adsorbents such as activated carbons, mineral adsorbents, synthetic polymeric adsorbents, ion exchange resins, lignin and lignocellulosic materials, adsorbents based on polysaccharides and others [66]. Among the available adsorbents Amberlite XAD adsorbents are widely used in the concentration of polyphenols [68]. Zhang et al. [69] reported the isolation and structural characterization of 10 phenolic compounds from strawberry extracts using a combination of Amberlite XAD-16 and C18 columns, HPLC-UV, and nuclear magnetic resonance spectroscopy methods.

3.2.2 Countercurrent-chromatography

Countercurrent chromatography is a technique widely used in the purification of natural products [70]. Countercurrent chromatography is a liquid–liquid partition chromatography process in which both the mobile phase and the stationary phase are liquid [70]. The main advantage of countercurrent chromatography, when compared to equivalent techniques such as low pressure liquid chromatography, is that there are no adsorption losses in the stationary phase [70]. The range of selectivity offered by chromatographic resins is equivalent to the range of selectivity offered by the different solvent systems [70].

Several studies have shown the importance of countercurrent chromatography for the purification of bioactive compounds from strawberry. The compound 2,5-dimethyl-4-hydroxy-3[2H]-furanone 6′O-malonyl-β-d-glucopyranoside was isolated from a strawberry glycosidic extract (Fragaria × *ananassa*, cv. Senga Sengana) by countercurrent chromatography [71]. Peonidin-3-glucoside and malvidin-3-glucoside were obtained from grapes in a single step, while in a second step, cyanidin-3-glucoside was isolated [72]. In another research, the separation of anthocyanin monomers of high purity from mulberry fruits was developed [73].

3.2.3 Two-phase aqueous system

Two-phase aqueous system is a liquid–liquid fractionation technique that is usually formed by mixing two polymers in aqueous media, for example, polyethylene glycol and dextran or maltodextrin, or by a polymer and a salt, such as polyethylene glycol and salts of phosphates, citrates or sulphates [74–76]. This method has advantages over other purification techniques due to a comparatively low consumption of energy and time, as well as the possibility to be designed for a continuous operation. Moreover, two-phase aqueous systems are effective for many types of substances, especially for the concentration and purification of bioactive compounds [74, 75]. Several studies have demonstrated the suitability of this technique for the purification of bioactive compounds such as phenolic compounds from fig fruits (*Ficus carica* L.) [76], or the purification of gallic acid from natural matrices with ionic liquids [77]. Furthermore, two-phase aqueous system has been applied for the purification of polyphenols from a model solution of gallic acid and three real samples of red and white wine, and orange juice in combination with macro and micro extractors [78]. Polyphenols have been also extracted from *Aronia melanocarpa* berries, using ultrasound-assisted extraction in combination with the two-phase aqueous system [79].

4. Stabilization of biomass by obtaining bioenergy and bioproducts

4.1 Biogas production

Anaerobic digestion is a microbiological process, in absence of oxygen, where organic matter is progressively degraded by an heterogeneous bacterial population to methane (55–70%) and carbon dioxide (30–45%) [80]. Anaerobic digestion presents some fundamental advantages such as the possibility of working at high rates of organic load, and the produced methane can be used as an energy source due to its heating value (35,793 kJ/m^3, at 1 atm, 0°C), which equals to 1 kg of raw coal or 0.76 kg of standard coal [3, 11]. The use of biogas for energy supply reduces deforestation, soil erosion and environmental pollution [81, 82]. Also, it can improve the energy efficiency of various production processes due to the energetic contribution that provides [82]. In addition, a wet waste called digestate, which is a mixture of partially degraded organic matter, microbial biomass, and inorganic compounds, is produced during biomethanization and could be used as a base for fertilizers or organic amendments [82, 83].

Several studies on anaerobic digestion of strawberries extrudate have been carried out. The results of one these studies reveal that strawberries extrudates have a high level of anaerobic biodegradability (90% in VS, (total volatile solids)) and that a substantial amount of methane can be obtained in this way (312 mL CH_4 STP/g added VS) (STP: standard temperature and pressure conditions, i.e. 0°C, 1 atm) at an organic loading rate range of 2.04 to 3.51 kg VS/m^3·d [84]. In another study of anaerobic digestion of strawberry waste from supermarkets, using an organic loading rate of 0.55–4.4 (g/L·d), the experimental biogas and methane yields were 0.588 and 0.231 L/g, respectively [85]. It has been observed that sometimes it is necessary to co-digest strawberry extrudate with a substrate that provides alkalinity, such as sewage sludge [86, 87]. Co-digestion studies of strawberry extrudate with other substrates such as fish waste [3] and glycerol [83] have also been studied. Anaerobic digestion of strawberry extrudate is a promising technique but it should be further studied since low alkalinity of the extrudate together with formation of inhibitory compounds caused by the extraction process could negatively affect the digestion process.

4.2 Compost production

Composting has been proposed for a long time as a quite cheap option for agricultural waste management [2]. Composting has also been proposed as a post-treatment for the produced digestate after anaerobic digestion [88]. Composting is the bio-oxidative conversion of organic waste into an organic amendment. According to Gutiérrez et al. (2017) [2], the cost of composting varies in a wide range from $40 to $500 per throughput ton depending on the technology. Composting costs vary widely depending on the type of operation, which ranges from the most simple ones, such as opening windrows, to more complex procedures like in-vessel aerobic composting that allows smell emissions to be controlled and prevents environmental pollution [2]. The great disadvantage is that a considerable amount of offensive odors can be emitted during the process due to the generation of volatile organic compounds [89]. Other disadvantages are the long process time and the necessity of a proper monitoring [90]. Co-composting of a waste mixture containing strawberry extrudate, fish waste, sewage sludge and bulking agent has been successfully proven [2, 89].

4.3 Bioethanol production

Bioethanol is one of the most produced alcohols from the fermentation of sugars found in fruits and vegetables [7, 91–93]. Theoretically, any organic product with a high content of sugars and starch, such as strawberry extrudate, may be susceptible to obtaining bioethanol [91]. Inedible sources from the strawberry extrudate such as lignocellulosic biomass, which mainly comprises cellulose, hemicellulose, and lignin, can be hydrolysed to produce a mixture of pentoses and hexoses that can be transformed into bioethanol [94]. Bioethanol from agro-waste, such as strawberry extrudate, could be a promising technology that involves four processes, pre-treatment, enzymatic hydrolysis, fermentation and distillation, this final step is crucial for the process to be economically viable on a commercial scale due to high energy consumption in the form of steam to increase the yield of bioethanol production when lignocellulose materials are used as raw material [93]. These processes have several challenges and limitations, such as the efficient pre-treatment process to eliminate lignin from the lignocellulosic agro-residues. The proper pre-treatment process can increase the concentrations of fermentable sugars after enzymatic hydrolysis, thus improving the efficiency of the entire process [92].

4.4 Bioplastics production

Fossil fuel depletion, global warming, and problems of pollution of the environment that provoke plastics in its life cycle are encouraging the development of biodegradable plastics [95, 96]. Agri-food waste are usually rich in many useful substances such as lipids, polysaccharides, and aromatics, which could be used for the manufacture of biodegradable polymeric materials. Bioplastics already play an important role in the sectors of packaging, agriculture, consumer electronics and motoring, but still have a very low share in the total production of plastics. Currently, about 1% of the annual tons of plastic are bioplastics [97]. Examples of such bioplastics are exopolysaccharides, polycaprolactone, polybutylene succinate, polybutylene adipate terephthalate, polyhydroxyalkanoates or polyhydroxybutyrates [97, 98]. For obtaining bioplastics from agri-food waste, the waste must be treated to extract or isolate specific macromolecules, such as cellulose, lignin, suberin, starch, or monomers, such as vegetable oils, tannins and terpenes [96, 99]. A study conducted on the production of bioplastics from Murta fruit extract, that is a native Chilean berry, showed the feasibility of using berries for bioplastic production [8].

4.5 Biochar production

Biochar is the solid carbonaceous residue produced through organic waste and used as a soil improver [100, 101]. Biochar is produced through several types of methods such as pyrolysis, torrefaction or hydrothermal carbonization [100, 101].

There are no studies reported in the literature dealing with the production of biochar from strawberry extrudates. However, the above-mentioned techniques (pyrolysis, torrefaction and hydrothermal carbonization) could be potentially applied for this substrate. Several studies have been carried out on the hydrothermal carbonization of other agri-food waste, such as olive cuttings and olive pulp [102]; grape marc [103]; olive mill waste, canned artichoke and orange waste [104].

5. Conclusions

This chapter has reviewed up-to-date literature on the bioactive compounds contained in strawberries, which have an important health and market value. Different extraction and purification techniques to obtain valuable compounds from strawberry extrudate have been reviewed and analyzed. The reviewed techniques present different advantages and drawbacks that were analyzed to facilitate the selection of the most suitable process for each valorisation scenario. Finally, different stabilization options for the biomass remaining after extraction have also been reviewed. Stabilization is required to avoid severe environmental impacts, and additionally could be an economically beneficial aid for balancing the cost of the extraction of high value-added compounds. As usually for any waste management option, selection of the best extraction, purification and stabilization technique for the strawberry extruded is a tailor-made solution for each situation.

Acknowledgements

This research was funded by the Spanish Ministry of Economy and Competitiveness by Project CTM2017-83870R. We acknowledge support of the publication fee by the CSIC Open Access Publication Support Initiative through its Unit of Information Resources for Research (URICI).

Conflict of interest

The authors declare no conflict of interest.

Author details

Juan Cubero-Cardoso[1], Antonio Serrano[1,2], Ángeles Trujillo-Reyes[1], Denys K. Villa-Gómez[2], Rafael Borja[1] and Fernando G. Fermoso[1*]

1 Instituto de la Grasa (C.S.I.C.), Sevilla, Spain

2 School of Civil Engineering, The University of Queensland, St Lucia, QLD, Australia

*Address all correspondence to: fgfermoso@ig.csic.es

References

[1] FAO. FAOSTAT: Food and agriculture organization of the united nations [Internet]. 2020 [cited 2020 Jul 17]. Available from: http://www.fao.org/faostat/es/#data/QC/visualize

[2] Gutiérrez MC, Serrano A, Siles JA, Chica AF, Martín MA. Centralized management of sewage sludge and agro-industrial waste through co-composting. Journal of Environmental Management. 2017 Jul;196:387-93.

[3] Serrano A, Siles JA, Chica AF, Martín MÁ. Agri-food waste valorization through anaerobic co-digestion: Fish and strawberry residues. Journal of Cleaner Production. 2013 Sep;54:125-32.

[4] Aharonov-Nadborny R, Raviv M, Graber ER. Soil spreading of liquid olive mill processing wastes impacts leaching of adsorbed terbuthylazine. Chemosphere. 2016;156(Supplement C): 220-7.

[5] Giampieri F, Tulipani S, Alvarez-Suarez JM, Quiles JL, Mezzetti B, Battino M. The strawberry: Composition, nutritional quality, and impact on human health. Nutrition. 2012;28(1):9-19.

[6] Dias MI, Barros L, Sousa MJ, Oliveira MBPP, Santos-Buelga C, Ferreira ICFR. Enhancement of nutritional and bioactive compounds by in vitro culture of wild *Fragaria vesca* L. vegetative parts. Food Chemistry. 2017;235:212-9.

[7] Dias MOS, Ensinas A V., Nebra SA, Maciel Filho R, Rossell CE V, Maciel MRW. Production of bioethanol and other bio-based materials from sugarcane bagasse: Integration to conventional bioethanol production process. Chemical Engineering Research and Design. 2009 Sep 1;87(9):1206-16.

[8] López de Dicastillo C, Bustos F, Guarda A, Galotto MJ. Cross-linked methyl cellulose films with murta fruit extract for antioxidant and antimicrobial active food packaging. Food Hydrocolloids. 2016;60(Supplement C):335-44.

[9] Ariza MT, Reboredo-Rodríguez P, Mazzoni L, Forbes-Hernández TY, Giampieri F, Afrin S, et al. Strawberry achenes are an important source of bioactive compounds for human health. International Journal of Molecular Sciences. 2016;17(7):1103.

[10] Wijngaard H, Hossain MB, Rai DK, Brunton N. Techniques to extract bioactive compounds from food by-products of plant origin. Food Research International. 2012;46(2): 505-13.

[11] Fermoso FG, Serrano A, Alonso-Fariñas B, Fernández-Bolaños J, Borja R, Rodríguez-Gutiérrez G. Valuable Compound Extraction, Anaerobic Digestion, and Composting: A Leading Biorefinery Approach for Agricultural Wastes. Journal of Agricultural and Food Chemistry American Chemical Society; Aug 15, 2018 p. 8451-68.

[12] Pérez-Loredo MG, Hernández- De Jesús L, Barragán-Huerta BE. Extracción de compuestos bioactivos de pitaya roja (*Stenocereus stellatus*) aplicando pretratamientos con microondas, ultrasonido y enzimáticos. Agrociencia. 2017;51(2):135-51.

[13] Rajendran K, Drielak E, Sudarshan Varma V, Muthusamy S, Kumar G. Updates on the pretreatment of lignocellulosic feedstocks for bioenergy production–a review. Vol. 8, Biomass Conversion and Biorefinery. Springer Verlag; 2018. p. 471-83.

[14] Machineni L. Lignocellulosic biofuel production: review of alternatives. Biomass Conversion and Biorefinery. Springer Verlag; 2019.

[15] Rodríguez-Gutiérrez G, Cardoso JCJC, Rubio-Senent F, Serrano A, Borja R, Fernández-Bolaños J, et al. Thermally-treated strawberry extrudate: A rich source of antioxidant phenols and sugars. Innovative Food Science and Emerging Technologies. 2019 Jan 1;51:186-93.

[16] Basu A, Nguyen A, Betts NM, Lyons TJ. Strawberry As a Functional Food: An Evidence-Based Review. Critical Reviews in Food Science and Nutrition. 2014;54(6):790-806.

[17] Sójka M, Klimczak E, Macierzyński J, Kołodziejczyk K. Nutrient and polyphenolic composition of industrial strawberry press cake. European Food Research and Technology. 2013;237(6):995-1007.

[18] Kafkas E, Koşar M, Paydaş S, Kafkas S, Başer KHC. Quality characteristics of strawberry genotypes at different maturation stages. Food Chemistry. 2007;100(3):1229-36.

[19] Tulipani S, Romandini S, Suarez JMA, Capocasa F, Mezzetti B, Busco F, et al. Folate content in different strawberry genotypes and folate status in healthy subjects after strawberry consumption. BioFactors. 2008;34(1):47-55.

[20] da Silva Pinto M, Lajolo FMFM, Genovese MIMI, Pinto MS, Lajolo FMFM, Genovese MIMI. Bioactive compounds and quantification of total ellagic acid in strawberries (Fragaria x ananassa Duch.). Food Chemistry. 2008;107(4):1629-35.

[21] Ochoa M CI, Ayala A AA. Los Flavonoides: Apuntes Generales y su Aplicación en la Industria de Alimentos. Revista Ingeniería y Competitividad; Vol 6, Núm 2. 2004;

[22] Robards K, Prenzler PD, Tucker G, Swatsitang P, Glover W. Phenolic compounds and their role in oxidative processes in fruits. Food Chemistry. 1999;66(4):401-36.

[23] Böhm H. G. Mazza und E. Miniati: Anthocyanins in Fruits, Vegetables and Grains. 362 Seiten, zahlr. Abb. und Tab. CRC Press, Boca Raton, Ann Arbor, London, Tokyo 1993. Preis: 144.— £. Vol. 38, Food / Nahrung. CRC press; 1994. 343-343 p.

[24] da Silva FL, Escribano-Bailón MT, Pérez Alonso JJ, Rivas-Gonzalo JC, Santos-Buelga C. Anthocyanin pigments in strawberry. LWT - Food Science and Technology. 2007;40(2):374-82.

[25] da Silva Pinto M, Lajolo FM, Genovese MI. Bioactive compounds and quantification of total ellagic acid in strawberries (Fragaria x ananassa Duch.). Food Chemistry. 2008;107(4):1629-35.

[26] Aaby K, Skrede G, Wrolstad RE. Phenolic composition and antioxidant activities in flesh and achenes of strawberries (fragaria ananassa). Journal of Agricultural and Food Chemistry. 2005;53(10):4032-40.

[27] Aaby K, Ekeberg D, Skrede G. Characterization of phenolic compounds in strawberry (Fragaria x ananassa) fruits by different HPLC detectors and contribution of individual compounds to total antioxidant capacity. Journal of Agricultural and Food Chemistry. 2007;55(11):4395-406.

[28] Klopotek Y, Otto K, Böhm V. Processing strawberries to different products alters contents of vitamin C, total phenolics, total anthocyanins, and antioxidant capacity. Journal of Agricultural and Food Chemistry. 2005;53(14):5640-6.

[29] Garzón GA. Anthocyanins As Natural Colorants And Bioactive Compounds: A Review. Acta Biológica Colombiana. 2008;13(3):27-36.

[30] Shaik-Dashagirisaheb YB, Varvara G, Murmura G, Saggini A, Caraffa A, Antinolfi P, et al. Inhibitor effect of antioxidant flavanoids quercetin, and capsaicin in mast cell inflammation. European Journal of Inflammation. 2013;357(2):353-7.

[31] Hannum SM. Potential Impact of Strawberries on Human Health: A Review of the Science. Critical Reviews in Food Science and Nutrition. 2004;44(1):1-17.

[32] Kashket S. Sweetness and flavor enhanced compositions and method of preparing such compositions. 1990 Apr 7;

[33] Porras-Loaiza A p., López-Malo A. Importancia de los grupos fenólicos en los alimentos. Temas Selectos de Ingeniería de Alimentos. 2009;3(1):121-34.

[34] Santos-Buelga C, Scalbert A. Proanthocyanidins and tannin like compounds nature, occurrence, dietary intake and effects on nutrition and health. Journal of the Science of Food and Agriculture. 2000;80(May):1094-117.

[35] Saucedo-Pompa S, Rojas-Molina R, Aguilera-Carbó AF, Saenz-Galindo A, GarzaB H de La, Diana Jasso-Cantú, et al. Edible Films based on canadelilla wax to improve the shelr life and quality of hass avocado. Food Research International. 2009;42(4):511-515.

[36] Joseph ALII, DiNardo JC. Methods of preparing and using botanical antioxidant compositions. Google Patents; 2017.

[37] Ma WL, Zhao X, Lin ZY, Mohammed MOA, Zhang ZF, Liu LY, et al. A survey of parabens in commercial pharmaceuticals from China and its implications for human exposure. Environment International. 2016;95(Supplement C):30-5.

[38] Azmir J, Zaidul ISM, Rahman MM, Sharif KM, Mohamed A, Sahena F, et al. Techniques for extraction of bioactive compounds from plant materials: A review. Journal of Food Engineering. 2013;117(4):426-36.

[39] Cowan MM. Plant Products as Antimocrobial Agents, Clinical Microbioligy Reviews, 565, 568, 570. Departement of Microbiology, Miami University. Oxford, Ohio; 1999.

[40] Zapata LM, Heredia AM, Quinteros CF, Malleret AD, Clemente G, Cárcel JA. Optimización de la extracción de antocianinas de arándanos. Ciencia, Docencia y Tecnología. 2014;25(49):166-92.

[41] Zainal-Abidin MH, Hayyan M, Hayyan A, Jayakumar NS. New horizons in the extraction of bioactive compounds using deep eutectic solvents: A review. Analytica Chimica Acta. 2017;979:1-23.

[42] Garcia-Viguera C, Zafrilla P, Tomás-Barberán FA. The use of acetone as an extraction solvent for anthocyanins from strawberry fruit. Phytochemical Analysis. 1998;9(6):274-7.

[43] Kajdžanoska M, Petreska J, Stefova M. Comparison of different extraction solvent mixtures for characterization of phenolic compounds in strawberries. Journal of Agricultural and Food Chemistry. 2011 May 25;59(10): 5272-8.

[44] Rincón B, Rodríguez-Gutiérrez G, Bujalance L, Fernández-Bolaños J, Borja R. Influence of a steam-explosion pre-treatment on the methane yield and kinetics of anaerobic digestion of two-phase olive mil solid waste or alperujo. Process Safety and Environmental Protection. 2016;102(Supplement C): 361-9.

[45] Struck S, Plaza M, Turner C, Rohm H. Berry pomace - A review of processing and chemical analysis of its polyphenols. International Journal

of Food Science and Technology. 2016;51(6):1305-18.

[46] Jain S, Jain S, Wolf IT, Lee J, Tong YW. A comprehensive review on operating parameters and different pretreatment methodologies for anaerobic digestion of municipal solid waste. Renewable and Sustainable Energy Reviews. 2015;52:142-54.

[47] Hendriks ATWM, Zeeman G. Pretreatments to enhance the digestibility of lignocellulosic biomass. Bioresource Technology. 2009;100(1):10-8.

[48] Serrano A, Fermoso FG, Alonso-FariñasB, Rodríguez-GutierrezG, Fernandez-Bolaños J, Borja R. Olive mill solid waste biorefinery: High-temperature thermal pre-treatment for phenol recovery and biomethanization. Journal of Cleaner Production. 2017 Apr 1;148:314-23.

[49] Mrabet A, Jiménez-Araujo A, Fernández-Bolaños J, Rubio-Senent F, Lama-Muñoz A, Sindic M, et al. Antioxidant phenolic extracts obtained from secondary Tunisian date varieties (*Phoenix dactylifera* L.) by hydrothermal treatments. Food Chemistry. 2016 Apr 1;196(Supplement C):917-24.

[50] Flores E. Extracción de Antioxidantes de las Bayas del Sauco (*Sambucus nigra L. subsp*. peruviana) con Ultrasonido, Microondas, Enzimas y Maceración para la Obtención de Zumos Funcionales. Informacion Tecnologica. 2017;28(1):121-32.

[51] Liazid A, Guerrero RF, Cantos E, PalmaM, BarrosoCG. Microwaveassisted extraction of anthocyanins from grape skins. Food Chemistry. 2011;124(3):1238-43.

[52] Bélafi-Bakó K, Cserjési P, Beszédes S, Csanádi Z, Hodúr C. Berry Pectins: Microwave-Assisted Extraction and Rheological Properties. Food and Bioprocess Technology. 2012;5(3):1100-5.

[53] Ballard TS, Mallikarjunan P, Zhou K, O'Keefe S. Microwave-assisted extraction of phenolic antioxidant compounds from peanut skins. Food Chemistry. 2010;120(4):1185-92.

[54] Shouqin Z, Jim X, Changzheng W. Note: Effect of high hydrostatic pressure on extraction of flavonoids in propolis. Food Science and Technology International. 2005;11(3):213-6.

[55] Corrales M, Toepfl S, Butz P, Knorr D, Tauscher B. Extraction of anthocyanins from grape by-products assisted by ultrasonics, high hydrostatic pressure or pulsed electric fields: A comparison. Innovative Food Science and Emerging Technologies. 2008;9(1):85-91.

[56] Patras A, Brunton NP, Da Pieve S, Butler F. Impact of high pressure processing on total antioxidant activity, phenolic, ascorbic acid, anthocyanin content and colour of strawberry and blackberry purées. Innovative Food Science and Emerging Technologies. 2009 Jul;10(3):308-13.

[57] Cao X, Zhang Y, Zhang F, Wang Y, Yi J, Liao X. Effects of high hydrostatic pressure on enzymes, phenolic compounds, anthocyanins, polymeric color and color of strawberry pulps. Journal of the Science of Food and Agriculture. 2011;91(5):877-85.

[58] Marszałek K, Mitek M, Skąpska S. The effect of thermal pasteurization and high pressure processing at cold and mild temperatures on the chemical composition, microbial and enzyme activity in strawberry purée. Innovative Food Science and Emerging Technologies. 2015;27(Supplement C): 48-56.

[59] Yan L-G, He L, Xi J. High intensity pulsed electric field as an innovative

technique for extraction of bioactive compounds—A review. Critical Reviews in Food Science and Nutrition. 2017;57(13):2877-88.

[60] Parniakov O, Lebovka NI, Van Hecke E, Vorobiev E. Pulsed Electric Field Assisted Pressure Extraction and Solvent Extraction from Mushroom (Agaricus Bisporus). Food and Bioprocess Technology. 2013;7(1):174-83.

[61] Neuman E, Schaefer-Ridder M, Wang Y, Hofschneider PH. Gene transfer into mouselyoma celles by electroporation in high electric fields. Embo J. 1982 Jul 1;1(7):1841-.

[62] Angersbach A, Heinz V, Knorr D. Effects of pulsed electric fields on cell membranes in real food systems. Innovative Food Science & Emerging Technologies. 2000;1(2):135-49.

[63] Odriozola-Serrano I, Soliva-Fortuny R, Martín-Belloso O. Impact of high-intensity pulsed electric fields variables on vitamin C, anthocyanins and antioxidant capacity of strawberry juice. LWT - Food Science and Technology. 2009;42(1):93-100.

[64] Odriozola-Serrano I, Soliva-Fortuny R, Martín-Belloso O. Phenolic acids, flavonoids, vitamin C and antioxidant capacity of strawberry juices processed by high-intensity pulsed electric fields or heat treatments. European Food Research and Technology. 2008;228(2):239-48.

[65] Pan Z, Qu W, Ma H, Atungulu GG, McHugh TH. Continuous and pulsed ultrasound-assisted extractions of antioxidants from pomegranate peel. Ultrasonics Sonochemistry. 2012;19(2):365-72.

[66] Soto ML, Moure A, Domínguez H, Parajó JC. Recovery, concentration and purification of phenolic compounds by adsorption: A review. Journal of Food Engineering. 2011;105(1):1-27.

[67] Qiu N, Guo S, Chang Y. Study upon kinetic process of apple juice adsorption de-coloration by using adsorbent resin. Journal of Food Engineering. 2007;81(1):243-9.

[68] Ahmad A, Siddique JA, Laskar MA, Kumar R, Mohd-Setapar SH, Khatoon A, et al. New generation Amberlite XAD resin for the removal of metal ions: A review. Journal of Environmental Sciences (China). 2015;31(Supplement C):104-23.

[69] Zhang Y, Seeram NP, Lee R, Feng L, Heber D. Isolation and identification of strawberry phenolics with antioxidant and human cancer cell antiproliferative properties. Journal of Agricultural and Food Chemistry. 2008;56(3):670-5.

[70] Valls J, Millán S, Martí MP, Borràs E, Arola L. Advanced separation methods of food anthocyanins, isoflavones and flavanols. Journal of Chromatography A. 2009;1216(43):7143-72.

[71] Roscher R, Herderich M, Steffen JP, Schreier P, Schwab W. 2,5-dimethyl-4-hydroxy-3[2H]-furanone 6'O-malonyl-beta-D- glucopyranoside in strawberry fruits. Phytochemistry. 1996;43(1):155-9.

[72] Renault J-H, Thepenier P, Zeches-hanrot M, Le Men-Olivier L, Durand A, Foucault A, et al. Preparative separation of anthocyanins by gradient elution centrifugal partition chromatography. Journal of Chromatography A. 1997;763(1):345-52.

[73] Chen Y, Du F, Wang W, Li Q, Zheng D, Zhang W, et al. Large-scale isolation of high-purity anthocyanin monomers from mulberry fruits by combined chromatographic techniques. Journal of Separation Science. 2017 Sep 1;40(17):3506-12.

[74] Iqbal M, Tao Y, Xie S, Zhu Y, Chen D, Wang X, et al. Aqueous two-phase system (ATPS): an overview and

advances in its applications. Biological Procedures Online. 2016;18(1):1-18.

[75] Molino JVD, Viana Marques D de A, Júnior AP, Mazzola PG, Gatti MSV. Different types of aqueous two-phase systems for biomolecule and bioparticle extraction and purification. Biotechnology Progress. 2013;29(6):1343-53.

[76] Feng YC, Li WL, He FM, Kong TT, Huang XW, Gao ZH, et al. Aqueous Two-Phase System as an Effective Tool for Purification of Phenolic Compounds from Fig Fruits (*Ficus carica* L.). Separation Science and Technology (Philadelphia). 2015;50(12):1785-93.

[77] Cláudio AFM, Ferreira AM, Freire CSR, Silvestre AJD, Freire MG, Coutinho JAP. Optimization of the gallic acid extraction using ionic-liquid-based aqueous two-phase systems. Separation and Purification Technology. 2012;97(Supplement C):142-9.

[78] Tušek AJ, Šalić A, Zelić B. Mathematical modelling of polyphenol extraction by aqueous two-phase system in continuously operated macro- and micro-extractors. Separation Science and Technology (Philadelphia). 2017;52(5):864-75.

[79] Xu YY, Qiu Y, Ren H, Ju DH, Jia HL. Optimization of ultrasound-assisted aqueous two-phase system extraction of polyphenolic compounds from *Aronia melanocarpa* pomace by response surface methodology. Preparative Biochemistry and Biotechnology. 2017;47(3):312-21.

[80] Monge O, Certucha Barragn MT, Almendariz Tapi FJ. Microbial Biomass in Batch and Continuous System. In: Biomass Now - Sustainable Growth and Use. InTech; 2013.

[81] Pérez I, Garfi M, Cadena E, Ferrer I. Technical, economic and environmental assessment of household biogas digester for rural communities. Renewable Energy. 2014;62(Supplement C):7.

[82] Bożym M, Florczak I, Zdanowska P, Wojdalski J, Klimkiewicz M. An analysis of metal concentrations in food wastes for biogas production. Renewable Energy. 2015 May 1;77:467-72.

[83] Serrano A, Siles JA, Chica AF, Martin MA. Improvement of mesophilic anaerobic co-digestion of agri-food waste by addition of glycerol. Journal of Environmental Management. 2014 Jul;140:76-82.

[84] Siles JA, Serrano A, Martín A, Martín MA. Biomethanization of waste derived from strawberry processing: Advantages of pretreatment. Journal of Cleaner Production. 2013 Mar 1;42: 190-7.

[85] Arhoun B, Gomez-Lahoz C, Abdala-Diaz RT, Rodriguez-Maroto JM, Garcia-Herruzo F, Vereda-Alonso C. Production of biogas from co-digestion of livestock and agricultural residues: A case study. Journal of Environmental Science and Health - Part A Toxic/Hazardous Substances and Environmental Engineering. 2017;52(9):856-61.

[86] Serrano A, Siles JA, Gutiérrez MC, Martín MÁ. Improvement of the biomethanization of sewage sludge by thermal pre-treatment and co-digestion with strawberry extrudate. Journal of Cleaner Production. 2015 Mar 1;90: 25-33.

[87] Serrano A, Siles JA, Chica AF, Martín MÁ. Anaerobic co-digestion of sewage sludge and strawberry extrudate under mesophilic conditions. Environmental Technology (United Kingdom). 2014 Dec 2;35(23):2920-7.

[88] Vaneeckhaute C, Lebuf V, Michels E, Belia E, Vanrolleghem PA, Tack FMG, et al. Nutrient Recovery from Digestate: Systematic Technology

Review and Product Classification. Waste and Biomass Valorization. 2017;8(1):21-40.

[89] Toledo M, Gutiérrez MC, Siles JA, García-Olmo J, Martín MA. Chemometric analysis and NIR spectroscopy to evaluate odorous impact during the composting of different raw materials. Journal of Cleaner Production. 2017 Nov 20;167:154-62.

[90] Parthiba Karthikeyan O, Trably E, Mehariya S, Bernet N, Wong JWC, Carrere H. Pretreatment of food waste for methane and hydrogen recovery: A review. Bioresource Technology. 2018 Feb 1;249:1025-39.

[91] Goh CS, Tan KT, Lee KT, Bhatia S. Bio-ethanol from lignocellulose: Status, perspectives and challenges in Malaysia. Bioresource Technology. 2010;101(13):4834-41.

[92] Guerrero AB, Ballesteros I, Ballesteros M. The potential of agricultural banana waste for bioethanol production. Fuel. 2018;213(Supplement C):176-85.

[93] Gupta A, Verma JP. Sustainable bio-ethanol production from agro-residues: A review. Renewable and Sustainable Energy Reviews. 2015;41(Supplement C):550-67.

[94] Tan KT, Lee KT, Mohamed AR. Role of energy policy in renewable energy accomplishment: The case of second-generation bioethanol. Energy Policy. 2008;36(9):3360-5.

[95] Heredia-Guerrero JA, Athanassiou A. Editorial: Non-polysaccharide Plant Polymeric Materials. Vol. 3, Frontiers in Materials. Frontiers Media SA; 2016.

[96] Heredia-Guerrero JA, Heredia A, Domínguez E, Cingolani R, Bayer IS, Athanassiou A, et al. Cutin from agro-waste as a raw material for the production of bioplastics. Journal of Experimental Botany. 2017 Nov 9;68(19):5401-10.

[97] Rujnić-Sokele M, Pilipović A. Challenges and opportunities of biodegradable plastics: A mini review. Waste Management & Research. 2017;35(2):132-40.

[98] Yaradoddi J, Patil V, Ganachari S, Banapurmath N, Hunashyal A, Shettar A, et al. Biodegradable Plastic Production From Fruit Waste Material and Its Sustainable Use for Green Applications. International Journal of Pharmaceutical Research & Allied Sciences. 2016;5(4):56-66.

[99] Gandini A, Lacerda TM, Carvalho AJF, Trovatti E. Progress of Polymers from Renewable Resources: Furans, Vegetable Oils, and Polysaccharides. Chemical Reviews. 2016;116(3):1637-69.

[100] de la Rosa JM, Rosado M, Paneque M, Miller AZ, Knicker H. Effects of aging under field conditions on biochar structure and composition: Implications for biochar stability in soils. Science of the Total Environment. 2018;613-614(Supplement C):969-76.

[101] Oliveira FR, Patel AK, Jaisi DP, Adhikari S, Lu H, Khanal SK. Environmental application of biochar: Current status and perspectives. Bioresource Technology. 2017;246(Supplement C):110-22.

[102] Volpe M, Fiori L. From olive waste to solid biofuel through hydrothermal carbonisation: The role of temperature and solid load on secondary char formation and hydrochar energy properties. Journal of Analytical and Applied Pyrolysis. 2017;124 (Supplement C):63-72.

[103] Basso D, Patuzzi F, Weiss-Hortala E, Rada EC, Castello D, Baratieri M, et al. Waste to biofuel through hydrothermal carbonization. Venice 2014: V

International Symposium on Energy from Biomass and Waste. 2014;47(Part A):114-21.

[104] Benavente V, Calabuig E, Fullana A. Upgrading of moist ageo-industrial wastes by hydrothermal carbonization. Journal of Analytical and Applied Pyrolysis. 2015;113 (Supplement C):89-98.

Saccharomyces

BIOTECHNOLOGY HANDBOOKS

Series Editors: Tony Atkinson and Roger F. Sherwood

PHLS Centre for Applied Microbiology and Research
Division of Biotechnology
Salisbury, Wiltshire, England

Volume 1 *PENICILLIUM* AND *ACREMONIUM*
Edited by John F. Peberdy

Volume 2 *BACILLUS*
Edited by Colin R. Harwood

Volume 3 CLOSTRIDIA
Edited by Nigel P. Minton and David J. Clarke

Volume 4 *SACCHAROMYCES*
Edited by Michael F. Tuite and Stephen G. Oliver

A Continuation Order Plan is available for this series. A continuation order will bring delivery of each new volume immediately upon publication. Volumes are billed only upon actual shipment. For further information please contact the publisher.

Saccharomyces

Edited by

Michael F. Tuite

The University of Kent, Canterbury
Kent, England

and

Stephen G. Oliver

Manchester Biotechnology Centre, UMIST
Manchester, England

Plenum Press • New York and London

Library of Congress Cataloging-in-Publication Data

Saccharomyces / edited by Michael F. Tuite and Stephen G. Oliver.
p. cm. -- (Biotechnology handbooks ; v. 4)
Includes bibliographical references and index.
ISBN 0-306-43634-5
1. Saccharomyces--Biotechnology. 2. Saccharomyces. I. Tuite, Michael F. II. Oliver, S. G. (Stephen George), 1949- . III. Series.
TP248.27.Y43S22 1991
660'.62--dc20 90-19442
CIP

ISBN 0-306-43634-5

A Division of Plenum Publishing Corporation
233 Spring Street, New York, N.Y. 10013

Printed in the United States of America

Contributors

J. R. Dickinson • School of Pure and Applied Biology, University of Wales College of Cardiff, Cardiff CF1 3TL, Wales

M. J. Dobson • Department of Botany, University of Nottingham, Nottingham NG7 2RD, England

A. J. Kingsman • Department of Biochemistry, University of Oxford, Oxford OX1 3QU, England

S. M. Kingsman • Department of Biochemistry, University of Oxford, Oxford OX1 3QU, England

C. Kreutzfeldt • Institut fur Pharmakologie und Toxikologie, Philipps-Universitat Marburg, Lahnberge, 3550 Marburg, Federal Republic of Germany

T. M. Matthews • Department of Chemical Engineering, University of Manchester, Institute for Science and Technology, Manchester M60 1QD, England

E. J. Mellor • Department of Biochemistry, University of Oxford, Oxford OX1 3QU, England

Stephen G. Oliver • Manchester Biotechnology Centre, University of Manchester, Institute for Science and Technology, Manchester M60 1QD, England

Michael F. Tuite • Biological Laboratory, University of Kent, Canterbury, Kent CT2 7NJ, England

C. Webb • Department of Chemical Engineering, University of Manchester, Institute for Science and Technology, Manchester M60 1QD, England

R. B. Wickner • National Institutes of Health, Bethesda, Maryland 20892

W. Witt • Institut fur Pharmakologie und Toxikologie, Philipps-Universitat Marburg, Lahnberge, 3550 Marburg, Federal Republic of Germany

Preface

In this volume we aim to present an easy-to-read account of the genus *Saccharomyces* that we hope will be of value to all students and researchers wishing to exploit this important genus, be it for academic or commercial purposes. Individual chapters have been commissioned to cover specific aspects of the biology of *Saccharomyces* species: growth, genetics, and metabolism, with the emphasis on methodology. Basic principles are discussed without an over-detailed, step-by-step breakdown of specific techniques, and lengthy discussions of standard molecular, biological, and biochemical techniques (e.g., polyacrylamide gel electrophoresis, protein purification, DNA sequencing) have been avoided. We hope the volume will provide a quick reference to the current status of a wide range of *Saccharomyces*-specific methodologies without focusing exclusively on recent developments in molecular techniques which can be found in the ever increasing numbers of "cloning manuals." By necessity, much of what is described in this volume concentrates on one particular species of *Saccharomyces,* namely *Saccharomyces cerevisiae.* This is not just a reflection of the authors' interests, but indicates the extent to which this simple eukaryote has been studied by biologists from all walks of life, for all sorts of reasons. If this volume can provide a broader knowledge base to the experienced yeast researcher, or ease the path of someone just starting work with *Saccharomyces,* then we will have achieved our aim.

We are grateful to all the authors and to Ken Derham at Plenum Press for their patience during the preparation of this volume.

Michael F. Tuite

University of Kent, Canterbury

Stephen G. Oliver

University of Manchester

Contents

Chapter 6

Chapter 7

Introduction 1

MICHAEL F. TUITE and STEPHEN G. OLIVER

Fungi have been exploited by mankind for many thousands of years, with perhaps the earliest recorded examples being the use of yeast to make ethanol, a process known to the Sumerians and the Babylonians before 6000 B.C. In the ensuing 8000 years one particular genus of yeast, namely, *Saccharomyces*, has played a central role in the commercial exploitation of fungi by mankind. These facultative anaerobes utilize the Embden–Meyerhof pathway to convert sugars to pyruvic acid, with each molecule of pyruvic acid then being reductively decarboxylated to give rise to one molecule each of ethanol and carbon dioxide. This simple, efficient way of fermenting glucose to ethanol and carbon dioxide has provided the foundation for two of our major food industries, brewing and baking (Fig. 1).

Yet from its earliest foundations yeast "biotechnology" has received little attention from "Biotechnologists," be they from biological or chemical disciplines. The development of brewing and baking strains with improved fermentation characteristics has occurred largely on an empirical basis using technologies that have remained essentially unaltered for millennia. Such traditional approaches do have their limitations, although one has to admit that they have been successful in the development of effective strains for use in many traditional processes. However, it has only been with the advent of modern molecular genetic techniques that yeast biotechnology has begun to take on a new face in which the organism itself has been subjected to the trickery that can profoundly change its genetic constitution. Even 10 years ago the idea that yeasts would be used commercially to produce effective vaccines against hepatitis B virus can hardly have been contemplated, yet it is a testimony to the flexibility and manipulability of these organisms that this dream is now a reality.

MICHAEL F. TUITE • Biological Laboratory, University of Kent, Canterbury, Kent CT2 7NJ, England. STEPHEN G. OLIVER • Manchester Biotechnology Centre, University of Manchester, Institute for Science and Technology, Manchester M60 1QD, England.

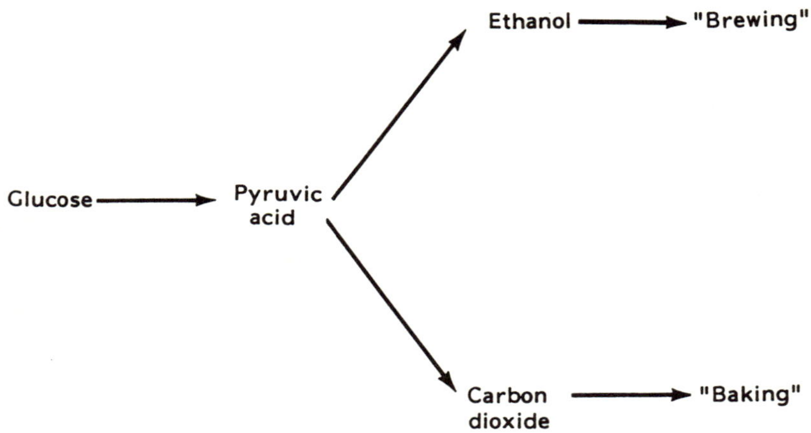

Figure 1. The exploitation of yeast metabolism.

Yeasts have also become important model systems for basic research into the biology of the eukaryotic cell, with one particular species, *S. cerevisiae,* being at the forefront of this research. In the early 20th century detailed studies by Warburg, Crabtree, and others into the fermentative ability of this species revealed a number of important aspects of respiration and fermentation, particularly the regulation of metabolism by glucose. Yet it was not until the pioneering work of Winge and Lindegren in the 1940s that *S. cerevisiae* became the subject of genetic research. Once the limitations of a homothallic lifestyle had been overcome (by mutation), genetic analysis became routine and increasingly came to be used to complement biochemical studies. More recently, with the development of powerful molecular genetic tools, a remarkable upsurge in the exploitation of *S. cerevisiae* in fundamental research has occurred, which has relied to a larger extent on the already accumulated knowledge of the biochemistry, physiology, and genetics of the species. Today, much of our understanding about eukaryotic gene structure and function, cytoskeletal organization, and the cell cycle comes from studies with this simplest of eukaryotes.

For anyone who has not worked with this genus before (and it is largely for such people that this book has been written) *Saccharomyces* yeasts appear not unlike bacterial—they are unicellular and grow rapidly on simple, well-defined media with a population doubling time under 2 hr. Yet they have many of the fundamental characteristics of a higher eukaryotic cell: a nuclear membrane, cytoplasmic organelles such as mitochondria, receptor and second messenger systems, and so on. The ability to rationally manipulate all aspects of gene expression by *in vitro* genetic techniques offers *S. cerevisiae* a unique place among eukaryotic model systems.

To allow rational genetic manipulation of the biosynthetic capabilities of any organism does not just require an ability to clone a gene into the said organism, but at the very least requires a detailed understanding of the organism's life cycle, culture requirements, and metabolism. In this book we aim to provide a series of chapters, each dealing with a fundamental aspect of the biology of *Saccharomyces*, but with the emphasis on methodology and current academic and commercial exploitation of members of this genus. It is not our intention, however, to provide a state-of-the-art "methods" book with highly detailed protocols, but rather to provide an overview of the wide range of traditional and modern methodologies available to yeast researchers, with references to original publications in which the fine details can be found. Lengthy discussions of standard molecular and biochemical techniques are also avoided since these can be found in a number of widely available standard texts.

Structural Biochemistry 2

C. KREUTZFELDT and W. WITT

1. INTRODUCTION

Saccharomyces cerevisiae is a unicellular eukaryote in which a combination of biochemical, genetic, and molecular biological techniques have been used in the study of cell structure, organelle biogenesis and function, and the regulation of growth and mating. This chapter provides an overview of the biochemistry of the structural components of the yeast cell.

2. BASIC MACROMOLECULES IN *SACCHAROMYCES*

2.1. Proteins and Peptides

The architecture of the yeast cell has long attracted the interest of cell biologists, and consequently, many proteins and peptides have been analyzed in respect to their role in regulation and maintenance of cell structure and metabolism. A number of topics have been selected for detailed consideration in this chapter: tubulin and actin are the main structural proteins in the yeast cell; the ribosomal proteins have been physically characterized, but their individual functions are largely unknown; mating factors are peptide hormones which trigger physiological and morphological changes. The discussion of glycoproteins concentrates on the processing and sorting of proteins and the formation of the cell wall.

2.1.1. Tubulin

During the cell cycle, cytoplasmic and nuclear microtubules are formed. This process has been extensively studied by electron microscopy

C. KREUTZFELDT • Phase GmbH, D-2410, Mölln, Federal Republic of Germany.
W. WITT • Institut fur Pharmakologie und Toxikologie, Philipps-Universitat Marburg, Lahnberge, 3550 Marburg, Federal Republic of Germany.

(for reviews, see Byers, 1981; King and Hyams, 1982). Tubulin, the major component of the microtubule proteins, has been purified from *S. cerevisiae* (Clayton *et al.*, 1979; Kilmartin, 1981) and these preparations have been shown to copolymerize with brain tubulin and to self-assemble *in vitro* to form microtubules. Two-dimensional electrophoretic analysis suggests that yeast tubulin, like brain tubulin, contains two subunits, alpha and beta, with mol. wt. 55,000 and 52,000, respectively. Besides similarities in molecular weight and polymerization between yeast and brain tubulins, there are differences concerning the number of protofilaments of microtubles formed *in vitro,* the sensitivity to certain drugs, and the speed of *in vitro* depolymerization (Kilmartin, 1981). Although cDNA clones for alpha and beta chicken brain tubulin did not hybridize to yeast DNA under stringent conditions (Cleveland *et al.*, 1980), it was nevertheless possible to use such probes to isolate the yeast beta tubulin gene *TUB2* (Neff *et al.*, 1983). The amino acid sequence predicted from the DNA sequence of *TUB2* gives rise to a protein of mol. wt. 51,073 (457 residues) which shows sequence homology of more than 70% with chicken beta tubulin.

Recently, the yeast alpha-tubulin genes *TUB1* and *TUB3* have been sequenced (Schatz *et al.*, 1986a). *TUB1* and *TUB3* give rise to slightly different polypeptides with mol. wt. 49,701 and 49,694, respectively. Both alpha-tubulins are highly homologous to those from other species. Yeast microtubules have been shown to contain both alpha-tubulins. Null mutation experiments have led to the conclusion that both tubulins essentially fulfill common qualitative functions (Schatz *et al.*, 1986b).

2.1.2. Actin

Yeast actin has an intracellular concentration of about 0.5 mg/ml cell volume (Greer and Schekman, 1982a) and its physical and physiological properties have been studied *in vitro*. The protein has been purified from the cytosol of *S. cerevisiae* by column chromatography to yield a polypeptide of mol. wt. 43,000 (Greer and Schekman, 1982a). Inhibition of DNaseI activity can be used as an assay to identify fractions containing actin.

KCl and Mg^{2+} induce polymerization of actin to produce microfilaments of 7 nm diameter which can be decorated with heavy muscle meromyosin (HMM). HMM ATPase activity has been found to be stimulated by yeast actin. Phalloidin binds to yeast actin filaments and protects them from depolymerization induced by KI. While cytochalasin B decreases the stability of yeast and muscle actin to equal extents, a number of differences between these two actins have been identified. In contrast to muscle actin, the polymerization of yeast actin is strongly affected by low concentrations of Ca^{2+} (Greer and Schekman, 1982b) by increasing the critical polymerization concentration. This effect is rapidly reversed by complexing Ca^{2+} with EGTA or by addition of Mg^{2+}. It has been suggested that Ca^{2+} interacts directly with yeast actin.

In yeast there is a single gene for actin and this has been cloned and sequenced (Gallwitz and Sures, 1980; Ng and Abelson, 1980). The predicted amino acid sequence consists of 374 residues and is about 90% homologous to mammalian cytoplasmic gamma actin and *Physarum* actin. Disruption of the yeast actin gene has confirmed that it encodes an essential function (Shortle *et al.,* 1982). Conditionally expressed actin gene mutations have been constructed and these have revealed that actin is involved in the organization, assembly, and function of the yeast cell surface (Shortle *et al.,* 1984; Novick and Botstein, 1985; see also Section 4.3).

2.1.3. Ribosomal Proteins

As in other eukaryotes, the yeast cytoplasmic ribosome consists of a large (60S) and a small subunit (40S). The precise number of the yeast ribosomal proteins (r-proteins) found in each subunit is still under contention, since different results have been obtained depending on conditions of preparation and the two-dimensional (2-D) polyacrylamide gel electrophoresis (PAGE) system used. However, there is agreement that the small subunit of *Saccharomyces* contains 30 ± 5 and the large one 40 ± 5 proteins (Warner, 1982). The r-proteins comprise about 15% of total cell proteins. The numbering of r-proteins largely depends on the 2-D PAGE system used for resolution. Therefore, different nomenclatures exist in the literature. Most of them designate the proteins derived from the small subunit with the prefix S to a number and those from the large one with a L. In a comparative study, Otaka and Osawa (1981) have correlated the most common nomenclatures for r-proteins of *S. cerevisiae*. In more recent publications, the nomenclature of Kruiswijk *et al.* (1978) is preferred.

Almost all r-proteins have been purified by preparative column chromatography. Their amino acid composition and, in some cases, the N-terminal amino acid sequence have been determined (Itoh *et al.,* 1979; Otaka *et al.,* 1982, 1983) by protein chemical methods. Like r-proteins from other sources, most yeast r-proteins are basic and only a few are acidic (S2, S5, S19, S26, L1, L24, L44/45; nomenclature of Kruiswijk *et al.,* 1978).

Several phospho-r-proteins (S2, S10, L44/45; nomenclature of Kruiswijk *et al.,* 1978) have been identified in *Saccharomyces*. S10 seems to be the analog of mammalian S6 (Zinker and Warner, 1976; Leer *et al.,* 1982).

Resistance to trichodermin, cycloheximide, and cryptopleurine has been found to be conferred by altered ribosomal proteins L3, L29, and rp59, respectively, and this phenomenon has been used for molecular cloning of these proteins (for literature see below).

A considerable number of genes coding for yeast r-proteins S7, S10, S11, S16A, S24, S31, S33, L3*(tcm1)*, L4, L16, L17A, L25, L29*(cyh2)*, L32, L34, L36, L46, rp24, rp28, rp29, rp39, rp51, rp52, rp58, rp59*(cry1)*, and rp63 have been cloned, and several have also been sequenced (the mutations conferring resistance to the antibiotics mentioned are given in parentheses). These investigations permit the following conclusions about r-protein gene organization in yeast: (1) the r-protein genes are normally not clustered, (2) some r-proteins are encoded by two unlinked genes, and (3) many, but not all, r-protein genes contain introns (Leer *et al.*, 1984, 1985a; Warner *et al.*, 1985; Larkin and Woolford, 1983, and literature cited therein). Current research is directed to understanding how the synthesis of about 70 different r-proteins is coordinately regulated. There are indications that this occurs posttranscriptionally, but the autogenous regulation seems to be rather different from that elucidated for *Escherichia coli* (Warner *et al.*, 1986; Planta *et al.*, 1986; Gourse *et al.*, 1986; Nomura *et al.*, 1984). However, according to present knowledge, regulation mechanisms that act on the transcriptional level cannot be excluded. Comparison of DNA sequences of various yeast r-protein genes has revealed two different conserved sequences located within a region from about −200 to −500 upstream of the polypeptide initiation codon (Teem *et al.*, 1984; Leer *et al.*, 1985b). These sequences, called HOMOL1 [consensus sequence: AACATC(CT)(GA)T(GA)CA] and ribosomal protein gene (RPG) box [consensus sequence: ACCCATACAT(TC)(TA)], have been exclusively found in upstream regions of r-protein genes (for the two known exceptions, see below), but in some of them the HOMOL1 or RPG box is missing. If both HOMOL1 and RPG box are present, the latter is located 3′ to the HOMOL1 box. The genes coding for the r-proteins S33 and L3 seem not to be flanked by any of these boxes. HOMOL1 and RPG box have been suggested to be involved in the control of r-protein gene transcription as enhancer-like structures (Leer *et al.*, 1985b). This view has been strengthened by Woudt *et al.* (1986), who have shown that deletion of one or both conserved upstream sequences (RPG1, RPG2) of ribosomal protein gene L25 strongly affects the transcription of this gene.

5′ upstream sequences which are closely related to HOMOL1 and RPG have been also identified in the *TEF1* and *TEF2* genes, respectively, which both code for yeast elongation factor EF1-alpha (Huet *et al.*, 1985). A protein factor named translational upstream factor (TUF) detected in a partially purified protein fraction has been shown to bind with varying affinities to HOMOL1 and RPG boxes of several ribosomal protein genes, as well as to the related sequences of the *TEF1* and *TEF2* genes (Huet *et al.*, 1985; Vignais *et al.*, 1987). The TUF protein may thus be involved in the coordinate regulation of those genes which code for the proteins of the translational machinery.

2.1.4. Mating Factors

Haploid *S. cerevisiae* cells constitutively secrete peptide pheromone into the medium. They are known as alpha- and a-factors and are produced by cells of alpha- and a-mating type, respectively. These factors, which act only on cells of the opposite mating type, induce the mating process (Thorner, 1981; Sprague *et al.*, 1983), finally yielding alpha/a diploids insensitive to either pheromone. Alpha-factor, which is now commercially available (e.g., Sigma, St. Louis), is commonly used for the synchronization of cell cultures (see also Section 6). The pheromone peptides have been purified and their amino acid sequences have been established:

a-factor:
NH_2-Tyr-Ile-Ile-Lys-Gly-Val-Phe-Trp-Asp-Pro-Ala-(S-Far)Cys-OCH_3

or

NH_2-Tyr-Ile-Ile-Lys-Gly-Leu-Phe-Trp-Asp-Pro-Ala-(S-Far)Cys-OCH_3

alpha-factor:

NH_2-Trp-His-Trp-Leu-Gln-Leu-Lys-Pro-Gly-Gln-Pro-Met-Tyr-COOH

(S-Far)Cys-OCH_3 denotes S-farnesyl cysteine methyl ester.Cyx denotes a cysteine residue modified by an as yet unkonwn group (Stötzler *et al.*, 1976; Andregg *et al.*, 1988). Most likely, there are two genes in a-cells each of them coding for a different a-factor. Thus two different factors are secreted by a single haploid strain (Thorner, 1981; Sprague *et al.*, 1983). The N-terminal Trp residue is missing in some alpha-factor preparations, possibly owing to action of an aminopeptidase. However, the remaining peptide exhibits full biological activity. Almost any other modification, including oxidation of the Met residue, results in severe reduction of biological activity.

The response of target cells to pheromone (arrest of growth at G1, elongation yielding pear-shaped cells) can be directly assayed under the microscope (Duntze *et al.*, 1973). The action of the pheromone is only transient, since both factors are rapidly degraded by proteases of the cells of opposite mating type. The rate of degradation has been estimated to be about 10^7 alpha-factor molecules/min at a concentration of 10^6–10^7 cells/ml (25°C). The usual concentration of pheromones in the

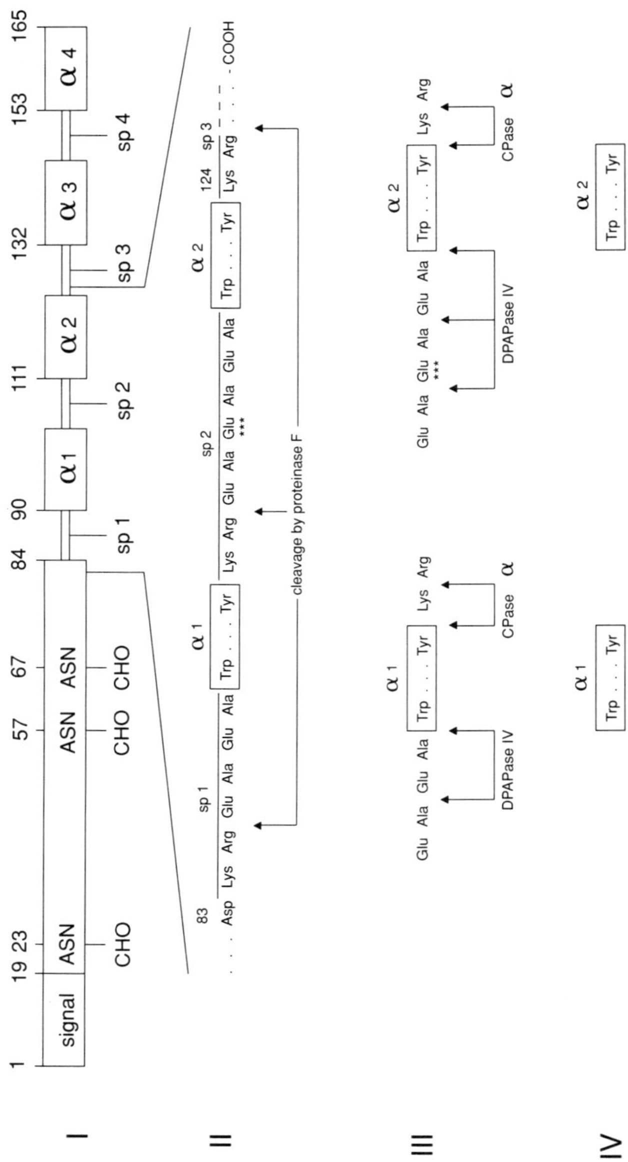
I
1
19
23
57
67
84
90
111
132
153
165
signal
ASN
ASN
ASN
CHO
CHO
CHO
α 1
α 2
α 3
α 4
sp 1
sp 2
sp 3
sp 4
II
83
Asp
Lys Arg Glu Ala Glu Ala
sp 1
Trp . . . Tyr
Lys Arg Glu Ala Glu Ala Glu Ala
sp 2
***
124
Lys Arg
sp 3
- COOH
cleavage by proteinase F
III
Glu Ala Glu Ala
DPAPase IV
CPase α
Lys Arg
IV
α 1
α 2

Figure 1. Proposed scheme of alpha-factor maturation. This model has emerged from studies by different groups (Kurjan and Herskowitz, 1982; Julius *et al.*, 1983, 1984; Achstetter and Wolf, 1985a). Since alpha-factor maturation comprises glycosylation and secretion, we refer also to Sections 2.1.5 and 3.3. Transcription of the alpha-factor gene finally gives rise to an mRNA which codes for a large polypeptide containing a leader sequence with a terminal signal sequence and four tandemly arranged sequences, each consisting of a spacer sequence (sp 1, etc.) followed by the alpha-factor amino acid sequence (stage I). The prepro-alpha-factor is suggested to be cotranslationally inserted into the endoplasmic reticulum where the polypeptide backbone (mol. wt. 18,600) is core-glycosylated at sites indicated (CHO). Core glycosylation increases the mol. wt. to 26,000, which is consistent with the addition of three $(GlcNAc)_2$ $(Man)_9$ units. Unlike other secreted proteins (e.g., invertase), the signal sequence is not removed at this stage. It has been suggested that the signal sequence is required to immobilize the prepro-alpha-factor on membrane structures to ensure proper processing. There are some indications that in the Golgi bodies, mannose-rich outer chains are polymerized onto the core oligosaccharides. Also, proteolytic processing seems to begin at this stage and continues in the secretory vesicles which appear later in the secretory pathway. The first proteolytic cleavages in the prepro-alpha-factor molecule are performed, most likely, by proteinase F which has been found to be associated with membranes (stage II). This enzyme recognizes the first two residues (Lys Arg) at the N-termini of each of the spacer sequences. Spacer sequence 2 is identical to that of spacers 3 and 4 except that the residue marked by "***" is replaced by Asp in spacers 3 and 4 (the amino acid sequence of the last two tandem sequences is not shown; the processing of alpha 3 and 4 happens exactly as shown for alpha 1 and 2). Proteinase F is believed to be the product of the *KEX2* gene, since mutants carrying the *kex2* mutation are devoid of this enzyme activity and defective in production of active alpha-factor and killer toxin. After proteinase F action, the four pro-alpha-factor molecules are further processed by enzymatic removal of additional residues. The N-terminal residues are cleaved off step by step most likely by dipeptidyl aminopeptidase IV (DPAPase) and those at the carboxyl terminus by carboxypeptidase alpha (CPase; stage III), finally yielding four mature alpha-factor molecules (stage IV; for complete amino acid sequence of alpha-factor, see text). Both enzymes have been found to be attached to membranes. It should be noted that these processing intermediates cannot be detected in wild-type cells but only in mutants, i.e., those belonging to the *sec* family. Recently, synthesis and glycosylation of prepro-alpha-factor have been performed in homologous yeast cell-free extracts (Rothblatt and Meyer, 1986a; Waters and Blobel, 1986).

medium is in the range from 10^{-7} to 10^{-8} M. Investigations using mutants (*sst1* or *bar1*) deficient in degradation of alpha-factor have shown that the ratio of pheromone concentration and titer of wild-type responding cells is important for the biological effects elicited.

The genes coding for the a- and alpha-factor have been sequenced (Brake *et al.,* 1985; Kurjan and Herskowitz, 1982). One alpha-factor gene codes for a large prepropolypeptide (165 residues) from which four copies of mature alpha-factor are excised during protein processing (Fig. 1). Another alpha-factor gene codes for a precursor (120 amino acids) that contains two copies of mature alpha-factor (Singh *et al.,* 1983).

Receptors for pheromones on haploid cells have been inferred from kinetic and genetic investigations (Jenness *et al.,* 1983). The products of the *STE2* and *STE3* genes are believed to be constituents of the receptors for alpha-factor and a-factor, respectively. This suggestion has recently been further confirmed (Nakayama *et al.,* 1987; Hagen *et al.,* 1986, and references therein). The number of alpha-factor binding sites and the dissociation constant have been estimated to be 8000 sites/cell and 6×10^{-9} M, respectively (Jenness *et al.,* 1986).

2.1.5. Glycoproteins

Many of the yeast proteins undergo posttranslational modification by covalent binding of nonprotein groups, e.g., phosphorylation (McDonough and Mahler, 1982; Castellanos and Mazon, 1985) and acylation (Wen and Schlesinger, 1984; Towler and Glaser, 1986). In addition, about 3% of the membrane protein and less than 2% of the total cell protein are glycosylated (Tanner, 1984).

Generally, the carbohydrate moiety of yeast glycoproteins is less variable than those of mammalian cells. Oligosaccharide chains bound to serine or threonine (*O*-linked) are not longer than five carbohydrate residues. Those bound to asparagine (*N*-linked) are composed of a large number of mannose units forming branched oligosaccharides, which are attached via two *N*-acetylglucosamine residues to the protein backbone. Synthesis and structure of glycoproteins in yeast and related topics are covered in reviews (Tanner, 1984; Schekman, 1985; Tanner and Lehle, 1987).

The synthesis mechanism of the *O*-linked oligosaccharides in yeast and other fungi differs from that of mammalian cells in that the first mannosyl residue directly attached to serine or threonine is transferred from dolichyl-P-Man in the endoplasmic reticulum (ER). GDP-Man is the donor for the outer carbohydrate units. However, the synthesis of the *N*-linked carbohydrate chains proceeds in essentially the same way as in mammalian cells. Briefly, a core oligosaccharide of the composition $(GlcNAc)_2Man_9Glc_3$ linked to dolichol via a pyrophosphate bridge is transferred to a target asparagine residue of a protein chain inserted into the rough ER. Target asparagines are those in a -Asn-X-Ser/Thr motif. The large part of this oligosaccharide is assembled by the subsequent transfer of two GlcNAc and five Man residues from nucleotide donors to dolicholphosphate. This heptasaccharide is then elongated by attaching four more Man and three Glc units from dolicholphosphate-Man or -Glc as donors. Three Glc and one Man residue are cut off rapidly (Byrd *et al.*, 1982) and up to seven mannosyl units may be transferred after transport of the glycoprotein to the Golgi to yield the modified core oligosaccharide of yeast mannoproteins (Tanner and Lehle, 1987). An outer chain consisting of 50 or more Man residues may be attached to the core region of secretory glycoproteins.

The structure of the carbohydrate moiety of these *S. cerevisiae* high-mannose mannoproteins is shown in Fig. 2 (Tanner and Lehle, 1987). The assembly of carbohydrate in yeast glycoproteins have been analyzed using numerous mutants defective in glycoprotein processing (designated *mnn*), and structural details have been elucidated by selective chemical and enzymatic degradations and by nuclear magnetic resonance (NMR) spectroscopy (see, e.g., Byrd *et al.*, 1982; Tsai *et al.*, 1984; Trimble and Atkinson, 1986, and references therein). This topic is also covered in reviews (Ballou, 1982; Catley, 1983; Tanner and Lehle, 1987).

Many glycoproteins with enzymatic properties have been described (Tanner and Lehle, 1987). A few of them have been analyzed in detail and are discussed below.

Invertase occurs in two mature forms in yeast: one is secreted and glycosylated; the second, a minor form, is found in the cytoplasm and is not glycosylated. A single invertase structural gene (*SUC2*) specifies two mRNA transcripts for both forms (Carlson and Botstein, 1982). The larger transcript includes a coding sequence for the signal peptide that determines the secretion of the glycosylated form of invertase (Carlson *et al.*, 1983). *Saccharomyces cerevisiae* external invertase is composed of two indistinguishable subunits each of mol. wt. about 60,000. Six core-sized oligosaccharide chains and three polymannose chains ranging in size to over 50 mannosyl residues are attached to each of the polypeptides, resulting in a mol. wt. of about 135,000 in a single subunit (Trimble *et al.*, 1983, and references therein). Multiple forms of external invertase have been separated by isoelectric focusing. The different charges of the glycoproteins have been found to be mainly due to variable contents of mannosylphosphodiester groups in the polymannose outer chains (Frevert and Ballou, 1982). The higher functional stability of the secreted invertase compared to the internal form has been attributed to the carbohydrate moiety, which facilitates refolding of the denaturated polypeptide and the formation of enzymatically active oligomers (Chu *et al.*, 1985).

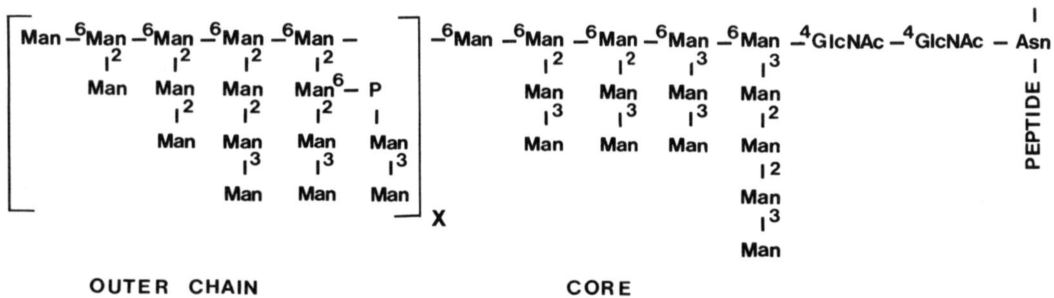

Figure 2. Structure of N-linked carbohydrate chains of mannoproteins, according to Tanner and Lehle (1987). The linkages between both GlcNAc residues, and between GlcNAc and Man are of β type. All other Man residues are (α1–6)-, (α1–2)-, or (α1–3)-linked, as indicated here.

In contrast to invertase, yeast carboxypeptidase Y, a vacuolar glycoprotein with a mol. wt. of about 60,000 and a carbohydrate content of about 17%, contains only core oligomannose units with an average chain length of 13 mannosyl residues (Hasilik and Tanner, 1978). Some of these chains are phosphorylated with all of the phosphate diesterified (Hashimoto *et al.*, 1981) or in phosphomonoester groups (Schwaiger *et al.*, 1982).

Only three of the four oligosaccharide chains in the native enzyme are suspectible to hydrolysis by endoglycosidase H, an enzyme that splits the chitibiosyl group attached to asparagine in glycoproteins (Tarentino *et al.*, 1974). The fourth group located at Asn becomes accessible after denaturation of the glycoprotein with SDS. Only seven of the nine chains of native yeast external invertase are readily released by endoglycosidase H. In both enzymes, all of the carbohydrate-bound phosphate is removed with the accessible oligosaccharides. These findings have been explained by the hypothesis that some of the carbohydrate chains in both enzymes become inaccessible during glycoprotein maturation to mannose elongation and phosphate addition (Trimble *et al.*, 1983).

2.2. Polysaccharides, Structural Mannoprotein, and Polyphosphate

Chemically heterogenous compounds are discussed together in this section because of their structural and functional relationship. Glycogen and polyphosphate function as storage or reserve substances in the yeast cell; mannan (now preferentially called mannoprotein), glucan, and chitin are the main structural constituents of the cell wall.

Polyphosphate is composed of inorganic phosphate in pyrophosphate linkages with chain lengths of three to several hundred. Polyphosphate is found mainly in the vacuole (Urech *et al.*, 1978; Wiemken *et al.*, 1979), but has been also detected on the outside of the yeast plasma membrane by toluidine-binding (Tijssen *et al.*, 1981). In addition to its function as phosphate reserve in yeast cells, polyphosphate seems to be involved in the sequestration of basic amino acids in the vacuole (Dürr *et al.*, 1979).

Glycogen has about the same structure as in mammalian cells (see Catley, 1983, and references therein). The involvement of this storage polysaccharide in the carbohydrate metabolism is discussed in Chapter 3 of this volume.

Glucan is one of the main components of the yeast cell wall. Depending on the extraction procedure and the yeast strains used, different classes of this heterogenous group of polysaccharides have been described. The major alkali-insoluble component consists of (β 1-3)-linked D-glycopyranose units with occasional branching via (β 1-6)-

linked glucose. A mol. wt. of at least 240,000 has been found. A minor component with a minimum mol. wt. of 22,000 contains mainly glucose in (β 1-6)-linkages. A polysaccharide with an extended (β 1-3)-glucose chain having a low degree of branching and side chains with (1-3)- and (1-6)-linked glucose residues constitutes the main component of the alkali-soluble glucan fraction (see Duffus *et al.,* 1982, and Catley, 1983, for reviews and references).

Glucan synthetase catalyzing the formation of (1-3)-β-D-glucan from UDP-glucose is probably localized at the inner surface of the plasma membrane (Shematek *et al.,* 1980). Shematek and Cabib (1980) have proposed a complex model for the regulation of this enzyme: GTP and a heat-labile "supernatant factor" which is phosphorylated by a kinase activate, while Mg^{2+} inhibits the glucan synthesis. Mg^{2+} is a cofactor for endogenous phosphatases, which dephosphorylate the "supernatant factor."

A variety of activating as well as inhibitory and so far uncharacterized endogenous factors have been extracted from different yeast cell fractions (Guillen *et al.,* 1985).

Chitin, a linear polysaccharide of (β 1-4)-linked GlcNAc, is found only in the primary septum and in the bud scars of the cell wall. The synthesis also proceeds at the inner surface of the plasma membrane and is strictly regulated by the cell cycle events (Cabib *et al.,* 1984). Two chitin synthases (called I and II) have been found in yeast (Orlean, 1987). Chitin synthase I is present in the cell as a zymogen, which can be converted to the active form by a variety of proteases including trypsin and yeast protease B (Duran and Cabib, 1978), but the responsible protease activating *in vivo* remains to be characterized. Chitin synthase I has been isolated by trapping it in the soluble chitin formed in the reaction with UDP-GlcNAc as donor (Kang *et al.,* 1984). Chitin synthase II is highest in the logarithmic growth phase and can be measured without protease activation. It has been proposed that chitin synthase II is responsible for chitin formation *in vivo,* while chitin synthase I may normally be inactive (Orlean, 1987). The insertion of the enzyme into the plasma membrane bilayer (Sentandreu *et al.,* 1984) and the restriction of the active enzyme(s) to certain sites in the membrane (Cabib *et al.,* 1984) have been also discussed as possible regulatory factors of chitin deposition and, consequently, of septum formation.

Mannoprotein is another major structural component of the yeast cell wall. The bulk mannoprotein is a heterogenous group of proteoglycans with different molecular weights and carbohydrate contents. More than 60 bands have been separated by SDS-PAGE of SDS-extracts purified from *S. cerevisiae* cell walls (Valentin *et al.,* 1984). The protein content of these compounds usually does not exceed 10%. The carbohydrate moiety of cell wall mannoprotein is well analyzed (see Cabib *et al.,*

1982; Ballou, 1982; Catley, 1983; Tanner and Lehle, 1987, for reviews and references). It is synthesized via the same lipid intermediates as the mannoprotein enzymes described earlier. Branched polysaccharide chains consisting of up to 150 Man units are linked to Asn and short oligomannan chains to serine or threonine.

In contrast to some mannoprotein enzymes, the protein moieties of the structural mannoproteins are largely unknown. Only a few electrophoretically homogenous compounds have been isolated. A fraction called "structural cell wall mannoprotein" has been extracted from *S. cerevisiae mnn9* mutant cell wall (Frevert and Ballou, 1985). This uniformly sized (mol. wt. 180,000) mannoprotein consists of 88% carbohydrate, which is evenly distributed between asparagine and hydroxyamino acids. The protein moiety shows an unusually high proline content of 11%. Attempts to further characterize the protein moiety and to prove its homogeneity have failed since it has not been possible to remove the carbohydrate completely without degradation of the protein. Another glycoprotein with mol. wt. 33,000 which contains only one *N*-linked core-sized oligosaccharide has been released from the cell wall by treatment with zymolyase (Pastor *et al.,* 1984), and its secretory pathway has been analyzed by serological procedures (Sanz *et al.,* 1987). An *O*-glycosylated glycoprotein with mol. wt. 22,000 has been extracted from the surface of intact cells with mercaptoethanol. Evidence has been presented that this is a mating-type-specific a-cell agglutinin (Orlean *et al.,* 1986).

2.3. Lipids

The lipid content of yeast cells can vary considerably as a consequence of a variety of influences. Nevertheless, yeast strains have been classified in respect to the percentage of lipids in the dry weight as low-lipid strains (<5%), medium (5–15%), and high-lipid strains (>15%). *Saccharomyces cerevisiae* strains belong to the medium and high-lipid group (Rattray *et al.,* 1975).

The function and synthesis of lipids as membrane constituents and the phospholipid composition of *S. cerevisiae* cells and subcellular membrane fractions have been reviewed (Henry, 1982; Prasad, 1985). A more general overview on yeast lipids (Rattray *et al.,* 1975) also includes lipid-related substances and minor compounds. In a comparative study, the lipid composition of 30 species of yeast has been analyzed (Kaneko *et al.,* 1976). The efficiency of different lipid extraction methods suitable for yeast cells has been compared. Initial heating in 80% ethanol has been recommended for complete extraction of polar lipids and inactivation of lipases (Hanson and Lester, 1980).

An array of glycerophospholipids typical for eukaryotic cells is found in the group of the polar lipids, comprising phosphatidylcholine, -ethanolamine, -monomethylethanolamine, dimethylethanolamine, -glycerol and -inositol, cardiolipin, phosphatidic acid, and lyso compounds. An unusual high content of phosphatidylinositol (almost 40% of the complete phospholipid fraction) has been detected in certain mutant strains (Henry, 1982). Phosphatidylinositol biphosphate is also present in yeast cells, and evidence has been presented that *RAS*-encoded GTP-binding proteins are involved in regulation of the release of inositol monophosphate and triphosphate from inositol phospholipids (Kaibuchi *et al.*, 1986).

Sphingolipids are represented in yeast cells by a group of inositol phosphoceramides with the structure mannose-(inositolphosphate)$_{1\text{-}2}$-ceramide. *In vitro* studies have shown that transfer of phosphoinositol from phosphatidylinositol is involved in the biosynthesis of these compounds (see Becker and Lester, 1980, for references). C-16 and C-18 fatty acids predominate in the pattern of yeast polar lipids, oleic acid (18:1) being the major component in many yeasts (Rattray *et al.*, 1975). Yeast cells are capable of fatty acid elongation and, consequently, the occurrence of fatty acids with chain lengths of up to C-30 has been reported (Welch and Burlingame, 1973).

As in all typical eukaryotic cells, phospholipids represent the major membrane lipid constituents, together with free sterols. Ergosterol is the main sterol component in yeast cells (Rattray *et al.*, 1975). Sterol esters are one of the main constituents of lipid particles. These particles also contain triglycerides and small amounts of free fatty acids, phospholipids, and protein. They function as lipid storage vesicles which may serve as a reservoir of building blocks for membrane lipids (Clausen *et al.*, 1974). A significant increase in the content of unsaturated free fatty acids in baker's yeast cells at the end of the growth phase in batch culture has been described. Based on the finding that valyl-tRNA synthetase was inhibited by unsaturated free fatty acids, a regulatory role of these compounds in growth control and protein synthesis has been suggested (Black, 1985).

The lipid content and the composition of membrane lipids are more variable in yeast than in any other eukaryotic organism. Factors influencing lipid composition include carbon source and other nutrients, temperature, and lipid precursors in the medium. Growth at anaerobic conditions prevents the synthesis of ergosterol and unsaturated fatty acids. A growing number of mutants defective in synthesis of fatty acids, phospholipids, and sterols has been isolated (see Henry, 1982, and Prasad, 1985, for references). Since reliable procedures for isolation of organelle membranes (e.g., mitochondrial membranes and plasma membrane) are available, the yeast cell represents an excellent subject for studying the influence of individual lipid components on membrane structure, function, and biogenesis. Examples of recent work in this field are the investigation of the influence of sterols on the lipid phase transitions by use of fluorescence polarization (Low *et al.*, 1985), the correlation of lipid synthesis and invertase secretion (Letts and Dawes, 1983), and the effect of plasma membrane phospholipid unsaturation on the general amino acid permease (Calderbank *et al.*, 1985). The involvement of lipids in transport processes and some aspects of the coordination of membrane phospholipid synthesis have been reviewed (Henry *et al.*, 1984; Prasad and Rose, 1986).

2.4. Ribonucleic Acids

In this section we refer to nuclear encoded mRNA, rRNA, and tRNA. Nuclear DNA and related topics are covered in Chapter 9 of this volume. Some aspects of the mitochondrial nucleic acids are discussed in Section 3.4.

2.4.1. mRNA

As in other eukaryotes, yeast mRNAs are synthesized as precursor molecules which are posttranscriptionally processed to yield finally mature translatable mRNAs. The monocistronic coding region of mature yeast mRNA is flanked by noncoding sequences, as in other eukaryotes. At the 5′ end there is a leader sequence with a terminal cap structure and at the 3′ end most mRNA species carry a poly(A) tail of variable length. The functional significance of leader and trailer sequences of eukaryotic mRNA has been reviewed by Baralle (1983) and it will not be discussed here.

Introns appear to be comparatively rare in yeast protein-encoding genes, except for ribosomal protein genes (see also Section 2.1.3). Introns have also been detected in the actin, Matal, *TUB1, TUB3, KIN28,* and *COX4* genes (Gallwitz and Sures, 1980; Ng and Abelson, 1980; Miller, 1984; Schatz *et al.*, 1986a; Simon *et al.*, 1986; Schneider and Guarente, 1987). It has been found that the mechanism of yeast pre-mRNA splicing parallels that in the mammalian system (reviewed by Padgett *et al.*, 1985, 1986), since formation of lariat RNAs could be demonstrated during splicing of pre-mRNAs whose mature forms code for the yeast ribosomal proteins L29 and rp51A and for actin (Newman *et al.*, 1985; Rodriguez *et al.*, 1984; Domdey *et al.*, 1984). For lariat RNA formation, a branching site within the intron is important besides the conserved 5′ terminal splice junction. In *Saccharomyces* this branching site (UACUAAC), which is located about 20–60 nucleotides upstream of the 3′ splice site, is highly conserved and has been found in all introns sequenced thus far (Langford and Gallwitz, 1983). In mammals, however, this branch site does not reveal such a high degree of conservation.

There is considerable evidence that U1 snRNPs participate in splicing in the mammalian system (Mattaj, 1984). In *S. cerevisiae,* there are at least 24 unique snRNAs (Guthrie, 1986). Some of them are clearly nonessential for growth, in contrast to, at least, the giant LSR1 RNA (1175 nucleotides). The LSR1 RNA, which contains homologies to vertebrate U2 and U4–U6 snRNAs (Guthrie, 1986; Ares, 1986, and references cited therein), has also been suggested to be the large trimethyl-capped RNA component in the yeast spliceosome (Ares, 1986; Pikielny *et al.,* 1986; Brody and Abelson, 1985). Besides the LSR1 RNA, three different snRNAs of 160–215 nt in length have been reported also to be present in the splice complex (Pikielny *et al.,* 1986).

The importance of splicing in gene regulation is currently under study, and Warner *et al.* (1985) have suggested that, in *Saccharomyces,* certain ribosomal proteins may regulate their own synthesis by binding to pre-mRNA molecules and inhibiting the splice reaction.

Mature mRNA molecules appear mainly in the cytosolic fraction, but have also been detected in the rough ER and in the mitochondrial fractions, respectively (Swida *et al.,* 1982; Ades and Butow, 1980). In the latter fraction, cytoplasmic polysomes have been found to be associated with the mitochondrial outer membrane (Ades and Butow, 1980). The ER-bound mRNA has been demonstrated to be translatable *in vitro* (Swida *et al.,* 1982).

The half-lives of mature mRNAs have been estimated by use of the temperature-sensitive mutant *rna1* in which the transport of RNA from the nucleus to the cytoplasm is blocked at the restrictive temperature (Koch and Friesen, 1979; Chia and McLaughlin, 1979). A typical experimental design permitting an estimate of mRNA half-lives is described next:

The cellular proteins of *rna1* are pulse-labeled (e.g., 3 min, ^{35}S-methionine) after shift to the nonpermissive temperature. An equal amount of long-term labeled cells (^{3}H-methionine) grown at the permissive temperature is added to the pulse-labeled cells for purpose of internal standardization. Subsequently, the proteins are extracted, submitted to 2-D PAGE, and the radioactivity (^{35}S:^{3}H ratio) of single spots is determined. This experiment is carried out several times, whereby the temporal interval between temperature shift and start of the pulse-labeling procedure is progressively increased. It can be observed that the *de novo* synthesis of individual proteins decreases with time. For calculation of the mRNA half-lives, it is assumed that the synthesis rate of individual proteins is proportional to the amount of fuctional mRNA.

The half-lives deduced from these studies have been found to be within the range from 3.5 min to more than 70 min, with an average value around 20 min. In similar experiments, Gorenstein and Warner

Table I. Half-Lives of mRNAs Coding for Ribosomal Proteins as Measured by Gorenstein and Warner (1976)[a]

Ribosomal protein	Half-life time (min)
S7	10.5
L5	11.1
L6	9.6
rp18	11.1
S13	8.7
rp28	8.5
rp29	11.7
S12	12.2
L22	12.1
S28	8.7
S27	10.2

[a] Whenever possible, the ribosomal proteins are designated according to the nomenclature of Kruiswijk *et al.* (1978).

(1976) have measured the half-lives of mRNAs of several ribosomal proteins (Table I).

2.4.2. rRNA

Each yeast ribosome contains four molecules of rRNA. The small ribosomal subunit contains a 17S (18S) rRNA and the large subunit has three RNA molecules with a size of 5S, 26S (25S), and 5.8S (the values given in parentheses are alternatively used in the literature). The latter is associated with the 26S (25S) rRNA by hydrogen bonding. The 26S (25S) and 17S (18S) rRNA of *Saccharomyces* have already been sequenced (Veldman *et al.*, 1981a; Georgiev *et al.*, 1981; Rubtsov *et al.*, 1980), and the 5S and 5.8S rRNA sequences of different yeasts and other organisms have been compiled (Erdmann *et al.*, 1985). Synthesis and maturation of rRNAs are quite similar in all eukaryotes. For *S. cerevisiae* these processes have been reviewed by Warner (1982). Hadjiolov (1985) has made a detailed survey of the nucleolar events and ribosome biogenesis in the eukaryotic kingdom. We provide only a short summary of the basic processes that have been elucidated for *S. cerevisiae*.

The rDNA is organized in approximately 100–140 repeat units located on chromosome XII. Transcription of a repeat unit yields two transcripts. One transcript is the 5S rRNA, which does not undergo posttranscriptional alterations. The other is a 37S (35S) transcript, which contains the sequences for 17S (18S), 5.8S, and 26S (25S) rRNA. This precursor is methylated (about 60 methyl groups per molecule). Termi-

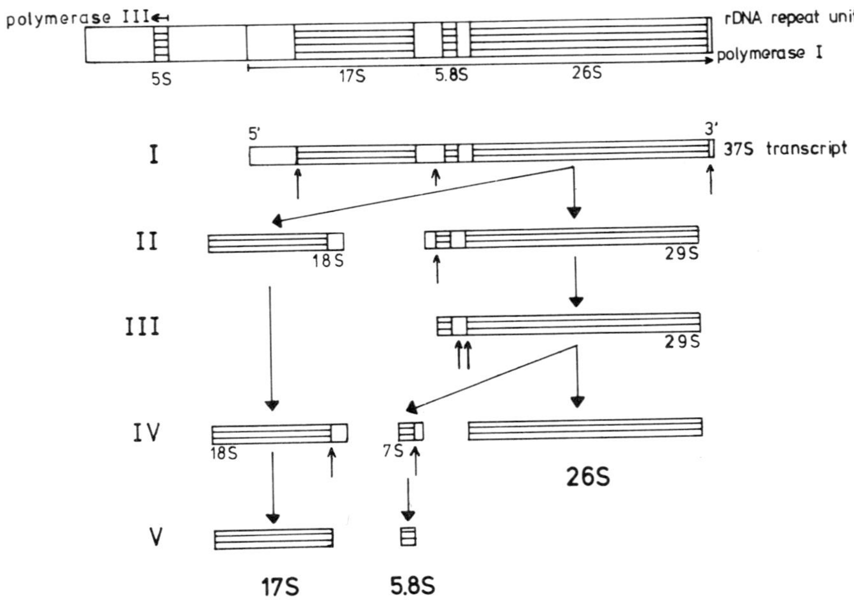

Figure 3. Processing of the 37S rRNA primary transcript yielding 26S, 17S, and 5.8S rRNA. This model has been suggested by Veldman *et al.* (1981b). Although it fits the available experimental data, direct evidence is missing. Stages I–V represent a time scale. Arrows in upward direction indicate cleavage sites.

nal and internal spacer sequences are cleaved off step by step. Thus intermediates are formed as depicted in Fig. 3. The processing occurs in the nucleus, but the 18S (20S) precursor, which is already packaged into a 40S ribonucleoprotein complex, is trimmed at the 3′ end in the cytoplasm.

2.4.3. tRNA

In *Saccharomyces* there are about 400 tRNA genes which appear to be randomly distributed throughout the yeast genome (reviewed by Guthrie and Abelson, 1982). Mutations yielding suppressor tRNAs have been observed in *Saccharomyces* (Guthrie and Abelson, 1982; Sherman, 1982). The sequences of numerous *Saccharomyces tRNA* species are known and are periodically published (Sprinzl *et al.*, 1985).

The tRNA genes code for precursors that are posttranscriptionally processed. This comprises trimming at the 3′ and 5′ termini, including addition of CCA to the 3′ end, base modifications, and, in some cases, intron removal (Guthrie and Abelson, 1982, and references therein). Approximately 40 yeast tRNA genes contain introns. These are 14–60 nucleotides in length and, in contrast to mRNA introns, do not have conserved sequences at their boundaries. However, the intron location is conserved since it always starts one nucleotide to the 3′ side of the anticodon. An endonuclease and a ligase which are required for the tRNA splicing have been partially purified (Peebles *et al.*, 1983; Greer *et al.*, 1983). The following splicing mechanism has been proposed: The endonuclease, which recognizes the intron sequence, precisely excises it to yield tRNA halves, one with a 2′,3′ cyclic phosphate and the other with a 5′ hydroxyl terminus at the cleavage sites. Prior to enzymatic ligation of the tRNA halves, phosphorylation and activation by adenylation of the 5′ terminus and opening of the 2′,3′ cyclic phosphate terminus to the 2′ phosphate are essential intermediate steps. This splicing mechanism is similar to that found in the wheat germ system but it is different from that in mammals (for a comparison see Filipowicz and Gross, 1984; Cech, 1983; Padgett *et al.*, 1986).

3. CELL STRUCTURE

Yeast cells contain a set of subcellular compartments typical of eukaryotic cells. Morphologically, or at least functionally, defined organelles include the nucleus, mitochondria, endoplasmic reticulum, Golgi complex, secretory vesicles, and vacuoles. The cell envelope consists of a rigid cell wall separated from the plasma membrane by the periplasmic space. Not all of the subcellular organelles are completely independent from each other, but are specialized structures derived from an extended intramembranous system (see, e.g., Matile and Wiemken, 1976). Nevertheless, isolation procedures for intact organelles or at least membrane fragments exist (see Section 5.1).

In the subsequent discussion of the cell organelles, the nucleus has been omitted. However, this topic is covered in Sections 4.2 and 4.3.

3.1. Cell Envelope

The arrangement of the structures in the yeast cell envelope is schematically shown in Fig. 4. The plasma membrane forms a diffusion barrier for most low-molecular-weight substances between cytoplasm and the culture medium. Some enzymes secreted by the yeast cells, typically glycoproteins with hydrolase activities, do not penetrate the cell wall and are trapped between plasma membrane and cell wall in the periplasmic space. The cell wall is mostly composed of mannoprotein and glucan. Although apparently rigid in structure, it participates in a number of morphological changes, including budding, mating, and spo-

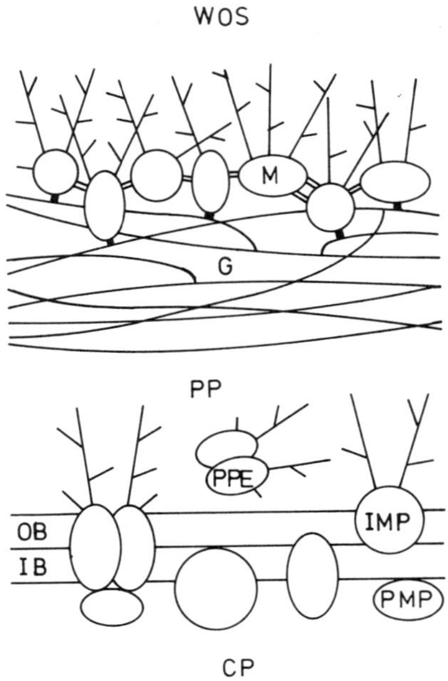

Figure 4. Scheme of the yeast cell surface, according to Zlotnik *et al.* (1984) and Schekman and Novick (1982). CP, cytoplasm; IB, OB, inner and outer halves, respectively, of the plasma membrane bilayer; PMP, peripheral membrane protein; IMP, integral membrane protein; PPE, periplasmic enzyme; PP, periplasmic space; G, glucan; M, mannoprotein; WOS, wall outer surface; (=) S-S thioester bond; (−) covalent bond.

rulation. The carbohydrate moieties of mannoprotein on its surface determine the immunochemical properties of the yeast cell.

3.1.1. Plasma Membrane

The plasma membrane is presented in Fig. 4 in accordance with the current membrane model. Intrinsic membrane proteins are inserted in a lipid bilayer consisting mostly of phospholipids and sterols. Extrinsic proteins (interacting with membrane lipids and proteins by polar binding) cover part of the bilayer surface. The carbohydrate moieties of membrane-bound glycoproteins extend only from the external surface of the membrane.

The phospholipid composition of the plasma membrane is similar to the phospholipid pattern found in complete yeast lipid extracts (see Henry, 1982, for reviews and references), while the sterol content seems to be higher than in intracellular membranes (Rank *et al.*, 1978). The lipid composition of yeast cells and their membranes exhibits great variation and can easily be manipulated. Therefore, yeasts are an ideal model for studying the properties and functions of lipids in eukaryotic cell membranes and are now being increasingly used for this purpose. Mainly the plasma membrane and mitochondria have been analyzed in this respect. Among other questions, lipid protein interactions, effects of lipid fluidity and composition on membrane enzymes and transport processes, and the fluidity and other physical properties of the lipid matrix itself in relation to fatty acid composition and sterol content have been investigated. These topics will not be discussed in detail and, therefore, the reader is referred to the reviews of Prasad (1985) and Prasad and Rose (1986).

The proteins of the plasma membrane have been analyzed by high-resolution 2D-PAGE to find marker polypeptides (Robertson *et al.*, 1980) or to study the membrane assembly in relation to secretory processes (Tschopp *et al.*, 1984). The efficiency of different detergents to solubilize plasma membrane proteins has been compared (Navarette and Serrano, 1983). Currently, only a few enzymes have been released in a functional state from the plasma membrane and identified in the pattern of the membrane polypeptides. Some of them are listed in Table II together with other enzymes assumed to be associated with the plasma membrane of *S. cerevisiae*. The numerous permeases of the yeast cell are not included. This topic is discussed in Chapter 3 of this volume.

It is important to realize that the procedures for plasma membrane

Table II. Enzymes Associated with the Plasma Membrane of *S. cerevisiae*

	Analytical procedure	Reference
Chitin synthetase	Purification by entrapment in the reaction product	Kang *et al.*, 1984
Glucan synthetase	Activity test	Shematek *et al.*, 1980
Proton-translocating ATPase	Solubilization, purification, reconstitution	Malpartida and Serrano, 1981a
Phosphorylated ATPase-intermediate	Solubilization and purification	Malpartida and Serrano, 1981b
Tyrosine kinase	Activity test and separation of phosphoproteins	Castellanos and Mazon, 1985
Adenylate cyclase	Activity test, inhibition by α-factor	Liao and Thorner, 1980
Phospholipase B and lysophospholipase	Solubilization and purification	Witt *et al.*, 1984
NAD(P)H ferricyanide reductase, flavins	Activity test, difference spectroscopy	Ramirez *et al.*, 1984
Lipid-mediated glycosylation of proteins	Activity test	Welten-Verstegen *et al.*, 1980

separation are not perfect and that contamination from organelle membranes cannot be excluded, leading to erroneous results in respect to subcellular distribution of enzyme activities. One of the reasons is that the endoplasmic reticulum is closely associated with the plasma membrane and connections between both membranes may exist (Schekman and Novick, 1982).

Plasma membrane fractions isolated by different procedures may vary greatly in protein and lipid composition (see Section 5). This may be due to artifacts of preparation, but it has also been explained by the assumption that certain membrane domains with characteristic properties coexist in the plasma membrane (Rank *et al.,* 1978). Membrane fractions with different protein patterns and densities have been separated (Tschopp and Schekman, 1983). Evidence for a laterally heterologous distribution of membrane components has also been gained by freeze–fracture electron microscopy, e.g., particles detectable on the cytoplasmic surface of the membrane bilayer arranged in semicrystalline arrays when the cells are grown to stationary phase (Steere *et al.,* 1980; Kramer *et al.,* 1978).

A growing field of research is evaluation of the processes involved in the plasma membrane assembly using mutant strains defective in lipid biosynthesis or secretion (e.g., Tschopp *et al.,* 1984; Ramirez *et al.,* 1983; Ferro-Novick *et al.,* 1984; Klig *et al.,* 1985; Letts and Henry, 1985). Both topics have been reviewed (Henry, 1982; Schekman and Novick, 1982; Novick, 1985; Schekman, 1985). Apparently, cell surface proteins in the plasma membrane, as well as in the cell wall, are transported to the cell periphery by the same or very similar mechanisms as secreted proteins (see Novick, 1985, for references).

3.1.2. Cell Wall

The cell wall structure and the synthesis and assembly of its constituents as a simple and versatile model for morphogenesis have been studied intensively. A variety of methods have been used for this purpose, including separation and chemical analysis of wall constituents; serological procedures; binding of dyes, antibodies, and lectins; selective degradation by enzymes; cell wall mutants; fungicides and other antibiotics (e.g., tunicamycin); and electron microscopic procedures. These topics have been reviewed (Ballou, 1982; Cabib *et al.,* 1982, 1984; Schekman and Novick, 1982; Sentandreu *et al.,* 1984; Catley, 1983).

Glucan and mannoprotein, the main structural constituents of the *S. cerevisiae* cell wall, are found in roughly equal amounts. Chitin is almost entirely restricted to the bud scar (Cabib *et al.,* 1982, 1984) and does not amount to more than about 4% of the cell wall material (Roberts *et al.,* 1983). Lipid contents between 3% and 9% have been reported (see Rattray *et al.,* 1975, for review and references). Glucan is regarded as the main structural determinant component of the wall (Cabib *et al.*, 1982; Catley, 1983). The shape of (β 1-3)glucan has been described as a long, twisted chain with a hydrophilic side due to hydroxyl groups and a hydrophobic side due to methine groups. The interaction of the hydrophobic surfaces is leading to a double helix with very high rigidity (Ballou, 1982). At least a part of the cell wall glucan seems to be arranged in a microfibrillar network. Chitin is polymorphic and exists in three crystalline forms called α, β, and γ-chitin. α-Chitin, the form with maximal hydrophobic binding, has been detected in yeast bud scars (see Ballou, 1982; Cabib *et al.*, 1982; Catley, 1983, for details and references). Mannoproteins do not seem to form any crystalline or fibrillar structures in the yeast cell wall (Cabib *et al.*, 1982).

The precise arrangement of these components in the cell wall of *Saccharomyces* is not known, but the available experimental evidence has been combined in a model (Zlotnik *et al.*, 1984): the outer layer of the wall consists of mannoproteins linked to each other by S-S or thioester bonds and hydrophobic interaction, and linked by covalent bonds of unknown nature to glucan, which forms the inner cell wall layer. Both layers are not completely separated; mannoproteins may penetrate the glucan layer to a certain, but unknown extent. Selective degradation with proteases has shown that the mannoproteins in the external layer determine the cell wall porosity (Zlotnik *et al.*, 1984).

3.2. Vacuoles

In contrast to other organelles such as mitochondria, vacuoles do not represent a clearly separated organelle fraction, but are part of an intramembranous system including the endoplasmic reticulum and other intracellular spaces separated by membranes from the cytoplasm. Therefore, it has been recommended that the term vacuole be used in a purely descriptive sense (Matile and Wiemken, 1976; Matile, 1978).

Form and size of the vacuole may change considerably during the cell cycle. Large vacuoles, which represent by far the most voluminous organelles of the yeast cell, shrink and fragment into small vesicles at the stage of bud initiation. These vesicles fuse to form one or two large vacuoles when the budding process proceeds (Wiemken *et al.*, 1970). At this stage, relatively pure and undamaged vacuoles can be isolated whose functions and structural properties have been investigated.

Many proteases, including carboxypeptidase Y and proteases A and B, and several other hydrolases, have been detected in the yeast vacuole. It has been claimed that many of these enzymes are exclusively located in the vacuolar space or at least absent from the cytoplasm (Wiemken *et al.*, 1979). The characteristic content of proteases and other hydrolases has led to the suggestion that vacuoles play a similar role in yeast cells to that of lysosomes in animal cells (see Matile, 1978, for references).

The second principal function of yeast vacuoles is storage of metabolites. One-fourth of the amino acid pool and an even higher percentage of arginine has been found in the vacuolar fraction of *S. cerevisiae* protoplasts (Wiemken *et al.,* 1979). The role of vacuolar polyphosphate in trapping basic amino acids has already been mentioned in Section 2.2. Polyphosphate has been found to be located exclusively in the vacuolar sap of protoplasts and has been used as a marker during the preparation of vacuoles (Wiemken *et al.,* 1979).

Right-side-out vacuolar membranes have been used to demonstrate an active transport of arginine driven by an electrochemical potential of protons generated by ATP hydrolysis. A potential of 180 mV, with contribution of 1.7 pH units, interior acid, has been determined. The Mg^{2+}-activated, proton-translocating ATPase of the vacuolar membrane has clearly different properties compared to the mitochondrial and plasma membrane ATPases and thus may also serve as vacuolar marker enzyme (Kakinuma *et al.,* 1981, and references therein).

The chemical composition of the vacuolar membrane (tonoplast) has also been studied in some detail. In comparative investigations of tonoplast and plasma membrane from *S. cerevisiae,* a higher content of phospholipids and unsaturated fatty acids has been found along with a lower sterol content in the vacuolar membrane. Differences in the protein—especially the glycoprotein—pattern have also been described. These findings have been discussed in relation to the higher "elasticity" of the tonoplast in comparison to the plasma membrane (Kramer *et al.,* 1978; Niedermeyer, 1976; Schwencke, 1977). Regions of the vacuolar membrane depleted of intramembranous particles have been detected by electron microscopy in freeze-fractured tonoplasts isolated from cells entering the stationary phase. The lipids of these vacuoles have been extracted and analyzed by differential scanning calorimetry, and an endothermic transition around the growth temperature has been found. These results have been interpreted as indication for patches of gel phase lipid in the tonoplast, which exclude intramembranous particles (Moeller *et al.,* 1981).

The transversally asymmetrical distribution of tonoplast carbohydrate, probably mannan chains of membrane glycoproteins, has been analyzed by concanavalin A binding. Carbohydrate is bound to the membrane surface opposite to the cytoplasm in both tonoplast and plasma membrane (Boller *et al.,* 1976).

3.3. Endoplasmic Reticulum and Secretory and Endocytic Apparatus

When thin sections of whole wild-type *Saccharomyces* cells are inspected electron microscopically for secretory organelles, vesicles that obviously function in secretion can be unequivocally detected in the growth region (see also Section 4.3). The Golgi seems to be totally absent, and normally only a few profiles may account for the ER. Our present knowledge about the secretory apparatus and secretion in yeast comes almost exclusively from studies with several conditionally lethal, temperature-sensitive mutants of the *sec* family defective in different steps of the secretory pathway, but not in overall protein synthesis (reviewed by Schekman, 1985; Novick, 1985; Ferro-Novick, 1985).

Tracing of the intracellular level of secretory enzymes (e.g., invertase, acidic phosphatase) has revealed that *sec* mutants can be divided into two classes. Mutants of both classes are blocked in protein export, but members of class A accumulate intracellularly active secretory enzymes whereas those of class B do not. Later it was found that class A and class B mutants are blocked in secretion at late and early steps, respectively.

The organelles and the order of their participation in secretion have been elucidated by double mutant analysis, by histochemical means, and by inspection of the phenotype of different *sec* mutants at the ultrastructural level. The latter studies have revealed that due to the block in secretion, certain mutants accumulate ER, others Golgi structures, and again others vesicles (for electron micrographs, see Novick *et al.*, 1980, 1983; Esmon *et al.*, 1981). It is now firmly established that in *Saccharomyces* the general pathway of secretion is in the order ER → Golgi → vesicles → cell surface, and the functions of ER and Golgi in glycosylation are well documented (see also Section 2.1.5). This pathway is similar to that in mammals (Walter and Lingappa, 1986). However, in contrast to the mammalian ER, the existence of a receptor for the signal recognition protein (Meyer, 1982) has not been proved for yeast ER. In a homologous yeast cell-free system, the prepro-alpha-factor has been demonstrated to cross ER membranes uncoupled from translation in an ATP-dependent fashion, whereas other secretory proteins tested did not reveal this behavior (Rothblatt and Meyer, 1986b; Waters and Blobel, 1986; for review, see Zimmermann and Meyer, 1986). Whether post-translational translocation also happens *in vivo* remains to be established.

Later steps in secretion have also been studied in cell-free systems. Glycosylation and transfer of invertase from ER to Golgi have been performed *in vitro* (Haselbeck and Schekman, 1986).

Recently, evidence has been obtained that endocytosis occurs in yeast (Riezman, 1985; Makarow, 1985a,b; for a recent review, see Riezman *et al.*, 1986). Riezman (1985) has shown that the fluorescent dye lucifer yellow carbohydrazide (LY) is taken up from the medium and accumulated in the vacuole of *S. cerevisiae*. This happens in an energy- and temperature-dependent, but not saturable, fashion. Certain *sec* mutants (class A) which accumulates vesicles have been demonstrated to be defective in these functions. Thus gene products required in the late steps of secretion may also be essential for endocytosis.

Two mutants defective in the endocytic pathway have been isolated

(Chvatchko *et al.*, 1986). The *end1* mutant is defective in the internalization of LY and alpha-factor, whereas *end2*, most probably, is unable to transport the endocytic contents into the vacuole. These studies and those from another laboratory suggest that alpha-factor uptake occurs via receptor-mediated endocytosis (Jenness and Spartick, 1986).

Makarow (1985a) has reported the internalization of enveloped viruses into spheroplasts. Moreover, bacterial alpha-amylase and FITC-dextran, which are taken up even by intact cells in a temperature-dependent fashion, have been shown to be transported into vacuoles (Makarow, 1985b). Studies with FITC-dextran have revealed that lack of ATP does not inhibit internalization of substrate but inhibits its transfer from an internal intermediate compartment into the vacuole. This intermediate compartment exhibits parallels to the mammalian endosomes (Makarow and Nevalainen, 1987).

Coated vesicle and clathrin triskelions have been identified in yeast by electron microscopy but their intracellular source is still unknown (Mueller and Branton, 1984). Moreover, clathrin seems not to be required for cell growth and protein secretion (Payne and Schekman, 1985).

3.4. Mitochondria

S. cerevisiae is a facultative anaerobe, which means it possesses the property to switch from respiration to fermentation, and vice versa. Mitochondria of fully respiring cells (aerobic growth on a nonfermentable carbon source) are rich in cristae structures. Under these conditions, 50 or more mitochondria per cell have been observed (Stevens, 1981). Glucose-repressed cells contain only a few mitochondria which are poor in cristae and whose respiratory enzymes exhibit only low activity (Stevens, 1981). A large number of respiratory-deficient mutants (petites), which are viable when grown on a fermentable carbon source, are known. These mutants have stimulated much research on yeast mitochondrial genetics, gene expression, and organelle biogenesis (for reviews, see Stevens, 1981; Dujon, 1981; Rosamond, 1982; Schatz and Butow, 1983; Hay *et al.*, 1984; Douglas and Takeda, 1985; Trivedi *et al.*, 1985; see also the volume *Mitochondria '83*, in which Yaffe, 1983, has given a summary of protein import into mitochondria). We will restrict ourselves to a description of the major mitochondrial components. Any further information is enclosed in the reviews cited.

Mitochondria are formed by an inner and outer membrane. The former may be strongly evaginated into the matrix space to form cristae. Both membranes are rich in phospholipids (Henry, 1982). Cardiolipin (diphosphatidylglycerol) is prominent in the inner membrane (Prasad, 1985). The sterol:phospholipid ratio is about 1:30 in both membranes (Bottema and Parks, 1980).

Many proteins and enzymes are associated with the mitochondrial membranes (Dujon, 1981). Some of them are used as marker enzymes allowing discrimination between outer and inner mitochondrial membrane (see also Table III). Most mitochondrial proteins (approximately 95%) are encoded by nuclear genes, synthesized on free cytoplasmic polysomes, and finally imported into mitochondria (a detailed descrip-

Table III. Marker Enzymes for Yeast Subcellular Compartments

Organelle or compartment	Marker	References for assays
Cytoplasm	Glucose-6-P dehydrogenase	Kato *et al.*, 1979
Vacuole		
Membrane	α-Mannosidase	van der Wilden *et al.*, 1973
Vacuolar sap	Protease A	Wiemken *et al.*, 1979
	Protease B	
	Carboxypeptidase Y	
Endoplasmic reticulum	NADPH: cytochrome c	Schatz and Klima, 1964
	Oxidoreductase	Dow *et al.*, 1981
	UDP-GlcNAc-*N*-acetylglucosaminyl dolichylphosphate transfer reactions	Lehle and Tanner, 1978
Golgi	Mannoprotein outer-chain oligosaccharide mannosyltransferase	Nakajima and Ballou, 1975
Nucleus		
Nucleoplasm	DNA-dependent RNA polymerase	Jerome and Jaehning, 1986
	Aspartate transcarbamoylase	Nagy *et al.*, 1982
Nuclear envelope	Identification by electron microscopy	Mann and Mecke, 1982
Mitochondria		
Matrix space	Aconitase	Racker, 1950
	Fumarase	
Intermembrane space	Cytochrome c	Djavadi-Ohaniance *et al.*, 1978
	Peroxidase	
	Flavocytochrome b_2	Appleby and Morton, 1959
Inner membrane	Cytochrome c oxidase	Mason *et al.*, 1973
Outer membrane	Kynurenine hydroxylase	Bandlow, 1972
	Porin	Daum *et al.*, 1982
Plasma membrane	Vanadate-sensitive Mg^{2+}-ATPase	Borst-Pauwels and Peters, 1981
		Tschopp and Scheckman, 1983
Cell wall		
Periplasmic space	Invertase	Goldstein and Lampen, 1975
Secretory vesicles	Acidic phosphatase	Haguenauer-Tsapis and Hinnen, 1984

tion of this process is given by Schatz and Butow, 1983; Yaffe, 1983; Hay *et al.*, 1984; Douglas and Takeda, 1985; the latter two references contain tables with imported proteins studied thus far).

Mitochondria have their own translational machinery. Synthesis of mitochondrial proteins with high fidelity has been demonstrated *in vitro* with isolated mitochondria (McKee *et al.*, 1984). The components of the mitochondrial protein synthesis apparatus coded by mitochondrial genes are the two rRNAs (15S and 21S, constituents of the small and large ribosomal subunits, respectively), a complete set of tRNAs (25 genes), and a few mRNA species. The mitochondrial ribosomal proteins, which are all coded by nuclear genes (except for the [*var1*] polypeptide, which seems to be associated with the small ribosomal subunit), have been analyzed (Faye and Sor, 1977). The small ribosomal subunit (37S) contains 33 proteins (mol. wt. ranging from 9500 to 60,000) and the large subunit (50S) has 38 proteins (mol. wt. ranging from 10,000 to 41,000). The ribosomal proteins derived from mitochondria are less basic than those from the cytoplasm.

The circular mitochondrial DNA (75,000 base pairs, 25 μm in circumference), which is extremely rich in AT base pairs (82%), codes for only a few polypeptides, certainly no more than 15: subunits I, II, and III of cytochrome c oxidase; subunits 6 and 9 of the ATPase complex; apocytochrome b; [*var1*] polypeptide and at least eight polypeptides as inferred from genetic data, e.g., maturases required for intron removal. The physical identification of two maturases coded by the cob-box region of the mitochondrial genome has been reported (Jacq *et al.*, 1984; Guiso *et al.*, 1984).

4. LIFE CYCLE

Saccharomyces is able to grow vegetatively as either a haploid or diploid and many industrial strains have a polyploid or aneuploid chromosome constitution. In the normal sexual cycle a diploid cell results from a fusion of two haploid cells of opposite mating type (alpha and *a*). This diploid cell, which is now heterozygous with respect to mating type (alpha/*a*), has lost the ability to mate and neither secretes mating factors nor responds to them, but it undergoes meiosis when exposed to certain starvation conditions (e.g., lack of nitrogen) and forms an ascus which normally contains four haploid spores, two of each mating type. Transfer of the spores to nutrient medium results in germination and rapid fusion within the ascus to form two diploid alpha/*a*-cells. Frequently, single germinating spores from heterothallic strains escape from the ascus and give rise to colonies of vegetatively growing haploid cells when kept isolated from cells of opposite mating type. The vegetative growth is characterized by budding and mitosis.

Cell cycle events and those just mentioned are accompanied by changes in cell structure which are described below.

4.1. Mating

For mating, haploid cells have to be conditioned by mating factors. In response to the action of mating factors, an increase of agglutination can be microscopically observed, normally about 30–60 min after mixing the cells of opposite mating type. Increase of agglutination happens not only in G1 of the cell cycle. By about 90–120 min, large cell aggregates are formed from which single unbudded cells arrested at G1 can be released by sonication. In addition, some free elongated pear-shaped cells (shmoos) appear. The occurrence of shmoos on solid medium has been used as a bioassay to estimate the concentration of pheromone in the medium (Duntze *et al.*, 1970; Betz *et al.*, 1977). Mating happens only between cells arrested in G1. This behavior of mating cells requires alterations mainly at the cell surface (reviewed by Sprague *et al.*, 1983; Thorner, 1981; Betz *et al.*, 1981) which are summarized below.

In most strains the drastic increase of agglutination is a function inducible by the opposite mating factor. However, there are also strains that are constitutive in this respect when grown at 23°–26°C, but behave like inducible strains when grown around 36°C. From such constitutive strains, sex-specific agglutination substances have been extracted (Tohoyama *et al.*, 1979; Yamaguchi *et al.*, 1984). Obviously, these agglutination substances, which turned out to be glycoproteins, are only found in a functional state on the cell surface of haploids and zygotes, but not on diploids and spores (Tohoyama *et al.*, 1979; Tohoyama and Yanagishima, 1981). The agglutination substance extracted from a-cells could mask alpha-cells, thus inhibiting their normal property to agglutinate with a-cells; the opposite is also true. The biological activity of the sex-specific agglutination substance derived from a-cells can be abolished by mercaptoethanol, which has little effect on the agglutination substance derived from alpha-cells.

Increased agglutination as a result of pheromone induction requires the presence of both a carbon and a nitrogen source and it is also dependent on RNA and protein synthesis, but, interestingly, not on the intact glycosylation pathway, since inhibition of glycosylation by drugs (e.g., tunicamycin) does not inhibit alpha-factor-induced agglutination. In response to mating factor, the synthesis of cell wall carbohydrates changes such that the glucan/mannan ratio increases. Furthermore, shorter oligosaccharide side chains are more frequently found in cell wall mannan (Lipke *et al.*, 1976).

Cells treated with mating factors become elongated and form tips which function as a conjugation tube. The cell wall of this tip accumulates chitin, a minor component of the vegetative cell wall. Accordingly, the chitin synthetase activity of the plasma membrane fraction from alpha-factor-treated a-cells increases (Schekman and Brawley, 1979). However, inhibition of chitin synthetase by Polyoxin D (500 μg/ml) fails to block morphogenesis (Lipke *et al.*, 1976). Bud scars—wall structures rich in chitin—seem to influence the mating efficiency since cells with a high number of scars show reduced mating potential. The reason may be that tip formation, like bud initiation, is impeded by too many scars.

In electron micrographs, the cell wall appears diffuse after alpha-factor exposure and vesicles have been shown to accumulate in the tip region (Lipke *et al.*, 1976; Betz *et al.*, 1981). These vesicles are believed to transport lytic enzymes capable of digesting the plasma membrane and cell wall between conjugating cells, thus allowing cytoplasmic continuity. Very little is known about the nature of these enzymes and their regulation. Endo- and exo-1,3-β-glucanases, which might digest the cell wall, have been reported to be produced in *Saccharomyces* constitutively, but mutants deficient in the exoglucanase activity have not lost the ability to mate (Del Rey *et al.*, 1979).

The biochemical processes involved in nuclear fusion are largely unknown. It has been observed that nuclear fusion is initiated at the spindle pole bodies (SPB), which eventually yield a single, large SPB.

Regulation of the cell recognition process is also not fully understood. It has been proposed that the mating sites, which seem to be determined by the cytoplasmic microtubules, have to be in proper orientation to each other. This would strongly diminish the number of possible mating partners within mating aggregates (Sprague *et al.*, 1983) and would explain that mating between more than two cells of opposite mating type rarely occurs (approximately one per 10^4 zygotes formed).

The intracellular signal triggered by the pheromones is another problem that needs further investigation. It is known that the intracellular level of cyclic AMP (cAMP) is lowered most likely due to the inhibition of the membrane-bound cAMP synthetase (Liao and Thorner, 1980). Any further discussion is beyond the scope of this chapter. We therefore refer to the review by Thorner (1982).

4.2. Meiosis and Sporulation

When diploid alpha/a-cells are shifted to nitrogen-free acetate medium (sporulation medium), they undergo meiosis and form ascospores (for media and procedures, see Haber and Halvorson, 1975). The cytology and the metabolic processes occurring during meiosis have been reviewed by Byers (1981) and by Esposito and Klapholz (1981). When cells are shifted to sporulation medium, they will complete their cell cycle to G1 stage and enter the meiotic pathway. As the sporulation medium seems to depress bud growth, a considerable asynchrony of spore formation is caused by daughter cells derived from those mother cells which have started budding just at the time of medium shift. Asynchrony can be reduced by the use of stationary cells in G1 stage; however, this may diminish the efficiency of sporulation.

At the ultrastructural level, the progress of meiosis has been extensively studied (e.g., Guth *et al.,* 1972; Byers and Goetsch, 1975a). In general, the early events of yeast meiosis are similar to those in other eukaryotes (Fig. 5). Before cells enter the meiotic prophase, the DNA is replicated (premeiotic S-phase) about 4–10 hr after medium shift. By the time of DNA replication, the SPB, which is localized at the nuclear envelope, is duplicated yet remains paired throughout the meiotic prophase (Fig. 5a,b; Moens and Rapport, 1971). Prior to the onset of premeiotic S-phase, dense material, termed dense body (DB; Fig. 5b), appears in the nucleus. In the DB, central elements of synaptonemal complexes (SC) are frequently visible. At leptotene, unpaired lateral elements and in zygotene short SCs become detectable (Zickler and Olson, 1975). At pachytene, the SCs reach maximum length and DB structures have vanished or have shrunken (Fig. 5c). At this stage, the yeast karyotype is accessible to analysis (the yeast karyotype has been also resolved by gel electrophoresis; see Schwartz and Cantor, 1984). SC inspection normally reveals approximately 17 bivalents (in diploids), which coincides with the number of known genetic linkage groups

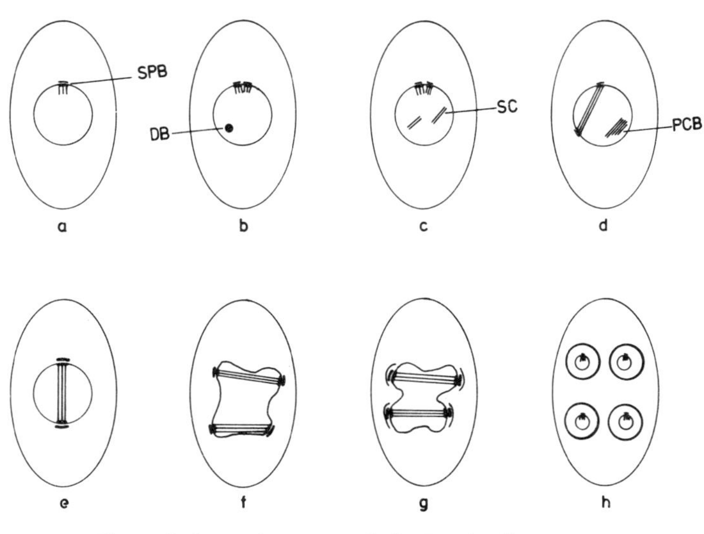

Figure 5. Stages in yeast meiosis. For details, see text.

(Byers and Goetsch, 1975a). Classical condensation of chromosomes has never been unequivocally demonstrated in either meiosis or mitosis.

In diplotene, a single polycomplex body (PCB) is formed and individual SCs are no longer visible (Fig. 5d). At about this time, the SPBs separate and a spindle is formed which spreads between the SPBs. When the spindle is maximally extended, the SPBs are in opposition and their cytoplasmic sides—the outer plaques—enlarge (Fig. 5e) and the PCB is no longer detectable. This also indicates the final stage of meiosis I. Cells stained for DNA appear binucleated under the light microscope.

Subsequently, each of the SPBs duplicates. The SPB pairs separate by moving along the still intact nuclear envelope, which has now gained a lobed shape (Moens and Rapport, 1971). Two spindles are formed each running between two SPBs which are finally located in opposition (Fig. 5f,g). The prospore wall develops by coalescence of vesicles at the cytoplasmic surface of the outer plaques of the SPBs (Fig. 5g). The prospore wall initially clings to the nuclear membrane during enlargement but later both structures are found separated. The space between the developing prospore wall and the nuclear envelope fills up with cytoplasmic material, including mitochondrial particles. When the second nuclear division is complete, the spindles disintegrate, the prospore wall closes, and individual ascospores become visible. The wall growth continues by deposition of material between the two unit membranes of the spore wall (Fig. 5h) and the inner layer of the spore wall becomes the plasmalemma of the ascospore. The wall of the mature ascospore contains glucan and mannan. Studies with mutants auxotrophic for glucosamine suggest that the outer layer of the ascosporal wall is enriched in glucosamine. This is probably deposited as chitin and may be the cause of the hydrophobicity of the spore surface. In contrast to normal spores, such mutants are sensitive to β-glucanases (Ballou *et al.*, 1977).

When cells have entered the meiotic pathway, only a single mitochondrion per cell is detectable by electron microscopy (Stevens, 1981). When the meiosis is complete, each spore within an ascus normally contains a single mitochondrion. However, this does not represent a complete packing of the organelle into the spores. Quantitative evaluation of thin sections has revealed that, on a volume basis, only 52% of the mitochondrial material, 32% of the cytoplasm, 10% of the vacuole(s), and 96% of the nucleus are enveloped into spores; the rest remains in the ascal epiplasm (Brewer and Fangman, 1980). Since the spores occupy about 30% of the total ascal space, only the cytoplasm is randomly enveloped and all other organelles mentioned are nonrandomly enveloped.

4.3. Vegetative Growth

In nutrient medium *Saccharomyces* cells grow by budding. The emergence of a bud indicates that a cell has entered a new division cycle.

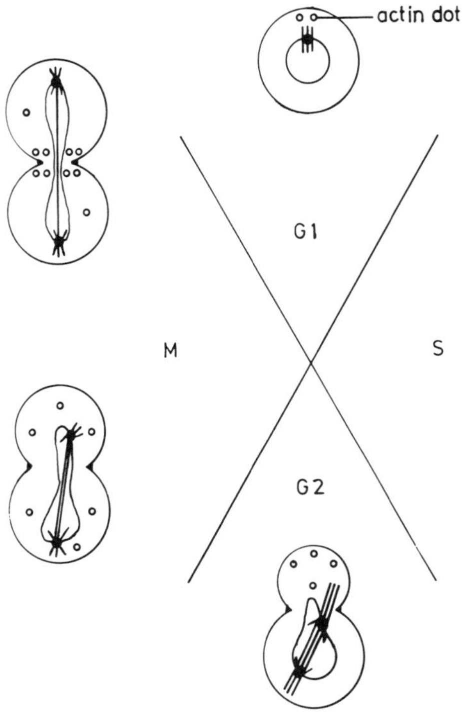

Figure 6. The cell cycle of *Saccharomyces cerevisiae*. For details, see text.

During the cell cycle, there is an increase in cellular components, including duplication of genetic material. Increase in total cell mass and cell volume is almost exclusively due to bud growth. The segregation of chromosomes happens mitotically within an intact nuclear membrane (Fig. 6). After completion of cytokinesis and formation of wall septa, cell separation gives rise to a daughter cell with all the cellular components and functions necessary for vegetative growth. An ordered sequence of morphological changes during mitotic division has been defined by extensive electron microscopic studies (Byers and Goetsch, 1975b; Peterson and Ris, 1976; Byers, 1981). In the following discussion of changes in cell structure during cell cycle, we lay stress on the involvement of microtubules and actin filaments whose role has been investigated by immunochemical methods in mutant and wild-type strains of *S. cerevisiae* and *S. uvarum** (Kilmartin and Adams, 1984; Adams and Pringle, 1984; Novick and Botstein, 1985).

*Taxonomists agree that *S. cerevisiae* and *S. uvarum* are identical. Kilmartin and Adams (1984), however, have emphasized that their *S. uvarum* strain was more suitable for the cytoimmmunochemical studies because of its more rodlike shape compared to *S. cerevisiae*.

Budding occurs only in some yeasts, but not in other eukaryotes. A bud is formed by a localized outgrowth of the cell wall at an obviously nonrandom site (see below). Electron microscopy has revealed that the SPB embedded in the nuclear envelope is duplicated as soon as a bud emerges, but remains paired until the bud has grown out to about one-third in diameter of the mother cell (reviewed by Byers, 1981; King and Hyams, 1982). Two subsets of microtubules emanate from the SPBs, designated as (1) nucleoplasmic and (2) cytoplasmic microtubules according to their subcellular distribution. The cytoplasmic microtubules are orientated to the budding region of the cell wall, which overlays the paired SPBs. The cytoplasmic microtubules may be involved in the localization of vesicles detectable near the budding site. These vesicle are thought to be engaged in the remodeling of the cell wall. The nucleoplasmic microtubules form the mitotic spindle. In haploid cells, initially 5–10 continuous microtubules interconnect the now separated SPBs. In addition, 17 discontinuous nucleoplasmic microtubules (per SPB) are detectable. In diploids 34 discontinuous microtubules per SPB are seen while the same number of continuous microtubules as in haploids (5–10) are observed (King and Hyams, 1982; Peterson and Ris, 1976). The number of the discontinuous microtubules is coincident with the number of chromosomes. Therefore, it is generally assumed that the discontinuous microtubules attach to the centromeres of the chromosomes, although there is no direct ultrastructural evidence for this. The number of continuous microtubules decreases as the cell proceeds in its division cycle until only one microtubule along with the nuclear axis with a length of about 8 μm is left (King *et al.*, 1982).

A filamentous ring in the neck region develops soon after bud emergence. It persists for most of the cell cycle but finally disappears at cytokinesis (Byers and Goetsch, 1976). The function and chemical nature of these filaments (10 nm in diameter) are still unknown. Figure 6 shows schematically the localization of the SPBs, the orientation of the mitotic spindle, and the movement of the nucleus during bud enlargement, based on electron microscopic evidence.

Immunofluorescence studies have extended knowledge on the localization of tubulin and actin during the cell cycle (Kilmartin and Adams, 1984; Adams and Pringle, 1984; Novick and Botstein, 1985). Immunofluorescent labeling of tubulin revealed that the SPBs nucleate microtubules throughout the cell cycle. In the period during spindle completion, and before movement of the elongated spindle into the enlarged bud, long bundles of cytoplasmic microtubules emanate from both SPBs. One bundle of cytoplasmic microtubules always extends toward or into the bud.

When actin is labeled by fluorescent antibodies or fluorescent phalloidin conjugate (which binds to F-actin with high affinity), a ring of actin dots on apparently unbudded cells, and around the base of small buds, becomes visible besides faintly stained filaments obviously connected with the dots. The development of the actin dot ring is not dependent on the presence of the chitin ring which is formed at the budding site of the mother cell and remains on the mother cell as a bud scar after cell separation (see also Section 3.1.2). During spindle elongation, actin dots are found primarily in the tip region of the growing bud. The bud tip has been shown to be the growth region (Adams and Pringle, 1984). When the spindle is fully elongated, actin dots are almost equally distributed between mother cell and bud. In addition, filamentous structures have been observed. By the time of cytokinesis, actin dots appear to be concentrated in the neck region on both mother cell and bud. Actin dots in other parts of the cell are also visible (see also scheme depicted in Fig. 6). The actin dot rings which appear at the beginning and end of the cell cycle most likely do not correspond to the filamentous ring in the neck observed electron microscopically (see above), since the filamentous ring has been found to be incompletely developed or totally absent in those stages when the actin dot rings have been identified. There seem to be no pronounced interactions between microtubules and actin filaments as judged by double fluorescent labeling, but it cannot fully be ruled out that thin interacting filaments have not been resolved by this method.

The role of actin during the cell cycle is not fully understood, but there is some evidence that actin filaments may take part in the localization of secretory vesicles into the growth region of the bud. Temperature-sensitive actin mutants are partially blocked in secrection of invertase at the restricted temperature. The secreted form of invertase is known to be transported to the cell surface in vesicles (see also Section 2.1.5). Similarly, the deposit of chitin (bud scar) may also result from the participation of actin filaments in directing the flow of vesicle containing chitin synthetase (zymogen) to the budding site. Vesicular transport of chitin synthetase (zymogen) was suggested several years ago to be responsible for the incorporation of chitin into the growth zone of alpha-factor-treated a-cells (Schekman and Brawley, 1979; see also Section 4.1). This view is further supported by the finding that the actin mutants mentioned earlier fail to deposit chitin in a localized way at the nonpermissive temperature; instead chitin appears to be distributed over most of the cell surface (Novick and Botstein, 1985). Other interpretations are possible, but there is strong evidence that actin plays a major role in polarized growth and cell surface organization. Further investigations are necessary to confirm the suggested functions.

5. CELL FRACTIONATION

The rigid cell wall of yeast is the main obstacle to the preparation of cell-free extracts and organelles. The disintegration of the wall can be achieved mechanically or by enzymatic digestion of its constituents. The drastic treatment required to disrupt yeast cells by mechanical means results in severe artifacts; organelles are damaged, and membrane fragments of different organelles may fuse or form enclosures. These effects may interfere with fractionation and purification of these organelles and membranes.

The enzymatic digestion of wall components, while cell integrity is maintained by an appropriate osmotic stabilizer, results in the formation of protoplasts. Digestive juice from *Helix pomatia* and extracellular enzymes from bacteria and fungi have been widely used for this purpose in most cases. Several enzyme preparations are commercially available (e.g., Helicase, Reactifs IBF; Zymolyase-100T, Kirin Brewery; Novozym 234, Novo Industri; Lyticase, Sigma). The utilization of some other purified and crude enzymes for cell wall digestion in several yeasts has been compared (Mann and Jeffery, 1986). Methods for preparation of protoplasts and applications for studying cell wall regeneration, fusion, and transformation have been reviewed (Peberdy, 1979; Freeman and Peberby, 1983).

5.1. Isolation of Organelles

A large number of protocols have been published aiming at the separation of several membranous organelles from the same cell homogenate in parallel (e.g., Martinez *et al.*, 1984) or the purification of only one organelle or membrane fraction. In many cases the fractionation scheme has been optimized simply to obtain maximal yield of membrane-associated functions, e.g., enzymes with high specific activities.

5.1.1. Mitochondria

Mechanical as well as enzymatic digestion of the cell wall has been used to isolate mitochondria from *S. cerevisiae*. Cell disruption with a Braun homogenizer MSK has been recommended for small-scale preparations (Guerin *et al.*, 1979; Trembath and Tzagoloff, 1979). Up to 30 strains can be processed simultanously (Trembath and Tzagoloff, 1979). For the large-scale preparation of mitochondria, cells have been homogenized by continuous grinding in a Dyno mill (Guerin *et al.*, 1979) or frozen in liquid nitrogen and subsequently disrupted in a Waring blendor at top speed (Trembath and Tzagoloff, 1979). Mechanical procedures, hand shaking with glass beads, and careful use of a Dyno mill are also recommended for preparation of pure mitochondria (without contamination of nuclear DNA) for transcriptional studies (Rickwood and Hayes, 1984).

Mitochondria of higher quality with respect to respiratory control,

Q_{O_2} value, and membrane integrity are usually obtained by enzymatic digestion of the cell wall and gentle mechanical desintegration of the protoplast membrane (Guerin *et al.*, 1979). Snail gut juice or Zymolyase has been used (Guerin *et al.*, 1979; Daum *et al.*, 1982) to prepare the protoplasts, and the resulting cell homogenate is usually fractionated by differential centrifugation; mitochondria are pelleted and washed at around 10,000 *g*. Discontinuous density gradient centrifugation has been used for further purification (Riezman *et al.*, 1983). The separation of inner and outer mitochondrial membranes has been achieved by linear density gradient centrifugation after homogenization with a Dounce homogenizer or sonication (Daum *et al.*, 1982; Riezman *et al.*, 1983).

5.1.2. Plasma Membrane

Mechanical disruption (in most cases with glass beads in a Braun cell homogenizer MSK) and enzymatic digestion have also been used as first steps in plasma membrane preparation. After differential centrifugation of the cell homogenate, separation of plasma membrane fragments is usually achieved by discontinuous or continuous density gradient centrifugation. Mitochondrial membranes can be removed by aggregation at their isoelectric point at pH 4.5. Protoplasts are directly subjected to osmotical lysis or are stabilized first by coating with concanavalin A. In the latter case, plasma membrane sheets are obtained, which can be separated by low-speed centrifugation. These sheets form vesicles upon removal of concanavalin A. These procedures have been modified and combined in many ways (see Goffeau and Slayman, 1981, for discussion and references).

In another approach, protoplasts that naturally carry negative surface charges at acidic pH have been attached to cationic silica microbeads (Schmidt *et al.*, 1983). After saturation of surplus positive charges by polyacrylic acid and subsequent osmotic lysis at pH 8, microbead-coated plasma membranes are separated by low-speed centrifugation (1000 *g*) from the lysate. This procedure yields at least 85% of the plasma membrane and the fraction appears to be essentially free from contaminating organelle membranes. The plasma membrane ATPase is recovered in highly active condition.

Two highly purified plasma membrane fractions, differing in their protein pattern and slightly in density, have been separated in the following way: after cell wall digestion, concanavalin A coating, lysis in the presence of DNase and RNase, alpha-methylmannoside treatment, and elimination of mitochondrial contamination on a sucrose gradient, the membranes are sonicated at high salt concentration to remove tightly bound ER from the plasma membranes. Finally, gradient centrifugation using a shallow Renografin gradient results in separation of two fractions identified as plasma membranes by marker enzymes and surface labeling (Tschopp and Schekman, 1983).

5.1.3. Vacuoles

Procedures for preparation of vacuoles inevitably start with enzymatic digestion of the cell wall because mechanical disruption of intact cells results in fragmentation of the vacuolar membranes. Isoosmotic lysis of protoplasts by exposure to polybases (e.g., DEAE dextran) has been used by Wiemken and co-workers (e.g., Wiemken *et al.*, 1979) to release intact vacuoles, which retain enclosed enzymes and low-molecular-weight compounds. Purification of the vacuolar fraction is usually achieved by several density gradient centrifugations (Wiemken *et al.*, 1979). Isoosmotic rupture of protoplasts by DEAE dextran has been also used to avoid contamination by vacuolar proteases in preparations of cytosol and mitochondria (Schwencke *et al.*, 1983).

5.1.4. Nucleus

Protocols for preparation of nuclei from yeast cells generally include osmotic lysis and gentle mechanical homogenization of protoplasts in the presence of detergents (e.g., Silver *et al.*, 1984) or low concentrations of the gradient medium (e.g., Nagy *et al.*, 1982; Schultz, 1978) used for the subsequent continuous or discontinuous density gradient centrifugation (e.g., Percoll). Yeast nuclei obtained in these or similar ways retain the activity of the different RNA polymerases and can be used for *in vitro* transcription experiments (e.g., Ide, 1981; Jerome and Jaehning, 1986). Nuclear membranes have been isolated by suspending purified nuclei with a tight-fitting Dounce homogenizer, extensive treatment with RNase and DNase, and density gradient centrifugation on Ficoll (Mann and Mecke, 1982).

5.1.5. Microsomes

The microsomal fraction is usually isolated from yeast cells by cell wall digestion followed by differential and density gradient centrifugation (see Käppeli *et al.*, 1982, for references), but other procedures have also been reported. Precipitation of microsomes by addition of $CaCl_2$ has been used with a recovery of cytochrome P-450 greater than 80% (Käppeli *et al.*, 1982). Initial cell homogenization in a French Press and isolation of rough and smooth endoplasmic reticulum by repeated sucrose density gradient centrifugation have been described (Swida *et al.*, 1982).

5.2. Identification of Yeast Cell Subfractions

The isolation of cellular membranous organelles is the main procedure to analyze the subcellular distribution of cell constituents and functions. This requires identification of cell fractions and evaluation of the extent of contamination by other organelles. The activities of enzymes being exclusively bound to certain organelle membranes or enclosed therein (marker enzymes) have been tested for this purpose. Marker enzymes should be used very carefully for several reasons: enzymes selected as markers for contaminating fractions may be inactivated or washed out during the purification procedure; the localization of some enzymes frequently used as markers is not precisely known; no marker enzyme has been found to be present only in the plasma membrane and the nuclear envelope; soluble enzymes may be unspecifically attached to membranes. Nevertheless, a list of marker enzymes suitable for yeast cell fractionation is presented in Table III. It is partly based on a similar list compiled by Schekman (1982). References are included in which the assay conditions are described.

In addition, subcellular fractionation can be followed by other procedures; e.g., typical membrane invaginations detectable by electron microscopy have been used to identify plasma membranes (Kramer *et al.*, 1978) and the purification of the same fraction has been evaluated by iodination of proteins at the surface of protoplasts (Tschopp and Schekman, 1983).

5.3. Autolytic Enzymes in Yeast Cells

Another obstacle in analyzing yeast cell constituents in general, and especially in lengthy cell fractionations, is the occurrence of a large number of hydrolytic enzymes capable of digesting intra- and extracellular compounds (Babayan and Bezrukov, 1985). A variety of proteases has been found (Achstetter and Wolf, 1985b; Wolf, 1986). They are not restricted to the vacuole, but seem to occur in nearly all cell compartments. Membrane phospholipids are hydrolyzed to fatty acids as soon as the yeast is homogenized (see Witt *et al.*, 1984, for references). Cell wall-degrading enzymes—mostly glucanases, but also chitinases and mannanases—have been found in many yeast species (Fleet, 1984).

The older literature on degradation of enzymes and other proteins during their isolation from yeast by endogenous proteases has been reviewed and discussed (Pringle, 1975). Pringle (1975) includes a list of known cases of proteolytic artifacts in yeast studies and a useful chapter on antiartifact methodology. The problem of proteolytic digestion can now be handled by using different protease inhibitors which became commercially available in large number in recent years; e.g., phenylmethyl sulfonylfluoride, *p*-hydroxymercuribenzoate, pepstatin, and chymostatin have been used together to prevent proteolytic artifacts in a study on intracellular protein processing and sorting (Mechler *et al.*, 1982).

6. METHODS OF SYNCHRONIZATION

Synchronously dividing cells are desirable when studying cell cycle or related phenomena. There are different methods to induce cell synchrony:

6.1. Imposition of a Cell Cycle Block

The basic premise of this method is that due to the blocking procedure, the cells of a culture reach a common stage in the cell cycle. From this point, synchronous growth continues for two or three generations when the block is released. The simplest way to impose a cell cycle block is to starve the cells. The cells then arrest at G1 but continue division when transferred to fresh medium (Mitchison and Carter, 1975). G1 arrest may also be achieved by inhibiting DNA synthesis, by exposing cells with mating type *a* to alpha-factor (this is more convenient than the opposite procedure, since alpha-factor is easier to purify and is also commercially available), and by use of certain cell-division-cycle (*cdc*) mutants, e.g., *cdc28* (Bücking-Throm *et al.,* 1973; Elliott and McLaughlin, 1983; Mitchison and Carter, 1975). Dozens of *cdc* mutants are now available so that blockage at distinct stages is possible. A critical survey of these mutants is given by Pringle and Hartwell (1981).

6.2. Centrifugational Methods

Cells at different stages of the cell cycle can be selected by centrifugational methods from an asynchronously growing culture. Separated cell cycle fractions may be used directly for analysis or may serve as starting material for a culture to be grown synchronously.

During the cell cycle, the density of cells changes periodically (Schulman, 1978). In *Saccharomyces* the maximum density is reached shortly after bud emergence and thereafter drops to a minimum during cytokinesis (Shulman, 1978). High- and low-density cells can be separated by isopycnic equilibrium centrifugation in density gradients prepared with Ludox, Renografin, or Percoll (Mitchison and Carter, 1975; Shulman, 1978; Baldwin and Kubitschek, 1984).

Since cell size also varies during the cell cycle, sorting according to size yields cell cycle fractions. Centrifugal elutration has been successfully applied for this purpose (Gordon and Elliott, 1977). In principle, this procedure is simply a counterflow centrifugation method. At constant rotor speed, a cell suspension is pumped opposite to the direction of the centrifugal forces through the rotor. Fractionation is carried out by incrementally increasing the flow rate.

Velocity sedimentation in sucrose gradients can be applied to obtain small cells which are normally at the start of the cell cycle. The advantage is that separations can be carried out at low speed in a simple laboratory centrifuge equipped with a swinging bucket rotor. If large amounts of cells have to be processed, the method can be scaled up in a zonal rotor (Mitchison and Carter, 1975).

REFERENCES

Achstetter, T., and Wolf, D. H., 1985a, Hormone processing and membrane bound proteinases in yeast, *EMBO J.* **4:**173–177.

Achstetter, T., and Wolf, D. H., 1985b, Proteinases, proteolysis and biological control in the yeast *Saccharomyces cerevisiae, Yeast* **1:**139–157.

Adams, A. E. M., and Pringle, J. R., 1984, Relationship of actin and tubulin distribution to bud growth in wild-type and morphogenetic-mutant *Saccharomyces cerevisiae, J. Cell Biol.* **98:**934–945.

Ades, I., and Butow, R., 1980, The products of mitochondria-bound cytoplasmic polysomes in yeast, *J. Biol. Chem.* **255:**9918–9924.

Andregg, R. J., Beta, R., Carr, S. A., Crabb, J. W., and Dontze, W., 1988, Structure of *Saccharomyces cerevisiae* mating hormone a-factor, *J. Biol. Chem.* **263:**18236–18240.

Appleby, C. A., and Morton, R. K., 1959, Lactic dehydrogenase and cytochrome b_2 of baker's yeast, *Biochem. J.* **71:**492–499.

Ares, M., 1986, U2 RNA from yeast is unexspectedly large and contains homology to vertebrate U4, U5, and U6 small nuclear RNAs, *Cell* **47:**49–59.

Babayan, T. L., and Bezrukov, M. G., 1985, Autolysis in yeast, *Acta Biotechnol.* **5:**129–136.

Baldwin, W. W., and Kubitschek, H. E., 1984, Bouyant density variation during the cell cycle of *Saccharomyces cerevisiae, J. Bacteriol.* **158:**701–704.

Ballou, C. E., 1982, Yeast cell wall and cell surface, in: *The Molecular Biology of the Yeast Saccharomyces, Metabolism and Gene Expression* (J. N. Strathern, E. W. Jones, and J. R. Broach, eds.), Cold Spring Harbor Monograph Series, Cold Spring Harbor, NY, pp. 335–360.

Ballou, C. E., Maitra, S. K., Walter, J. W., and Whelan, W. L., 1977, Developmental defects associated with glucosamine auxotrophie in *Saccharomyces cerevisiae, Proc. Natl. Acad. Sci. USA* **74:**4351–4355.

Bandlow, W., 1972, Membrane separation and biogenesis of the outer membrane of yeast mitochondria, *Biochim. Biophys. Acta* **282:**105–122.

Baralle, F. E., 1983, The functional significance of leader and trailer sequences in eucaryotic mRNAs, *Int. Rev. Cytol.* **81:**71–106.

Becker, G. W., and Lester, R. L., 1980, Biosynthesis of phospoinositol-containing sphingolipids from phosphatidylinositol by a membrane preparation from *Saccharomyces cerevisiae, J. Bacteriol.* **142:**747–754.

Betz, R., MacKay, V. L., and Duntze, W., 1977, a-Factor from *S. cerevisiae:* Partial characterization of a mating hormone produced by cells of mating type a, *J. Bacteriol.* **132:**462–472.

Betz, R., Manney, T. R., and Duntze, W., 1981, Hormonal control of gametogenesis in the yeast *Saccharomyces cerevisiae, Gamete Res.* **4:**571–584.

Black, S., 1985, Inhibition of the reductive activation of valyl-tRNA synthetase from yeast by unsaturated fatty acids and associated observations on newly found lipophilic substances from yeast, *J. Biol. Chem.* **260:**433–439.

Boller, T., Dürr, M., and Wiemken, A., 1976, Asymmetric distribution of concanavalin A binding sites on yeast plasmalemma and vacuolar membrane, *Arch. Microbiol.* **109:**115–118.

Borst-Pauwels, G. W. F. H., and Peters, P. H. J., 1981, Factors affecting the inhibition of yeast plasma membrane ATPase by vanadate, *Biochim. Biophys. Acta* **642:**173–181.

Bottema, C. K., and Parks, L. W., 1980, Sterol analysis of the inner and outer mitochondrial membranes in yeast, *Lipids* **15:**987–992.

Brake, A., Brenner, C., Najarian, R., Laybourn, P., and Merryweather, J., 1985, Structure of genes encoding precursors of the yeast peptide mating pheromone a-factor, in: *Transport and Secretion of proteins* (M.-J. Gething, ed.), Cold Spring Harbor Laboratory, Cold Spring Harbor, NY, pp. 103–108.

Brewer, B. J., and Fangman, W. L., 1980, Preferential inclusion of extrachromosomal genetic elements in yeast meiotic spores, *Proc. Natl. Acad. Sci. USA* **77:**5380–5384.

Brody, E., and Abelson, J., 1985, The "spliceosome": Yeast premessenger RNA associates with a 40S complex in a splicing dependent reaction, *Science* **228:**963–967.

Bücking-Throm, E., Duntze, W., Hartwell, L. H., and Manney, T. R., 1973, Reversible arrest of haploid yeast cells at the initiation of DNA synthesis by a diffusible sex factor, *Exp. Cell Res.* **76:**99–110.

Byers, B., 1981, Cytology of the yeast life cycle, in: *The Molecular Biology of the Yeast Saccharomyces, Life Cycle and Inheritance* (J. N. Strathern, E. W. Jones, and J. R. Broach, eds.), Cold Spring Harbor Monograph Series, Cold Spring Harbor, NY, pp. 59–97.

Byers, B., and Goetsch, L., 1975a, Electron microscopic observations on the meiotic karyotype of diploid and tetraploid *Saccharomyces cerevisiae, Proc. Natl. Acad. Sci. USA* **72:**5056–5060.

Byers, B., and Goetsch, L., 1975b, Behaviour of spindles and spindle plaques in the cell cycle and conjugation of *Saccharomyces cerevisiae, J. Bacteriol.* **124:**511–523.

Byers, B., and Goetsch, L., 1976, A highly ordered ring of membrane-associated filaments in budding yeast, *J. Cell Biol.* **69:**717–721.

Byrd, J. C., Tarentino, A. L., Maley, F., Atkinson, P. H., and Trimble, R. B., 1982, Glycoprotein synthesis in yeast: Identification of $Man_8GlcNAc_2$ as an essential intermediate in oligosaccharide processing, *J. Biol. Chem.* **257:**14657–14666.

Cabib, E., Roberts, R., and Bowers, B., 1982, Synthesis of the yeast cell wall and its regulation, *Annu. Rev. Biochem.* **51:**763–793.

Cabib, E., Kang, M. S., Bowers, B., Elango, N., Mattia, E., Slater, M. L., and Au-Young, J., 1984, Chitin synthesis in yeast, a vectorial process in the plasma membrane, *FEMS Symp.* **27:**91–100.

Calderbank, J., Keenan, M. H. J., and Rose, A. H., 1985, Plasma-membrane phospholipid unsaturation affects expression of the general amino-acid permease in *Saccharomyces cerevisiae* Y185, *J. Gen. Microbiol.* **131:**57–65.

Carlson, M., and Botstein, D., 1982, Two differentially regulated mRNAs with different 5′ ends encode secreted and intracellular forms of yeast invertase, *Cell* **28:**145–154.

Carlson, M., Taussig, R., Kustu, S., and Botstein, D., 1983, The secreted form of invertase in *Saccharomyces cerevisiae* is synthesized from mRNA encoding a signal sequence, *Mol. Cell. Biol.* **3:**439–447.

Castellanos, R. M. P., and Mazon, M. J., 1985, Identification of phosphotyrosine in yeast proteins and of a protein tyrosine kinase associated with the plasma membrane, *J. Biol. Chem.* **260:**8240–8242.

Catley, B. J., 1983, Regulation of yeast and fungal polysaccharides excluding chitin and cellulose, *Prog. Indust. Microbiol.* **18:**129–200.

Cech, T. R., 1983, RNA splicing: Three themes with variations, *Cell* **34:**713–716.

Chia, L.-L., and McLaughlin, C., 1979, The half-life of mRNA in *Saccharomyces cerevisiae, Mol. Gen. Genet.* **170:**137–144.

Chu, F. K., Takase, K., Guarino, D., and Maley, F., 1985, Diverse properties of external and internal forms of yeast invertase derived from the same gene, *Biochemistry* **24:**6125–6132.

Chvatchko, Y., Howald, I., and Riezman, H., 1986, Two yeast mutants defective in endocytosis are defective in pheromone response, *Cell* **46:**355–364.

Clausen, M. K., Christiansen, K., Jensen, P. K., and Behnke, O., 1974, Isolation of lipid particles from baker's yeast, *FEBS Lett.* **43:**176–179.

Clayton, L., Pogson, C. I., and Gull, K., 1979, Microtubule proteins in yeast *Saccharomyces cerevisiae, FEBS Lett.* **106:**67–70.

Cleveland, D. W., Lopata, M. A., MacDonald, R. J., Cowan, N. J., Rutter, W. J., and Kirschner, M. W., 1980, Number and evolutionary conservation of alpha- and beta-tubulin and cytoplasmic beta- and gamma-actin genes using specific cloned cDNA probes, *Cell* **20:**95–105.

Daum, G., Böhni, P. C., and Schatz, G., 1982, Import of proteins into mitochondria: Cytochrome b and cyrochrome c peroxidase are located in the intermembrane space of yeast mitochondria, *J. Biol. Chem.* **257:**13028–13033.

Del Rey, F., Santos, T., Garcia-Acha, I., and Nombela, C., 1979, Synthesis of 1,3-β-glucanases in *Saccharomyces cerevisiae* during the mitotic cycle, mating and sporulation, *J. Bateriol.* **139:**924–931.

Djavadi-Ohaniance, L., Rudin, Y., and Schatz, G., 1978, Identification of enzymatically inactive apocytochrome c peroxidase in aerobically grown *Saccharomyces cerevisiae, J. Biol. Chem.* **253:**4402–4407.

Domdey, H., Apostol, B., Lin, R.-J., Newmann, A., Brody, E., and Abelson, J., 1984, Lariat structures are *in vivo* intermediates in yeast pre mRNA splicing, *Cell* **39:**611–621.

Douglas, M., and Takeda, M., 1985, Nuclear genes encoding mitochondrial proteins in yeast, *Trends Biochem. Sci.* **10:**192–194.

Dow, J. M., Carron, R. R., and Villa, V. D., 1981, Role of membranes of mycelial *Mucor rouxii* in synthesis and secretion of cell wall matrix polymers, *J. Bacteriol.* **145:**272–279.

Duffus, J. H., Levi, C., and Manners, D. J., 1982, Yeast cell-wall glucans, *Adv. Microb. Physiol.* **23:**151–181.

Dujon, B., 1981, Mitochondrial genetics and functions, in: *The Molecular Biology of the Yeast Saccharomyces, Life Cycle and Inheritance* (J. N. Strathern, E. W. Jones, and J. R. Broach, eds.), Cold Spring Harbor Monograph Series, Cold Spring Harbor, NY, pp. 505–634.

Duntze, W., MacKay, V., and Manney, T. R., 1970, *Saccharomyces cerevisiae:* A diffusible sex factor, *Science* **168:**1472–1473.

Duntze, W., Stötzler, D., Bücking-Throm, E., and Kalbitzer, S., 1973, Purification and partial characterization of alpha-factor, a mating-type specific inhibitor of cell reproduction from *Saccharomyces cerevisiae, Eur. J. Biochem.* **35:**357–365.

Duran, A., and Cabib, E., 1978, Solubilization and partial purification of yeast chitin synthase: Confirmation of the zymogenic nature of the enzyme, *J. Biol. Chem.* **253:**4419–4425.

Dürr, M., Urech, K., Boller, T., Wiemken, A., Schwencke, J., and Nagy, M., 1979, Sequestration of arginine by polyphosphate in vacuoles of yeast (*Saccharomyces cerevisiae*), *Arch. Microbiol.* **121:**169–175.

Elliott, S. G., and McLaughlin, C. S., 1983, The yeast cell cycle: Coordination of growth and division rates, *Prog. Nucl. Acid Res. Mol. Biol.* **28:**143–176.

Erdmann, V. A., Wolters, J., Huysmans, E., and DeWachter, R., 1985, Collection of published 5S, 5.8S and 4.5S ribosomal RNA sequences, *Nucl. Acids Res.* **13** (Suppl.):r105–r153.

Esmon, B., Novick, P., and Schekman, R., 1981, Compartmentalized assembly of oligosaccharides on exported glycoproteins in yeast, *Cell* **25:**451–460.

Esposito, R. E., and Klapholz, S., 1981, Meiosis and Ascospore development, in: *The Molecular Biology of the Yeast* Saccharomyces, *Life Cycle and Inheritance* (J. N. Strathern, E. W. Jones, and J. R. Broach, eds.), Cold Spring Harbor Monograph Series, Cold Spring Harbor, NY, pp. 211–287.

Faye, G., and Sor, F., 1977, Analysis of mitochondrial ribosomal proteins of *Saccharomyces cerevisiae* by two dimensional polyacrylamide gel electrophoresis, *Mol. Gen. Genet.* **155:**27–35.

Ferro-Novick, S., 1985, Genetic approaches to the study of protein targeting across a lipid bilayer, *Trends Biochem. Sci.* **10:**425–427.

Ferro-Novick, S., Novick, P., Field, C., and Schekman, R., 1984, Yeast secretory mutants that block the formation of active cell surface enzymes, *J. Cell Biol.* **98:**35–43.

Filipowicz, W., and Gross, H. J., 1984, RNA ligation in eukaryotes, *Trends Biochem. Sci.* **9:**68–71.

Fleet, G. H., 1984, The occurrence and function of endogenous wall-degrading enzymes in yeasts, *FEMS Symp.* **27:**227–238.

Freeman, R. F., and Peberby, J. F., 1983, Protoplast fusion in yeasts, in: *Yeast genetics. Fundamental and applied aspects* (J. F. T. Spencer, D. M. Spencer, and A. R. W. Smith, eds.), Springer, New York, pp. 243–253.

Frevert, J., and Ballou, C. E., 1982, Yeast invertase polymorphism is correlated with variable states of oligosaccharide chain phosphorylation, *Proc. Natl. Acad. Sci. USA* **79:**6147–6150.

Frevert, J., and Ballou, C. E., 1985, *Saccharomyces cerevisiae* structural cell wall mannoprotein, *Biochemistry* **24:**753–759.

Gallwitz, D., and Sures, R., 1980, Structure of a split gene: Complete nucleotide sequence of the actin gene in *Saccharomyces cerevisiae, Proc. Natl. Acad. Sci. USA* **77:**2546–2550.

Georgiev, O. I., Nikolaev, N., Hadjiolov, A. A., Skryabin, K. G., Zakharyev, V. M., and Bayev, A. A., 1981, The structure of the yeast ribosomal RNA genes. 4. Complete sequence of the 25S rRNA gene from *Saccharomyces cerevisiae, Nucl. Acids Res.* **9:**6953–6973.

Goffeau, A., and Slayman, C. W., 1981, The proton-translocating ATPase of the fungal plasma membrane, *Biochim. Biophys. Acta* **639:**197–223.

Goldstein, A., and Lampen, J. O., 1975, β-D-Fructofuranoside fructohydrolase from yeast, *Meth. Enzymol.* **42:**504–511.

Gordon, C. N., and Elliott, S. G., 1977, Fractionation of *Saccharomyces cerevisiae* cell populations by centrifugal elutration, *J. Bacteriol.* **129:**97–100.

Gorenstein, C., and Warner, J. R., 1976, Coordinate regulation of the synthesis of eukaryotic ribosomal proteins, *Proc. Natl. Acad. Sci. USA* **73:**1547–1551.

Gourse, R. L., Sharrock, R. A., and Nomura, M., 1986, Control of ribosome synthesis in *Escherichia coli,* in: *Structure, Function and Genetics of Ribosomes* (B. Hardesty and G. Kramer, eds.), Springer-Verlag, New York, pp. 766–788.

Greer, C., and Schekman, R., 1982a, Actin from *Saccharomyces cerevisiae, Mol. Cell. Biol.* **2:**1270–1278.

Greer, C., and Schekman, R., 1982b, Calcium control of *Saccharomyces cerevisiae* actin assembly, *Mol. Cell. Biol.* **2:**1279–1286.

Greer, C. L., Peebles, C. L., Gegenheimer, P., and Abelson, J., 1983, Mechanism of action of a yeast RNA ligase in tRNA splicing, *Cell* **32:**537–546.

Guerin, B., Labbe, P., and Somlo, M., 1979, Preparation of yeast mitochondria (*Saccharomyces cerevisiae*) with good P/O and respiratory control ratios, *Meth. Enzymol.* **55:**149–159.

Guillen, A., Leal, F., Andaluz, E., and Larriba, G., 1985, Endogenous factors that modula .e yeast glucan synthetase in cell-free extracts, *Biochim. Biophys. Acta* **842:**151–161.

Guiso, N., Dreyfus, M., Siffert, O., Danchin, A., Spyridakis, A., Gargouri, A., Claisse, M., and Slonimski, P. P., 1984, Antibodies against synthetic oligopeptides allow identification of the mRNA-maturase encoded by the second intron of the yeast cob–box gene, *EMBO J.* **3:**1769–1772.

Guth, E., Hasimoto, T., and Conti, S. F., 1972, Morphogenesis of ascospores in *Saccharomyces cerevisiae, J. Bacteriol.* **109:**869–880.

Guthrie, C., 1986, Finding functions for small nuclear RNAs in yeast, *Trends Biochem. Sci.* **11:**430–434.

Guthrie, C., and Abelson, J., 1982, Organization of tRNA genes in *Saccharomyces cerevisiae,* in: *The Molecular Biology of the Yeast Saccharomyces, Metabolism and Gene Expression* (J. N. Strathern, E. W. Jones, and J. R. Broach, eds.), Cold Spring Harbor Monograph Series, Cold Spring Harbor, NY, pp. 487–528.

Haber, J. E., and Halvorson, H. O., 1975, Methods in sporulation and germination of yeasts, in: *Methods in Cell Biology,* Vol. 11 (D. M. Prescott, ed.), Academic Press, New York, pp. 46–69.

Hadjiolov, A. A., 1985, *The Nucleolus and Ribosome Biogenesis,* Springer-Verlag, Vienna, New York, pp. 1–263.

Hagen, D. C., McCaffrey, G., and Sprague, G. F., 1986, Evidence the yeast *STE3* gene encodes a receptor for the peptide pheromone a-factor: Gene sequence and implications for the structure of the presumed receptor, *Proc. Natl. Acad. Sci. USA* **83:**1418–1422.

Haguenauer-Tsapis, R., and Hinnen, A., 1984, A deletion that induces the signal peptidase cleavage site impairs processing, glycosylation, and secretion of cell surface yeast acid phosphatase, *Mol. Cell. Biol.* **4:**2668–2675.

Hanson, B. A., and Lester, R. L., 1980, The extraction of inositol-containing phospholipids and phosphatidylcholine from *Saccharomyces cerevisiae* and *Neurospora crassa, J. Lipid Res.* **21:**309–315.

Haselbeck, A., and Schekman, R., 1986, Interorganelle transfer and glycosylation of yeast invertase *in vitro, Proc. Natl. Acad. Sci. USA* **83:**2017–2021.

Hashimoto, C., Cohen, R. E., Zhang, W.-J., and Ballou, C. E., 1981, Glycoprotein chains on yeast carboxypeptidase Y are phosphorylated, *Proc. Natl. Acad. Sci. USA* **78:**2244–2248.

Hasilik, A., and Tanner, W., 1978, Carbohydrate moiety of carboxypeptidase Y and perturbation of its biosynthesis, *Eur. J. Biochem.* **91:**567–575.

Hay, R., Böhni, P., and Gasser, S., 1984, How mitochondria import proteins, *Biochim. Biophys. Acta* **779:**65–87.

Henry, S. A., 1982, Membrane lipids of yeast: Biochemical and genetic studies, in: *The Molecular Biology of the Yeast* Saccharomyces, *Metabolism and Gene Expression* (J. N. Strathern, E. W. Jones, and J. R. Broach, eds.), Cold Spring Harbor Monograph Series, Cold Spring Harbor, NY, pp. 101–158.

Henry, S. A., Klig, L. S., and Loewy, B. S., 1984, The genetic regulation and coordination of biosynthetic pathways in yeast: Amino acid and phospholipid synthesis, *Annu. Rev. Genet.* **18:**207–31.

Huet, J., Cottrelle, P., Cool, M., Vignais, M.-L., Thiele, D., Marck, C., Buhler, J. M., Sentenac, A., and Fromageot, P., 1985, A general upstream binding factor for genes of the yeast translational apparatus, *EMBO J.* **4:**3539–3547.

Ide, G. J., 1981, Nucleoside 5′,-[γ-S]triphosphates will initiate transcription in isolated yeast nuclei, *Biochemistry* **20:**2633–2638.

Itoh, T., Higo, K., and Otaka, E., 1979, Isolation and characterization of twenty-three ribosomal proteins from large subunits of yeast, *Biochemistry* **18:**5787–5793.

Jacq, C., Banroques, J., Becam, A. M., Slonimski, P. P. Guiso, N., and Danchin, A., 1984, Antibodies against a fused "lacZ-yeast mitochondrial intron" gene product allow identification of the mRNA maturase encoded by the fourth intron of the yeast cob-box gene, *EMBO J.* **3:**1567–1572.

Jenness, D. D., and Spatrick, P., 1986, Down regulation of the alpha-factor peromone in *Saccharomyces cerevisiae, Cell* **46:**345–353.

Jenness, D., Burkholder, A. C., and Hartwell, H., 1983, Binding of alpha-factor pheromone to yeast *a*-cells: Chemical and genetic evidence for an alpha-factor receptor, *Cell* **35:**521–529.

Jenness, D. D., Burkholder, A. C., and Hartwell, L. H., 1986, Binding of alpha-factor pheromone to *Saccharomyces cerevisiae* a-cells: Dissociation constant and number of binding sites, *Mol. Cell. Biol.* **6:**318–320.

Jerome, J. F., and Jaehning, J. A., 1986, mRNA transcription in nuclei isolated from *Saccharomyces cerevisiae, Mol. Cell. Biol.* **6:**1633–1639.

Julius, D., Blair, L., Brake, A., Sprague, G., and Thorner, J., 1983, Yeast alpha factor is processed from a large precursor polypepide: The essential role of a membrane-bound dipeptidyl aminopeptidase, *Cell* **32:**839–852.

Julius, D., Schekman, R., and Thorner, J., 1984, Glycosylation and processing of prepro-alpha-factor through the yeast secretory pathway, *Cell* **36:**309–318.

Kaibuchi, K., Miyajima, A., Arai, K.-I., and Matsumoto, K., 1986, Possible involvement of *RAS*-encoded proteins in glucose-induced inositolphospholipid turnover in *Saccharomyces cerevisiae, Proc. Natl. Acad. Sci. USA* **83:**8172–8176.

Kakinuma, Y., Ohsumi, Y., and Anraku, Y., 1981, Properties of H^+-translocating adenosine triphosphatase in vacuolar membranes of *Saccharomyces cerevisiae, J. Biol. Chem.* **256:**10859–10863.

Kaneko, H., Hosohara, M., Tanaka, M., and Itoh, T., 1976, Lipid composition of 30 species of yeast, *Lipids* **11:**837–844.

Kang, M. S., Elago, N., Mattia, E., Au-Young, J., Robbins, P. W., and Cabib, E., 1984, Isolation of chitin synthase from *Saccharomyces cerevisiae,* purification of an enzyme by entrapment in the reaction product, *J. Biol. Chem.* **259:**14966–14972.

Käppeli, O., Sauer, M., and Fiechter, A., 1982, Convenient procedure for the isolation of highly enriched cytochrome P-450–containing microsomal fraction from *Candida tropicalis, Anal. Biochem.* **126:**179–182.

Kato, N., Sahm, H., Schuette, H., and Wagner, F., 1979, Purification and properties of glucose-6-phosphate dehydrogenase and 6-phosphogluconate dehydrogenase from a methanol-utilizing yeast, *Candida boidinii, Biochem. Biophys. Acta* **566:**1–11.

Kilmartin, J. V., 1981, Purification of yeast tubulin by self-assembly *in vitro, Biochemistry* **20:**3629–3633.

Kilmartin, J. V., and Adams, A. E. M., 1984, Structural rearrangements of tubulin and actin during the cell cycle of the yeast *Saccharomyces, J. Cell Biol.* **98:**922–933.

King, S. M., and Hyams, J. S., 1982, The mitotic spindle of *Saccharomyces cerevisiae:* Assembly, structure, and function, *Micron* **13:**93–117.

King, S. M., Hyams, J. S., and Luba, A., 1982, Absence of microtubule sliding and an analysis of spindle formation from the yeast *Saccharomyces cerevisiae, J. Cell. Biol.* **94:**341–349.

Klig, L. S., Homann, M. J., Carman, G. M., and Henry, S. A., 1985, Coordinate regulation of phospholipid biosynthesis in *Saccharomyces cerevisiae:* Pleiotropically constitutive opil mutant, *J. Bacteriol.* **162:**1135–1141.

Koch, H., and Friesen, J. D., 1979, Individual messenger RNA half lives in *Saccharomyces cerevisiae, Mol. Gen. Genet.* **170:**129–135.

Kramer, R., Kopp, F., Niedermeyer, W., and Fuhrmann, G. F., 1978, Comparative studies of the structure and composition of the plasmalemma and the tonoplast in *Saccharomyces cerevisiae, Biochim. Biophys. Acta* **507:**369–380.

Kruiswijk, T., Planta, R. J., and Mager, W. H., 1978, Quantitative analysis of the protein composition of yeast ribosomes, *Eur. J. Biochem.* **83:**245–252.

Kurjan, J., and Herskowitz, I., 1982, Structure of a yeast pheromone gene (MFalpha): A putative alpha-factor precursor contains four tandem copies of mature alpha-factor, *Cell* **30:**933–943.

Langford, C. J., and Gallwitz, D., 1983, Evidence for an intron-contained sequence required for the splicing of yeast RNA polymerase II transcripts, *Cell* **33:**519–527.

Larkin, J. C., and Woolford, J. L., 1983, Molecular cloning and analysis of the *CRY1* gene: A yeast ribosomal protein gene, *Nucl. Acids Res.* **11:**403–420.

Leer, R. J., van Raamsdonk-Duin, M. C., Molenaar, M. T., Cohen, L. H., Mager, W., and Planta, R. J., 1982, The structure of the gene coding for the phosphorylated ribosomal protein S10 in yeast, *Nucl. Acids Res.* **10:**5869–5878.

Leer, R. J., van Raamsdonk-Duin, M. C., Hagendoorn, J. M., Mager, W. H., and Planta, R. J., 1984, Structural comparison of yeast ribosomal protein genes, *Nucl. Acids Res.* **12:**6685–6700.

Leer, R. J., van Raamsdonk-Duin, M. C., Kraakman, P., Mager, W. H., and Planta, R. J., 1985a, The genes of yeast ribosomal proteins S24 and L46 are adjacent and divergently transcribed, *Nucl. Acids Res.* **13:**701–709.

Leer, R. J., Van Ramsdonk-Duin, M. M. C., Mager, W. H., and Planta, R. J., 1985b, Conserved sequences upstream of yeast ribosomal protein genes, *Curr. Gen.* **9:**273–277.

Lehle, L., and Tanner, W., 1978, Glycosyltransfer from dolichylphosphate sugars to endogenous and exogenous glycoprotein acceptors in yeast, *Eur. J. Biochem.* **83:**563–570.

Letts, V. A., and Dawes, I. W., 1983, Temperature-sensitive *Saccharomyces cerevisiae* mutant defective in lipid biosynthesis, *J. Bacteriol.* **156:**212–221.

Letts, V. A., and Henry, S. A., 1985, Regulation of phospholipid synthesis in phosphatidyl serine synthase-deficient (*chol*) mutants of *Saccharomyces cerevisiae, J. Bacteriol.* **163:**560–567.

Liao, H., and Thorner, J., 1980, Yeast mating pheromone alpha-factor inhibits adenylate cyclase, *Proc. Natl. Acad. Sci. USA* **77:**1898–1902.

Lipke, P. N., Taylor, A., and Ballou, C. E., 1976, Morphogenic effects of α-factor on *Saccharomyces cerevisiae* a cells, *J. Bacteriol.* **127:**610–618.

Low, C., Rodriguez, R. J., and Parks, L. W., 1985, Modulation of yeast plasma membrane composition of a yeast sterol auxotroph as a function of exogenous sterol, *Arch. Biochem. Biophys* **240:**530–538.

Makarow, M., 1985a, Endocytosis in *Saccharomyces cerevisiae:* Internalization of enveloped virus into sheroplasts, *EMBO J.* **4:**1855–1860.

Makarow, M., 1985b, Endocytosis in *Saccharomyces cerevisiae:* Internalization of α-amylase and fluorescent dextran into cells, *EMBO J.* **4:**1861–1866.

Makarow, M., and Nevalainen, T., 1987, Transport of a fluorescent macromolecule via endosomes to a vacuole in *Saccharomyces cerevisiae, J. Cell Biol.* **104:**67–75.

Malpartida, F., and Serrano, R., 1981a, Reconstitution of the proton-translocating adenosine triphosphatase of yeast plasma membranes, *J. Biol. Chem.* **256:**4157–4177.

Malpartida, F., and Serrano, R., 1981b, Phosphorylated intermediate of the ATPase from the plasma membrane of yeast. *Eur. J. Biochem.* **116:**413–417.

Mann, W., and Jeffery, J., 1986, Yeasts in molecular biology. Spheroplast preparation with *Candida utilis, Schizosaccharomyces pombe* and *Saccharomyces cerevisiae, Biosci. Rep.* **6:**597–602.

Mann, K., and Mecke, D., 1982, The isolation of *Saccharomyces cerevisiae* nuclear membranes with nuclease and high-salt treatment, *Biochim. Biophys. Acta* **687:**57–62.

Martinez, J. P., Murqui, A., Flores, A., and Sentandreu, R., 1984, Subcellular fractionation of actively growing protoplasts of *Saccharomyces cerevisiae, Biochim. Biophys. Acta* **805:**59–71.

Mason, T. L., Poyton, R. O., Wharton, D. C., and Schatz, B., 1973, Cytochrome c oxidase from baker's yeast. I. Isolation and properties, *J. Biol. Chem.* **248:**1346–1354.

Matile, P., 1978, Biochemistry and function of vacuoles, *Annu. Rev. Plant Physiol.* **29:**193–223.

Matile, P., and Wiemken, A., 1976, Interaction between cytoplasm and vacuole, in: *Transport in Plants,* Vol. 3 (C. R. Stocking and U. Heber, eds.), Springer, Berlin, pp. 255–287.

Mattaj, I. W., 1984, SnRNAs: From gene architecture to RNA processing, *Trends Biochem. Sci.* **9:**435–437.

McDonough, J. P., and Mahler, H. P., 1982, Covalent phosphorylation of the Mg^{2+}-dependent ATPase of yeast plasma membrane, *J. Biol. Chem.* **257:**14579–14581.

McKee, E. E., McEwen, J. E., and Poyton, R. O., 1984, Mitochondrial gene expression in *Saccharomyces cerevisiae.* II. Fidelity of translation in isolated mitochondria from wild type and respiratory-deficient mutant cells, *J. Biol. Chem.* **259:**9332–9338.

Mechler, B., Müller, M., Müller, H., Meussdoerffer, F., and Wolf, D. H., 1982, *In vivo* biosynthesis of the vacuolar proteinases A and B in the yeast *Saccharomyces cerevisiae, J. Biol. Chem.* **257:**11203–11206.

Meyer, D. I., 1982, The signal hypothesis—A working model, *Trends Biochem. Sci.* **7:**320–321.

Miller, A. M., 1984, The yeast Mata1 gene contains two introns, *EMBO J.* **3:**1061–1065.

Mitchison, J. M., and Carter, B. L. A., 1975, Cell cycle analysis, in: *Methods in Cell Biology,* Vol. 11 (D. M. Prescott, ed.), Academic Press, New York, pp. 201–219.

Moeller, C. H., Mudd, J. B., and Thomson, W. W., 1981, Lipid phase separations and intramembranous particle movements in the yeast tonoplast, *Biochim. Biophys. Acta* **643:**376–386.

Moens, P. B., and Rapport, E., 1971, Spindles, spindle plaques, and meiosis in the yeast *Saccharomyces cerevisiae* (Hansen), *J. Cell Biol.* **50:**344–361.

Mueller, S. C., and Branton, D., 1984, Identification of coated vesicles in *Saccharomyces cerevisiae, J. Cell Biol.* **98:**341–346.

Nagy, M., Laporte, J., Penverne, B., and Herne, G., 1982, Nuclear localization of aspartate transcarbamoylase in *Saccharomyces cerevisiae, J. Cell Biol.* **92:**790–794.

Nakajima, T., and Ballou, C. E., 1975, Yeast manno-protein biosynthesis: Solubilization and selective assay of four mannosyltransferases, *Proc. Natl. Acad. Sci. USA* **72:**3912–3916.

Nakayama, N., Miyajima, A., and Arai, K., 1987, Common signal transduction system by STE2 and STE3 in haploid cells of *Saccharomyces cerevisiae:* Autocrine cell-cycle arrest results from forced expression of STE2, *EMBO J.* **6:**249–254.

Navarrete, R., and Serrano, R., 1983, Solubilization of yeast plasma membranes and mitochondria by different types of nondenaturating detergents, *Biochim. Biophys. Acta* **728:**403–408.

Neff, N. F., Thomas, J. H., Grisafi, P., and Botstein, D., 1983, Isolation of the β-tubulin gene from yeast and demonstration of its essential function *in vivo, Cell* **33:**211–215.

Newman, A. J., Lin, R., Cheng, S., and Abelson, J., 1985, Molecular consequences of specific intron mutations on yeast mRNA splicing *in vivo* and *in vitro, Cell* **42:**335–344.

Ng, R., and Abelson, J., 1980, Isolation of the gene for actin in *Saccharomyces cerevisiae, Proc. Natl. Acad. Sci. USA* **77:**3912–3916.

Niedermeyer, W., 1976, The elasticity of the yeast cell tonoplast related to its ultrastructure and chemical composition. II. Chemical and cytochemical investigations, *Cytobiology* **13:**380–393.

Nomura, M., Gourse, R., and Baughman, G., 1984, Regulation of the synthesis of ribosomes and ribosomal components *Annu. Rev. Biochem.* **53:**75–117.

Novick, P., 1985, Intracellular transport mutants of yeast, *Trends Biochem. Sci.* **10:**432–434.

Novick, P., and Botstein, D., 1985, Phenotypic analysis of temperature-sensitive yeast actin mutants, *Cell* **40:**405–416.

Novick, P., Field, C., and Schekman, R., 1980, Identification of 23 complementation groups required for post-translational events in the yeast secretory pathway, *Cell* **21:**205–215.

Novick, P., Ferro, S., and Schekman, R., 1981, Order of events in the yeast secretory pathway, *Cell* **25:**461–469.

Orlean, P., 1987, Two chitin synthases in *Saccharomyces cerevisiae, J. Biol. Chem.* **262:**5732–5739.

Orlean, P., Ammer, H., Watzele, M., and Tanner, W., 1986, Synthesis of an O-glycosylated cell surface protein induced in yeast by α-factor, *Proc. Natl. Acad. Sci. USA* **83:**6263–6266.

Otaka, E., and Osawa, S., 1981, Yeast ribosomal proteins. V. Correlation of several nomenclatures and proposal of a standard nomenclature, *Mol. Gen. Genet.* **181:**176–182.

Otaka, E., Higo, K., and Osawa, S., 1982, Isolation of seventeen proteins and amino-terminal amino sequences of eight proteins from cytoplasmic ribosomes of yeast, *Biochemistry* **21:**4545–4550.

Otaka, E., Higo, K., and Itoh, T., 1983, Yeast ribosomal proteins. VII. Cytoplasmic ribosomal proteins from *Schizosaccharomyces pombe, Mol. Gen. Genet.* **191:**519–524.

Padgett, R. A., Grabowski, P. J., Konarska, M. M., and Sharp, P., 1985, Splicing messenger RNA precursors: Branch sites and lariat RNAs, *Trends Biochem. Sci.* **10:**154–157.

Padgett, R. A., Grabowski, P. J., Konarska, M. M., Seiler, S., and Sharp, P., 1986, Splicing of messenger RNA and RNA precursors, *Annu. Rev. Biochem.* **55:**1119–1150.

Pastor, F. I. J., Valentin, E., Herrero, E., and Sentandreu, R., 1984, Structure of the *Saccharomyces cerevisiae* cell wall. Mannoproteins released by zymolyase and their contribution to wall architecture, *Biochim. Biophys. Acta* **802:**292–300.

Payne, G. S., and Schekman, R., 1985, A test of clathrin function in protein secretion and cell growth, *Science* **230:**1009–1014.

Peberdy, J. F., 1979, Fungal protoplasts: Isolation, reversion, and fusion, *Annu. Rev. Microbiol.* **33:**21–39.

Peebles, C. L., Gegenheimer, P., and Abelson, J., 1983, Precise excision of intervening sequences from precursor tRNAs by a membrane-associated yeast endonuclease, *Cell* **32:**525–536.

Peterson, J. B., and Ris, H., 1976, Electron-microscopic study of the spindle and chromosome movement in the yeast *Saccharomyces cerevisiae, J. Cell Sci.* **22:**219–242.

Pikielny, C. W., Rymond, B. C., and Rosbash, M., 1986, Electrophoresis of ribonucleoproteins reveals an ordered assembly pathway of yeast splicing complexes, *Nature* **324:**341–345.

Planta, R. J., Mager, W. H., Leer, R. J., Woudt, L. P., Raue, H. A., and El-Baradi, T. T. A. L., 1986, Structure and expression of ribosomal protein genes in yeast, in: *Structure, Function, and Genetic of Ribosomes* (B. Hardesty and G. Kramer, eds.), Springer-Verlag, New York, pp. 699–718.

Prasad, R., 1985, Lipids in the structure and function of yeast membrane, *Adv. Lipid Res.* **21:**187–242.

Prasad, R., and Rose, A. H., 1986, Involvement of lipids in solute transport in yeasts, *Yeast* **2:**205–220.

Pringle, J. R., 1975, Methods for avoiding proteolytic artefacts in studies of enzymes and other proteins from yeast, *Meth. Cell Biol.* **12:**149–184.

Pringle, J. R., and Hartwell, L. H., 1981, The *Saccharomyces cerevisiae* cell cycle, in: *The Molecular Biology of the Yeast Saccharomyces, Life Cycle and Inheritance* (J. N. Strathern, E. W. Jones, and J. R. Broach, eds.), Cold Spring Harbor Monograph Series, Cold Spring Harbor, NY, pp. 97–142.

Racker, E., 1950, Spectrophotometric measurements of the enzymatic formation of fumaric and cis-aconitic acids, *Biochim. Biophys. Acta* **4:**211–214.

Ramirez, R. M., Ishida-Schick, T., Krilowicz, B. L., Leish, B. A., and Atkinson, K. D., 1983, Plasma membrane expansion terminates in *Saccharomyces cerevisiae* secretion-defective mutants while phospholipid synthesis continues, *J. Bacteriol.* **154:**1276–1283.

Ramirez, J. M., Gallego, G. G., and Serrano, R., 1984, Electron transfer constituents in plasma membrane fractions of *Avena sativa* and *Saccharomyces cerevisiae, Plant Sci. Lett.* **34:**103–110.

Rank, G. H., Gerlach, J. H., Robertson, A. J., and Van Hoeven, R. P., 1978, High viscosity vesicles of yeast separated at pH 4 have surface glycoproteins, *Nature* **273:**682–684.

Rattray, J. B. M., Schibeci, A., and Kidby, D. K., 1975, Lipids of yeast, *Bacteriol. Rev.* **39:**197–231.

Rickwood, D., and Hayes, A., 1984, An evaluation of methods used to prepare yeast mitochondria for transcriptional studies, *Prep. Biochem.* **14:**163–171.

Riezman, H., 1985, Endocytosis in yeast: Several of the yeast secretory mutants are defective in endocytosis, *Cell* **40:**1001–1009.

Riezman, H., Hay, R., Gasser, S., Daum, G., Schneider, G., Witte, C., and Schatz, G., 1983, The outer membrane of yeast mitochondria: Isolation of outside-out sealed vesicles, *EMBO J.* **2:**1105–1111.

Riezman, H., Chvatchko, Y., and Dulic, V., 1986, Endocytosis in yeast, *Trends Biochem. Sci.* **11:**325–328.

Roberts, R. L., Bowers, B., Slater, M. L., and Cabib, E., 1983, Chitin synthesis and localization in cell division cycle mutants of *S. cerevisiae, Mol. Cell. Biol.* **3:**922–930.

Robertson, A. J., Gerlach, J. H., Rank. G. H., and Fowke, L. C., 1980, Yeast cell wall, membrane, and soluble marker polypeptides identified by comparative two-dimensional electrophoresis, *Can. J. Biochem.* **58:**567–572.

Rodriguez, J. R., Pikielny, C. W., and Rosbash, M., 1984, *In vivo* characterization of yeast mRNA processing intermediates, *Cell* **39:**603–610.

Rosamond, J., 1982, The molecular biology of the mitochondrion, *Biochem. J.* **202:**1–8.

Rothblatt, J. A., and Meyer, D. I., 1986a, Secretion in yeast: Reconstitution of the translocation and glycosylation of α-factor and invertase in a homologous cell-free system, *Cell* **44:**619–628.

Rothblatt, J. A., and Meyer, D. E., 1986b, Secretion in yeast: Translocation and glycosylation of prepro-α-factor *in vitro* can occur via an ATP-dependent post-translational mechanism, *EMBO J.* **5:**1031–1036.

Rubtsov, P. M., Musakhanov, M. M., Zakharyev, V. M., Krayev, A. S., Skryabin, K. G., and Bayev, A. A., 1980, The structure of the yeast ribisomal RNA genes. I. The complete nucleotide sequence of the 18S ribosomal RNA gene from *Saccharomyces cerevisiae, Nucl. Acids Res.* **8:**5779–5794.

Sanz, P., Herrero, E., and Sentandreu, R., 1987, Secretory pattern of a major integral mannoprotein of the yeast cell wall, *Biochim. Biophys. Acta* **924:**193–203.

Schatz, G., and Butow, R. A., 1983, How are proteins imported into mitochondria? *Cell* **32:**316–318.

Schatz, G., and Klima, J., 1964, Triphosphopyridine nucleotide: Cytochrome c reductase of *Saccharomyces cerevisiae:* a microsomal enzyme, *Biochim. Biophys. Acta* **81:**448–461.

Schatz, P. J., Pillus, L., Grisafi, P., Solomon, F., and Botstein, D., 1986a, Two functional α-tubulin genes of the yeast *Saccharomyces cerevisiae* encode divergent proteins, *Mol. Cell Biol.* **6:**3711–3721.

Schatz, P. J., Solomon, F., and Botstein, D., 1986b, Genetically essential and non-essential α-tubulin genes specify functionally interchangeable proteins. *Mol. Cell Biol.* **6:**3722–3733.

Schekman, R., 1982, Biochemical markers for yeast organelles, in: *The Molecular Biology of the Yeast Saccharomyces, Metabolism and Gene Expression* (J. N. Strathern, E. W. Jones, and

J. R. Broach, eds.), Cold Spring Harbor Monograph Series, Cold Spring Harbor, NY, pp. 651–652.
Schekman, R., 1985, Protein localization and membrane traffic in yeast, *Annu. Rev. Cell Biol.* **1**:167–195.
Schekman, R., and Brawley, V. L., 1979, Localized deposition of chitin on the yeast cell surface in response to pheromone, *Proc. Natl. Acad. Sci. USA* **76**:645–649.
Schekman, R., and Novick, P., 1982, The secretory process and yeast cell surface assembly, in: *The Molecular Biology of the Yeast Saccharomyces, Metabolism and Gene Expression* (J. N. Strathern, E. W. Jones, and J. R. Broach, eds.), Cold Spring Harbor Monograph Series, Cold Spring Harbor, NY, pp. 361–393.
Schmidt, R., Ackermann, R., Kratky, Z., Wassermann, B., and Jacobson, B., 1983, Fast and efficient purification of yeast plasma membranes using cationic silica microbeads, *Biochim. Biophys. Acta* **732**:421–427.
Schneider, J. C., and Guarente, L., 1987, The untranslated leader of nuclear COX4 gene of *Saccharomyces cerevisiae* contains an intron, *Nucl. Acids Res.* **15**:3515–3529.
Schultz, L. D., 1978, Transcriptional role of yeast deoxyribonucleic acid dependent ribonucleic acid polymerase III, *Biochemistry* **17**:750–758.
Schwaiger, H., Hasilik, A., von Figura, K. Wiemken, A., and Tanner, W., 1982, Carbohydrate-free carboxypeptidase Y is transferred into the lysosome-like yeast vacuole, *Biochem. Biophys. Res. Commun.* **104**:950–956.
Schwartz, D. C., and Cantor, C. R., 1984, Separation of yeast chromosome-sized DNAs by pulse field gradient gel electrophoresis, *Cell* **37**:67–75.
Schwencke, J., 1977, Characteristics and integration of the yeast vacuole with cellular function, *Physiol. Veg.* **15**:491–517.
Schwenke, J., Canut, H., and Flores, A., 1983, Simultaneous isolation of the yeast cytosol and well-preserved mitochondria with negligible contamination by vacuolar proteinases, *FEBS Lett.* **156**:274–280.
Sentandreu, R., Herrero, E., and Elorza, M. V., 1984, The assembly of wall polymers in yeast, *FEMS Symp.* **27**:51–61.
Shematek, E. M., and Cabib, E., 1980, Biosynthesis of the yeast cell wall. II. Regulation of β-(1->3) glucan synthetase by ATP and GTP, *J. Biol. Chem.* **255**:895–902.
Shematek, E. M., Braatz, J. A., and Cabib, E., 1980, Biosynthesis of the yeast cell wall: I. Preparation and properties of β-(1->3) glucan synthetase, *J. Biol. Chem.* **255**:888–894.
Sherman, F., 1982, Suppression in the yeast *Saccharomyces cerevisiae,* in: *The Molecular Biology of the Yeast Saccharomyces, Metabolism and Gene Expression* (J. N. Strathern, E. W. Jones, and J. R. Broach, eds.), Cold Spring Harbor Monograph Series, Cold Spring Harbor, NY, pp. 463–486.
Shortle, D., Harber, J., and Botstein, D., 1982, Lethal disruption of the yeast actin gene by DNA transformation, *Science* **217**:371–373.
Shortle, D., Novick, P., and Botstein, D., 1984, Construction and genetic characterization of temperature-sensitive alleles of the yeast actin gene, *Proc. Natl. Acad. Sci. USA* **81**:4889–4893.
Shulman, R. W., 1978, Yeast cell selection and synchrony: Density gradients and mating factor block, in: *Methods in Cell Biology, Vol. 20* (D. M. Prescott, ed.), Academic Press, New York, pp. 43–45.
Silver, P. A., Keegan, L. P., and Ptashne, M., 1984, Amino terminus of the yeast GAL4 gene product is sufficient for nuclear localization, *Proc. Natl. Acad. Sci. USA* **81**:5951–5955.
Simon, M., Seraphin, B., and Faye, G., 1986, KIN28, a yeast split gene coding for a putative protein kinase homologous to CDC28, *EMBO J.* **5**:2697–2701.
Singh, A., Chen, E. Y., Lugovoy, J. M., Chang, C. N., Hitzman, R. A., and Seeburg, P. H., 1983, *Saccharomyces cerevisiae* contains two discrete genes coding for the α-factor pheromone, *Nucl. Acids Res.* **11**:4049–4063.

Sprague, G. F., Blair, L. C., and Thorner, J., 1983, Cell interactions and regulation of cell type in the yeast *Saccharomyces cerevisiae, Annu. Rev. Microbiol.* **37:**623–660.

Sprinzl, M., Moll, J., Meissner, F., and Hartmann, T., 1985, Compilation of tRNA sequences, *Nucl. Acids Res.* **13** (Suppl.):r1–r50.

Steere, R. L., and Erbe, E. F., and Moseley, J. M., 1980, Prefracture and cold-fracture images of yeast plasma membranes, *J. Cell Biol.* **86:**113–122.

Stevens, B., 1981, Mitochondrial structure, in: *The Molecular Biology of the Yeast Saccharomyces, Life Cycle and Inheritance* (J. N. Strathern, E. W. Jones, and J. R. Broach, eds.), Cold Spring Harbor Monograph Series, Cold Spring Harbor, NY, pp. 471–504.

Stötzler, D., Kiltz, H.-H., and Duntze, W., 1976.,Primary structure of α-factor peptides from *Saccharomyces cerevisiae, Eur. J. Biochem.* **69:**397–400.

Swida, U., Kreutzfeldt, C., Ramezani-Rad, M., and Käufer, N., 1982, Isolation and characterization of rough and smooth endoplasmic reticulum from *Saccharomyces cerevisiae, FEMS Microbiol. Lett.* **15:**313–318.

Tanner, W., 1984, Regulation of glycoprotein synthesis in yeast by mating pheromones, in: *Structure, Function, and Biosynthesis of Plant Cell Walls, Proc. Annu. Symp. Bot., 7th,* (M. W. Dugger and S. Bartinicki-Garcia, eds.), Waverly Press, Baltimore, pp. 20–32.

Tanner, W., and Lehle, L., 1987, Protein glycosylation in yeast, *Biochim. Biophys. Acta* **906:**81–99.

Tarentino, A. L., Plummer, T. H., and Maley, F., 1974, The release of intact oligosaccharides from specific glycoproteins by endo-β-*N*-acetylglucosaminidase H, *J. Biol. Chem.* **249:**818–824.

Teem, J. L. Abovich, N., Käufer, N. F., Schwindinger, W. F., Warner, J. R., Levy, A., Wooldford, J., Leer, R. J., van Ramsdonk-Duin, M. M. C., Mager, W. H., Planta, R. J., Schultz, L., Friesen, J. D., Fried, H., and Rosbash, M., 1984, A comparison of yeast ribosomal protein gene DNA sequences, *Nucl. Acids Res.* **12:**8295–8312.

Thorner, J., 1981, Pheromonal regulation of development in *Saccharomyces cerevisiae,* in: *The Molecular Biology of the Yeast Saccharomyces, Life Cycle and Inheritance* (J. N. Strathern, E. W. Jones, and J. R. Borach, eds.), Cold Spring Harbor Monograph Series, Cold Spring Harbor, NY, pp. 143–180.

Thorner, J., 1982, An essential role of cyclic AMP in growth control: The case of yeast, *Cell* **30:**5–6.

Tijssen, J. P. F., Beekes, H. W., and van Steveninck, J., 1981, Localization of polyphosphates at the outside of the yeast cell plasma membrane, *Biochim. Biophys. Acta* **649:**529–532.

Tohoyama, H., and Yanagishima, N., 1981, Changes in the sexual agglutination ability during the formation of vegetative cells from spores in *Saccharomyces cerevisiae, Mol. Gen. Genet.* **183:**205–208.

Tohoyama, H., Yoshida, M. H. K., and Yanagishima, N., 1979, Regulation of the production of the agglutination substances responsible for sexual agglutination in *Saccharomyces cerevisiae:* Changes associated with conjugation and temperature shift, *Mol. Gen. Genet.* **174:**269–280.

Towler, D., and Glaser, L., 1986, Protein fatty acid acylation: Enzymatic synthesis of an N-myristoylglycyl peptide, *Proc. Natl. Acad. Sci. USA* **83:**2812–2816.

Trembath, M. K., and Tzagoloff, A., 1979, Large- and small-scale preparations of yeast mitochondria, *Meth. Enzymol.* **55:**160–170.

Trimble, R. B., and Atkinson, P. H., 1986, Structure of yeast external invertase $Man_{8\text{-}14}GlcNAc$ processing intermediates by 500-megahertz 'H NMR spectroscopy, *J. Biol. Chem.* **261:**9815–9824.

Trimble, R. B., Maley, F., and Chu, F. L., 1983, Glycoprotein biosynthesis in yeast: Protein conformation affects processing of high mannose oligosaccharides and carboxypeptidase Y and invertase, *J. Biol. Chem.* **258:**2562–2567.

Trivedi, A., Fantin, D. J., and Tustanoff, E. R., 1985, *Saccharomyces cerevisiae* as a model eukaryote for studies on mitochondriogenesis, *Microbiol. Sci.* **2**:10–13.

Tsai, P. -K., Frevert, J., and Ballou, C. E., 1984, Carbohydrate structure of *Saccharomyces cerevisiae* mnn9 mannoprotein, *J. Biol. Chem.* **259**:3805–3811.

Tschopp, J., and Schekman, R., 1983, Two distinct subfractions in isolated *Saccharomyces cerevisiae* plasma membranes, *J. Bacteriol.* **156**:222–229.

Tschopp, J., Esmon, P. C., and Schekman, R., 1984, Defective plasma membrane assembly in yeast secretory mutants, *J. Bacteriol.* **160**:966–970.

Urech, K., Dürr, M., Boller, T., and Wiemken, A., 1978, Localization of polyphosphate in vacuoles of *Saccharomyces cerevisiae, Arch. Microbiol.* **116**:275–278.

Valentin, E., Herrero, E., Pastor, F. I. J., and Sentandreu, R., 1984, Solubilization and analysis of mannoprotein molecules from the cell wall of *Saccharomyces cerevisiae, J. Gen. Microbiol.* **130**:1419–1428.

Van der Wilden, W., Matile, P., Schellenberg, M., Meyer, J., and Wiemken, A., 1973, Vacuolar membranes: Isolation from yeast cells, *Z. Naturforsch.* **28c**:416–421.

Veldman, G. M., Klootwijk, J., de Regt, V. C. H. F., and Planta, R. J., 1981a, The primary and secondary structure of yeast 26S rRNA, *Nucl. Acids Res.* **9**:6936–6952.

Veldman, G. M., Klootwijk, J. van Heerikhuizen, H., and Planta, R. J., 1981b, The nucleotide sequence of the intergenic region between the 5.8S and 26S rRNA genes of the yeast ribosomal RNA operon. Possible implications for the interactions between 5.8S and 26S rRNA and the processing of the primary transript, *Nucl. Acids Res.* **9**:4847–4862.

Vignais, M-L., Woudt, L. P., Wassenar, G. M., Mager, W. H., Sentenac, A., and Planta, R. J., 1987, Specific binding of TUF factor to upstream activation sites of yeast ribosomal protein genes, *EMBO J.* **6**:1451–1457.

Walter, P., and Lingappa, V. R., 1986, Mechanism of protein translocation across endoplasmic reticulum membrane, *Annu. Rev. Cell Biol.* **2**:499–516.

Warner, J. R., 1982, The yeast ribosome: Structure, function, and synthesis, in: *The Molecular Biology of the Yeast Saccharomyces, Metabolism and Gene Expression* (J. N. Strathern, E. W. Jones, and J. R. Broach, eds.), Cold Spring Harbor Monograph Series, Cold Spring Harbor, NY, pp. 529–560.

Warner, J. R., Mitra, G., Schwindinger, W. F., Studeng, M., and Fried, H. W., 1985, *Saccharomyces cerevisiae* coordinates accumulation of yeast ribosomal proteins by modulating mRNA splicing, translational initiation, and protein turnover, *Mol. Cell Biol.* **5**:1512–1521.

Warner, J. R., Elion, E. A., Dabeva, M. D., and Schwindinger, W. F., 1986, The ribosomal genes of yeast and their regulation, in: *Structure, Function, and Genetic of Ribosomes* (B. Hardesty and G. Kramer, eds.), Springer-Verlag, New York, pp. 719–732.

Waters, M. G., and Blobel, G., 1986, Secretory protein translocation in a yeast cell-free system can occur posttranslationally and requires ATP hydrolysis, *J. Cell Biol.* **102**:1543–1550.

Welch, J. W., and Burlingame, A. L., 1973, Very long chain fatty acids in yeast, *J. Bacteriol.* **115**:464–466.

Welten-Verstegen, G. W., Boer, P., and Stern-Parve, E. P., 1980, Lipid-mediated glycosylation of endogenous proteins in isolated plasma membranes of *Saccharomyces cerevisiae, J. Bacteriol.* **141**:342–349.

Wen, D., and Schlesinger, M. J., 1984, Fatty acid-acylated proteins in secretory mutants of *Saccharomyces cerevisiae, Mol. Cell Biol.* **4**:688–694.

Wiemken, A., Matile, P., and Moor, H., 1970, Vacuolar dynamics in synchronously budding yeast, *Arch. Microbiol.* **70**:89–103.

Wiemken, A., Schellenberg, M., and Urech, K., 1979, Vacuoles: The sole compartments of digestive enzymes in yeast (*Saccharomyces cerevisiae*), *Arch. Microbiol.* **123**:23–35.

Witt, W., Schweingruber, M. E., and Mertsching, A., 1984, Phospholipase B from the plasma membrane of *Saccharomyces cerevisiae:* Separation of two forms with different carbohydrate content, *Biochim. Biophys. Acta* **795:**108–116.

Wolf, D. H., 1986, Cellular control in the eukaryotic cell through action of proteinases: The yeast *Saccharomyces cerevisiae* as a model organism, *Microbiol. Sci.* **3:**107–114.

Woudt, L. P., Smit, A. B., Mager, W. H., and Planta, R. J., 1986, Conserved sequence elements upstream of the gene encoding yeast ribosomal protein L25 are involved in transcription activation, *EMBO J.* **5:**1037–1040.

Yaffe, M., 1983, Import of proteins into mitochondria, a survey, in: *Mitochondria '83, Nucleo–mitochrondrial Interactions* (R. J. Schweijen, K. Wolf, and F. Kaudewitz, eds.), Walter deGruyter Verlag, Berlin, pp. 47–55.

Yamaguchi, M., Yoshida, K., and Yanagishima, N., 1984, Isolation and biochemical and biological characterization of an a-mating-type-specific glycoprotein for sexual agglutination from the cytoplasm of a-cells in the yeast *Saccharomyces cerevisiae, Arch. Microbiol.* **140:**113–119.

Zickler, D., and Olson, L. W., 1975, The synaptonemal complex and the spindle plaque during meiosis in yeast, *Chromosoma* **50:**1–23.

Zimmerman, R., and Meyer, D. I., 1986, 1986: A year of new insights into how proteins cross membranes, *Trends Biochem. Sci.* **11:**512–515.

Zinker, S., and Warner, J. R., 1976, The ribosomal proteins of *Saccharomyces cerevisiae,* phosphorylated and exchangeable proteins, *J. Biol. Chem.* **251:**1799–1807.

Zlotnik, H., Fernandez, M. P., Bowers, B., and Cabib, E., 1984, *Saccharomyces cerevisiae* mannoproteins form in external cell wall layer that determines wall porosity, *J. Bacteriol.* **159:**1018–1026.

Metabolism and Biosynthesis 3

J. R. DICKINSON

1. INTRODUCTION

A recent Symposium on Yeast Cell Biology was opened with the statement "In ten years yeast has changed from *E[scherichia] coli* with a nucleus to the universal cell" (Hartwell, 1985). This bold assertion underlines the current importance of yeast as an experimental system. In terms of our understanding of metabolism and biosynthesis, the extent of our knowledge of *Saccharomyces cerevisiae* is surpassed only in the case of *E. coli.* Nevertheless, despite this mass of information we are still ignorant of the actual steps in a number of primary metabolic pathways. In addition, many important details about the regulation of individual pathways (the means whereby the cell integrates the many separate pathways to effect growth and proliferation) remain undiscovered. Furthermore, the greater structural complexity of the eukaryotic yeast cell (compared with the prokaryotic *E. coli*) means that compartmentalization of metabolites is a further complication, the significance of which is now starting to be seriously considered by more and more researchers.

At this point it is appropriate to give a note of caution. The newcomer to yeast metabolism is urged not to survey the general schemes of metabolic pathways that appear in biochemistry textbooks and as wallcharts. These are only compilations, the information having been derived from numerous different systems. Not all of these pathways exist in yeast. Furthermore, yeast does some things differently from *E. coli* (the model system on which many metabolic pathways are based). For example, the biosynthesis of lysine is accomplished by entirely different pathways in yeast and bacteria. The reviews on carbohydrate metabolism by Fraenkel (1982) and on amino acid and nucleotide biosynthesis by Jones and Fink (1982) are excellent starting points for newcomer and expert alike.

J. R. DICKINSON • School of Pure and Applied Biology, University of Wales College of Cardiff, Cardiff CF1 3TL Wales.

The genetics of *S. cerevisiae* are better worked out than for any other eukaryote. In addition, the organism is amenable to genetic manipulation. As a result, metabolic questions can be examined far more extensively and in greater detail than would otherwise be possible. For example, we are not just limited to examining the effects of individual mutations. Using "conventional" yeast genetic methods (described elsewhere in this volume) it is possible to combine different mutations into a single cell. The potential for examining control of a metabolic pathway then becomes apparent especially where gene–enzyme relationships in a pathway are known. Comparisons between haploids, homozygous, and heterozygous diploids, and the use of cloned genes maintained in the yeast cell at defined copy number, enable an understanding of importance of gene dosage in various aspects of metabolic control.

Probably the best-known feature of *Saccharomyces* is its facultative nature; that is, it can be grown aerobically or anaerobically. Thus, the effects of a mutation affecting glycolysis can be studied by growing the cells aerobically on (say) ethanol, whereas if one wants to examine the consequences of defects in oxidative (mitochondrial) functions, the cells can be grown by fermentation of glucose. This convenient feature of yeast has been exploited fully by those interested in the metabolism, structure, and biogenesis of mitochondria. Undoubtedly our knowledge of mitochondria would be greatly diminished without work done on yeast.

2. OUTLINE OF CARBOHYDRATE METABOLISM

Carbohydrate metabolism in *Saccharomyces* has been reviewed by Fraenkel (1982). Two of the most pertinent points made in that review are: (1) strains of *Saccharomyces* used for biochemical and genetic study grow best by fermentation, even aerobically, and (2) the enzymes of glycolysis represent 30–65% (depending on estimate) of soluble protein: thus they are a major component of total cellular protein. There are five physiological phenomena in *Saccharomyces* which need to be considered every time we grow a culture. These are catabolite repression (usually only glucose repression is considered), catabolite inactivation, the induction of respiratory enzymes by oxygen, the Pasteur effect, and catabolite conversion.

2.1. Catabolite Repression

Catabolite repression is the lowering of the levels of cellular functions involved in respiration by a carbon source such as glucose. In batch culture at the initially high substrate concentrations growth will be accomplished by fermentation. As the substrate is consumed, its concentration falls and the cells become derepressed; i.e., induction of respiratory enzymes and components of electron transport occurs. It has been shown that the hexokinase PII isozyme is bifunctional and has, in addition to its "normal" catalytic function, a regulatory domain for triggering glucose repression (Entian and Frohlich, 1984; Entian *et al.*, 1984; Frohlich *et al.*, 1984).

Other sugars are less repressing than glucose, and the means whereby the cell monitors the concentration and identity of the carbon source have still to be worked out. However, some progress should result from the finding that the *SNF1* gene product is a protein kinase (Celenza and Carlson, 1986). Upon glucose deprivation the *SNF1* function is required in the expression of glucose-repressible genes involved in the metabolism of sucrose, galactose, maltose, melibiose, and nonfermentable carbon sources. It is now known that *SNF1* is the same gene as *CCR1* (also known as *CAT1*) (Celenza and Carlson, 1986). Further clues are likely to come from the identification of *u*pstream *a*ctivation *s*ites (UAS) of glucose-repressible genes. These sequences have been identified by deletion analysis, deletion within the UAS of a structural gene resulting in inability to derepress synthesis of the respective enzyme. This allows identification of a formal genetic model in which derepression by a positive effector binding to DNA permits transcription. Clearly, a complete biochemical description of the phenomenon of catabolite repression is still some way off but now seems a realistic possibility.

2.2. Catabolite Inactivation

Catabolite inactivation refers to the following sequence of events: Addition of glucose or a metabolically related sugar to cells growing on acetate or ethanol results in a loss of activity of the key gluconeogenic enzymes fructose 1,6-bisphosphatase, cytoplasmic malate dehydrogenase, and phosphoenolpyruvate carboxykinase (Gancedo, 1971; Witt *et al.*, 1966; Ferguson *et al.*, 1967; Haarasilta and Oura, 1975). The advantages of such a mechanism to the cell are obvious: it no longer needs to perform gluconeogenesis in the presence of glucose. Glycolysis using the newly added glucose with simultaneous gluconeogenesis from the original acetate (or ethanol) would be metabolically futile. In the case of fructose 1,6-bisphosphatase inactivation occurs in two stages. The first stage is rapid, with a decrease of activity of 60% occurring within about 3 min; this is accomplished by phosphorylation of the enzyme (Tortora *et al.*, 1981; Muller and Holzer, 1981; Mazon *et al.*, 1982a). The second stage, a proteolytic degradation, takes about 1 hr (Funayama *et al.*, 1980; Tortora *et al.*, 1981). It has been proposed that addition of glucose causes an increase in adenosine 3′:5′-cyclic monophosphate (cAMP) and that the phosphorylation occurs by the action of a cAMP-dependent protein kinase (Purwin *et al.*, 1982; Van der Platt and Van Solingen, 1974; Tortora *et al.*, 1982; Mazon *et al.*, 1982b). A study of the effect of glucose on the three gluconeogenic enzymes in a mutant deficient in the regulatory subunit of cAMP-dependent protein kinase (*bcy1* mutation), and in a double mutant that also carried the *cyr1* mutation (*CYR1* is the gene coding for adenylate cyclase, the enzyme responsible for the synthesis of cAMP from ATP), confirmed the involvement of cAMP in catabolite inactivation but indicated that other mechanisms are also involved (Tortora *et al.*, 1984).

2.3. Effect of Oxygen

The formation of intact functional mitochondria with fully active respiratory enzymes requires molecular oxygen. Despite its importance, virtually nothing is known about the way this occurs. It is well known that utilization of substrates such as glycerol, ethanol, pyruvate, and lactate occurs oxidatively using the tricarboxylic acid (TCA) cycle and mitochondrial electron transport. Growth on acetate requires the glyoxylate cycle in addition. However, it is often not appreciated that TCA cycle enzyme activities are still present in anaerobic conditions. TCA cycle activity is needed for the provision of intermediates for biosynthesis. Reference to schemes of metabolic pathways will confirm that, for example, the TCA cycle intermediate 2-oxoglutarate is essential in the biosynthesis of glutamate, glutamine, proline, lysine, and arginine (Jones and Fink, 1982). Furthermore, glutamate, which is derived from 2-oxoglutarate (via NADP-dependent glutamate dehydrogenase), is recognized as a nitrogen source in *Saccharomyces* and is itself used in the biosynthesis of aspartate, and hence threonine, methionine, and cysteine, and also tyrosine, phenylalanine, serine, leucine, isoleucine, and histidine. Clearly, we are now in the realms of nitrogen anabolic pathways. Thus, although carbon and nitrogen metabolism are frequently considered as separate aspects (e.g., in reviews and by researchers in their investigations), it is apparent that this division is artificial, made only for convenience, and that the yeast cell does not recognize such a separation. Accordingly, after the introduction to nitrogen metabolism (Section 3), this chapter will take an holistic view to emphasize the unity of cellular metabolism.

2.4. The Pasteur Effect

The Pasteur effect is the control exerted by respiration over glycolytic rate. When cells are growing fermentatively under anaerobic conditions, the rate of glucose utilization will be high. A switch to aerobic conditions results in the onset of respiration with a consequent rapid decrease in the rate of fermentation (glycolysis). The advantage of such a mechanism to the cell is explained in terms of the far greater yield of ATP available from complete oxidation of its carbon source *via* respiration compared with glycolysis. Canonical wisdom states that the Pasteur effect is mediated primarily by the allosteric regulation of phosphofructokinase, which is considered to be the chief "pacemaker" enzyme of glycolysis. Among the components that have been shown to stimulate phosphofructokinase are AMP, ADP, inorganic phosphate, and fructose-6-phosphate (fructose-6-P). ATP and citrate are known to inhibit the enzyme. Thus when respiration is proceeding at high rates, oxidative phosphorylation will be taking place, producing large amounts of ATP and consuming ADP, AMP, and phosphate, and increased levels of TCA cycle activity will give rise to higher concentrations of citrate than would occur during anaerobic glycolysis. This will result in inhibition of phosphofructokinase (by ATP and citrate) and reduced stimulation from AMP, ADP, and phosphate. Conversely, when mitochondrial activity is low, the ATP concentration will be low, so there will be little or no inhibition of phosphofructokinase. In addition, ADP and AMP concentrations will be high, thus stimulating phosphofructokinase and thereby increasing glycolytic flux. Notice also that phosphofructokinase is stimulated by its substrate (fructose-6-P).

This view of the Pasteur effect cannot be the complete story, however. Just one of the questions in this area is the difficulty in reconciling *in vivo* glycolytic flux through phosphofructokinase with *in vitro* kinetic evidence. A typical intracellular concentration of fructose-6-P is 0.2 mM whereas maximal activity of phosphofructokinase (required to produce the *in vivo* glycolytic fluxes observed is only attained *in vitro* as substrate concentrations of around 5 mM. Fructose-2,6-bisphosphate, not the product of the phosphofructokinase (fructose-6-P 1-kinase) reaction, but produced by fructose-6-P 2-kinase, is a potent stimulator of phosphofructokinase (Clifton and Fraenkel, 1983). It seems likely, therefore, that this compound is also important in regulating phosphofructokinase activity *in vivo*. Studies in wild-type strains and in mutants affected in aspects of glycolysis have revealed there are also other controls operating on the overall glycolytic rate. Some of these are mentioned in subsequent sections of this chapter.

2.5. Catabolite Conversion

The term *catabolite conversion* (Fröhlich and Entian, 1982) is not generally well known. This effect is seen in the conversion of enolase isoenzymes after addition of glucose to derepressed cells. Enolase is the first gluconeogenic enzyme that is also required for glycolysis, and it has been demonstrated that conversion of gluconeogenic enolase I to glycolytic enolase II is an additional regulatory mechanism to prevent gluconeogenesis in the presence of glucose. A further control accompanying the coordinated decrease in activity of enolase I and increase in activity of enolase II is the simultaneous repression of isoenzyme I and induction of isoenzyme II (Entian *et al.*, 1987).

Table I. List of *Saccharomyces* Genes Involved in Carbon Metabolism That Have Been Cloned[a]

Gene designation	Gene product	Reference
ADH1 (ADC1)	(Fermentative) alcohol dehydrogenase I	Williamson *et al.* (1980); Young *et al.* (1982)
ADH2 (ADR2)	(Oxidative) alcohol dehydrogenase II	Young *et al.* (1982)
ADH3 (ADM)	(Mitochondrial) alcohol dehydrogenase	Young *et al.* (1982)
ADR1	[Regulator of alcohol dehydrogenase II synthesis]	Denis and Young (1983)
ADR6	Required for derepression of alcohol dehydrogenase II	Taguchi and Young (1987)
CHO1	Phosphatidyl serine synthase	Letts *et al.* (1983)
CIT1 (GLU3)	Citrate synthase	Suissa *et al.* (1984)
ENO1	Enolase I	Holland *et al.* (1981)
ENO2	Enolase II	Holland *et al.* (1981)
FAS1	Fatty acid synthetase	Kuziora *et al.* (1983); Roberts and Schewizer (1984)
FAS2	Fatty acid synthetase	
FBP1	Fructose bisphosphatase	Sedivy and Fraenkel (1985)
GAL1	Galactokinase	St. John and Davis (1979); Miyajima *et al.* (1984)
GAL2	Membrane-bound subunit of galactose permease	Tschopp *et al.* (1986)
GAL4 (gal81)	[Transcriptional activator of *GAL1, GAL2, GAL7, GAL10,* and *MEL1*]	Laughon and Gesteland (1982); Johnston and Hopper (1982)
GAL7	Galactose-1-phosphate uridyltransferase	St. John and Davis (1979)
GAL10	UDP-glucose-4-epimerase	St. John and Davis (1979)
GAL80	[Regulator of *GAL4* gene product activity]	Nogi *et al.* (1984)
GCR1	Positive transcriptional regulator of enolase and glyceraldehyde-3-P-dehydrogenase genes	Holland *et al.* (1987)
GLK1	Glucokinase	Walsh *et al.* (1983)
GPM1	Phosphoglycerate mutase	Kawasaki and Fraenkel (1982)
HEX2	[Needed for glucose repression, regulates hexokinase PII synthesis]	Niederacher and Entian (1987)

Table I. (*Continued*)

Gene designation	Gene product	Reference
HXK1	Hexokinase PI (A)	Walsh *et al.* (1983); Entian *et al.* (1984)
HXK2 (HEX1)	Hexokinase PII (B)	Walsh *et al.* (1983); Frohlich *et al.* (1984)
INO1	Inositol-1-phosphate synthase	Klig and Henry (1984)
LPD1	Lipoamide dehydrogenase	Roy and Dawes (1987)
MAL11	Maltose permease	
MAL12	Maltase	Charron *et al.* (1986)
MAL13	Positive regulator of *MAL12* gene	
MAL2-8C		Rodicio and Zimmerman (1985a)
MAL3	[Regulator of α-glucosidase synthesis]	Michels and Needleman (1983)
MAL4		Rodicio and Zimmerman (1985b)
MAL61	Maltose permease	
MAL62	Maltase	Dubin *et al.* (1986)
MAL63	Positive regulator of *MAL12* gene	
PDC1	Pyruvate decarboxylase	Schmitt *et al.* (1983)
PFK1	Phosphofructokinase α subunit	Clifton and Fraenkel (1982); Heinisch (1986)
PFK2	Phosphofructokinase β (regulatory?) subunit	Heinisch (1986)
PGI1	Phosphoglucose isomerase	Kawasaki and Fraenkel (1982); Aguilera and Zimmerman (1986)
PGK1	Phosphoglycerate kinase	Hitzeman *et al.* (1980); Dobson *et al.* (1982); Kawasaki and Fraenkel (1982)
PGM1 (GPM1)	Phosphoglycerate mutase	Rodicio and Heinisch (1987)
PIS1	Phosphatidyl inositol synthase	Nikawa and Yamashita (1984)
PYK1 (CDC19)	Pyruvate kinase	Kawasaki and Fraenkel (1982); Burke *et al.* (1983)
SNF1	[Derepressor of many glucose-repressible genes]	Celenza and Carlson (1984)
SNF3	"High-affinity" glucose transporter	Bisson *et al.* (1987)
SUC2	Invertase	Hohmann and Zimmermann (1986)
TDH1	Glyceraldehyde-3-P-dehydrogenase I	Holland and Holland (1980); McAlister and Holland (1985)
TDH2	Glyceraldehyde-3-P-dehydrogenase II	
TDH3	Glyceraldehyde-3-P-dehydrogenase III	
TPI1	Triose phosphate isomerase	Kawasaki and Fraenkel (1982)

[a]Former or alternative gene designations are given in parentheses.

2.6. Molecular Genetics of Carbon Metabolism

The *S. cerevisiae* glycolysis genes and other genes involved in carbon metabolism that have been cloned are shown in Table I.

3. OUTLINE OF NITROGEN METABOLISM

Nitrogen catabolism has been reviewed by Cooper (1982a). Amino acid and nucleotide biosynthesis have been reviewed by Jones and Fink (1982) and by Henry *et al.* (1984). Many compounds will serve as nitrogen source for *S. cerevisiae* though some are obviously better than others (see Cooper, 1982a).

3.1. Amino Acid Biosynthesis

The pathways of amino acid biosynthesis are under complex regulation at both genetic and metabolic levels. The so-called general control of amino acid biosynthesis is involved in the coordinate repression and derepression of 30 or more enzymes in different pathways. Derepression occurs upon amino acid starvation, repression takes place in conditions of medium sufficiency. Each biosynthetic pathway is also characterized by its own specific control mechanisms which operate with respect to the end products and/or intermediates of the particular pathway. The two levels of control coexist. The general control does not derepress the synthesis of every enzyme in every pathway. For example, all of the enzymes of the tryptophan biosynthetic pathway except phosphoribosyl-anthranilate isomerase are under general control, whereas in the leucine pathway only α-isopropylmalate synthase is under general control (Hsu *et al.*, 1982). So far, the biosynthetic pathways for arginine, histidine, isoleucine, leucine, lysine, tryptophan, and valine have been shown to be regulated by the general control mechanism (Henry *et al.*, 1984).

Cloning of the structural genes encoding the amino acid biosynthetic enzymes has been a major impetus toward unraveling the mechanism of general control. Other useful genetic approaches have been the isolation of mutants affected in general control of amino acid biosynthesis and the subsequent cloning of the regulatory genes involved. Biochemical and physiological studies have involved starving a given amino acid auxotroph for the requisite amino acid or provision of an analog to a wild-type strain that will cause inhibition of an enzyme in a particular biosynthetic pathway. The end result is the same, for the cell becomes depleted of the end product of the pathway (the amino acid) and subsequently derepresses the enzymes of the pathway along with all other enzymes under general control.

Analysis of pathway-specific regulation, which might be thought to be a simpler problem, is not always straightforward. Obviously, study of those pathways subject to general control will be more difficult than study of those which are not. An additional complication occurs when an amino acid can serve as a nitrogen source. Another possible snag (the consideration of which has general application and is not limited solely to questions concerning amino acid metabolism) involves the frequently made assumption that provision of an intermediate to a mutant unable to synthesize a particular compound will result in a pattern of metabolism similar to that occurring in the wild type. Clearly, this approach takes no account of the possible regulatory effects on either the immediate pathway concerned (e.g., feedback inhibition or controls of transcription or translation of the remaining intact enzymes in that pathway), effects on other related pathways (e.g., where cross-pathway controls are exerted), or, assuming none of the foregoing gross disturbances occur, the metabolic inbalance resulting from abnormal concentrations of an intermediate in the wrong subcellular compartment.

3.2. Proteases

Saccharomyces cerevisiae possesses an extensive range of proteases. An excellent review on the synthesis and function of these enzymes has been published by Jones (1984). Table II summarizes our current knowledge of the properties, roles, cellular location, and genetics of these important enzymes. More and more yeast proteases are being discovered each year, though the physiological role of many of these remains obscure. Wolf has commented that the "conventional" view of proteinase action of a limited number of proteases being responsible for the degradation of many intracellular proteins may have to be replaced by the idea of many proteases being responsible for degradation, with each protease cutting only a few cellular proteins (Wolf, 1982). If this view is correct, then biochemical explanations of the specificity of protease action are simplified, but the problem of the cellular regulation of many proteases becomes more complex.

Proteases are considered to be involved in a wide variety of functions both general and specific in *S. cerevisiae*. These include general protein turnover and degradation, sporulation, protein degradation in the cell cycle, catabolite inactivation, degradation of NAD-dependent glutamate dehydrogenase upon nitrogen starvation and NADP-dependent glutamate dehydrogenase following carbon starvation, zymogen processing, formation and destruction of α-factor, removal of the products of missense and nonsense mutations and heterologous gene expression, "killer" toxins, and secretion processes (see Jones, 1984, for an extensive list of references). The major outstanding problems in this area are that there

Table II. Proteases of *Saccharomyces cerevisiae*[a]

Enzyme	Characteristics	Glycosylation	Cellular location	Cellular role(s)	Structural gene	Other genes
Proteinase A	Acid endoproteinase, 1 × 41–80 kDa	+	Vacuole	N starvation–induced protein protein degradation; protein turnover in stationary phase	*PRA1*	*PEP2-PEP15*, *ABM6*, *ABM8*
Proteinase B	Serine sulfhydryl endoproteinase, 1 × 31-44 kDa	+	Vacuole	N starvation–induced protein degradation; spore dispersal	*PRB1*	*PEP2-PEP8*, *PEP11-PEP16*, *PRB2-PRB4*, *RAD6*, *PSO2*
Carboxypeptidase Y	Serine sulfhydryl exopeptidase, 1 × 59-63 kDa	+	Vacuole	N starvation–induced protein degradation; metabolism of peptides	*PRC1*	*PEP1-PEP16*, *PEP21*, *ABM6*, *ABM8*
Aminopeptidase I (LAPIV)	Exopeptidase (Zn^{2+}) 600 kDa (12 × 53 kd)	+	Vacuole	Metabolism of peptides	*LAP4*?	*PEP4*
Mitochondrial protease	Metallo endoproteinase, 115 kDa	n.d.	Mitochondrial matrix	Processing of imported precursors to mitochondrial proteins	Unknown, but nuclear	
Dipeptidyl aminopeptidase A	Cleaves X-Pro-Y and X-Ala-Y to yield X-Pro and X-Ala; heat stable	n.d.	Membrane	Processing of pheromone precursors (α-factor)	*STE13*?	
Dipeptidyl aminopeptidase B	Cleaves X-Pro-Y and X-Ala-Y to yield	n.d.	Membrane	Unknown		

	X-Pro and X-ala; heat labile					
Lysine-arginine-cleaving endopeptidase	Cleave C-terminal to Lys-Arg and Arg-Arg pairs	n.d.	Membrane	Processing of α-factor and killer toxin precursors	*KEX2*?	
Proteinase M	Cleaves oligopeptides with blocked α-amino groups	n.d.	Membrane	Unknown		
Proteinase P	As for proteinase M; cleavage catalyzed may be the same as that catalyzed by pheromone peptidase	n.d.	Membrane	Unknown		
Pheromone peptidase	Cleaves α-factor between Leu-6 and Lys-7 (see proteinase P)	n.d.	Membrane? or secreted into medium	Recovery from α-factor–induced cell cycle arrest	*SST1 (BAR1)*?	
Aminopeptidase II (LAPI)	Exopeptidase (Me^{2+}), 85 kDa	n.d.	Periplasm and/or cytoplasm?	Unknown	*LAP1*?	*RAD1*?, *PSO2*?, *RAD6*?
Aminopeptidase II (LAPII)	Exopeptidase (Me^{2+}), 85 kDa	n.d.	Periplasm and/or cytoplasm?	Unknown	*LAP2*?	*RAD1*?, *PSO2*?, *RAD6*?
Leucine aminopeptidase III (LAPIII)	Exopeptidase, 94 kDa	n.d.	n.d.	Unknown	*LAP3*?	
Aminopeptidase Co	Exopeptidase (Co^{2+}), 100 kDa	n.d.	n.d.	Unknown		
Carboxypeptidase S	Exopeptidase (Zn^{2+})	n.d.	n.d.	Metabolism of peptides	*CPS1*? *(DUT1?)*	

[a]Reproduced, with permission, from the Annual Review of Genetics, Volume 18, © 1984 by Annual Reviews Inc.

are proteases with no known function and cellular processes which obviously involve proteolytic action where the protease concerned has not been identified. Much of this is due to methodology. Proteases are generally characterized in terms of their activity toward particular classes of substrates and with respect to their pH optima. The substrates used in protease assays are frequently either radioactively labeled peptides or proteins coupled to chromogenic moieties. The reaction is terminated by addition of cold trichloroacetic acid. Enzyme activity is measured in terms of the amount of radioactivity or chromophore that has been rendered trichloroacetic acid–soluble (unconsumed substrate remains trichloroacetic acid–precipitable). Other methods (e.g., that described by Saheki and Holzer, 1975) simply rely on colorimetric estimation (Biuret, Folin, etc.) of trichloroacetic acid-soluble peptides released by protease action. It is thus apparent why it is difficult to tie known *in vitro* activities to observed *in vivo* proteolytic events. Genetic attempts at answering these questions have not, in general, been productive. A more useful approach might be to first determine the precise subcellular location of the process presumed to be controlled by protease action and then to investigate which protease(s) are active there. Subcellular fractionation coupled with cytochemical techniques would seem to be appropriate. One can then address the question "Does a given protease attack the protein or peptide of interest *in vitro* and is its mode of catalysis consistent with events observed *in vivo*?" Finally, one would need to demonstrate that the kinetics of *in vivo* regulation of the protease activity correlates with the kinetics of the original cellular process under investigation.

3.3. Nucleotide Metabolism

The biochemistry and genetics of nucleotide biosynthesis have been comprehensively reviewed by Jones and Fink (1982). This section will deal solely with those aspects of nucleotide metabolism that deserve special mention because of physiological, biochemical, or genetic consequences of naturally occurring features or mutations in nucleotide metabolism and the way we handle yeast as a result of these. In addition, some recent findings will be mentioned especially with respect to the metabolism of the cyclic nucleotide cAMP.

3.3.1. Purine Nucleotides

De novo synthesis of the purine nucleotides AMP and GMP starts with phosphoribosyl pyrophosphate. A sequence of 10 reactions which involves the products of *ADE4, ADE5, ADE8, ADE6, ADE7, ADE2, ADE1,* and *ADE13* genes, phosphoribosylaminoimidazolecarboxamide formyltransferase, and IMP cyclohydrolase leads to the formation of

IMP where the AMP and GMP pathways branch (Jones and Fink, 1982). AMP and GMP can also be formed from supplied adenine and guanine by the action of adenine phosphoribosyltransferase and guanine phosphoribosyltransferase. In addition, IMP can be formed from hypoxanthine. *ade1* and *ade2* mutations are frequently used genetic markers because of a strong red coloration upon adenine starvation which enables convenient marker assignment in such strains. The pigment is believed to derive from the intermediate phosphoribosylaminoimidazole which accumulates in cells bearing *ade1* and *ade2* mutations.

3.3.2. Pyrimidine Nucleotides

De novo synthesis of pyrimidine nucleotides proceeds from glutamine through six enzyme-catalyzed steps to UMP. After UMP the details are very vague. Figure 1 outlines the pathways that might exist in *S. cerevisiae.* The implications for the paucity of our knowledge in this area are quite serious. For example, the compound hydroxyurea (HU) is frequently used as an inhibitor of DNA synthesis. It is often stated that HU is presumed to inhibit DNA replication by inhibiting the enzyme ribonucleoside diphosphate reductase, thus preventing the synthesis of (the DNA precursors) deoxyribonucleoside triphosphates, but its definite mode of action and any possible side effects are not known.

UMP can be derived from exogenous uracil via uracil permease and uracil phosphoribosyl transferase or from uridine, in which case uptake is by uridine permease whence the uridine may be metabolized via either uridine kinase to UMP or uridine ribohydrolase, which results in uracil. Cytosine can be taken up by cytosine permease with subsequent deamination to uracil, and cytidine can be taken up by cytidine permease with cytidine deaminase catalyzing its conversion to uridine (Cooper, 1982b; Jones and Fink, 1982).

Yeasts, in common with most other fungi, lack thymidine kinase (Grivell and Jackson, 1968), so it is not possible to specifically label thymine nucleotides or DNA in yeast by the use of radioactively labeled thymine or thymidine. Consequently we have to employ indirect methods for this. Typical approaches for isotopic labeling of DNA in *S. cerevisiae* include those of Williamson (1965), Hartwell (1967), and Hatzfeld (1973), in which a less specific precursor such as adenine or uracil is used to label total nucleic acid, followed by elimination of the radioactivity due to RNA by solubilizing in alkali (e.g., IM NaOH); RNAase could also be used. For measurements of RNA an estimate of total nucleic acid is usually sufficient as RNA is roughly 98% of total cellular nucleic acid (Hartwell, 1967).

Disturbance of the levels of thymine nucleotides can have profound genetic effects on the cell. Such effects are achieved by inhibition of

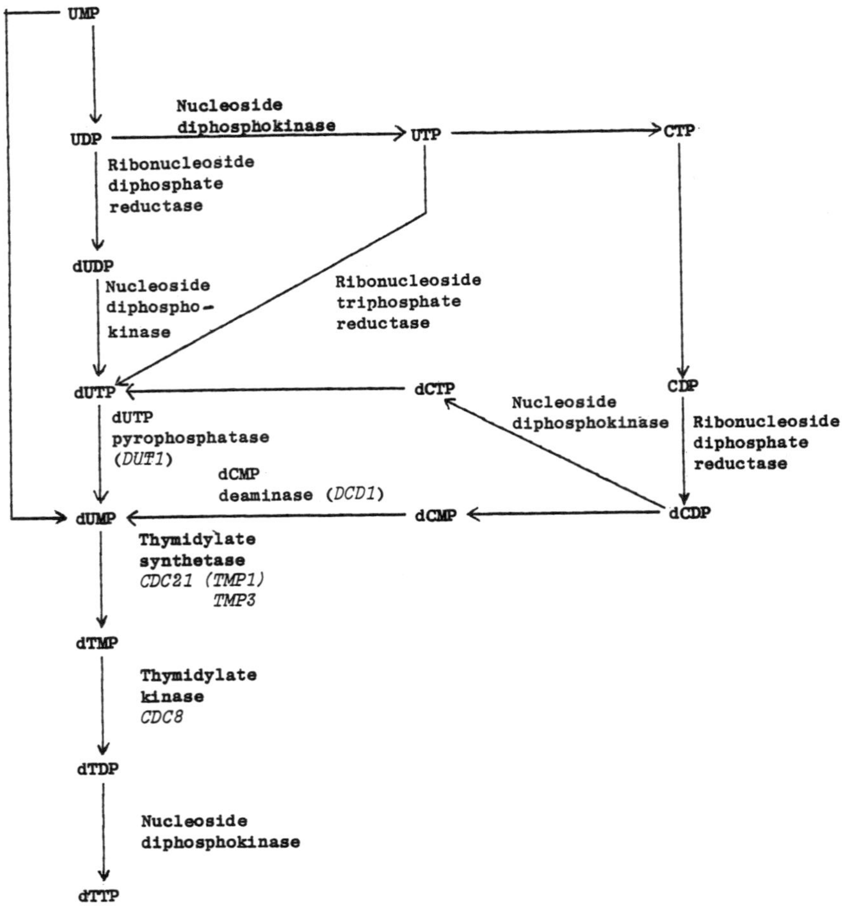

Figure 1. The possible pathways of pyrimidine nucleotide biosynthesis in *S. cerevisiae*. Enzymes are named and genes assigned where they have been definitely identified.

dTMP synthesis using either a thymidylate synthetase mutant (*tmp1*), folate antagonists to block the methylation of dUMP to dTMP, or 5-fluoro dUMP. All three methods produce starvation for thymine nucleotides. This causes gene conversion and crossing over, unequal sister chromatid crossing over, and mating-type switching (Kunz *et al.*, 1980, 1984, 1985; Barclay *et al.*, 1982; Kunz and Haynes, 1982; Eckardt *et al.*, 1983). The earlier isolation of *tup1* mutants (which incorporate dTMP) had shown that while some alleles produce pleiotropic defects, there was a general effect on mating efficiency. Strains that were *MATα tup1* had reduced efficiency of mating with *MATa* strains but had higher than normal mating efficiency with *MATα* strains such that they mated almost equally well with *a* and α tester strains (Wickner, 1974).

3.3.3. Cyclic Nucleotides

Cyclic nucleotides have attracted a lot of attention recently. cAMP has been implicated in the regulation of a number of processes in *S. cerevisiae,* including the cell cycle, sporulation, mating, and catabolite inactivation (discussed in Section 2.2.). cAMP is formed from ATP by adenylate cyclase. It is degraded by the enzyme phosphodiesterase to produce 5′AMP. cAMP acts *via* cAMP-dependent protein kinase, which is composed of catalytic and cAMP-binding (regulatory) subunits. Binding of cAMP to the regulatory subunit causes dissociation of the subunits, thus rendering the catalytic subunit active.

The involvement of cAMP in mating has been shown by the fact that the mating pheromone α-factor inhibits adenylate cyclase *in vitro* and causes arrest at "start" and that exogenous cAMP can reduce α-factor–induced G1 arrest. Furthermore, *ste5* mutants which do not mate have an α-factor–insensitive adenylate cyclase (Liao and Thorner, 1980). Isobutylmethylxanthine inhibits phosphodiesterase *in vitro* and antagonizes α-factor arrest (Liao and Thorner, 1981). Temperature-sensitive *cyr1* mutants (defective in adenylate cyclase) cannot mate at the restrictive temperature (Matsumoto *et al.,* 1983a). Evidence for a cAMP regulatory system controlling cell cycle progression comes from the fact that *cyr1* mutants arrest as class II "start" mutants in the absence of cAMP (Matsumoto *et al.,* 1982). The cell division cycle mutation *cdc35* which causes cell cycle arrest at start is allelic to *cyr1* (Casperson *et al.,* 1985). The mutations *cyr1* and *CYR3* (the latter results in an altered regulatory subunit of cAMP-dependent protein kinase requiring elevated levels of cAMP for dissociation), are suppressed by the mutation *bcy1,* which results in a cAMP-independent protein kinase due to deficiency of the regulatory subunit (Matsumoto *et al.,* 1982). The probably involvement of cAMP in initiation of sporulation is shown by the fact that diploids homozygous for *cyr1* or *CYR3* can sporulate in rich media when wild-type strains would not (Matsumoto *et al.,* 1983b) and that glucose repression of sporulation can be alleviated by cAMP (Tsuboi *et al.,* 1972; Tsuboi and Yanagishima, 1973). Other elements of the cAMP regulatory system that have been discovered include the following mutants; the defects these mutations confer are given in parentheses: *cyr2* (altered catalytic subunit of cAMP-dependent protein kinase) (Uno *et al.,* 1984); *IAC* (increased adenylate cyclase) (Uno *et al.,* 1983); *pde1* (reduced activity of "high K_m form" of cAMP phosphodiesterase) (Uno *et al.,* 1983), and *ppd1* (deficient in phosphoprotein phosphatase, which dephosphorylates phosphorylated proteins—i.e., works in the reverse direction of protein kinase) (Matsumoto *et al.,* 1985). In addition to these mutants which were identified by classical mutational analyses, direct cloning has indicated there are three different genes (*TPK1, TPK2,* and *TPK3*) encoding the catalytic subunits of cAMP-dependent protein kinase (Toda *et al.*, 1987a) and a low K_m ("high affinity") cAMP phosphodiesterase encoded by *PDE2* (Sass *et al.*, 1986).

Further interest has been generated by the discovery that yeast has *RAS* genes. Two such genes have been identified (*RAS1* and *RAS2*) which have structural and functional homologies to mammalian *ras* oncogenes. Cells containing disruptions in both *RAS* genes do not proliferate and have lowered cAMP. *RAS2 val19,* an allele with a missense mutation analogous to one that activates the transforming potential of mammalian *ras* genes, results in increased cAMP levels within the cell and renders the cell constitutive for proliferation (Toda *et al.*, 1985; Kataoka *et al.*, 1985; Broek *et al.*, 1985). The *bcy1* mutation suppresses *ras1 ras2* mutants but *RAS2 val19* does not suppress *cyr1* (Toda *et al.*, 1985), thus the *CYR1* and *BCY1* gene products must be dependent on the action of the *RAS* proteins. The mammalian adenylate cyclase is membrane bound and has a hormone receptor which upon binding hormone stimulates GTP binding of the G-protein subunit, rendering it active and stimulating the catalytic subunit. Slow hydrolysis of the GTP to GDP by the GTPase activity of the G protein returns it to the inactive form. Yeast *RAS* and human *ras* genes having "oncogenic" alterations result in a loss of GTPase activity (Toda *et al.*, 1985). Other elements that have to be incorporated into our understanding of this complex system are the gene products of *CDC25,* which regulates adenylate cyclase possibly by regulating guanine nucleotide bound to *RAS* proteins (Robinson *et al.*, 1987; Broek *et al.*, 1987); *RAM1,* which is required for processing of *RAS* proteins (Powers *et al.*, 1986); *RHO1* and *RHO2* (*Ras homologs*) and the various *SRA* (*suppressor of ras*) genes (Cannon *et al.*, 1986). The situation may not be as complicated as it first seems, as several of these apparently different genes may be identical as in the case of *SRA4* (= *CYR1* = *CDC35*), *sra1* (= *BCY1* = *REG1*), and *SRA3* = *CYR2*. Thus, once again the biochemical details are incomplete; nevertheless this area is currently under intense investigation in many laboratories throughout the world. The information that is sought will enable an understanding of the integrative network of controls involving mating, proliferation, sporulation, and aspects of carbon metabolism.

3.4. Molecular Genetics of Nitrogen Metabolism

Table III lists those genes of *S. cerevisiae* involved in nitrogen catabolism and anabolism that have been cloned.

4. TRANSPORT OF SUBSTRATES INTO THE CELL

Transport processes in *S. cerevisiae* have been reviewed by Cooper (1982b). There have been two ways of attempting to identify the transpor-

Table III. List of *Saccharomyces* Genes Involved in Nitrogen Anabolism and Catabolism[a]

Gene designation	Gene product	Reference
ADE1	Phosphoribosylaminoimidazole succinocarboxyamide synthetase	Crowley and Kaback (1984)
ADE2	Phosphoribosylaminoimidazole carboxylase	Sasnauskas *et al.* (1982)
ADE4	Glutamine phosphoribosylpyrophosphate amidotransferase	Mantsala and Zalkin (1984)
ARG3	Ornithine carbamoyltransferase	Crabeel *et al.* (1981)
ARG4	Arginosuccinate lyase	Hsiao and Carbon (1979)
ARGRI (ARG80)	[Specific regulation of arginine metabolism]	
ARGRII (ARG81)		Dubois *et al.* (1987)
ARGRIII (ARG82)		Dubois and Messenguy (1985)
ARO1A	3-Enol pyruvoylshikimate-5-phosphate synthase	
ARO1B	Shikimate kinase	Larimer *et al.* (1983)
ARO1C	3-Dehydroquinate synthase	
ARO1D	Shikimate dehydrogenase	
ARO1E	3-Dehydroquinate dehydratase	
ARO3	Deoxy-D-arabino-D-heptulosonate phosphate synthase	Teshiba *et al.* (1986)
ARO7	Chorismate mutase	Ball *et al.* (1986)
CAN1	Arginine permease	Broach *et al.* (1979)
CAR1	Arginase	Jauniaux *et al.* (1982); Sumrada and Cooper (1982)
CAR2	Ornithine transaminase	Jauniaux *et al.* (1984); Sumrada and Cooper (1985)
CDC8	Thymidylate kinase	Sclafani and Fangman (1984)
CPA1	Carbamoylphosphate synthetase	Messenguy *et al.* (1983)
CPA2	Carbamoylphosphate synthetase	Lusty and Jeffrey (1982); Messenguy *et al.* (1983)
CYR1	Adenylate cyclase	Masson *et al.* (1984); Casperson *et al.* (1985)
DAL2	Allantoicase	Yoo *et al.* (1985)
DAL3	Ureidoglycolate hydrase	
DAL5	[Component of allantoate transport system]	Rai *et al.* (1987)
DCD1	dCMP deaminase	McIntosh and Haynes (1986)
DFR1	Dihydrofolate reductase	McIntosh *et al.* (1986)
DUR1,2	Urea amidolyase	Genbauffe and Cooper (1986)
DUT1	dUTP pyrophosphatase	McIntosh *et al.* (1986)
FUR4 (URA-P)	Uracil permease	Chevallier (1982)
GCD4	Negative transcriptional regulator of *GCN4*	Skvirsky *et al.* (1986)

(*continued*)

Table III. (*Continued*)

Gene designation	Gene product	Reference
GCN1 (NDR1) *GCN2 (AAS1)* *GCN3 (AAS2)* *GCN4 (AAS3)* *GCN5*	[Involved in general control of amino acid biosynthesis]	Hinnebusch and Fink (1983b)
HIS1	ATP phosphoribosyltransferase	Hinnebusch and Fink (1983a)
HIS3	Imidazoleglycerolphosphate dehydratase	Struhl *et al.* (1979)
HIS4A	Phosphoribosyl-AMP cyclohydrolase	
HIS4B	Phosphoribosyl-ATP-pyrophosphohydrolase	Donahue *et al.* (1982)
HIS4C	Histidinol dehydrogenase	
HIS5	Histidinol phosphate aminotransferase	Harashima *et al.* (1981); Satoshi *et al.* (1981)
ILV1	Threonine deaminase	Petersen *et al.* (1983); Holmberg (1984)
ILV2	Acetohydroxyacid synthase	Holmberg (1984); Polaina (1984)
ILV3	Dihydroxyacid dehydratase	Holmberg (1984); Polaina (1984)
ILV4	Acetohydroxyacid isomeroreductase	Holmberg (1984)
ILV5	Acetohydroxyacid isomeroreductase	Polaina (1984)
KEX2	Endopeptidase processes prepro-α-factor	David *et al.* (1984)
LEU1	α-Isopropylmalate dehydratase	Hsu *et al.* (1982); Hsu and Schimmel (1984)
LEU2	β-Isopropylmalate dehydratase	Ratzkin and Carbon (1977)
LEU3	[Regulator of *LEU1* and *LEU2* expression]	Brisco *et al.* (1987)
LEU4	α-Isopropylmalate synthase	Chang *et al.* (1984)
LEU5	[Nonessential regulator of leucine biosynthetic pathway]	Drain and Schimmel (1986)
LYS2	2-Aminoadipate reductase	Eibel and Philippsen (1983)
MET2	Homoserine acetyltransferase	Langin *et al.* (1986)
MET3	ATP: sulfate adenylyl transferase	Cherest *et al.* (1985)
MET6	Homocysteine methyltransferase	Csaikl and Csaikl (1986)
MET25	Homocysteine synthase	Sangsoda *et al.* (1985)
PDE2	"High affinity" cAMP phosphodiesterase	Sass *et al.* (1986)
PPR1	[Regulator of expression of *URA1* and *URA3* products]	Losson and Lacroute (1982)
PPR2	Regulator of *URA4* gene expression	Hubert *et al.* (1983)
PRA1 (PEP4)	Vacuolar aspartyl protease	Woolford *et al.* (1986); Ammerer *et al.* (1986)

Table III. (*Continued*)

Gene designation	Gene product	Reference
PUT2	Pyrroline-5-carboxylate dehydrogenase	Brandriss (1983)
RNR2	Ribonucleotide reductase (small subunit)	Elledge and Davis (1987)
SPE1	Ornithine decarboxylase	Fonzi and Sypherd (1985)
SPT2 (SPM2)	[Negative regulator of Ty-controlled gene expression]	Roeder *et al.* (1985)
TMP1	Thymidylate synthetase	Taylor *et al.* (1982)
TMP2	Dihydrofolate reductase	Nath and Baptist (1984)
TRP1	Phosphoribosylanthranilate isomerase	Struhl *et al.* (1979)
TRP2	Anthranilate synthase	
TRP3	Anthranilate synthase: (indole-3-glycerol phosphate synthase	Aebi *et al.* (1982)
TRP4	Anthranilate phosphoribosyltransferase	Niederberger *et al.* (1984)
TRP5	Tryptophan synthase	Zalkin and Yanofsky (1982)
URA1	Dihidroorotate dehydrogenase	Guerry-Kopecko and Wickner (1980)
URA2B	Aspartate carbamoyltransferase	Souciet *et al.* (1982)
URA2C (CPU1)	Carbamoylphosphate synthetase	
URA3	Orotidine-5′-phosphate decarboxylase	Bach *et al.* (1979)
URA6	Uridine monophosphokinase	Liljelund and Lacroute (1986)

[a]Former or alternative gene designations are given in parentheses.

ters and their specificity. Biochemical studies center on identification of the kinetics of substrate uptake. Usually Michaelis–Menten kinetics are assumed and the data obtained are fitted to a model employing equations of this type. Competitive inhibition by substrates is adduced to have demonstrated that the two substrates are transported by the same carrier. Genetic methods for probing transport use mutants defective in uptake of one substrate to test whether uptake of other substrates is also affected. Neither approach is satisfactory alone. Both biochemical and genetic studies are required with measurements made under appropriate physiological conditions.

The dual biochemical and genetic approach is elegantly demonstrated in a series of experiments by Bisson and Fraenkel (1983a,b, 1984), who used both wild-type strains and strains with one or more of the following mutations—*hxk1* (lacking hexokinase A or PI), *hxk2* (lacking hexokinase B or P-II), and *glk1* (lacking glucokinase)—to study glucose and fructose uptake. They used ^{14}C-labeled glucose, fructose, mannitol, and 6-deoxy-D-glucose (an analog of glucose that is taken up by the cell but not metabolized). After incubation of cells and labeled sugar, uptake was terminated by addition of a large excess of cold water followed by rapid filtration to collect the cells and further washing with cold water. Radioactivity due to hexose uptake was measured by liquid scintillation counting. These workers demonstrated that there are both low- and high-affinity mechanisms for glucose uptake. The high-affinity uptake system was low when cells were grown in high concentrations (100 mM) of glucose and high when cells were in low concentrations (0.5 mM) of glucose, in stationary phase of growth on glucose, and in other carbon sources such as galactose and lactate plus glycerol or ethanol. The activity of the high-affinity transport system is dependent on the activity of the three glucose phosphorylating enzymes. Hence, triple kinase mutants (*hxk1 hxk2 glk1*) lacking all three kinase activities do not show high-affinity uptake of either glucose or fructose (Bisson and Fraenkel, 1983a), whereas the *hxk1 hxk2* double mutant lacks only high-affinity fructose uptake but retains high-affinity glucose uptake.

A number of important questions concerning the mechanism of hexose uptake remain unanswered (Bisson and Fraenkel, 1984). It is not apparent from this work whether the high- and low-affinity uptake systems are different systems or part of the same system. Nor is it clear whether the uptake of glucose is an energy-linked process or involves facilitated diffusion. Experiments using analogs of glucose such as 6-deoxyglucose indicate that both uptake systems involve facilitated diffusion (Kotyk, 1967; Cirillo, 1968; Heredia *et al.*, 1968; Bisson and Fraenkel, 1983b, 1984). However, the problem with studying "native" glucose arises from the fact that it is very rapidly phosphorylated either while entering the cell or immediately thereafter. Bisson and Fraenkel have suggested that the hexokinases and/or glucokinase may be involved in derepression of a component of the transporter system, possibly the membrane carrier itself (Bisson and Fraenkel, 1984).

One possible way to resolve this problem might be to use ^{13}C nuclear magnetic resonance (NMR). ^{13}C NMR spectroscopy has already been usefully employed to study the uptake and metabolism of glucose by *S. cerevisiae in vivo* (den Hollander *et al.*, 1979). One of the advantages of this approach is that it is possible to discriminate between anomeric forms of sugars. Figure 2 shows a typical experiment, and in this case preferential use of the β-anomer of glucose is shown. Using NMR to study glucose transport mutants should elucidate this long-standing mystery. Until recently glucose transport mutants had not been obtained, however it has now been shown that the *SNF3* gene is required for high-affinity glucose transport (Bisson *et al.*, 1987). Notice that the glucose transport mutant was originally isolated as a *snf* (*s*ucrose *n*on-*f*ermenting) mutant; i.e. the phenotype is more subtle than one might have guessed. This general point is considered again later (Section 5.1). The use of NMR has potential in many other areas of yeast central metabolism, as outlined in Section 5.2.

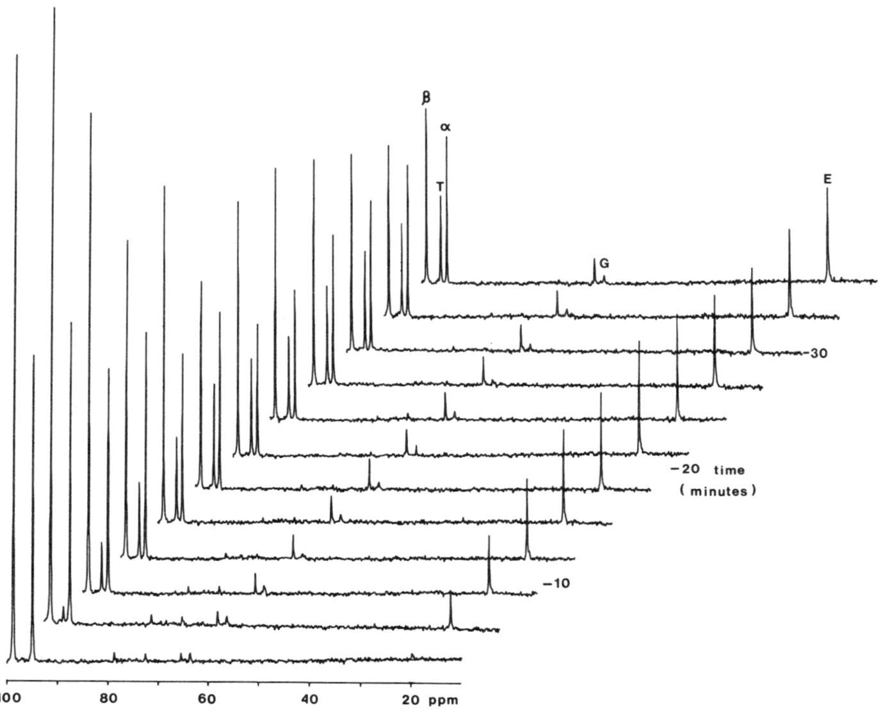

Figure 2. Time course of utilization of [1-^{13}C] glucose by *Saccharomyces cerevisiae. In vivo* time-elapsed proton decoupled NMR spectra are shown. Each spectrum is the result of 1000 pulses. α, C-1 of α-D-glucose; β, C-1 of β-D-glucose; T, C-1, 1′ of trehalose; E, C-1 of ethanol; G, C-1 of glycerol.

If we wish to examine carefully the rate of synthesis of macromolecules (e.g., protein or nucleic acid), the only reliable way is to use a pulse labeling method with an appropriate precursor (in these cases an amino acid and, say, uridine, respectively). Thus, we rely on transport (uptake) of the labeled precursor even though we are not directly concerned with the transport process itself. Such "uptake and incorporation" studies make certain assumptions which need careful examination in each specific case. The basic assumption is that the amount of precursor incorporated (into protein or nucleic acid) during the short period of the pulse is a measure of the rate of synthesis of the macromolecule. The precursor chosen should be present in low concentrations in the cell (said to have a "small pool"). Pool sizes vary in different strains and in different media. The presence of multiple pools should be considered. It can often be shown that the cell has a "metabolic pool" which is supplied by endogenous biosynthesis and uptake from the exogenous label and is used for synthesis of the macromolecule. In addition, there may be two kinds of sequestered pool: a "storage pool" due to uptake of exogenous precursor and possibly also supplied from the metabolic pool and an "exchangeable pool" involved in exchange of exogenous and endogenous precursor. With a small metabolic pool incorporation will be a linear function with time. If the pool is large the rate of incorporation will increase until the pool reaches a constant specific activity. The terminology employed here is that of Creanor and Mitchison (1982), whose precise work demonstrates the considerations just outlined.

The importance of precursor pool sizes cannot be overemphasized, especially in relation to the cell cycle. The amount of radiolabel incorporated may not be a true measure of macromolecular biosynthesis through the cell cycle for a variety of reasons. For example, the size of the metabolic pool may vary through the cell cycle. The contributions from exogenous and endogenous precursor may vary through the cell cycle (i.e., transport of the precursor may be cell cycle modulated) and the various storage pools may vary in their contributions to biosynthesis through the cell cycle. It thus becomes important to determine whether changes in specific activity of the metabolic pool occur. It is virtually impossible in most cases to extract the metabolic pool on its own. The use of a mutant auxotrophic for the precursor to be used is an obvious way of abolishing its endogenous synthesis, but this has other inherent problems discussed previously (see Section 3.1). It is possible, however, to arrange conditions in which a storage pool is stable and hence its effect on incorporation into macromolecules can be ignored. That such conditions exist (arrived at by trial and error) can be demonstrated by a pulse-chase type of experiment. Uptake, measured as total cellular radioactivity, should cease immediately following the chase or transfer to fresh medium. Incorporation into macromolecules from the metabolic pool may continue increasing for a given period of time before it too becomes constant. The difference between total radioactivity and incorporated radioactivity is a measure of the size of the storage pool(s).

5. REGULATION OF METABOLISM

5.1. Genetic Studies

From the preceding material in this chapter the reader will already have recognized this author's opinion as to the power of a combined biochemical–genetic approach to questions of metabolism and its regulation. The genetic studies of recent years involving the isolation of mutants, examination of their physiology and biochemistry, the cloning of genes, and subsequent analyses of gene expression have provided knowledge on regulation at the genetic level and have also furnished important information on pathways and their control at the biochemical level. For example, in the pathway of arginine biosynthesis gene–enzyme relationships have been established for the complete pathway, regulatory mutations have been investigated, and the principal properties and intracellular location of all the enzymes are known (Jones and Fink, 1982; Henry *et al.,* 1984).

Genetic approaches have not been without their shortcomings, however. For example, the usefulness of having mutants for each stage of glycolysis has long been recognised. Prior to the isolation of aldolase (*fba1*) mutants (Lobo, 1984), it had been suggested that aldolase mutants might have unexpected properties, and evidence of isozymes of aldolase was cited in explanation for the lack of such mutants (Fraenkel, 1982). However, the single-gene mutants were apparently obtained readily and were typical of the majority of glycolysis mutants in that they were unable to grow on glucose and showed growth inhibition by glucose when growing on ethanol. In other words, it is not always clear why attempts to obtain particular mutants fail even when one knows beforehand the biochemical lesion that is sought and one might be able to predict some phenotypes on the basis of this.

The biochemical–genetic approach for analyzing the control over glycolysis by phosphofructokinase has produced some startling results. It now seems clear that *S. cerevisiae* possesses two phosphofructokinase isoenzymes controlled by at least five genes. The gene *PFK1* codes for the catalytic subunit of the familiar soluble allosteric enzyme. *PFK2* is a common element coding for both the regulatory subunit of the soluble enzyme and also a particulate phosphofructokinase. Mutants lacking the particulate enzyme define at least three other genes *PFK3, PFK4,* and *PFK5* (Parmer *et al.,* 1984). Phosphofructokinase mutants have been isolated in a number of different laboratories (Clifton *et al.,* 1978; Ciriacy and Breitenbach, 1979; Navon *et al.,* 1979; Clifton and Fraenkel, 1982; Lobo and Maitra, 1982, 1983; Breitenbach-Schmitt *et al.,* 1984a; Parmar *et al.,* 1984). This has led to confusion on the behalf of some authors as to whether the mutation designated *pfk1* by Fraenkel's group should be regarded as the same as the *pfk1* or the *pfk2* mutation of Maitra's group (Breitenbach-Schmitt *et al.,* 1984a). Such distinctions need not concern us here; the confusion is more apparent than real. Maitra's group have investigated this area extensively and provided convincing evidence dispelling any confusion (Nadkarni *et al.,* 1984).

The discovery of a particulate phosphofructokinase is quite unusual because all the other glycolytic enzymes of yeast are soluble. But an even more unexpected finding is that of Breitenbach-Schmitt *et al.* (1984b), who have presented evidence for the existence of a "bypass" pathway parallel to the phosphofructokinase–aldolase reaction sequence. These workers examined the metabolism of *pfk* mutants growing on glucose by analyzing the fate of ^{14}C-labeled glucose. The "classical" view of glycolysis states that only C-3 and C-4 of glucose should contribute to carbon dioxide production. However, by using U-^{14}C- glucose and 3,4-^{14}C-glucose it was shown that in wild-type cells only two-thirds of the carbon atoms released as CO_2 derive from the C-3 and C-4 positions of glucose. The remainder derive from other C atoms in glucose. In *pfk1* or *pfk2* mutants all six positions of glucose contributed equally to CO_2 production (Breitenbach-Schmitt *et al.,* 1984a). These findings suggest that soluble phosphofructokinase catalyzes only part of the total flow of carbon in glycolysis and that the enzyme is dispensable in glycolysis. Mutants were isolated in three different genes (*BYP1, BYP2, BYP3*) which block glycolysis in single *pfk1* and *pfk2* mutants (Breitenbach-Schmitt *et al.,* 1984b). In the *byp1* mutant CO_2 was derived almost exclusively from the C-3 and C-4 positions of glucose (i.e., according to the "classical" pattern of glycolysis). From accompanying enzyme assays and identification of accumulating metabolites in *pfk* and *byp* mutants it appears that the nonoxidative portion of the pentose phosphate cycle has a significant role in hexose fermentation. This is in contrast to use of the oxidative portion of the cycle *via* glucose-6-P dehydrogenase, which is known to make a very small contribution (around 2.5%) to carbon flux from glucose (Lagunas and Gancedo, 1973). The implications of this work are clearly very important. No doubt many of us think we know all there is to be known about a pathway as fundamental as glycolysis in yeast. It is interesting to speculate whether the *BYP1, BYP2,* and *BYP3* genes identified by the Darmstadt group are the same as the *PFK3, PFK4,* and *PFK5* genes described by workers in Bombay.

Genetic methods have been slow in yielding biochemical information on cell cycle control in yeast. Approximately 80 genes are known to be involved in cell cycle progression, and much has been learned in physiological and genetic terms especially of the temporal order of gene expression and dependency relations between the largely unidentified gene products involved (Pringle and Hartwell, 1981). A great number of cell cycle genes have been both cloned and sequenced. However, the biochemical defect in specific cell cycle mutants is known in only a very few cases. Those mutations where the biochemical defect is known are *alg1* (protein *N*-glycosylation) (Klebl *et al.,* 1984); *bcy1** (regulatory subunit of cAMP-dependent protein kinase) (Matsumoto *et al.,* 1983a; Toda *et al.,* 1987b); *cdc7** (Patterson *et al.,* 1986); *cdc8** (thymidylate kinase) (Jong *et al.,* 1984; Sclafani and Fangman, 1984); *cdc9** (DNA ligase) (Johnston and Nasmyth, 1978; Barker *et al.,* 1985); *cdc19* (*pyk1*)* (pyruvate kinase) (Kawasaki, 1979; Kawasaki and Fraenkel, 1982; Burke *et al.,* 1983); *cdc21* (*tmp1*)* (thymidylate synthetase) (Game, 1976; Taylor *et al.,* 1982, 1987); *cdc28** (protein kinase) (Reed *et al.,* 1985); *cdc30* (*pgi2*)* (phosphoglucose isomerase I) (Dickinson and Williams, 1987, unpublished results); *cdc35* (*cyr1*)* (adenylate cyclase) (Masson *et al.,* 1984; Casperson *et al.,* 1985);

cyr2 (catalytic subunit of cAMP-dependent protein kinase) (Uno *et al.*, 1984); *CYR3* (affects the regulatory subunit of cAMP-dependent protein kinase) (Uno *et al.*, 1982); *pde1** *pde2** ("low affinity" and "high affinity" cAMP phospodiesterases, respectively) (Sass *et al.*, 1986; Toda *et al.*, 1987a); *ppd1* (*glc1*) (phosphoprotein phosphatase) (Uno *et al.*, 1983); and *top2** (DNA topoisomerase II) (DiNardo *et al.*, 1984). (The asterisk denotes where the mutation has been shown to affect the structural gene.) Clearly, biochemical understanding of regulation and control of the cell cycle lags behind genetic and physiological descriptions. This is especially true in the case of mutants originally isolated as *c*ell *d*ivision *c*ycle (*cdc*) mutants. This gap in knowledge has become a pressing problem in this area and hinders attempts to understand the integration of metabolism involved in the cell's control of growth and cell division.

Attempts at identification of *CDC* gene products have proceeded in a somewhat piecemeal fashion. Those *cdc* mutants affected in DNA replication are relatively easy to work out because one knows which overall process to inspect. By an ordered sequence of experiments in which one examines whether initiation of DNA synthesis occurs, if ligation is effected, or if DNA replication is terminated, it is possible to pinpoint the defect. Similarly, we might guess that mutants affected in mitosis should yield those with defective tubulins, as has been the case in other eukaryotic systems. However, in the majority of cell cycle mutants there is no such convenient "pointer" as DNA replication or mitosis. The essential problem remains: How do we start to guess which biochemical function (from among the thousands of possibilities within the cell) is affected in a particular mutant?

It has now become a fairly standard procedure to clone the gene of interest and then to sequence it. The base sequence can then be compared with that of other cloned genes to inspect for homologies and can also be used to predict the primary structure (amino acid sequence) of the protein encoded. In some cases this exercise allows identification of the gene product on the basis of convincing homology with known proteins that have been sequenced previously. One should be prepared to expect that this approach may be rather limited because an insufficient number of proteins have been sequenced. The essential problem remains: we need a technique capable of "seeing" the whole metabolism in the cell.

5.2. Physiological and Biochemical Studies

Generally physiological and biochemical methods do not make interesting reading and so this section will concentrate on selected newer and novel approaches.

When, late in 1981, I turned my attention to the metabolism of sporulation, it seemed that despite previous studies on many separate and diverse aspects of the phenomenon, there was no clear and integrated view of the metabolic changes associated with the initiation of sporulation and the subsequent events of the process itself. Indeed, many of the previous biochemical studies had presented conflicting views. *In vivo* ^{13}C NMR was presented as an holistic technique capable of providing the overview of metabolism that was required (Dickinson *et al.*, 1983). By reference to this particular problem (the metabolism of sporulation), we can see the utility of the NMR method to other aspects of central intermediary metabolism. It is, of course, nondestructive and noninvasive.

To those unfamiliar with NMR it is a technique that has the reputation of being far too insensitive to be any use in *in vivo* analysis of metabolism. It is true, for example, that the low (1.1%) relative natural abundance of ^{13}C (the nucleus of carbon actually observed in the NMR experiment) means that we need to employ artificially enriched substrates for most analyses of carbon metabolism. Nevertheless, there are cellular components we can observe due to natural abundance of ^{13}C. These are obviously those compounds present in large amounts within the cell, one example being the storage carbohydrate trehalose, which is present in quite high amounts under most conditions. Another popular misconception about NMR for this type of study is that one needs a superpowerful spectrometer. Nothing could be further from the truth. A considerable amount of work has been done using very powerful machines, sometimes with so-called "wide-bore" magnets. (A "wide-bore" magnet allows the use of a wide sample tube, thus enabling a greater quantity of cells to be analyzed than would otherwise be possible.) Perfectly adequate spectra can be obtained from yeast in normal-sized (10 mm) tubes in machines of modest field strength.

Sporulation in the presence of [2-^{13}C] acetate was studied by ^{13}C NMR spectroscopy. Immediately after transfer to sporulation medium the major metabolite produced was glutamate. From the labeling pattern observed it was concluded that both the tricarboxylic acid cycle and the glyoxylate cycle were operating. At about 4 hr net glutamate synthesis ceased and trehalose synthesis (gluconeogenesis) was initiated *via* glyoxylate cycle-derived intermediates. Subsequently, beginning at 4 hr the synthesis of large amounts of fatty acid began, concomitant with onset of operation of the pentose phosphate pathway. The involvement of the pentose phosphate pathway in sporulation was previously unsuspected and was discovered on the basis of the distribution of ^{13}C-labeled carbon atoms in trehalose. We now know that the NADPH provided by the pentose phosphate reaction sequence is used to fuel the fatty acid biosynthesis (Dickinson and Hewlins, 1988).

These NMR studies provided some important new clues. For example, when cells enter sporulation there is a shift in the operation of the

TCA cycle leading to glutamate accumulation. An increase of glutamate during sporulation had been reported previously (Rousseau, 1972), but it was believed that amino acids were derived from proteolysis (Tingle *et al.*, 1973). Clearly, the ^{13}C NMR studies gave a fresh interpretation of this, namely, that initiation of sporulation involves a reduction in the activity of the 2-oxoglutarate dehydrogenase complex. This has been confirmed (Dickinson *et al.*, 1985). A further consequence of this would be to reduce GTP synthesis *via* succinyl CoA synthetase because of the reduced carbon flux through this portion of the TCA cycle after the down-regulation of the 2-oxoglutarate dehydrogenase. Freese *et al.*, (1985) have confirmed that GTP levels always fall upon initiation of sporulation. Furthermore, *spd* mutants that are derepressed for sporulation (i.e., sporulate in rich media) and have greatly reduced sensitivity to NH_4^+ repression of sporulation (Dawes, 1975; Vezinhet *et al.*, 1979) have reduced levels of 2-oxoglutarate dehydrogenase activity.

The use of ^{13}C-enriched substrates allows a view of the whole of carbon metabolism that is proceeding. By analyzing the labeling pattern in a particular intermediate it is possible to work out the relative contribution of each pathway involving that intermediate. This is something that cannot usually be done using radioactive labeling without recourse to very sophisticated chemistry as well as tedious separation techniques. We can get information on phosphorylated intermediates, energy metabolism, and phospholipids by ^{31}P NMR. (^{31}P has 100% natural abundance, so artificially enriched substrates are not required here.) The use of ^{31}P NMR is well demonstrated in a study of wild-type and glycolysis mutants of *S. cerevisiae* (Navon *et al.*, 1979). The mutant strains were shown to accumulate characteristic sugar phosphates upon incubation with glucose. An additional piece of information that can be gained from ^{31}P NMR spectra is the intracellular pH of the environment of individual metabolites (Moon and Richards, 1973). Intracellular pH is conveniently determined on the basis of the chemical shifts of inorganic phosphate and/or other phosphosphorylated intermediates. It is even possible to observe the different pH values of different organelles. Obviously this type of information can be particularly valuable when attempting to correlate enzyme activities (measured *in vitro*) with *in vivo* metabolic fluxes. Analogous studies using ^{15}N labeling can be devised to reveal aspects of cellular nitrogen metabolism.

Another new technique is the application of membrane inlet mass spectrometry to measurements of dissolved gas concentrations. This allows a considerable improvement over former methods which have usually relied on gas phase measurements. It is obviously preferable to be able to make measurements in the actual cellular environment. The principles of operation and design of such equipment have been described (Lloyd and Scott, 1983). The technique has been used to observe the Pasteur effect in yeast (Lloyd *et al.*, 1983a) (as shown in Fig. 3), and to study the effects of NH_4^+ on glycolysis and respiration (Lloyd *et al.*, 1983b). The instrumentation has much to commend it, including the facility to make continuous measurements of numerous gases with a single probe and the possibility of monitoring many fermentations simultaneously. In Cardiff we have built a probe suitable for use in an NMR tube giving the possibility of simultaneous monitoring of dissolved gas metabolism along with *in vivo* NMR analysis of metabolic pathways. An additional feature of the instrumentation is its reliability over long periods, which, with its proven use in batch cultures and chemostats, would seem to be purpose-built for industrial use in computer-controlled fermentation processes.

An interesting report appeared describing the construction of a complete yeast glycolysis system *in vitro* (Welch and Scopes, 1985). Individually purified enzymes from *S. cerevisiae* were combined in proportions and at concentrations comparable to those existing *in vivo*. This cell-free glycolysis system produced ethanol and carbon dioxide from glucose. The authors point out that this demonstrates the there is nothing in principle that prevents the use of purified enzymes for other pathwaylike transformations or possibly the use of semipurified preparations that allow production of more valuable products. There also seems to be considerable potential here for analyzing control of a metabolic pathway (at the level of enzyme activity only).

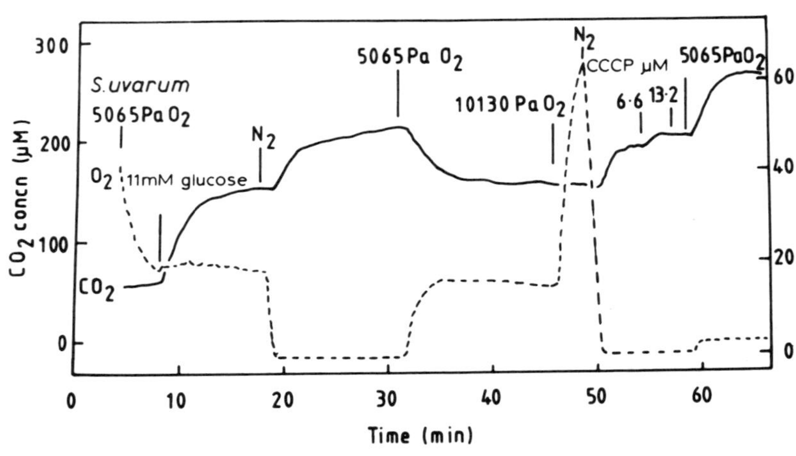

Figure 3. The Pasteur effect in *Saccharomyces uvarum* observed by mass spectrometry. The cells were continuously stirred in the reaction vessel. At the marker lines the composition of the gas stream entering the open system was altered, or additions were made as shown. ------ Dissolved O_2 concentration; ——dissolved CO_2 concentration. (Reprinted by permission from *Biochemical Journal,* **212:** *749–754*. Copyright © 1983 the Biochemical Society, London.)

5.3. Further Aspects of Regulation

Thus far we have examined metabolism as a self-contained entity, but three additional types of control must be considered: mating-type effects, the cell cycle, and sporulation.

5.3.1. Mating-Type Effects

Mating-type effect (a/α suppression) was observed in overproducing mutants of *S. cerevisiae* affected in arginine catabolism and urea utilization. The *cargA*$^+$ O^h, cargB$^+$ O^h, and *durO^h* mutations resulted in massive overproduction of arginase, ornithine transminase, and urea carboxylase-allophanate hydrolase (respectively) in haploid strains of either mating type, and in diploids homozygous at the mating-type locus. However, mating-type heterozygosity (i.e., in $\frac{\text{MAT}a}{\text{MAT}\alpha}$ diploids) caused suppression of the otherwise constitutive overproduction (Dubois *et al.*, 1978; Lemoine *et al.*, 1978; Deschamps and Wiame, 1979). A less dramatic overproduction of iso-2-cytochrome c involving a similar mating-type–regulated mutation, *CYC7-H2*, has been demonstrated by Rothstein and Sherman (1980). In the case of the *CYC7-H2* mutation the transposable element Ty1 was found to be inserted in the 5′ region of the gene coding for iso-2-cytochrome c (Errede *et al.*, 1980a,b). A Ty element was also found to be inserted in the *cargA* region in the *cargA*$^+$ O^h mutant (Jauniaux *et al.*, 1981). Taguchi *et al.* (1984) have reported on Ty-mediated overproduction of the glucose-repressible alcohol dehydrogenase isozyme ADHII. In this case the mating-type effect was greatest in the presence of glucose, a repressing carbon source. Thus, Ty-mediated gene expression is regulated by mating-type heterozygosity and carbon catabolite control.

5.3.2. The Cell Cycle

It is not the purpose of this section to consider the cell cycle *per se*, but we must take account of it. Any observations made on a population of asynchronous cells will necessarily be time-averaged. This is not merely an academic abstraction. As each cell progresses through the processes involved in growth, DNA replication, mitosis, and cell division, many aspects of temporal regulation of metabolism are invoked. Whether this temporal regulation reflects the circadian clock or a cellular clock, such as size control in *S. cerevisiae* (Sudbery *et al.*, 1980), is unimportant in this context. It now seems clear that the overall pattern of protein synthesis throughout the cell cycle is exponential and only 2% of proteins show periodic or modulated patterns of synthesis, and these are not among the abundant proteins (Lorincz *et al.,* 1982). It is presumed these proteins include regulators. Furthermore, in most cases where enzyme activity has been analyzed carefully, it follows a linear pattern with a rate change once per cycle (Creanor *et al.,* 1983). However, exceptions are known, and these appear to involve enzymes and proteins that are temporally and functionally linked to DNA replication. These include enzymes producing DNA precursors and the histones (Storms *et al.,* 1984; Moll and Wintersberger, 1976). Thus if we are in the business of observing, preparing, or attempting to increase the yield of proteins that are cell cycle modulated (or their metabolic products), then we may choose to ignore the cell cycle at our cost.

If it became essential to make synchronous cultures in production-scale quantities, it should be possible to obtain reasonable synchrony by means of an appropriate fed-batch culture rendering it analogous to laboratory-scale "starvation–refeeding" regimes.

5.3.3. Sporulation

The switch from vegetative growth and mitosis to the alternative development pathway of meiosis and sporulation that can occur in diploid strains of *S. cerevisiae* is under complex regulation (Esposito and Klapholz, 1981). Cells must be heterozygous at the mating-type locus. The switch can take place only during the G1 phase of the cell cycle and is also governed by nutritional conditions. Induction of sporulation requires a shiftdown from a rich growth medium to one containing a poorly utilized carbon source such as acetate, usually in the absence of nitrogen. NH_4^+ in particular is a strong repressor of sporulation. Thus the process is simultaneously subject to mating-type control, cell cycle control, and carbon- and nitrogen-source regulation. Hence, in a consideration of sporulation we can start to appreciate the significance to the cell of the array of controls that can be exercised over metabolic processes in terms of its life cycle.

6. CONCLUDING REMARKS

In this chapter I have outlined some approaches that have provided us with our current knowledge of metabolism and biosynthesis. Metabolism in *S. cerevisiae* is regulated at many different levels, with interactions occurring between levels. There are a number of conspicuous gaps in our knowledge, including the nature of intermediates in pathways as well as many more details concerning regulation. This biochemical information is essential, for without it exploitation of the organism must remain empirical rather than logical. For example, the use of a cloned gene in construction of overproducing strains or for other aspects of process optimization will be seriously hindered where the biochemical identity of a gene product remains unknown. Without a complete biochemical description of the pathway involved, including regulation, the likely result of such an exercise would be the shifting of a "bottleneck" from one part of a pathway to another. Thus biochemistry becomes an even more important foundation to our understanding of the many features of this fascinating organism.

ACKNOWLEDGMENT. I am grateful to Professor P. K. Maitra for helpful comments.

REFERENCES

Aebi, M., Niederberger, P., and Hütter, R., 1982, Isolation of the *TRP2* and the *TRP3* genes of *Saccharomyces cerevisiae* by functional complementation in yeast, *Curr. Genet.* **5:**39–46.

Aguilera, A., and Zimmerman, F. K., 1986, Isolation and molecular analysis of the phosphoglucose isomerase structural gene of *Saccharomyces cerevisiae, Mol. Gen. Genet.* **202:**83–89.

Ammerer, G., Hunter, C. P., Rothman, J. H., Saari, G. C., Valls, L. A., and Stevens, T. H., 1986, *PEP4* gene of *Saccharomyces cerevisiae* encodes proteinase A, a vacuolar enzyme required for processing of vacuolar precursors, *Mol. Cell. Biol.* **6:**2490–2499.

Bach, M-L., Lacroute, F., and Botstein, D., 1979, Evidence for transcriptional regulation of OMP-decarboxylase in yeast by hybridization of mRNA to the yeast structural gene cloned in *E. coli., Proc. Natl. Acad. Sci. USA.* **76:**386–390.

Ball, S. G., Wickner, R. B., Cottarel, G., Schaus, M., and Tirtiaux, C., 1986, Molecular cloning and characterisation of *ARO7-OSM2* a single yeast gene necessary for chorismate mutase activity and growth in hypertonic medium, *Mol. Gen. Genet.* **205:**326–330.

Barclay, B. J., Kunz, B. A., Little, J. G., and Haynes, R. H., 1982, Genetic and biochemical consequences of thymidylate stress, *Can. J. Biochem.* **60:**172–194.

Barker, D. G., White, J. H. M., and Johnston, L. H., 1985, The nucleotide sequence of the DNA ligase gene (*CDC9*) from *Saccharomyces cerevisiae:* A gene which is cell cycle regulated and induced in response to DNA damage, *Nucl. Acid Res.* **13:**8323–8337.

Bisson, L. F., and Fraenkel, D. G., 1983a, Involvement of kinases in glucose and fructose uptake by *Saccharomyces cerevisiae, Proc. Natl. Acad. Sci. USA.* **80:**1730–1734.

Bisson, L. F., and Fraenkel, D. G., 1983b, Transport of 6-deoxyglucose in *Saccharomyces cerevisiae, J. Bacteriol.* **155:**995–1000.

Bisson, L. F., and Fraenkel, D. G., 1984, Expression of kinase-dependent glucose uptake in *Saccharomyces cerevisiae, J. Bacteriol.* **159:**1013–1017.

Bisson, L. F., Neigeborn, L., Carlson, M., and Fraenkel, D. G., 1987, The *SNF3* gene is required for high affinity glucose transport in *Saccharomyces cerevisiae, J. Bacteriol.* **169:**1656–1662.

Brandriss, M. C., 1983, Proline utilisation in *Saccharomyces cerevisiae:* Analysis of the cloned *PUT2* gene, *Mol. Cell. Biol.* **3:**1846–1856.

Breitenbach-Schmitt, I., Heinisch, J., Schmitt, H. D., and Zimmerman, F. K., 1984a, Yeast mutants without phosphofructokinase activity can still perform glycolysis and alcohol fermentation, *Mol. Gen. Genet.* **195:**530–535.

Breitenbach-Schmitt, I., Schmitt, H. D., Heinisch, J., and Zimmerman, F. K., 1984b, Genetic and physiological evidence for the existence of a second glycolytic pathway in yeast parallel to the phosphofructokinase–aldolase reaction sequence, *Mol. Gen. Genet.* **195:**536–540.

Brisco, P. R. G., Cunningham, T. S., and Kohlhaw, G. B., 1987, Cloning, disruption, and chromosomal mapping of yeast *LEU3,* a putative regulatory gene, *Genetics* **115:**91–99.

Broach, J. R., Strathern, J. N., and Hicks, J. B., 1979, Transformation of yeast: development of a hybrid cloning vector and isolation of the *CAN1* gene, *Gene* **8:**121–133.

Broek, D., Samiy, N., Fasano, O., Fujiyama, A., Tamanoi, F., Northup, J., and Wigler, M., 1985, Differential activation of yeast adenylate cyclase by wild-type and mutant *RAS* proteins, *Cell* **41:**763–769.

Broek, D., Toda, T., Michaeli, T., Levin, L., Birchmeier, C., Zoller, M., Powers, S., and Wigler, M., 1987, The *S. cerevisiae CDC25* gene product regulates the *RAS*/adenylate cyclase pathway, *Cell* **48:**789–799.

Burke, R. L., Tekamp-Olson, P., and Najarian, R., 1983, Isolation characterization and sequencing of the pyruvate kinase gene of *Saccharomyces cerevisiae, J. Biol. Chem.* **258:**2193–2201.

Cannon, J. F., Gibbs, J. B., and Tatchell, K., 1986, Suppressors of the *ras2* mutation of *Saccharomyces cerevisiae, Genetics* **113:**247–264.

Casperson, G. F., Walker, N., and Bourne, H. R., 1985, Isolation of the gene coding for adenylate cyclase in *Saccharomyces cerevisiae, Proc. Natl. Acad. Sci. USA.* **82:**5060–5063.

Celenza, J. L., and Carlson, M., 1984, Cloning and genetic mapping of *SNF-1* a gene required for expression of glucose repressible genes in *Saccharomyces cerevisiae, Mol. Cell. Biol.* **4:**49–53.

Celenza, J. L., and Carlson, M., 1986, A yeast gene that is essential for release from glucose repression encodes a protein kinase, *Science* **233:**1175–1180.

Chang, L. F. L., Cunningham, T. S., Gatzek, P. R., Chen, W. J., and Kohlhaw, G. B., 1984, Cloning and characterization of yeast *LEU4,* one of 2 genes responsible for α-isopropylamalate synthesis, *Genetics* **108:**91–106.

Charron, M. J., Dubin, R. A., and Michels, C. A., 1986, Structural and functional analysis of the *MAL1* locus of *Saccharomyces cerevisiae, Mol. Cell. Biol.* **6:**3891–3899.

Cherest, H., Nguyen, N. T., and Surdin-Kerjan, Y., 1985, Transcriptional regulation of the *MET3* gene of *Saccharomyces cerevisiae, Gene* **34:**269–281.

Chevallier, M-R., 1982, Cloning and transcriptional control of a eukaryotic permease gene, *Mol. Cell. Biol.* **2:**977–984.

Ciriacy, M., and Breitenbach, I., 1979, Physiological effects of seven different blocks in glycolysis in *Saccharomyces cerevisiae, J. Bacteriol.* **139:**152–168.

Cirillo, V. P., 1968, Relationship between sugar structure and competition for the sugar transport system in bakers' yeast, *J. Bacteriol.* **95:**603–611.

Clifton, D., and Fraenkel, D. G., 1982, Mutant studies of yeast phosphofructokinase, *Biochemistry* **21:**1935–1942.

Clifton, D., and Fraenkel, D. G., 1983, Fructose 2,6-bisphosphate and fructose-6-P 2-kinase in *Saccharomyces cerevisiae* in relation to metabolic state in wild type and fructose-6-P 1-kinase mutant strains, *J. Biol. Chem.* **258:**9245–9249.

Clifton, D., Weinstock, S. G., and Fraenkel, D. B., 1978, Glycolysis mutants in *Saccharomyces cerevisiae, Genetics* **88:**1–11.

Cooper, T. G., 1982a, Nitrogen metabolism in *Saccharomyces cerevisiae,* in: *The Molecular Biology of the Yeast* Saccharomyces. *Metabolism and Gene Expression* (J. N. Strathern, E. W. Jones, and J. R. Broach, eds.), Cold Spring Harbor Laboratory, Cold Spring Harbor, NY, pp. 39–99.

Cooper, T. G., 1982b, Transport in *Saccharomyces cerevisiae,* in: *The Molecular Biology of the Yeast* Saccharomyces. *Metabolism and Gene Expression* (J. N. Strathern, E. W. Jones, and J.

R. Broach, eds.), Cold Spring Harbor Laboratory, Cold Spring Harbor, New York, pp. 399–461.

Crabeel, M., Messenguy, F., Lacroute, F., and Glansdorff, 1981, Cloning *ARG3*, the gene for ornithine carbamoyltransferase from *S. cerevisiae*. Expression in *E. coli* requires secondary mutations. Production of plasmid β-lactamase in yeast, *Proc. Natl. Acad. Sci. USA* **78:**5026–5030.

Creanor, J., and Mitchison, J. M., 1982, Patterns of protein synthesis during the cell cycle of the fission yeast *Schizosaccharomyces pombe*, *J. Cell. Sci.* **58:**263–285.

Creanor, J., Elliott, S. G., Bisset, Y. C., and Mitchison, J. M., 1983, Absence of step changes in activity of certain enzymes during the cell cycle of budding and fission yeasts in synchronous cultures, *J. Cell. Sci.* **61:**339–349.

Crowley, J. C., and Kaback, D. B., 1984, Molecular cloning of chromosome 1 DNA from *Saccharomyces cerevisiae:* Isolation of the *ADE1* gene, *J. Bacteriol.* **159:**413–417.

Csaikl, V., and Csaikl, F., 1986, Molecular cloning and characterisation of the *MET6* gene of *Saccharomyces cerevisiae*, *Gene* **46:**207–214.

David, J., Brake, E., Blair, L., Kunisawa, R., and Thorner, J., 1984, Isolation of the putative structural gene for the lysine-arginine-cleaving endopeptidase required for processing of yeast prepro-α-factor, *Cell* **37:**1075–1090.

Dawes, I. W., 1975, Study of cell development using derepressed mutations, *Nature* **255:**707–708.

den Hollander, J. A., Brown, T. R., Ugurbil, K., and Shulman, R. G., 1979, ^{13}C nuclear magnetic resonance studies of anaerobic glycolysis in suspensions of yeast cells, *Proc. Natl. Acad. Sci. USA.* **76:**6069–6100.

Denis, C. L., and Young, E. T., 1983, Isolation and characterization of the positive regulatory gene *ADR1* from *Saccharomyces cerevisiae*, *Mol. Cell. Biol.* **3:**360–370.

Deschamps, J., and Wiame, J. M., 1979, Mating-type effect on *cis* mutations leading to constitutivity of ornithine transminase in diploid cells of *Saccharomyces cerevisiae*, *Genetics* **92:**749–758.

Dickinson, J. R., and Hewlins, M. J. E., 1988, A study of the role of the hexose monophosphate pathway with respect to fatty acid biosynthesis in sporulation of *Saccharomyces cerevisiae*, *J. Gen. Microbiol.* **134:**333–337.

Dickinson, J. R., and Williams, A. S., 1987, The *cdc30* mutation in *Saccharomyces cerevisiae* results in a temperature-sensitive isoenzyme of phosphoglucose isomerase, *J. Gen. Microbiol.* **133:**135–140.

Dickinson, J. R., Dawes, I. W., Boyd, A. S. F., and Baxter, R. L., 1983, ^{13}C NMR studies of acetate metabolism during sporulation of *Saccharomyces cerevisiae*, *Proc. Natl. Acad. Sci. USA* **80:**5847–5851.

Dickinson, J. R., Ambler, R. P., and Dawes, I. W., 1985, Abnormal amino acid metabolism in mutants of *Saccharomyces cerevisiae* affected in the initiation of sporulation, *Eur. J. Biochem.* **148:**405–406.

DiNardo, S., Voelkel, K., and Sternglanz, R., 1984, DNA topoisomerase II mutant of *Saccharomyces cerevisiae:* Topoisomerase II is required for segregation of daughter molecules at the termination of DNA replication, *Proc. Natl. Acad. Sci. USA* **61:**2616–2620.

Dobson, M. J., Tuite, M. F., Roberts, N. A., Kingsman, A. J., Kingsman, S. M., Perkins, R. E., Conroy, S. C., Dunbar, B., and Fothergill, L. A., 1982, Conservation of high efficiency promoter sequences in *Saccharomyces cerevisiae*, *Nucl. Acid Res.* **10:**2625–2637.

Donahue, T. F., Farabaugh, P. J., and Fink, G. R., 1982, The nucleotide sequence of the *His4* region of yeast, *Gene* **16:**47–59.

Drain, P., and Schimmel, P., 1986, Yeast *LEU5* is a *PET*-like gene that is not essential for leucine biosynthesis, *Mol. Gen. Genet.* **204:**397–403.

Dubin, R. A., Perkins, E. L., Needleman, R. B., and Michels, C. A., 1986, Identification of a second *trans*-acting gene controlling maltose fermentation in *Saccharomyces carlsbergensis, Mol. Cell. Biol.* **6:**2757–2765.

Dubois, E., and Messenguy, F., 1985, Isolation and characterization of the yeast *ARGRIII* gene involved in regulating both anabolism and catabolism of arginine, *Mol. Gen. Genet.* **198:**283–289.

Dubois, E., Hiernaux, D., Grenson, M., and Wiame, J. M., 1978, Specific induction of catabolism and its relation to repression of biosynthesis in arginine metabolism of *Saccharomyces cerevisiae, J. Mol. Biol.* **122:**383–406.

Dubois, E., Bercy, J., and Messenguy, F., 1987, Characterisation of two genes *ARGRI* and *ARGRIII* required for specific regulation of arginine metabolism in yeast, *Mol. Gen. Genet.* **207:**142–148.

Eckerdt, F., Kunz, B. A., and Haynes, R. H., 1983, Variation of mutation and recombination frequencies over a range of thymidylate concentrations in a diploid thymidylate auxotroph of yeast, *Curr. Genet.* **7:**399–402.

Eibel, H., and Philippsen, P., 1983, Identification of the cloned *Saccharomyces cerevisiae LYS2* gene by an integrative transformation approach, *Mol. Gen. Genet.* **191:**66–73.

Elledge, S. J., and Davis, R. W., 1987, Identification and isolation of the gene encoding the small subunit of ribonucleotide reductase from *Saccharomyces cerevisiae:* DNA damage-inducible gene required for mitotic viability, *Mol. Cell. Biol.* **7:**2783–2793.

Entian, K-D., and Frohlich, K-U., 1984, *Saccharomyces cerevisiae* mutants provide evidence of hexokinase PII as a bifunctional enzyme with catalytic and regulatory domains for triggering carbon catabolite repression, *J. Bacteriol.* **158:**29–35.

Entian, K-D., Kopetzki, E., Frohlich, K. U., and Mecke, D., 1984, Cloning of hexokinase isoenzyme PI from *Saccharomyces cerevisiae.* PI transformants confirm the unique role of hexokinase isoenzyme PII for glucose repression in yeasts, *Mol. Gen. Genet.* **198:**50–54.

Entian, K-D., Meurer, B., Köhler, H., Mann, K-H., and Mecke, D., 1987, Studies on the regulation of enolases and compartmentation of cytosolic enzymes in *Saccharomyces cerevisiae, Biochem. Biophys. Acta* **923:**214–221.

Errede, B., Cardillo, T. S., Sherman, F., Dubois, E., Deschamps, J., and Wiame, J. M., 1980a, Mating signals control expression of mutations resulting from insertion of a transposable repetitive element adjacent to diverse yeast genes, *Cell* **22:**427–436.

Errede, B., Cardillo, T. S., Wever, G., and Sherman, F., 1980b, Studies on transposable elements in yeast. 1. ROAM mutations causing increased expression of yeast genes: their activation by signals directed toward conjugation functions and their formation by insertion of Ty1 repetitive elements, *Cold Spring Harbor Symp. Quant. Biol.* **45:**593–602.

Esposito, R. E., and Kalapholz, S., 1981, Meiosis and ascospore development, in: *The Molecular Biology of the Yeast* Saccharomyces. *Life Cycle and Inheritance* (J. N. Strathern, E. W. Jones, and J. R. Broach, eds.), Cold Spring Harbor Laboratory, Cold Spring Harbor, NY, pp. 211–287.

Ferguson, J. J., Boll, M., and Holzer, H., 1967, Yeast malate dehydrogenase: Enzyme inactivation in catabolite repression, *Eur. J. Biochem.* **1:**21–25.

Fonzi, W. A., and Sypherd, P. S., 1985, Expression of the gene for ornithine decarboxylase of *Saccharomyces cerevisiae* in *Escherichia coli, Mol. Cell. Biol.* **5:**161–166.

Fraenkel, D. G., 1982, Carbohydrate metabolism, in: *The Molecular Biology of the Yeast* Saccharomyces. *Metabolism and Gene Expression* (J. N. Strathern, E. W. Jones, and J. R. Broach, eds.), Cold Spring Harbor Laboratory, Cold Spring Harbor, NY, pp. 1–37.

Freese, E. B., Olempska-Beer, Z., Hartig, A., and Freese, E., 1985, Initiation of meiosis and sporulation of *Saccharomyces cerevisiae* by sulfur or guanine deprivation, *Dev. Biol.* **102:**438–451.

Frohlich, K-U., and Entian, K-D., 1982, Regulation of gluconeogenesis in the yeast *Sac-*

charomyces cerevisiae. Evidence for conversion of enolase isoenzymes, *FEBS Lett.* **139:**164–166.
Frohlich, K-U., Entian, K-D., and Mecke, D., 1984, Cloning and restriction analysis of the hexokinase PII gene of the yeast *Saccharomyces cerevisiae, Mol. Gen. Genet.* **194:**144–148.
Funayama, S., Gancedo, J. M., and Gancedo, C., 1980, Turnover of yeast fructose bisphosphatase in different metabolic conditions, *Eur. J. Biochem.* **109:**61–66.
Game, J. C., 1976, Yeast cell cycle mutant *cdc21* is a temperature-sensitive thymidylate auxotroph, *Mol. Gen. Genet.* **146:**313–315.
Gancedo, C., 1971, Inactivation of fructose-1,6-diphosphatase by glucose in yeast, *J. Bacteriol.* **107:**401–405.
Genbauffe, F. S., and Cooper, T. G., 1986, Induction and repression of the urea amidolyase gene in *Saccharomyces cerevisiae, Mol. Cell. Biol.* **6:**3954–3964.
Grivell, A. R., and Jackson, J. F., 1968, Thymidine kinase: Evidence for its absence from *Neurospora crassa* and some other microorganisms and the relevance of this to the specific labelling of deoxyribonucleic acid, *J. Gen. Microbiol.* **54:**307–317.
Guerry-Kopecko, P., and Wickner, R. B., 1980, Cloning of the *URA1* gene of *Saccharomyces cerevisiae, J. Bacteriol.* **143:**1530–1533.
Haarasilta, S., and Oura, E., 1975, On the activity and regulation of anaplerotic and gluconeogenic enzymes during the growth processes of baker's yeast, *Eur. J. Biochem.* **52:**1–7.
Harashima, S., Sidhu, R. S., Tohe, A., and Oshima Y., 1981, Cloning of the *HIS5* gene of *Saccharomyces cerevisiae* by yeast transformation, *Gene* **16:**1–3.
Hartwell, L. H., 1967, Macromolecular synthesis in temperature-sensitive mutants of yeast, *J. Bacteriol.* **93:**1662–1670.
Hartwell, L. H., 1985, Cetus–UCLA Symposium on Yeast Cell Biology, Keystone, CO.
Hatzfeld, J., 1973, DNA labelling and its assay in yeast, *Biochim. Biophys. Acta* **299:**34–42.
Heinisch, J., 1986, Isolation and characterisation of the two structural genes coding for phosphofructokinase in yeast, *Mol. Gen. Genet.* **202:**75–82.
Henry, S. A., Klig, L. S., and Loewy, B. S., 1984, The genetic regulation and coordination of biosynthetic pathways in yeast: Amino acid and phospholipid synthesis, *Annu. Rev. Genet.* **18:**207–231.
Heredia, C. F., Sols, A., and DelaFuente, 1968, Specificity of the constitutive hexose transport in yeast, *Eur. J. Biochem.* **5:**324–329.
Hinnebusch, A. G., and Fink. G. R., 1983a, Repeated DNA sequences upstream from *HIS1* also occur at several other co-regulated genes in *Saccharomyces cerevisiae, J. Biol. Chem.* **258:**5238–5247.
Hinnebusch, A. G., and Fink, G. R., 1983b, Positive regulation in the general amino acid control of *Saccharomyces cerevisiae, Proc. Natl. Acad. Sci. USA* **80:**5374–5378.
Hitzeman, R. A., Clarke, L., and Carbon, J., 1980, Isolation and characterisation of the yeast 3-phosphoglycerokinase gene (*PGK*) by an immunological screening technique, *J. Biol. Chem.* **255:**12,073–12,080.
Hohmann, S., and Zimmermann, F. K., 1986, Cloning and expression on a multicopy vector of five invertase genes of *Saccharomyces cerevisiae, Curr. Genet.* **11:**217–225.
Holland, J. P., and Holland, M. J., 1980, Structural comparisons of two nontandemly repeated yeast glyceraldehyde-3-P dehydrogenase genes, *J. Biol. Chem.* **255:**2596–2605.
Holland, M. J., Holland, J. P., Thill, G. P., and Jackson, K. A., 1981, The primary structure of two yeast enolase genes. Homology between the 5′ noncoding flanking regions of yeast enolase and glyceraldehyde-3-phosphate dehydrogenase genes, *J. Biol. Chem.* **256:**1385–1395.
Holland, M. J., Yokoi, T., Holland, J. P., Myambo, K., and Innis, M. A., 1987, The *GCR1* gene encodes a positive transcriptional regulator of the enolase and glyceraldehyde-3-

phosphate dehydrogenase gene families in *Saccharomyces cerevisiae, Mol. Cell. Biol.* **7:**813–820.

Holmberg, S., 1984, Genetic improvement of brewer's yeast, *Trends Biotechnol.* **2:**98–102.

Hsiao, C-L., and Carbon, J., 1979, High frequency transformation of yeast by plasmids containing the cloned yeast *ARG4* gene, *Proc. Natl. Acad. Sci. USA* **76:**3829–3833.

Hsu, Y. P., and Schimmel, P., 1984, Yeast *Leu1* repression of mRNA levels and relationship of 5′ noncoding region to that of *Leu2, J. Biol. Chem.* **259:**3714–3719.

Hsu, Y. P., Kohlhaw, G., and Niederberger, P., 1982, Evidence that α-isopropylmalate synthase of *Saccharomyces cerevisiae* is under the "general" control of amino acid biosynthesis, *J. Bacteriol.* **150:**969–972.

Hubert, J. B., Guyonvarch, A., Kammerer, B., Exinger, Fl., Liljelund, P., and Lacroute, F., 1983, Complete sequence of an eukaryotic regulatory gene, *EMBO J.* **2:**2071–2073.

Jauniaux, J. C., Dubois, E., Crabeel, M., and Wiame, J. M., 1981, DNA and RNA analysis of the arginase mutants in *Saccharomyces cerevisiae, Arch. Int. Physiol. Biochem.* **89:**B111–B112.

Jauniaux, J. C., Dubois, E., Vissers, S., Crabeel, M., and Wiame, J. A., 1982, Molecular cloning, DNA structure and RNA analysis of the arginase gene in *Saccharomyces cerevisiae.* A study of *cis*-dominant regulatory mutations, *EMBO J.* **1:**1125–1131.

Jauniaux, J. C., Degols, G., Vissers, S., and Wiame, J. M., 1984, Regulation of L-ornithine transaminase synthesis in *Saccharomyces cerevisiae:* Gene cloning, RNA and *cis*-dominant regulatory mutations analysis. Abstract of paper presented at *Twelfth International Conference on Yeast Genetics and Molecular Biology*), Edinburgh, U. K., p. 168.

Johnston, L. H., and Nasmyth, K. A., 1978, *Saccharomyces cerevisiae* cell cycle mutant *cdc9* is defective in DNA ligase, *Nature* **274:**891–893.

Johnston, S. A., and Hopper, J. E., 1982, Isolation of the yeast regulatory gene *GAL4* and analysis of its dosage effects on the galactose-melibiose regulon, *Proc. Natl. Acad. Sci. USA* **79:**6971–6975.

Jones, E. W., 1984, The synthesis and function of proteases in *Saccharomyces:* Genetic approaches, *Annu. Rev. Genet.* **18:**233–270.

Jones, E. W., and Fink, G. R., 1982, Regulation of amino acid and nucleotide biosynthesis in yeast, in: *The Molecular Biology of the Yeast* Saccharomyces. *Metabolism and Gene Expression* (J. N. Strathern, E. W. Jones, and J. R. Broach, eds.), Cold Spring Harbor Laboratory, Cold Spring Harbor, NY, pp. 181–299.

Jong, A. Y. S., Kuo, C. L., and Campbell, J. L., 1984, The *CDC8* gene of yeast encodes thymidylate kinase, *J. Biol. Chem.* **259:**1052–1059.

Kataoka, T., Powers, S., Cameron, S., Fasano, O., Goldfarg, M., Broach, J., and Wigler, M., 1985, Functional homology of mammalian and yeast *RAS* genes, *Cell* **40:**19–26.

Kawasaki, G. H., 1979, The *cdc19* mutation of *Saccharomyces cerevisiae* is a temperature-sensitive defect for the glycolytic enzyme, pyruvate kinase, in: *The Molecular Biology of Yeast* (Abstract of paper presented at Cold Spring Harbor Meeting, 1979), Cold Spring Harbor Laboratory, Cold Spring Harbor, NY, p. 35.

Kawasaki, G. H., and Fraenkel, D. G., 1982, Cloning of yeast glycolysis genes by complementation, *Biochem. Biophys. Res. Commun.* **108:**1107–1112.

Klebl, F., Huffaker, T., and Tanner, W., 1984, A temperature-sensitive *N*-glycosylation mutant of *S. cerevisiae* that behaves like a cell-cycle mutant, *Exp. Cell Res.* **150:**309–313.

Klig, L. S., and Henry, S. A., 1984, Isolation of the yeast *INO1* gene: located on an autonomously replicating plasmid the gene is fully regulated, *Proc. Natl. Acad. Sci. USA* **81:**3816–3820.

Kotyk, A., 1967, Properties of the sugar carrier in bakers' yeast II, Specificity of transport, *Folia Microbiol.* **12:**121–131.

Kunz, B. A., and Haynes, R. H., 1982, DNA repair and the genetic effects of thymidylate stress in yeast, *Mutat. Res.* **93:**353–375.

Kunz, B. A., Barclay, B. J., Game, J. C., Little, J. G., and Haynes, R. H., 1980, Induction of mitotic recombination in yeast by starvation for thymine nucleotides, *Proc. Natl. Acad. Sci. USA* **77**:6057–6061.

Kunz, B. A., Taylor, G. R., Konforti, B., Glickman, B. W., and Haynes, R. H., 1984, Inhibition of thymidylate biosynthesis induces mitotic unequal sister chromatid recombination in *Saccharomyces cerevisiae, Curr. Genet.* **8**:211–217.

Kunz, B. A., Taylor, G. R., and Haynes, R. H., 1985, Mating-type switching in yeast is induced by thymine nucleotide depletion, *Mol. Gen. Genet.* **199**:540–542.

Kuziora, M. A., Chalmers, J. H. Jr., Douglas, M. S., Hitzeman, R. A., Mattick, J. S., and Wakil, S. J., 1983, Molecular cloning of fatty acid synthetase genes from *Saccharomyces cerevisiae, J. Biol. Chem.* **258**:11648–11653.

Lagunas, R., and Gancedo, J. M., 1973, Reduced pyridine nucleotides balance in glucose-growing *Saccharomyces cerevisiae, Eur. J. Biochem.* **37**:90–94.

Langin, T., Faugeron, G., Goyon, C., Nicolas, A., and Rossignol, J-L., 1986, The *MET2* gene of *Saccharomyces cerevisiae:* Molecular cloning and nucleotide sequence, *Gene* **49**:283–293.

Larimer, F. W., Morse, C. C., and Beck, A. K., 1983, Isolation of the *ARO1* cluster gene of *Saccharomyces cerevisiae, Mol. Cell. Biol.* **3**:1609–1614.

Laughon, A., and Gesteland, R. F., 1982, Isolation and preliminary characterization of the *GAL4* gene, a positive regulator of transcription in yeast, *Proc. Natl. Acad. Sci. USA* **79**:6827–6831.

Lemoine, Y., Dubois, E., and Wiame, J. M., 1978, The regulation of urea amidolyase of *Saccharomyces cerevisiae, Mol. Gen. Genet.* **166**:251–258.

Letts, V. A., Klig, L. S., Bae-Lee, M., Carman, G. M., and Henry, S. A., 1983, Isolation of the yeast structural gene for the membrane-associated enzyme phosphatidyl serine synthetase, *Proc. Natl. Acad. Sci. USA* **80**:7279–7283.

Liao, H., and Thorner, J., 1980, Yeast mating pheromone α-factor inhibits adenylate cyclase, *Proc. Natl. Acad. Sci. USA* **77**:1898–1902.

Liao, H., and Thorner, J., 1981, Adenosine 3′,5′ phosphate phosphodiesterase and pheromone response in the yeast *Saccharomyces cerevisiae, J. Bacteriol.* **148**:919–925.

Liljelund, P., and Lacroute, F., 1986, Genetic characterisation and isolation of the *Saccharomyces cerevisiae* gene coding for uridine monophosphokinase, *Mol. Gen. Genet.* **205**:74–81.

Lloyd, D., and Scott, R. I., 1983, Direct measurement of dissolved gases in microbiological systems using membrane inlet mass spectrometry, *J. Microbiol. Meth.* **1**:313–328.

Lloyd, D., Kristensen, B., and Degn, H., 1983a, Glycolysis and respiration in yeasts. The Pasteur effect studied by mass spectrometry, *Biochem. J.* **212**:749–754.

Lloyd, D., Kristensen, B., and Degn, H., 1983b, Glycolysis and respiration in yeasts: The effect of ammonium ions studied by mass spectrometry, *J. Gen. Microbiol.* **129**:2125–2127.

Lobo, Z., 1984, *Saccharomyces cerevisiae* aldolase mutants, *J. Bacteriol.* **160**:222–226.

Lobo, Z., and Maitra, P. K., 1982, A particulate phosphofrutokinase from yeast, *FEBS Lett.* **137**:279–282.

Lobo, Z., and Maitra, P. K., 1983, Phosphofructokinase mutants of yeast, *J. Biol. Chem.* **258**:1444–1449.

Lorincz, A. T., Miller, M. J., Xuong, N-H., and Geiduschek, E. P., 1982, Identification of proteins whose synthesis is modulated during the cell cycle of *Saccharomyces cerevisiae, Mol. Cell. Biol.* **2**:1532–1549.

Losson, R., and Lacroute, F., 1982, Cloning of a eukaryote regulatory gene, *Mol. Gen. Genet.* **184**:394–399.

Lusty, C. J., and Jeffrey, L. U., 1982, Cloning of a yeast gene coding for arginine-specific carbamoyl-phosphate synthetase, *Proc. Natl. Acad. Sci. USA* **79**:2240–2244.

Mantsala, P., and Zalkin, H., 1984, Nucleotide sequence of *Saccharomyces cerevisiae ADE4* encoding glutamine phosphoribosylpyrophosphate amidotransferase, *J. Biol. Chem.* **259**:8478–8484.

Masson, P., Jacquemin, J. M., and Culot, M., 1984, Molecular cloning of the *tsm0185* gene responsible for adenylate cyclase activity in *Saccharomyces cerevisiae, Ann. Microbiol. (Inst. Pasteur)* **135:**343–351.

Matsumoto, K., Uno, I., Oshima, Y., and Ishikawa, T., 1982, Isolation and characterization of yeast mutants deficient in adenylate cyclase and cAMP-dependent protein kinase, *Proc. Natl. Acad. Sci. USA* **79:**2355–2359.

Matsumoto, K., Uno, I., and Ishikawa, T., 1983a, Control of cell division in *Saccharomyces cerevisiae* mutants defective in adenylate cyclase and cyclic AMP-dependent protein kinase, *Exp. Cell. Res.* **146:**151–161.

Matsumoto, K., Uno, I., and Ishikawa, T., 1983b, Initiation of meiosis in yeast mutants defective in adenylate cyclase and cAMP-dependent protein kinase, *Cell* **32:**417–423.

Matsumoto, K., Uno, I., Kato, K., and Ishikawa, T., 1985, Isolation and characterization of a phosphoprotein phosphatase-deficient mutant in yeast, Yeast **1:**25–38.

Mazon, M. J., Gancedo, J. M., and Gancedo, C., 1982a, Inactivation of yeast fructose-1,6-bisphosphatase. *In vivo* phosphorylation of the enzyme, *J. Biol. Chem.* **257:**1128–1130.

Mazon, M. J., Gancedo, J. M., and Gancedo, C., 1982b, Phosphorylation and inactivation of yeast fructose-bisphosphatase *in vivo* by glucose and proton ionophores. A possible role for cAMP, *Eur. J. Biochem.* **127:**605–608.

McAlister, L., and Holland, M. J., 1985, Isolation and characterisation of yeast strains carrying mutations in the glyceraldehyde-3-phosphate dehydrogenase genes, *J. Biol. Chem.* **260:**15013–15018.

McIntosh, E. M., and Haynes, R. H., 1986, Sequence and expression of the dCMP deaminase gene (*DCD1*) of *Saccharomyces cerevisiae, Mol. Cell. Biol.* **6:**1711–1721.

McIntosh, E. M., Gadsden, M. H., and Haynes, R. H., 1986, Transcription of genes encoding enzymes involved in DNA synthesis during the cell cycle of *Saccharomyces cerevisiae, Mol. Gen. Genet.* **204:**363–366.

Messenguy, F., Feller, A., Crabeel, M., and Pierard, A., 1983, Control mechanisms acting at the transcriptional and post-transcriptional levels are involved in the synthesis of the arginine pathway carbamoyl-phosphate synthase of yeast, *EMBO J.* **2:**1249–1254.

Michels, C. A., and Needleman, R. B., 1983, A genetic and physical analysis of the *MAL1* and *MAL3* genes in standard strains of *Saccharomyces cerevisiae, Mol. Gen. Genet.* **191:**225–230.

Miyajima, A., Nakayama, N., Miyajima, I., Arai, N., Okayama, H., and Arai, K., 1984, Analysis of full-length cDNA clones carrying *GAL1* of *Saccharomyces cereviaiae*—A model system for cDNA expression, *Nucl. Acid Res.* **12:**6397–6414.

Moll, R., and Wintersberger, E., 1976, Synthesis of yeast histones in the cell cycle, *Proc. Natl. Acad. Sci. USA* **73:**1863–1867.

Moon, R. B., and Richards, J. H., 1973, Determination of intracellular pH by P magnetic resonance, *J. Biol. Chem.* **248:**7276–7278.

Muller, D., and Holzer, H., 1981, Regulation of fructose-1,6-bisphosphatase in yeast by phosphorylation/dephosphorylation, *Biochem. Biophys. Res. Commun.* **103:**926–933.

Nadkarni, M., Parmar, L., Lobo, Z., and Maitra, P. K., 1984, Mutations in the regulatory subunit of soluble phosphofructokinase from yeast, *FEBS Lett.* **175:**294–298.

Nath, K., and Baptist, E. W., 1984, Cloning of a yeast dihydrofolate reductase gene in *E. coli, Curr. Genet.* **8:**265–270.

Navon, G., Shulman, R. G., Yamane, T., Eccleshall, T. R., Lam, K-B., Baronofsky, J. J., and Marmur, J., 1979, Phosphorus-31 nuclear magnetic resonance studies of wild-type and glycolytic pathway mutants of *Saccharomyces cerevisiae, Biochemistry* **18:**4487–4499.

Niederacher, D., and Entian, K-D., 1987, Isolation and characterisation of the regulatory *HEX2* gene necessary for glucose repression in yeast, *Mol. Gen. Genet.* **206:**505–509.

Niederberger, P., Furter, R., Prantl, F., Braus, G., and Huetter, R., 1984, Cloning of yeast *TRP* genes, *Riv. Biol.* **77:**600–603.

Nikawa, J., and Yamashita, S., 1984, Molecular cloning of the gene encoding CDPdiacylglycerol-inositol 3-phosphatidyl transferase in *Saccharomyces cerevisiae, Eur. J. Biochem.* **143:**251–256.

Nogi, Y., Shimada, H., Matsuzaki, Y., Hashimoto, H., and Fukasawa, T., 1984, Regulation and expression of the galactose gene cluster in *Saccharomyces cerevisiae*. 2. The isolation and dosage effect of the regulatory gene *GAL80, Mol. Gen. Genet.* **195:**29–34.

Parmar, L., Lobo, Z., Nadkarni, M., and Maitra, P. K., 1984, Multiple genes control particulate phosphofructokinase of yeast, *Mol. Gen. Genet.* **197:**515–516.

Patterson, M., Sclafani, R. A., Fangman, W. L., and Rosamond, J., 1986, Molecular characterisation of the cell cyle gene *CDC7* from *Saccharomyces cerevisiae, Mol. Cell. Biol.* **6:**1590–1598.

Petersen, J. G. L., Holmberg, S., Nilsson-Tillgren, T., and Kielland-Brandt, M. C., 1983, Molecular cloning and characterization of the threonine deaminase (*ILV1*) gene of *Saccharomyces cerevisiae, Carlsberg Res. Commun.* **48:**149–160.

Polaina, J., 1984, Cloning of the *ILV2, ILV3* and *ILV5* genes of *Saccharomyces cerevisiae, Carlsberg Res. Commun.* **49:**577–584.

Powers, S., Michaelis, S., Broek, D., Anna-A., S. S., Field, J., Herskowitz, I., and Wigler, M., 1986, *RAM,* a gene of yeast required for a functional modification of *RAS* proteins and for production of mating pheromone a-factor, *Cell* **47:**413–422.

Pringle, J. R., and Hartwell, L. H., 1981, The *Saccharomyces cerevisiae* cell cycle, in: *The Molecular Biology of the Yeast* Saccharomyces. *Life Cycle and Inheritance* (J. N. Strathern, E. W. Jones, and J. R. Broach, eds.), Cold Spring Harbor Laboratory, Cold Spring Harbor, NY, pp. 97–142.

Purwin, C., Leidig, F., and Holzer, H., 1982, Cyclic AMP-dependent phosphorylation of fructose-1,6-bisphosphatase in yeast, *Biochem. Biophys. Res. Commun.* **107:**1482–1489.

Rai, R., Genbauffe, F. S., Lea, H. Z., and Cooper, T. G., 1987, Transcriptional regulation of the *DAL5* gene in *S. cerevisiae, J. Bacteriol.* **169:**3521–3524.

Ratzkin, B., and Carbon, J., 1977, Functional expression of cloned yeast DNA in *E. coli, Proc. Natl. Acad. Sci. USA* **74:**487–491.

Reed, S. I., Hadwiger, J. A., and Lorincz, A. T., 1985, Protein kinase activity of the yeast cell division cycle gene *CDC28, Proc. Natl. Acad. Sci. USA* **82:**4055–4059.

Roberts, L. M., and Schewizer, E., 1984, Molecular cloning of the yeast fatty acid synthetase genes *FAS1* and *FAS2*. Illustrating the structure of the *FAS1* cluster gene by transcription mapping and transcription studies, *Mol. Gen. Genet.* **194:**457–465.

Robinson, L. C., Gibbs, J. B., Marshall, M. S., Sigal, I. S., and Tatchell, K., 1987, CDC25: A component of the *RAS*-adenylate cyclase pathway in *Saccharomyces cerevisiae, Science* **235:**1218–1221.

Rodicio, R., and Heinisch, J., 1987, Isolation of the yeast phosphoglyceromutase gene and construction of deletion mutants, *Mol. Gen. Genet.* **206:**133–140.

Rodicio, R., and Zimmerman, F. K., 1985a, Cloning of maltase regulatory genes in *Saccharomyces cerevisiae*. I. Isolation of the *MAL2-8C* regulatory gene, *Curr. Genet.* **9:**539–545.

Rodicio, R., and Zimmerman, F. K., 1985b, Cloning of maltase regulatory genes in *Saccharomyces cerevisiae*. 2. Isolation of the *MAL4* regulatory gene, *Curr. Genet.* **9:**547–551.

Roeder, G. S., Beard, C., Smith, M., and Keranen, S., 1985, Isolation and characterisation of the *SPT2* gene, a negative regulator of Ty-controlled yeast gene expression, *Mol. Cell. Biol.* **5:**1543–1553.

Rothstein, R. J., and Sherman, F., 1980, Dependence on mating type for the overproduction of iso-2-cytochrome c in the yeast mutant *CYC7-H2, Genetics* **94:**891–898.

Rousseau, P., 1972, Germination of yeast spores in *Saccharomyces cerevisiae,* Ph.D. thesis, University of Wisconsin, Madison.

Roy, D. J., and Dawes, I. W., 1987, Cloning and characterisation of the gene encoding lipoamide dehydrogenase in *Saccharomyces cerevisiae, J. Gen. Microbiol.* **133:**925–933.

Saheki, T., and Holzer, H., 1975, Proteolytic activities in yeast, *Biochim. Biophys. Acta* **384:**203–214.

Sangsoda, S., Cherest, H., and Surdin-Kerjan, Y., 1984, The expression of the *MET25* gene of *Saccharomyces cerevisiae* is regulated transcriptionally, *Mol. Gen. Genet.* **200:**407–414.

Sasnauskas, K., Gedminieno, G., Naktinis, V., Citavicius, D., and Janulaitis, A., 1982, Cloning of the *ADE2* gene of *Saccharomyces cerevisiae, DOK1. Adad. Nauk. SSSR* **263:**224–227.

Sass, P., Field, J., Nikawa, J., Toda, T., and Wigler, M., 1986, Cloning and characterisation of the high-affinity cAMP phosphodiesterase of *Saccharomyces cerevisiae, Proc. Natl. Acad. Sci. USA.* **83:**9303–9307.

Satoshi, H., Sidhu, R. S., Tohe, A., and Oshima, Y., 1981, Cloning of the *HIS5* gene of *Saccharomyces cerevisiae* by yeast transformation, *Gene* **16:**335–342.

Schmitt, H. D., Ciriacy, M., and Zimmerman, F. K., 1983, The synthesis of yeast pyruvate decarboxylase is regulated by large variations in the messenger RNA level, *Mol. Gen. Genet.* **192:**247–252.

Sclafani, R. A., and Fangman, W. L., 1984, Yeast gene *CDC8* encodes thymidylate kinase and is complemented by herpes thymidine kinase gene TK, *Proc. Natl. Acad. Sci. USA* **81:**5821–5825.

Sedivy, J. M., and Fraenkel, D. G., 1985, Fructose bisphosphatase of *Saccharomyces cerevisiae.* Cloning, disruption and regulation of the *FBP1* structural gene, *J. Mol. Biol.* **186:**307–319.

Skvirsky, R. C., Greenberg, M. L., Myers, P. L., and Greer, H., 1986, A new negative control gene for amino acid biosynthesis in *Saccharomyces cerevisiae, Curr. Genet.* **10:**495–501.

Souciet, J. L., Hubert, J. C., and Lacroute, F., 1982, Cloning and restriction mapping of the yeast *URA2* gene coding for the carbamoyl phosphate synthetase aspartate transcarbamoylase complex, *Mol. Gen. Genet.* **186:**385–390.

St.John, T. P., and Davis, R. W., 1979, Isolation of glactose inducible DNA sequences from *Saccharomyces cerevisiae* by differential plaque filter hydridization, *Cell* **16:**443–452.

Storms, R. K., Ord, R. W., Greenwood, M. T., Mirdamadi, B., Chu, F. K., and Belfort, M., 1984, Cell cycle dependent expression of thymidylate synthase in *Saccharomyces cerevisiae, Mol. Cell. Biol.* **4:**2858–2864.

Struhl, K., Stinchcomb, D. T., Scherer, S., and David, R. W., 1979, High-frequency transformation of yeast: Autonomous replication of hybrid DNA molecules, *Proc. Natl. Acad. Sci. USA* **76:**1035–1039.

Sudbery, P. E., Goodey, A. R., and Carter, B. L. A., 1980, Genes which control cell proliferation in the yeast *Saccharomyces cerevisiae, Nature* **288:**401–404.

Suissa, M., Suda, K., and Schatz, G., 1984, Isolation of the nuclear *Saccharomyces cerevisiae* gene for citrate synthase and 15 other mitochondrial proteins by a new screening method, *EMBO J.* **3:**1773–1782.

Sumrada, R. A., and Cooper, T. G., 1982, Isolation of the *CAR1* gene from *Saccharomyces cerevisiae* and analysis of its expression, *Mol. Cell. Biol.* **2:**1514–1523.

Sumrada, R. A., and Cooper, T. G., 1985, Target sites for positive and negative regulatory elements controlling expression of an inducible eukaryotic gene, *UCLA Symp. Mol. Cell. Biol. New Ser.* **30:**291–301.

Taguchi, A. K. W., and Young, E. T., 1987, The cloning and mapping of *ADR6,* a gene required for sporulation and for expression of the alcohol dehydrogenase II isoenzyme from *Saccharomyces cerevisiae, Genetics* **116:**531–540.

Taguchi, A. K. W., Ciriacy, M., and Young, E. T., 1984, Carbon source dependence of transposable element-associated gene activation in *Saccharomyces cerevisiae, Mol. Cell. Biol.* **4:**61–68.

Taylor, G. R., Barclay, B. J., Storms, R. K., Friesen, J. D., and Haynes, R. M., 1982, Isolation of the thymidylate synthetase gene (*TMP1*) by genetic complementation in yeast, *Mol. Cell. Biol.* **2:**437–442.

Taylor, G. R., Lagosky, P. A., Storms, R. K., and Haynes, R. H., 1987, Molecular characterisation of the cell cycle regulated thymidylate synthase gene of *Saccharomyces cerevisiae, J. Biol. Chem.* **262:**5298–5307.

Teshiba, S., Furter, R., Niederberger, P., Braus, G., Paravicini, G., and Hütter, R., 1986, Cloning of the *ARO3* gene of *Saccharomyces cerevisiae* and its regulation, *Mol. Gen. Genet.* **205:**353–357.

Tingle, M., Singh Klar, A. J., Henry, S. A., and Halvorson, H. O., 1973, Ascospore formation in yeast, in: *Microbial Differentiation, 23rd Symposium of the Society for General Microbiology* (J. M. Ashworth and J. E. Smith, eds.), Cambridge University Press, Cambridge, pp. 209–243.

Toda, T., Uno, I., Ishikawa, T., Powers, S., Kataoka, T., Broek, D., Cameron, S., Broach, J., Matsumoto, K., and Wigler, M., 1985, In yeast, *RAS* proteins are controlling elements of adenylate cyclase, *Cell* **40:**27–36.

Toda, T., Cameron, S., Sass, P., Zoller, M., and Wigler, M., 1987a, Three different genes in *S. cerevisiae* encode the catalytic subunits of the cAMP-dependent protein kinase, *Cell* **50:**277–287.

Toda, T., Cameron, S., Sass, P., Zoller, M., Scott, J. D., McMullen, B., Hurwitz, M., Krebs, E. G., and Wigler, M., 1987b, Cloning and characterisation of *BCY1,* a locus encoding a regulatory subunit of the cAMP dependent protein kinase in yeast, *Mol. Cell Biol.* **7:**1371–1377.

Tortora, P., Birtel, M., Lenz, A. G., and Holzer, H., 1981, Glucose-dependent metabolic interconversion of fructose-1,6-bisphosphatase in yeast, *Biochem. Biophys. Res. Commun.* **100:**688–695.

Tortora, P., Burlini, N., Hanozet, G. M., and Guerritore, A., 1982, Effect of caffeine on glucose-induced inactivation of gluconeogenetic enzymes in *Saccharomyces cerevisiae.* A possible role of cyclic AMP, *Eur. J. Biochem.* **126:**617–622.

Tortora, P., Burlini, N., Caspani, G., and Guerretore, A., 1984, Studies on glucose-induced inactivation of gluconeogenetic enzymes in adenylate cyclase and cAMP-dependent protein kinase mutants, *Eur. J. Biochem.* **145:**543–548.

Tschopp, J. F., Emr, S. D., Field, C., and Schekman, R., 1986, *GAL2* codes for a membrane-bound subunit of the galactose permease in *Saccharomyces cerevisiae, J. Bacteriol.* **166:**313–318.

Tsuboi, M., and Yanagishima, N., 1973, Effects of cAMP, theophilline and caffeine on the glucose repression of sporulation in *Saccharomyces cerevisiae, Arch. Mikrobiol.* **93:**1–12.

Tsuboi, M., Kamisaka, S., and Yanagishima, N., 1972, Effects of cyclic 3′-5′-adenosine monophosphate on the sporulation of *Saccharomyces cerevisiae, Plant Cell Physiol.* **13:**585–588.

Uno, I., Matsumoto, K., and Ishikawa, T., 1982, Characterization of cAMP requiring yeast mutants altered in the regulatory subunit of protein kinase, *J. Biol. Chem.* **257:**14110–14115.

Uno, I., Matsumoto, K., and Ishikawa, T., 1983, Characterization of a cyclic nucleotide phosphodiesterase-deficient mutant in yeast, *J. Biol. Chem.* **258:**3539–3542.

Uno, I., Matsumoto, K., Adachi, K., and Ishikawa, T., 1984, Characterization of cyclic AMP requiring yeast mutants altered in the catalytic subunit of protein kinase, *J. Biol. Chem.* **259:**12508–12513.

Van der Plaat, J. B., and Van Solingen, P., 1974, Cyclic 3′,5′-adenosine monophosphate stimulates trehalose degradation in baker's yeast, *Biochem. Biophys. Res. Commun.* **56:**580–587.

Vezinhet, F., Kinnaird, J. H., and Dawes, I. W., 1979, The physiology of mutants derepressed for sporulation in *Saccharomyces cerevisiae, J. Gen. Microbiol.* **115:**391–402.

Walsh, R. B., Kawasaki, G., and Fraenkel, D. G., 1983, Cloning of genes that complement yeast hexokinase and glucokinase mutants, *J. Bacteriol.* **154:**1002–1004.

Welch, P., and Scopes, R. K., 1985, Studies on cell-free metabolism: Ethanol production by a yeast glycolytic system reconstituted from purified enzymes, *J. Biotechnol.* **2:**257–273.

Wickner, R. B., 1974, Mutants of *Saccharomyces cerevisiae* that incorporate deoxythymidine-5′-monophosphate into deoxyribonucleic acid *in vivo*, *J. Bacteriol.* **117**:252–260.

Williamson, D. H., 1965, The timing of DNA synthesis in the cell cycle of *Saccharomyces cerevisiae*, *J. Cell. Biol.* **25**:517–528.

Williamson, V. M., Benetzen, J., Young, E. T., Nasmyth, K., and Hall, B. J., 1980, Isolation of the structural gene for alcohol dehydrogenase by genetic complementation in yeast, *Nature* **283**:214–216.

Witt, I., Kronau, R., and Holzer, H., 1966, Repression von alkoholdehydrogenase, isocitratlyase und malatsynthase in hefe durch glucose, *Biochim. Biophys. Acta* **118**:522–537.

Wolf, D. H., 1982, Proteinase action *in vitro* versus proteinase function *in vivo:* Mutants shed light on intracellular proteolysis in yeast, *Trends Biochem. Sci.* **7**:35–37.

Woolford, C. A., Daniels, L. B., Park, F. J., Jones, E. W., Van Arsdell, J. N., and Innis, M. A., 1986, The *PEP4* gene encodes an aspartyl protease implicated in the posttranslational regulation of *Saccharomyces cerevisiae* vacuolar hydrolases, *Mol. Cell. Biol.* **6**:2500–2510.

Yoo, H-S., Genbauffe, F. S., and Cooper, T. G., 1985, Identification of the ureidoglycolate hydrolase gene in the *DAL* gene cluster of *Saccharomyces cerevisiae, Mol. Cell Biol.* **5**:2279–2288.

Young, E. T., Williamson, V. M., Taguchi, A. K. W., Smith, M., Sledziewski, A., Russel, D., Osterman, J., Denis, C. L., Cox, D., and Beier, D., 1982, The alcohol dehydrogenase genes of the yeast *Saccharomyces cerevisiae* isolation, structure and regulation, *Basic Life Sci.* **19**:335–361.

Zalkin, H., and Yanofsky, C., 1982, Yeast gene *TRP5:* Structure, function, regulation, *J. Biol. Chem.* **257**:1491–1450.

Methods in Classical Genetics 4

R. B. WICKNER

1. INTRODUCTION

This chapter deals with the classical genetics of *Saccharomyces cerevisiae* largely from the methods viewpoint. Some of the interesting areas that have been studied in this way will be mentioned, particularly in the section on cytoplasmic genetics, which is the author's particular interest.

Many previous reviews have dealt with the subjects treated here, but the reader is especially referred to the two volumes entitled *The Molecular Biology of the Yeast* Saccharomyces (Strathern *et al.*, 1981), in which the results of studies of yeast genetics are summarized. Reviews on nearly all aspects of yeast genetics and molecular biology by leading workers in the field can be found in these volumes.

2. LIFE CYCLE OF *S. CEREVISIAE*

While most wild strains are diploid and most strains used for fermentation are diploid or of even higher ploidy, most laboratory strains are haploid. Yeast of any ploidy up to at least tetraploid can be easily constructed and relatively stably maintained. The usual life cycle is shown diagramatically in Fig. 1.

Haploid cells can grow mitotically indefinitely with a doubling time as low as 90 min on rich medium. But yeast has two sexes, or mating types, called *a* and α. They are not called male and female or + and − because the mating process is symmetrical. Insofar as is known, all the genetic information from both parents, both chromosomal and cytoplasmic (nonchromosomal), is contributed to the zygotes. When cells of opposite mating type come in contact on rich medium, each arrests the cell

*This chapter was completed in 1985.

R. B. WICKNER • National Institutes of Health, Bethesda, Maryland 20892.

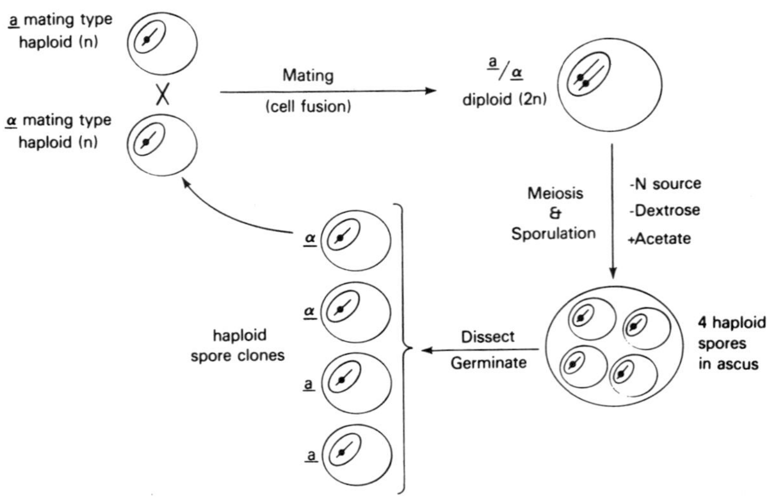

Figure 1. Life cycle of *S. cerevisiae.* Mating of cells of opposite mating type occurs under growth conditions. Diploids can grow mitotically indefinitely or, under the conditions indicated, be induced to undergo meiosis and spore formation. Mating type segregates 2 *a* : 2 α in the usual crosses. It is determined by alleles at a locus on chromosome III.

cycle of the other near the start point via a constitutively produced polypeptide mating pheromone. That produced by α cells is called α hormone and has the sequence (Trp)-His-Trp-Leu-Gln-Leu-Lys-Pro-Gly-Gln-Pro-Met)Tyr. That produced by *a* cells is called *a* hormone and has the sequence Lys-Gly-Val-Phe-Trp-Ala-Asx-Pro. Interestingly, the α pheromone, that has a sex-related function in yeast, is similar in sequence to the human hypothalamic hormone, gonadotropin-releasing hormone (Glu-His-Trp-Ser-Tyr-Gly-Leu-Arg-Pro-Gly-NH_2), and can act in place of the latter (Loumaye *et al.,* 1982).

The expressed mating type of a yeast strain depends on what information is present at the *MAT* locus on chromosome III. While most laboratory strains stably maintain their mating type, most wild-type strains rapidly alternate between *a* and α. This phenomenon is due to the presence of extra copies of *a* and α mating-type information at sites on chromosome III distant from the *MAT* locus. These sequences are frequently transposed to the *MAT* locus with elimination of the sequences previously residing there (Herskowitz and Oshima, 1981). Expression of the extra copies of *a* and α information is repressed. The HO gene, necessary for the transposition reaction, is missing or defective in most laboratory strains, so that these strains have a stable mating type. This control of cell "differentiation" by DNA rearrangement or transposition is analogous to similar events in higher eukaryotes. Yeast also has a transposon, called Ty1, whose properties resemble both bacterial transposons and RNA tumor viruses.

The haploid cells of opposite mating type, arrested near their start points, agglutinate in a specific reaction and fuse to form a transient heterokaryon, that is, a single cell with two nuclei. In the normal mating process, this is followed quickly by karyogamy (nuclear fusion) to produce a diploid cell (the zygote). The zygotes can be recognized under the microscope and directly picked out of a mating mixture with a micromanipulator needle. The first bud of a zygote usually emerges from the equatorial region of the dumbbell-shaped zygote cell, giving an easily recognized appearance. More commonly, one selects prototrophic diploid cells from the mating mixture using the complementary nutritional markers of the parents.

The diploid cells formed can grow indefinitely by mitosis and, at any point, be induced to undergo meiosis and spore formation by alteration of the culture conditions. Specifically, acetate is used as the carbon source (usually 1% potassium acetate), and little or no dextrose (0.05% or less) or nitrogen source (0.1% yeast extract or less) is included. Meiosis and sporulation are inhibited by the usual levels of dextrose or nitrogen source, and these processes, which generate diverse offspring in place of the uniform offspring resulting from mitotic reproduction, can be viewed as a response to adversity. The whole process takes 12–24 hr in liquid, but 24–48 hr on plates, and each diploid nucleus undergoes a typical meiosis producing four haploid nuclei, each of which is encapsulated in a spore coat. These four spores remain within what was the diploid cell's wall. That wall is called the ascus (= sac), and the spores are called ascospores. For this reason, *S. cerevisiae* is a member of the *Ascomycetes.* The four spores, or the four haploid "spore" clones they produce on germination, are called a tetrad.

The nutritional control of the mitosis–meiosis decision is mediated by the adenylate cyclase-cAMP-dependent protein kinase system (Matsumoto *et al.,* 1982, 1983a,b; Uno *et al.,* 1982; reviewed by Matsumoto *et al.,* 1985). Specifically, cyclic AMP is needed to begin the cell cycle and acts at a point in G_1 even before the point of action of α-hormone. On the other hand, cyclic AMP must be absent in order for cells to enter meiosis.

This mitosis–meiosis decision for yeast is quite similar to a growth-differentiation decision in higher eukaryotes. Perhaps it should thus not be surprising that yeast has proteins homologous to the mammalian "Ras" oncogenes whose overproduction or malfunction in mammalian cells results in cancer (Gallwitz *et al.,* 1983; DeFeo-Jones *et al.,* 1983). These yeast *RAS* proteins are involved in controlling cAMP production and thus in modulating the growth-differentiation decision in yeast (Toda *et al.*, 1985; Tatchell *et al.*, 1985). They are GTP-dependent stimulators of adenylate cyclase (Uno *et al.*, personal communication; Toda *et al.*, 1985).

The four ascospores from each diploid cell are routinely dissected free from each other using a needle attached to a micromanipulator while viewing the process under the microscope. As will be described in detail in Section 3, the information available by being able to analyze all four products of each single meiotic event (tetrad analysis) is sufficiently valuable to justify the effort involved in mastering the technique of tetrad dissection. This process is carried out under a magnification of around 250×. In order to have plenty of room between the objective and the surface of the agar slab on which the dissection is done (Fig. 2), it is desirable to use a 10× or 15× objective with a 25× or 20× or 15× pair of eyepieces. The stage has to be moved around a lot, and it is useful to be able to return to a place where you were previously dissecting. This can be done without taking your eyes away from the eyepieces if the stage has handles such that one turn = 1 mm. This is the case for the stage designed by Fogel and Mortimer, available from Lawrence Precision Machine in Hayward, California. A micromanipulator attached directly to the stage of the microscope reduces the effects of nearby vibrations and reduces the likelihood of accidentally moving the needle out of the field. A large variety of micromanipulators are available, and price, ease of repair, frequency of need of repair, and ease of learning are all worth considering. Those for which the needle is controlled in two or three dimensions by the motions of a single lever are probably the easiest on which to learn,

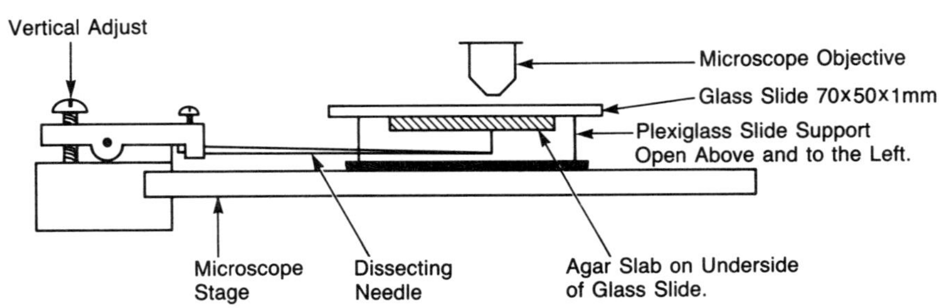

Figure 2. A dissecting apparatus. Schematic cross-section through a dissecting apparatus of the Mortimer–Fogel type. The sporulated culture is streaked on a thin agar slab, which rests on a large glass slide. The slide with agar slab is inverted and placed on a rectangular plexiglass holder which is open above (for the agar slab) and to the left (to admit the dissecting needle). The needle's down motion is controlled by the apparatus to the left attached to the stage of the microscope. The other two dimensions are controlled by moving the plexiglass slide holder.

but are significantly more expensive and somewhat more complex to repair.

The four spores of each tetrad are dissected apart and placed in a defined pattern on a thin slab of agar or agar containing rich medium (YPAD). This can be done (1) directly on a Petri dish inverted over the dissecting needle (Fogel *et al.,* 1981) or (2) on a slab of agar on a glass plate. In the latter case, the slab must then be transferred to a YPAD plate for germination and growth. The spore clones can be then [case (1)] directly replicated to various media for testing or [(1) or (2)] picked and inoculated onto a YPAD master plate for growing and testing. The author uses a master plate and places the colonies in an array made by gently pressing the surface of the agar with a 48-point inoculator [available from West Coast Scientific, Inc., Emeryville, CA (catalog No. WC48)], suitable for inoculating half the wells of a 96-well microtiter plate (e.g., Falcon catalog No. 3040). This facilitates multiple matings done in analyzing the tetrads, for example, mating-type and complementation tests. The 48-prong inoculator is dipped into an H_2O suspension of the strain(s) used in the mating, and then the adhering cells are simultaneously delivered to each of the 48 spots on a fresh YPAD replica of the master plate.

All the markers in the cross are generally analyzed (you'll be sorry later if you don't) by replica-plating to suitable media and by doing complementation tests. Mating-type testing is done by mating YPAD replicas of the master plate separately with known auxotrophic *a* and α tester strains. The formation of prototrophs able to grow on SD (synthetic dextrose) plates shows that mating occurred. Simple methods for carrying out large numbers of complementation tests have been described (e.g., Wickner, 1979). For example, a mixture of *a* and α *his1 ade1 lys1 met4* strains can be mated with the spore clones. This one plate can then be replica-plated to -his, -ade, -lys, and -met plates to score these four markers.

After all the spore clones have been suitably analyzed, these strains can become parents for further crosses.

3. TETRAD ANALYSIS

If the haploid parents of a cross differ in some trait determined by a single chromosomal gene, the diploids formed by mating will be heterozygous at that locus. The mechanics of meiosis (Fig. 3A, 3B, and 3C) then require that two of the four nuclei formed will have the genotype of one parent and two will have the genotype of the other parent. This is called 2 : 2 segregation (or 2+ : 2−, etc.). Deviations from 2 : 2 segregation can be due to polygenic control of the trait examined or to gene

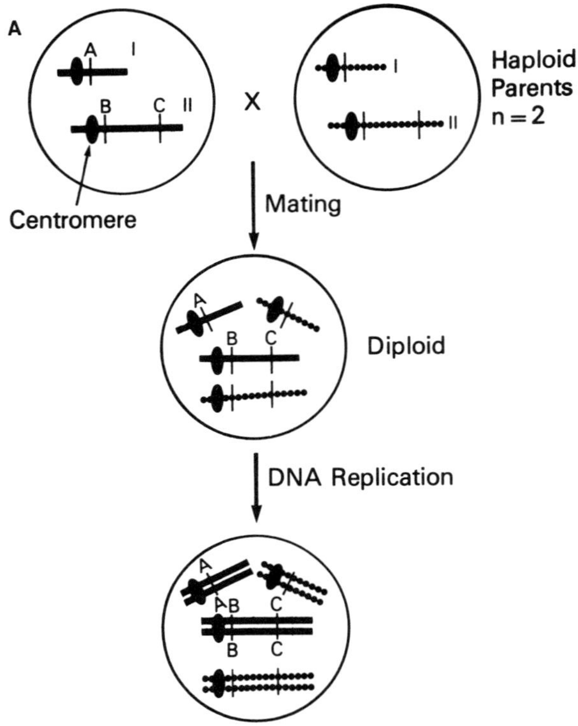

Figure 3. The mechanics of meiosis. (A) Haploid ($n = 2$) cells of opposite mating-type mate. One parent has marker A on chromosome I and markers B and C on chromosome II. Markers A and B are close to their respective centromeres. The diploid is induced to undergo meiosis and begins by replicating its DNA. (B) Homologous chromosomes pair and recombination occurs. (Following recombination, if cells are returned to growth medium, recombinant diploid cells can be detected that have not completed sporulation.) Meiosis proceeds. The first meiotic division is shown here. Homologous chromosomes separate. (C) The second meiotic division then occurs without intervening DNA replication so that haploid progeny are produced. A single recombination between markers B and C has produced a tetratype ascus with respect to these two genes. A and B are nonparental ditype with respect to each other because of their chromosomes' assortment at division I. Note that a crossover centromere distal to marker A did not alter the distribution of marker A in the haploid spores.

conversion, mitotic recombination during growth of the diploid, cytoplasmic inheritance, or aneuploidy (see below).

The centromere is the site on a chromosome that is attached to the mitotic or meiotic apparatus and by which a chromosome is pulled to one or the other daughter nucleus. Thus, it is essential to remember that the factor determining the segregation of a marker into the daughter nuclei is the connection of the copies of that marker to their respective

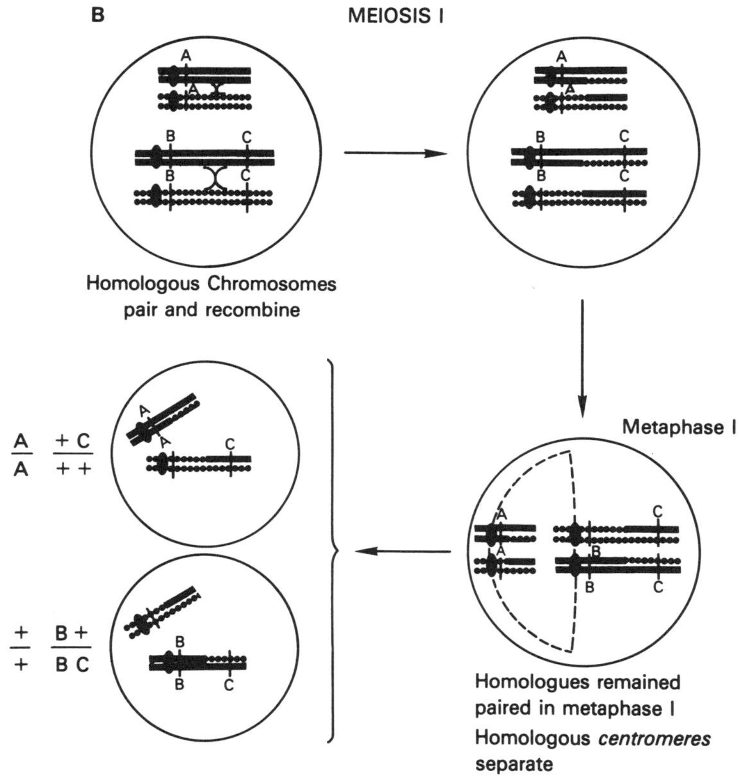

Figure 3. *(continued)*

centromeres. Recombinations occurring between a marker and its centromere can change that connection, but recombinations occurring centromere-distal to the marker do not affect that connection.

If the parents in a cross differ in two single chromosomal genes (e.g., A B X + +) and if these two genes are completely unlinked to each other or to their respective centromeres, then the distributions of mutant and wild-type alleles of the two genes are random with respect to each other. All the possible relative segregation patterns are as follows:

AB	AB	AB	A+	A+	A+
AB	A+	A+	AB	AB	A+
++	+B	++	+B	++	+B
++	++	+B	++	+B	+B
PD	T	T	T	T	NPD

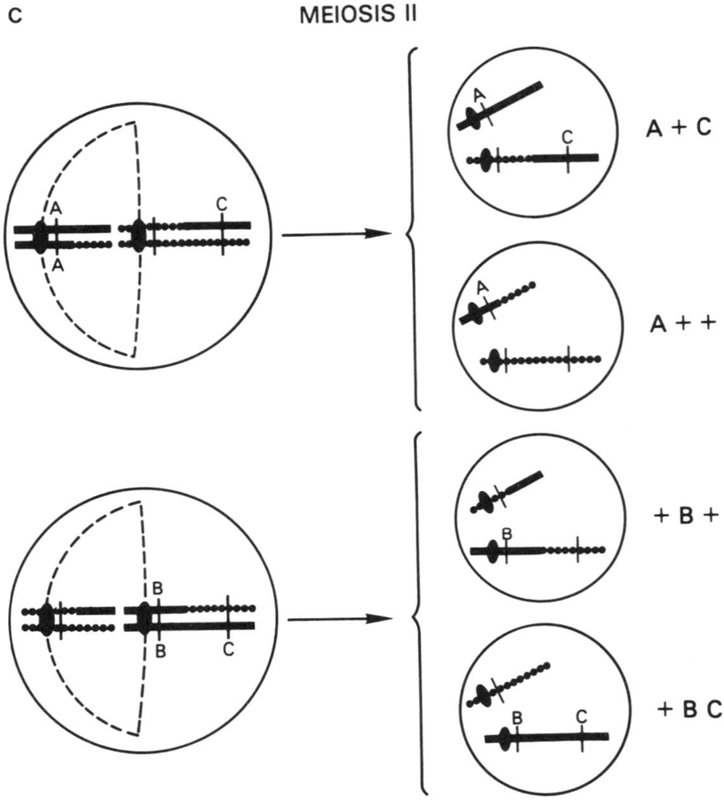

Figure 3. *(continued)*

The first of these tetrad types has only two types of spore clones, namely, A B and + +. It is thus called a ditype tetrad. Since these two spore types have the same genotypes as the parent strains with respect to these two markers, this is called a parental ditype (PD) tetrad. The last tetrad also has only two types of spore clone genotypes with respect to markers A and B, namely, A+ and +B. Since both of these are recombinant genotypes when compared to the parents, such a tetrad is called a nonparental ditype (NPD) tetrad with respect to these two markers. Each of the other tetrads has four genotypes, two parental (A B and + +) and two recombinant (A+ and +B), and these are thus called tetratype (T) tetrads. As can be seen, the ratio of PD : NPD : T tetrads is 1 : 1 : 4 in this case. As will be shown, deviations from this ratio indicate either (1) linkage of A and B, in which case PD > NPD, or (2) linkage of both A and B to their respective centromere, in which case PD : NPD : T :: 1 : 1 : < 4.

If the two markers B and C are very closely linked to each other, most of the tetrads will be PD. As can be seen in Fig. 3, a single crossover between B and C changes the PD into a T ascus. The crossover shown is between strands (chromatids) 2 and 3, but the reader can satisfy himself that the same result would be found for a single crossover between any pair of nonsister chromatids (1 and 3, 2 and 3, 1 and 4, or 2 and 4).

For a second recombination event between B and C, the result depends on which pair of chromatids are involved relative to the first crossover. In Fig. 4C, the second exchange involves the pair of chromatids not involved in the first. Only this kind of double crossover (four-strand double crossover) results in an NPD tetrad. (Conversely, observing an NPD tetrad implies that at least two crossovers have occurred between B and C, assuming they are on the same chromosome; see

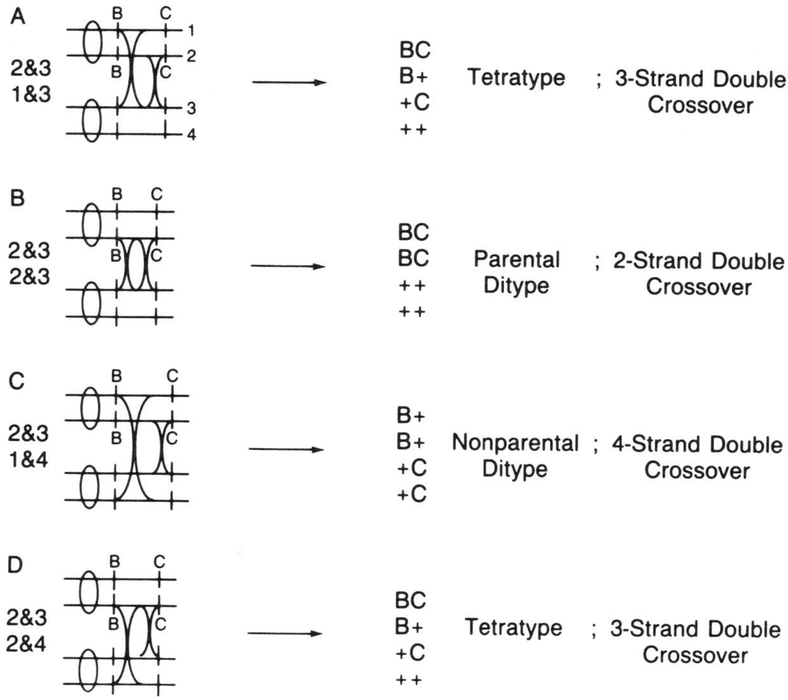

Figure 4. Double crossovers. Given that one crossover in a particular interval involves a particular pair of strands (2 and 3 in this diagram), the second crossover in that interval can involve the same pair of strands (result: PD tetrad), the two strands not involved in the first crossover (result: NPD tetrad), or (for the other two possibilities) one strand involved in the first crossover and one not involved (results: T tetrads). Thus, one double crossover in four would be expected to result in an NPD tetrade.

below.) The other possible second exchanges are 1 and 3 or 2 and 4, producing tetratype tetrads (three-strand double crossover), and 2 and 3, reversing the effect of the first crossover producing a parental ditype (two-strand double crossover). To arrive at a simple formula for genetic distances, one assumes that no more than two crossovers occur in a particular (short) interval and uses the facts that (1) the NPD tetrads are one-fourth of the total double crossovers and (2) among the T tetrads are half of the double crossover tetrads, a number equal to twice the NPD tetrads.

Thus, in the interval between B and C, the total crossovers =

$$1 \times \text{single crossovers} + 2 \times \text{double crossovers} =$$
$$1 \times (\text{T} - 2\ \text{NPD}) + 2 \times (4\ \text{NPD}) = \text{T} + 6\ \text{NPD}$$

Since one crossover produces only half recombinant and half parental spores, the percent recombination (i.e., the percent of progeny that are recombinant in the limiting case of very short distances) = 100 × total crossovers/2 (total tetrads) = 50 (T + 6 NPD)/(PD + NPD + T) (Perkins, 1949). The units are centiMorgans (cM).

This formula was derived assuming that only one or two crossovers occur in the B–C interval. It is thus increasingly inaccurate with increasing genetic distance. More accurate formulae have been published (Snow, 1979), as has a curve relating cM to linearly additive map units (Mortimer and Schild, 1980; Ma and Mortimer, 1983).

If two markers are on different chromosomes, but each is quite close to its respective centromere, the two markers will be found to produce fewer tetratype asci than expected. Each marker will be seen to segregate from its wild-type allele at the first meiotic division. This is called first-division segregation and is illustrated in Fig. 5. Note that for the two centromere-linked genes A and B in Fig. 5, the crossover (which will only rarely come between A and its centromere or between B and its centromere) has no effect on their segregation relative to each other. There are thus only two ways A and B are likely to segregate, depending only on which centromeres go to which daughter cell at the *first* meiotic division, namely, PD and NPD (Fig. 5). If A, for example, is known to be essentially *at* its centromere, then B's distance from its centromere can be calculated simply by the formula $\text{cM} = \dfrac{\text{T} \times 100}{2\ (\text{total tetrads})}$. Centromere linkage can, by definition, only be detected in the range where single crossovers (tetratype tetrads) are relatively rare, i.e., up to about 30 cM from the centromere. Since direct linkage of two markers on the same chromosome is detected by a deficiency of *double* crossovers (NPD tetrads), it is detectable up to a separation of about 60 cM.

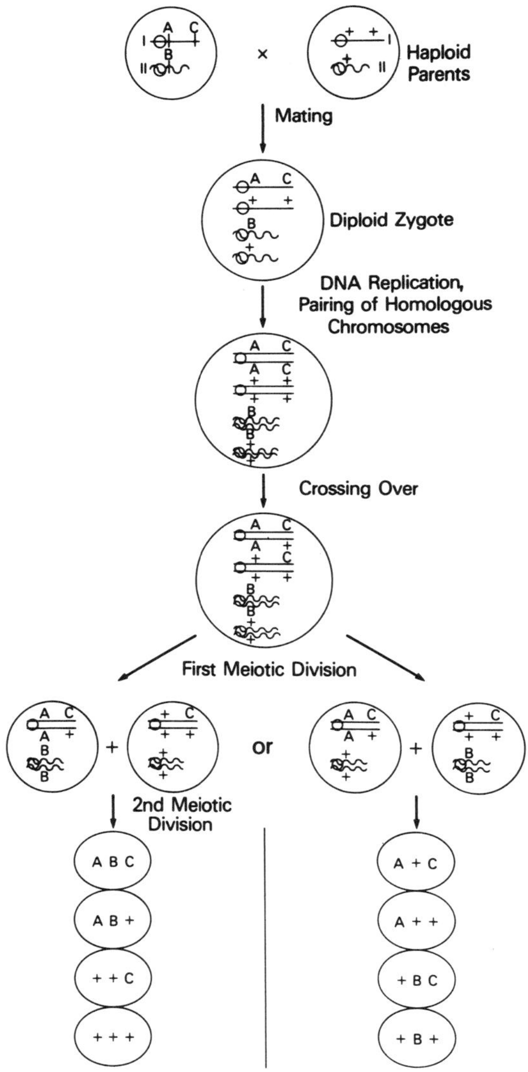

Figure 5. Centromere linkage. The events here are the same as in Fig. 3. Markers A and B, unlike marker C, are unaffected by the crossing-over stage because they are very close to their respective centromeres. The segregation of A and B relative to each other depends only on whether the A's and B's go to the same pole in the first meiotic division (on the left) to produce a PD tetrad with respect to A and B, or the A's go with the B^+'s and the B's go with the A^+'s (on the right) to produce an NPD tetrad. Tetratype tetrads for A and B can occur only if, as shown here for marker C, there is at least one crossover between A and its centromere or between B and its centromere.

The reader who wants to become familiar with tetrad analysis should (1) spend some time with pencil and paper working out the consequences of various crossovers and segregations as shown in the figures for this section and (2) apply the analysis described here to some actual data such as that shown in Appendix II.

4. ANEUPLOIDY

An aneuploid strain is one whose chromosome number is not a multiple of the basic haploid number, *n*. An otherwise haploid strain carrying one extra chromosome III, for example, is said to be "disomic for chromosome III." An otherwise diploid strain carrying three chromo-

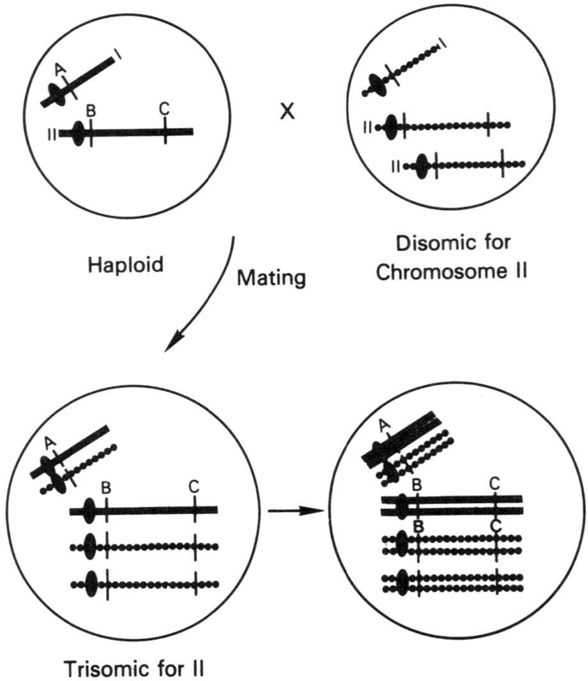

Figure 6. An extra chromosome II, as shown here, results in deviation of segregation from the normal 2+ : 2− pattern for markers on that chromosome as long as two of the chromosomes have the dominant allele and one has the recessive allele. This phenomenon is used in certain mapping methods (see text). A single crossover between B and C is shown here. Centromere-linked genes on chromosome II (like B) segregate 4+ : 0 and 2+ : 2B. Centromere-distant genes on chromosome II (like C) segregate mostly 3+ : 1C with some 4+ : 0 and 2+ : 2C tetrads (see text).

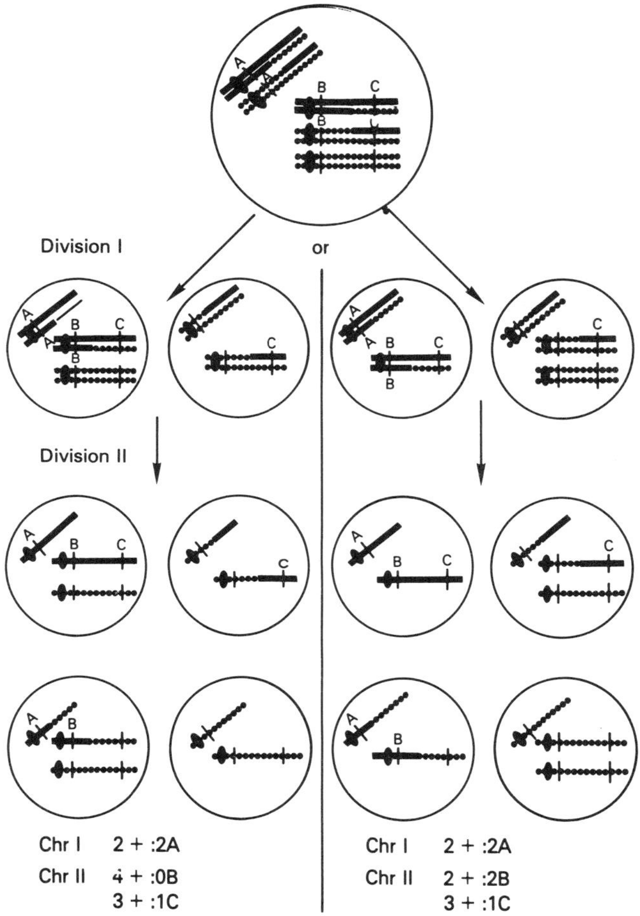

Figure 6. *(continued)*

somes III is said to be "trisomic for chromosome III." An otherwise diploid strain that lacks one of the pair of chromosomes II, for example, is said to be "monosomic for chromosome II."

Yeast is quite tolerant of extra chromosomes, and a single extra chromosome may be relatively stably propagated in mitosis. Disomy for chromosome XI is particularly widespread among laboratory strains. However, many single extra chromosomes are unstable, and multiple extra chromosomes generally cause slow growth or no growth. This selects for loss of some of the extra chromosomes.

We shall examine first the effect of one extra chromosome on meiotic segregation and later describe mapping methods based on the use of aneuploid strains.

As shown in Fig. 6, mating a normal haploid with markers A, B, and

C (A B C) with a strain disomic for a chromosome carrying the wild-type alleles of markers B and C (+ ++/++), results in a trisomic strain of genotype A/+ BC/++/++. The marker, A, on the chromosome for which an extra copy is not present segregates 2 A : 2 + as for a normal cross. However, the extra copies of chromosome II carrying the wild-type alleles of B and C mask the mutant alleles in many of the spore clones. For marker B, which is close to its centromere, a mixture of 4 + : 0 (two-thirds of all tetrads) and 2 + : 2 B (one-third) is observed. This ratio is expected since one of the three chromosomes II is presumably chosen at random to segregate by itself to one of the daughter cells at the first meiotic division. That will be the chromosome carrying B (the recessive mutation) one-third of the time and will produce a 2 + : 2 B tetrad. The other outcomes will produce 4 + : 0 tetrads. For marker C, which is assumed to be far from its centromere, a mixture of 2 + : 2 C, 3 + : 1 C, and 4 + : 0 tetrads will be produced in the ratio of 1 : 10 : 4 (see Table I). A detailed discussion of these ratios and other quantitative

Table I. Theoretical Distribution of Tetrad and Spore Types from −/+/+

		Tetrad types				Spore types	Frequency
	Frequency	2/3	1/3				
	Phenotype	4+:0	2+:2−				
Tightly centromere-linked gene[a]		*trp1*/+	*trp1*			+ =	1/3
		trp1/+	*trp1*			+/+ =	1/6
		+	+/+			*trp1* =	1/6
		+	+/+			*trp1*/+ =	1/3
	Frequency	4/15	1/15	8/15	2/15	Spore types	Frequency
	Phenotype	4+:0	2+:2−	3+:1−	3+:1−		
Centromere-unlinked gene[b]		*rna1*/+	*rna1*	*rna1*	*rna1*/*rna1*	+ =	10/30
		rna1/+	*rna1*	*rna1*/+	+/+	+/+ =	6/30
		+	+/+	+/+	+	*rna1* =	5/30
		+	+/+	+	+	*rna1*/+ =	8/30
						rna1/*rna1* =	1/30

[a]The segregation of *trp1* is independent of recombination events because of its tight centromere linkage. At the first meiotic division, one of the three homologs will migrate alone to one pole. If this is the one carrying the *trp1* mutation, a 2 *trp1* : 2 +/+ tetrad will result. If it is one of the two chromosomes carrying *trp*+, a 2 *trp1*/+ : 2 + tetrad will result. Therefore, the ratio of the former to the latter tetrad type is 1 : 2, as shown.

[b]We assume that *rna1* is so far from its centromere that many crossovers occur between it and its centromere and the distribution of the two *rna1* mutant alleles among the six chromatids is random. There are 15 possible distinct arrangements. If they are equally likely, the indicated distribution results. For intermediate distances, the calculations of Shaffer *et al.* (1971) and Riley and Manney (1978) predict intermediate frequencies of the −/− segregants. Our calculation assumes completely trivalent pairing. Any bivalent–univalent pairing will decrease the overall frequency of −/− segregants.

aspects has been published (Shaffer *et al.*, 1971; Culbertson and Henry, 1973; Riley and Manney, 1978).

This phenomenon is the basis for certain mapping methods (see Section 6.3).

5. MUTANT INDUCTION AND ISOLATION

5.1. Mutagenesis

Saccharomyces cerevisiae can be mutagenized by any of the agents useful in bacteria, e.g., ethylmethanesulfonic acid (EMS) (Lindegren *et al.*, 1965), nitrosoguanidine, UV or X irradiation, and frameshift mutagens (reviewed by Mortimer and Manney, 1971), or using *in vitro* methods to specifically modify a gene of interest (see Chapter 5). One must generally strike a balance between cell killing and mutation induction. Usually at very high percent killing, the yield of desired mutants actually falls as a percent of survivors. We usually use 4% EMS in 0.1 M potassium phosphate, pH 7.6, at 20°C for 2 hr. This gives about 50% killing with our strains, but strain variation is to be expected.

What is to be done with the mutagenized cells depends on what kind of mutation is sought. If it is necessary to know that one's mutations are independent, then the mutagenized cells must be divided before growth, either by plating out or by growing cells in many separate tubes if mutation expression prior to exposure to selection conditions is necessary.

5.2. Mutant Isolation

Selection for nongrowth under some conditions can be made using nystatin (Snow, 1966) or inositol-less death (Henry *et al.*, 1975) in a manner quite like penicillin selection for *Escherichia coli*. This can be used for selecting auxotrophs or other mutants whose growth arrest is not lethal. Often temperature-sensitive or cold-sensitive mutants in essential macromolecular processes are isolated by plating mutagenized cells on rich medium at the permissive temperature, replica-plating to a plate at the nonpermissive temperature, and then subscreening clones that did not grow for some more specific trait. Examples include (1) cell cycle mutants isolated by subscreening for a uniform bud size of initially asynchronous cells whose growth was arrested by shifting to the nonpermissive temperature (reviewed by Pringle and Hartwell, 1981) and (2) protein, RNA, or DNA synthesis mutants screened by rapid measurements of differential arrest of macromolecular synthesis at the nonpermissive temperature.

Selection of new mutants as revertants of other mutations is a standard method for studying the interactions of genes or their products.

Revertants can be intragenic (including true reversions to wild type) or extragenic. Any mutation that reverses a mutant phenotype without eliminating the original mutation is called a suppressor. Extragenic suppressors can be informational suppressors, such as amber, ocher, or missense tRNA suppressors, or functional suppressors. Informational suppressors are generally allele-specific, suppressing only certain alleles of a particular gene, and locus nonspecific, suppressing some mutant alleles of each of many different genes. Functional suppressors may suppress all mutant alleles of a given locus by supplying an alternate means of accomplishing what the original gene normally does. In other cases, a functional suppressor may be a mutation in a gene whose product normally interacts with the original mutant protein and reverses the original mutant phenotype by stabilizing that protein.

Methods of mutant isolation are extremely diverse and depend too much on the type of mutant to be isolated, to be listed in detail here. Suffice it to say that essentially all methods used with any microorganism can be used with *S. cerevisiae.*

6. GENETIC MAPPING METHODS

Genetic mapping is a useful way of characterizing a mutant or set of mutants. Its primary value is in identifying the mutated gene by a method independent of the phenotype and the actual possession of the mutant strain itself. It demonstrates that the gene being studied is distinct from all genes previously mapped to other locations—an otherwise impractical task. As the *S. cerevisiae* map is becoming denser, mapping, with increasing frequency, leads to the finding of unexpected identify of the newly mapped gene and a gene previously identified and mapped by another phenotype. This is commonplace in *E. coli* work. Mapping can facilitate later efforts to clone a gene if it is found near an already cloned gene. Since yeast chromosomes have been separated by gel electrophoresis, we can also expect to have chromosome-specific banks in the near future.

6.1. Meiotic Mapping

The primary mapping method is simply meiotic mapping by tetrad analysis. Experience with all other methods has led to pitfalls in even the most expert hands. Thus, while other methods generally lead more quickly to the assignment of a chromosome, only meiotic mapping is reliable for the final conclusion. In view of this, some workers have constructed sets of strains with large numbers of markers scattered at intervals throughout the genome. These are then simply crossed to the new mutants, tetrads are dissected and analyzed, and the results are examined for meiotic linkage (Contopoulou and Mortimer, cited in Mortimer and Schild, 1981; Gaber *et al.*, 1983). The only limitation of this method is that not quite all of the yeast genome has been sufficiently mapped with standard markers to assure that a location will be found in this way, but experience shows that for over 80% of mutants one is successful.

Gene order can be derived assuming that distances between genes are additive. If the A–B distance is 10 cM, the A–C distance is 6 cM, and the C–B distance is 4 cM, the gene order should be ACB. As a check, one can analyze individual tetrads. For short distances like this, double crossovers should be rare. For most tetrads, at least two of the markers A, B, and C should be in the PD (parental ditype) configuration. The third marker may be tetratype with respect to the other two if a single crossover has occurred. A crossover in the A–C interval in the example cited would produce tetrads of the following type:

A	C	B
+	C	B
A	+	+
+	+	+

A crossover between C and B would yield:

A	C	B
A	C	+
+	+	B
+	+	+

But a tetrad of the following type

A	C	B
A	+	B
+	C	+
+	+	+

would be very rare relative to the other two types because it requires two crossovers (one between A and C, one between C and B) *if* ACB is the correct gene order. Table II shows the rare tetrad type for each of the three possible gene orders of the markers in a cross of the type A B C X + + +.

Table II. Ordering Genes by Examination of Individual Tetrads

Parents: A B C × + + +

Gene order	Rare tetrad type		
A B C	A B C		
	A + C		
	+ B +		
	+ + +		
A C B	A C B		A B C
	A + B	=	A B +
	+ C +		+ + C
	+ + +		+ + +
C A B	C A B		A B C
	C + B	=	+ B C
	+ A +		A + +
	+ + +		+ + +

6.2. Mitotic Recombination

Mitotic recombination has been used similarly to survey chromosome arms for the location of a gene (e.g., Mortimer and Hawthorne, 1973). Here, one constructs diploids of the type $a^-b^-c^-x^-/+ + + +$, where a^-, b^-, and c^- are standard markers and x^- is the unmapped gene. Mitotic recombination is induced by UV or other mutagen treatment.

A correlation between homozygosis of one of the standard markers and of the unmapped markers means that they are located on the same chromosome arm. Thus, one isolates a^- and b^- and c^- clones and tests whether one of these classes has become x^-. If, for example, most b^- clones are x^-, then x is on the some chromosome arm as b. To understand this, one must look at the details of the consequences of mitotic recombination.

A recombination between nonsister chromatids (Fig. 7) produces two possible outcomes depending on how the product nonsister chromatids segregate into the daughter cells. In 50% of cases the two daughter cells have each become homozygous, one for certain alleles on one homolog (C in the figure) and one for those on the other homolog (+). It is important to keep in mind here, as in meiotic recombination, that only the configuration of markers centromere-*distal* to the site of the recombination event is changed. Thus, marker B does not become homozygous in Fig. 7 because it is centromere-proximal to the crossover site. The other 50% of cases produces one daughter cell that is unchanged from the parent

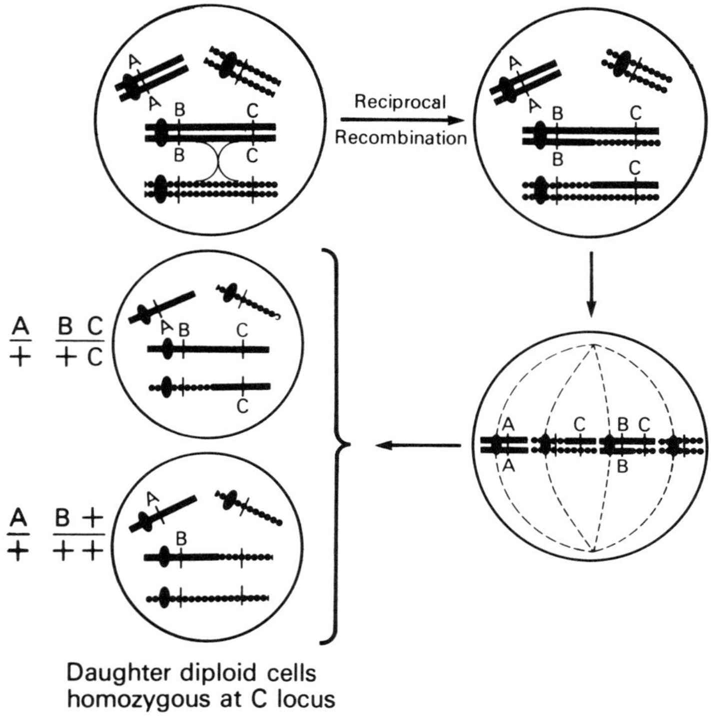

Figure 7. Mitotic recombination can result in homozygosis of genes centromere-distal to the site of the recombination event.

diploid cell and one that is still heterozygous for all markers, but with the linkage relationships reversed at the site of the recombination.

The use of mitotic recombination as a mapping method relies on the fact that, in Fig. 7, for example, marker C signals the homozygosis of all markers distal to marker C. Mitotic gene conversion is an important exception, so that a number of mitotic recombinants for each marker must be examined.

6.3. Aneuploidy Methods

A general method used to locate many of the genes now on the *S. cerevisiae* map relies on the effects of extra chromosomes on the pattern of segregation. As can be seen in Fig. 6, if a normal haploid carrying markers on various chromosomes is crossed with a strain carrying an extra chromosome, markers on the chromosome for which an extra copy is present will not always segregate 2+ : 2−. Instead, the tetrads will be a mixture of 2+ : 2−, 3+ : 1−, and 4+ : 0 ("aneuploid segregation"). The proportions of these tetrad types vary according to the distance from its centromere of the marker in the +/+/− configuration. For a marker very close to its centromere, two-thirds of the tetrads are 4+ : 0 and one-third are 2+ : 2−. For a marker far from its centromere, the majority of tetrads are 3+ : 1−, and a smaller number of 2+ : 2−, 4+ : 0 and even a few 1+ : 3− tetrads will be observed. In contrast, markers on other chromosomes will be segregating 2+ : 2−.

Use of this phenomenon for mapping new genes relies on the assumption that the extra chromosome(s) is an entire chromosome and not just one arm or other fragment of a chromosome. This is almost always true, but because exceptions are known, a chromosome assignment must always be checked by testing for meiotic linkage.

One approach used is to have, along with the unmapped marker, many standard markers (Mortimer and Hawthorne, 1973). Strains of ploidy varying around $n + n/2$ can be generated by sporulating a triploid strain. Most of the spore clones from a triploid are inviable, but the few survivors are multiply aneuploid and can be mated with a series of strains each carrying the unmapped and several standard markers. If, in a particular cross, the unmapped marker segregates 2+ : 2−, then it is eliminated (provisionally) from each chromosome whose standard marker showed "aneuploid segregation" (4+ : 0, 3+ : 1−, and 2+ : 2−) in that cross. Contrariwise, if the unmapped marker showed aneuploid segregation in the cross, then it is eliminated from each chromosome whose standard marker showed 2+ : 2− segregation.

A variation of this method incorporates the standard markers into the triploid strain, eliminating the need to construct a lot of strains (Wickner, 1979). Another variation uses a set of fairly stable multiply aneuploid strains whose extra chromosomes are known (Hilger *et al.*, 1982). Another variation uses complementing alleles at the locus to be mapped to specifically select a strain that is aneuploid only for the chromosome on which the locus to be mapped resides (Hilger and Mortimer, 1980).

6.4. The *spo11* Mapping Method (Klapholz and Esposito, 1982)

The *spo11* gene is necessary for meiotic recombination. Diploids that are homozygous for a defect in this gene undergo a defective meiosis, most of whose products are inviable. Most of those which are viable are haploids but contain an assortment of parental *unrecombined* chromosomes. This phenomenon can be used to map a gene. A *spo11 can1* (or *cyh2*) strain carrying the mutation to be mapped is mated with a series of *spo11* CAN^+ CYH^+ tester strains carrying markers on known chromosomes, and the diploids are sporulated. The rare viable products of meiosis are selected using the *can1* (recessive resistance to canavanine) or *cyh2* (recessive resistance to cycloheximide) marker. These are then scored for the unmapped marker and the standard marker. In this type of cross, because of the *spo11* defect, two markers anywhere on the same chromosome show nearly perfect linkage to each other even if they would be unlinked in a normal cross. Analysis of about 200 spore clones usually suffices to locate on which chromosome a marker is located. A clone of a gene or other chromosomal DNA sequence can likewise be located by putting it in a *URA3*-carrying integrating vector (e.g., YIp5), integrating it by homology with the insert and then mapping the *URA3* gene on the vector. The *spo11* strains all carry a nonreverting allele of *ura3* for this purpose. Note that it must first be checked that the plasmid integrated by its homology with the unmapped gene. The most commonly used *ura3* hosts are not deleted for *URA3*, and integration can also occur there.

6.5. Chromosome-Loss Methods

The loss of chromosomes from a diploid strain can be induced by growth of the strain in the presence of methyl benzimidazole-2-y1-carbamate (MCB) (Wood, 1982). Likewise, treating a *rad52/rad52* diploid strain with x-rays or methylmethanesulfonate (MMS) induces loss of chromosomes (Mortimer *et al.*, 1981). Diploid strains homozygous for *chl1* (*ch*romosome *l*oss) undergo spontaneous, nonrandom chromosome loss that could be used for mapping, but there are some chromosomes which are virtually never lost, limiting the usefulness of this method (Liras *et al.*, 1978). Diploids homozygous for the temperature-sensitive cell division cycle mutants *cdc6* or *cdc14* undergo frequent loss of chromosomes when maintained for several hours at the nonpermissive temperature (Kawasaki, 1979). All chromosomes tested are lost with some frequency, and this method has been used to map several genes. The logic here is the same as for the *spo11* method, except that the degree of haploidy achieved is not usually complete. If, for example, the recessive allele of the unmapped marker is on the same chromosome as a particular standard marker, these two markers will show virtually complete linkage.

These chromosome-loss methods and the *spo11* method represent a substantial advance in mapping methods. They should make genetic mapping a routine part of the characterization of any new mutant.

6.6. 2-μm DNA Mapping (Falco and Botstein, 1983)

If an integration vector carries a part of the 2-μm DNA plasmid that does not include its origin of replication, it remains unable to replicate unless it integrates into the chromosomal DNA. Such a vector can be used to map a cloned sequence. The vector integrates at the chromosomal site homologous to the cloned sequence. For some as yet unknown reason, this makes the chromosomal arm carrying the integrated segment of 2-μm DNA unstable. When the haploid strain carrying the 2-μm DNA integrated at the site of the unmapped cloned sequence is mated with a normal multiply marked haploid strain, the 2-μm DNA-carrying chromosome arm is lost at rather high frequency. This can occur by mitotic recombination or by chromosome loss. This can be used to map the cloned sequence even if one does not have a mutant for the gene cloned. Of course, as described earlier, the *spo11* method can also be used for this purpose.

Loss of the entire chromosome represents only a minority of these events, but this minority class of events makes it possible to map a mutation located on the opposite side of the centromere from the standard marker for that chromosome.

6.7. Overall Mapping Strategies

All mutants to be mapped should first be tested for centromere linkage. For those which are not centromere-linked, two overall strategies are feasible. As the yeast map is becoming more well defined, with fewer uncharted areas, the brute force meiotic mapping approach with multiply marked strains (see Section 6.1) is increasingly attractive for a typical mutant. For mutations selected as suppressors of other mutations, this is more difficult because it involves reconstructing many strains to introduce the suppressed mutant. Of course, any method used will involve some of this.

The other alternative for initial mapping is to use one of the methods described earlier to locate on which chromosome the mutant is located.

For mapping cloned fragments, the preceding method (Gaber *et al.*'s strains are all *leu2*$^-$), the *spo11* method (Klapholz and Esposito's strains are all *ura3*$^-$), and the 2-μm mapping method (Falco and Botstein, 1983) have all been used with success.

6.8. The Genetic Map

Genetic linkage studies have defined 17 chromosomes for *S. cerevisiae* (Fig. 8; reviewed by Mortimer and Schild, 1985). Of these, chromosomes I–XVI each have many genes and have even been demonstrated biochemically as distinct DNA molecules (Carle and Olson, 1985). Chromosome XVII is defined by a single tightly centromere-linked gene, *KRB1*, whose unusual genetic properties suggest the possibility of an unusual structure (Wickner and Leibowitz, 1977; Wickner *et al.*, 1983).

The classes of mutants available include those affecting nearly all aspects of the yeast life cycle and metabolism (see examples in Table III). The primary source for standard markers for mapping purposes is the Yeast Genetic Stock Center (YGSC), Department of Biophysics and Medical Physics, University of California, Berkeley, California 94720, supervised by R. K. Mortimer, whose work has established most of the map. While many specialized mutant types are available from YGSC, some others must be obtained from individual investigators. YGSC also carries special strains for mapping, as hosts for transformation, and as tools for special types of genetic study.

7. NON-MENDELIAN GENETICS

Most phenotypic differences between yeast strains are determined by chromosomal genes. If two haploid strains, A and B, differ in some trait that is determined by a single chromosomal gene difference, then mating these two strains and inducing meiosis in the diploid strain formed generally produces two spore clones with the phenotype of one parent and two with the phenotype of the other parent, that is, 2 : 2 segregation in meiosis.

A nonchromosomal gene, however, if present in high copy number in one parent and absent from the other parent, will be present in high copy number in the diploid cells formed by mating the two parents, and all four of the spores formed by meiosis will get one or more copies of this gene. Thus, instead of the 2+ : 2− segregation seen with the chromosomally determined phenotypic difference, a phenotypic difference between the parents determined by a nonchromosomal gene generally produces 4+ : 0 segregation in meiosis, with all of the spore clones showing the phenotype determined by possession of the nonchromosomal gene. Often even a low copy number plasmid will be inherited by all the offspring because of a specific segregation mechanism. On the other hand, some high-copy-number plasmids can be unstably inherited because of the absence of a proper segregation mechanism. This is the case for many 2-μm DNA-based and ARS-containing cloning vectors (Kikuchi, 1983).

A second method to distinguish chromosomal from nonchromosomal genes is mitotic segregation. The diploid strain formed between A and B, if grown mitotically, will generally not show segregation of the parental phenotypic difference if that difference is chromosomally determined. Both chromosomal alleles are transmitted to both daughter cells at each generation, and the dominant allele will control the phenotype. If A has a nonchromosomal genome and B lacks that genome completely, the A × B diploids will all have the phenotype of strain A. But if A and B both have

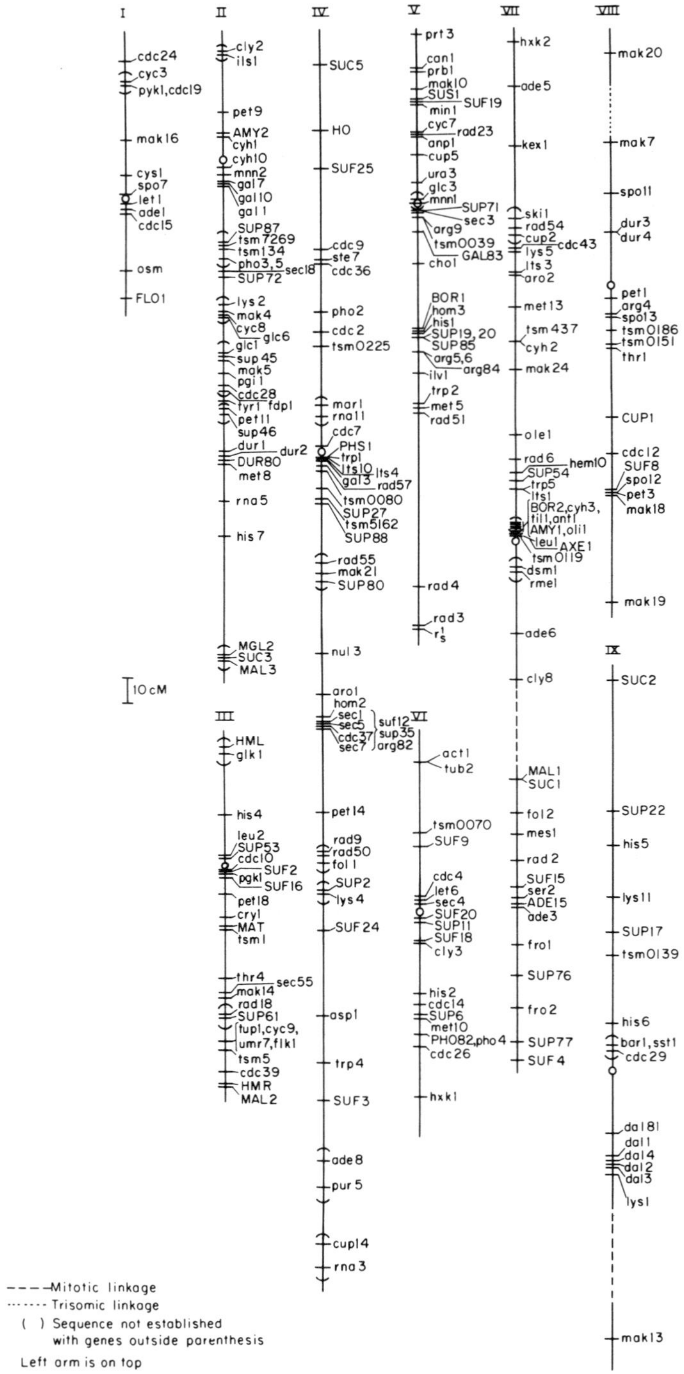

Figure 8. The genetic map of *S. cerevisiae* (reviewed by Mortimer and Schild, 1985).

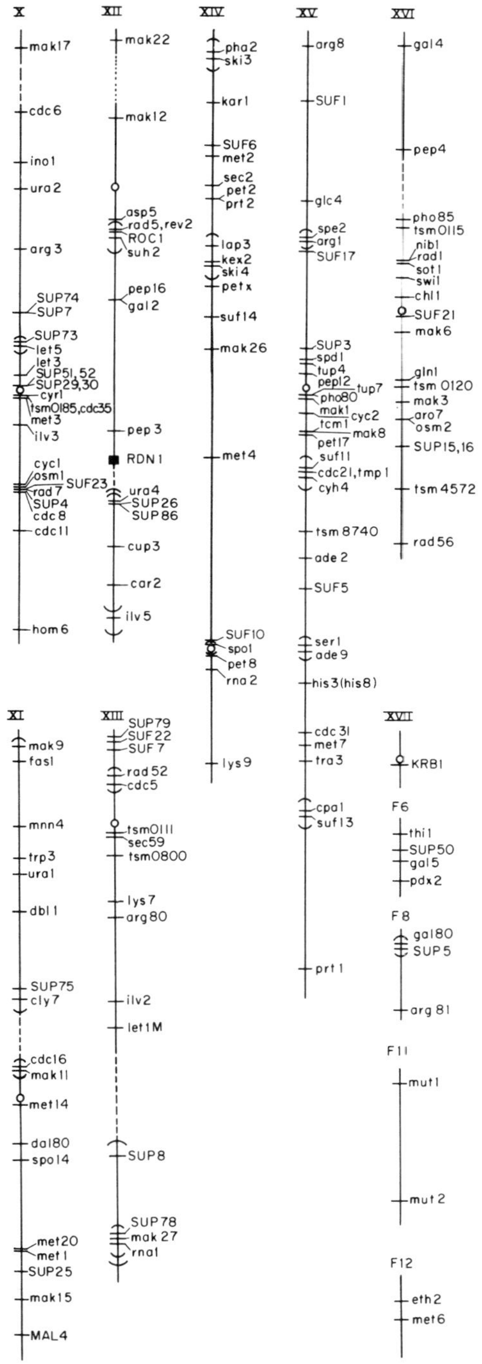

Figure 8. *(continued)*

Table III. Examples of Types of Yeast Mutants Available[a]

Auxotrophs—adenine, arginine, histidine, leucine, tryptophan, tyrosine, phenylalanine, threonine, homoserine, aspartate, isoleucine–valine, lysine, methionine, uracil, inositol, thiamine, cysteine, fatty acids, dTMP	Chromosomal mutants affecting mitochondrial function (petite)
Fermentation—sucrose, maltose, melibiose, α-methylglucoside, galactose, glycolytic pathway	Acid and alkaline phosphatase structural and regulatory mutants
General control of amino acid biosynthesis	Recombinant-deficient and chemical mutagen-sensitive
Amber and ocher suppressible alleles and suppressors	Radiation-sensitive cAMP metabolism
Nitrogen assimilation	Sporulation-deficient
Mannan mutants	Steroid synthesis
"Killer" (dsRNA replication and expression) mutants (nuclear and cytoplasmic)	Secretion-defective
Permeability mutants	Heme synthesis
Drug resistance (nuclear and mitochrondrial)	Nuclear fusion-defective
Proteinase-deficient mutants	Nucleotide synthesis
Mitochrondrial genome mutants	Sterile (nonmating)
	Mating-type interconversion system
	Protein synthesis
	Cell division cycle
	Cell lysis, osmotic-sensitive RNA synthesis
	Flocculation, smooth colony
	Low temperature-sensitive
	Glycogen and glucosamine accumulation

[a]Many of these strains of available from the YGSC. Others must be obtained from the individual investigators. YGSC publishes a useful catalog of strains that includes helpful hints, a map, references, and other information.

the cytoplasmic genome, and that genome is differently marked in A than in B, then, on mitotic growth of the A × B diploids, eventually some cells will get only the genome from A while others will get only that from B, and the two phenotypes will be seen to *segregate in mitosis*. This phenomenon is due to the absence of a regular segregation mechanism and to the statistical fluctuations in the copies of the genome from A vs. the copies from B. Although mitotic recombination does produce segregation (homozygosis) of chromosomal heteroalleles, this process is rare and produces only a fraction of a percent of homozygous progeny in the population. Mitotic segregation of nonchromosomal genes generally results in virtually the entire population becoming homozygous. The native 2-μm DNA does have a regular segregation mechanism, and so differently marked molecules do not necessarily segregate even after extremely prolonged growth (Livingston, 1977).

The third method for distinguishing chromosomal from nonchromosomal genes uses a mutant described by Conde and Fink (1976) which is defective for nuclear fusion. Normally, when two haploid cells mate, the cellular fusion is quickly followed by nuclear fusion to form a diploid cell. But if one of the parent cells carries the *kar1* mutation, its nuclei are somehow not able to fuse efficiently. The cells proceed to divide, and since the two nuclei did not fuse, they are separated into separate daughter cells. These daughter cells thus each have the same haploid nuclei as the two parents, but since cytoplasmic mixing occurred at the binucleate (or heterokaryon) stage, a cytoplasmic genome present in only one of the parent cells is present in both daughter cells (Fig. 9). The transfer of a phenotype, in a cross of this kind (called "cytoduction" or a "*kar* cross") from cells with one nucleus to cells with the other nucleus, is evidence that the phenotype is determined by a nonchromosomal gene.

A fourth method relies on the fact that most nonchromosomal genomes can be eliminated (or "cured") by one or another growth condition

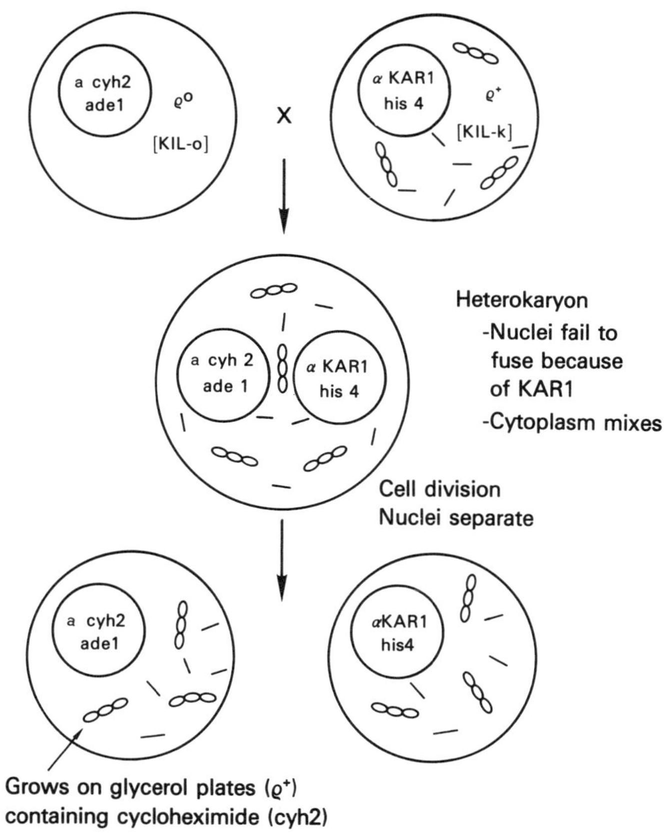

Figure 9. The *kar1-1* mutation results in failure of nuclei to fuse after cell mating. The cytoplasm of both parents mix at the heterokaryon stage, and the daughter cells, each carrying one of the parental nuclei, have a mixture of the parental cytoplasms.

or chemical treatment. The mitochondrial DNA (ρ) is eliminated by growth in the presence of 50 μg/ml of ethidium bromide (Goldring *et al.*, 1970). L-A and M double-stranded RNAs (dsRNAs) are eliminated by growth at elevated temperature (37–39°C) (Wickner, 1974; Sommer and Wickner, 1982), and M is eliminated by growth in the presence of low concentrations of cycloheximide (Fink and Styles, 1972). [PSI] (see below) is eliminated by growth in high osmotic strength media (Singh *et al.*, 1979) or in the presence of 5 mM guanidine hydrochloride (Tuite *et al.*, 1981). The curing method of distinguishing chromosomal from non-chromosomal genes is limited by the fact that curing methods are initially hard to find and are not available for all known cytoplasmic elements.

These four methods of distinguishing chromosomal from non-chromosomal traits are summarized in Table IV.

The molecules and phenotypes showing non-Mendelian (non-chromosomal) inheritance in *S. cerevisiae* include mitochondrial DNA (respiration and resistance to choramphenicol, erythromycin, antimycin, and oligomycin), various dsRNAs [the killer phenotypes, [HOK], [NEX], [EXL] (defined in Table VI)], 2-μm DNA (see also Chapter 5), [PSI] (increased ocher suppression by chromosomal ocher suppressors), [URE3] (ability to use ureidosuccinate to bypass a *ura2* mutation in the presence of ammonia), and 20S RNA (production of 20S RNA on transfer to acetate medium). The genomes responsible for [URE3] and 20S RNA are not yet known. Recent evidence indicates that [PSI] is carried on

Table IV. Chromosomal vs. Nonchromosomal Traits

Method	Result for: Single chromosomal gene	Result for: Nonchromosomal gene	Exceptions
Meiotic segregation	2+ : 2−	4+ : 0	[URE3]
Mitotic segregation	+/− heterozygosity stably maintained[a]	+ and − segregate[b]	If "−" means no genome, the "+" is stable
Cytoduction (*kar* crosses)	Not transferred[c]	Efficiently transferred	2-μm DNA is transferred with 50% efficiency
Curing	Rarely mutated	Mutated at close to 100% efficiency	No method known for URE3, 20S RNA, T, W, and L-(BC)

[a]Rarely, mitotic recombination results in separation of + and − alleles.
[b]Mitotic segregation requires that the parents each have the same cytoplasmic genome, but that that genome be marked differently, "+" in one parent and "−" in the other.
[c]Rarely, chromosomal markers are transferred.

3-μm DNA, the ribosomal DNA repeat sequence present as a covalent circle in most strains. It is presumably generated by recombination between two tandemly repeated copies of the chromosomal ribosomal DNA repeats. We will discuss some aspects of the mitochondrial and killer systems.

7.1. The Mitochondrial Genome

The mitochondrion is a complex organelle whose main function is to produce ATP by oxidative phosphorylation. Of the many mitochondrial proteins, it is estimated that 95% are encoded by nuclear genes (called *pet* genes), and only 5% by the mitochondrial DNA. Mitochondrial DNA (mitDNA or ρ) is a double-stranded circular molecule about 70 kb in length. Each haploid cell has about 50 copies of mitDNA. Growth of ρ^+ cells in the presence of high levels of ethidium or acridines results in rapid and complete loss of mitDNA to produce ρ° cells. $\rho^+ \times \rho^\circ$ crosses yield 4 ρ^+ : 0 segregation in meiosis. Oligomycin blocks respiration by acting on the mitochondrial ATPase complex, two subunits of which are encoded by mitochondrial DNA. Oligomycin-resistant (O^R) mutants include mutants in each of these two mitochondrial genes. In a cross of the type $O^R \times O^S$ (oligomycin-sensitive), the diploids formed show rapid segregation of O^R from O^S. ρ can also be transferred by *kar* crosses (cytoduction). Mating a ρ^+ *kar1* strain and a ρ° strain results in ρ^+ cytoductants.

The mitochondrial genome has been studied in great detail (for excellent reviews, see Dujon, 1981, and Slonimski *et al.*, 1982). Hundreds of mutants have been analyzed, and the entire genome has been sequenced. Figure 10 shows a map of the mitochondrial DNA, and Table V lists the genes. The mitochondrion has its own protein synthesis apparatus with different ribosomes, tRNA, and other elements, quite distinct from the usual cytoplasmic ribosome system. The large and small rRNA are coded by mitDNA, as are the tRNAs used, but the ribosomal proteins, tRNA charging enzymes, and other protein synthesis factors are coded by nuclear genes. The mitochondrial ribosomes are insensitive to cycloheximide, which acts on the cytoplasmic (nuclear-coded) ribosomes, but they are sensitive to many of the antibiotics that affect bacterial ribosomes, such as chloramphenicol, erythromycin, and spiramycin. These antibiotics were very useful in distinguishing mitochondrially synthesized from cytoplasmically synthesized proteins. Also, mutants resistant to these drugs played a large role in the early studies of mitochondrial genetics.

While many mature proteins synthesized in the cytoplasm are imported into the mitochondria, as far as is known, no nuclear-encoded messenger RNAs (mRNAs) are translated on mitochondrial ribosomes,

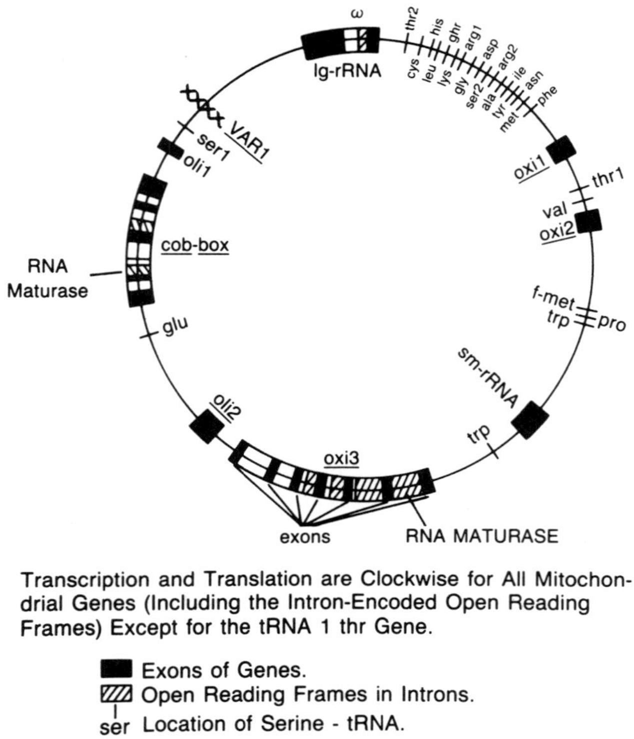

Figure 10. The mitochondrial genome is about 50 kb. The genes are described in more detail in Table V.

and all proteins encoded by mitDNA are translated in the mitochondria. Examination of the sequences encoding those proteins has revealed that mitochondria have a slightly different genetic code than the code that chromosomal genes use. The major difference is that the UGA codon, read as a stop codon by the cytoplasmic ribosomes, is read as tryptophan by the mitochondrial ribosomes. Human mitochondria do the same thing. Also, at least one CUA codon is translated as threonine in mitochondria instead of leucine.

Another striking aspect of mitochondrial DNA is the presence of intervening sequences in three mitochondrial genes: the large rRNA gene, the structural gene for cytochrome b ("*cob-box*" gene), and the gene for subunit I of cytochrome oxidase (*oxi3*). Genetic studies, combined with sequencing and analysis of transcripts and protein products, have

Table V. Mitochondrial Genes

Gene name	Gene product	Phenotype of mutants	Special features
oxi1	Subunit II of cytochrome oxidase	Respiratory-deficient	No intervening sequences
oxi2	Subunit III of cytochrome oxidase	Respiratory-deficient	No intervening sequences
oxi3	Subunit I of cytochrome	Respiratory-deficient	5 introns; long open reading frames in first 4 introns; different strains have more or fewer introns
"*cob-box*"	Cytochrome b	Respiratory-deficient resistance to diuron, antimycin, mucidin, or funiculoson	5 introns; 2 with long open reading frames; ORFs needed to excise intron
oli1	Subunit 9 of ATPase	Phosphorylation-deficient resistance to oligomycin, venturicidin, or ossamycin	No intervening sequences
oli2	Subunit 6 of ATPase	Phosphorylation-deficient resistance to oligomycin	No intervening sequences
var1	Mitoribosomal polypeptide	Variable size of mitoribosomal protein	Part of *var1* product is coded by *var1* locus; part (?) nuclear
"lg-rRNA"	21S rRNA	Resistance to chloramphenicol, erythromycin, or spiromycin; polarity of recombination (ω)	An intervening sequence; (ω^+) induces preferential inheritance of itself
"sm-rRNA"	15S rRNA	Resistance to paromomycin	
"tRNA"	Transfer RNAs	Deficient in mitochondrial protein synthesis	UGA = tryptophan instead of amber
rep1 *rep2* *rep3*	Origins of replication	Extreme suppressiveness when in a ρ^- mutant	Each has a similar 300-bp sequence; but *rep*° ρ^- mutants can still replicate

revealed functions for some of these intervening sequences (introns). Other reading frames have been found in the intron in the large rRNA gene, in two of the five *cob-box* introns, and in four of the seven introns of the *oxi3* gene.

Frameshift and chain-termination mutants in the open reading frames of the second intron of the *cob-box* gene result in failure to excise that very intron from the primary gene transcript. These same intron mutants in the *cob-box* gene fail to express the *oxi3* gene properly, owing to a failure to properly excise introns from the *oxi3* primary transcript. These results indicate that the protein encoded by an intron of the cytochrome b structural gene (*cob-box* gene) plays a necessary role in the excision of introns in both the *cob-box* gene and the *oxi3* gene. An *oxi3* intron also appears to encode an "RNA maturase."

The intron in the gene for the large rRNA is present in some strains (called ω) and absent in others (called w^-). In crosses of the type $\omega^+ \times \omega^-$, almost all of the progeny are ω^+; that is, they have the intervening sequence. This "selfish DNA" role for the intervening sequence is vaguely reminiscent of the "selfish RNA" role that Cech (1983) has suggested for a self-excising intron in the *Tetrahymena* rDNA gene.

Before we leave the mitochondria, mention should be made of the ρ^- or petite mutation, by which Ephrussi *et al.* (1949) first recognized the mitochondrial genome. If a single ρ^+ colony is replated, around 1% of the subclones will be found to be unable to grow using ethanol or glycerol as a carbon source; that is, they will be unable to respire, although they can use dextrose. Most of these ρ^- mutants are due to extensive deletions of mitDNA, leaving only a small sequence which is duplicated in tandem to produce a molecule of about the same size as normal mitDNA.

7.2. The Killer Systems

Certain strains of *S. cerevisiae* secrete a protein toxin that kills other strains, but secreting strains are immune to the action of the toxin. Ability to secrete the toxin (K^+ phenotype) and to resist its lethal effects (R^+ phenotype) is determined by a linear dsRNA called M. In fact, there are several distinct killer toxins (K_1, K_2, etc.) determined by distinct M dsRNAs (M_1, M_2, etc.). The M_1 dsRNA has been shown to code for the protein precursor to the K_1 toxin (for reviews of the killer systems, see Wickner, 1983, 1985, and Tipper and Bostian, 1984. (See Fig. 11.)

In addition to M_1 or M_2, various yeast strains carry a variety of other dsRNAs (Table VI). L-A dsRNA is the only other variety known to be essential for the killer phenomenon, although all killer strains also have the minor species, W. L-A codes for the major coat protein of the intracellular viruslike particles (VLP) in which both L-A and M are found.

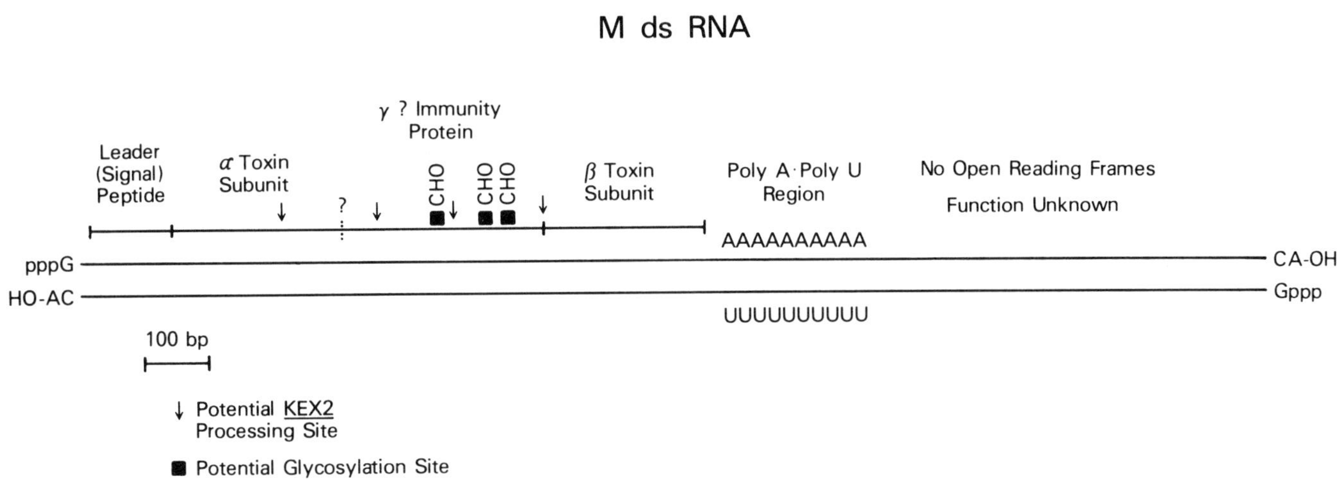

Figure 11. Most of the sequence of M_1 dsRNA is known from a combination of DNA and RNA sequencing.

Table VI. Components of the *S. cerevisiae* Killer Systems

dsRNAs	
M_1	A 1.8-kb dsRNA that codes for polpeptide toxin and a resistance function. Strains carrying M_1 have the genotype [KIL-k_1]. Secretion of an active K_1 toxin is the K_1^+ phenotype. Resistance to the K_1 toxin is the R_1^+ phenotype.
M_2	A 1.5-kb dsRNA that codes for a second toxin which kills M-o- and M_1-containing cells. M_2, [KIL-k_2], K_2^+, and R_2^+ are analogous to M_1, [KIL-k_1], K_1^+, and R_1^+.
L-A	A 4.5-kb dsRNA that is noninfectious and stably maintained like a plasmid, but encapsidated like a virus. It encodes its major capsid protein (81 kDa). The genes [EXL], [NEX], and [HOK] (see below) are present in various combinations on various forms of L-A, denoted L-A-E ([EXL] alone), L-A-HN ([HOK] and [NEX]; found in all K_1 killers), L-A-HE ([HOK] and [EXL]), or L-A-H ([HOK] only; found in certain K_2 strains). L-A depends on MAK3, MAK10, and PET18 for replication.
L-B and L-C	4.5-kb dsRNAs unrelated to L-A and present in VLPs with different major proteins. L-B and L-A or L-C and L-A are compatible. L-B and L-C show some sequence homology. L-(BC) means an L dsRNA of the L-B or L-C type, which, because it has not been fully characterized, could be either.
T and W	2.76-kb and 2.25-kb minor dsRNAs. They do not cross-hybridize with each other, other dsRNAs, or cell DNA. They are cytoplasmically inherited. The copy number of T and W is induced 10-fold by growth at 37°C.
Cytoplasmic genes	
[HOK]	*h*elper *o*f *k*iller. This non-Mendelian gene supplies the helper function needed by M_1 for replication in a wild-type strain. It is located on certain forms of L-A dsRNA, namely, L-A-HN, L-A-HE, and L-A-H. To test for [HOK], the combination of L-A-E + M_1 is introduced from a *ski2* strain into the *SKI*$^+$ M-o strain to be tested. L-A-E cannot support M_1 replication in a *SKI*$^+$ host, but if [HOK] is present, M_1 will be maintained
[EXL]	*excl*uder of M_2 dsRNA. This non-Mendelian trait prevents the replication of [KIL-k_2] if [NEX] is absent, but not if [NEX] is present. [EXL] is located on certain forms of L-A, namely, L-A-E and L-A-HE.
[NEX]	M_2 *nonex*cludable by [EXL], but does not prevent exclusion of M_2 by strains carrying M_1. [NEX] is located on L-A-HN, the form of L-A found in wild-type K_1 killer strains.
[KIL-b]	A frequently encountered natural variant of [KIL-k_1] in which M_1 does not need many of the *MAK* gene products, the copy number of M_1 is higher than usual, the strain is a superkiller, and the degree of repression of L-A copy number by M_1 is greater than usual.
[KIL-d]	A mutant form of [KIL-k_1] in which haploid strains unstably maintain and incompletely express M_1, but diploid strains stably maintain and normally express M_1.
[KIL-sk]	A mutant form of M_1 dsRNA in which toxin stability is increased compared to the parent, resulting in the superkiller phenotype.

Table VI. (*Continued*)

[KIL-i]	A mutant form of M_1 dsRNA in which no active toxin is produced, but cells are immune to toxin action.
[KIL-s]	Deletion mutants of M_1 which prevent the replication of (suppress) the parent M_1 dsRNA. These mutants are analogous to defective interfering particles of animal viruses.
Chromosomal genes	
MAK	*ma*intenance of [*K*IL-k_1]. *MAK* genes comprise at lease 29 chromosomal genes necessary to maintain M_1 dsRNA. Mutants carrying the recessive alleles, *mak*, are K^-R^- M-o. At least some of the genes are also required for the maintenance of M_2. *MAK3, MAK10,* and *PET18* are needed by all forms of L-A. *MAK27* is needed by L-A-E.
clo	A complex chromosomal defect resulting in loss of L-B or L-C.
SKI	*s*uper*ki*ller. Mutants carrying the recessive alleles of *ski2, ski3, ski4, ski6, ski7,* or *ski8* produce more killer toxin and have increased copy number of M_1, M_2, L-A, and L-(BC). ski^- mutations eliminate the need for some of the *MAK* genes and prevent the exclusion of M_2 by L-A-HN in *mkt* strains. The *ski5* mutant has none of these traits and lacks a cell surface protease that normally degrades the toxin.
MKT	*m*aintenance of [KIL-k_2] in the presence of [NEX]. Strains having a recessive allele, *mkt* (about 80% of laboratory strains), cannot maintain [KIL-k_2] at 30°C if [NEX] (L-A-HN) is present.
MKS	*mkt* suppressor. Like most ski^- mutations, mutations in *mks1, mks2,* or MKS50 prevent exclusion of M_2 by L-A-HN in *mkt1* hosts, but do not show other characteristics of ski^- mutations.
KEX	*k*iller *ex*pression. Two chromosomal genes needed to process the toxin precursor. Mutants are K^-R^+. The *kex* mutants also do not properly process the α-hormone precursor, and many other secreted proteins are affected.
REX	*r*esistance *ex*pression. One chromosomal gene needed to express M_1-determined resistance to the toxin.
KRE	*k*iller *re*sistant. Three chromosomal genes needed for normal toxin action on sensitive cells. *KRE1* affects β(1–6)glucan, the normal toxin cell wall receptor. *KRE2* also affects toxin binding, but *KRE3* does not.
SEC	*sec*retion. Chromosomal genes for general protein secretion; mutants in these genes are conditional lethal and also affect killer toxin secretion.

These VLPs are found only inside the cell and are not infectious, so that their status as viruses is questionable. However, their structural resemblance to virus particles and the existence of a particle RNA polymerase producing full-length, message strand, single-stranded RNA (ssRNA) is certainly reminiscent of a virus. L-B and L-C determine different VLP major coat proteins from that determined by L-A, but in normal strains, M requires L-A, and neither L-B nor L-C will suffice.

The T and W dsRNAs are usually present at very low copy number, but when cells are grown at 37°C, their copy number increases 10-fold or more to reach copy numbers similar to that of M. T and W are not present in the same VLPs as are the L-A, L-B, or L-C molecules, but where they are located in the cell is not yet known.

While the study of most viruses and plasmids has centered on the viral or plasmid genome itself, the study of the killer systems has been somewhat more global, with equal emphasis on the chromosomal genes involved in replication, expression, and regulation of the dsRNA genomes. Table VI summarizes the chromosomal genes involved in the killer systems.

Those involved in expression of the killer phenotype are the *kex1* and *kex2* genes and the *sec* genes. The *kex1* and *kex2* genes are relatively specific for the killer toxin, although *kex2* mutations also affect secretion of α pheromone and prevent meiotic spore formation. *KEX2* is now known to encode a protease that cleaves after a pair of basic residues. The *sec* mutants are defective in various stages of general protein secretion and are all temperature-sensitive (ts) for growth. The secretion of the killer toxin apparently follows the general protein secretory pathway, but, for some reason, needs two extra genes, *kex1* and *kex2*.

Resistance to the toxin produced by M dsRNA requires, in addition, the *REX1* product. The *KRE* products are involved in normal toxin action. *kre1* mutants lack the toxin receptor—β(1-6)glucan—in the cell wall.

The mak^- mutants (*ma*intenance of *k*iller) lose M_1 dsRNA. Thirty *MAK* genes have been identified and mapped. Biochemical components shown by these mutations to be involved in M replication are the polyamines spermidine and spermine, ribosomal protein L3 (*MAK8*), and topoisomerase I (*MAK1*). How these components are involved in the replication process is unknown. Some *MAK* genes are essential for cell growth (*MAK16, PET18, MAK30*); one (*PET18*) is essential as well for mitDNA replication. The *MAK3, MAK10,* and *PET18* genes are needed for L-A replication as well as M replication. Since M requires L-A, perhaps M's requirement for these three chromosomal genes is via its requirement for L-A. The *MKT* genes are needed for M_2 (but not M_1) if L-A-HN, a particular form of L-A, is present. As may be seen in Table VII, different dsRNAs have very different requirements for chromosomal genes for their replication.

The SKI gene products lower the levels of L-(BC), L-A, and M dsRNAs. The ski^- mutants were first identified by their *s*uper*ki*ller phenotypes, but they have several other phenotypes. The ski^- mutations suppress many mak^- mutations as well as the mkt^- mutations, and if M is present, ski^- mutants are cold-sensitive (cs) for growth. This may be due to the elevated amounts of M dsRNA utilizing all of some cell component needed for growth at the low temperature.

Table VII. Different Chromosomal Gene Requirements for dsRNAs

dsRNAs	*MAK3*	*MAK10*	*PET18*[a]	*MAK1*	*MAK8*	*SPE2*	*SPE10*	*MAK27*	*MKT1*	*CLO*
M_1	+[b]	+	+	+	+	+	+	+	+	–
M_2	+	+	+	+	+	+	+	+	+	–
L-A-HN	+	+	+	–	–	–	–	–	–	–
L-A-HE	+	+	+	–	–	–	–	–	–	–
L-A-E	+	+	+	–	–	–	+	+	–	–
L-B	–[c]	–	–	–	–[c]		–[c]	–[c]	–[c]	+
L-C		–	–	–						+
T	–	–	–	–	–			–	–	–
W	–	–	–	–	–			–	–	–

[a]Mitochondrial DNA also requires *PET18* for its replication.
[b]+ = gene required for replication or maintenance; – = gene not required for replication or maintenance.
[c]Requirement determined for an undifferentiated member of the L-B and L-C class of molecules [L-(BC)].

A number of dsRNA exclusion phenomena have been described. M_1 excludes M_2; L-A-E (a form of L-A) excludes M_2 if L-A-HN is absent; L-A-HN excludes M_2 if the host is *mkt*$^-$; and deletion mutants of M_1 (S dsRNAs) exclude the wild-type M_1 dsRNA. Another interaction among the dsRNAs is the ability of M_1 or M_2 to lower the copy number of L-A 10-fold or more. The basis of these interactions is unclear but could involve competition among dsRNAs for replication proteins.

7.3. 2-μm DNA (Reviewed by Broach, 1981)

2-μm DNA is a 6300-bp, circular, double-stranded molecule present in most *Saccharomyces* strains at 50–100 copies per cell. It has a pair of 599 inverted repeat sequences, frequent recombination between which is catalyzed by a 2-μm DNA-coded, site-specific recombinase called FLP. Thus, two intracellular forms of 2-μm DNA are in equilibrium. 2-μm DNA also encodes two complementable replication functions and a site (called STB) needed for proper segregation. Crossing strains with and without 2-μm DNA results in 4 : 0 segregation, but 2-μm DNA is only transmitted at a 50% rate by cytoduction. The DNA is found in histone-containing chromatin, and each molecule replicates exactly once per cell cycle (in S phase), making it seem likely that 2-μm DNA is actually located in the nucleus. 2-μm DNA acts like a multicopy minichromosome, with an origin of replication, a centromere, chromatin structure, and cell-cycle–controlled replication. It has no essential function for the cell. 2-μm DNA can be cured by transformation with pJDB219, a *LEU2*-carrying vector which is itself unstable. After pJDB219 is lost, the cells remain healthy.

The origin of replication of 2-μm DNA is used in YEp vectors (see Chapter 5).

8. APPENDIX I: MEDIA

To prepare solid media, the ingredients listed are autoclaved together for 15 min, except as indicated below, and, after cooling to 50–60°C, poured into Petri plates. One liter of medium is autoclaved in a 2-liter flask. About 30 plates are poured per liter of medium. Plates are left to dry at room temperature for a few days and are then stored at 4°C in the plastic bags that the empty Petri dishes came in. The plates are good indefinitely unless they dry out or become contaminated.

YPAD-Rich Medium

1% yeast extract

2% peptone

2% dextrose

2% agar

0.04% adenine sulfate (prevents *ade1* and *ade2* mutants from accumulating red pigment and dying)

SD Synthetic Minimal Medium with Dextrose

0.67% yeast nitrogen base without amino acids (available from Difco or from Gibco)

2% dextrose

2% agar

YPAG Glycerol Medium—Only Respiration-Competent Cells Can Grow

1% yeast extract

2% peptone

3% vol/vol glycerol

2% agar

0.04% adenine sulfate

H Synthetic Complete Medium

0.04% adenine sulfate

0.67% yeast nitrogen base without amino acids

2% dextrose

2% agar

10 ml/liter 1 M K_2HPO_4

50 ml/liter of appropriate mix

Mix Composition

Uracil	0.5 g/liter
L-Tryptophan	0.5 g/liter
L-Histine	0.5 g/liter
L-Arginine	0.5 g/liter
L-Methionine	0.5 g/liter
L-Tyrosine	0.7 g/liter
L-Isoleucine	0.7 g/liter
L-Lysine	0.7 g/liter
L-Phenylalanine	1.2 g/liter
L-Valine	3.0 g/liter

After autoclaving, add 5 ml of sterile 5% L-threonine, 5 ml of 2% sodium aspartate, and 5 ml of 1% L-leucine per liter of medium.

Mixes with the above composition lacking one or more nutrients are prepared and kept at room temperature. The author uses H-ade, H-ura, H-trp, H-his, H-arg, H-tyr, H-lys, H-phe, H-ilv (lacking isoleucine *and* valine), and H-MAT (lacking methionine, aspartate, and threonine). The H-MAT plates make scoring of most methionine, threonine, and aspartate auxotrophs more clear-cut than if only the single amino acid was left out.

Sporulation Medium

1% potassium acetate
0.1% yeast extract
0.05% dextrose
2% agar

9. APPENDIX II: SAMPLE TETRAD DATA

These data were collected from a cross of the following parent strains:

$$1653\text{-}12C = MAT\ \alpha\ trp1\ [\text{KIL-k}_1]$$
$$\times$$
$$2011\text{-}33A = MAT\ a\ mak21\ aro1D\ pet19\ ade8\ rna3\ [\text{KIL-o}]$$

mak21 results in loss of M_1 dsRNA and thus the K_1^- phenotype. *aro1D* results in a requirement for both tryptophan and phenylalanine, while *trp1* strains need only tryptophan. To score *trp1,* all segregants were mated with *a trp1* and α *trp1* tester strains. The formation of prototrophic diploids is indicated by a "+" in the appropriate column and implies that the spore clone was $TRP1^+$. The *pet19* segregants are unable to use glycerol as a carbon source and thus cannot grow on YPG plates. The *ade8* mutation makes cells auxotrophic for adenine. The *rna3* mutation results in temperature sensitivity for cell growth. 37°C is the nonpermissive temperature, and growth on YPAD plates at 37°C is indicated under the heading "37°." Mating type is scored by mating the spore clones on YPAD plates with standard *a ura4* and α *ura4* tester strains. Ability to mate with the *a ura4* strain is seen as the formation of prototrophic diploids (a "+" in the column headed "× *a* tester") and implies that the spore clone tested has the α mating type.

With the data shown here, the reader can construct a partial map of chromosome IV. What can be said about the location of the centromere? Of the mating type locus? Hint: the data have been arranged in a rather suggestive way. Look at the segregation of individual tetrads.

Tetrad number		−*trp*	K_1	−*phe*	YPG	−*ade*	37°	× *a* tester	× α tester	× *a trpl* tester	× α *trp1* tester
1	A	−	−	+	+	−	+	+	−	−	−
	B	−	−	−	−	+	−	−	+	−	+
	C	+	+	+	+	+	+	+	+	+	+
	D	−	+	−	−	−	−	−	+	−	−
2	A	−	−	−	−	−	−	−	+	−	+
	B	−	−	+	+	+	+	+	−	−	−
	C	−	+	−	−	−	−	−	+	−	+
	D	−	+	+	+	+	+	+	−	−	−
3	A	+	−	+	−	+	+	−	+	−	+
	B	−	−	−	+	+	−	+	−	−	−
	C	−	+	−	−	−	−	+	−	−	−
	D	+	+	+	+	−	+	−	+	−	+
4	A	−	−	−	−	−	−	−	+	−	−
	B	+	−	+	+	+	+	+	−	+	−
	C	−	+	−	−	−	+	+	−	+	−
	D	−	+	+	+	+	−	−	+	−	−
5	A	−	−	+	−	−	−	−	+	−	−
	B	−	−	−	+	−	−	+	−	+	−
	C	−	+	−	−	+	+	−	+	−	−
	D	+	+	+	+	+	+	+	+	+	+
6	A	−	+	−	−	−	−	−	+	−	+
	B	−	−	−	−	+	+	−	+	−	+
	C	−	−	+	+	−	−	+	−	−	−
	D	−	+	+	+	+	+	+	−	−	−
7	A	−	−	−	+	−	+	−	+	−	+
	B	−	−	−	−	+	−	+	−	−	−
	C	+	+	+	−	−	−	−	+	−	+
	D	−	+	+	+	+	+	+	−	−	−
8	A	−	−	−	−	−	−	−	+	−	+
	B	−	+	−	−	−	−	−	+	−	+
	C	−	−	+	+	+	+	+	−	−	−
	D	−	+	+	+	+	+	+	−	−	−
9	A	−	−	−	−	−	+	−	+	−	−
	B	+	−	+	+	−	−	+	−	+	−
	C	−	+	−	−	+	+	−	+	−	−
	D	+	+	+	+	+	−	+	+	+	+
10	A	−	−	−	−	+	+	+	−	−	−
	B	+	−	+	+	−	+	−	+	−	+
	C	−	+	+	+	+	−	+	−	−	−
	D	−	+	−	−	−	−	−	+	−	+

(continued)

Tetrad number		−*trp*	K_1	−*phe*	YPG	−*ade*	37°	× *a* tester	× α tester	× *a trpl* tester	× α *trp1* tester
11	A	−	−	−	−	−	−	−	+	−	−
	B	+	+	+	−	−	+	+	−	+	−
	C	−	−	−	+	+	−	+	−	+	−
	D	−	+	+	+	+	+	−	+	−	−
12	A	−	−	−	−	−	−	+	−	−	−
	B	+	−	+	+	+	+	−	+	−	+
	C	−	+	−	−	−	−	+	−	−	−
	D	+	+	+	+	+	+	−	+	−	+
13	A	−	−	−	−	−	−	+	−	+	−
	B	−	−	+	+	−	+	−	+	−	−
	C	−	+	−	−	+	−	+	−	−	−
	D	+	+	+	+	+	+	−	+	−	+
14	A	−	−	+	−	−	−	+	−	−	−
	B	−	+	−	−	+	+	−	+	−	+
	C	+	−	+	+	−	+	+	−	+	−
	D	−	+	−	+	+	−	−	+	−	−
15	A	+	−	+	+	+	+	+	+	+	+
	B	−	−	−	−	+	+	−	+	−	+
	C	−	+	−	−	−	−	−	+	−	−
	D	−	+	+	+	−	−	+	−	−	−
16	A	−	−	−	−	−	−	−	+	−	+
	B	−	−	−	−	+	+	+	−	−	−
	C	+	+	+	+	−	−	−	+	−	+
	D	−	+	+	+	+	+	+	−	−	−
17	A	−	−	−	−	−	−	−	+	−	+
	B	−	−	−	+	+	+	−	+	−	+
	C	−	+	+	+	−	−	+	−	−	−
	D	−	+	+	−	+	+	+	−	−	−
18	A	−	−	−	−	−	+	+	−	+	−
	B	−	−	−	−	+	−	+	−	−	−
	C	+	+	+	+	+	+	+	+	+	+
	D	−	+	+	+	−	−	−	+	−	−
19	A	−	−	−	−	+	+	+	−	−	−
	B	−	−	−	+	−	−	−	+	−	+
	C	−	+	+	−	+	+	+	−	−	−
	D	+	+	+	+	−	−	−	+	−	+
20	A	−	−	−	−	−	−	−	+	−	+
	B	−	−	−	−	+	+	−	+	−	−
	C	−	+	+	+	−	−	+	−	−	−
	D	+	+	+	+	+	+	+	+	+	+
21	A	+	−	−	−	−	−	−	+	−	+
	B	−	−	−	+	+	−	−	+	−	+
	C	−	+	+	−	−	+	+	−	−	−
	D	−	+	+	+	+	+	+	−	−	−
22	A	−	−	−	−	+	+	+	−	−	−

Tetrad number		−*trp*	K_1	−*phe*	YPG	−*ade*	37°	× *a* tester	× α tester	× *a trpl* tester	× α *trp1* tester
	B	+	+	+	−	+	+	−	+	−	+
	C	+	−	+	+	−	−	−	+	−	+
	D	−	+	−	+	−	−	+	−	−	−
23	A	+	−	+	+	+	+	−	+	−	+
	B	−	−	+	−	+	−	+	−	−	−
	C	−	+	−	−	−	−	+	−	+	−
	D	−	+	−	+	−	+	−	+	−	−
24	A	−	−	−	−	+	+	−	+	−	+
	B	−	−	−	+	−	+	−	+	−	+
	C	−	+	+	−	−	−	+	−	−	−
	D	−	+	+	+	+	−	+	−	−	−
25	A	−	−	−	−	−	+	−	+	−	−
	B	+	+	+	−	−	−	+	−	+	−
	C	−	−	−	+	+	+	−	+	−	+
	D	−	+	+	+	+	−	+	−	−	−
26	A	−	−	−	−	−	−	+	−	+	−
	B	−	+	+	−	+	+	−	+	−	−
	C	−	−	−	+	−	−	+	−	−	−
	D	+	+	+	+	+	+	−	+	−	+
27	A	−	−	−	−	+	−	−	+	−	+
	B	+	+	+	−	−	+	+	−	+	−
	C	−	−	−	+	−	−	+	−	−	−
	D	−	+	+	+	+	+	−	+	−	−
28	A	−	−	−	−	+	+	−	+	−	+
	B	−	−	−	−	−	−	+	−	−	−
	C	−	+	+	+	+	−	−	+	−	−
	D	+	+	+	+	−	+	+	−	+	−
29	A	−	+	−	−	−	+	+	−	+	−
	B	−	+	−	−	+	+	−	+	−	−
	C	+	−	+	+	−	−	+	−	+	−
	D	−	+	+	+	+	−	−	+	−	−
30	A	−	−	−	−	−	+	−	+	−	−
	B	−	−	−	+	−	−	+	−	+	−
	C	+	+	+	+	+	−	+	+	+	+
	D	−	+	+	−	+	+	−	+	−	−
31	A	−	+	−	−	−	−	+	−	−	−
	B	−	−	−	−	+	−	−	+	−	+
	C	+	−	+	+	−	+	−	+	−	+
	D	−	+	+	+	+	+	+	−	−	−
32	A	+	−	+	+	−	−	−	+	−	+
	B	−	−	−	−	−	−	+	−	+	−
	C	−	+	−	−	+	+	−	+	−	−
	D	−	+	+	+	+	+	+	−	−	−

(continued)

Tetrad number		−*trp*	K_1	−*phe*	YPG	−*ade*	37°	× *a* tester	× α tester	× *a trpl* tester	× α *trp1* tester
33	A	+	−	+	+	+	+	+	−	+	−
	B	−	+	−	−	−	−	+	−	+	−
	C	−	−	−	+	+	+	−	+	−	−
	D	−	+	+	−	−	−	−	+	−	−
34	A	−	−	−	−	−	+	−	+	−	−
	B	+	+	+	−	+	+	−	+	−	+
	C	−	−	−	+	+	−	+	−	−	−
	D	+	+	+	+	−	−	+	−	+	−
35	A	+	−	+	+	+	+	+	+	+	+
	B	−	−	−	−	−	−	−	+	−	−
	C	−	+	−	+	−	−	+	−	−	−
	D	+	+	+	−	+	+	−	+	−	+
36	A	−	−	−	−	+	+	+	−	+	−
	B	−	+	−	−	−	−	−	+	−	−
	C	+	+	+	+	−	−	+	−	+	−
	D	−	−	+	+	+	+	−	+	−	−

REFERENCES

Broach, J. R., 1981, The yeast plasmid 2 μ circle, in: *The Molecular Biology of the Yeast Saccharomyces: Life Cycle and Inheritance* (J. N. Strathern, E. W. Jones, and J. R. Broach, eds.), Cold Spring Harbor Laboratory, Cold Spring Harbor, NY, pp. 445–470.

Carle, G. F., and Olson, M. V., 1985, An electrophoretic karyotype for yeast, *Proc. Natl. Acad. Sci. USA* **82:**3756–3760.

Cech, T. R., 1983, RNA splicing: Three themes with variations, *Cell* **34:**713–716.

Conde, J., and Fink, G. R., 1976, A mutant of *Saccharomyces cerevisiae* defective for nuclear fusion, *Proc. Natl. Acad. Sci. USA* **73:**3651–3654.

Culbertson, M. R., and Henry, S. A., 1973, Genetic analysis of hybrid strains trisomic for the chromosome containing a fatty acid synthetase gene complex (*fas1*) in yeast, *Genetics* **75:**441–458.

DeFeo-Jones, D., Scolnick, E. M., Koller, R., and Dhar, R., 1983, *ras*-Related gene sequences identified and isolated from *Saccharomyces cerevisiae, Nature* **306:**707–709.

Dujon, B., 1981, Mitochondrial genetics and functions, in: *The Molecular Biology of the Yeast Saccharomyces: Life Cycle and Inheritance* (J. N. Strathern, E. W. Jones, and J. R. Broach, eds.), Cold Spring Harbor Laboratory, Cold Spring Harbor, NY, pp. 505–635.

Ephrussi, B., Hottinguer, H., and Tavlitzki, J., 1949, Action de l'acriflavine sur les levures. II. Etude genetique du mutant "petite colonie," *Ann. Inst. Pasteur (Paris)* **76:**419.

Falco, S. C., and Botstein, D., 1983, A rapid chromosome mapping method for cloned fragments of yeast DNA, *Genetics* **105:**857–872.

Fink, G. R., and Styles, C. A., 1972, Curing of a killer factor in *Saccharomyces cerevisiae, Proc. Natl. Acad. Sci. USA* **69:**2846–2849.

Fogel, S., Mortimer, R. K., and Lusnak, K., 1981, Mechanism of meiotic gene conversion or "wanderings on a foreign strand," in: *The Molecular Biology of the Yeast Saccharomyces: Life Cycle and Inheritance* (J. N. Strathern, E. W. Jones, and J. R. Broach, eds.), Cold Spring Harbor Laboratory, Cold Spring Harbor, NY, pp. 289–339.

Gaber, R., Mathison, L., Edelman, I., and Culbertson, M. R., 1983, Frameshift suppression in *Saccharomyces cerevisiae.* VI. Complete genetic map of twenty-five suppressor genes, *Genetics* **103:**389–407.

Gallwitz, D., Donath, C., and Sander, C., 1983, A yeast gene encoding a protein homologous to the human C-has/bas protooncogene product, *Nature* **306:**704–707.

Goldring, E. S., Grossman, L. E., Krupnick, D., Cryer, D. R., and Marmur, J., 1970, The petite mutation of yeast: Loss of mitochondrial DNA during induction of petites with ethidium bromide, *J. Mol. Biol.* **52:**323–335.

Henry, S. A., Donahue, T., and Culbertson, M., 1975, Selection of spontaneous mutants by inositol starvation in *Saccharomyces cerevisiae, Mol. Gen. Genet.* **143:**5–11.

Herskowitz, I., and Oshima, Y., 1981, Control of cell type in *Saccharomyces cerevisiae:* Mating type and mating-type interconversion. in: *The Molecular Biology of the Yeast Saccharomyces: Life Cycle and Inheritance* (J. N., Strathern, E. W. Jones, and J. R. Broach, eds.), Cold Spring Harbor Laboratory, Cold Spring Harbor, NY, pp. 181–209.

Hilger, F., and Mortimer, R. K., 1980, Genetic mapping of *arg1* and *arg8* in *Saccharomyces cerevisiae* by trisomic analysis combined with interallelic complementation, *J. Bacteriol.* **141:**270–274.

Hilger, F., Prevot, M., and Mortimer, R. K., 1982, Genetic mapping of *arg, cpa, car,* and *tsm* genes in *Saccharomyces cerevisiae* by trisomic analysis, *Curr. Genet.* **6:**93–98.

Kawasaki, G., 1979, Karyotypic instability and carbon source effect in cell cycle mutants of *Saccharomyces cerevisiae,* Ph.D. thesis, University of Washington, Seattle.

Kikuchi, Y., 1983, Yeast plasmid requires a cis-acting locus and two plasmid proteins for its stable maintenance, *Cell* **35:**487–493.

Klapholz, S., and Esposito, R. E., 1982, A new mapping method employing a meiotic rec^- mutant of yeast, *Genetics* **100:**387–412.

Lindegren, G., Hwang, Y. L., Oshima, Y., and Lindegren, C. C., 1965, Genetical mutants induced by ethyl methane sulfonate in *Saccharomyces, Can. J. Genet. Cytol.* **7:**491–499.

Liras, P., McCusker, J., Mascioli, S., and Haber, J. E., 1978, Characterization of a mutation in yeast causing nonrandom chromosome loss during mitosis, *Genetics* **88:**651–671.

Livingston, D. M., 1977, Inheritance of 2 μ mDNA plasmid from *Saccharomyces, Genetics* **86:**73–84.

Loumaye, E., Thorner, J., and Catt, K. J., 1982, Yeast mating pheromone activates mammalian gonadotrophs: Evolutionary conservation of a reproductive hormone? *Science* **218:**1323–1325.

Ma, C., and Mortimer, R. K., 1983, Empirical equation that can be used to determine genetic map distances from tetrad data, *Mol. Cell. Biol.* **3:**1886–1887.

Matsumoto, K., Uno, I., Oshima, Y., and Ishikawa, T., 1982, Isolation and characterization of yeast mutants deficient in adenylate cyclase and cAMP-dependent protein kinase, *Proc. Natl. Acad. Sci. USA* **79:**2355–2359.

Matsumoto, K., Uno, I., and Ishikawa, T., 1983a, Initiation of meiosis in yeast mutants defective in adenylate cyclase and cyclic AMP-dependent protein kinase, *Cell* **32:**417–432.

Matsumoto, K., Uno, I., and Ishikawa, T., 1983b, Control of cell division in *Saccharomyces cerevisiae* mutants defective in adenylate cyclase and cAMP-dependent protein kinase, *Exp. Cell Res.* **146:**151–161.

Matsumoto, K., Uno, I., and Ishikawa, T., 1985, Genetic analysis of the role of cAMP in yeast, *Yeast* **1:**15–24.

Mortimer, R. K., and Hawthorne, D. C., 1973, Genetic mapping in *Saccharomyces.* IV. Mapping of temperature-sensitive genes and use of disomic strains in localizing genes, *Genetics* **74:**33–54.

Mortimer, R. K., and Manney, T. R., 1971, Mutation induction in yeast, in: *Chemical Mutagens,* Vol. 1 (A. Hollaender, ed.), Plenum Press, New York, pp. 289–310.

Mortimer, R. K., and Schild, D., 1980, Genetic map of *Saccharomyces cerevisiae, Microbiol. Rev.* **44:**519–571.

Mortimer, R. K., and Schild, D., 1981, Genetic mapping in *Saccharomyces cerevisiae,* in: *The Molecular Biology of the Yeast Saccharomyces: Life Cycle and Inheritance* (J. N. Strathern, E. W. Jones, and J. R. Broach, eds.), Cold Spring Harbor Laboratory, Cold Spring Harbor, NY, pp. 11–26.

Mortimer, R. K., and Schild, D., 1985, Genetic map of *Saccharomyces cerevisiae, Microbiol. Rev.* **49:**519–571.

Mortimer, R. K., Contopoulou, R., and Schild, D., 1981, Mitotic chromosome loss in a radiation-sensitive strain of the yeast *Saccharomyces cerevisiae, Proc. Natl. Acad. Sci. USA* **78:**5778–5782.

Perkins, D. D., 1949, Biochemical mutants in the smut fungus *Ustilago maydis, Genetics* **34:**607–626.

Pringle, J. R., and Hartwell, L. H., 1981, The *Saccharomyces cerevisiae* cell cycle, in: *The Molecular Biology of the Yeast Saccharomyces: Life Cycle and Inheritance* (J. N. Strathern, E. W. Jones, and J. R. Broach, eds.), Cold Spring Harbor Laboratory, Cold Spring Harbor, NY: pp. 97–142.

Riley, M. I., and Manney, T. R., 1978, Tetraploid strains of *Saccharomyces cerevisiae* that are trisomic for chromosome III, *Genetics* **89:**667–684.

Shaffer, B., Brearley, R., Littlewood, R., and Fink, G. R., 1971, A stable aneuploid of *Saccharomyces cerevisiae, Genetics* **67:**483–495.

Singh, A., Helms, C., and Sherman, F., 1979, Mutation of the non-Mendelian suppressor, ψ^+, in yeast by hypertonic media, *Proc. Natl. Acad. Sci. USA* **76:**1952–1956.

Slonimski, P. P., Borst, P., and Attardi, G. (eds.), 1982, *Mitochondrial Genes,* Cold Spring Harbor Laboratory, Cold Spring Harbor, NY.

Snow, R., 1966, An enrichment method for auxotrophic yeast mutants using the antibiotic "nystatin," *Nature* **211:**206–207.

Snow, R., 1979, Maximum likelihood estimation of linkage and interference from tetrad data, *Genetics* **92:**231–245.

Sommer, S. S., and Wickner, R. B., 1982, Co-curing of plasmids affecting killer dsRNAs of *Saccharomyces cerevisiae:* [HOK], [NEX], and the abundance of L are related and further evidence that M_1 requires L, *J. Bacteriol.* **150:**545–551.

Strathern, J. N., Jones, E. W., and Broach, J. R., eds., 1981, *The Molecular Biology of the Yeast Saccharomyces,* 2 vols., Cold Spring Harbor Laboratory, Cold Spring Harbor, NY.

Tatchell, K., Robison, L. C., and Breitenbach, M., 1985, *RAS2* of *Saccharomyces cerevisiae* is required for gluconeogenic growth and proper response to nutrient limitation, *Proc. Natl. Acad. Sci. USA* **82:**3785–3789.

Tipper, D. J., and Bostian, K. A., 1984, Double-stranded RNA killer systems in yeasts, *Microbiol. Rev.* **48:**125–156.

Toda, T., Uno, I., Ishikawa, T., Powers, S., Kataoka, T., Broek, D., Cameron, S., Broach, J., Matsumoto, K., and Wigler, M., 1985, In yeast, RAS proteins are controlling elements of adenylate cyclase, *Cell* **40:**27–36.

Tuite, M. F., Mundy, C. R., and Cox, B. S., 1981, Agents that cause a high frequency of genetic changes from [psi$^+$] to [psi$^-$] in *Saccharomyces cerevisiae, Genetics* **98:**691–711.

Uno, I., Matsumoto, K., and Ishikawa, T., 1982, Characterization of cyclic AMP-requiring yeast mutants altered in the regulatory subunit of protein kinase, *J. Biol. Chem.* **257:**14110–14115.

Wickner, R. B., 1974, "Killer character" of *Saccharomyces cerevisiae:* Curing by growth at elevated temperature, *J. Bacteriol.* **117:**1356–1357.

Wickner, R. B., 1979, Mapping chromosomal genes of *Saccharomyces cerevisiae* using an improved genetic mapping method, *Genetics* **92:**803–821.

Wickner, R. B., 1983, Genetic control of replication of the double-stranded RNA segments of the killer systems in *Saccharomyces cerevisiae, Arch. Biochem. Biophys.* **222:**1–11.

Wickner, R. B., 1985, Killer systems in yeast, in: *Current Topics in Medical Mycology,* Vol. 1 (M. R. McGinnis, ed.), Springer-Verlag, New York (in press).

Wickner, R. B., and Leibowitz, M. J., 1977, Dominant chromosomal mutation bypassing chromosomal genes needed for killer RNA plasmid replication in yeast, *Genetics* **87:**453–469.

Wickner, R. B., Boutelet, F., and Hilger, F., 1983, Evidence for a new chromosome in *Saccharomyces cerevisiae, Mol. Cell. Biol.* **3:**415–420.

Wood, J. S., 1982, Mitotic chromosome loss induced by methyl benzimidazole-2-yl-carbamate as a rapid mapping method in *Saccharomyces cerevisiae, Mol. Cell Biol.* **2:**1080–1087.

Recombinant DNA Techniques 5

A. J. KINGSMAN, E. J. MELLOR, M. J. DOBSON, and S. M. KINGSMAN

1. INTRODUCTION

The aim of this chapter is to introduce the procedures currently used to isolate and manipulate yeast genes. We restrict our discussion to those procedures that are, at least in part, peculiar to the yeast *Saccharomyces cerevisiae.* We do not deal with procedures that are common to all other organisms, such as cDNA bank construction, hybrid-selected or arrested translation systems, nucleic acid probe strategies, or radioimmune detection strategies, as these have been covered in a variety of general texts (e.g., Maniatis *et al.*, 1982; Glover, 1985).

One of the earliest procedures for the isolation of yeast genes used an *Escherichia coli* complementation system (Ratzkin and Carbon, 1977; Struhl *et al.,* 1976). A genome representative collection of yeast DNA fragments was inserted into an *E. coli* plasmid and then used to transform a variety of *E. coli* auxotrophs. Some of the *E. coli* mutations were complemented by yeast DNA and most of these fragments were found to contain the corresponding yeast gene. In this way the yeast *LEU2* gene, which encodes isopropylmalate dehydrogenase, was isolated by complementation of an *E. coli leuB* mutation (Ratzkin and Carbon, 1977) and the yeast *TRP1* gene, encoding phosphoribosyl anthranilate isomerase, was isolated by complementation of an *E. coli trpC* mutation (Struhl *et al.,* 1976). This procedure was unreliable because it depended on expression of the yeast genes in *E. coli* via the fortuitous presence in the yeast DNA of sequences that resemble *E. coli* expression signals. In fact, it was successful in only about 25% of cases. However, it did provide the first cloned selectable yeast genes and led directly to the development of yeast transformation techniques that used these genes as selectable markers. With the advent of yeast transformation and plasmid

A. J. KINGSMAN, E. J. MELLOR, M. J. DOBSON, and S. M. KINGSMAN • Department of Biochemistry, University of Oxford, Oxford OX1 3QU, England.

vectors, yeast gene isolation is no longer achieved by *E. coli* complementation, but rather by direct yeast complementation.

2. YEAST TRANSFORMATION

Two major procedures are used for the introduction of DNA into yeast cells. In the first (Hinnen *et al.*, 1978; Beggs, 1978) exponentially growing cells are treated with an enzyme, either glusulase, zymolyase, or lyticase, to remove the cell wall (see Chapter 9). The resulting spheroplasts are then exposed to DNA in the presence of Ca^{2+} ions followed by treatment with polyethylene glycol of molecular weight about 4000 Da (PEG 4000). The spheroplasts are then embedded in a regeneration agar that usually imposes selection for a selectable marker and acts as a physical support for the regeneration of the wall. The mechanism of DNA uptake is not understood, but it may involve DNA entrapment during PEG-mediated spheroplast fusion. However, it is worth noting that diploids are not created at high frequencies ($<5\%$) (M. J. Dobson, unpublished data) during the transformation process with haploid strains. Using this technique with anautonomous plasmid vector, transformation frequencies as high as $10^6/\mu g$ DNA can be obtained.

The second procedure (Ito *et al.*, 1983) does not require the removal of the cell wall. Whole cells are subjected to prolonged (2–8 hr) exposure to Li^+ ions at 4°C followed by treatment with DNA and PEG 4000. Like the *E. coli* techniques, the cells are then given a heat shock and plated directly onto the surface of selective plates. This procedure gives transformation frequencies about 100-fold lower than the spheroplast procedure, but it does have the advantage that transformants grow on the surface of the selection plates rather than embedded in them. This makes subsequent processing (e.g., replica plating) easier. The reduction in transformation frequencies with the Li^+ technique is not so pronounced (i.e., 10-fold) for linear DNA fragments (see Section 6).

3. PLASMID VECTORS

Almost all yeast plasmid vectors are, in fact, both yeast and *E. coli* vectors. They usually comprise all or part of an *E. coli* vector such as pBR322 and a yeast relication and selection system. This arrangement allows *E. coli* to be used as the "preparative" organism for the various manipulations associated with recombinant DNA procedures. Yeast autonomous plasmid vectors are based on two types of replication system. The first is dependent on chromosomal DNA fragments that are able to replicate autonomously (Struhl *et al.*, 1979; Stinchcomb *et al.*, 1979;

Kingsman *et al.*, 1979). These are the autonomously replicating sequences (ARSs) and may be chromosomal origins of replication (Williamson, 1985). The second makes use of sequences derived from the endogenous yeast 2 μ plasmid (e.g., Beggs, 1978; Struhl *et al.*, 1979; Broach, 1982).

3.1. ARS-Based Plasmids

A commonly used set of ARS-based plasmids is based on the plasmid YRp7 (Struhl *et al.*, 1979). This is a pBR322 derivative containing a 1.45-kb EcoRI fragment from the yeast genome. This fragment contains the *TRP1* gene, which can be used as a selectable marker, and the *ARS1* sequence. YRp7 transforms yeast at frequencies up to 10^5/μg DNA; it is maintained in the nucleus but, like all recombinant ARS plasmids, it is extremely unstable (Kingsman *et al.*, 1979). Under selective conditions only 20–60% of cells contain plasmid, and under nonselective conditions more than 95% of cells lose the plasmid after 15 generations in haploids but in diploids ARS plasmids are more stable. Interestingly, in those cells that contain plasmid, copy numbers are high (20–80/cell). The instability appears to be due to a segregation defect that results in a bias toward segregation to mother cells (Murray and Szostak, 1983a). The instability of ARS-based plasmids can be overcome by the addition of a segment of DNA from a yeast centromere (Clarke and Carbon, 1980). These ARS–CEN plasmids also transform at frequencies round 10^5/μg DNA, but in contrast to the ARS plasmids, they are almost entirely stable and they are maintained at a copy number of 1–2/cell. Examples of ARS and ARS–CEN plasmids are given in Fig. 2 and a comprehensive list can be found in Parent *et al.* (1985).

ARS plasmids are of limited usefulness because of their instability although there are some circumstances where this can be exploited to identify a plasmid-encoded function by coloss with a selectable marker. ARS–CEN plasmids are favored when a low-copy-number, stable system is required without the use of an integrative vector (Williamson, 1985) (see Section 6).

3.2. 2-μm-Based Plasmids

The 2-μm plasmid is present in most laboratory strains of *S. cerevisiae* at copy numbers from 20 to 200/cell (Broach, 1982). It is a circle of 6318 bp and contains two inverted repeats of 599 bp (Fig. 1). 2-μm replication and maintenance are dependent on two *cis*-acting and two *trans*-acting functions (Jayaram *et al.*, 1983). The *cis*-acting functions are the origin of replication and a series of 62-bp repeated sequences, termed *REP3* or *STB*, that seem to be involved in the fidelity of plasmid

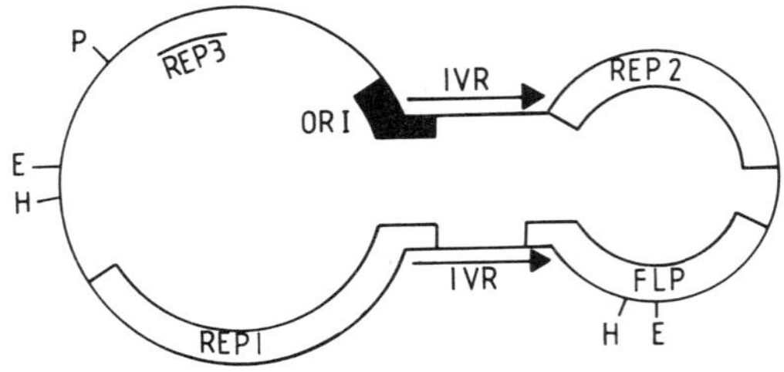

Figure 1. Organization of the 2-μm plasmid. The A-form of the plasmid is shown. Recombination across the inverted repeats would invert one side of the "dumbbell" with respect to the other to produce the B-form. Open boxes = large open reading frames. Closed box = origin of replication. IVR = inverted repeat. Restriction sites: H = *Hind*III; E = *Eco*RI; P = *Pst*I.

segregation. The *trans*-acting functions are encoded by *REP1* and *REP2*. It is not clear, at present, whether these genes control replication, segregation, or both. A third major coding sequence, *FLP*, encodes a recombinase that mediates site-specific recombination between the inverted repeats. 2-μm-based vectors may contain all four essential functions, in which case they are able to replicate in the absence of an endogenous 2 μ plasmid (i.e., in a 2-μm° strain), or they may contain only the *cis*-acting functions, in which case stable maintenance is dependent on the complementing *REP1* and *REP2* genes of the endogenous plasmid in a 2-μm$^+$ strain.

2-μm-based vectors are capable of transforming yeast at frequencies as high as 10^6/μg DNA and maintaining copy numbers as high as 400/cell (Dobson *et al.*, unpublished data). However, the precise characteristics of any vector vary with the particular construction used. Copy numbers as low as 5–10 are not uncommon. An example of a commonly used 2-μm-based plasmid is given in Fig. 2 and a comprehensive list of 2-μm-based vectors can be found in Parent *et al.* (1985).

3.3. Yeast Gene Isolation Using Plasmid Vectors

Most yeast genes have been isolated by selecting, from a genome representative plasmid pool, a molecule that is able to complement the defect in a known yeast mutant (e.g., Nasmyth and Reed, 1980). In the case of, for example, an auxotrophic mutation, selection is direct and requires only that growth is obtained on minimal medium. Alternatively, in the case of more complex phenotypes, such as those conferred by the mating-type genes, it may be necessary to obtain sufficient yeast transformants, using the plasmid selectable marker, to ensure representation of the entire genome and then screen these transformants for the desired phenotype (Nasmyth and Tatchell, 1980). The numbers involved can

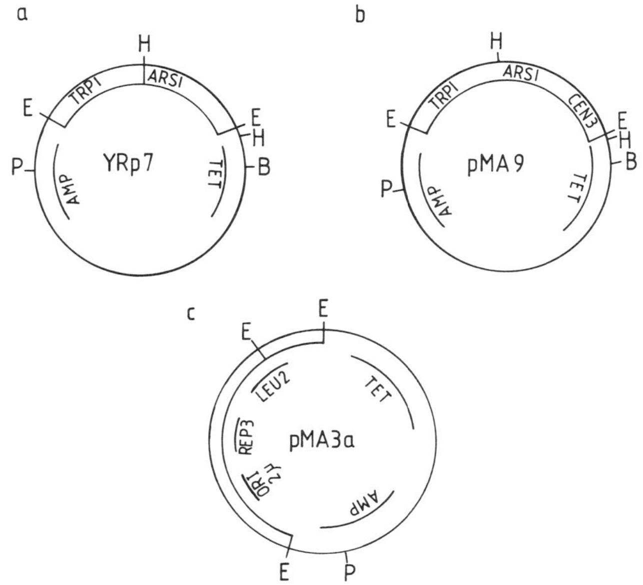

Figure 2. (a) The ARS plasmid YRp7. (b) The ARS–CEN plasmid pMA9. (c) The 2-μm-based plasmid pMA3a. Open box = yeast DNA. Thin line = pBR322 DNA with ampicillin and tetracycline resistance markers shown. Restriction sites: H = *Hind*III; E = *Eco*RI; B = *Bam*HI; P = *Pst*I.

easily be calculated from considerations of the genome size and the size of the fragments inserted into the vector (Clarke and Carbon, 1976). Plasmid systems are favored for this approach because they give high frequencies of transformation, and, more important, because they contain bacterial plasmid replicons they can be retrieved from the initial yeast transformant by preparing DNA from the transformant and using it to transform *E. coli,* selecting for a bacterial marker (e.g., ampicillin resistance). The bacterial clone can then be propagated to produce large amounts of the yeast DNA fragment.

The vast array of mutations (Mortimer and Schild, 1985) and this very simple gene isolation procedure maintain yeast as an attractive molecular genetic system.

3.4. Site-Directed Mutagenesis in Yeast Using Single-Stranded Plasmids

It is often useful to be able to isolate a cloned gene as a single-stranded DNA (ssDNA) as well as a double-stranded DNA (dsDNA). A number of plasmids that can be propagated in *E. coli* as either ssDNA or dsDNA which can also replicate in *S. cerevisiae* as dsDNA have been developed by incorporating the replication and packaging signals from filamentous bacteriophage such as f1, into the plasmid (Cesareni and Murray, 1987). The demonstration that *S. cerevisiae* can be efficiently transformed with ssDNA plasmids (Singh *et al.*, 1982) has subsequently led to the use of such plasmids for carrying out site-directed mutagenesis of cloned genes directly in yeast (Burke and Olson, 1986; Walder and Walder, 1986). In this method yeast is transformed directly with a heteroduplex molecule comprised of the ssDNA plasmid, carrying the gene of choice, and a short oligonucleotide, containing the desired nucleotide alteration, annealed to the ssDNA plasmid. No prior *in vitro* synthesis of the complementary DNA strand is necessary. A further development of strategies for site-directed mutagenesis directly in yeast has come from the work of Moerschell *et al.* (1988), who have shown that one can mutate genomic sequences by transforming yeast directly with the mutant oligonucleotide. Mutants can be generated with oligonucleotides as short as 20 nucleotides and with amounts as low as 100 μg, but this has the drawback that a direct selection method is required for isolating the generated mutant.

4. MINICHROMOSOME VECTORS

Linear autonomous vectors can be created by assembling an *ARS*, a *CEN*, a selectable marker, and a *TEL* (telemere) (Murray and Szostak, 1983b) (Fig. 3). These vectors resemble minichromosomes, although they are substantially less stable than authentic chromosomes. However, they appear to become more stable with increased size, making them particularly suitable for the introduction of large DNA fragments into yeast. In this context, the development of yeast artificial chromosome (YAC) vectors (see Section 5) has provided the opportunity for assembling whole metabolic pathways where multiple genes need to be used.

5. YAC CLONING VECTORS

The cloning of very large DNA fragments is of central importance to the analysis of the genomes of higher eukaryotes, including the pro-

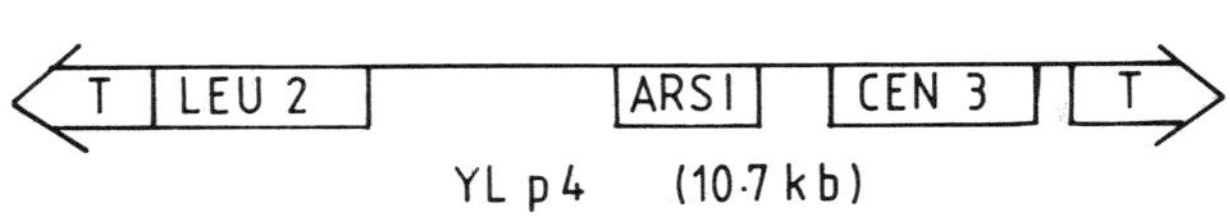

Figure 3. The yeast minichromosome YLp4. T = *Tetrahymena* telomeres. Open boxes = yeast DNA. Thin line = pBR322 DNA.

posed DNA sequencing of the human genome. Many genes from such organisms, e.g., the human factor VIII gene and the *Drosophila* bithorax genes, span enormous distances. These large genes have previously had to be cloned as a series of overlapping cut fragments of less than 40 kb in order to meet the size constraints that packaging in bacteriophage imposes. This is both laborious and unsatisfactory, in that some very large loci may never be reliably reconstituted. However, Burke *et al.* (1987) have developed a method using YACs to clone DNA fragments up to 500 kb. This new technique is illustrated in Fig. 4.

The circular plasmid vector pYAC2 may be propagated readily in *E. coli*. Circular plasmid molecules, isolated from the bacterium, can be cleaved into two fragments that constitute chromosome arms. The left arm contains all the structural and functional requirements of a yeast chromosome: a centromere, an *ARS* element, but only one telomere. The right arm contains only a telomere. Each arm carries a different yeast gene as a selectable marker; *TRP1* for the left arm and *URA3* for

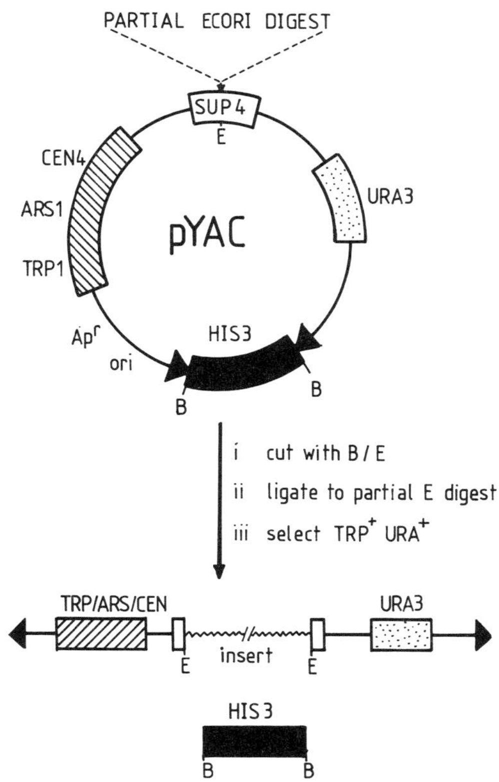

Figure 4. YAC cloning vector pYAC. Restriction sites: B = *Bam*HI; E = *Eco*RI. For detailed description of the cloning procedure, see text.

the right. The cleavage site that separates the two arms lies within the yeast *SUP4* gene which encodes a tyrosine-inserting, ocher suppressor tRNA. Large (>100 kb) fragments of heterologous DNA may be ligated to the two arms. Transformation of yeast with the ligation mix yields about 100 transformant colonies per microgram of DNA. By selecting for *TRP1* and *URA3* simultaneously, only molecules that have both arms are obtained. The presence of an insert is easily detected in a host strain carrying an *ade2*-ocher mutation. In YACs without an insert, *SUP4* will be functional and transformant colonies white. In YACs carrying heterologous DNA, *SUP4* is inactivated, the *ade2*-ocher allele is not suppressed, and transformant colonies are red. The recombinant YAC molecules appear to be stably maintained without structural rearrangement. This, in addition to their use in analyzing large genomes, YAC vectors have potential for introducing whole new metabolic pathways into *S. cerevisiae* to greatly extend its catabolic and anabolic capabilities.

6. INTEGRATIVE SYSTEMS

The first experiment that resulted in the successful transformation of *S. cerevisiae* used the simplest of integrative systems (Hinnen *et al.*, 1978). The plasmid used was a bacterial plasmid containing the yeast *LEU2* gene and the recipient strain contained a double mutation in *LEU2* to reduce the frequency of spontaneous reversion to LEU^+. When LEU^+ transformants were obtained, it was found that, in most cases, the entire plasmid had integrated into chromosome III via homologous recombination across the *LEU2* regions (Fig. 5). A variety of inte-

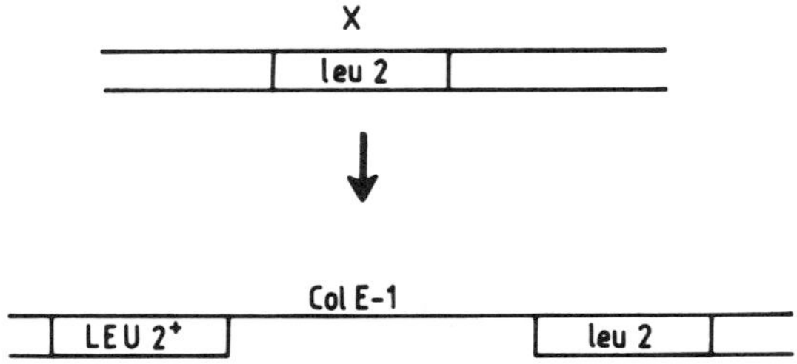

Figure 5. Integration of plasmid pYE(*LEU2*)10 into the yeast genome. Open box = yeast DNA. Thin line = ColE1 DNA.

grative strategies have arisen from these initial experiments. Together they make yeast unique among eukaryotes in that chromosomal genes can be accurately replaced by any derivative with almost total reliability.

6.1. Targeted Integration

The observation that a double-strand break in a plasmid-borne yeast DNA fragment is highly recombinogenic (Orr-Weaver *et al.*, 1981) has led to a system for directing integration of a DNA fragment to a specific chromosomal site. If a plasmid contains two chromosomal yeast DNA fragments and is used to transform yeast, integration will occur at the chromosomal location of either sequence with more or less equal frequency. However, if a double-strand break is produced, probably by cleavage with a restriction enzyme, in one of those fragments, then integration at the corresponding chromosomal locus will be stimulated at least 100-fold (Fig. 6). Integration can therefore be targeted to a site by double-strand cleavage. Either transformation procedure can be used. A minor problem with this technique is that integration of tandem arrays is relatively common. This can be detected easily by Southern blots (Southern, 1975) and can be diminished by using reduced amounts of DNA. This technique, like the simple integrative vector system described ear-

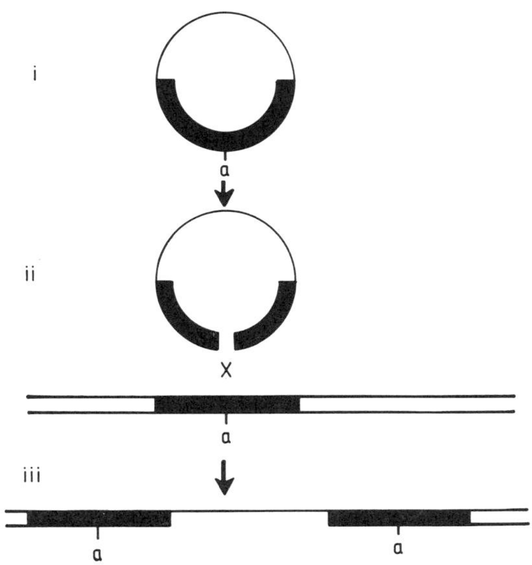

Figure 6. Targeted integration. Closed box = yeast DNA. Thin line = bacterial vector. a = any restriction site in the yeast DNA. A cloned piece of yeast DNA containing restriction site a (i) is cleaved at a (ii) and then used to transform yeast. Integration occurs at the point of cleavage to generate a duplication of the cloned yeast fragment with the components of that duplication separated by the bacterial plasmid sequences (iii).

lier, results in a duplication of the target DNA with the duplicates separated by the remainder of the plasmid (Fig. 6).

6.2. Allelic Rescue

Having cloned a wild-type yeast gene by complementation, for example, it is frequently the case that DNA sequence analysis of mutant alleles of that gene will be most illuminating. Strategies have been devised, therefore, to isolate mutant alleles by using targeted integration of a plasmid vector at a site close to the mutation using homologous recom-

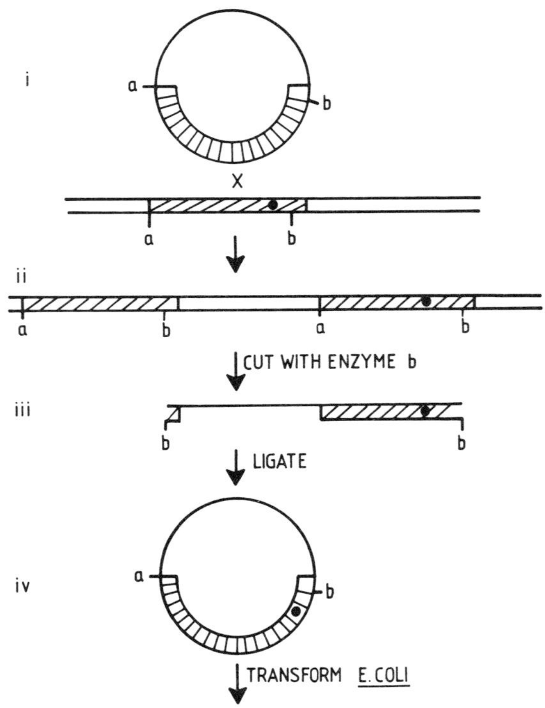

Figure 7. Allelic rescue I. Hatched box = cloned yeast DNA. Open box = yeast genome. Thin line = bacterial plasmid. Black spot = mutation. a and b = restriction sites within the cloned yeast DNA. A wild-type gene inserted into a bacterial vector is used to transform a yeast strain that is mutant at the corresponding locus (i). Integration occurs by recombination across the yeast homologous sequences to generate a duplication of the cloned fragment flanking the bacterial vector sequences (ii). Total DNA is then prepared and digested with enzyme a and enzyme b in separate reactions. Digestion with b is shown (iii). Depending on which side of the mutation the original integration event occurred, the reaction with either a or b will generate a fragment containing the mutant allele linked to bacterial vector sequences (iii). The cut DNA is then ligated and used to transform *E. coli* selecting for a bacterial plasmic marker (iv). Transformants are then screened, for example, for plasmids that are unable to transform the mutant yeast strain to wild type.

bination between the wild-type and mutant genes. The resulting structure has the plasmid sequences flanked by one mutant and one wild-type gene (Fig. 7), the precise configuration depending on the crossover point. If DNA is prepared from the resulting transformant and then cleaved with an appropriate restriction fragment, the plasmid and mutant allele sequences will be on the same fragment (Stiles *et al.,* 1981). Ligation followed by transformation of *E. coli* will give bacterial transformants containing plasmids carrying the mutant allele. Plasmids can be screened for their inability to correct the mutation in yeast.

A more elegant technique that guarantees that the mutant allele is recovered exploits the ability of yeast cells to carry out gap repair during integration of a linearized plasmid (Fig. 6) (Orr-Weaver *et al.,* 1982). In this procedure integration of a plasmid bearing a selectable marker is directed to the mutant gene by the presence of a double-strand break in the plasmid-borne wild-type gene. However, the wild-type gene is not just broken, it is also gapped at the break by using either one or more restriction enzymes or a restriction enzyme and an exonuclease. The gap is designed to run over the region of the mutation. When the plasmid integrates, the gap is repaired using the mutant gene as template. The resulting structure has the plasmid vector sequences flanked on both

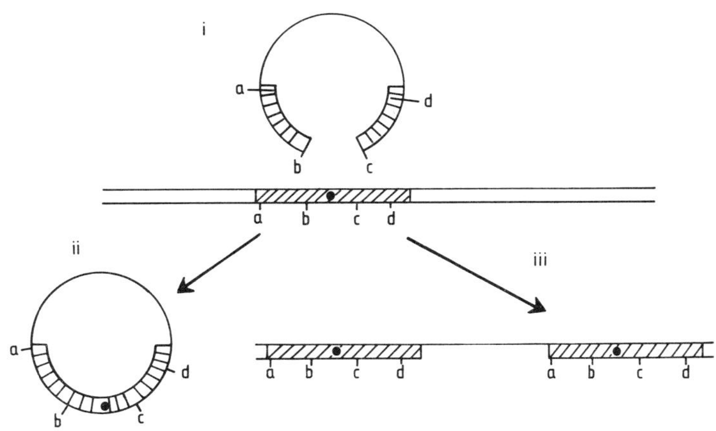

Figure 8. Allelic rescue by gap repair. Hatched box = cloned yeast DNA. Open box = yeast genome. Thin line = bacterial vector DNA. Black spot = mutation. a, b, c, and d are restriction sites in the cloned yeast DNA. A gap is created in the cloned DNA over the region of the desired mutation by cleavage at b and c (i). The resulting linear molecule is used to transform a yeast strain carrying the desired mutant allele (i). Gap repair may occur directly (ii) and if the gapped plasmid is potentially an autonomous yeast plasmid the mutant allele will be transferred to and maintained on this autonomous molecule, or gap repair may occur as a function of integration (iii). However, unlike the situation in Fig. 7, digestion with any of the enzymes a, b, c, or d will give a fragment that carries the mutant allele. If the digested DNA is ligated and used to transform *E. coli,* all transformants will contain the desired allele.

sides by a mutant gene. Appropriate cleavage of the transformant DNA followed by ligation and *E. coli* transformation results in recovery of the mutant allele with 100% efficiency.

The gap repair strategy can also be used to "transfer" a mutant allele to an autonomous plasmid that has been gapped (Orr-Weaver and Szostak, 1983; Orr-Weaver *et al.*, 1982). If an unstable ARS plasmid is used, the gapped plasmid and therefore its selectable marker can only survive by gap repair, either via integration or not. Transformants with autonomous repaired plasmids are easily identified, as the selectable marker will show marked instability (Fig. 8). This also has the advantage that recovery of the plasmid carrying the mutant allele is achieved by transformation of *E. coli* directly with total yeast DNA preparations.

6.3. Gene Disruption

Using a cloned yeast gene to disrupt its chromosomal counterpart is an important procedure. It can be used to assess the phenotype of a null mutation or to confirm that a cloned DNA fragment confers a previously identified phenotype. A simple procedure is to insert a fragment, generated by restriction enzymes or exonucleases, that comes from within the cloned gene, into an integrative vector (Shortle *et al.*, 1982). Inte-

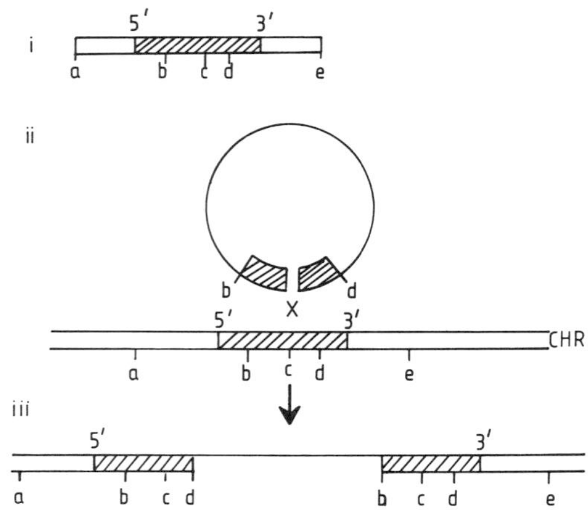

Figure 9. Gene disruption. Open and hatched boxes = yeast DNA. Hatched box = a gene within the cloned fragment a–e (i). Thin line = bacterial vector sequences. CHR = chromosome. a, b, c, d, and e = restriction sites in the cloned yeast DNA. The chromosomal copy of the hatched gene is to be disrupted. The internal fragment b–d is subcloned by inserting it into a bacterial vector. It is then cleaved at c and used to transform a yeast strain (ii). Integration occurs at c to create a duplication of fragment b–d but with the 5′ and 3′ ends of the hatched gene separated by one copy of b–d and the bacterial vector (iii).

gration is then directed into the chromosomal copy of the gene by making a double-strand break in the fragment (Fig. 9). This creates a duplication of the cloned fragment in the genome, but the 5′ and 3′ ends of the gene are separated by the vector sequences. The gene will therefore be nonfunctional. The only major disadvantage of this technique is that the disruption will be unstable because of the duplication.

A more satisfactory procedure involves replacement of the genomic copy of a gene by mutated derivative. The general name for this is "gene transplacement" (Scherer and Davis, 1979). For example, a small deletion may be introduced into the cloned gene using restriction enzymes and an exonuclease. The mutated derivative is then inserted into an integrative vector and forced to recombine with its genomic complement. Integration can occur on either side of the deletion to produce a duplication of the gene-flanking vector sequences. One copy of the gene will be wild type, the other will carry the deletion. The transformant will also bear the phenotype determined by the selectable marker. A second recombination event on the opposite side of the deletion to the integrative event will resolve the duplication, eliminate the selectable marker, and leave the mutant derivative in place of the wild-type gene (Fig. 10). Scoring for loss of the selectable marker permits identification of clones that have undergone a second recombination event. The position of that event with respect to the deletion can be determined either by the phenotype confered by mutant alleles or by appropriate Southern hybridization experiments. *URA3* is frequently used as a selectable marker for this type of approach as its loss can be selected by growth on 5-fluoroorotic acid (Boeke *et al.*, 1984). The disadvantages of this technique are that two steps are required and that 5-fluorooritic acid is very expensive.

A particularly elegant and convenient disruption procedure makes use of the recombinogenic properties of "ends" to insert a selectable marker into a gene (Rothstein, 1983). The key feature of this approach is that the marker provides both selection and disruption. A selectable marker is first inserted into a cloned gene *in vitro*. This can be a simple insertion, where a marker fragment is inserted at a suitable restriction site, or it can be a replacement, where the marker fragment replaces a region of the cloned gene. In the former this would create an insertion mutation; in the latter, an insertion/deletion mutation. A relatively large fragment comprising ends that are homologous with the target gene flanking the selectable marker is then isolated, usually from an agarose gel. This fragment is then used to transform a recipient selecting for the marker. Recombination with the chromosomal copy takes place across the ends of the fragment and so the chromosomal copy is replaced, in one step, by a derivative that is disrupted. This disruption links a selectable marker to the gene in question and it is nonrevertable (Fig. 11).

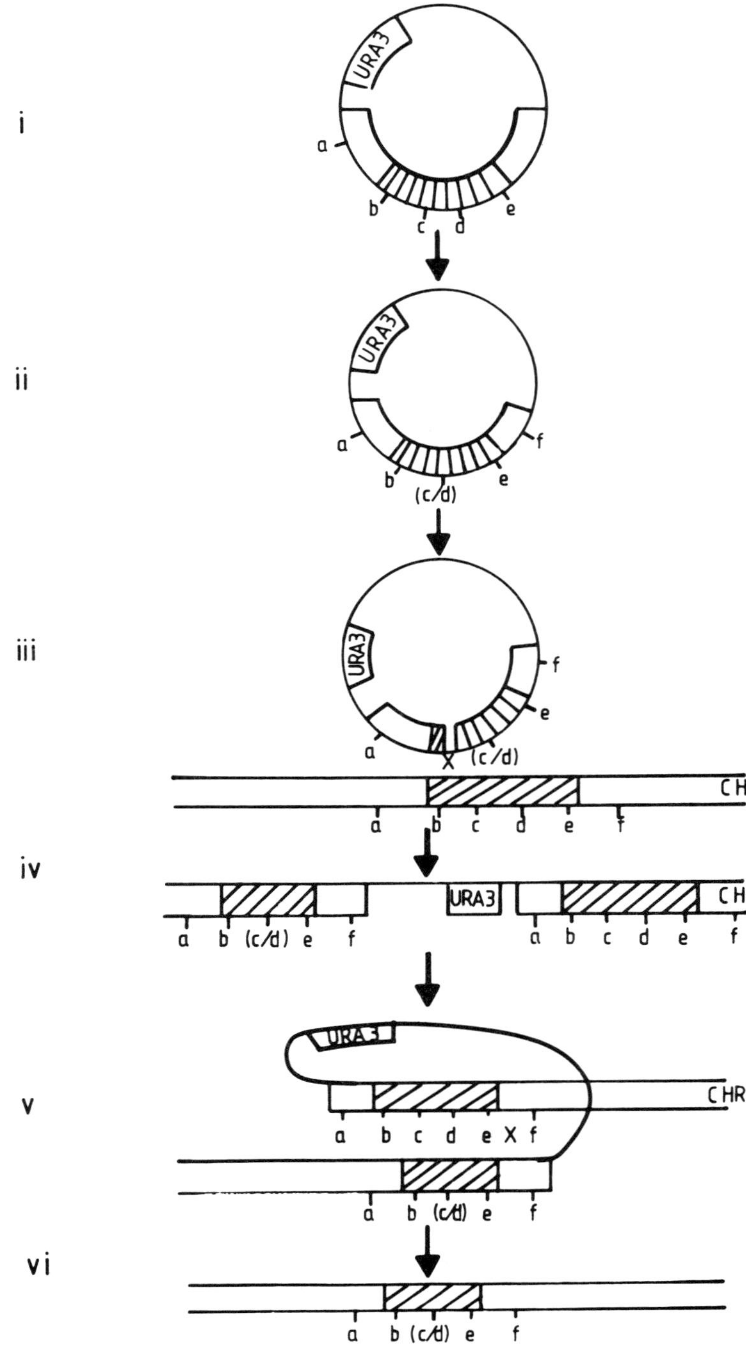

Figure 10. Gene transplacement. Open and hatched boxes = yeast DNA. Hatched box = a gene present on a cloned fragment. URA3 = *URA3* gene as the selectable marker. Thin line = bacterial vector sequences. CHR = chromosome. a, b, c, d, e, and f = restriction sites within the cloned yeast DNA. The hatched gene in the genome of a yeast strain is to be

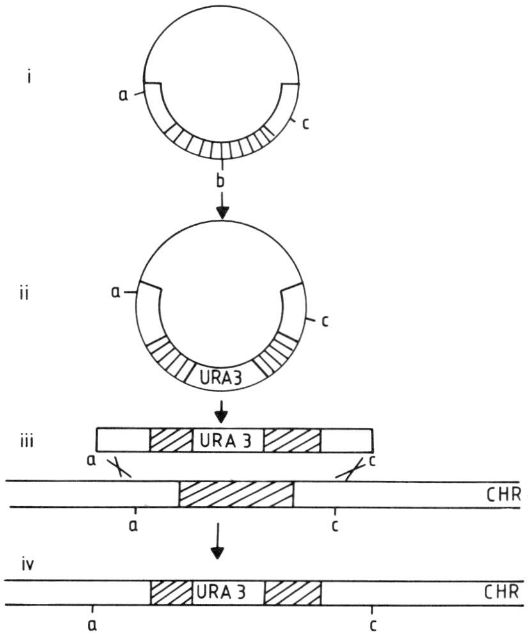

Figure 11. Marker-mediated gene disruption. Hatched and open boxes = yeast DNA. Hatched box = gene present on a cloned yeast DNA fragment. Thin line = bacterial vector sequences. CHR = chromosome. URA3 = the *URA3* gene as a selectable marker. a, b, and c are restriction sites within the cloned yeast DNA fragment. A disrupted version of the hatched gene is to replace its genomic copy. The gene is present on an integrative vector (i). It is disrupted by insertion of the *URA3* marker at b (ii). The resulting plasmid is cleaved at a and c and the fragment a–c is purified. This fragment is used to transform a yeast strain selecting for URA$^+$. Recombination occurs at a and c (iii) to replace the genomic copy of the gene by a disrupted derivative (iv).

6.4. Refinements of Transplacement

The procedure described in Section 6.3 is a convenient way to achieve gross disruption of a yeast gene. However, one of the most common modern experimental strategies in yeast molecular biology is to make relatively small, subtle changes to a gene and then ask if that change

---

replaced by a deleted version. The hatched gene is present on a cloned fragment in an integrative vector carrying *URA3* as the selectable marker (i). The deleted derivative of the gene is created by removing fragment c–d (ii). The plasmid containing the deletion is then cleaved at b and used to transform a yeast strain to URA$^+$. Integration occurs at b (iii) to create a duplication, one component of which carries the mutation and the other does not (iv). Excision of the vector sequences and one component of the duplication by recombination (v) can be selected on 5-fluoroorotic acid. Depending on which side of the mutation the excision event occurs, either the wild-type or the mutant allele will be left in the genome. The mutant allele is shown (vi). This can be tested either by the resulting phenotype or by Southern hybridization (see text).

affects function. The mutations can vary from single nucleotide changes to relatively small deletions. Clearly, it is common for there to be no easy way to select for replacement of a wild-type gene by a mutant gene, as there is when a selectable marker is the mutagenic agent. This is a particular problem when, as is usually the case, multiple versions of the mutant gene need to be analyzed.

A variety of conceptually similar tricks have been devised to solve these problems. They all involve marking the wild-type or resident gene in such a way that its loss is selectable. One of the first procedures to be described uses a recessive drug resistance marker (Struhl, 1983). First, the strain to be used is made resistant to cycloheximide by either crossing in or isolating a mutation at the *cyh2* locus. The gene to be analyzed is then disrupted, by a method described in the previous section, with the wild-type *CYH2* gene. This strain, now sensitive to cycloheximide, is the starting strain for all further transplacements. Mutant derivatives of the

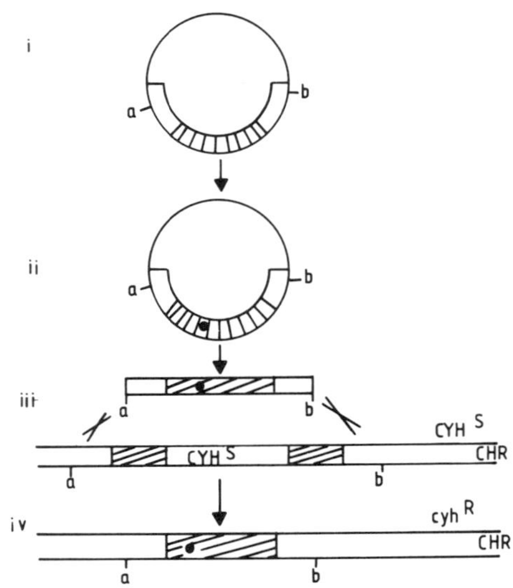

Figure 12. Selected transplacement. Open and hatched boxes = yeast DNA. Hatched box = a gene present on a cloned yeast DNA fragment. Thin line = bacterial vector sequences. CHR = chromosome. CYH = *CYH2* gene. Black spot = mutation. a and b = restriction sites within the cloned yeast DNA fragment. The hatched gene in the genome is to be replaced by derivative containing a mutation. The hatched gene is present on a cloned fragment in a bacterial vector (i). A mutation in that gene is created *in vitro* (ii). The resulting plasmid is cut at a and b and the fragment a–b is purified and used to transform a yeast strain selecting for cycloheximide resistance (iii). This strain has already undergone a mating to introduce a *cyh*^{r} mutation at the *CYN*2 locus and then a gene disruption at the hatched locus using the wild-type *CYH*^{s} gene. The strain is therefore sensitive to cycloheximide by virtue of this insertion at the hatched gene. When fragment a–b recombines with its disrupted genomic counterpart, it replaces it and the transformant becomes cycloheximide resistant (iv).

gene to be analyzed are introduced into this strain by transformation. They replace the *CYH2* marked resident gene by recombination and the cell becomes resistant to cycloheximide. The transplacement can therefore be selected. (See Fig. 12.) Other similar strategies have been devised using a suppressable canavanine resistance marker (Nasmyth, 1985) or the ability to select *ura3* mutants on 5-fluoroorotic acid (Rothstein, 1985).

A less sophisticated and perhaps more convenient technique makes use of the high frequency with which cotransformation of two DNA molecules occurs in yeast (Rudolph *et al.*, 1985). In this procedure a mutant DNA fragment is used to replace the resident gene directly. Yeast cells that have taken up DNA in the transformation process are chosen by selecting for a marker present on a plasmid that is mixed with the mutant fragment. Typically between 1 and 10% of plasmid-bearing transformants also carry the transplacement. If the resident gene is marked either by the phenotype that it confers or by any other marker that can be scored, then transformants carrying the transplacement are easily detected.

7. SELECTION SYSTEMS

Many of the procedures described in this chapter depend on the use of easily selectable markers. Auxotrophic markers are usually used for laboratory strains as they are very easy and cheap to use. However, there is increasing interest in manipulating industrial strains that may not have appropriate mutations because of their history, or because they are of dubious ploidy, or because the process for which they are used precludes any other genetic background. A typical example would be strains used in the food or brewing industries. In these cases it is necessary to mark any DNA molecule that is to be used with a dominant selectable marker. Two systems are most commonly used. The first is the copper-thionein gene of *S. cerevisiae, CUP1,* which confers resistance to copper (Butt *et al.*, 1984). This gene can be included in any of the types of vector described and selection is achieved by plating on medium containing copper (Henderson *et al.*, 1985). The second commonly exploited system uses a bacterial transposon (Tn601) gene encoding aminoglycoside phosphotransferase (Jiminez and Davies, 1980). This enzyme inactivates the antibiotic G418 which is active against yeast cells. G418 resistance is most effective when the gene is expressed from a yeast promoter (see Chapter 7). The only disadvantage of the system is that relatively high concentrations of G418 are required and it is expensive.

ACKNOWLEDGMENTS. We thank Frank Caddick for excellent artwork and photography. M.J.D. is a MRC Senior Research Fellow. Work in the Oxford laboratory was supported by MRC, SERC, AFRC, the Royal Society of London, the Nuffield Foundation, Celltech Ltd. and Delta Biotechnology Ltd.

REFERENCES

Beggs, J. D. 1978, Transformation of yeast by a replicating hybrid plasmid, *Nature*, **275:**104–109.

Boeke, J. D., Lacroute, F., and Fink, G. R. 1984, A positive selection for mutants lacking orotidine-5′-phosphate decarboxylase activity in yeast: 5-Fluoro-orotic acid resistance, *Mol. Gen. Genet.* **197:**345–346.

Broach, J. R. 1982, The yeast plasmid 2 μ circle, *Cell* **28:**203–204.

Burke, D. T., and Olson, M. V. 1986, Oligo-directed mutagenesis of *E. coli* and yeast by simple co-transformation of the primer and template, *DNA* **5:**325–332.

Burke, D. T., Carle, G. F., and Olsen, M. V. 1987, Cloning of large segments of exogenous DNA into yeast by means of artificial yeast chromosome vectors, *Science* **236:**806–813.

Butt, T. R., Sternberg, E. J., Gorman, J. A., Clark, P., Hamer, D., Rosenberg, M., and Crooke, S. T. 1984, Copper metallothionein of yeast: Structure of the gene and regulation of expression, *Proc. Natl. Acad. Sci. USA* **81:**3332–3336.

Cesareni, G., and Murray, J. A. H. 1987, Plasmid vectors carrying the replication origin of filamentous single-stranded phages, in: *Genetic Engineering*, Vol. 9 (J. K. Setlow, ed.), Plenum Press, New York, pp. 137–144.

Clarke, L., and Carbon, J. 1976, A colony bank containing synthetic Col E1 hybrid plasmids representative of the entire *E. coli* genome, *Cell* **9:**91–99.

Clarke, L., and Carbon, J. 1980, Isolation of a yeast centromere and construction of functional small circular chromosomes, *Nature* **257:**504–509.

Glover, D. M. 1985, *DNA Cloning. Vol. I/II. A Practical Approach,* IRL Press, Oxford.

Henderson, R. C. A., Cox, B. S., and Tubb, R. 1985, The transformation of brewing yeasts with a plasmid containing the gene for copper resistance, *Curr. Genet.* **9:**133–138.

Hinnen, A., Hicks, J. B., and Fink, G. R. 1978, Transformation of yeast, *Proc. Natl. Acad. Sci. USA* **75:**1929–1933.

Ito, H., Fukuda, Y., Murata, K., and Kimura, A. 1983, Transformation of intact yeast cells treated with alkali cations, *J. Bacteriol.* **153:**163–168.

Jayaram, M., Li, Y.-Y., and Broach, J. R. 1983, The yeast plasmid 2 μ circle encodes components required for its high copy number propagation, *Cell* **34:**95–104.

Jiminez, A., and Davies, J. 1980, Expression of a transposable antibiotic resistance element in *Saccharomyces, Nature* **287:**869–871.

Kingsman, A. J., Clarke, L., Mortimer, R. K., and Carbon, J. 1979, Replication in *Saccharomyces cerevisiae* of plasmid pBR313 carrying DNA from the yeast *TRP1* region, *Gene* **7:**141–152.

Maniatis, T., Fritsch, E. F., and Sambrook, J. 1982, *Molecular Cloning. A Laboratory Manual,* Cold Spring Harbor Laboratory, Cold Spring Harbor, NY.

Moerschell, R. P., Tsunasawa, S., and Sherman, F. 1988, Transformation of yeast with synthetic oligonucleotides, *Proc. Natl. Acad. Sci. USA* **85:**524–528.

Mortimer, R. K., and Schild, D. 1985, Genetic map of *Saccharomyces cerevisiae,* edition 9, *Microbiol. Rev.* **49:**181–213.

Murray, A. W., and Szostak, J. W. 1983a, Pedigree analysis of plasmid segregation in yeast, *Cell* **34:**961–970.

Murray, A. W., and Szostak, J. W. 1983b, Construction of artificial chromosomes in yeast, *Nature* **305:**189–193.

Nasmyth, K. 1985, At least 1400 base pairs of 5′-flanking DNA is required for the correct expression of the HO gene in yeast, *Cell* **42:**213–223.

Nasmyth, K., and Reed, S. I. 1980, Isolation of genes by complementation in yeast: Molecular cloning of a cell-cycle gene, *Proc. Natl. Acad. Sci. USA* **77:**2119–2123.

Nasmyth, K., and Tatchell, K. 1980, The structure of transposable yeast mating-type loci, *Cell* **19:**753–764.

Orr-Weaver, T. L., and Szostak, W. J. 1983, Yeast recombination: The association between double-strand gap repair and crossing-over, *Proc. Natl. Acad. Sci. USA* **80:**4417–4421.

Orr-Weaver, T. L., Szostak, J. W., and Rothstein, R. J. 1981, Yeast transformation: A model system for the study of recombination, *Proc. Natl. Acad. Sci. USA* **78:**6354–6358.

Orr-Weaver, T. L., Szostak, J. W., and Rothstein, R. J. 1982, Genetic applications of yeast transformations with linear and gapped plasmids. In: *Methods in Enzymology*, Vol. 101 (R. Wu, L. Grossman, and K. Moldave, eds.), Academic Press, New York, pp. 228–245.

Parent, S. A., Fenimore, C. M., and Bostian, K. A. 1985, Vector systems for the expression, analysis and cloning of DNA sequences in *S. cerevisiae, Yeast* **1:**83–138.

Ratzkin, B., and Carbon, J. 1977, Functional expression of cloned yeast DNA in *E. coli, Proc. Natl. Acad. Sci. USA* **74:**487–491.

Rothstein, R. J. 1983, One-step gene disruption in yeast, in: *Methods in Enzymology*, Vol. 101 (R. Wu, L. Grossman, and K. Moldave, eds.), Academic Press, New York, pp. 202–211.

Rothstein, R. J. 1985, Cloning in yeast, in: *DNA Cloning. Vol. II. A Practical Approach* (D. M. Glover, ed.), IRL Press, Oxford, pp. 45–66.

Rudolph, H., Koenig-Rauseo, I., and Hinnen, A. 1985, One-step gene replacement in yeast by cotransformation, *Gene* **36:**87–95.

Scherer, S., and Davis, R. W. 1979, Replacement of chromosome segments with altered DNA sequences constructed *in vitro, Proc. Natl. Acad. Sci. USA* **76:**4951–4955.

Shortle, D., Haber, J. E., and Botstein, D. 1982, Lethal disruption of the yeast actin gene by integrative DNA transformation, *Science* **217:**371–373.

Singh, H., Bieker, J. J., and Dumas, L. B. 1982, Genetic transformation of *Saccharomyces cerevisiae* with single-stranded circular DNA vectors, *Gene* **20:**441–449.

Southern, E. M. 1975, Detection of specific sequences among DNA fragments separated by gel electrophoresis, *J. Mol. Biol.* **98:**503–517.

Stiles, J. I., Szostak, J. W., Young, A. T., Wu, R., Consaul, S., and Sherman, F. 1981, DNA sequence of a mutation in the leader region of the yeast iso-1-cytochrome mRNA, *Cell* **25:**277–284.

Stinchcomb, D. T., Struhl, K., and Davis, R. W. 1979, Isolation and characterisation of a yeast chromosomal replicator, *Nature* **282:**39–43.

Struhl, K. 1983, The new yeast genetics, *Nature* **305:**391–396.

Struhl, K., Cameron, J. R., and Davis, R. W. 1976, Functional genetic expression of eukaryotic DNA in *E. coli, Proc. Natl. Acad. Sci. USA* **73:**1471–1475.

Struhl, K., Stinchcomb, D. T., Scherer, S., and Davis, R. W. 1979, High frequency transformation of yeast; autonomous replication of hybrid molecules, *Proc. Natl. Acad. Sci. USA* **76:**1035–1039.

Walder, R. Y., and Walder, J. A. 1986, Oligodeoxynucleotide-directed mutagenesis using the yeast transformation system, *Gene* **42:**133–139.

Williamson, D. H. 1985, The yeast ARS element, six years on: A progress report, *Yeast* **1:**1–14.

Expression of Heterologous Genes

6

MICHAEL F. TUITE

1. INTRODUCTION

With the advent of recombinant DNA technology in the mid-1970s a number of shrewd academics quickly realized the commercial potential of these tools in producing proteins of high commercial value, which, until that time, had proven both difficult and expensive to isolate in quantities sufficient for clinical trials. While *Escherichia coli* as a host organism was an obvious choice in the early stages of commercial genetic engineering (Harris, 1983), it soon became apparent that it was far from the ideal host. It did not make sense, for example, to expect a eukaryotic gene product to be expressed in a prokaryotic environment *and* be correctly folded and modified at the posttranslational level. So by 1980 the search was underway to find a more suitable host and the budding yeast *Saccharomyces cerevisiae* immediately presented itself as an attractive alternative. It is a host that is widely accepted for human consumption by both the consumer and the regulatory authorities, has no associated toxins, and, with the development of an effective plasmid-based gene cloning system (see Chapter 5), its exploitation as a host for producing high-value proteins has progressed rapidly.

Inserting a heterologous gene into a high-copy-number vector and introducing it by transformation into yeast is a relatively straightforward exercise, but such a manipulation does not guarantee even minimal expression of the heterologous gene. To ensure successful and efficient transcription of the heterologous gene it must be inserted into a plasmid immediately adjacent to a DNA sequence that can effectively bind yeast RNA polymerase II, i.e., a yeast promoter. Once transcribed into mRNA, the next hurdle is to get yeast ribosomes to efficiently initiate upon, and to

MICHAEL F. TUITE • Biological Laboratory, University of Kent, Canterbury, Kent CT2 7NJ, England.

accurately translate, the mRNA into the encoded amino acid sequence. Yet success at this second stage of expression still does not guarantee the ultimate goal, namely, to produce high levels of the fully matured heterologous gene product. For example, many eukaryotic proteins need, to a greater or lesser extent, a degree of posttranslational modification or processing to ensure biological activity, and we must also ensure that the polypeptide, once synthesized, is not then rapidly degraded by host proteolytic activities. We must also minimize or circumvent potential deleterious effects that the heterologous polypeptide may present to the general welfare of the yeast host by coaxing the yeast to secrete the polypeptide out into the surrounding growth medium. Finally, we must ensure that the plasmid system being exploited (be it high copy number or low copy number) is stably maintained both under short-term batch growth conditions and in the more long-term continuous culture systems, particularly where high-level synthesis of the foreign polypeptide is desirable.

The aim of this chapter is to provide answers, where possible, to a number of questions normally raised when one first sets out to explore the possibility of expressing a heterologous gene in yeast, namely;

1. What are the best plasmid constructions for introducing a heterologous gene into yeast?
2. How can one ensure that the gene will be transcribed and its corresponding mRNA translated in its new environment?
3. What steps can one take to ensure maximum yield of the gene product?
4. What advantages and disadvantages (if any) are associated with using this microorganism as a host?

Table I can be used as a simple guide to the basic manipulations

Table I. Expression of a Heterologous Gene in Yeast: A Step-by-Step Guide

Step 1	Clone heterologous gene into an appropriate *expression vector*
Step 2	Introduce vector into yeast via *transformation*
Step 3	*Transcription* of the gene and transport of mRNA from the nucleoplasm to the cytoplasm
Step 4	*Initiation of translation* of mRNA
Step 5	*Elongation* of polypeptide chain
Step 6	*Posttranslational modification* of product
Step 7a or	*Intracellular accumulation* of protein
Step 7b	*Removal of signal sequence* during transfer into endoplasmic reticulum, followed by
Step 8	*Secretion* into surrouding medium

required to achieve successful expression of a heterologous gene in yeast, which will be discussed in detail in the following sections.

2. INTRODUCTION OF HETEROLOGOUS DNA INTO YEAST

2.1. Basic Cloning Technology

The routes that can be adopted in cloning and manipulating both prokaryotic and eukaryotic genes have been well documented in a number of "laboratory manuals" (e.g., Maniatis *et al.*, 1982; Dillon *et al.*, 1985). Similarly, the structure and fundamental characteristics of basic cloning vectors, together with transformation systems for introducing cloned genes into yeast, have been described elsewhere in this volume (Chapter 5) and will not be dealt with in detail here.

2.2. Basic Vector Design

A plasmid for introducing heterologous DNA into yeast should (1) be directly selectable in yeast, (2) be replicated autonomously in yeast, (3) possess a high copy number, (4) be able to replicate in *E. coli,* and (5) be directly selectable in *E. coli.* To suit these properties such a plasmid should be either of the YRp or the YEp category, depending on the yeast origin of replication required (see Chapter 5). What are the options available for the various plasmid encoded sequences? Particularly, which sequences have proven useful to date?

2.2.1. Selectable Markers

The majority of basic yeast cloning plasmids carry wild-type genes that are capable of functionally complementing corresponding auxotrophic mutations in the host strain (Table II). However, one might want to introduce vectors into genetically undefined strains (e.g., polyploid sterile brewing strains) or into another *Saccharomyces* species for which there is no appropriate complementable genetic marker. In addition, in large-scale batch culture, where one wants to maintain selective pressure for retention of the plasmid, it is prohibitively expensive to use defined medium lacking the required nutrient. These needs have led to the development of a series of dominant selectable markers (Table II) that require little or no manipulation of the host genotype and can be selected for in rich growth medium.

With the economic and regulatory constraints placed on the use of antibiotic markers, the use of the *CUP1* gene as a selectable marker offers great potential. A multicopy plasmid carrying the *CUP1* gene can

Table II. Selectable Markers That Can Be Used in Yeast Cloning and Expression Vectors

Selectable gene	Host genotype	Selection medium[a]	Reference
Auxotrophic selection			
LEU2	*leu2*	CDM-leu	Kingsman *et al.*, this volume
HIS3	*his3*	CDM-his	Kingsman *et al.*, this volume
URA3	*ura3*	CDM-ura	Kingsman *et al.*, this volume
TRP1	*trp1*	CDM-trp	Kingsman *et al.*, this volume
Dominant selection			
CUP1	Wild type	URM + $CuSO_4$	Henderson *et al.*, 1985
HSV-TK (thymidine kinase)	Wild type	URM + thymidine + amethopterin + sulfanilamide	McNeil and Freisen, 1981; Zhu *et al.*, 1984
R-dhfr (*E. coli* dihydrofolate reductase)	Wild type	URM + sulfanilamide + amethopterin	Miyajima *et al.*, 1984
M-dhfr (mouse dihydrofolate reductase)	Wild type	URM + sulfanilamide + methotrexate	Zhu *et al.*, 1985
hph	Wild type	URM + hygromycin B	Kaster *et al.*, 1984
Km^r	Wild type	URM + G418	Webster and Dickson, 1983
M1 killer toxin immunity gene	K^-: killer sensitive	URM pH 4.6	Bussey and Meaden, 1985

[a]CDM = complete defined medium; URM = undefined rich medium.

be used to directly select for transformants in copper-sensitive brewing strains (Henderson *et al.*, 1985) by plating directly into regeneration medium containing appropriate concentrations of $CuSO_4$, usually 0.5–1 mM. Selection can then be maintained by growing transformants in the presence of this cheap, relatively nontoxic salt.

2.2.2. Origins of Replication

For a plasmid to be able to replicate autonomously in yeast it must contain an origin of DNA replication. Such sequences have been obtained from several different sources: (1) the endogenous 2-μm circle plasmid which carries a single origin of replication (designated ORI), (2) sequences from yeast chromosomal DNA, (3) sequences from yeast mitochondrial DNA, and (4) sequences from heterologous DNA.

In cases 2–4 these sequences are defined purely on the criteria that they confer the ability of autonomous replication in yeast and are referred to as *a*utonomous *r*eplicating *s*equences (ARS) (see Williamson, 1985). Plasmids using an homologous ARS or the 2-μm ORI have generally been incorporated in vectors for introducing foreign genes into yeast. ARS-based plasmids have a high copy number, in the order of 30–50 copies per cell, as do the majority of ORI-based vectors (but see Section 6.1). A major problem with the majority of 2-μm ARS plasmids is that they are highly unstable even when selective pressure is maintained (see Section 6.1).

Almost invariably the *E. coli* origin of replication is of the colE1 type derived from pBR322 or one of its close relatives (pBR325, pBR328, pAT153, etc.). Similarly, the selectable marker for transfer and manipulation in *E. coli* is generally the ampicillin-resistance and/or the tetracycline-resistance determinant derived from the pBR family of plasmid vectors.

3. TRANSCRIPTION OF HETEROLOGOUS GENES

It is unusual to observe transcription of a heterologous gene in yeast unless that gene is linked to a yeast promoter. In addition, a number of other considerations have to be taken into account to ensure that transcription is both efficient and accurate.

3.1. The Problem of Introns

Many of the genes from eukaryotes whose expression is sought in yeast naturally contain noncoding introns. While *S. cerevisiae* has few genes that contain introns, e.g., genes for actin and ribosomal proteins, recent evidence has indicated that yeast has a strict requirement for a sequence located *within* the intron in order to facilitate splicing of the intron from the transcribed RNA (Langford and Gallwitz, 1983). This sequence (5′ TACTAACA 3′) is located between 20 and 55 nucleotides upstream of the 3′ splice site of all yeast introns and is absent from most introns of higher eukaryotic genes. Consequently, although yeast is clearly capable of transcribing eukaryotic, DNA sequences, it is unable to excise the introns posttranscriptionally (Beggs *et al.*, 1980; Langford *et al.*, 1983; Handa *et al.*, 1985) unless they fortuitously contain this sequence (Woudt *et al.*, 1985). A further problem can arise if the heterologous intron contains a sequence capable of functioning as a signal for transcription termination in yeast, as observed by Beggs *et al.* (1980) in their efforts to express a genomic (intron-containing) copy of the rabbit β-globin gene in yeast.

The problem posed by heterologous introns can be effectively overcome by using cDNA copies of the gene of interest (which therefore lacks introns), and since the normal route for cloning higher eukaryotic genes is via cDNA, this generally presents no great difficulty.

3.2. Encoded Pre–Pro Sequences

If a protein whose expression is sought in yeast is a naturally secreted product, then it will be synthesized with a signal or "pre"-sequence that will be encoded by a cDNA copy of the gene. In addition, some of these proteins have additional amino-terminal or carboxy-terminal "pro"-sequences that are removed posttranslationally in order to activate the mature product in yeast; these prosequences are also encoded by the gene. Do these additional sequences affect the transcription and ultimate expression of the gene product? Detailed analysis of the expression of the bovine chymosin gene in yeast (Mellor *et al.*, 1983), a naturally secreted product with an additional prosequence, clearly indicates they can do so. Using a complete cDNA copy of the chymosin gene, variants were constructed in which the presequence had been replaced by a methionine codon (to give met-prochymosin) or in which both the pre- and prosequence had been replaced by a methionine codon (to give met-chymosin) (Fig. 1). The expression of these two variants was compared to that achieved with the intact preprochymosin gene, and significant differences were found with maximum levels being achieved with the met-prochymosin gene (5% total cell protein) and minimal levels with the met-chymosin gene (<0.1% total cell protein). Preprochymosin was expressed at levels up to 1% of total cell protein.

The need to retain the presequence to allow efficient expression of a heterologous gene in yeast was further confirmed by the efforts of Edens *et al.* (1984) to express the gene coding for the sweet-tasting protein thaumatin. They observed 20-fold higher levels of expression if the encoded presequence was retained over that observed for an engineered met-thaumatin gene. Presumably, in the case of both chymosin and thaumatin, retention of the presequence facilitates its translation on the ribosomes of the yeast endoplasmic reticulum where the presequence is removed after passage through the membrane.

In contrast to these observations, however, Hitzeman *et al.* (1983b) noted that expression of the human interferon genes IFNα1 and IFNα2 was significantly *less* if the associated presequences were retained, but their data were based on levels of biologically active IFN detectable in or outside the yeast cell rather than direct measurements of the levels of

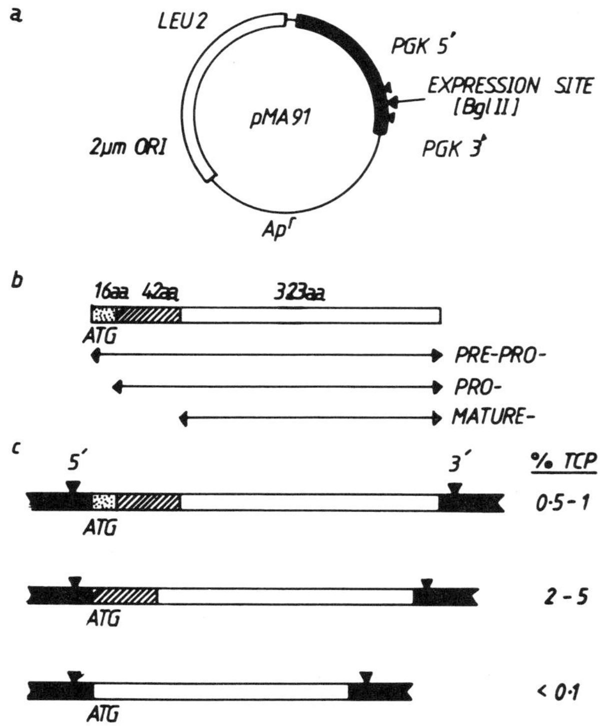

Figure 1. The expression of different derivatives of calf chymosin in yeast: the influence of the pre- and prosequences on levels of expression. (a) The yeast expression plasmid pMA91 (Mellor *et al.*, 1983) that contains the *LEU2* and 2-µm circle origin of replication (open box) and the 5′ promoter and 3′ terminator sequences of the gene encoding phosphoglycerate kinase (*PGK;* closed box). The heterologous gene is "sandwiched" between these control sequences by insertion into the unique BglII "expression site." ▼ indicates mRNA start and polyadenylation sites. (b) Naturally occurring derivatives of calf chymosin: the precursor polypeptide preprochymosin, which during secretion loses the 16-amino-acid signal peptide (stippled box) to form the zymogen prochymosin. At low pH the 42-amino-acid prosequence (hatched box) is removed via an autocatalytic cleavage process to produce the 323-amino-acid mature chymosin polypeptide that has an aspartyl proteinase activity. (c) Levels of expression of the three different derivatives of chymosin in yeast, represented as percentage of total cell protein in transformed strains (% TCP). For details of construction of the DNA fragments, see Mellor *et al.* (1983).

IFN-related proteins, and, as the authors suggest, pre-IFN could lack biological activity.

These data suggest that if one wants to express a naturally secreted heterologous protein intracellularly in yeast, care must be taken to examine the levels of expression obtained both with the native full-length gene and with a truncated version lacking any potential amino-terminal pre- and/or prosequences.

3.3. Transcriptional Promoters

Optimal transcription of a heterologous gene in yeast requires that it be coupled to a yeast transcriptional promoter. Yeast promoters contain many of the canonical sequences that are common to higher eukaryotic promoters, although the locations of these sequences, namely the "TATA" box and the "CAAT" box, with respect to the mRNA start site, are much more heterogeneous than observed in higher eukaryotes (for review, see Struhl, 1986). Yeast genes also contain upstream *cis*-acting elements (upstream activator sequences; UAS) that modulate transcription and appear to be analogous to higher eukaryotic "enhancer" elements (Guarente, 1984). The basic features of a yeast transcriptional promoter and translation sequence requirements are shown in Fig. 2.

Given the similarity of transcription signals between yeast and higher eukaryotes, it is somewhat surprising that *S. cerevisiae* is largely unable to efficiently and accurately initiate transcription on a heterologous eukaryotic promoter. Examples of aberrant initiation on a heterologous promoter include the rabbit β-globin gene (Beggs *et al.*, 1980), the *Dictyostelium* discoidin I gene (Jellinghaus *et al.*, 1982), the gene from *Drosophila* capable of complementing the yeast *ade8* mutation (Henikoff and Furlong, 1983), the *Drosophila hsp70* heat shock gene (Nicholson and

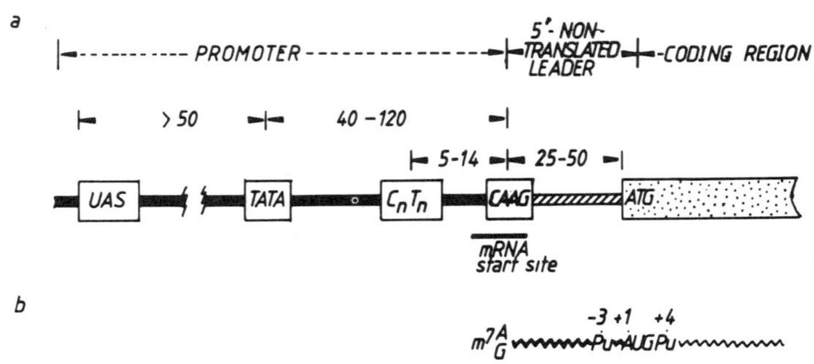

Figure 2. The basic sequence features of an efficient yeast transcriptional promoter and its corresponding mRNA. (a) The promoter region contains the following elements: UAS = upstream activating sequence; TATA = consensus sequence common to all eukaryotic promoters and implicated in controlling the fidelity of transcription initiation; C_nT_n = pyrimidine (cytosine + thymine) rich tract between 15 and 30 nucleotides in length; CAAG = consensus sequence at or adjacent to which mRNA start site maps; ATG = initiation of translation codon. The approximate distance in nucleotides between these elements in indicated. (b) The 5′ end of the corresponding mRNA $m^7{}^{A}_{G}$ = methylated cap; Pu = purine; AUG = initiation of translation codon.

Moran, 1984), and the bean phaseolin gene (Cramer *et al.*, 1985). Even the promoters of a number of genes from the fission yeast *Schizosaccharomyces pombe* are not correctly recognized by *S. cerevisiae* RNA polymerase II (Russell, 1983; Losson and Lacroute, 1983).

There are a number of examples of bacterial 5′ noncoding sequences functioning as promoters in yeast, e.g., those associated with the *bla* and *lac* genes (Breunig *et al.*, 1982), but in both these cases the precise 5′ ends of the mRNA transcripts synthesized in yeast were not mapped and it is therefore not completely clear whether yeast RNA polymerase II initiates at or near the *E. coli* promoter.

There are at least three reported examples of yeast RNA polymerase II successfully recognizing the natural transcription initiation site of a heterologous gene, namely, for the maize storage protein genes coding for the 19-kDa and 21-kDa polypeptides (Langridge *et al.*, 1984), the French bean phaseolin gene (Cramer *et al.*, 1985), and the soybean leghemoglobin Lbc_3 gene (Jensen *et al.*, 1986).

It is therefore unwise to expect that the natural promoter of a

Table III. Homologous Promoter Sequences Used to Direct the Expression of Foreign Genes in Yeast

Gene	Gene product	Efficiency[a]	Reference
ADH1	Alcohol dehydrogenase I	High	Hitzman *et al.* (1981); Ammerer (1983)
ENO	Enolase	High	Innis *et al.* (1985)
PGK	3-Phosphoglycerate kinase	High	Tuite *et al.* (1982); Hitzman *et al.* (1983a)
TPI	Triose phosphate isomerase	High	Smith *et al.* (1985); Zhu *et al.* (1985)
GAP	Glyceraldehyde-3-phosphate dehydrogenase	High	Urdea *et al.* (1983)
GAL1,10	Galatokinase/galactose epimerase	Medium	Goff *et al.* (1984); Stepien *et al.* (1983)
PHO5	Alkaline phosphatase	Medium	Miyanohara *et al.* (1983); Rosenberg *et al.* (1984)
MFα1	α-Mating factor	Medium	Bitter *et al.* (1984); Singh *et al.* (1984)
ARG3	Ornithine carbamoyl transferase	Low	Cabezon *et al.* (1984)
TRP1	N(5′-phosphoribosyl)-anthianilate isomerase	Low	Dobson *et al.* (1983)
TRP5	Tryptophan synthase	Low	Miyajima *et al.* (1984)
URA3	Orotidine-5′-phosphate decarboxylase	Low	Smith *et al.* (1985)
ACT1	Actin	Low	Zhu *et al.* (1985)
CYC1	Iso-1-cytochrome C	Low	Gritz and Davies (1983)

[a]Defined by relative mRNA concentration; high $> 1\%$ total mRNA; medium, 0.1–1.0% total mRNA; low $< 0.1\%$ total mRNA.

heterologous gene will function efficiently in yeast. Consequently, attention has turned to cloning and identifying yeast promoters and coupling them to heterologous gene sequences. As shown in Table III, a large number of yeast promoters have been used for this purpose.

When transcribing a heterologous gene in yeast using a yeast promoter, two strategies are available: (1) that a promoter lacking any adjacent coding sequence be fused directly to the intact coding sequence to be expressed, i.e., to construct a *transcriptional fusion*—this results in the synthesis of a native gene product; or (2) that a promoter, together with a few amino-terminal codons of its corresponding gene, be fused in the correct translational reading frame to the heterologous gene sequence, i.e., to construct a *translational fusion*—this will result in the synthesis of a hybrid or fusion protein which may or may not retain the required biological activity.

Almost invariably the option of a transcriptional fusion is chosen, but to be able to engineer such a "promoter" fragment one must have detailed structural and functional information with regard to DNA sequences important for both transcriptional efficiency and regulation. In this context perhaps the efficient phosphoglycerate kinase (*PGK*) promoter is one of the better-understood promoters (Stanway *et al.*, 1987). As shown in Fig. 3, a series of *PGK*-promoter fragments have been isolated and fused to the mature human IFNα2 sequence and levels of

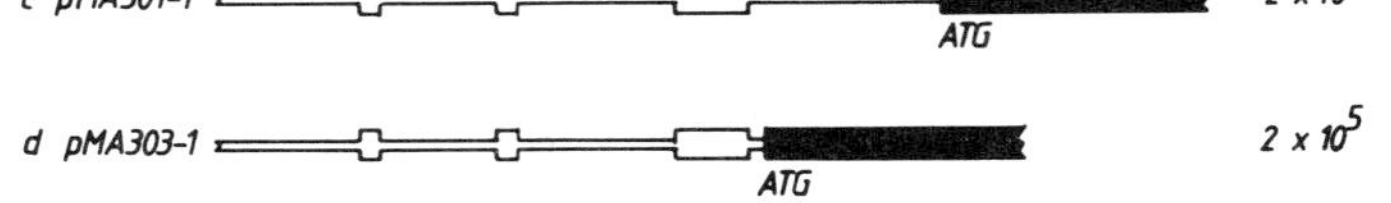

Figure 3. Transcriptional and translational fusion vectors based on the yeast phosphoglycerate kinase gene (PGK): a comparison of levels of expression of the human interferon IFNα2 gene. (a) cPGK: gives the basic structure of the intact *PGK* gene, giving nucleotide positions relative to the ATG codon. T = TATA box; CT = pyrimidine-rich tract. Hatched region = PGK coding sequence. (b) pMA230-1: a translational fusion with an intact human α2 interferon (1FNα2) gene (closed box). The hybrid protein contains 12 amino acids of the yeast PGK protein. The levels of 1FNα2-PGK expression are shown in terms of molecules per transformed cell. (c) pMA301-1: a transcriptional fusion containing all *PGK* 5′ noncoding DNA up to −1 followed by the intact human 1FNα2 gene. (d) pMA303-1: a transcriptional fusion containing all *PGK* 5′ noncoding DNA up to −48, followed by the intact human 1FNα2 gene. (Data taken from Kingsman and Kingsman, 1983.)

expression of biologically active interferon assayed in transformed yeasts. The optimum levels are achieved when a translational fusion is employed (pMA230), although a transcriptional fusion vector that carries all essential promoter sequences also directs comparatively high levels of interferon synthesis. Removal of promoter sequences near the coding sequence results in a significant reductions in expression levels. Analogous experiments using *PGK*-promoter fragments that have deletions at the distal side of the promoter fragment have shown that a UAS is necessary for optimum promoter activity (Ogden *et al.*, 1986). It is important, therefore, that the location of the functional elements of the chosen promoter be identified and precisely located.

As detailed in Table III, yeast promoters can vary considerably in their efficiencies. The majority of heterologous gene expression studies to date have used the efficient promoters, particularly those associated with glycolytic genes (*PGK, ENO, TPI, GAP*), to ensure that maximal levels of the heterologous mRNA are synthesized.

3.4. Termination of Transcription

Once initiated, transcription of the adjacent heterologous gene occurs unheeded provided no introns are present (see Section 3.1). The next problem is how to get RNA polymerase II to terminate transcription once the required heterologous gene has been transcribed. Yeast genes contain DNA sequences in their 3′ noncoding regions that apparently signal not only transcription termination, but also polyadenylation (Zaret and Sherman, 1982, 1984) since the majority of yeast mRNAs have defined 3′ ends with poly(A) tails (McLaughlin *et al.*, 1973). The nature of these "termination" sequences is still not clear. Zaret and Sherman (1982) have identified the sequence TTT-TAG(T)-TTT in the 3′ ends of a number of (but not all) yeast genes and shown that loss of this sequence from the *CYC1* gene results in a dramatic drop in the levels of *CYC1* mRNA coupled with an increase in size of the mRNA by 1 kb. In contrast, Henikoff and Cohen (1984) have identified a sequence (T_5ATA) that can act autonomously as a partial termination signal, but which requires additional downstream sequences for optimal functioning. Again, this sequence is not found in all yeast genes. Few yeast genes contain the polyadenylation signal (AATAAA) identified in higher eukaryotic genes (Proudfoot and Brownlee, 1976).

An effective termination/polyadenylation sequence 3′ to the heterologous gene would therefore seem essential for efficient production of the mRNA transcript, particularly when one is dealing with a circular template. Consequently, the majority of yeast expression vectors are constructed such that a homologous 3′ noncoding sequence follows the heterologous gene and a number of such sequences have been so employed: *TRP1* (Hitzeman *et al.*, 1981, 1983a), *ADH1* (Urdea *et al.*, 1983), *SUC2* (Goff *et al.*, 1984), *GAP* (Rosenberg *et al.*, 1984), *ARG3* (Cabezon *et al.*, 1984), 2-μm *FLP* (Hitzeman *et al.*, 1983b), *PGK* (Mellor *et al.*, 1983), *TRP5* (Miyajima *et al.*, 1985), CYC1 (Zhu *et al.*, 1985), MFα1 (Brake *et al.*, 1984; Bitter *et al.*, 1984), *TPI* (Thim *et al.*, 1986). In all cases these 3′ sequences ensured efficient termination of transcription irrespective of "strength" of the promoter to which they were coupled.

Although it was originally suggested that the presence of a homologous 3′ termination sequence in the immediate vicinity of the foreign gene was an absolute requirement for efficient expression of the gene (Hitzeman *et al.*, 1983b), it is now clear that heterologous 3′ noncoding sequences often (albeit fortuitously) contain effective transcriptional termination signals [e.g., human IFNα2 (Tuite *et al.*, 1982); human γ-IFN (Derynck *et al.*, 1983)], while Goff *et al.* (1984) have demonstrated that the addition of an extra 1000–2000 nucleotides on the 3′ end of calf chymosin mRNA (synthesized using the yeast *GAL1* promoter) is no less stable than the same mRNA whose synthesis was terminated by a *SUC2* terminator sequence and gave rise to a much shorter, more "authentic" transcript.

We know very little about the nucleotide sequences important for termination of transcription and polyadenylation of mRNAs in yeast, and it is not clear as to the necessity for such sequences immediately adjacent to the heterologous gene. Transcription must, however, be terminated effectively at some point during expression of the foreign gene, which leads to the general recommendation that an identified transcriptional terminator should be included in a yeast expression vector.

3.5. Regulated Promoters

A number of heterologous gene products synthesized at high levels in *S. cerevisiae* have a detrimental effect on cell growth; some may even be toxic. Strong yeast promoters that are constitutive (e.g., *ADH1*) thus present a problem when used to direct the synthesis of such a toxic product; namely, how does one achieve high intracellular levels of the product without impairing growth or viability of the host?

One possible way to alleviate this problem is to use a promoter whose expression can be effectively regulated up or down by changes in growth conditions, be they media modifications or changes in temperature. The first demonstration of the ability to regulate heterologous gene expression in yeast was with the *PGK* promoter coupled to the human IFNα2 gene (Tuite *et al.*, 1982). PGK is one of a number of glycolytic enzymes whose synthesis is coordinately regulated by the carbon source; when cells are grown on a fermentable carbon source such as glucose, enzyme levels can be 100-fold higher than when they are grown on a nonfermentable source such as acetate (Maitra and Lobo, 1971). This regulation occurs at the level of transcription (Holland and Holland, 1978). The synthesis of human IFNα2 coupled to the *PGK* promoter showed similar regulation, with the intracellular levels of IFNα2 being some 20-fold higher on glucose medium than on acetate medium (Tuite *et al.*, 1982). However, the level of IFNα2 in acetate-grown cells (approximately 8×10^4 molecules/cell) was still quite high and it would be more desirable if this "uninduced" level were lower.

Currently three tightly regulated yeast promoters are being used widely in expression plasmids.

3.5.1. GAL1,7,10

The promoters of these three genes, encoding the galactose-utilizing enzymes in yeast, are tightly repressed in cells grown in the presence of glucose, but are induced approximately 1000-fold when glucose is replaced by galactose (St. John and Davis, 1981; Johnston and Davis, 1984). This activation by galactose can be achieved at relatively high cell densities, thus making these promoters particularly attractive for utilizing in tightly regulated expression plasmids. To date, only the *GAL1* and *GAL10* promoters have been used for this purpose (e.g., Stepien *et al.*, 1983; Smith *et al.*, 1985; Kataoaka *et al.*, 1985; Kornbluth *et al.*, 1987). The potential of utilizing the *GAL* promoters is limited somewhat by the level of the endogenous *GAL4*-encoded protein, which becomes rate-limiting when the cell contains multiple copies of an induced *GAL* promoter (Johnston and Hopper, 1982); the GAL4 protein is a positive regulatory factor required for the activation of *GAL* promoters. A concomitant regulated overexpression of the GAL4 protein may overcome this problem.

3.5.2. PHO5

This gene encodes the yeast-repressible acid phosphatase and its transcription is tightly repressed in medium containing inorganic phosphate. *PHO5* transcription can be activated by transferring cells to medium depleted of inorganic phosphate (Bostian *et al.*, 1980), but this means of regulating heterologous gene expression cannot be efficiently carried out during large-scale fermentations. Consequently, Kramer *et al.* (1984) have used a host strain carrying mutations in the *PHO80* and *PHO4* regulatory genes such that at 35°C the *PHO5* promoter is not activated even in low phosphate medium, but when cells are switched to 23°C the promoter is activated in medium either containing or lacking inorganic phosphate. This simple temperature switch has been used to successfully regulate the expression of human leukocyte interferon (Kramer *et al.*, 1984).

3.5.3. α-Factor

The expression of mating-type-specific genes, including the gene coding for α-factor, i.e., *MFα1*, is tightly regulated by the products of the *SIR* genes. Brake *et al.* (1984) made use of a conditional (temperature-sensitive) *sir3* mutant to develop a highly regulated promoter system for expressing foreign genes in yeast based on the *MFα1* promoter. They coupled the gene for human epidermal growth factor (hEGF) to the *MFα1* promoter and introduced the plasmid into a *sir3* mutant. If these transformed cells were grown at 36°C, the *MFα1* promoter was not functional, but when the transformed cells were shifted down to the permissive temperature (24°C), a 10^5-fold increase in expression of hEGF was observed after approximately six generations of growth. An added advantage to using the *MFα1* system is that the foreign proteins are secreted if they are provided with the *MFα1* signal sequence (Emr *et al.,* 1983; Brake *et al.,* 1984, and see Section 5.3).

A number of promoters are thus available which, in conjunction with appropriate host mutations, provide highly regulated transcription systems for heterologous gene expression in yeast.

4. TRANSLATION OF HETEROLOGOUS mRNAs

Once transcription of the heterologous gene has been efficiently and accurately achieved, the heterologous mRNA must then be recognized and translated by yeast ribosomes. As with transcription, there are a number of sequence features that many yeast mRNAs have in common (Fig. 2), and these must be taken into consideration if one is to achieve efficient and accurate translation of a heterologous mRNA.

4.1. Codon Bias

The choice of synonymous codons in both prokaryotic and eukaryotic organisms is strongly biased, and in this respect *Saccharomyces* is no exception (Bennetzen and Hall, 1982; Sharp *et al.,* 1986; and Table IV). This nonrandom choice of synonymous codons appears to be directly related to the availability of tRNA isoacceptors within the cell (Ikemura, 1982). Furthermore, genes that are strongly expressed show a much greater bias in their codon usage than genes that are expressed at lower levels; this is believed (but not formally proven) to reflect differences in the rate of translation of these mRNAs. The concentration of charged cognate tRNA governs the step time required to add amino acids opposite each codon, and rapid translation would therefore be favored by the use of triplets for high-abundance tRNAs. Analysis of

Table IV. Relative Synonymous Codon Usage in Highly Expressed Saccharomyces Genes[a]

	U		C		A		G	
U	UUU	0.42	UCU	3.17	UAU	0.26	UGU	1.78
	UUC	1.58	UCC	2.17	UAC	1.74	UGC	0.22
	UUA	0.80	UCA	0.23	UAA	2.61	UGA	0.24
	UUG	4.50	UCG	0.09	UAG	0.16	UGG	1.00
C	CUU	0.13	CCU	0.50	CAU	0.52	CGU	0.64
	CUC	0.02	CCC	0.03	CAC	1.48	CGC	0.01
	CUA	0.42	CCA	3.44	CAA	1.89	CGA	0.00
	CUG	0.13	CCG	0.03	CAG	0.11	CGG	0.00
A	AUU	1.36	ACU	1.94	AAU	0.28	AGU	0.17
	AUC	1.58	ACC	1.78	AAC	1.72	AGC	0.16
	AUA	0.06	ACA	0.22	AAA	0.38	AGA	5.20
	AUG	1.00	ACG	0.06	AAG	1.62	AGG	0.14
G	GUU	2.18	GCU	2.72	GAU	0.84	GGU	3.80
	GUC	1.65	GCC	1.13	GAC	1.16	GGC	0.15
	GUA	0.04	GCA	0.12	GAA	1.83	GGA	0.02
	GUG	0.13	GCG	0.04	GAG	0.17	GGG	0.03

[a]The data are taken from Sharp *et al.* (1986), who analyzed the synonymous codon usage in 32 *Saccharomyces* genes showing a codon bias index characteristic of a highly expressed gene, i.e., greater than 0.6.

110 yeast genes (Sharp *et al.*, 1986) reveals that while yeast is capable of translating all 61 amino acid–specifying codons, there are a number of codons that are very rarely used in yeast mRNAs, e.g., CGA/CGG,arg.

Codon bias in higher eukaryotic mRNAs is markedly different from that observed for yeast mRNAs, yet it is clear that yeast can efficiently translate a wide range of heterologous mRNAs. There are at least three *reported* instances of heterologous genes, successfully transcribed in yeast, yet whose mRNAs are not, for whatever reason, translated into protein: human synthetic proinsulin (Stepien *et al.*, 1983), rat growth hormone (Ammerer, 1983), and the Drosophilia *hsp70* gene (Nicholson and Moran, 1984). If one determines the codon bias index (Bennetzen and Hall, 1982) for heterologous mRNAs successfully translated, it is clear that this appears to have minimal influence. For example, the codon bias index for hepatitis B surface antigen is 0.02 (Bitter and Egan, 1984), while highly expressed yeast genes such as *ADH1* and *PGK* have indices of 0.90 or greater.

With the advanced technology of oligonucleotide synthesis it has now become feasible to examine directly the influence of codon bias by chemically synthesizing a gene using only those codons that are used at high frequency in yeast mRNAs. This indeed has been done already; for example, Urdea *et al.* (1983) synthesized a gene coding for human epidermal growth factor, a 53-amino polypeptide, using only these codons found in highly expressed yeast genes. Not surprisingly, this synthetic gene was expressed in yeast when linked to the yeast *GAP* promoter although the levels of expression reported (30 μg/liter) were low. Levels of expression with the native gene for hEGF were not reported. Similarly, Bitter *et al.* (1984) have reported the expression of synthetic genes for β-endorphin and α-interferon in yeast. Bitter and Egan (1984) constructed a partly synthetic Hepatitis B surface antigen (HBsAg) gene in which the first 30 codons were chemically synthesized utilizing "ideal" yeast codons while the remainder of the gene contained native sequences. While this hybrid gene was expressed 10- to 15-fold more efficiently than the native HBsAg gene, this is probably the result of concommitant modification of the 5′ noncoding region of the HBsAg mRNA (see Section 4.2).

From the preceding discussion it is clear that there is little evidence that codon bias plays an important role in determining the steady-state levels of a heterologous protein in yeast although it may influence the fidelity with which a heterologous mRNA is translated. On the other hand, successful expression in yeast of two synthetic genes constructed using codons preferred by *E. coli,* namely, those for human growth hormone (Tokunaga *et al.,* 1985) and human proinsulin (Stepien *et al.,* 1983), has been achieved.

4.2. 5′-Untranslated mRNA Leader

The translation initiation codon AUG is invariably preceded by a 5′ untranslated leader sequence of up to 100 bases in length. In eukaryotes, including yeast, this includes a methylated "cap" at the 5′ end of the mRNA. The 40S subunit of the ribosome is believed to recognize the 5′ cap and subsequently migrate along the mRNA until it encounters the first AUG codon (Kozak, 1981, 1984). Analysis of the nucleotide sequences of cloned yeast genes encoding abundant proteins reveals that the occurrence of bases in the 5′ untranslated leader of these genes is not random (Table V). In particular, there is a strong bias *for* A residues (51.9%) and *against* G residues (4.8%). In addition, these 5′ leader regions are comparatively short for highly expressed genes, being on average 41 nucleotides (Table V).

Are those features of the 5′ noncoding region important for translation of heterologous mRNAs in yeast? The answer is almost certainly yes. For example, Kniskern *et al.* (1986), by replacing the heterologous 5′ nontranslated sequence of hepatitis B core antigen (HBcAg) with an authentic yeast (GAP) 5′ sequence, increased the levels of detectable protein by almost two orders of magnitude without significantly changing the steady-state levels of RNA. Similarly, Bitter and Egan (1984) replaced the 5′ nontranslated region of the HBsAg gene (which includes

Table V. Nucleotide Composition of 5′ Untranslated mRNA Leader Sequences in 10 Highly Expressed Yeast Genes

Gene	Length of 5′ leader (nucleotides)	No. of nucleotides A	U	G	C
GAP11	82	39	19	6	18
GAP491	44	24	8	3	9
GAP63	53	30	10	2	11
ENO8	36	20	7	1	8
ENO46	41	21	6	2	12
PGK1	39	20	12	2	5
TPI1	20	13	3	0	4
PYK1	33	19	4	1	9
ADH[a]					
I	37	16	9	2	10
II	27	12	7	1	7
AVE	41	51.9%	20.6%	4.8%	22.6%

[a]There are two major *ADH1* transcripts.

10 G residues and only six A residues in the 25 nucleotides immediately preceding the translational initiation codon) with a chemically synthesized fragment containing 27 nucleotides immediately 5′ to the initiation codon of the efficiently expressed yeast *GAP491* gene (containing 17 A residues and only one G residue). A 10- to 15-fold increase in the levels of HBsAg synthesized was achieved by this manipulation.

The work of Sherman and his colleagues (Baim *et al.*, 1985; Laz *et al.*, 1988) using *in vitro* modifications of the yeast *CYC1* mRNA have confirmed that sequences 5′ to the AUG will inhibit translation if they contain a high proportion of G residues or if they can form hairpin loop structures.

Sequences within the 5′ untranslated leader therefore have profound effects on expression of both homoglous and heterologus mRNAs in yeast and must be taken into consideration when engineering a gene for high-level expression in yeast.

4.3. AUG Context

In higher eukaryotic mRNAs Kozak (1984, 1986) has shown that the nucleotide context within which the initiator AUG codon is found can have a profound effect on translational efficiency of the corresponding mRNA, and that the nucleotides immediately flanking the AUG codon are not random. A recent analysis of 131 yeast mRNA sequences clearly indicates a nonrandom distribution of nucleotides around the AUG codon (Cigan and Donahue, 1987). In particular, the −3 position is occupied by an adenine residue in 75% of the mRNAs examined. However, the bias in nucleotide distribution around the yeast AUG codon (5′-AAA*AUG*UCU-3′; Cigan and Donahue, 1987) is different from the higher eukaryotic consensus (5′-CACC*AUG*G-3′; Kozak, 1984, 1986). The nucleotides adjacent to the AUG may be important in dictating the efficiency with which a particular AUG codon is recognized as a "stop" signal for the migrating 40S ribosomal subunit (Kozak, 1981). The importance of the nucleotide at −3 for maximal translational efficiency has been confirmed by Baim *et al.* (1985), with *CYC1* mRNA with a purine, preferably A, at the −3 position being required for highest efficiencies of translation of this mRNA. The sequence context is not absolutely required for translation, however.

It therefore seems necessary to have the AUG initiator codon of a heterologous mRNA in a favorable nucleotide environment if one is to achieve maximum efficiency of translation of that mRNA, namely, to have a purine (preferably A) at the −3 position and preceded by a 5′ untranslated leader that is less than 50 nucleotides in length, rich in A residues, and lacking any G residues or stable secondary structure potential.

5. POSTTRANSLATIONAL MODIFICATIONS

Once a heterologous mRNA has been successfully translated into a polypeptide, subsequent modification is often necessary to produce a mature and functional polypeptide. This is particularly true when the polypeptide is a secretory or membrane protein which is subjected to three principal types of modification: formation of disulfide bonds, specific proteolytic cleavages, and addition of carbohydrate residues. While little is known about how correct disulfide bond formation is catalyzed in yeast, a great deal is known about the latter two, particularly the steps one takes to ensure they take place.

5.1. Yeast Secretion Pathway

High intracellular levels of expression of naturally secreted heterologous proteins in yeast might not be desirable for a number of reasons: the high levels might prove toxic to the host cell, the product may accumulate as an insoluble aggregate, or the product may not be fully posttranslationally processed. In principle, getting yeast to secrete proteins into the surrounding medium should overcome these problems and furthermore should ease the downstream processing requirements since a secreted protein will be contaminated with few other proteins.

Yeast is capable of efficiently secreting a number of its own polypeptides, provided the nascent polypeptide has a short (15–30 amino acid) hydrophobic amino-terminal region or "signal sequence." This signal sequence facilitates contact with the membrane of the endoplasmic reticulum (ER). The nascent polypeptide is then cotranslationally translocated across the membrane of the ER at which point the signal peptide is cleaved by a membrane-bound protease. The secretory pathway then involves a series of membrane-bound structures that transfer the mature polypeptides to their site of release at the plasma membrane.

As in higher eukaryotes, protein secretion in yeast is often linked to protein glycosylation (for review, see Tanner and Lehle, 1987; and see Chapter 2). Initially *N*-linked asparagine core glycosylation of protein occurs probably during translocation of the polypeptide through the membrane of the ER. Core oligosaccharides are then extended in a Golgi-like organelle by addition of outer-chain sugar branches. Finally, fully glycosylated proteins are packaged into vesicles which are transported to the growing bud where they fuse with the plasma membrane and discharge their contents into the periplasmic space (Schekman and Novick, 1982).

The yeast cell surface consists of at least three layers, the cell wall, a periplasm, and a plasma membrane (Ballou, 1982). Most of the enzymes secreted naturally by yeast are located in the periplasm or in the cell wall; examples include invertase, acid phosphatase, and L-asparaginase. Some smaller nonglycosylated proteins, such as α-factor and killer toxin, are efficiently secreted through the cell wall and into the surrounding culture medium, yet the total amount of protein yeast cells secrete into the medium represents less than 1% of the total cell-associated protein. The factors that determine periplasmic or extracellular location of yeast proteins are unknown although, as in other organisms, they are all synthesized with a cleavable hydrophobic amino-terminal signal sequence (see Section 5.3).

For the reasons just outlined it would be most appropriate to engineer yeast to secrete the foreign protein out into the medium rather than into the cell-associated periplasmic space. This approach has been undertaken with remarkable success using both homologous and heterologous signal sequences.

5.2. Use of Heterologous Signal Sequences

As catalogued by Perlman and Halvorson (1983), signal sequences found in both prokaryotes and eukaryotes, although varying in length and amino acid composition, all possess a hydrophobic core of approximately 12 amino acids followed by a three-amino-acid peptidase recognition sequence: A-X-B, with A being a large aliphatic amino acid (such as leu, val, or ile) as are those found at B (principally ala, gly, and ser).

The most frequently occurring peptidase recognition sequence found is *ala X ala*.

Reasoning that secretion signals may be functionally similar in all eukaryotes, a number of groups have examined the ability of heterologous signal sequences to direct the secretion of foreign proteins from yeast, with varying degrees of success (Table VI). The results obtained clearly demonstrate that heterologous signal sequences can be recognized and processed in the yeast secretory pathway, but their applicability is far from satisfactory because (1) the efficiency of secretion is relatively low and, with the exception of mouse α-amylase (Thomsen, 1983) and *Aspergillus awamori* glucoamylase (Innis *et al.*, 1985), the majority of the protein remains associated with the cell, and (2) a number of aberrantly processed proteins are often synthesized, both intracellularly and extracellularly, where the signal peptide has only been partially removed. For example, Hitzeman *et al.* (1983b) have shown that, of the IFNα2 secreted from yeast using its associated signal sequence, only 64% was properly processed, with the remaining 36% containing an additional three amino acids at the amino terminus. Intracellular forms of IFNα2 contained, in addition, a third species with an additional 14 amino terminal amino acids. That is not to say yeast cannot correctly process heterologous signal sequences—Sato *et al.* (1986) have shown that the 15-amino-acid signal of human salivary α-amylase is cleaved correctly

Table VI. Heterologous Signal Sequences Examined in Yeast: Location of the Gene Product

<table>
<tr><th rowspan="2">Gene</th><th rowspan="2">Reference</th><th colspan="3">Location of gene product (% total)</th></tr>
<tr><th>Intracellular</th><th>Periplasmic</th><th>In medium</th></tr>
<tr><td>Human IFNα1</td><td>Hitzeman et al., 1983b</td><td>96</td><td colspan="2">4</td></tr>
<tr><td>Human IFNα2</td><td>Hitzeman et al., 1983b</td><td>92–98</td><td colspan="2">2–8</td></tr>
<tr><td>Human IFNγ</td><td>Hitzeman et al., 1983b</td><td>79–80</td><td colspan="2">10–21</td></tr>
<tr><td>Wheat α-amylase</td><td>Rothstein et al., 1984</td><td colspan="2">40–70</td><td>30–60</td></tr>
<tr><td>Human immunoglobulin λ chain</td><td>Wood et al., 1985</td><td colspan="2">60–90</td><td>10–40</td></tr>
<tr><td>Human immunoglobulin μ chain</td><td>Wood et al., 1985</td><td colspan="2">85–95</td><td>5–15</td></tr>
<tr><td>Mouse α-amylase</td><td>Thomsen, 1983</td><td colspan="2">10</td><td>90</td></tr>
<tr><td>Human salivary α-amylase</td><td>Nakamura et al., 1986</td><td>34</td><td>21</td><td>45</td></tr>
<tr><td>French bean phaseolin</td><td>Cramer et al., 1985</td><td colspan="2">100</td><td>0</td></tr>
<tr><td>Chicken egg white lysozyme</td><td>Oberto and Davison, 1985</td><td colspan="2">33</td><td>66</td></tr>
<tr><td>A. awamori glucoamylase</td><td>Innis et al., 1985</td><td colspan="2">10</td><td>90</td></tr>
</table>

and the N-terminal glutamine residue correctly modified into pyroglutamate.

Because of the unpredictability of heterologous signal sequence function in yeast it is therefore more appropriate to couple signal sequences from native yeast genes onto a heterologous coding sequence.

5.3. Homologous Signal Sequences

For optimal product purification it would be advisable to use a signal sequence for a truly extracellular protein. The major candidate for this is the signal sequence of α-factor.

The α-factor locus (*MFα1*) (Fig. 4) encodes a 165-amino-acid prepropolyprotein precursor containing a 22-amino-acid hydrophobic, amino-terminal signal sequence. Only the mature 13-amino-acid peptide with α-factor activity is secreted from the cell. *MFα1*-based secretion vectors generally retain both the signal sequence and the adjacent 61 amino acid prosequence (Fig. 4) and have been shown to efficiently direct the secretion of a number of foreign proteins, for example, β-

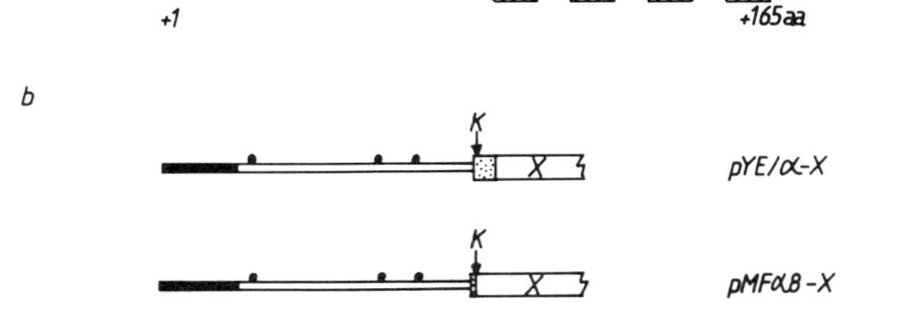

Figure 4. Secretion vectors based on the yeast α-factor gene *MFαl*. (a) Schematic representation of the *MFα1* locus and the α-factor precursor polypeptide showing the postulated 20-amino-acid signal sequence (closed box), the 61-amino-acid prosegment (open box), the spacer peptides (stippled box), and the α-factor peptides (hatched box). ● = *N*-glycosylation sites (asn x thr); K = *KEX2*-encoded endopeptidase site (lys arg ↓). For details of processing see Chapter 2. (b) Examples of two secretion vectors utilizing fusions with the *MFα1* gene and heterologous gene X. pYE/α-X (Bitter *et al.,* 1984) contains the signal peptide, the prosegment, and the first spacer peptide (arg glu ala glu ala). The first amino acid of X occupies the same position in the hybrid protein as does the first amino acid of the first α-factor peptide in the native precursor. pMFα8-X (Miyajima *et al.,* 1985) contains just the lys arg of the first spacer peptide, thus retaining the *KEX2* processing site. In both vectors the *MFα1* promoter is used to direct transcription.

endorphin (Bitter *et al.*, 1984), α-interferon (Bitter *et al.*, 1984; Singh *et al.*, 1984), human epidermal growth factor (Brake *et al.*, 1984), bovine prochymosin (Smith *et al.*, 1985), mouse interleukin-2 (Miyajima *et al.*, 1985), rat carboxypeptidase A (Cardell *et al.*, 1985), and somatomedin-C (Ernst, 1986). Although such constructs efficiently secrete mature foreign proteins into the medium, both aberrant amino terminal processing and internal proteolytic cleavages are often observed. In particular, incomplete removal of the (glu-ala) repeats from the amino terminal region, due to rate-limiting quantities of the dipeptidyl aminopeptidase encoded by the *STE13* gene (Julius *et al.*, 1983), has been observed (Brake *et al.*, 1984; Bitter *et al.*, 1984; Thim *et al.*, 1986). This problem can be overcome by simultaneously introducing into the transformed cell extra copies of the *STE13* gene either on the same plasmid or directly integrated into the chromosome. Brake *et al.* (1984) have demonstrated that removal of the (glu-ala) repeats from the vector by *in vitro* mutagenesis results in all the foreign proteins being secreted with authentic amino terminal sequences—this construct retained the lys-arg *KEX2* protease-cleavage sequence immediately adjacent to the coding sequence and was therefore sufficient for the necessary amino-terminal processing.

A second, more serious problem is internal proteolytic cleavage of the secreted foreign proteins with, in the extreme case of β-endorphin (Bitter *et al.*, 1984), no complete mature protein being found outside the cell. These cleavages occur exclusively on the carboxyl side of certain (but not all) lys residues within the protein. The extent to which these internal cleavages occur is not determined by adjacent residues, but is rather determined by the conformation-dependent accessibility of susceptible bonds in the foreign protein. It is clear that not all proteins are equally susceptible to this proteolytic cleavage; for example, 95% of α-interferon secreted from yeast using the *MFα1* signal remains intact (Bitter *et al.*, 1984). The enzyme responsible for this internal cleavage is almost certainly the endopeptidase encoded by the *KEX2* gene, and since the enzyme is required for secretion *per se*, it is difficult to envisage how one might overcome this potentially deleterious activity. Using a *kex2* mutant is clearly not suitable although perhaps the activity of the enzyme can be modified by *in vitro* mutagenic techniques, particularly since the *KEX2* gene has been isolated (Julius *et al.*, 1984).

Much work is still required in optimizing yeast secretion vectors both at the level of signal sequence design and in identifying genetic and physiological parameters important for achieving maximal secretion. Yeast signal sequences other than *MFα1* are currently under investigation, particularly the signal for the yeast killer toxin (Skipper *et al.*, 1985) and for the extracellular glucoamylase gene (*STA1*, sometimes designated *DEX1*) that has been isolated from *Saccharomyces diastaticus* (Yamashita *et al.*, 1985). The use of the signal sequences from the *PHO5* and *SUC2* genes (coding for repressible acid phosphatase and invertase, respectively) is somewhat limited in that their gene products are not truly extracellular, but rather are secreted into the periplasmic space and thus remain, to a large extent, cell-associated.

The growth phase of the culture is important if one is trying to achieve optimal levels of secretion. A number of reports have indicated that maximal levels of secretion are achieved in stationary phase cultures; Brake *et al.* (1984) for example reported that only 33% of human epidermal growth factor (linked to the *MFα1* signal) was secreted from logarithmically growing cells, yet 24 hr *after* the cells reached stationary phase less than 1% of hEGF remained intracellular, with the recombinant product representing greater than 90% of the secreted proteins in the culture supernatant. The physiological or biochemical basis for these differences is unknown but clearly requires more extensive investigation, as does the observation of Rothstein *et al.* (1984) that levels of secreted wheat α-amylase are much higher in rich medium, although in this case secretion was directed by a heterologous secretion signal.

In yeast there seems to be no size limitation on the proteins that can be effectively secreted from the cell; for example, Schultz *et al.* (1987) have reported the secretion of the 400-kDa major envelope glycoprotein (gp350) of Epstein–Barr virus.

5.4. Posttranslational Modification and Secretion

Protein secretion in yeast, as in other eukaryotic cells, appears to be tightly coupled with posttranslational modification of the nascent polypeptide. Apart from cleavage of the amino-terminal signal sequence by a "signal peptidase," at least four additional modifications can occur to protein naturally secreted from yeast:

1. Addition of a core oligosaccharide to asparagine residues via an *N*-linked glycosylation event, generally at the sequence asn-X-ser/thr (Lehle and Bause, 1984)
2. Formation of an *O*-glycosidic linkage to the seryl or threonyl residues
3. Covalent attachment of long-chain fatty acid
4. Formation of disulfide bridges between cysteine residues

The addition of core oligosaccharides occurs in the endoplasmic reticulum as does the fatty acid acylation, although these two modifications are not coupled. Oligosaccharide side chains are extended in the Golgi-like apparatus prior to transfer to the periplasmic space. Little, if anything, is known about disulfide bond formation in yeast—there is currently no evidence for the presence of a protein disulfide isomerase activity (Freedman, 1984) although at least one naturally secreted protein from yeast (the killer toxin) has at least three disulfide bridges (Bostian *et al.*, 1984).

Use of the antibiotic tunicamycin allows one to determine whether or not a protein synthesized in yeast is subject to *N*-linked glycosylation since this antibiotic is a specific inhibitor of such reactions and can act *in vivo*. For example, Wood *et al.* (1985) used tunicamycin to demonstrate that a significant proportion of immunoglobulin μ chain produced in yeast is *N*-glycosylated. Alternatively, putative glycoproteins can be deglycosylated *in vitro* using either endoglycosidase H or trifluoromethane sulfonic acid.

To ensure the biological activity of a secreted heterologous protein posttranslational modification must be both efficient and correct. A number of investigations have reported the accumulation of nonauthentic forms of several heterologous proteins secreted from yeast that are immunologically related to the authentic product; for example, human salivary α-amylase (Sato *et al.*, 1986), Epstein–Barr virus major envelope glycoprotein gp350 (Schultz *et al.*, 1987), and human T-cell leukemia virus type I envelope protein (Kuga *et al.*, 1986) are all found to be synthesized in more than one form due to incomplete glycosylation of the proteins. There is also evidence that yeast can fail to add any carbohydrate to a naturally glycosylated protein, e.g., hepatitis B surface antigen (Hitzeman *et al.*, 1983a).

The core structure of yeast oligosaccharides is identical to that present on mammalian glycoproteins (Ballou, 1982) but atypically the outer chains are mannose-rich. It is therefore to be expected that the glycosylation patterns of recombinant mammalian glycoproteins will not, by and large, be authentic.

5.5. Other Posttranslational Modification Events

Viral and mammalian proteins can also require other posttranslational modifications in addition to glycosylation and disulfide bond formation to generate biological activity, and recent evidence suggests that yeast can carry out several of the required modifications; for example, phosphorylation and myristoylation ($p60^{v\text{-}src}$; Kornbluth *et al.*, 1987) and amino terminal acetylation (human Cu, Zn superoxide dismutase; Hallewell *et al.*, 1987) have been directly demonstrated in yeast. In addition, yeast, like other eukaryotes, has an enzyme (methionine aminopeptidase) that can cleave off the N-terminal methionine residue cotranslationally. This enzyme is inhibited by the presence of a charged residue in the penultimate residue (Huang *et al.*, 1987), and therefore the possibility exists that the N-terminal methionine will not be removed from a heterologous protein which may effect its biological activity.

6. MAXIMIZING PRODUCT YIELD

Previous sections have dealt with the question "How can one get a heterologous gene expressed authentically in yeast?" Once initial expression has been achieved, attention is then often turned toward maximizing the levels of recoverable, fully processed, biologically active protein. Potentially, given a high-copy-number plasmid (50–100 copies/cell) carrying an efficiently transcribed gene, one should obtain levels in excess of 50% total cell protein. This potential synthesis output has indeed been achieved with homologous gene products; for example, if the entire *PGK1* coding sequence (coding for the abundant glycolytic enzyme phosphoglycerate kinase) and corresponding transcription signals are introduced into yeast on an autonomous high-copy-number YEp vector, more than 50% of the accumulated total cell protein is phosphoglycerate kinase (PGK) (Fig. 5). This compares to a normal level of 1–2% in nontransformed wild-type yeast cells (Mellor *et al.*, 1985). Yet replacing the PGK coding region with a heterologous PGK coding sequence,

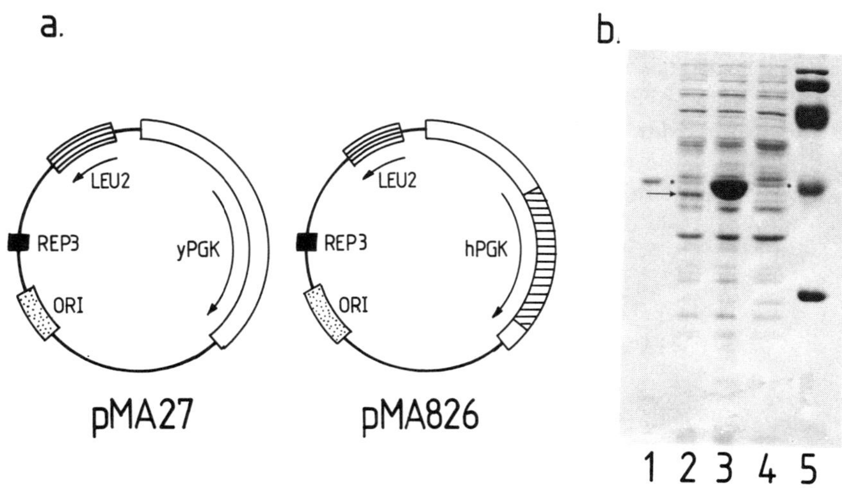

Figure 5. Overexpression of homologous and heterologous phosphoglycerate kinases (PGK) in *S. cerevisiae* transformants. (a) The structure of expression plasmids for high-level synthesis of yeast PGK (pMA27) and human PGK (pMA826). In each case the PGK coding sequence is transcribed from the yeast *PGK* promoter and both plasmids contain the *LEU2* selectable marker and the 2-μm plasmid origin of replication (ORI) and *REP3* (*STB*) sequences. (b) One-dimensional SDS polyacrylamide gel stained with Coomassie blue to detect proteins. Lane 1: Pure yeast PGK. Lane 2: Total soluble protein from a pMA826 transformant (human PGK is indicated by the arrow). Lane 3: Total soluble protein from a pMA27 transformant. Lane 4: Total soluble protein from an untransformed strain. Lane 5: Molecular weight markers. Yeast PGK is indicated by *.

namely, that coding for human PGK, does not result in 50% of total cell protein as human PGK, but rather only 1–2% total cell protein (Fig. 5).

The discrepancy between homologous and heterologous gene expression in yeast is a cause for some concern, and Table VII indicates the possible reasons for such a discrepancy and the consequences each would have on the copy number of the plasmid and the steady-state levels of the mRNA and protein. All three parameters can be assayed routinely (see Chapter 9).

The possible causes for these discrepancies in expression levels in the *PGK*-based expression system have been investigated in detail (Mellor *et al.*, 1985; Chen *et al.*, 1984). The major observation is that there is at least a 10-fold reduction in the steady-state levels of the heterologous mRNA compared to *PGK* mRNA. This indicates there is either a reduction in transcription initiation and/or elongation on the heterologous gene, or the heterologous mRNA has a short half-life. The recent evidence that the *PGK* gene of yeast contains a transcriptional "enhancer" element *within* its coding region (and is thus missing from the heterologous expression plasmid; Mellor *et al.*, 1987) suggests a possible reason for these results; namely, transcription initiation from the *PGK* 5′ promoter region is not optimal in the absence of coding and/or 3′ sequences. It should be stressed that this may be unique to the *PGK* gene and this comparison is indeed further complicated by the relatively long half-life of both PGK mRNA and protein (Chen *et al.*, 1984). It does, however, illustrate that placing a heterologous gene between efficient 5′ and 3′ transcription signals may not be sufficient to guarantee maximal transcription of the heterologous gene.

The efficiency of gene expression initiation is not the only para-

Table VII. Possible Reasons for Differences in Efficiency of Expression of Homologous vs. Heterologous Genes in Yeast

	Consequences		
	Plasmid	Steady state	
Cause	Copy no.	mRNA levels	Protein levels
1. Decreased plasmid copy number	Low	Low	Low
2. Structural instability of plasmid	High	Low	Low
3. Increased protein turnover	High	High	Low
4. Decreased translation of mRNA	High	High	Low
5. Decreased transcription of gene	High	Low	Low
6. Increased mRNA turnover	High	Low	Low

meter to which careful attention must be paid. Two other parameters should also be considered: the stability of the recombinant plasmid and the genotype of the host strain into which the recombinant plasmid has been introduced.

6.1. Stability of Recombinant Plasmids in Yeast

Recombinant plasmids, present at high copy number in yeast, show varying degrees of stability both in the absence and, to a lesser extent, in the presence of selective pressure. There are two types of plasmid instability: *segregational instability*, giving rise to plasmid-free cells, and *structural instability*, which generates plasmids, structurally related to the parental plasmid, but having either deletions, insertions, or rearrangements. The majority of studies to date have concentrated on the former phenomenon and can be monitored by the loss of a plasmid-encoded phenotypic marker, e.g., auxotrophic requirement. It is, however, important to be aware of the latter class of instability, because it can lead to either loss or reduction in the levels of expression of the plasmid-borne heterologous gene even though that plasmid may still retain the "scorable" phenotypic marker used to monitor plasmid retention.

The parameters that influence the stability of a recombinant plasmid in yeast can be broadly subdivided as follows:

1. The source of the origin of replication
2. The copy number of the plasmid
3. The mechanism for partitioning the plasmid during mitotic cell division
4. The nature of the gene product whose synthesis is being directed by the heterologous gene present on the plasmid
5. The genotype of the host strain employed

The majority of recombinant plasmids utilized in heterologous gene expression systems carry either the 2-μm circle origin of replication (ORI) or a homologous ARS sequence (see Section 2.2.2). It is generally accepted that plasmids utilizing the 2-μm circle sequence show greater stability (Futcher and Cox, 1984).

6.1.1. Stability of 2-μm-Based Vectors

The stability of the 2-μm plasmid is generally attributed to (1) its ability to increase its copy number when its copy number is low (Futcher, 1986), and (2) an efficient, plasmid-encoded partitioning system (Jayaram *et al.*, 1983; Kikuchi, 1983) that ensures equal distribution of plasmid molecules to mother and daughter cells after mitosis. The latter partition system operates without modulating the copy number of the plasmid and consists of two *trans*-acting functions (encoded by *REP1* and *REP2*) and two *cis*-acting functions (*REP3* and ORI). Invariably manipulation of 2-μm sequences during the construction of a chimaeric yeast plasmid leads to disruption or total loss of one or more of these functions. The consequences of such losses on general segregational stability of a constructed plasmid are summarized in Table VIII.

2-μm-based vectors lacking the ORI sequence clearly cannot be used to transform yeast unless an alternative origin of replication (i.e., an ARS sequence) is also present on the plasmid. Removal of any one of the other *REP* components can significantly reduce the stability of the plasmid without impairing its transforming ability (Table VIII). For the two *trans*-acting functions (*REP1* and *REP2*) absence of one or both of these sequences from the constructed plasmid can be overcome by using a host strain that already carries the native 2-μm plasmid which can supply the missing *trans*-acting functions. Note that the majority of laboratory yeast strains are [cir$^+$]).

Loss of *REP3* from the vector has more serious consequences on plasmid stability; a *REP3*-defective plasmid carrying the 2-μm *ORI* is as unstable as an ARS-containing vector. The stabilizing activity of *REP3* is independent of its distance from, or orientation with respect to, the *ORI* sequence (Kikuchi, 1983). The stabilizing property of *REP3* is not restricted to the 2-μm ORI, since it will also stabilize a vector carrying an ARS (Kikuchi, 1983). Thus any vector, to be maximally stable and retain high copy number, should contain the *REP3* sequence, with the *REP1* and *REP2* functions being supplied by the same plasmid or via the endogenous 2-μm population. The origin of the ORI/ARS sequence *per*

Table VIII. Summary of the Stability Characteristics of a Range of Recombinant Yeast Plasmids

Plasmid	2-μm REP function present	% Plasmid-bearing cells (selective)	Segregation bias[a]	Mitotic stability (nonselective)
YRp [cir$^+$]	None	10–30%	M	Very low
YRp [cir^o]	None	10–30%	M	Very low
YEp	None	n.d.	M	Very low
YEp [cir$^+$]	REP3	>90%	None	High
YEP [cir^o]	REP3	n.d.	M	Very low
YCp + ARS	None	>90%	None	High
YRp	REP3	n.d.	None	High
YEp [cir$^{+/o}$]	REP1, REP2, REP3	n.d.	n.d.	Moderate

n.d. = not determined.
[a] M = segregation bias toward mother cell.

se seems less important for stability. No 2-μm-based vector has yet been described that retains the high degree of segregational stability of the unmodified 2-μm circle under nonselective conditions, clearly indicating that other, as yet unidentified, parameters are involved.

6.1.2. ARS-Based Plasmids

ARS-based plasmids are generally unstable even under selective conditions. For example, in a population transformed with a pBR322-*TRP1-ARS1* plasmid (YRp7) and grown under selective conditions (for TRP^+ phenotype) less than 20% of the cells actually carry the plasmid (Struhl *et al.*, 1979; Kingsman *et al.*, 1979). During growth under non-selective conditions there is an exponential loss of the plasmid from the population such that after only 10 generations less than 1% of the cells still retain the plasmid (Struhl *et al.*, 1979; Kingsman *et al.*, 1979). The use of ARS-based plasmids should therefore be avoided when constructing expression vectors since applying selective pressure clearly does not guarantee plasmid retention.

Why ARS-based plasmids are so unstable is still a point of contention (see Williamson, 1985). Since their copy number is as high as, if not higher than, 2-μm-based plasmids, their instability appears to be due to an inability to efficiently segregate during mitotic growth, with plasmid invariably segregating to the parental (mother) cell (Murray and Szostak, 1983).

The degree of segregational instability of both ARS and 2-μm ORI-based plasmids can be greatly reduced by the incorporation of a *CEN* sequence (*CEN* = a DNA fragment from a yeast centromere; Clarke and Carbon, 1980), but this is at the expense of lowering the copy number to one to two plasmid molecules per cell.

6.1.3. Instabilities Induced by Overexpression of a Cloned Gene

The overexpression of a cloned gene in yeast, as in *E. coli* (Carrier *et al.*, 1983) can lead to a reduction in cell growth rate. Consequently, a cell in which either the expression plasmid has been lost or in which it has undergone a structural rearrangement such that the cloned gene is no longer expressed will have a selective growth advantage over a cell carrying a functional expression system. Cells with such a selective growth advantage quickly overtake and predominate in a culture. In the absence of selection for a plasmid-borne marker such as *LEU2*, the majority of "nonexpressors" would be plasmid-free, while if such selection was applied for plasmid retention the observed instability would arise by structural rearrangement. The former is shown in Fig. 6; while the standard cloning plasmid pMA3a and a derivative carrying and overexpressing

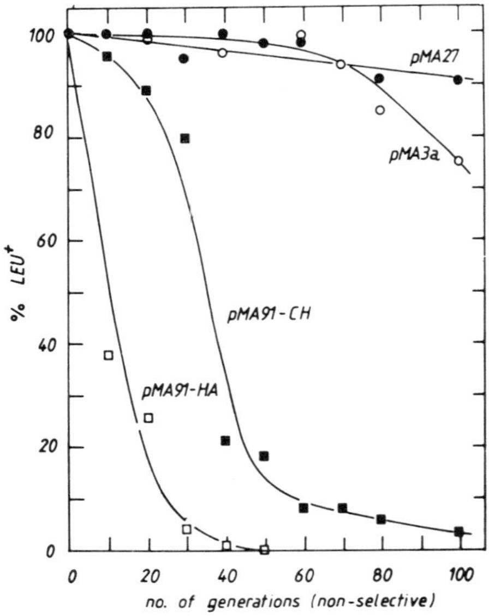

Figure 6. Relative segregational stabilities of expression plasmids in yeast during nonselective growth. All plasmids are based on the high-copy-number *LEU2*/2-μm origin plasmid pMA3a that is relatively stable during nonselective growth. Plasmid loss is monitored by loss of the plasmid-encoded *LEU2* marker following growth through a number of generations in medium containing leucine. pMA3a (○); pMA27 (●) = pMA3a carrying the intact yeast *PGK* gene; pMA91-CH (■) = pMA27 with the *PGK* coding sequence replaced by the calf prochymosin gene (for details see Fig. 1); pMA91-HA (□) = pMA27 with the *PGK1* coding sequence replaced by the hemagglutinin gene of influenza virus. (Data from K. Walters and M. F. Tuite, unpublished.)

the homologous *PGK* gene (see Fig. 5) are fairly stable under nonselective growth conditions, a derivative in which the homologous PGK *coding* sequence has been replaced either by the calf prochymosin gene (pMA91-CH) or by the influenza HA gene (pMA91-HA) is lost rapidly from the culture (M. F. Tuite, and K. Walters, unpublished data). If selection is retained (via the *LEU2* gene), then both pMA91-CH and pMA91-HA undergo structural rearrangements at high frequency (K. Walters, Akhamaloka, and M. F. Tuite, unpublished data).

A range of structural rearrangements have been noted by us to occur with expression plasmids when selective pressure is maintained, and they appear to generally arise via a homologous recombination event between plasmid-borne sequences and chromosomal sequences. For an expression vector this invariably involves the 5′ and/or 3′ sequences of the promoter. The use of strains blocked in mitotic recombination (e.g., *rad52;* Malone and Esposito, 1980) should reduce or even eliminate such structural instability, as would the use of hosts in which the potential substrates for homologous recombination had been deleted from the chromosome by gene disruption (see Chapter 5).

homologous recombination had been deleted from the chromosome by gene disruption (see Chapter 5).

There is as yet no fully effective means of eliminating the growth disadvantage of actively expressing cells, although by using a tightly regulated promoter system (see Section 3.5) expression during the active growth phase of the culture, i.e., exponential phase, can be turned off and switched on at a point late in the growth cycle as the cells enter stationary phase where biomass is almost maximum and cells have finished dividing, giving little chance for segregational loss to occur. An alternative approach would be to use an effective secretion system that ensures no deleterious intracellular buildup of the heterologous gene product (see Section 5.3).

Another consequence of introducing a high-copy-number plasmid into yeast is that their replication may impose a heavy energetic drain on the host cell. One option would be to use plasmids whose copy number can be switched from low to high during growth. No plasmids analogous to the temperature-sensitive runaway, plasmid replication mutants described in *E. coli* (Uhlin *et al.*, 1979; Yasuda and Takagi, 1983) have yet been described for yeast.

6.1.4. Use of Integrated Expression Systems

The pioneering work of Hinnen *et al.* (1978) demonstrated that heterologous DNA sequences can be stably integrated into the yeast genome and integrating might offer an alternative approach to ensure the mitotic stability of an expression system. The technology for integrating plasmid DNA sequences into yeast chromosomes is discussed at length in Chapter 5. The feasibility of this approach for heterologous gene expression has been demonstrated by Hitzeman *et al.* (1984), who integrated a *TRP1* vector, lacking an *ARS* but carrying the human IFNα1 gene coupled to the yeast *ADH1* promoter. They found that individual TRP^+ transformants obtained showed great variability in levels of expression of IFNα1 and this was a consequence of varying numbers of the plasmid tandemly integrated into the genome. All the transformants obtained were stable during nonselective growth, yet in no case were the levels of IFNα1 expression obtained ($<0.2\%$ total cell protein) comparable with the levels achieved with the same expression system located on an autonomous high-copy *ARS*-based plasmid (1–2%). Hitzeman *et al.* (1984) also reported similar findings for an integrated "expression unit" containing the *PGK* promoter coupled to the HBsAg gene and the *TRP1* terminator.

The use of integrated expression systems therefore is a promising alternative to the more traditional autonomous plasmid-based systems which have been the main focus of this chapter. They have the important advantage of long-term stability during nonselective growth conditions (Zhu *et al.*, 1986), but have the disadvantage that the levels of expression achieved are somewhat lower than achieved with an autonomous system. Yet, there may be situations where one wants to manipulate an endogenous biosynthetic pathway without requiring large amounts of the gene product, but where stability is of paramount importance.

6.2. Host Strains for Heterologous Gene Expression

Relatively few laboratory yeast strains have been used as hosts for expressing heterologous genes and generally contain one, or more, of the required auxotrophic mutations (*leu2, trp1, ura3, his3*) to allow selection and maintenance of the expression vector, but little else in the way of potentially beneficial mutations. Little attempt has been made to systematically modify the host strain characteristics to see if one can create a more suitable host for heterologous gene expression. Modifications to the host strain genotype that influence either plasmid copy number, plasmid integrity, plasmid segregation, or heterologous protein stability would be of immense value and, as will be discussed in the following Sections, such modifications can be made. The host strain as well as the expression system should therefore be carefully considered in parallel at the onset of a heterologous gene expression program.

6.2.1. Plasmid Stability and Copy Number

The stable propagation of a plasmid population in *S. cerevisiae* requires DNA replication functions, copy number control, and an effective mitotic segregation mechanism to be operational at all phases of the growth cycle. To what extent do plasmids in yeast rely on host-encoded functions, as opposed to autonomy in these functions? The detailed studies by Broach and others (see Section 6.1.1) have shown that the 2-μm circle encodes functions that allow DNA replication, that regulate copy number, and that ensure effective partition at mitosis. Consequently, the 2-μm circle in a very stable plasmid maintaining an effective copy number of approximately 50–100 molecules/cell. Yet, of course, replication does require host-encoded polymerases, and plasmid DNA replication, like chromosomal DNA replication, is under strict, nuclear-encoded cell cycle control (Zakian *et al.*, 1979).

The involvement of host-encoded functions in maintenance of the 2-μm circle population has recently been confirmed by the isolation of a mutation (*map1*) that causes a significant reduction in the copy number of the 2-μm circle (Kikuchi and Toh-E, 1986). This mutation also markedly decreases the stability of both YEp and YCp plasmids.

The ploidy of the host strain can have a marked effect on the segregational stability of yeast chimaeric plasmids (Mead *et al.*, 1986; Spalding and Tuite, 1989). For example, the rate of loss of a *TRP1-ARS1* plasmid during nonselective growth is dramatically reduced as the ploidy of the host strain is increased from haploid to tetraploid (Spalding and Tuite, 1989). In addition to the increase in segregational stability, the copy number of both chimeric plasmids and the endogenous 2-μm circle is also affected by the ploidy of the host strain; for example, a tetraploid strain has four times as many plasmid molecules per plasmid-bearing cell as a haploid strain (Mead *et al.*, 1986; Spalding and Tuite, 1989). The constitution of the mating-type system may also play an as yet undefined role in determining YRp plasmid copy number (Gullov and Friis, 1985).

Another approach to optimizing plasmid stability is to incorporate onto the expression plasmid a gene whose product is essential for cell growth and where the endogenous chromosomal copy of that gene has been inactivated. This means that the recombinant strain can be grown in a cheap complex medium. Loison *et al.* (1986) have recently described just such a system employing a host strain carrying the *ura3 fur1* mutations. Such strains are normally inviable, but can be grown provided they contain a plasmid with a functional *URA3* (oritidine-5′-phosphate decarboxylase) gene. Such an "autoselective" vector can be used to express high levels of human α-antitrypsin in various complex media (Loison *et al.*, 1986). In our hands, such vectors, based on a range of essential genes, show high levels of structural instability (A. Spalding *et al.*, unpublished). One possible way around this problem is to also introduce the *rad52* mutation into the host strain by gene disruption techniques and thus significantly reduce the likelihood of both inter- and intramolecular recombination events (see Section 6.1.3).

6.2.2. Protein Degradation

Saccharomyces cerevisiae contains more than 40 proteolytic enzymes (Achstetter and Wolf, 1985) and all the compartments of the cell represent possible locations for proteinases and possible sites for proteolysis. A number of these proteinases have been implicated in the degradation of aberrant and/or nonfunctional proteins, namely, proteinases yscA and yscB and carboxypeptidases yscY and yscS (Achstetter and Wolf, 1985). One might predict that mutant strains that lack one or more of these proteolytic enzymes might prove suitable hosts for the stable accumulation of heterologous proteins. Such mutants exist and are listed in Table IX, yet few data have yet appeared to confirm this notion. While a number of laboratories have carried out expression work in *pep4-3* strains (e.g., Mellor *et al.*, 1983; Smith *et al.*, 1985; Miyajima *et al.*, 1985), no dramatic enhancement of yields has yet been reported by use of these

Table IX. Mutations of *S. cerevisiae* That Result in Lack of One or More Proteolytic Enzymes

Mutation	Deficiency
pra1	Deficient in proteinase yscA, a vacuolar carboxylic endopeptidase
prb1	Deficient in proteinase yscB, a vacuolar serine sulfydral endopeptidase
pre1	Deficient in carboxypeptidase yscY, a vacuolar serine sulfydral exopeptidase
pep4-3	Defective in a variety of vacuolar hydrolase activities; accumulates precursor molecules
kex2	Defective in proteinase yscF, a metallo sulfydral endopeptidase (Ca^{2+}); inability to process α-factor and killer factor
cps1	Defective in carboxypeptidase yscS, a vacuolar metallo exopeptidase (Zn^{2+})

strains. The increasing tendency to secrete heterologous proteins from yeast (see Section 5) to a large extent overcomes any problems with protein degradation that arise as a consequence of intracellular accumulation.

6.2.3. Supersecretion Mutants

Irrespective of the strength of the promoter utilized in a secretion vector, there is apparently a limitation in the amount of protein the yeast secretion pathway can successfully process and efficiently secrete. While the rate-limiting step(s) have yet to be identified, Smith *et al.* (1985) have increased the capacity of the yeast secretion system by isolating host mutants that secrete up to 10 times the amount of calf prochymosin compared to the wild-type host even though these so-called "supersecreting" mutants synthesize the same *total* amount of prochymosin. Two genes have been identified—*SSC1* and *SSC2*—and *scc1 scc2* double mutants proved to be more efficient secreters than either *ssc1* or *ssc2* single mutants. These host genetic modifications will be of immense value in optimizing secretion of heterologous proteins from yeast.

7. SUMMARY AND PROSPECTS

The yeast *S. cerevisiae* is an ideal biotechnological resource made all the more useful in the modern biotechnology era by the wide range of gene manipulation techniques to which it can be subjected. There can be little argument that *S. cerevisiae* is an (almost) ideal host in which to express heterologous genes and largely satisfies a wide range of necessary properties that are needed for universal acceptability in the recombinant DNA industry. Table X lists the necessary properties and compares yeast's performance with the other three "popular" host organisms in the recombinant DNA industry, namely, *E. coli, Bacillus subtilis,* and mammalian cells. While it is clear that the latter probably represents the ideal (primarily because the majority of genes we seek to express are generally of mammalian origin), a large number of hurdles remain to be overcome in their development as cost-effective, large-scale producers of heterologous gene products, not the least being transfer of culture technology to the 1000-liter industrial fermentor.

Yeast has the greatest advantage in its wide acceptability as an industrial production strain, and the ease with which it can be genetically manipulated. While currently achievable levels of heterologous gene expression are generally lower than the maximal levels achievable in *E. coli,* this is due largely to our poor understanding of what DNA signals are required in an expression plasmid for optimal levels of transcription of the cloned gene and the critical parameters for efficient translation of its mRNA. The recent reports of high levels of expression of two recombinant proteins, hepatitis B core antigen (Kniskern *et al.,* 1986) and

Table X. Comparison of Yeast with Other Hosts Utilized in Heterologous Gene Expression Systems

Properties examined	Yeast	*E. coli*	*Bacillus*	Mammalian
1. Acceptibility as production strain	+++	±	++	++/±[a]
2. Science of fermentation	+++	+	++	+
3. Posttranslational modification of eukaryotic proteins	++	–	–	+++
4. Effective extracellular secretion systems	++	+	+++	?
5. Levels of heterologous gene expression	++	+++	?	?
6. Stability of heterologous proteins	++[b]	++[b]	?	?
7. Inducible expression systems	++	+++	–	++
8. Plasmid stability and integrity	++	++	+	?
9. Positive selection systems	+++	+++	+++	+++
10. Genetic modification of host	+++	+++	+++	±
11. Production of soluble heterologous protein	++	+	?	?
12. Ease of cell disruption	±	+++	+++	++
13. Viral contamination and induced cell lysis	+++	±	±	+
14. mRNA splicing	+[c]	–	–	+++[d]

[a]± if tumour or semitransformed cell lines used.
[b]If appropriate host mutants are used.
[c]Only its own introns.
[d]All except yeast.

human Cu/Zn superoxide dismutase (Hallewell *et al.*, 1987), do suggest that there is no *a priori* reason why high expression levels cannot be achieved in *S. cerevisiae.* The major selling point for yeast over *E. coli* is the availability of effective secretion systems that can give high extracellular levels of posttranslationally modified mammalian gene products.

More than 80 different heterologous genes have now been reported as being successfully expressed in *S. cerevisiae* using the strategies outlined in this chapter. It would, however, be dishonest to suggest that there are *no* problems involved in expressing heterologous genes in yeast. Some of these have already been touched on here; for example, the inability of yeast to efficiently and correctly splice out introns from higher eukaryotic mRNAs that lack the critical "TACTAAC" box, while an inconvenience, is readily overcome by using standard cDNA cloning techniques. On the other hand, plasmid stability is still a major problem, particularly since regulatory authorities are not enamoured by the inclusion of "selective agents" in industrial-scale fermentations, especially if the gene product is for human therapeutic use. The development of a truly stable, plasmid-based expression system, which can be maintained in the absence of an exogenously supplied selective agent, is eagerly awaited. Stability in this context is so often viewed purely from the segregational point of view, yet structural stability has to be given equal attention; a segregationally stable plasmid may still be subject to intramolecular rearrangements that reduce, even eliminate, heterologous gene expression. Stable integration of the expression system into the host chromosome clearly offers one route which has yet to be fully explored, although the problem here may well be achieving an appropriate copy number of the basic expression system. It is clear from the work of Smith *et al.* (1985) that integration of the transcriptional unit into the chromosome can lead to an increase in the efficiency of secretion irrespective of the copy number.

Major advances await to be made in developing a suitable host strain and require a greater understanding of host–plasmid and host–product interactions to improve both plasmid and product stability. The power of traditional genetics appears to have been largely ignored at the expense of recombinant genetics within this context. Another area that has been overlooked is the fidelity of heterologous gene expression in yeast, particularly at the level of translation; can the yeast protein synthesis machinery efficiently translate mRNAs that contain codons for which the corresponding isoacceptor tRNA may be in very short supply without making an error? In addition to such projected missense errors, there is also evidence that a significant degree of error at the level of translational termination occurs in yeast (Tuite and McLaughlin, 1982; M. F. Tuite *et al.*, unpublished). A consequence of a termination error is the production of a polypeptide with a C-terminal extension due to translation of the normally nontranslated 3′ region. The isolation of mutants of yeast that show increased translational fidelity could well prove valuable in ensuring maximal fidelity of heterologous gene expression.

The proof that *S. cerevisiae* will be a commercially viable choice for expressing recombinant proteins comes from the recent licensing of a hepatitis B subunit vaccine to Merck, Sharp, and Dohme (USA) for human use. The HBsAg particles assembled in yeast (Valenzuela *et al.*, 1982; Hitzeman *et al.*, 1983a) have proven to be highly immunogenic and safe in both animals and humans (McAleer *et al.*, 1984; Scolnick *et al.*, 1984; Hollinger *et al.*, 1986). In addition, further steps have been taken to improve this yeast-derived vaccine by engineering a modified HBsAg particle containing a polyalbumin receptor to improve its efficacy (Valenzuela *et al.*, 1985a) and by including a second immunogen, that for herpes simplex virus surface antigen, to produce a potential polyvalent vaccine (Valenzuela *et al.*, 1985b). A very recent and exciting further development in this context is the exploitation of the yeast Ty-like virus particle to generate novel subunit vaccines (Adams *et al.*, 1987). The use of yeast to produce safe, efficacious subunit vaccines is but one of a number of the new developments in yeast biotechnology that have come from exploiting genetic engineering technology.

REFERENCES

Achstetter, T., and Wolf, D. H., 1985, Proteinases, proteolysis and biological control in the yeast *Saccharomyces cerevisiae, Yeast* **1:**139–157.

Adams, S. E., Dawson, K. M., Gull, K., Kingsman, S. M., and Kingsman, A. J., 1987, The expression of hybrid HIV:Ty virus-like particles in yeast, *Nature* **329:**68–70.

Ammerer, G., 1983, Expression of genes in yeast using the *ADC1* promoter, *Methods Enzymol.* **101:**192–201.

Baim, S. B., Goodhue, C. T., Pietras, D. F., Eustice, D. C., Labhard, M., Freidman, L. R., Hampsey, D. M., Stiles, J. I., and Sherman, F., 1985, Rules of translation: Studies with altered forms of the yeast CYC1 gene, in: *Sequence Specificity in Transcription and Translation: ICN-UCLA Symposium,* Alan R. Liss, pp. 351–362.

Ballou, C. E., 1982, Yeast cell wall and cell surface, in: *The Molecular Biology of the Yeast Saccharomyces.* Vol. 2: *Metabolism and Gene Expression* (J. Strathern, E. W. Jones, and J. R. Broach, eds.), Cold Spring Harbor Laboratory, Cold Spring Harbor, NY, pp. 335–360.

Beggs, J. D., van den Berg, J., van Ooyen, A., and Weissman, C., 1980, Abnormal expression of chromosomal rabbit β-globin gene in *Saccharomyces cerevisiae, Nature* **283:**835–840.

Bennetzen, J. L., and Hall, B. D., 1982, Codon selection in yeast, *J. Biol. Chem.* **257:**3026–3031.

Bitter, G. A., and Egan, K. M., 1984, Expression of heterologous genes in *Saccharomyces cerevisiae* from vectors utilising the glyceraldehyde-3-phosphate dehydrogenase gene promoter, *Gene* **32:**263–274.

Bitter, G. A., Chen, K. K., Banks, A. R., and Lai, P-H., 1984, Secretion of foreign proteins from *Saccharomyces cerevisiae* directed by α-factor gene fusions, *Proc. Natl. Acad. Sci. USA* **81:**5330–5334.

Bostian, K. A., Hemire, J. M., Cannon, L. E., and Halvorson, H. O., 1980, *In vitro* synthesis of repressible yeast acid phosphatase: Identification of multiple mRNAs and products, *Proc. Natl. Acad. Sci. USA* **77:**4504–4508.

Bostian, K. A., Elliott, Q., Bussey, H., Burn, V., Smith, A., and Tipper, D. J., 1984, Sequence of the preprotoxin dsRNA gene of type I killer yeast: Multiple processing events produce a two-component toxin, *Cell* **36:**741–751.

Brake, A. J., Merryweather, J. P., Coit, D. G., Heberlein, U. A., Masiarz, F. R., Mullenbach, G. T., Urdea, M. S., Valenzuela, P., and Barr, P. J., 1984, α-Factor-directed synthesis and secretion of mature foreign proteins in *Saccharomyces cerevisiae, Proc. Natl. Acad. Sci. USA* **81:**4642–4646.

Breunig, K. D., Malcedonski, V., and Hollenberg, C. P., 1982, Transcription of the bacterial β-lactamase gene in *Saccharomyces cerevisiae, Gene* **20:**1–10.

Bussey, H., and Meaden, P., 1985, Selection and stability of yeast transformants expressing cDNA of an M1 killer toxin-immunity gene, *Curr. Genet.* **9:**285–291.

Cabezon, T., de Wilde, M., Herion, P., Loriau, R., and Bollen, A. 1984, Expression of human α_1-antitrypsin cDNA in the yeast *Saccharomyces cerevisiae, Proc. Natl. Acad. Sci. USA* **81:**6594–6598.

Cardell, S. J., Craik, C. S., Hilvert, D., Urdea, M. S., and Rutter, W. J., 1985, Site-directed mutagenesis shows that tyrosine 248 of carboxy-peptidase A does not play a crucial role in catalysis, *Nature* **317:**551–556.

Carrier, M. J., Nugent, M. E., Tacon, W. C. A., and Primrose, S. B., 1983, High expression of cloned genes in *E. coli* and its consequences, *Trends Biotechnol.* **1:**109–113.

Chen, C. Y., Oppermann, H., and Hitzeman, R. A., 1984, Homologous and heterologous gene expression in the yeast *Saccharomyces cerevisiae, Nucleic Acids Res.* **12:**8951–8970.

Cigan, A. M., and Donahue, T. D., 1987, Sequence and structural features associated with translational initiator regions in yeast—a review, *Gene* **59:**1–18.

Clarke, L., and Carbon, J., 1980, Isolation of a yeast centromere and construction of functional small circular chromosomes, *Nature* **257:**504–509.

Cramer, J. H., Lea, K., and Slightom, J. L., 1985, Expression of phaseolin cDNA genes in yeast under control of natural plant DNA sequences, *Proc. Natl. Acad. Sci. USA* **82:**334–338.

Derynck, R., Singh, A., and Goeddel, D. V., 1983, Expression of human interferon-γ cDNA in yeast, *Nucleic Acids Res.* **11:**1819–1837.

Dillon, J-A. R., Nasim, A., and Nestmann, E. R. (eds.), 1985, *Recombinant DNA Methodology,* Wiley, New York.

Dobson, M. J., Tuite, M. F., Mellor, J., Roberts, N. A., King, R. M., Burke, D. C., Kingsman, A. J., and Kingsman, S. M., 1983, Expression in *Saccharomyces cerevisiae* of human interferon-alpha directed by the *TRP* 5′ region, *Nucleic Acids Res.* **11:**2287–2302.

Edens, L., Bom, I., Ledeboer, A. M., Maat, J., Toonen, M. Y., Visser, C., and Verrips, C. T., 1984, Synthesis and processing of the plant protein thaumatin in yeast, *Cell* **37:**629–633.

Emr, S. D., Schekman, R., Flessel, M. C., and Thorner, J., 1983, An *MFα1-SUC2* (α-factor-invertase) gene fusion for study of protein localization and gene expression in yeast, *Proc. Natl. Acad. Sci. USA* **80:**7080–7084.

Ernst, J. F., 1986, Improved secretion of heterologous proteins by *Saccharomyces cerevisiae:* Effects of promoter substitution in alpha-factor fusions, *DNA* **5:**483–491.

Freedman, R. B., 1984, Native disulphide bond formation in protein biosynthesis: Evidence for the role of protein disulphide isomerase, *Trends Biochem. Sci.* **6:**438–441.

Futcher, A. B., 1986, Copy number amplification of the 2μm circle plasmid of *Saccharomyces cerevisiae, J. Theoret. Biol.* **119:**197–204.

Futcher, A. B., and Cox, B. S., 1984, Copy number and the stability of 2μm circle-based artificial plasmids of *Saccharomyces cerevisiae, J. Bacteriol.* **157:**283–290.

Goff, C. G., Moir, D. T., Kohno, T., Gravins, T. C., Smith, R. A., Yamasaki, E., and Taunton-Rigby, A., 1984, Expression of calf prochymosin in *Saccharomyces cerevisiae, Gene* **27:**35–46.

Gritz, L., and Davies, J., 1983, Plasmid-encoded hygromycin B resistance: The sequence of hygromycin B phosphotransferase gene and its expression in *Escherichia coli* and *Saccharomyces cerevisiae, Gene* **25:**179–188.

Guarente, L., 1984, Yeast promoters: Positive and negative elements, *Cell* **36:**799–800.

Gullov, K., and Friis, J., 1985, Maintenance and copy number control of ARS1 plasmids in *Saccharomyces cerevisiae:* Evidence of a mating type effect, *Curr. Gent.* **10:**21–27.

Hallewell, R. A., Mills, R., Tekamp-Olson, P., Blacher, R., Rosenberg, R., Otting, F., Masiarz, F. R., and Scandella, C. J., 1987, Amino terminal acetylation of authentic human Cu, Zn superoxide dismutase produced in yeast, *Bio/Technology* **5:**363–366.

Handa, H., Mizumoto, K., Oda, K., Okamoto, T., and Fukasawa, T., 1985, Transcription of the human adenovirus E1a gene in *Saccharomyces cerevisiae, Gene* **33:**159–168.

Harris, T. J. R., 1983, Expression of eukaryotic genes in *E. coli,* in: *Genetic Engineering,* Vol. 4 (R. Williamson, ed.) Academic Press, New York, pp. 128–175.

Henderson, R. C. A., Cox, B. S., and Tubb, R., 1985, The transformation of brewing yeasts with a plasmid containing the gene for copper resistance, *Curr. Genet.* **9:**133–138.

Henikoff, S., and Cohen, E. H., 1984, Sequence responsible for transcription termination on a gene segment in *Saccharomyces cerevisiae, Mol. Cell. Biol.* **4:**1515–1520.

Henikoff, S., and Furlong, C. E., 1983, Sequence of a *Drosophila* DNA segment that functions in *Saccharomyces cerevisiae* with its regulation by a yeast promoter, *Nucleic Acids. Res.* **11:**789–800.

Hinnen, A., Hicks, J. B., and Fink, G. R., 1978, Transformation of yeast, *Proc. Natl. Acad. Sci. USA* **75:**1929–1933.

Hitzeman, R. A., Hagie, F. E., Levine, H. L., Goeddel, D. V., Ammerer, G., and Hall, B. D., 1981, Expression of a human gene for interferon in yeast, *Nature* **293:**717–722.

Hitzeman, R. A., Chen, C. Y., Hagie, F. E., Patzer, E. J., Liu, C-C., Estell, D. A., Miller, J. V., Yaffe, A., Keleid, D. G., Levinson, A. D., and Oppermann, H., 1983a, Expression of hepatitis B virus surface antigen in yeast, *Nucleic Acids Res.* **11:**2745–2763.

Hitzeman, R. A., Leung, D. W., Perry, L. J., Kohr, W. J., Levine, H. L., and Goeddel, D. V., 1983b, Secretion of human interferons by yeast, *Science* **219:**620–625.

Hitzeman, R. A., Chen, C. Y., Hagie, F. E., Lugovoy, J. M., and Singh, A., 1984, Yeast: An alternative organism for foreign protein production, in: *Recombinant DNA Products: Insulin, Interferon and Growth Hormone* (A. P. Bollon, ed.), CRC Press, Boca Raton, FL, pp. 47–65.

Holland, M. J., and Holland, J. P., 1978, Isolation and identification of yeast mRNAs coding for enolase, glyceraldehyde-3-phosphate dehydrogenase and phosphoglycerate kinase, *Biochemistry* **17:**4900–4907.

Hollinger, F. B., Troisi, C. L., and Pepe, P. E., 1986, Anti-HBs responses to vaccination with a human hepatitis B vaccine made by recombinant DNA technology in yeast, *J. Infect. Dis.* **153:**156–159.

Huang, S., Elliott, R. C., Liu, P-S., Koduri, R. K., Weickmann, J. L., Lee, J. H., Blair, L. C., Ghosh-Dastidar, P., Bradshaw, R. A., Bryan, K. M., Einarson, B., Kendall, R. L., Kolacz, K. H. Y., and Saito, K., 1987, Specificity of cotranslational amino-terminal processing of proteins in yeast, *Biochemistry* **26:**8242–8246.

Ikemura, T., 1982, Correlation between the abundance of yeast transfer RNAs and the occurrence of the respective codons in protein genes, *J. Mol. Biol.* **158:**573–597.

Innis, M. A., Holland, M. J., McCabe, P. C., Cole, G. E., Whittman, V. P., Tal, R., Watt, K. W. K., Gelfand, D. H., Holland, J. P., and Meade, J. H., 1985, Expression, glycosyla-

tion and secretion of an *Aspergillus* glucoamalylase by *Saccharomyces cerevisiae, Science* **228:**21–26.
Jayaram, M., Li, Y. Y., and Broach, J. R., 1983, The yeast plasmid 2μ circle encodes components required for its high copy number propagation, Cell **34:**95–104.
Jellinghaus, V., Schatzle, V., Schmid, W., and Roewekamp, W., 1982, Transcription of a *Dictyostelium* discoidin-I gene in yeast: Alternative promoter sites used in two different eukaryotic cells, *J. Mol. Biol.* **159:**623–636.
Jensen, E. O., Marcker, K. A., and Villadsen, T. S., 1986, Heme regulates the expression in *Saccharomyces cerevisiae* of chimaeric genes containing 5′-flanking soybean leghemoglobin sequences, *EMBO, J.* **5:**843–847.
Johnston, M., and David, R. W., 1984, Sequences that regulate the divergent *GAL1-GAL10* promoter of *Saccharomyces cerevisiae, Mol. Cell Biol.* **4:**1440–1448.
Johnston, S. A., and Hopper, J. E., 1982, Isolation of the yeast regulatory gene *GAL4* and analysis of its dosage effects on the galactose/melibiose regulon, *Proc. Natl. Acad. Sci. USA* **79:**6971–6975.
Julius, D., Blair, L., Brake, A., Sprague, G., and Thorner, J., 1983, Yeast α-factor is processed from a larger precursor polypeptide: The essential role of a membrane bound dipeptidyl aminopeptidase, *Cell* **32:**839–852.
Julius, D., Brake, A., Blair, L., Kunisawa, R., and Thorner, J., 1984, Isolation of the putative structural gene for the lys-arg cleavage endopeptidase required for processing of yeast prepro-α-factor, *Cell* **37:**1075–1089.
Kaster, K. R., Burgett, S. G., and Ingolia, T. D., 1984, Hygromycin B resistance as a dominant selectable marker in yeast, *Curr. Genet.* **8:**353–358.
Kataoka, T., Powers, S., Cameron, S., Fasano, O., Goldfarb, M., Broach, J., and Wigler, M., 1985, Functional homology of mammalian and yeast RAS genes, *Cell* **40:**19–26.
Kikuchi, Y., 1983, Yeast plasmid requires a cis-acting locus and two plasmid proteins for its stable maintenance, *Cell* **35:**487–493.
Kikuchi, Y., and Toh-E., A., 1986, A nuclear gene of *Saccharomyces cerevisiae* needed for stable maintenance of plasmids, *Mol. Cell Biol.* **6:**4053–4059.
Kingsman, S. M., and Kingsman, A. J., 1983, The production of interferon in bacteria and yeast, in: *Interferons, Society for General Microbiology Symposium,* Vol. 35 (D. C. Burke and A. Morris, eds.), Cambridge University Press, Cambridge, pp. 211–254.
Kingsman, A. J., Clarke, L., Mortimer, R. K., and Carbon, J., 1979, Replication in *Saccharomyces cerevisiae* of plasmid pBR313 carrying DNA from the yeast *TRP1* region, *Gene* **7:**141–152.
Kniskern, P. J., Hagopian, A. Montgomery, D. L., Burke, P., Dunn, N. R., Hofmann, K. J., Miller, W. J., and Ellis, R. W., 1986, Unusually high-level expression of a foreign gene (hepatitis B virus core antigen) in *Saccharomyces cerevisiae, Gene* **46:**135–141.
Kornbluth, S., Jove, R., and Hanafusa, H., 1987, Characterization of avian and viral p60^{src} proteins expressed in yeast, *Proc. Natl. Acad. Sci. USA* **84:**4455–4459.
Kozak, M., 1981, Possible role of flanking nucleotides in recognition of the AUG initiator codon by eukaryotic ribosomes, *Nucleic Acids. Res.* **9:**5233–5252.
Kozak, M., 1984, Compilation and analysis of sequences upstream from the translational start in eukaryotic mRNAs, *Nucleic Acids Res.* **12:**857–879.
Kozak, M., 1986, Point mutations define a sequence flanking the AUG initiator codon that modulates translation by eukaryotic ribosomes, *Cell* **44:**283–292.
Kramer, R. A., de Chiara, T. M., Schaber, M. D., and Hilliker, S., 1984, Regulated expression of a human interferon gene in yeast: Control by phosphate concentration or temperature, *Proc. Natl. Acad. Sci. USA* **81:**367–370.
Kuga, T., Hattori, S., Yoshida, M., and Taniguchi, T., 1986, Expression of human T-cell leukemia virus type I envelope protein in *Saccharomyces cerevisiae, Gene* **44:**337–340.

Langford, C., and Gallwitz, D., 1983, Evidence for an intron-contained sequence required for the splicing of yeast RNA polymerase II transcripts, *Cell* **33:**519–527.

Langford, C., Nellen, N., Niessing, J., and Gallwitz, D., 1983, Yeast is unable to excise foreign intervening sequences from hybrid gene transcripts, *Proc. Natl. Acad. Sci. USA* **80:**1496–1500.

Langridge, P., Eibel, H., Brown, J. W. S., and Feix, G., 1984, Transcription from maize storage protein gene promoters in yeast, *EMBO J.* **3:**2467–2471.

Laz, T., Clements, J., and Sherman, F., 1988, Structural features of mRNA that affect translation in yeast, in: *Genetics of Translation: New Approaches* (M. F. Tuite, M. Picard, and M. Bolotin-Fukuhara, eds.), Springer Verlag, New York, 1988.

Lehle, L., and Bause, E., 1984, Primary structural requirements from *N*- and *O*-glycosylation of yeast mannoproteins, *Biochim. Biophys. Acta* **799:**246–251.

Loison, G., Nguyen-Juilleret, M., Alouani, S., and Marquet, M., 1986, Plasmid-transformed *ura3 fur1* double mutants of *S. cerevisiae:* An autoselection system applicable to the production of foreign proteins, *Bio/Technology* **4:**433–437.

Losson, R., and Lacroute, F., 1983, Plasmids carrying the yeast OMP decarboxylase structural and regulatory genes: Transcription regulation in a foreign environment, *Cell* **32:**371–377.

Maitra, P. K., and Lobo, Z., 1971, A kinetic study of glycolytic enzyme synthesis on yeast, *J. Biol. Chem.* **246:**475–488.

Malone, R. E., and Esposito, R. E., 1980, The *RAD52* gene is required for homothallic interconversion of mating types and spontaneous mitotic recombination in yeast, *Proc. Natl. Acad. Sci. USA* **77:**503–507.

Maniatis, T., Fritsch, E. F., and Sambrook, J., 1982, *Molecular Cloning. A Laboratory Manual,* Cold Spring Harbor Laboratory, Cold Spring Harbor, NY.

McAleer, W. J., Buyak, E. B., Maigetter, R. Z., Wampter, D. E., Miller, W. J., and Hilleman, M. R., 1984, Human hepatitis B vaccine from recombinant yeast, *Nature* **307:** 178–180.

McLaughlin, C. S., Warner, J. R., Edmonds, M., Nakazato, H., and Vaughan, M. H., 1973, Polyadenylic acid sequences in yeast messenger ribonucleic acid, *J. Biol. Chem.* **248:**1466–1471.

McNeil, J. B., and Friesen, J. D., 1981, Expression of the herpes simplex virus thymidine kinase gene in *Saccharomyces cerevisiae, Mol. Gen. Genet.* **184:**386–393.

Mead, D. J., Gardner, D. C. J., and Oliver, S. G., 1986, Enhanced stability of a 2μ-based recombinant plasmid in diploid yeast, *Biotechnol. Lett.* **8:**391–396.

Mellor, J., Dobson, M. J., Roberts, N. A., Tuite, M. F., Emtage, J. S., Lowe, P. A., Patel, T., Kingsman, A. J., and Kingsman, S. M., 1983, Efficient synthesis of enzymatically active calf chymosin in *Saccharomyces cerevisiae, Gene* **24:**1–14.

Mellor, J., Dobson, M. J., Roberts, N. A., Kingsman, A. J., and Kingsman, S. M., 1985, Factors affecting heterologus gene expression in *Saccharomyces cerevisiae, Gene* **33:**215–226.

Mellor, J., Dobson, M. J., Kingsman, A. J., and Kingsman, S. M., 1987, A transcriptional activator is located in the coding region of the yeast *PGK* gene, *Nucleic Acids Res.* **15:**6243–6259.

Miyajima, A., Miyajima, I., Arai, K-I., and Arai, N., 1984, Expression of plasmid R388-encoded type II dihydrofolate reductase as a dominant selective marker in *Saccharomyces cerevisiae, Mol. Cell. Biol.* **4:**407–414.

Miyajima, A., Bond, M. W., Otsu, K., Arai, K-I., and Arai, N., 1985, Secretion of mature mouse interleukin-2 by *Saccharomyces cerevisiae:* Use of a general secretion vector containing promoter and leader sequences of the mating pheromone α-factor, *Gene* **37:**155–161.

Miyanohara, A., Toh-E., Nozaki, C., Hamada, F., Ohtomo, N., and Matsubara, K., 1983,

Expression of hepatitis B surface antigen gene in yeast, *Proc. Natl. Acad. Sci. USA* **80:**1–5.

Murray, A. W., and Szostak, J. W., 1983, Pedigree analysis of plasmid segregation in yeast, *Cell* **34:**911–917.

Nakamura, Y., Sto, T., Emi, M., Miyanohara, A., Nishide, T., and Matsubara, K., 1986, Expression of human salivary α-amylase gene in *Saccharomyces cerevisiae* and its secretion using the mammalian signal sequence, *Gene* **50:**239–245.

Nicholson, R. C., and Moran, L. A., 1984, Expression of a *Drosophila* heat-shock gene in cells of the yeast *Saccharomyces, Biosci. Rep.* **4:**963–972.

Oberto, J., and Davison, J., 1985, Expression of chicken egg white lysozyme by *Saccharomyces cerevisiae, Gene* **40:**57–65.

Ogden, J. E., Stanway, C., Kim, S., Mellor, J., Kingsman, A. J., and Kingsman, S. M., 1986, Efficient expression of the yeast *PGK* gene depends on an upstream activation sequence but does not require "TATA" sequences, *Mol. Cell Biol.* **6:**5335–4343.

Perlman, D., and Halvorson, H. O., 1983, A putative signal peptidase recognition site in eukaryotic and prokaryotic signal peptides, *J. Mol. Biol.* **167:**391–409.

Proudfoot, N. J., and Brownlee, G. G., 1976, 3′ non-coding region sequences in eukaryotic mRNA, *Nature* **263:**211–214.

Rosenberg, S., Barr, P. J., Najarian, R. C., and Hallewell, R. A., 1984, Synthesis in yeast of a functional oxidation-resistant mutant of human α_1-antitrypsin, *Nature* **312:**77–80.

Rothstein, S. J., Lazarus, C. M., Smith, W. E., Baulcombe, D. C., and Gatenby, A. A., 1984, Secretion of a wheat α-amylase expressed in yeast, *Nature* **308:**662–665.

Russell, P. R., 1983, Evolutionary divergence of the mRNA transcription initiation mechanism in yeast, *Nature* **301:**167–169.

Sato, T., Tsunasawa, S., Nakamura, Y., Emi, M., Sakiyama, F., and Matsubara, K., 1986, Expression of the human salivary α-amylase gene in yeast and characterisation of the secreted protein, *Gene* **50:**247–257.

Schekman, R., and Novick, P., 1982, The secretory process and yeast cell surface assembly, in: *The Molecular Biology of the Yeast Saccharomyces.* Vol. 2: *Metabolism and Gene Expression* (J. Strathern, E. W. Jones, and J. R. Broach, eds.), Cold Spring Harbor Laboratory, Cold Spring Harbor, NY, pp. 361–398.

Schultz, L. D., Tanner, J., Hofmann, K. J., Emini, E. A., Condra, J. H., Jones, R. E., Kieff, E., and Ellis, R. W., 1987, Expression and secretion in yeast of a 400-kDa envelope glycoprotein derived from Epstein–Barr virus, *Gene* **54:**113–123.

Scolnick, E. M., McLean, A. A., West, D. J., McAleer, W. J., Miller, W. J., and Buyak, E. B., 1984, Clinical evaluation in healthy adults of a hepatitis B vaccine made by recombinant DNA, *JAMA* **251:**2812–2815.

Sharp, P. M., Tuohy, T. M. F., and Mosurski, K. R., 1986, Codon usage in yeast: Cluster analysis clearly differentiates highly and lowly expressed genes, *Nucleic Acids Res.* **14:**5125–5143.

Singh, A., Lugovoy, J. M., Kohr, W. J., and Perry, L. J., 1984, Synthesis, secretion and processing of α-factor-interferon fusion proteins in yeast, *Nucleic Acids Res.* **12:**8927–8938.

Skipper, N., Sutherland, M., Davies, R. W., Kilburn, D., Miller, R. C., Warren, A., and Wong, R., 1985, Secretion of a bacterial cellulase by yeast, *Science* **230:**958–960.

Smith, R. A., Duncan, M. J., and Moir, D. T., 1985, Heterologous protein secretion from yeast, *Science* **229:**1219–1224.

Spalding, A., and Tuite, M. F., 1989, Host-plasmid interactions in *Saccharomyces cerevisiae:* effect of host-ploidy on plasmid stability and copy number, *J. Gen. Microbiol.* **135:**1037–1045.

Stanway, C., Kingsman, A. J., and Kingsman, S. M., 1987, The control of transcription in *Saccharomyces cerevisiae, Bioessays* **7:**62–67.

Stepien, P. P., Brousseau, R., Wu, R., Narang, S., and Thomas, D. Y., 1983, Synthesis of a human insulin gene. VI. Expression of the synthetic proinsulin gene in yeast, *Gene* **24:**289–297.

St. John, T. P., and Davis, R. W., 1981, The organisation and transcription of the galactose gene cluster of *Saccharomyces, J. Mol. Biol.* **152:**285–316.

Struhl, K., 1986, Yeast promoters, in: *Maximizing Gene Expression* (W. Reznikoff, W. Gold, and L. Gold, eds.), Butterworths, London, pp. 35–78.

Struhl, K., Stinchcomb, D. T., Scherer, S., and Davis, R. W., 1979, High-frequency transformation of yeast: Autonomous replication of hybrid DNA molecules, *Proc. Natl. Acad. Sci. USA* **76:**1035–1039.

Tanner, W., and Lehle, L., 1987, Protein glycosylation in yeast, *Biochim. Biophys. Acta* **906:**81–99.

Thim, L., Hansen, M. G., Norris, K., Hoegh, I., Boel, E., Forstrom, J., Ammerer, G., and Fiil, N. P., 1986, Secretion and processing of insulin precursors in yeast, *Proc. Natl. Acad. Sci. USA* **83:**6766–6770.

Thomsen, K. K., 1983, Mouse α-amylase synthesised by *Saccharomyces cerevisiae* is released into the culture medium, *Carlsberg Res. Commun.* **48:**545–555.

Tokunaga, T., Iwai, S., Gomi, H., Kodama, K., Ohtsuka, E., Ikehara, M., Chisaka, O., and Matsubara, K., 1985, Expression of a synthetic human growth hormone gene in yeast, *Gene* 39:117–120.

Tuite, M. F., and McLaughlin, C. S., 1982, Natural readthrough of a UGA termination codon in a yeast cell-free system: Evidence for involvement of both a mitochondrial and nuclear tRNA, *Mol. Cell. Biol.* **2:**490–497.

Tuite, M. F., Dobson, M. J., Roberts, N. A., King, R. M., Burke, D. C., Kingsman, S. M., and Kingsman, A. J., 1982, Regulated high efficiency expression of human interferon-alpha in *Saccharomyces cerevisiae, EMBO J.* **1:**603–608.

Uhlin, B. E., Molin, S., Gustafsson, P., and Nordstrom, K., 1979, Plasmids with temperature-dependent copy number for amplification of cloned genes and their products, *Gene* **6:**91–106.

Urdea, M. S., Merryweather, J. P., Mullenback, G. T., Coit, D., Hebertein, U., Valenzuela, P., and Barr, P. J., 1983, Chemical synthesis of a gene for human epidermal growth factor urogastrone and its expression in yeast, *Proc. Natl. Acad. Sci. USA* **80:**7461–7465.

Valenzuela, P., Medina, A., Rutter, W. J., Ammerer, G., and Hall, B. D., 1982, Synthesis and assembly of hepatitis B virus surface antigen particles in yeast, *Nature* **298:**347–350.

Valenzuela, P., Coit, D., and Kuo, C. H., 1985a, Synthesis and assembly in yeast of hepatitis B surface antigen particles containing the polyalbumin receptor, *Bio/Technology* **3:**317–322.

Valenzuela, P., Coit, D., Medina-Selby, M. A., Kuo, C. H., Nest, G. V., Burke, R. L., Urdea, M. S., and Graves, P. V., 1985b, Antigen engineering in yeast: synthesis and assembly of hybrid hepatitis B surface antigen-herpes simplex 1 gD particles, *Bio/Technology* **3:**323–328.

Webster, T. D., and Dickson, R. C., 1983, Direct selection of *Saccharomyces cerevisiae* resistant to the antibiotic G418 following transformation with a DNA vector carrying the kanamycin-resistance gene of Tn903, *Gene* **26:**243–252.

Williamson, D. H., 1985, The yeast *ARS* element, six years on: A progress report *Yeast* **1:**1–14.

Wood, C. R., Boss, M. A., Kenton, J. M., Calvert, J. E., Roberts, N. A., and Emtage, J. S., 1985, The synthesis and *in vivo* assembly of functional antibodies in yeast, *Nature* **314:**446–449.

Woudt, L-P., van den Heuvel, J., van Raamsdonk-Duin, M. M. C., and Mager, W. H., and Planta, R. J., 1985, Correct removal by splicing of a *Neurospora* intron in yeast, *Nucleic Acids Res.* **13:**7729–7739.

Yamashita, I., Suzuki, K., and Fukui, S., 1985, Nucleotide sequence of the extracellular glucoamylase gene *STA1* in the yeast *Saccharomyces diastaticus, J. Bacteriol.* **161:**567–573.

Yasuda, S., and Takagi, T., 1983, Overproduction of *Escherichia coli* replication proteins by use of runaway-replication plasmids, *J. Bacteriol.* **154:**1153–1161.

Zakian, V. A., Brewer, B. J., and Fangman, W. L., 1979, Replication of each copy of the yeast 2 micron DNA plasmid occurs during the S phase, *Cell* **17:**923–934.

Zaret, K. S., and Sherman, F., 1982, DNA sequences required for efficient transcription termination in yeast, *Cell* **28:**563–573.

Zaret, K. S., and Sherman, F., 1984, Mutationally altered 3′ ends of yeast *CYC1* mRNA affect transcript stability and translational efficiency, *J. Mol. Biol.* **177:**107–135.

Zhu, X-L., Ward, C., and Weissbach, A., 1984, Control of herpes simplex virus thymidine kinase gene expression in *Saccharomyces cerevisiae* by a yeast promoter sequence, *Mol. Gen. Genet.* **194:**31–41.

Zhu, J., Contreras, R., Gheysen, D., Ernst, J., and Fiers, W., 1985, A system for dominant transformation and plasmid amplification in *Saccharomyces cerevisiae, Bio/Technology* **3:**451–456.

Zhu, J., Contreras, R., and Fiers, W., 1986, Construction of stable laboratory and industrial yeast strains expressing a foreign gene by integrative transformation using a dominant selection scheme, *Gene* **50:**225–237.

"Classical" Yeast Biotechnology 7

STEPHEN G. OLIVER

1. HISTORY

Yeast was probably the first microorganism to be exploited by humans, and the brewing of alcoholic beverages and the leavening of bread using yeast represent the oldest, and still the largest, biotechnologies. The earliest records of the use of yeast date to before 6000 B.C. when the Sumerians and Babylonians exploited the organism, albeit unconsciously, in the production of beer (Corran, 1975). The use of yeast to leaven bread by the production of carbon dioxide was developed at a much later date, in Egypt at ca. 4000 B.C. The ancient Egyptians extended their bread-making technology to the production of an acidic beer called boza or boozah from a lightly baked "loaf" of germinated grain. The production of distilled liquor probably originated in either China or the Middle East and was common throughout the known world by the 14th century A.D.

2. BAKING

In the 18th and 19th centuries distiller's or brewer's yeast was supplied to the bakers, and it was not until the introduction of the Vienna process in 1846 that yeast was deliberately grown for use in baking (Oura *et al.*, 1979). Today, the provision of baker's yeast is a major industry, with some 1.5 million tonnes of fresh yeast produced annually. The same companies are frequently involved in supplying both the baking and distilling industries. Although the two processes demand yeast with rather different characteristics, the use of baker's yeast for the production of alcohol is extensive. For instance, virtually the whole of the Brazilian ethanol industry relies on Fleischmann's baking yeast for its fermentations.

STEPHEN G. OLIVER • Manchester Biotechnology Centre, University of Manchester, Institute for Science and Technology, Manchester M60 1QD, England.

The requirements for a good baker's yeast are that it should have a fast growth rate under aerobic conditions to reduce production time and that it should also have a high fermentation rate on bread dough. The development of suitable strains has been greatly facilitated by the fact that baker's yeasts, unlike brewing yeasts, are diploid organisms which produce a high proportion of viable spores (Hough, 1985).

The methods involved in the production of baker's yeast contrast greatly with those employed in alcohol fermentations since the rate of production and yield of biomass, rather than ethanol, must be optimized. Although continuous culture might appear the obvious way of achieving this, baker's yeast is usually produced in a fed-batch process. Thus the production of baker's yeast, unlike that of ethanol, is performed under aerobic conditions in substrate-limited conditions so as to maximize the conversion of assimilable carbon into biomass. Molasses is the most commonly used carbon source, although there is now some interest in exploiting starch products and whey as substrates. This would necessitate genetic manipulation of the yeast and is discussed more fully in Section 12. Feedback control systems for the addition of carbon substrate have been designed; however, nutrient is most commonly supplied to the culture according to a predetermined pattern. Frequently this involves a simple exponential increase in the rate of addition of molasses, but regimes in which the rate of feed is kept constant, or even reduced, as stationary phase is approached have been found to improve both the yield and keeping quality of the yeast (Burrows, 1979).

The sugars available to yeast in the fermentation of dough come from two sources. First there is the small amount of free sugars that the flour itself contains (Oura *et al.*, 1979; Burrows, 1979). In wheat flour these sugars amount to 1–2% by weight according to the particular type of flour; they consist of glucose, fructose, maltose, sucrose, and a variety of glucofructans. Typical figures for the sugar composition of wheat flour are given in Table I. The second source of fermentable sugar is the maltose produced by the hydrolysis of starch by wheat amylases which are activated during the making of dough. This maltose released from starch amounts to some 3% by weight (Burrows, 1979) and thus represents the principal source of fermentable carbon for the panary fermentation. Therefore, any good baker's yeast must be able to rapidly ferment maltose, yet there is no direct correlation between maltose fermentation and dough-raising ability (Lovgren and Hautera, 1977). Baker's yeast must also have high levels of invertase and related enzyme activities in order to hydrolyze sucrose and the glucofructans. Finally, although baker's yeast is produced in highly aerobic conditions, it is used in bread dough in circumstances in which oxygen is limiting or absent. For this reason the characteristics of a good baker's yeast must represent a compromise between the contrasting demands of production and use.

Table I. Sugar Content of Wheat Flour

Sugar	% w/w
Glucose	0.02
Fructose	0.04
Sucrose	0.26
Glu_1 -fru_2	0.40
Glu_1 -fru_3	0.26
Higher glucofructans	0.72
Maltose	0.12

3. BEER BREWING

The brewing of beer is an enormous enterprise and probably the largest biotechnological industry in the world. The amount of beer produced commercially each year in the world is ca. 10^{11} liters. For most of this production the principal source of fermentable sugars is malted barley, although a number of African beers use other cereals such as sorghum. In the United States it is common to supplement malt with other, cheaper, carbohydrate sources which may readily be converted to fermentable sugars since American malts have particularly high enzyme activity and are also rich in nitrogeneous compounds. The most common solid adjuncts used are maize and rice grits. Grits are produced by abrading cereal grains to remove the husk, embryo, and aleurone layer. The remaining endosperm, which is rich in starch, but contains little lipid or protein, is then milled to produce the grits. Liquid adjuncts may take the form of corn syrups (produced by the enzymic hydrolysis of grits) or solutions of cane sugar or invert sugar.

The basic process of beer brewing is outlined in Fig. 1. In the malting process, barley grains are germinated under controlled conditions such that enzymes are produced that begin the conversion of the starch and protein reserves of the seed's endosperm to fermentable sugars and amino acids which may later be used by yeast. This germination process is arrested by drying ("kilning") and the temperature employed at this stage determines the principal characteristics of the malt. If moderate temperatures are used, then the malt will be pale in color and high in enzyme activity. Kilning at high temperatures produces dark malts, suitable for ales and stouts, which are low in enzyme activity. American malts are usually of the former class and this permits the use of starch adjuncts described above, a practice forbidden by law in West Germany.

The next stage in the brewing process involves milling the malted barley into a coarse flour called grist. This is then steeped in water and incubated at ca. 65°C, a process known as mashing. The hydrolysis of

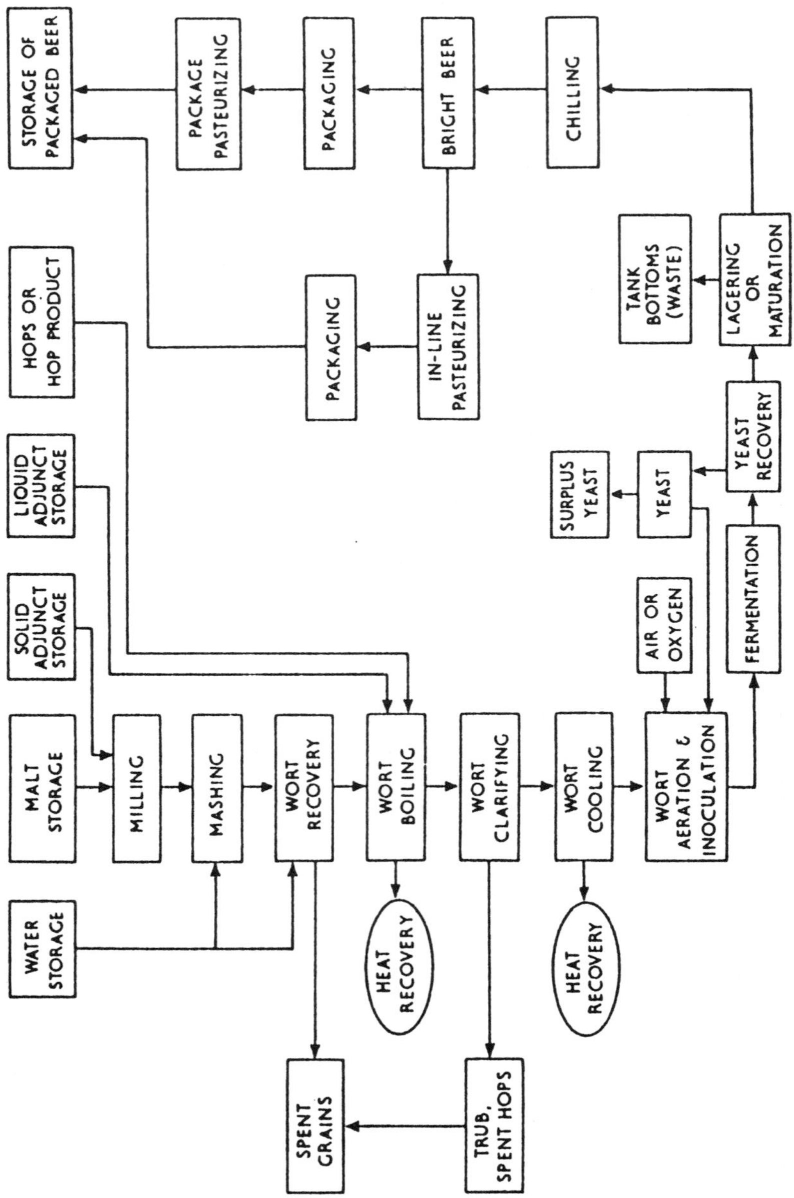

Figure 1. Flow diagram of the brewing process. (From Hough, 1985.)

starch to fermentable sugars continues in the mash and this is mainly due to the action of the barley α-amylase. This enzyme is highly thermotolerant and only 4% of its activity is lost during kilning (Preece, 1954). This contrasts markedly with β-amylase and endo-β-glucanase whose activities decline by 66% and 33%, respectively, as a result of kilning (Preece, 1954; Preece and Hoggan, 1957). There is also considerable conversion of protein and peptides to amino acids during mashing. This is largely due to the action of carboxypeptidase, which, again, is relatively heat tolerant (Mikola *et al.*, 1972). During mashing solid adjuncts to malt may be added. In order to aid the digestion of this supplementary starch, preparations of amylases and other carbohydrate-degrading enzymes, such as pullulanase, may be added. These commercial enzyme preparations are usually derived from filamentous fungi or bacteria, particularly *Aspergillus* or *Bacillus* species. Enzyme addition is especially prevalent in Western Europe (outside West Germany), where the use of dark malts means that there is insufficient enzyme activity to degrade additional starch.

At the end of the mash, the aqueous extract is separated from the spent grains and other solid material. The spent grains, which still contain about 20% protein by weight, are frequently used in cattle feed. The clear liquid, called wort or sweet wort (to emphasize its high content of fermentable sugars) is then boiled in a vessel called a wort kettle or copper (although it is now commonly constructed of stainless steel). The purposes of boiling the wort are manifold:

1. Inactivation of enzymes
2. Sterilization of the wort prior to inoculation with yeast
3. Precipitation of residual proteins
4. Reduction in pH through the precipitation of calcium phosphate (this favors yeast growth and militates against bacterial infections)
5. Distillation of volatile compounds which might impart off-flavors to the beer
6. Further development of the color originally imparted by the kilning of malt
7. Concentration of the wort which determines the starting gravity and therefore the ultimate alcohol content of the beer

Liquid adjuncts such as sugar solutions or corn syrup may be added to the wort kettle to provide a further source of fermentable carbon. Traditionally hops are added at this stage and boiling serves to isomerize the α-acids in the hops. However, hops are expensive and it is becoming increasingly common to use a process of postfermentation bittering, where hop extracts are added as a final step before conditioning of the beer. This practice is more efficient than the traditional method since bittering substances are frequently lost with the removal of the foam head after fermentation.

On completion of boiling, which may take up to 2 hr, the liquid wort is separated from spent hops and precipitated matter ("trub") either by filtration through a bed of spent hops or a stainless steel strainer or by use of a centrifuge or a whirlpool tank. The clarified wort is then cooled by passage through a heat exchanger and is finally ready for fermentation. (It is a chastening experience for a microbiologist to walk round a large modern brewery and see what a small part of the whole enterprise the fermentation vessels represent.)

Classically, two kinds of yeasts are used in the production of beer. Yeasts of the species *Saccharomyces cerevisiae* are used in top fermentations to produce the beers and ales traditionally drunk in the United Kingdom; the fermentation is conducted at a temperature of 15–22°C. Lagers are brewed at a low temperature, 8–15°C, using the bottom-fermenting yeast species, *S. uvarum.* This yeast was previously known as *S. carlsbergensis* in honor of the famous Danish beer maker. The distinction between these two supposed species of yeast has become increasingly blurred, and since *S. cerevisiae* and *S. uvarum* can readily interbreed to produce diploid yeasts which yield viable spores, it is probably best to regard them as varieties of the same species. For diagnostic purposes, *S. cerevisiae* and *S. uvarum* can easily be distinguished by their ability to use mellibiose (*S. uvarum* can and *S. cerevisiae* cannot). However, there are now bottom-fermenting ale yeasts and even, in the United Kingdom at least, top-fermenting lager yeasts.

The fermentation of wort to beer may be represented by the following equation (Hough, 1985):

$$\begin{array}{l}\text{Maltose + amino acid} \rightarrow \text{yeast + ethanol + } CO_2 \text{ + heat} \\ \quad\text{(from hydrolysis of barley starch)} \\ \quad 100\text{ g} + 0.5\text{ g} \rightarrow 5\text{ g} + 48.8\text{ g} + 46.8\text{ g} + 50\text{k Cal}\end{array}$$

The low amount of biomass produced, and the high amount of ethanol, contrasts markedly with the production of baker's yeast under aerobic conditions:

$$\begin{array}{l}\text{Sucrose (from molasses) + ammonia + oxygen} \rightarrow \text{yeast + water} \\ \quad\text{+ carbon dioxide + heat} \\ \quad 100\text{ g} \quad + \quad 5\text{ g} + 51\text{ g} \rightarrow 48\text{ g} + 35\text{ g} + 73\text{ g}\end{array}$$

In addition to ethanol, yeast may produce a number of higher alcohols, the so-called fusel alcohols. It is the reaction between acyl CoA moieties and those various alcohols which produce esters and fats; the former are important contributors to the flavor of beer. Most of these other alcohols are produced by the transamination of an amino acid, followed by decarboxylation and reduction:

$$(CH_3)_2(CH)_2NHCOOH \xrightarrow{\text{oxoacid amino acid}} (CH_3)_2CHCHO + CO_2$$

isobutyraldehyde

$$\downarrow NADH_2 \rightarrow NAD^+$$

$$(CH_3)_2CHCH_2OH$$

isobutanol

A mixture of isobutyraldehyde and methylglyoxal gives (Palamund and Hardwick, 1969) an earthy, musty aroma to the beer. Other important flavor compounds in beer include diketones and diacetyl. The latter imparts a "butterscotch" flavor and results from the chemical decarboxylation and oxidation of acetolactate excreted into the beer by yeast during the fermentation.

$$CH_3CO - \overset{OH}{\underset{CH_3}{C}} - COOH \rightarrow CH_3CO - \underset{CH_3}{C} = 0$$

acetetolactate   diacetyl

Sulfur compounds are other major flavor determinants in beer, the sulfur being derived either from sulfate or from the sulfur-containing amino acids methionine and cysteine. Two simple compounds that have very potent flavors are hydrogen sulfide (H_2S) and dimethyl sulfide [$(CH_3)_2S$].

Figures 2 and 3 contrast the course of a typical top fermentation

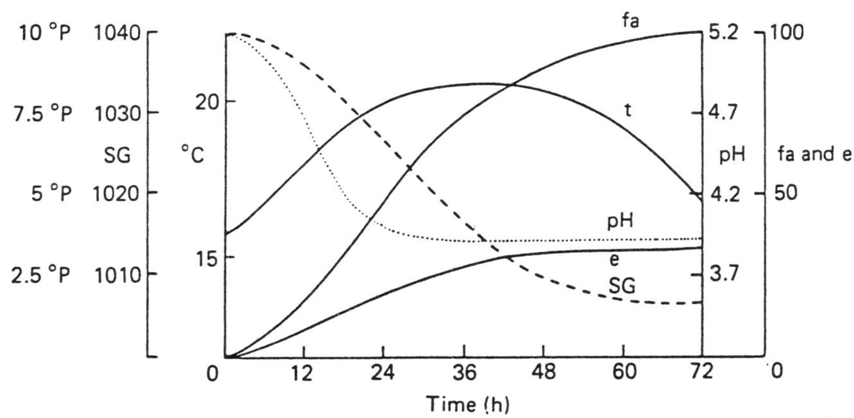

Figure 2. Time course of a top fermentation with an ale yeast. SG, specific gravity of wort; t, temperature; fa, fusel alcohol (mg l^{-1}); e, ester content (mg l^{-1}). (From Hough, 1985.)

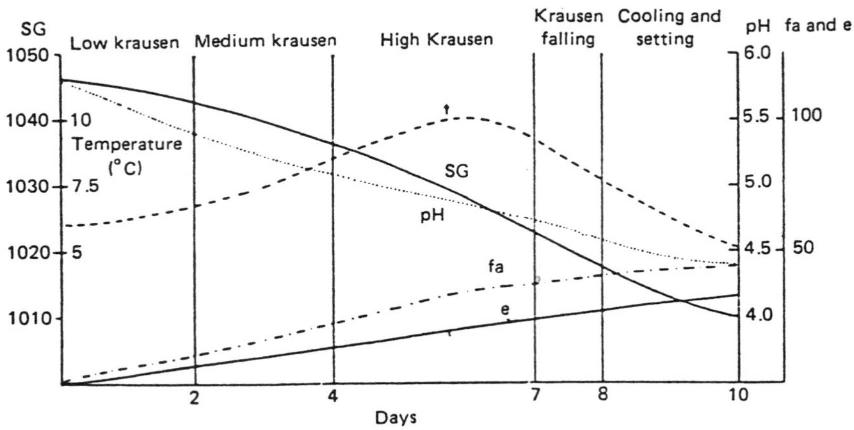

Figure 3. Time course of a bottom fermentation with a lager yeast. (Symbols as for Fig. 2.) (From Hough, 1985.)

with *S. cerevisiae* at a temperature of 15–20°C with that of a bottom fermentation by *S. uvarum* at 6–10°C. A major difference in the two processes is the time involved, the lager fermentation taking three times longer than the ale one. This is one reason why a conventionally brewed lager is more expensive than an equivalent ale. In the traditional ale fermentation, the yeast rises to the surface during the last 10–12 hr of the process and may be skimmed or sucked off. A yeasty foam also develops in the lager process; this is referred to as the "Krausen". In contrast to the ale fermentation, however, the Krausen falls to the base of the vessel toward the end of the lager process, taking the yeast with it.

In traditional breweries, a large variety of fermenter designs with different systems for the retrieval of yeast was employed; these had evocative names such as Yorkshire Stone Squares and Burton Unions. Even lager was traditionally brewed in open vessels. Today, closed fermentation vessels are the rule. In large, modern breweries these vessels are equipped with automatic cleaning systems and have controlled mixing and cooling devices. The most common design is the cylindroconical vessel (Fig. 4), where the flocculent yeast accumulates in the basal cone whose temperature may be controlled separately from that of the bulk of the vessel. Fermentation times are reduced in such vessels and little foam is formed and thus the amount of bittering substances lost during the fermentation is minimized. These two features make considerable contributions to the efficiency, and hence the profitability, of the brewing process.

Alcohol production is, of course, a growth-linked characteristic and therefore continuous culture commends itself as the optimal system of

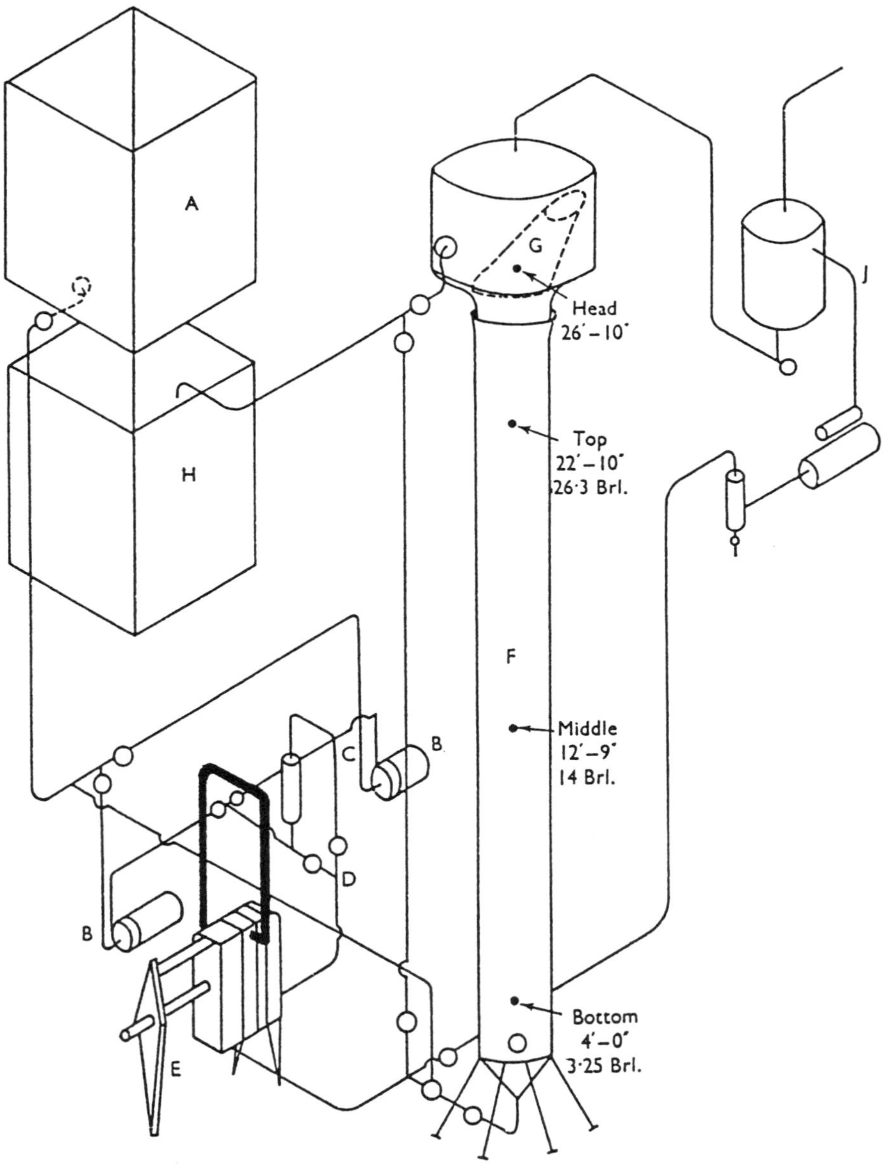

Figure 4. Production tower fermenter and ancillary equipment. A, wort collecting vessels; B, impeller-type pump; C, flowmeter; D, control valve; E, flash pasteurizer; F, tower; G, yeast separator; H, beer receiver; J, CO_2-collecting vessel. (From Ault *et al.*, 1969.)

fermentation. In the brewing industry, as in baker's yeast production, continuous culture has not caught on (at least, not in the Northern hemisphere). Attempts to introduce continuous beer production in Britain have failed, falling victim to problems of strain stability (Thorne, 1970; Hall, 1970) and infection. In one, now famous, case a continuous fermenter became infected with a strain of "killer" yeast (Maule and Thomas, 1973). Although saké brewers have guarded against this problem by developing production strains that are themselves killers, no such innovations have been made in British brewing practice. One suspects that the radical changes in working practices which continuous fermentation incurs had much to do with the system's demise in the United Kingdom.

In the Southern hemisphere, continuous culture has been more successful. A number of alcohol factories in Brazil employ a continuous fermentation system developed by the Alfa-Laval company (Carioca, 1984). The continuous brewing of beer has been very successful in New Zealand, where sophisticated systems of "cascade" fermentation (see Fig. 5), with a number of fermenters connected in series and recycling of yeast cells, have been employed (Coutts, 1961, 1966). A single-vessel, continuous system was employed in the United Kingdom; the pressure-cycle tower fermenters used (Ault *et al.*, 1969) were similar in concept, if not in scale, to the ill-fated ICI "pruteen" fermenter. A critical parameter for brewing beer in these tower fermenters is that a sufficient supply of oxygen to the wort is maintained. Although the alcoholic fermentation is an anaerobic process, small amounts of oxygen are required to sustain fatty acid and sterol biosynthesis by the yeast (Andreasen and Stier, 1953; 1954).

Freshly fermented beer is not ready for consumption and must undergo further treatment before it can be distributed and sold. In traditional brewing practice, postfermentation treatment, or conditioning, is a relatively simple process, but in modern beer manufacture, as in the other areas of biotechnology, downstream processing is assuming increasing importance.

In traditional cask conditioning of ale, freshly fermented beer is "racked off" into wooden or (nowadays) stainless steel casks. The beer is usually held in the cask under pressure and undergoes a secondary fermentation. The freshly fermented beer introduced into the cask should contain about 10^6 yeast cells ml^{-1} and these are responsible for the secondary fermentation. In traditional breweries, however, wild yeasts (particularly *Brettanomyces* species) also invaded the casks and imparted characteristic flavors to the beer. The secondary fermentation may rely on the presence of residual sugars in the fresh beer. Alternatively, the secondary fermentation may be primed by the addition of sugar syrups to the fresh beer. The use of caramel at this stage permits

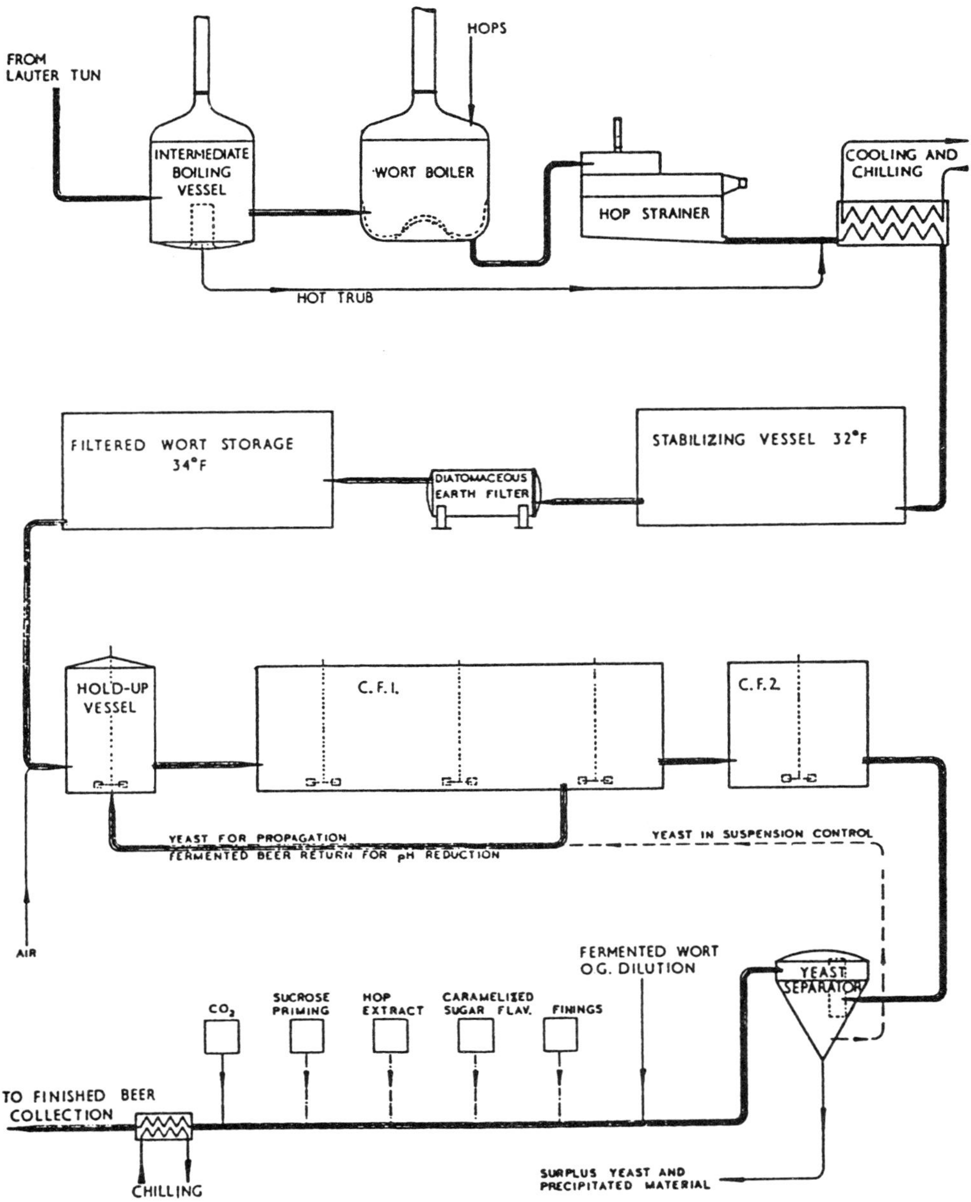

Figure 5. Cascade fermenter system for continuous beer production. (From Coutts, 1966.)

the color of the finished product to be controlled and the addition of isomerized hop extracts determines bitterness. Traditional lager production involves extensive postfermentation conditioning of the beer in enclosed lagering tanks which are held at 2°C for up to 9 months. Under these conditions the residual sugars are slowly fermented and the yeast settles out together with precipitated polyphenols (tannins) and polypeptides (proteins). This lagering yields a haze-free beer with a very long shelf life, but it is a very costly process.

Modern brewing practice has greatly reduced the time required for postfermentation conditioning of both ales and lagers. The primary fermentation is usually allowed to proceed to completion and then the beer is held at 14–18°C for 2 days (the diacetyl rest) to allow the volatilization of undesirable products of fermentation. The beer may then be chilled and carbonated artifically. Special treatments may be employed to prevent beer haze developing. Haze formation is catalyzed by heavy metals and is the result of polyphenols and polypeptides coming out of solution. Two types of haze may be distinguished, chill haze and permanent haze. Both appear on chilling the beer, but in the latter the aggregates formed do not redissolve on subsequent warming.

4. SAKÉ BREWING

Saké is often referred to as rice wine but, in fact, its production is more similar to that of beer than wine since starch is the primary carbohydrate source and must first be converted to sugars before yeast may undertake the alcoholic fermentation. The saké fermentation is not only very interesting, involving the action of two microorganisms in succession, but also gives rise to a beverage that has the highest alcohol content (ca. 25% vol./vol.) obtained by direct fermentation. It is believed that the original way of converting rice starch into fermentable sugars for saké fermentation was to chew steamed rice so that salivary amylases became mixed with it. The method of using filamentous fungi to degrade the starch was introduced from China at the beginning of the 5th century A.D. (Kodama and Yoshikawa, 1977). An outline of the modern process of saké brewing is given in Fig. 6.

The rice is "polished" before use by passage through a mill. This has the effect of increasing the relative carbohydrate content of the grains from ca. 72% to ca. 78% and results from a fall in the protein, lipid, and ash contents. Polishing increases the capacity of the grains to absorb water, which they do to the extent of 25–30% in the next step of the process—steeping. The steeped grains are then drained of excess water before steaming for 30–60 min. Steaming has a number of effects: it sterilizes the rice, permits further absorption of water, denatures pro-

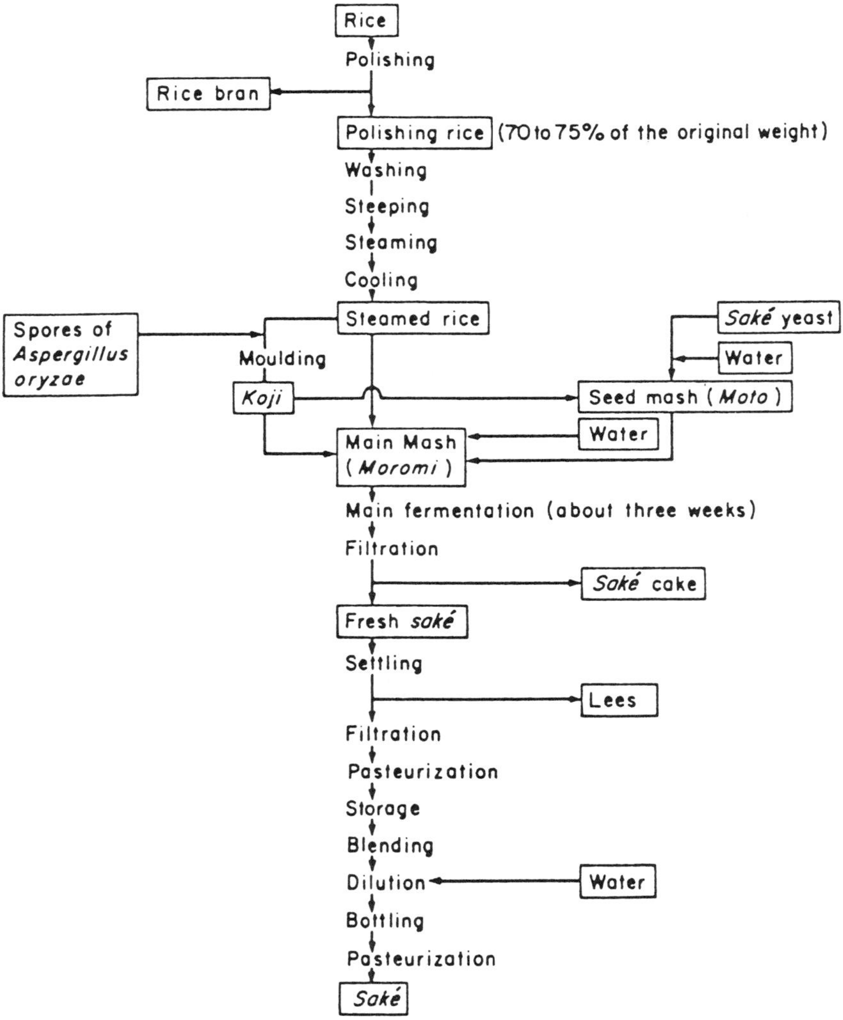

Figure 6. Flow diagram of the saké process. (From Kodama and Yoshikawa, 1977.)

teins, and converts starch to the α-form in which it is more susceptible to the action of the *Aspergillus* amylases.

In parallel with the pretreatment of the rice, the "koji" is prepared. Koji is a culture of *A. oryzae* grown on steamed rice (Murakami *et al.*, 1968). In all, about 50 kinds of enzymes have been reported as present in koji, but the most important to the saké process are α-amylase, amyloglucosidase, acid protease, and alkaline protease. The temperature of the koji fermentation influences the proportions of these enzymes, with high culture temperatures (ca. 42°C) favoring the amylases and lower ones (ca. 30°C) resulting in higher protease activity.

The final component required for sake fermentation is the yeast starter culture, or moto. There are two main methods by which this inoculum is prepared. In the traditional technique, referred to as ki-moto and yamahai-moto, the culture is naturally acidified by the action of lactic acid bacteria, which produce conditions under which the growth of saké yeast is selected over that of the wild yeasts previously found in the culture. Modern practice, sukujo-moto, demands that the culture is artificially acidified by adding lactic acid at the start of the saké fermentation, which is then pitched with pressed yeast grown in pure culture by procedures similar to those described for the generation of baker's yeast (see Section 2). A good saké yeast needs to exhibit tolerance to low pH, high osmotic pressure, and high ethanol concentrations. Saké producers are so convinced of the peculiar properties of their yeast that they have been led, in the past, to distinguish it by the specific name *Saccharomyces saké.* There is no doubt, however, that saké yeast is *Saccharomyces cerevisiae,* and its peculiar properties, especially in regard to ethanol tolerance, need to be carefully defined. Brown *et al.* (1981) compared the ethanol tolerance of the commercial saké yeast, Kyokai No. 7, with that of a laboratory strain, 5D-cyc. They found that the two yeasts were indistinguishable in terms of ethanol tolerance of net growth, but that fermentation by the saké yeast was more tolerant to the inhibitory effects of ethanol than was that of the laboratory strain (Fig. 7). The inhibition constant, K_i, for the effect of ethanol on fermentation rate was found to be 0.97 M for 5D-cyc but 1.32 M for Kyokai No. 7. The ability of saké yeast to withstand high ethanol concentrations during saké brewing may not simply be due to the properties of the yeast itself. Hayashida *et al.* (1975) have reported that a proteolipid produced by *Aspergillus oryzae* enhances the ethanol tolerance of saké yeast.

The main fermentation of the saké process, the "moromi," is carried out in a large, open vessel. The moromi is a fed-batch process in which successive additions of fresh substrate, in the form of steamed rice and koji, are made. In the moromi, the conversion of starch to fermentable sugars, by the action of *A. oryzae,* and the fermentation of those sugars to alcohol, by *S. cerevisiae,* occur simultaneously. This again distinguishes the brewing of saké from that of beer, where saccharification (malting and mashing) and fermentation occur sequentially. Since the final concentration of alcohol reached (ca. 20% vol./vol.) requires nearly 40% wt./vol. sugar to be supplied, it is evident that this can only be achieved by such a sequential process. Yeast would not be able to ferment such a high concentration of sugar if it were all provided at the start of the process.

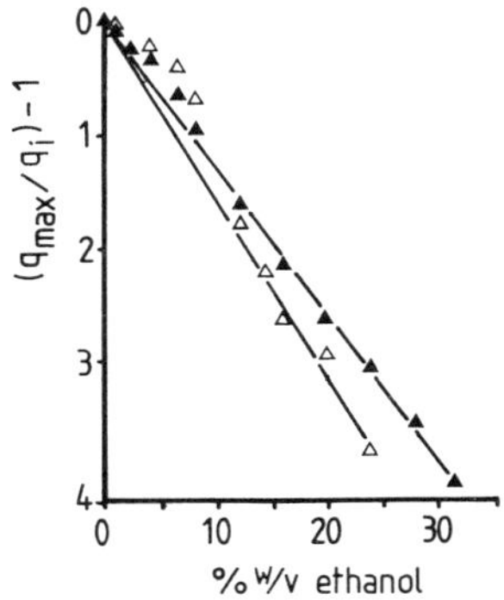

Figure 7. Effect of ethanol on fermentation in two yeast strains: plot for noncompetitive inhibition, —△—, a laboratory haploid strain of *S. cerevisiae;* —▲—, a commercial saké yeast. (From Brown *et al.,* 1981.)

The fermentation takes about 25 days to complete, and while the temperature varies considerably through the process, it never exceeds 18°C and at the early stages is as low as 7–8°C. It is often considered that this low temperature contributes to the ethanol tolerance of saké yeast. However, the effects of temperature on ethanol tolerance are complex and differ for growth and fermentation (Brown and Oliver, 1982a). A particularly low temperature is not necessarily beneficial.

At the end of the fermentation, when the alcohol content may be enhanced by the deliberate addition of ethanol diluted in a salts solution, the saké is harvested by filtration under pressure. The cake left behind on the filter, 'saké-kasu', contains starch, protein, and various enzymes as well as yeast cells. It is sold to the food industry for use in other processes such as pickle making. One tonne of polished rice yields 3000 liters of saké (at ca. 20% vol/vol ethanol) and 200–250 kg of saké-kasu. The saké is held at low temperature for 5–10 days, during which time it further matures owing to the action of enzymes both released from the *Aspergillus* and contained within the, now largely dead, yeast. It is important to avoid autolysis of the yeast cells at this stage. Much solid material settles out during storage and the saké is finally clarified by filtration through activated charcoal. A further period of maturation may then be allowed before pasteurization and bottling. Pasteurization was introduced into the saké process in the 16th century, thus considerably predating Louis Pasteur. Given the very high alcohol content of the product, it is perhaps surprising that it is needed at all. However, infections with highly ethanol-tolerant lactic-acid bacteria ("hiochi" bacteria) do occur and produce off-flavors which are mainly attributable to diacetyl formation (see Section 3). Deterioration of bottled saké can also result from photooxidation reactions, which are readily prevented by the use of opaque bottles.

5. WINE MAKING

The fermentation of grape must into wine is, at a superficial level at least, a much simpler process than the production of beer or saké. This is because the starting material contains fermentable sugars and there is no problem of converting nonfermentable polysaccharides into fermentable monomers. Grapes contain four principal sugars—glucose, fructose, arabinose, and rhamnose. In the unripe grapes, glucose is by far the most abundant sugar, but during the ripening process the proportion of fructose increases until it equals or even exceeds that of glucose (Lafon-Lafourcade, 1983). The total concentration of reducing sugars varies according to the strain of grapes employed as well as the conditions (particularly climatic) of their cultivation. However, the usual range is 120–150 g/liter, and of this total, the pentoses represent less than 1% (Vogt *et al.*, 1974). In addition to these monomers, polysaccharides are also present as pectins and gums. The principal acids present in grape must are tartaric and malic and their content varies from ca. 2 g/liter to 7 g/liter according to climate. This gives the must an average pH of 2.8–3.9. In hot, dry regions, such as Argentina, the content of these acids can be neglible. The citric acid content of grapes is 10% the level of malic or tartaric acids. Nitrogen is present to the extent of 0.1–1 g/liter and assumes a variety of forms, including free ammonia (3–10% of the total), amino acids (25–30%), and polypeptides (30–50%). The principal amino acids are arginine, glutamate, proline, and threonine and the proportion of total nitrogen present as amino acids is found to increase during ripening.

Grape must, therefore, contain the essential requirements for growth and fermentation by yeasts, and in traditional wine-making practice, the natural flora of the grape is relied upon to initiate the fermentation. A number of nutritional supplements will accelerate the process, however, and are permitted by the regulatory authorities. These supplements include phosphate, potassium, and thiamin (Lafon-Lafourcade, 1983). The latter, in addition to activating the fermentation, is also useful in preventing the accumulation of keto compounds which will combine with the SO_2 used to arrest the fermentation and prevent the microbial degradation of the wine. The requirement for small amounts of oxygen to permit fatty acid and sterol biosynthesis (Andreasen and Stier, 1953, 1954) for yeast growth may be fulfilled by a process known as "pumping over" (see Section 5.2), which is used exclusively for the production of red wines.

The performance of the yeasts present on the grapes is very variable and this frequently results in a long lag before the start of fermentation. In modern practice, therefore, it is usual to inoculate with ca. 10 g/hl of active dried yeast. The natural flora is often eliminated by sulfiting before this inoculum is added. Specially selected strains of *S. cerevisiae* or *S. bayanus* are used for this purpose and each winery jealously guards its own strains. As with *S. saké,* it is not clear that there is any genuine taxonomic distinction between *S. cerevisiae* and *S. bayanus.*

5.1. White Wine Production

White wine is produced from pure juice obtained from white grapes. The extraction of juice is performed rapidly to avoid oxidation and maceration of the skins. The use of perfectly sound grapes is preferred in order to avoid accidental contamination of the juice. The extraction process has four phases: crushing, draining, pressing, and clarification.

Crushing breaks the skins of the grapes and lets the juice run out of the pulp. Fragmentation or grinding of the grapes is avoided since this results in the production of large amounts of sediment. Vibrating screens are then used to drain off the juice since this accelerates the process and avoids oxidation. Following crushing, pressing permits extraction of the remaining juice from the pulp, and continuous, pneumatically driven presses are employed in large-scale operations. The problem with such equipment is that it may pulverize the press cake, leading to contamination with skin fragments. Oxidation can also be a problem at this stage. An alternate approach is to press intact grapes without crushing. This yields musts that are readily clarified and highly fermentable.

Clarification is the removal of suspended solids from the must and may be achieved by filtration, centrifugation, or, more traditionally, settling and decantation. Addition of sulfite to the settling tanks will prevent both fermentation by the indigenous flora and oxidative damage. Pectinolytic enzymes may be added at this stage to reduce the polysaccharide content and accelerate the process. Industrial enzyme preparations commonly contain cellulases, hemicellulases, xylanases, and proteases in addition to pectinases. Removal of protein from the must may also be achieved by addition of the absorbent clay bentonite. This procedure is familiar to biochemists who need to reduce nuclease activity in their preparations.

The clarified must is stored in casks or, these days, in large tanks before fermentation, which is carried out under anaerobic conditions at a temperature of 20°C. Fermentation is arrested by sulfiting, and this may be done early if it is desired to produce a sweet white wine. An alternate strategy is to add fresh must, often in concentrated form, to the wine at the time of bottling. Such a procedure is said to "mellow" a dry wine.

After fermentation, white wine should be rapidly clarified by filtration or centrifugation. Prolonged storage of the wine on the "lees" (yeast and other solid residues of fermentation) can cause the development of disagreeable odors due to the formation of sulfhydryl compounds. Nevertheless, some classes of wine are deliberately left on the lees to promote the malolactic fermentation.

5.2. Red Wine Production

The major distinction between the production of red and white wines is that the former requires maceration of the grapes while the latter avoids it. Maceration is the process whereby the must is left in prolonged contact with the solid parts of the grape—the skins, the seeds, and even, for some varieties, the stems. These impart color to the wine and also increase its tannin content. Crushing and destemming the grapes are often carried out simultaneously. Removal of stems gives the wine a more subtle, "refined" flavor. If the stems are left on the berries, the product has an astringent, "herby" taste.

After the crushing stage, production of red wine follows a course similar to that of the white varieties. However, the fermentation is carried out at a higher temperature, 25–30°C, since this facilitates color extraction. Even at this temperature, cooling of the fermenter will often be required since red wine fermentation is more rapid than that of white, as the must is rich in nitrogen and other "activating" substances. During the fermentation the practices of skin, seed, and pulp are carried to the top of the fermentation vessel by the rising bubbles of CO_2 and form a "cap" on the surface. The fermenter may be drained into a second vessel and the liquor "pumped over" into the first vessel again. This stimulates yeast growth by the addition of small quantities of oxygen and also promotes color extraction due to the passage of liquor through the cap. The liquor is finally drawn off into large casks where the conversion of residual sugars into alcohol is completed. This may also permit the malolactic fermentation. The wine is finally stabilized by the addition of SO_2, before a period of storage which may be prolonged to produce a mature red wine.

Storage of wine in barrels for the purposes of maturation involves decanting quarterly and the frequent addition of wine to replace evaporative losses. It is essential that the headspace of the barrel be kept full in order to prevent the growth of acetic acid bacteria. Many changes in the color and flavor of wine occur during maturation, some of which are acquired from the wooden casks themselves.

The last stage of the wine-making process before bottling is fining. In this clarification procedure, suspended impurities are precipitated by the addition of bentonite or protein (gelatin, globulin, casein, or egg white). The protein reacts with tannin to produce a precipitate which carries down other suspended particles. Filtration usually follows fining, although this stage is omitted for some varieties. The wine must then be bottled under conditions that prevent microbial contamination. In the hot-bottling process, this is achieved by heating the wine to 50°C and bottling at the same temperature. An alternative is to "flash"-pasteurize the wine at 80–85°C for several seconds immediately before filling the bottles.

6. STRAIN DEVELOPMENT

In all industrial processes involving microorganisms, considerable improvement in product yield or process efficiency may be achieved by genetic development of the production organism. This is true of yeast in the traditional biotechnological processes described here. Indeed, much unconscious selection of desirable traits must have been carried out by practitioners of the arts of brewing and bread making over the millennia in which they have been in use. Modern strain development has a wide range of techniques available for use with yeast, and I will not duplicate here the descriptions provided in other chapters in this volume but will, instead, provide some specific examples of their use. The endeavors of the geneticist in yeast strain development are often hampered by the fact that production strains have ill-defined genetics and are frequently unbalanced polyploids. The factors whose improvement is sought, such as flavor, flocculation ability, or alcohol tolerance, may themselves be poorly defined and controlled (or modified) by large numbers of genes. Our ignorance of the biochemical nature and genetic control of these traits may mean that, in some cases, the application of gene cloning techniques would be premature and wholly inappropriate. In other cases, such as the extension of substrate range, recombinant DNA technology holds great promise since it is hoped that new potentialities can be recruited in a specific manner that does not compromise the desirable traits the strain already possesses.

7. ETHANOL TOLERANCE

The life cycle of *S. cerevisiae* provides both haploid and diploid vegetative phases and permits isolation of all four products of meiosis. This makes it ideal for strain development by conventional genetic techniques (see Chapter 4 by R. B. Wickner). Such techniques usually involve the screening or selection of mutants on agar plates, and the geneticist designs conditions such that only the desired mutant will grow or not grow. For some characteristics such all-or-none selection or screening techniques are not practicable. This may be because the desired phenotype is determined by a large number of genes or because only relatively small, quantitative, improvements may be achieved. Both these problems are encountered when attempting to isolate yeast mutants with improved tolerance to ethanol.

Yeast is highly tolerant of ethanol when compared to other microorganisms, probably because of unconscious selection on the part of the producers of alcoholic beverages throughout history (Oliver, 1984). Thus only small improvements to the organism's tolerance are likely to be achieved in any strain improvement program. However, owing to the energy consumption profile of alcohol distillation, even relatively modest increases in final product concentration may be economically significant (Righelato, 1980). The likelihood that a large number of genes are involved in determining ethanol tolerance is indicated by the complexity of the effect that the alcohol has on its producer organism. Fermentation rate (Rubner, 1912), glucose consumption (Gray, 1941), biomass yield (Troyer, 1953), growth rate (Troyer, 1953; Holzberg *et al.,* 1967; Aiba *et al.,* 1968), and cell viability (Thomas *et al.,* 1978) have all been used as indicators of the relative tolerance of yeast strains to the toxic effects of ethanol. In fact, different inhibition constants may be determined for the inhibitory effect of ethanol on cell viability, true growth rate, and fermentation rate (Brown *et al.,* 1981). That this complexity at the phenotypic level is mirrored at the genotypic level is confirmed by the identification of both nuclear (Aguilera *et al.,* 1982; Sugden and Oliver, 1983) and cytoplasmic (Brown *et al.,* 1984) mutations which produce an ethanol-sensitive phenotype. The above description of the physiological and genetic characteristics of yeast ethanol tolerance makes it unlikely that conventional agar plate screening techniques will succeed in isolating tolerant mutants, and this has been found to be the case (Ismail and Ali, 1971a,b).

In this situation when only small quantitative increases are likely to be obtained as a result of multiple mutations, the use of continuous selection is indicated (Harder *et al.,* 1977; Dykhuizen and Hartl, 1983). However, the very complexity of the inhibitory effects of ethanol on yeast makes it difficult to design a suitable selection regime. For this reason, a system in which the intensity of selection was determined by the culture itself via a feedback control circuit was adopted by Brown and Oliver (1982b). The feedback selection system employed is diagrammed in Fig. 8. An infrared analyzer was used to monitor continuously the fermentation activity of the culture by determining the concentration of carbon dioxide in the exit gas. When this concentration equaled or exceeded the value set on the potentiometric controller, a relay closed. This switched on a peristaltic pump and introduced ethanol into the culture vessel. When the addition

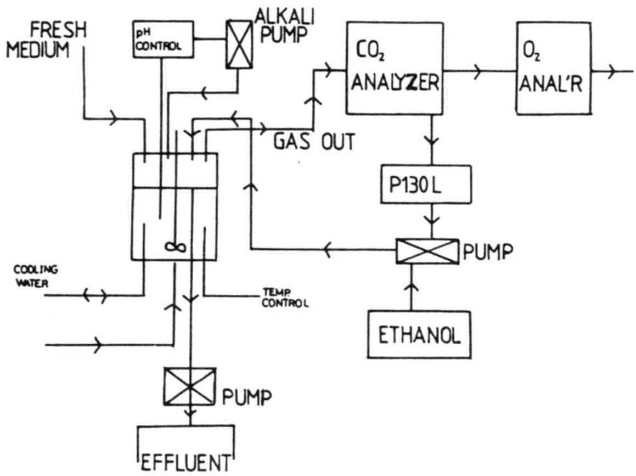

Figure 8. Continuous culture apparatus for feedback selection of ethanol-tolerant yeast mutants. (From Brown and Oliver, 1982b.)

of the alcohol had reduced the carbon dioxide concentration of the exit gas to below the value set by the controller, the relay opened and the ethanol pump was switched off. The ethanol was then gradually diluted out of the fermenter vessel and the rate of carbon dioxide production increased. When the carbon dioxide concentration in the exit gas exceeded the value set by the controller once more, the pump was switched on again. Therefore, the system was functionally analogous to a turbidostat. However, it was the rate of carbon dioxide production by the culture,

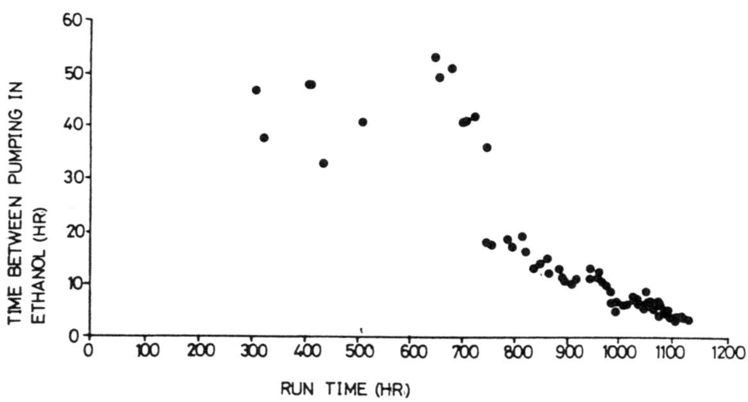

Figure 9. Improvement in ethanol tolerance of the culture. The frequency of switching of the ethanol pump in response to an increase in CO_2 concentration of the exit gas is shown. Each point represents a 7-hr moving average of the interval between operations of the pump. (From Brown and Oliver, 1982b.)

rather than its turbidity, which was held constant and the supply of inhibitor, rather than nutrients, which was regulatory.

The system indicated an improvement in the ethanol tolerance of the culture by an increase in the frequency with which the ethanol pump was switched on by the control system. The pump did not switch on at all for the first 12 days of the experiment. After that, it switched on occasionally until, 27 days (650 hr) into the experiment, there was a dramatic increase in the frequency with which the pump switched on (Fig. 9). This

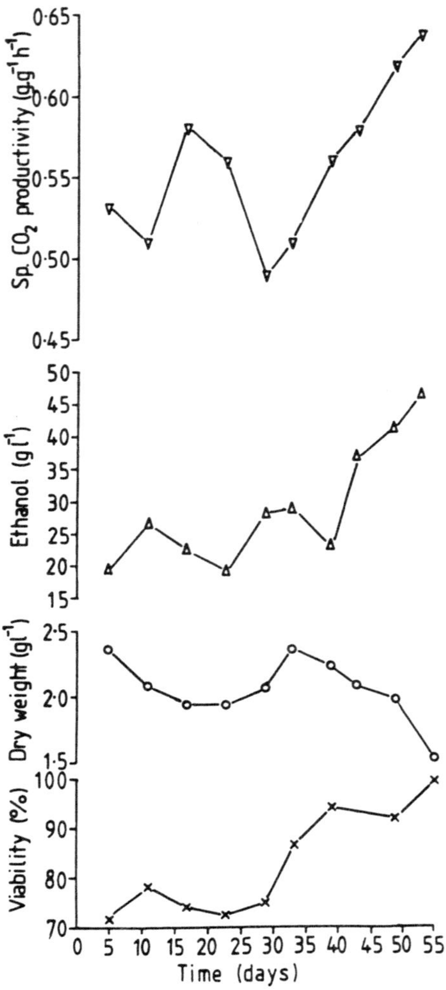

Figure 10. Changes in the performance of the culture during feedback continuous selection. x—x cell viability; ○—○, dry weight; △—△, ethanol concentration; ▽—▽, specific rate of production of CO_2. These parameters were measured every 2 days; the graph shows mean values for successive 6-day periods. (From Brown and Oliver, 1982b.)

increase in switching coincided with a number of changes in the physiological state of the culture (Fig. 10). The specific rate of carbon dioxide production and the concentration of ethanol in the growth medium increased, as did the viability of the culture. These changes were accompanied by a fall in the concentration of biomass in the culture vessel, as predicted by analogy with the turbidostat. The abruptness of these changes might be interpreted as being the result of some particular mutational event. This seems unlikely; mutants with a range of phenotypes were obtained and genetic analysis showed that these carried mutations in more than one gene.

In the first instance, mutants were isolated by selecting cells that formed colonies on 12% (wt./vol.) ethanol plates. This concentration of alcohol was lethal to the parent strain used in the experiment. The fermentation performance of these mutants was subsequently investigated and the best were found to ferment at twice the rate of the parent strain in the presence of 10% (wt./vol.) ethanol (Table II), were selected on the basis of an enhanced overal fermentation rate, the K_i of ethanol inhibition of fermentation being little changed from that in the wild type. Others, such as SB154 and SB155, had an overall fermentation rate no better that that of the wild-type strain, but the K_i of ethanol for fermentation inhibition was significantly increased. Mutants of the first class may find immediate application in the commercial production of ethanol. Indeed, SB160 has been shown to reduce significantly attenuation times in batch fermentation on a pilot-plant scale. However, it is likely that mutants of the second class will be more valuable in the long run if they are used as starting material for further strain development.

Although the mutants isolated could survive and ferment at enhanced rates in the presence of high concentrations of alcohol, the selection regime never exposed the culture to more than 5% (wt./vol.) ethanol. This illustrates the advantage of allowing the culture to determine the intensity of selection via the feedback system. If the selection had

Table II. Comparison of the Fermentation Performance of Mutants of *Saccharomyces uvarum* Isolated by Feed-Back Continuous Selection

Strain	Fermentation Rate (QCO_2; gg^{-1} h^{-1})	Ki (fermentation) (% w/v ethanol)
5d-cyc (wild type)	24.7	4.4
SB154	20.4	8.2
SB155	19.1	5.1
SB159	30.5	4.6
SB160	43.8	4.6

been imposed externally, then ethanol concentrations higher than 5% (wt./vol.) would certainly have been used.

8. FLOCCULATION

The ability of yeast to flocculate, that is, to form large aggregates of cells that will fall to the bottom of the culture vessel toward the end of the flocculation process, has long been considered a desirable trait for brewing yeasts to possess (Rose, 1984). Both the biochemistry and genetics of the process are complex and poorly understood. This situation has not been helped by the fact that flocculation ability has been bred out of many genetically defined laboratory strains, and it has only been in comparatively recent years that isogenic flocculent and nonflocculent pairs of haploid strains have been available (Russell *et al.*, 1980). A fuller understanding of the process of flocculation is required to enable its exploitation as a cheap, easy means of separating cells and culture liquor in modern biotechnological processes. It is a highly desirable trait, for instance, in strains used for ethanol production when continuous culture with cell recycle is employed (Cysewski and Wilke, 1976). If vectors containing a cloned flocculation gene could be constructed, they might provide a self-selecting system to ensure plasmid stability, or, at least, the preferential retention of plasmid-containing cells in continuous culture (J. R. Johnston and D. R. Berry, personal communication).

The physical process of flocculation is complex, and it appears that an initial input of mechanical energy is required to overcome the mutual repulsion between individual yeast cells (Stratford and Keenan, 1987). Once this barrier is overcome, it is evident that floc formation requires the presence of divalent cations, particularly Ca^{2+}. Most theories of flocculation view Ca^{2+} as providing a bridge between anionic groups on the walls of yeast cells (Mill, 1964).

The source of these groups is far from clear, and both proteins (Jayatissa and Rose, 1976; Beavan *et al.*, 1979; Stewart *et al.*, 1975) and polysaccharides (Stratford and Keenan, 1987; Lyons and Hough, 1970, 1971) have been suggested as candidates. A completely different class of hypothesis is that glycoproteins act to cross-link yeast cells in a manner analogous to that of plant lectins (Taylor and Orton, 1978; Miki *et al.*, 1982). The role of calcium would then be seen as promoting conformational changes in these lectinlike molecules. The idea that wall proteins are involved is supported by experiments demonstrating that treatment of whole cells with proteolytic enzymes can remove their flocculation ability (Nishihara *et al.*, 1977, 1982; Hodgson *et al.*, 1985). Moreover, Holmberg (1978) has identified cell wall proteins that are flocculation-specific by comparing extracts from isogenic flocculent and nonflocculent strains.

The investigation of the genetic control of flocculation has a confused history owing to the identification of a number of genes in different laboratories that subsequently proved to be allelic (Russell *et al.*, 1980). Currently, six genes are believed to be able, separately, to confer flocculation ability on yeast (Hodgson *et al.*, 1985; Johnston and Reader, 1983). Three of these genes are dominant and three recessive, although this may be conditioned by the genetic background. A nuclear gene designated *fsu1* (Holmberg and Kielland-Brandt, 1978) acts as a suppressor of flocculation ability and, in general, the cytoplasmic petite mutation prevents flocculation (Holmberg and Kielland-Brandt, 1978; Wilkie and Mudd, 1981). The gene *FLO1* has been mapped to chromosome I, some 37 cM (centimorgans) from *ADE1;* the genes previously identified as *FLO2* and *FLO4* are now considered to be alleles of *FLO1* (Russell *et al.*, 1980). The second dominant gene, *FLO5*, confers stronger flocculation ability than does *FLO1* and is not linked to the latter gene. Yamashita and Fukui (1983a and b) have identified a third dominant flocculence gene, *FLO8*, which maps to chromosome VIII, being linked to *ARG4*. This gene and its recessive allele, *flo8*, show an interesting behavior in respect to dominance. All *FLO8/FLO8* homozygotes are flocculent but *FLO8/flo8* heterozygotes are flocculent only in diploids that are homozygous at the mating-type locus (*a/a* or α/α diploids); *a*/α diploids do not flocculate. It may be that the recessive allele is the result of the insertion of the yeast Ty transposon adjacent to the gene, thus creating a ROAM mutation (Errede *et al.*, 1980) which puts *flo8* expression under mating-type control.

The fact that flocculation is a dominant character when expressed by the *FLO1*, *FLO4*, or *FLO8* gene should facilitate its transfer to industrial strains. Stewart (1981) attempted such a transfer via the technique of protoplast fusion. The fusion of a flocculent laboratory haploid to a nonflocculent brewing yeast (or *vice versa*) produced flocculent hybrids, but these hybrids failed to produce palatable beer. This illustrates a major problem with the technique of protoplast fusion. Complete genomes are combined and undesirable traits may be recruited into the hybrid in addition to the desired trait. Recombinant DNA technology offers the promise of converting nonflocculent strains to flocculent ones by the specific recruitment of a single gene. The announcement that a DNA fragment conferring flocculation ability has been cloned (Watari *et al.*, 1987) suggests that this promise will be fulfilled. However, the fact that the cloned fragment does not map to chromosome I, in spite of the fact that the gene bank was made from a *FLO1* strain, gives some cause for concern.

9. POLYSACCHARIDE UTILIZATION

The extension of the substrate range of *S. cerevisiae* to include polysaccharides such as starch and cellulose has excited much interest in recent years. In the brewing industry, amylolytic enzymes are added to the mash in the production of "lite" beers and to supplement the saccharification capacity of malt when starch adjuncts are used (see Section 3). More significant is probably the use of starch as a substrate for the production of industrial ethanol by fermentation. In the United States, the low-grade starch residues left from the enzymic processing of grain into high-fructose syrups offers a cheap substrate for ethanol production (Tubb, 1986). Enzymic conversion of starch to fermentable sugars has now largely replaced acid hydrolysis (Fogarty and Kelly, 1980), and if yeast could be given the ability to produce amylases itself, then the production of ethanol from starch would be much simplified. The cost of sufficient glucoamylase to saccharify 1 ton of starch is ca. $8 at 1986 prices (Schenberg Frascino and Da Costa, 1987). The cost savings would be increased to ca. $10/ton if yeast were able to process raw starch itself and thereby remove the need for precooking (Tubb, 1986). While these savings are fairly modest, the total sums involved are huge on an industry-wide basis. Further economic benefits might also accrue if the glucoamylase produced by the yeast could be retrieved and sold to corn syrup producers.

In Brazil, there are a number of advantages to the use of starch, rather than sugar cane, as a substrate for ethanol production. If cassava (*Manihot esculenta*) as well as sugar cane could be used for ethanol production, then a number of problems related to soil exhaustion and disease which stem from crop monoculture should be relieved (Carioca, 1984). Moreover, cassava can grow on poorer soils and over a greater climatic range than sugar cane. Its use would therefore release rich agricultural land for food production and permit the spread of the ethanol industry into poorer regions of Brazil where new sources of employment are badly needed. Although cassava is an annual crop, its use is less seasonal than sugar cane since roots may be stored in the soil. Thus cassava, whether used on its own or together with sugar cane, would relieve the seasonal fluctuations in ethanol production and provide some insulation against variations in the world price of sugar. It is estimated that 180 liters of ethanol can be obtained from 1 tonne of cassava using existing technologies (Carioca, 1984).

Since *S. cerevisiae* is unable to hydrolyze starch, new genes must be recruited into the organism to give it this ability. While this is an obvious candidate for the exploitation of recombinant DNA technology, a number of nonconventional genetic routes have been used to achieve this end.

10. RARE MATING

Most industrial strains of yeast are not capable of mating since they have a polyploid genetic constitution. However, mass matings of laboratory haploids with industrial strains produce occasional hybrid organisms as a result of rare-mating events (Gunge and Nakatomi, 1972; Spencer and Spencer, 1977). These matings probably result from a single mating-type transposition event at the *MAT* locus of one of the chromosomes III of the industrial strains since commercial lager yeasts, at least, appear to be triploid (Kielland-Brandt *et al.*, 1983). A major problem with the rare-mating technique, as with protoplast fusion (see Section 11), is to develop some way of selecting the hybrid organisms produced. A convenient method is to mate an auxotrophic haploid strain with a cytoplasmic petite mutant of the commercial yeast; prototrophic grande may then be selected on minimal medium plus glycerol (Spencer *et al.*, 1981).

Tubb and his colleagues (1981) exploited this technique to mate a starch-utilizing haploid strain of *S. diastaticus* with a commercial lager yeast. The initial hybrids were unsuitable for use in brewing since they gave a phenolic off-flavor due to the production of 4-vinylguaiacol. This undesirable characteristic was found to be determined by a single nuclear gene (*POF*), and it was found possible, in subsequent experiments, to isolate pof^- segregants (Goodey and Tubb, 1982). Nevertheless, the hybrid strain reduced the amount of dextrin in beer to only a limited extent since the *S. diastaticus* enzyme is unable to degrade the β-1,6 branch points in the starch molecule.

11. PROTOPLAST FUSION

Protoplast fusion is a more efficient way of producing hybrids between strains that do not normally mate since it does not rely on mating-type switching events. It is essential to have some way of selecting for the desired hybrid and against the parent strains, and the utility of the cytoplasmic petite mutation (which may be induced in polyploid commercial strains) has already been referred to. An alternate approach is the so-called "dead donor" technique in which the mating partner which carries no genetic markers is killed by UV irradiation or heat treatment prior to the fusion (Hockney and Freeman, 1980). Ferenczy and Kucsera (1985) have taken this technique one step further and developed a method whereby the hybrid products of a protoplast fusion event may be selected even when neither of the parental strains contains a genetic marker. This method, which I call the Lazarus technique, kills both parental strains by use of metabolic poisons. Poisons that attack a different class of enzymes are used for each parent, e.g., *N*-ethylmaleimide for one, and myconazole for the other. This results in the death of both parents, which may be brought back to life on fusion by phenotypic complementation. Each parent supplies a functional set of enzymes which the other lacks.

A number of workers have employed protoplast fusion to confer amylolytic activity on either brewery or distillery yeasts. Hockney and Freeman (1980) exploited both the dead-donor technique and the natural biotin auxotrophy of brewing strains to hybridize *S. diastaticus* and *S. cerevisiae.* Wilson *et al.* (1982) used complementation of auxotrophies between two haploid strains in the fusion of *Schwanniomyces alluvius* and *Saccharomyces uvarum.* Two groups of Brazilian researchers (Galembeck *et al.,* 1982; Echeverrigaray, 1983) have used protoplast fusion to form hybrids between the starch-degrading yeast *Lipomyces konoenkoae* and *S. cerevisiae.* Hybrids were selected as being able to grown on starch at 37°C since *S. cerevisiae* cannot utilize starch and *L. kononenkoae* is unable to survive at elevated temperatures. While successful fusants, capable of converting starch to ethanol, were obtained in all cases, the instability of the hybrids in the absence of selection was a universal problem.

12. RECOMBINANT DNA TECHNOLOGY

The techniques of gene cloning and transformation have been used to transfer genes encoding amylolytic and cellulolytic enzymes into yeast from a variety of sources. A problem arises when the recipient is a commercial yeast strain since the complementation of an auxotrophic marker by a yeast gene cloned in the vector may not be used as a way of selecting transformanants. What is required is a dominant selectable marker such as the drug resistances commonly used in the genetic manipulation of bacteria. Unfortunately, genes conferring resistance to drugs in yeast and other eukaryotes are frequently recessive since resistance is due to a structural alteration of the drug's target, rather than to destruction of the drug itself. An exception to this is resistance to the aminoglycoside antibiotic G418 which may be conferred upon yeast by the bacterial transposon Tn5 (Jimenez and Davies, 1980). A dominant selectable marker obtainable from yeast itself is the *CUP1* gene, which confers resistance to high levels of Cu^{2+} ions in a dosage-dependent manner (Fogel and Welch, 1982). Either of these genes may be incorporated into yeast cloning vectors and used to select transformants of commercial strains (Henderson *et al.,* 1985).

12.1. Cloning and Expression in Yeast of Genes Encoding Amylolytic Enzymes

A wide range of yeast species have the ability to utilize starch, but these do not include the varieties of *S. cerevisiae* used currently in the production of beer or industrial alcohol. In addition to these amylolytic yeasts, genes encoding suitable enzymes may also be sought in the filamentous fungi, bacteria, plants, and mammals.

Much work has been invested in the cloning and expression of amyloglucosidase genes from *S. diastaticus*. This amylolytic yeast is now regarded as a variety of *S. cerevisiae* and the transfer of the required genes from *S. diastaticus* has the twin advantages that there should be no barriers to their expression and no problems with the regulatory authorities. The difference between *S. cerevisiae* and *S. diastaticus* in their production of amylolytic enzymes appears to be mainly a function of gene regulation. *S. cerevisiae* produces an amyloglucosidase enzyme during sporulation, where it is believed to function in glycogen utilization (Colonna and Magee, 1978). In contrast, *S. diastaticus* produces an amylo-1,4-glucosidase throughout vegetative growth as the result of the action of one or more of a series of genes denoted *DEX* or *STA* (Erratt and Stewart, 1978). There is no evidence that the *S. diastaticus* enzyme has any ability to degrade past the β-1,6-linkage at the branch points of the starch molecule, and this severely limits its usefulness. Two groups have cloned genes from *S. diastaticus* which they separately call *DEX1* (Tamaki, 1978) and *STA1* (Meaden *et al.*, 1985); a comparison of their restriction maps indicates that they are probably allelic. In addition to its direct use in starch conversion, the cloned *DEX* gene may be used as a positively selectable marker in *S. cerevisiae* transformations and also has potential in secretion vectors.

A number of form-species of the fungus *Aspergillus* produce large amounts of extracellular amyloglucosidase. Industrial strains of *A. niger* or *A. awamori*, for instance, produce as much as 20 g/l^{-1} of the enzyme (Van Brunt, 1986). The successful expression of *S. cerevisiae* genes in *Escherichia coli* encouraged the belief that the expression of fungal genes in yeast should be relatively simple. This turned out not to be the case since the *Aspergillus* amyloglucosidase gene contains four introns, which *S. cerevisiae* is unable to splice, and this prevents expression of the genomic copy of the gene in yeast. The intron problem has been solved by two groups (Boel *et al.*, 1984; Innis *et al.*, 1985) by cloning a cDNA copy identified using an oligonucleotide probe. The cDNA gene may be expressed in yeast from a high-efficiency promoter, such as *ENO1* (Innis *et al.*, 1985). The product is glyosylated in yeast to a similar extent to the native enzyme, and more than 90% of the activity is found in the culture medium. The form of the *Aspergillus* enzyme produced is capable of degrading raw starch since it has debranching activity and thus offers a distinct advantage over the *S. diastaticus* enzyme. A merit of the latter, for the production of beer, if not industrial ethanol, is that it is readily inactivated by heat (Tubb, 1986).

The ubiquity of signal peptide sequences has permitted the expression in, and secretion from, yeast of α-amylase enzymes encoded by cDNA clones from either wheat (Rothstein *et al.*, 1984) or mouse (Thomsen, 1983). The coding sequence of the mouse α-amylase has also been fused to the signal peptide of the yeast pheromone (Astolfi Filho *et al.*, 1986). Very efficient secretion of the amylase is achieved by this route, although a direct comparison of the efficiency of the yeast and mouse signal sequences has yet to appear.

12.2. Cloning and Expression of Endoglucanase Genes in Yeast

Lignocellulosic agricultural wastes represent an enormous, and largely untapped, carbon and energy source for biotechnological processes. In Brazil alone more than 71 million tonnes of waste is produced every year from the cultivation of sugar cane, cassava, corn, rice, and soya. More than half of this waste is cellulose, and a number of genes encoding enzymes involved in cellulose degradation have been cloned and expressed in yeast (see Van Brunt, 1986). However, the accessibility of the cellulose is a problem both biochemically, due to its association with lignin, and physically. The latter problem is likely to favor filamentous organisms, rather than yeasts, for the initial processing of these agricultural wastes. However, the synthesis by yeast of one class of enzymes involved in cellulose degration, the endoglucanases, has important implications for traditional yeast biotechnologies such as beer brewing. Excess β-glucans in beer may lead to the formation of gels, hazes, and precipitates which compromise the quality of the product and lead to problems with its filterability (Cantwell *et al.*, 1986; Knowles *et al.*, 1986).

The construction of a yeast strain that is able to degrade β-glucans requires not only expression of the β-glucanase gene in yeast, but efficient secretion of the enzyme into the medium. A number of groups have cloned and expressed bacterial genes encoding endo-β-1,3-1,4-glucanase in yeast. Both *Bacillus subtilis* (Hinchliffe and Box, 1984; Cantwell *et al.*, 1986) and *Clostridium thermocellum* (Sacco *et al.*, 1984) have been used as sources of this gene, but there is no indication that the enzyme is excreted from *S. cerevisiae.* An alternative approach is to link the coding sequence of the bacterial gene to the signal sequence of a yeast gene that specifies a secreted product. This strategy was adopted by the Allelix group (Skipper *et al.*, 1985), who used the signal sequence of the yeast killer peptide to obtain the secretion of a bacterial endoglucanase specified by a gene cloned from *Cellulomonas fimi.*

The preceding account suggests that prokaryotic signal sequences are not functional in yeast. Eukaryotic ones, on the other hand, appear to be ubiquitous. Jackson *et al.* (1986) exploited the ability of the mouse α-amylase signal peptide to direct protein excretion in yeast (see Section 12.1) to achieve the excretion of the barley endo-β-1,3,-1,4-glucanase.

Two groups have cloned cDNA copies of two endo-β-1,4-glucanase genes, *egl1* and *egl2*, from the cellulolytic fungus *Trichoderma reesei* (Arsdell *et al.*, 1987; Pentilla *et al.*, 1987). Both these enzymes are glycoslyated and excreted by yeast, the native signal sequence being used. When expressed at high level using the yeast *PGK* promoter, the *Trichoderma* enzymes alter the morphology of the host yeast cells, making them larger and of irregular shape (Pentilla *et al.*, 1987)., This effect could be due to a weak effect of endoglucanases on the chitin of the yeast bud scars.

12.3. Cloning and Expression of Complete Metabolic Pathways: The Way Ahead

The use of recombinant DNA technology to improve the range and efficiency of yeast's activities in classical biotechnological processes has so far been limited to the recruitment of just one or two genes which specify novel activities. However, the technology currently available for yeast should permit the wholesale genetic engineering of the organism by addition of complete, novel metabolic pathways. A start in this direction has been made by the assembly of all the genes required for the complete yeast tryptophan biosynthetic pathway on a single plasmid (Niederberger *et al.*, 1984; Prasad *et al.*, 1987). It should be possible to extend this approach to recruit complete pathways from heterologous sources and also to construct novel pathways that do not presently exist in nature. The ability to construct artificial yeast chromosomes (Murray and Szostak, 1983) means that there is no obvious limit to the length of such pathways or to our conceptual horizons in yeast strain development.

REFERENCES

Aguilera, A., del Castillo, L., and Benitez, T., 1982, Alcohol and sucrose tolerant wine yeast strains, *Cien. Biol. (Portugal)* **7**:89–94.

Aiba, S., Shoda, M., and Nagatani, M., 1968, Kinetics of product inhibition in alcohol fermentation, *Biotechnol. Bioeng.* **10**:845–864.

Andreasen, A. A., and Stier, T. J. B., 1953, Anaerobic nutrition of *Saccharomyces cerevisiae*. I. Ergosterol requirement for growth in defined medium, *J. Cell. Comp. Physiol.* **41**:23–27.

Andreasen, A. A., and Stier, T. J. B., 1954, Anaerobic nutrition of *Saccharomyces cerevisiae*. II. Unsaturated fatty acid requirement for growth in defined medium, *J. Cell. Comp. Physiol.* **43**:271–277.

Arsdell, J. N., Kivok, S., Schweickart, V. L., Ladner, M. B., Gelfand, D. H., and Innis, M. A., 1987, Cloning, characterization and expression in *Saccharomyces cerevisiae* of endoglucanase I from *Trichoderma reesei*, *Bio/Technology* **5**:60–64.

Astolfi Filho, S., Galembeck, E. V., Faria, J. B., and Schenberg Frascino, A. C., 1986, Stable yeast transformants that secrete functional α-amylase encoded by cloned mouse pancreatic cDNA, *Bio/Technology* **4:**311–315.

Ault, R. G., Hampton, A. N., Newton, R., and Roberts, R. H., 1966. Biological and biochemical aspects of tower fermentation, *J. Inst. Brew.* **75:**260–277.

Beavan, M. J., Belk, D. M., Stewart, G. G., and Rose, A. H., 1979, Changes in electrophoretic mobility and lytic enzyme activity associated with development of flocculating in *Saccharomyces cerevisiae, Can. J. Microbiol.* **23:**68–74.

Boel, E., Hjort, I., Svensson, B., Norris, F., Norris, K. E., and Fiil, N. P., 1984, Glucoamylases E1 and E2 from *Aspergillus niger* are synthesised from two different but closely related mRNAs, *EMBO J.* **3:**1097–1102.

Brown, S. W., 1983, Ethanol tolerance in the yeast *Saccharomyces.* Ph.D. thesis, University of Kent at Canterbury.

Brown, S. W., and Oliver, S. G., 1982a, The effect of temperature on the ethanol tolerance of the yeast *Saccharomyces uvarum, Biotechnol. Lett.* **4:**269–274.

Brown, S. W., and Oliver, S. G., 1982b, Isolation of ethanol tolerant mutants of yeast by continuous selection, *Eur. J. Appl. Microbiol. Biotechnol.* **16:**119–122.

Brown, S. W., Oliver, S. G., Harrison, D. E. F., and Righelato, R. C., 1981, Ethanol inhibition of yeast growth and fermentation: Differences in the magnitude and complexity of the effect, *Eur. J. Appl. Microbiol. Biotechnol.* **11:**151–155.

Brown, S. W., Sugden, D. A., and Oliver, S. G., 1984, Ethanol production and tolerance in grade and petite yeast, *J. Chem. Technol. Biotechnol.* **34**(13)**:**116–120.

Burrows, S., 1979, Bakers' yeast, *Econ. Microbiol.* **4:**31–64.

Cantwell, B. A., Brazil, G., Murphy, N., and McConnell, 1986, Comparison of expression of the endo-β-1,3-1,4-glucanase gene from *Bracillus subtilis* in *Sacchromyces cerevisiae* from the *CYC1* and *ADH1* promoters, *Curr. Genet.* **11:**65–70.

Carioca, J. O. B., 1984, Potencial da biomassa, in: *Biomassa, Fundamentos and Aplicacoes technologias* (J. O. B. Carioca and H. L. Arora, eds.), Universidade Federal do Ceara e Ministerio do Interior, Fortazela, Brazil, pp. 65–77.

Colonna, W. J., and Magee, P. T., 1978, Glycogenolytic enzymes in sporulating yeast, *J. Bacteriol.* **134:**844–853.

Corran, H. S., 1975, *A History of Brewing,* David and Charles, London.

Coutts, M. W., 1961, British patent 872,391–872,400.

Coutts, M. W., 1966, The many facets of continuous fermentation, *Proc. Australian Section Inst. Brew. (Ninth Convention),* pp. 1–8, Institute of Brewing, Sydney.

Cysewski, G. R., and Wilke, C. R., 1976, Utilization of cellulosic materials through enzymatic hydrolysis. I. Fermentation of hydrolysate to ethanol and single-cell protein, *Biotechnol. Bioeng.* **18:**1297–1313.

Dykhuizen, D. E., and Hartl, D. L., 1983, Selection in chemostats, *Microbiol. Rev.* **47:**150–168.

Echeverrigaray, S. L., 1983, Estabilidade Genética e Heterose em Hybridos Interspecíficos de Leveduras. M.Sc. thesis, Universidade de Sao Paulo, Piracicaba, Brazil.

Erratt, J. A., and Stewart, G. G., 1978, Genetic and biochemical studies of yeast strains able to utilise dextrins, *J. Am. Soc. Brew. Chem.* **36:**151–161.

Errede, B., Cardillo, T. S., Wever, G., and Sherman, F., 1980, ROAM mutations causing increased expression of yeast genes: Their activation by signals directed toward conjugation functions and their formation by insertion of Ty1 repetitive elements, *Cell* **45:**593–602.

Ferenczy, L., and Kucsera, J., 1985, Gene transfer via chemically inactivated protoplasts ot yeasts, *Proc. 10th International Specialized Symposium on Yeasts (Varna),* p. 103, Institute of Molecular Biology, Sofia.

Fogarty, W. M., and Kelly, C. T., 1980, Amylases, amyloglucosidases and related glucanases, in: *Microbial Enzymes and Bioconversions* (A. H. Rose, ed.), Academic Press, New York, pp. 115–165.

Fogel, S., and Welch, J. W., 1982, Tandem gene amplification mediates copper resistance in yeast, *Proc. Natl. Acad. Sci. USA.* **79:**5342–5346.

Galembeck, E. V., Fernandes, B. L., Costa, S. O. P., and Schenberg Frascino, A. C., 1982, Fusion of protoplasts of different yeast genera: *Saccharomyces* and *Lipomyces, Microb. Genet. Bull.* p. 524.

Goodey, A. R., and Tubb, R. S.., 1982, Genetic and biochemical analysis of the ability of *Saccharomyces cerevisiae* to decarboxylate cinnamic acids, *J. Gen. Microbiol.* **128:**2615–2620.

Gray, W. D., 1941, Studies on alcohol tolerance of yeasts, *J. Bacteriol.* **42:**561–574.

Gunge, N., and Nakatomi, Y., 1972, Genetic mechanisms of rare matings of the yeast *Saccharomyces cerevisiae* heterozygous for mating type, *Genetics* **70:**41–58.

Hall, J. F., 1970, The use of Difco WLN agar for demonstration of the instability of strains of *Sacch. carlsbergensis, J. Inst. Brew.* **76:**522–523.

Harder, W., Kuenen, J. G., and Matin, A. A., 1977, Microbial selection in continuous culture, *J. Appl. Bacteriol.* **43:**1–24.

Hayashida, S., Der Fong, D., and Hongo, M., 1975, Mechanism of formation of high concentration alcohol in sake brewing. X. Physiological properties of yeast cells grown in proteolipid-supplemented medium, *Agr. Biol. Chem.* **39:**1025–1031.

Henderson, R. C. A., Cox, B. S., and Tubb, R. S., 1985, Transformation of brewing yeasts with a plasmid containing the gene for copper resistance, *Curr. Genet.* **9:**133–138.

Hinchliffe, E., and Box, W. G., 1984, Expression of the cloned endo-1,3-1,4-B-glucanase gene of *Bacillus subtilis* in *Saccharomyces cerevisiae, Curr. Genet.* **8:**471–475.

Hockney, R. C., and Freeman, R. F., 1980, Construction of polysaccharide-degrading brewing yeast by protoplast fusion, in: *Advances in Protoplast Research* (L. Ferenczy and G. Farkas, eds.), Pergamon Press, Oxford, pp. 139–144.

Hodgson, J. A., Berry, D. R., and Johnston, J. R., 1985, Discrimination by heat and proteinase treatments between flocculent phenotypes conferred on *Saccharomyces cerevisiae* by the gene *FLO1* and *FLO5, J. Gen. Microbiol.* **131:**3219–3227.

Holmberg, S., 1978, Isolation and characterisation of a polypeptide absent from non-flocculent mutants of *Saccharomyces cerevisiae, Carlsberg Res. Commun.* **43:**401–413.

Holmberg, S., and Kielland-Brandt, M. C., 1978, A mutant of *Saccharomyces cerevisiae* temperature sensitive for flocculation: Influence of oxygen and respiratory deficiency on flocculence, *Carlsberg Res. Commun.* **43:**401–413.

Holzberg, I., Finn, R. K., and Steinkraus, K. H., 1967, A kinetic study of the alcohol fermentation of grape juice, *Biotechnol. Bioeng.* **9:**413–427.

Hough, J. S., 1985, *The Biotechnology of Malting and Brewing,* Cambridge University Press, Cambridge.

Innis, M. A., Holland, M. J., McCabe, P. C., Cole, G. E., Wittmann, V. P., Tal, R., Watt, K. W. K., Gelfand, D. H., Holland, J. P., and Meade, J. H., 1985, Expression, glycosylation and secretion of an *Aspergillus* glucoamylase by *Saccharomyces cerevisiae, Science* **228:**21–26.

Ismail, A. A., and Ali, A. M. M., 1971a, Selection of high ethanol-yielding *Saccharomyces.* I. Ethanol tolerance and the effect of training in *Saccharomyces cerevisiae* Hansen, *Folia Microbiol.* **16:**350–354.

Ismail, A. A., and Ali, A. M. M., 1971b, selection of high ethanol-yielding *Saccharomyces.* II. Genetics of ethanol tolerance, *Folia Microbioll* **16:**350–354.

Jackson, E. A., Ballance, G. M., and Thomsen, K. K., 1986, Construction of a yeast vector directing the synthesis and release of barley (1-3,1-4)-β-glucanase, *Carlsberg Res. Commun.* **51:**445–458.

Jayatissa, P. M., and Rose, A. H., 1976, Role of wall phosphomannan in flocculation of *Saccharomyces cereivisiae*, *J. Gen. Microbiol.* **35:**61–68.

Jimenez, A., and Davies, J., 1980, Expression of a transposable antibiotic resistance element in *Saccharomyces*, *Nature* **287:**869–871.

Johnston, J. R., and Reader, H. P., 1983, Genetic control of flocculation, in: *Yeast Genetics: Fundamental and Applied Aspects* (J. F. T. Spencer, D. M. Spencer, and H. R. W. Smith, eds.), Springer Verlag, New York, pp. 205–224.

Kielland-Brandt, M. C., Nilsson-Tillgren, N., Petersen, J. G. L., Holmberg, S., and Gjermansen, C., 1983, Approaches to the genetic analysis and breeding of brewer's yeast, in: *Yeast Genetics. Fundamental and Applied Aspects* (J. F. T. Spencer, D. M. Spencer, and A. R. W. Smith, eds.), Academic Press, New York, pp. 421–455.

Knowles, J., Lehtinen, U., Nikkola, M., Pentilla, M., Suihko, M-L., Home, S., Vilpola, A., and Enari, T-M., 1986, Glucanolytic brewer's yeast, *Proc. 21st Eur. Brew. Conv. (Madrid)*, p. 37, Institute of Brewing, Madrid.

Kodama, K., and Yoshikawa, K., 1977, Saké, *Econ. Microbiol.* **1:**423–475.

Lafon-Lafourcade, S., 1983, Wine and brandy, *Biotechnology* **5:**81–163.

Lovgren, T., and Hautera, P., 1977, Maltose fermentation and leavening ability of bakers' yeast, *Eur. J. Appl. Microbiol.* **4:**37–43.

Lyons, T. P., and Hough, J. S., 1970, Flocculation of brewer's yeast, *J. Inst. Brew.* **76:**564–571.

Lyons, T. P., and Hough, J. S., 1971, Further evidence for the cross-bridging hypothesis for flocculation of brewer's yeast, *J. Inst. Brew.* **77:**300–305.

Maule, A. P., and Thomas, P. D., 1973, Strains of yeast lethal to brewery yeasts, *J. Inst. Brew.* **79:**137–141.

Meaden, P., Ogden, K., Bussey, H., and Tubb, R. S., 1985, A *DEX* gene conferring production of extracellular amyloglucosidase in yeast, *Gene* **34:**325–334.

Miki, B. L. A., Poon, N. H., James, A. P., and Seligy, V. L., 1982, Possible mechanism for flocculation interactions governed by gene *FLO1* in *Saccharomyces cerevisiae*, *J. Bacteriol.* **150:**878–889.

Mikola, J., Pietila, K., and Enari, T-M., 1972, Inactivation of malt peptidases during masking, *J. Inst. Brew.* **78:**384–388.

Mill, P. J., 1964, The nature of interactions between flocculent cells in the flocculation of *Saccharomyces cerevisiae*, *J. Gen. Microbiol.* **96:**165–174.

Murakami, H., Sagama, H., and Takase, S., 1968, Non-productivity of aflatoxin by Japanese industrial strains of *Aspergillus*. III. Common characteristics of aflatoxin-producing strains, *J. Gen. Appl. Microbiol. (Tokyo)* **14:**251–262.

Murray, A. W., and Szostak, J. W., 1983, Construction of artificial chromosomes in yeast, *Nature* **305:**189–193.

Niederberger, P., Aebi, M., Furter, R., Prantl, F., and Hutter, R., 1984, Expression of an artificial yeast *TRP*-gene cluster in yeast and *Escherichia coli*, *Mol. Gen. Genet.* **195:**481–486.

Nishihara, H., Toraya, T., and Fukui, S., 1977, Effect of chemical modification of cell surface components of a brewer's yeast on the floc-forming ability, *Arch. Microbiol.* **150:**890–899.

Nishihara, H., Toraya, T., and Fukui, S., 1982, Flocculation of all walls of brewer's yeast and effects of metal ions, protein denaturants and enzyme treatments, *Arch. Microbiol.* **131:**112–115.

Oliver, S. G., 1984, Biological limits to ethanol production, *Chem. Indust.* **14:**425–427.

Oura, E., Soumalainen, H., and Viskari, R., 1979, Breadmaking, *Econ. Microbiol.* **4:**88–146.

Palamund, S. R., and Hardwick, W. A., 1969, Studies on the relative flavour importance of some beer constituents, *Tech. Q. Master Brew. Assoc. Amer.* **6:**117–128.

Pentilla, M. E., Andre, L., Saloheimo, M., Lehtovaara, P., and Knowles, J. K. C., 1987,

Expression of *Trichoderma reesei* endoglucanases in the yeast *Saccharomyces cerevisiae*, *Yeast* **3:**175–187.
Prasad, R., Niederberger, P., and Hutter, R., 1987, Tryptophan accumulation in *Saccharomyces cerevisiae* under the influence of an artificial yeast *TRP* gene cluster, *Yeast* **3:**95–105.
Preece, I. A., 1954, *The Biochemistry of Brewing,* Oliver and Boyd, Edinburgh.
Preece, I. A., and Hoggan, J., 1957, Carbohydrate modification during malting, *Proceedings of the European Brewery Convention Congress,* Elsevier, London, pp. 72–83.
Righelato, R. C., 1980, Anaerobic fermentation: Alcohol production, *Phil. Trans. Roy. Soc. B* **295:**491–500.
Rose, A. H., 1984, Physiology of cell aggregation; flocculation by *Saccharomyces cerevisiae* as a model system, in: *Microbial Adhesion and Aggregation* (K. C. Marshall, ed.), Spring-Verlag, Berlin, pp. 323–335.
Rothstein, S. J., Lazurus, C. M., Smith, W. E., Baulcombe, D. C., and Gatenby, A. A., 1984, Secretion of a wheat α-amylase expressed in yeast, *Nature* **308:**662–665.
Rubner, M., 1912, The physiology and nutrition of yeast during alcoholic fermentation, *Arch. Physiol.* **1:**1–392.
Russell, I., Stewart, G. G., Reader, H. P., Johnston, J. R., and Martin, P. A., 1980, Revised nomenclature of genes that control yeast flocculation, *J. Inst. Brew.* **86:**120–121.
Sacco, M., Millet, J., and Aubert, J. P., 1984, Cloning and expression in *Saccharomyces cerevisiae* of a cellulase gene from *Clostridium thermocellum, Ann. Microbiol. Inst. Pasteur* **135A:**485–488.
Schenberg Frascino, A. C., and Da Costa, S. O. P ., 1987, Molecular approaches to Alcohol Biotechnology in Brazil, in: *Critical Reviews in Biotechnology,* Vol. 6 (G. G. Stewart and I. Russell, eds.), CRC Press, Boca Raton, FL, pp. 323–355.
Skipper, N., Sutherland, M., Davies, R. W., Kilburn, D., Miller, R. C., Jr., Warren, A., and Wong, R., 1985, Secretion of a bacterial cellulase from yeast, *Science* **230:**958–960.
Spencer, J. F. T., and Spencer, D. M., 1977, Hybridisation of non-sporulating strains of brewer's and distiller's yeasts, *J. Inst. Brew.* **83:**287–289.
Spencer, J. F. T., Land, P., and Spencer, D. M., 1981, The use of mitochondrial mutants in the isolation of hybrids obtained by protoplast fusion, *Mol. Gen. Genet.* **178:**651–654.
Stewart, G. G., 1981, The genetic manipulation of industrial yeast strains, *Can. J. Microbiol.* **27:**973–990.
Stewart, G. G., Russell, I., and Garrison, I. F., 1975, Some considerations on the flocculation characteristics of ale and lager yeast strains, *J. Int. Brew.* **81:**248–257.
Stratford, M., and Keenan, M. H. J., 1987, Yeast flocculation: Kinetics and collision theory, *Yeast* **3:**201–206.
Sugden, D. A., and Oliver, S. G., 1983, Reduced ethanol tolerance: One of the pleiotropic effects of the *pep4*.3 mutation in *Saccharomyces cerevisiae, Biotechnol. Lett.* **5:**419–422.
Tamaki, H., 1978, Genetic studies of ability to ferment starch in *Saccharomyces* gene polymorphism, *Mol. Gen. Genet.* **164:**205–209.
Taylor, N. W., and Orton, W. L., 1978, Aromatic compounds and sugars in flocculation of *Saccharomyces cerevisiae, J. Inst. Brew.* **84:**113–114.
Thomas, D. S., Hossack, J. A., and Rose, A. H., 1978, Plasma-membrane lipid composition and ethanol tolerance in *Saccharomyces cerevisiae, Arch. Microbiol.* **117:**239–245.
Thomsen, K. K., 1983, Mouse α-amylase synthesised by *Saccharomyces cerevisiae* is released into the culture medium, *Carlsberg Lab. Res. Commun.* **48:**545–555.
Thorne, R. S. W., 1970, Yeast mutation during continuous culture, *J. Inst. Brew.* **76:**555–563.
Troyer, J. R., 1953, A relation between cell multiplication and alcohol tolerance in yeasts, *Mycologia* **45:**20–39.

Tubb, R. S., 1986, Amylolytic yeasts for commercial applications, *Trends Biotechnol.* **4:**96–104.

Tubb, R. S., Brown, A. J. P., Searle, B. A., and Goodey, A. R., 1981, Development of new techniques for the genetic manipulation of brewing yeasts, in: *Current Developments in Yeast Research* (G. G. Stewart and I. Russell, eds.), Pergamon Press, Oxford, pp. 775–779.

Van Brunt, J., 1986, Fungi: The perfect host? *Bio/Technology* **4:**1057–1062.

Vogt, E., Jakob, L., Lemperle, E., and Weiss, E., 1974, *Wein,* Eugen Ulmer Verlag, Stuttgart.

Watari, J., Takata, Y., Nishikawa, N., and Kamada, K., 1987, Cloning of a gene controlling yeast flocculence, *Proceedings of the 21st European Brewing Convention (Madrid),* pp. 537–544, Institute of Brewing, Madrid.

Wilkie, D., and Mudd, R. C., 1981, Aspects of mitochondrial control of cell surface characteristics in *Saccharomyces cerevisiae,* in: *Advances in Biotechnology: Current Developments in Yeast Research* (G. G. Stewart and I. Russell, eds.), Pergamon Press, Toronto, pp. 345–349.

Wilson, J. J., Khachatourian, G. G., and Ingledew, W. M., 1982, Protoplast fusion in the yeast, *Saccharomyces alluvius, Mol. Gen. Genet.* **186:**95–100.

Yamashita, I., and Fukui, S., 1983a, Molecular cloning of a glucoamylase gene in the yeast *Saccharomyces, Agric. Biol. Chem.* **47:**2689–2692.

Yamashita, S., and Fukui, S., 1983b, Mating signals control expression of both starch fermentation genes and a novel flocculation gene *FLO8* in the yeast *Saccharomyces, Agric. Biol. Chem.* **47:**2889–2896.

Culture Systems 8

T. M. MATTHEWS and C. WEBB

1. INTRODUCTION

Many yeasts of the genus *Saccharomyces* have a long history of industrial usage where the substrates are generally either agricultural or processing wastes, or crops grown for the purpose. Principle examples of the production of food yeast are *S. uvarum,* which is recovered from malt wort in beer production, *S. cerevisiae* (baker's yeast), usually grown on molasses, and *S. fragilis,* which is cultured on cheese whey. Molasses, surplus grain, and sulfite waste liquor are the main raw materials used for fuel ethanol production, usually by fermentation with *S. uvarum* or *S. cerevisiae.* Cellulosic waste materials may also be used as a substrate after preliminary acid hydrolysis to break down long-chain polysaccharides to simple sugars. Many *Saccharomyces* species also play a large part in the production of alcoholic beverages, where substrates are usually provided by fruit, grain, or cereal crops.

The *Saccharomyces* yeasts can be grown in solutions of simple chemical substances but growth is more rapid and economically viable when the culture medium contains certain supplements.

Once a suitable culture medium satisfying the nutritional requirements of the *Saccharomyces* has been established, the physical and chemical conditions for growth must be optimized for maximum cell yield or growth rate. The aeration intensity, dissolved oxygen, and carbon dioxide concentrations in the medium can affect yeast growth and the temperature and pH can also influence the fermentation. In addition, high sugar and high ethanol concentrations can have inhibitory effects on yeast growth. The intensity of these effects is generally strain dependent; hence each parameter must be considered for each *Saccharomyces* species to be cultured.

T. M. MATTHEWS and C. WEBB • Department of Chemical Engineering, University of Manchester Institute for Science and Technology, Manchester M60 1QD, England

2. NUTRITIONAL REQUIREMENTS OF *SACCHAROMYCES*

2.1. Carbon

Saccharomyces species of yeast are able to ferment the hexose sugars of *d*-glucose, *d*-fructose, and *d*-mannose. The rate of fermentation of mannose is less than that of glucose, and with some species, fructose is fermented more slowly than glucose (Menzinsky, 1943).

The main sugar constituent of cane and beet molasses is sucrose, but as the rate of inversion of sucrose by *Saccharomyces* yeasts is far in excess of the rate of fermentation, a sucrose fermentation can be considered a glucose and fructose fermentation (Atkin *et al.*, 1946).

Maltose, the sugar present in wort, is not fermentable by all members of the genus. However, fermentation of maltose by brewer's and baker's yeasts will proceed and the fermentation rate may be increased by the addition of a small quantity of glucose or oxygen to the culture medium (Schultz *et al.*, 1940; Leibowitz and Hestrin, 1939).

Other sugars that can be used as a carbon source by some, but not all, *Saccharomyces* species include *d*-galactose, which can be fermented by brewer's and baker's yeasts, and lactose, which cannot, however, be fermented by *S. cerevisiae*.

All *l*-sugars and all pentoses can be considered unfermentable by *Saccharomyces* yeasts, and there is evidence that some pentoses can retard the rate of fermentation of hexoses by certain species (Sobotka *et al.*, 1936).

The carbohydrate present in some of the raw materials used industrially in *Saccharomyces* fermentation, and their hydrolysis products, are shown in Table I.

2.2. Nitrogen

Yeasts have a nitrogen content of around 10% of their dry weight; hence nitrogen is an important constituent of any growth medium. Many inorganic ammonium salts have been found to promote the growth of *Saccharomyces* species and the most efficient are the following:

Ammonium acetate
Ammonium carbonate
Ammonium bicarbonate
Ammonium lactate
Mono-, di-, and triammonium phosphate
Ammonium sulfate
Ammonium tartrate

Ammonium sulfate is the most widely used nitrogen source as it also provides a readily assimilable source of sulfur.

Table I. The Carbohydrate Content of Raw Materials Used in *Saccharomyces* Fermentation

Raw material	Carbohydrate present	Hydrolysis products
Cane sugar Beet sugar Molasses Cannery waste	Sucrose	Glucose Fructose
Grains Cassava Potatoes Jerusalem artichokes Sago	Starch	Glucose Maltose Maltotriose Higher-molecular-weight dextrins[a]
Wood and wood waste Agricultural waste Municipal waste	Cellulose	Glucose Mannose Galactose Xylose[a] Arabinose[a]
Whey	Lactose	Glucose Galactose

[a]Not fermentable by most *Saccharomyces* species.

Both baker's and brewer's yeasts are unable to assimilate nitrates, so nitrates are generally not suitable as nitrogen sources in culture media for the *Saccharomyces*.

Many amino acids provide suitable sources of nitrogen for yeast growth. In a survey (Schultz and Pomper, 1948) of 40 strains of yeast all classified as *Saccharomyces* and with an amino acid as the sole nitrogen source, seven amino acids were found to promote good growth in all 40 yeast strains and a further seven were found to support medium to good growth in around 90% of the yeasts tested. Five amino acids were found to be inadequate nitrogen sources for the growth of these by *Saccharomyces* species. The amino acids tested are listed in Table II.

A balanced amino acid mixture has been shown to be a more efficient nitrogen source than ammonium ions (Thorne, 1944; Hoggan, 1977). This could be due to the ability of yeasts to assimilate complete mixtures of amino acids without preliminary deamination. Amino acid takeup relieves the cell of a synthetic carbon demand which would otherwise have to be met from sugar metabolism (Jones *et al.*, 1969).

Urea can also be metabolized by *Saccharomyces* species as a nitrogen source and is used as such industrially in India and South America; however, for yeast growth comparable to that with ammonium sulfate as nitrogen source, the vitamin biotin is usually added to the culture medium to assist assimilation.

Table II. Suitability of Individual Amino Acids as Sole Nitrogen Source for Growth of *Saccharomyces* Strains

Group 1:	Good growth of all strains	Alanine
		Arginine
		Asparagine
		Aspartic acid
		Glutanic acid
		Leucine
		Valine
Group 2:	Good to moderate growth	Isoleuaine
		Methionine
		Phenylalanine
		Serine
		Tryptophan
Group 3:	Poor growth	Cystine
		Glycine
		Histidine
		Lysine
		Proline
		Threonine

2.3. Phosphorus

Phosphorus is essential for the growth of all yeasts, controlling the synthesis of lipids and carbohydrates and maintaining the integrity of the cell wall. Baker's and brewer's yeasts can grow on a medium without phosphorus for a short time but the phosphate reserves within the yeast cells are used for the growth (Markham and Bryne, 1968). Yeasts take up phosphate as the monovalent anion $H_2PO_4^-$ but the divalent anion HPO_4^{2-} is not absorbable (Rothstein, 1961). The amount of phosphate assimilated depends on the quantity supplied. Potassium dihydrogen phosphate is normally added in a concentration of around 0.6 mM/g cells to culture media for optimum fermentation rates.

2.4. Sulfur

Saccharomyces species can obtain the sulfur they require from inorganic sulfate, sulfite, or thiosulfite (Schultz and McManus, 1950), which are reduced to the amino acid methionine within the cell (Lewis and Wildenradt, 1969). Although methionine is the preferred sulfur source, ammonium sulfate is generally chosen for industrial fermentations on the basis of cost. Sulfur constitutes about 0.4% of the dry weight of yeast cells.

Table III. Trace Elements and the *Saccharomyces*

Group 1: macroelements (0.1–1 mM)	Group 2: microelements (0.1–100 μM)	Group 3: inhibitors (>100 μM)
Potassium	Cobalt	Silver
Magnesium	Boron	Arsenic
Calcium	Cadmium	Barium
Zinc	Chromium	Mercury
Iron	Copper	Lithium
Manganese	Iodine	Nickel
Chlorine	Molybdenum	Osmium
	Nickel	Lead
	Vanadium	Selenium
		Tellurium

2.5. Trace Elements

The trace elements required by *Saccharomyces* for active growth have been well documented and can be divided into three categories (Rose, 1976):

1. The macro elements, which are required in concentrations of 0.1–1.0 mM
2. The micro elements, which are required in concentrations of 0.1–100 μM
3. The inhibitors, which can adversely affect yeast growth when present in concentrations greater than 100 μM

The elements in these groups are listed in Table III. The presence of some of the elements in groups 1 and 2 in greater concentrations than those stated has been found to be inhibitory to cell growth in some cases. Growth media that contain organic compounds, for example, molasses and wort, may protect the yeast from the influence of otherwise inhibitory elements by precipitating them or forming complexes.

2.6. Growth Factors

Various growth factors are taken up during biosynthesis which relieve the cell of the need to synthesize the particular compound and hence save energy. These growth factors include vitamins, amino acids, nucleic acids, fatty acids, and sterols.

One key growth factor throughout the *Saccharomyces* is inositol, and a lack of this can lead to less efficient cell division (Smith, 1951) and morphological changes within the cell wall (Power and Challinor, 1969; Kirsop and Brown, 1972). Beet molasses contain sufficient concentrations of inositol for optimum growth of the yeast (White and Munns, 1950), but with other substrates a minimum of 2 mg/liter is required.

Biotin and pantothenate are essential for all strains of *Saccharomyces* (Williams *et al.,* 1940) and some strains also require thiamin, pyridoxine, *p*-aminobenzoic acid, niacin, folic acid, and riboflavin. The concentrations of these vitamins required for optimum growth rates are strain dependent, and with some species of the *Saccharomyces* the presence of thiamin in the absence of pyridoxine can reduce growth (Williams *et al.,* 1940; Gordon and Stewart, 1972).

Environmental conditions may also dictate a requirement for one or more growth-promoting substances; for example, ergosterol and unsaturated fatty acids are essential for yeast growth under anaerobic conditions (Nes *et al.,* 1978) and choline, carnitine, or leucine is required by some thermophilic yeast strains (Travassos and Cury, 1971).

Addition of all the necessary growth-promoting compounds to a culture medium can be costly, and hence lower-than-optimal growth rates become acceptable on consideration of the economics of the system. Alternatively, adequate growth-promoting factors can be provided by the addition of yeast extract to the fermentation medium.

3. PROCESS VARIABLES THAT INFLUENCE THE GROWTH OF *SACCHAROMYCES*

3.1. Hydrogen Ion Concentration

Most yeast of the genus *Saccharomyces* will grow at pH values in the range 2.4–8.6. The intracellular pH of *S. cerevisiae* is controlled within the range 5.8–6.3 and has been shown to be independent of external pH values ranging from 3 to 7 (Martiny, 1972). Yeast growth and fermentation rates do not appear to be affected by pH values varying between 3.5 and 6 in the bulk medium, and this is probably due to the tight control over intracellular pH.

When sucrose is used as a carbon source, the system is more pH sensitive than with glucose because yeast invertase activity is affected by low pH values.

Most laboratory-scale and industrial *Saccharomyces* fermentations are controlled at a pH between 4 and 5, which also helps reduce the risk of bacterial contamination within the reactor.

3.2. Temperature and Ethanol Inhibition Effects

Most strains of *Saccharomyces* will grow at any temperature between 0° and 40°C, though optimum temperature for maximum growth rate is strain dependent and generally in the range 28–35°C (Walsh and Martin, 1977). However, when considering cell yield rather than rate of growth, optimum temperatures tend to be somewhat lower, and in yeast manufacture a temperature of 26°C is maintained for the first 6 hr of the fermentation (Walter, 1940).

Growth of *S. cerevisiae* at temperatures near its tolerance limit (40°C) results in disruption of sterl and fatty acid synthesis (Thompson and Parks, 1974), destabilization of the plasma membrane, and a decrease in cell viability (Krouwel and Braber, 1979). It has also been observed that temperature has a marked effect on the mutation rate to respiratory deficiency in *S. cerevisiae,* which can produce significant quantities of such "petite" mutant cells at 40°C, though very few are detected at temperatures below 35°C (Jones *et al.*, 1981).

In fermentations for ethanol production, slightly higher temperatures are employed than for yeast cell production. Increasing the temperature of a fermentation of *S. cerevisiae* on a glucose-based medium from 30° to 39°C can result in a decrease in cell growth and an increase in ethanol productivity and eventual cell death at about 39°C. In general, the optimum temperature for ethanol production by *Saccharomyces* strains in 5–10°C higher than the optimum temperature for growth (White and Munns, 1951). However, ethanol produced during the fermentation has an inhibitory effect on yeast growth. In one study addition of ethanol to cultures of *S. cerevisiae* and *S. uvarum* growing exponentially at 23°C produced a 10% decrease in the growth rate at 2% (wt./vol.) ethanol, a 50% decrease in 6% (wt./vol.) ethanol, and a 90% decrease in 10% (wt./vol.) ethanol (Brown *et al.*, 1981). This inhibition of cell growth by high ethanol concentrations is greater at higher temperatures (Gray, 1941; Nagodawithana *et al.*, 1974). Table IV shows the

Table IV. Inhibitory Concentrations of Ethanol at Different Temperatures

Temperature (°C)	Ethanol (wt./vol.)
9	9.5
18	8.3
27	7.5
36	3.8

ethanol concentrations that prevented further fermentation by *S. cerevisiae* for a range of temperatures (Ranganathan and Bhat, 1958).

This effect is considered to be due to ethanol being produced within the cell at a higher rate than it can be transported through the cell membrane. This leads to enzyme inhibition and is followed by cell death (Nagodawithana *et al.,* 1974). For batch growth of the *Saccharomyces,* lower temperatures will produce a higher cell yield. In continuous culture or systems where ethanol is removed from the reactor, the ethanol inhibition effect is not so severe and higher temperatures can be employed. Heating culture media can be costly; so for economic reasons *Saccharomyces* cultures are generally maintained at temperatures between 25° and 30°C (Walter, 1940; Del Rosario *et al.,* 1979; Eroshin *et al.,* 1976).

3.3. Dissolved Oxygen and Substrate Inhibition Effects

Yeasts are unable to grow for more than four or five generations under fully anaerobic conditions (Jones *et al.,* 1981). Addition of air or oxygen to *Saccharomyces* fermentations is essential to maintain cell viability. Complete oxidation of the sugar to carbon dioxide and water will give optimum cell production. Oxygen acts as the final electron acceptor in oxidative phosphorylation, during yeast respiration. Under conditions of high dissolved oxygen concentrations, fermentation of the sugar to ethanol is inhibited. This effect, the Pasteur effect, was first noted by Pasteur in 1867 (Fiechter *et al.,* 1981).

The two extremes of respiration and fermentation of glucose can be represented by

$$C_6H_{12}O_6 \xrightarrow[\text{fully aerobic (respiration)}]{6O_2} 6CO_2 + 6H_2O \qquad G = -686 \text{ kcal}$$

$$C_6H_{12}O_6 \xrightarrow{\text{fully anaerobic (fermentation)}} 2C_2H_5OH + 2CO_2 \qquad G = -54 \text{ kcal}$$

Respiration releases more energy than fermentation and is therefore the preferred process. However, these two extremes are not realized in practice. Many *Saccharomyces* species are sensitive to glucose (Fiechter *et al.,* 1981) and their respiration is repressed in the presence of a concentration of glucose greater than 1.0 g/liter (Rickard and Hogan, 1978). Under such conditions the biomass yield will decrease and ethanol will be produced. This is known as the Crabtree effect or Contre–Pasteur effect. In a study of the Crabtree effect in various yeast strains, growing on a medium containing 30 g/liter glucose, seven of eight *Saccharomyces* species tested gave a positive Crabtree effect (de Deken, 1966). However,

Table V. Effect of Initial Sugar Concentration on the Specific Growth Rate of S. cerevisiae

Initial sugar concentration (% wt./vol.)	Specific growth rate (hr^{-1})
5	0.6
14	0.5
30	0.3

other research has concluded that the Crabtree effect should be considered a transient control mechanism and that a gradual derepression of respiration will occur (Barford *et al.*, 1980).

In the brewing industry the specific growth rate, viability, and yield of the *Saccharomyces* species employed have been found to increase with the level of oxygen concentration in the wort for levels of up to 20% saturation. Higher dissolved oxygen levels do not affect the fermentation, indicating that oxygen levels of around 20% of saturation are necessary for yeast cell maintenance and growth (Gordon and Stewart, 1972). More generally, the minimum level of oxygen required is strain dependent and also can be dependent on environmental conditions. Lower dissolved oxygen levels are required when unsaturated fatty acids and sterols are present in the culture media, but higher levels are needed if serine is present (Harding and Kirsop, 1979).

High concentrations of sugars also have an inhibitory effect on yeast growth though some *Saccharomyces* species are more sugar tolerant than others. In aerobic batch fermentations the specific growth rate of one strain of *S. cerevisiae* has been shown to fall as the initial sugar concentration is increased (Ghose and Tyagi, 1979b), as shown in Table V.

The specific growth rate of a more sugar-tolerant *S. uvarum* strain has been reported to be unaffected by sugar concentrations of up to 25% (wt./vol.) (Del Rosario *et al.*, 1979).

In the cultivation of *Saccharomyces* yeasts it is important that the initial sugar concentration in the culture medium not be so high as to prevent growth of the yeast.

3.4. Carbon Dioxide

Carbon dioxide, a by-product of yeast growth and ethanol production, is inhibitory to both these processes under aerobic and anaerobic conditions (Chen and Gutmanis, 1976; Kunkee and Ough, 1966). Carbon dioxide can affect the permeability and composition of yeast cell membranes and can also shift the equilibrium in carboxylation/ decarboxylation reactions in the metabolic pathways of the yeast. The inhibition effects are greater in high ethanol concentrations and at low pH values (Kunkee and Ough, 1966).

4. THE THEORY AND PRACTICE OF YEAST CULTURE SYSTEMS

4.1. Batch Systems

Historically *Saccharomyces* yeasts have been grown batchwise in submerged cultures, for the production of cells for food and fodder, and in order to ferment liquid media to alcoholic beverages. When a single-cell organism such as a *Saccharomyces* strain is grown in submerged culture, a plot of the logarithm of the dry weight of cells produced against time gives a characteristic curve dependent on strain and environmental conditions. A typical growth curve, as shown in Fig. 1, consists of three distinct sections: (A) the lag phase, (B) the exponential phase, (C) the stationary phase. A growth acceleration phase may be considered to follow the lag phase and precede the exponential phase, and similarly a growth retardation phase may be considered to occur before the stationary phase.

The *lag phase* represents the time period between inoculation of the culture medium with the organism and a measurable increase in cell

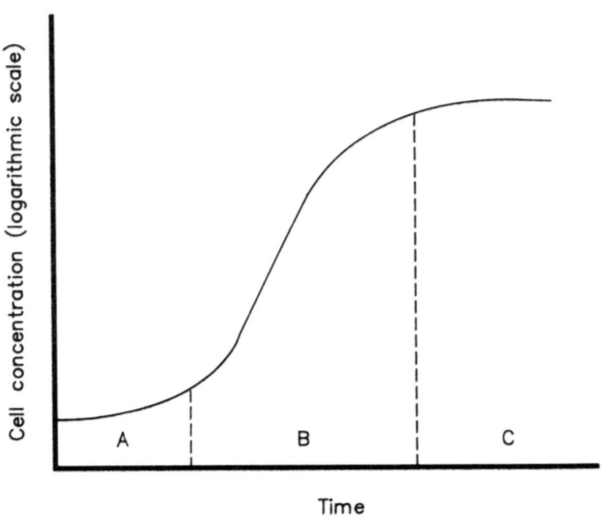

Figure 1. Idealized time course for yeast growth, showing (A) lag phase, (B) exponential phase, (C) stationary phase.

concentration. During this time the cells are adapting to their new environment. The lag phase can be shortened by using a large inoculum or an inoculum culture that is already growing exponentially under similar conditions. If the culture medium is near the optimum temperature for the yeast growth and contains all the essential nutritional requirements for the yeast, this will also decrease the apparent lag phase.

The *exponential phase* is the time period during which the specific growth rate (μ) is constant and at a maximum (μ_{max}) for the given strain and the environmental conditions. To achieve a constant specific growth rate individual cells must be budding at regular time intervals. The increase in cells per unit time is a product of the specific growth rate (μ) and the amount of cells present at a given time (x) and can be represented by

$$\frac{dx}{dt} = \mu_{max}\, x \qquad \text{or} \qquad x_o = x\, e^{(\mu_{max} t)} \tag{1}$$

where x_o is the dry mass of cells present at the start of the exponential growth phase.

Therefore,

$$\frac{x}{x_o} = e^{(\mu_{max} t)} \tag{2}$$

By taking logarithms of both sides of equation (2)

$$ln\, x - ln\, x_o = \mu_{max}\, t \tag{3}$$

Hence

$$\mu_{max} = \frac{ln\, x - ln\, x_o}{t} \tag{4}$$

The maximum specific growth rate (μ_{max}) is given by the gradient of section B of the growth curve (Fig. 1).

In general,

$$\mu_{max} = \frac{lnx_2 - lnx_1}{t_2 - t_1} \tag{5}$$

or

$$\mu_{max} = \frac{\log_{10} x_2 - \log_{10} x_1}{2.3\,(t_2 - t_1)} \tag{6}$$

The mean generation time or doubling time (g) of the yeast population can also be obtained from the growth curve or calculated as

$$\mu_{max} = \frac{\ln 2x - \ln x}{g} = \frac{\ln 2}{g}, \text{ i.e., } g = \frac{0.69}{\mu_{max}} \tag{7}$$

As the nutrients for yeast cell growth in the culture medium become limited and ethanol and carbon dioxide begin to accumulate, the increase in cells per unit time will no longer be given by μ_{max}. In 1950, Monod (Burrows, 1970) proposed a theory of cell growth based on two assumptions:

1. Microorganisms grow exponentially with a maximum specific growth rate μ_{max} until limited by some environmental condition to a lower specific growth rate μ.
2. The specific growth rate μ of any microbial culture is proportional to the limiting substrate concentration S.

Assumption (a) can be written

$$\frac{dx}{dt} = \mu x \tag{8}$$

Assumption (b) can be written

$$\mu = \mu_{max} \frac{S}{S + K_m} \tag{9}$$

where K_m is a constant numerically equal to the substrate concentrations at which the specific growth rate is half its maximum value, as shown in Fig. 2. Equation (8) is analogous to the Michaelis–Menten equation for enzymic reactions and indicates that when S is very large, $S/S + K_m$ tends to unity and μ tends to μ_{max} and when $S = 0$ then $\mu = 0$; i.e., if the limiting substrate has been consumed, no further growth will occur. Beyond this condition the accumulated cell mass therefore remains constant and growth enters the *stationary phase.*

The type and size of vessels employed for batch fermentation of *Saccharomyces* yeasts vary immensely. The extremes of the range are a sterile Erlenmeyer flask and an open vat with a volume of hundreds of cubic meters. A typical batch fermenter of the type that could be used under sterile conditions in a laboratory (a stirred tank reactor) is shown in Fig. 3. Temperatures, pH, DO (dissolved oxygen), and foam detecting probes may be inserted for monitoring and controlling the fermentation. Ports are available for the addition of alkali, air, antifoam, and the inoculum culture and for the removal of gases and sampling. Simple agitation can be achieved by a magnetic stirrer, but more commonly a

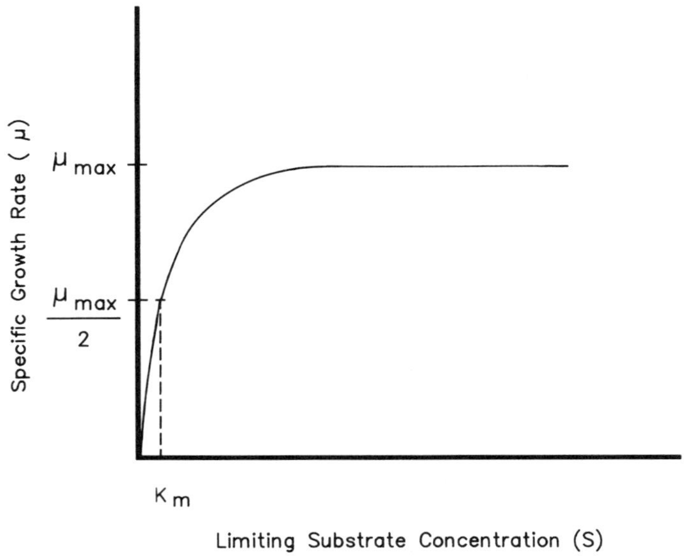

Figure 2. Idealized representation of the effect of limiting substrate concentration on specific growth rate.

mechanical stirrer is employed. More efficient agitation is achieved by means of shaped baffles.

The commercial manufacture of *Saccharomyces* yeasts (Pyke, 1957; Rhodes and Fletcher, 1966) usually begins with the transfer of a single cell of the preferred *Saccharomyces* strains to a nutrient agar slope. When grown, this culture is used to inoculate a flask of liquid medium containing around 6% (wt./vol.) sugar, which in turn will be transferred following a suitable growth period to a larger fermentation vessel containing a similar growth medium. Further transfers to progressively larger volumes of media take place at intervals of 12–24 hr and aeration is increased at each stage to maintain a steady dissolved oxygen concentration. The final production stage takes place when the production vessel containing around 200 m^3 medium is inoculated with yeast suspension to give an initial concentration in the reactor of around 20% wt./vol. biomass. The cells in this final inoculum are usually in the exponential phase of growth in order to minimize the lag phase. Throughout the entire process the pH is maintained below 4.5 to reduce the risk of bacterial contamination and above 3.5, since at lower pH values the color of the yeast produced can be affected and if sucrose is the carbon source, the yeast invertase activity may be affected. The temperature is initially maintained at 25°C but is allowed to rise gradually to 30°C by the end of the fermentation.

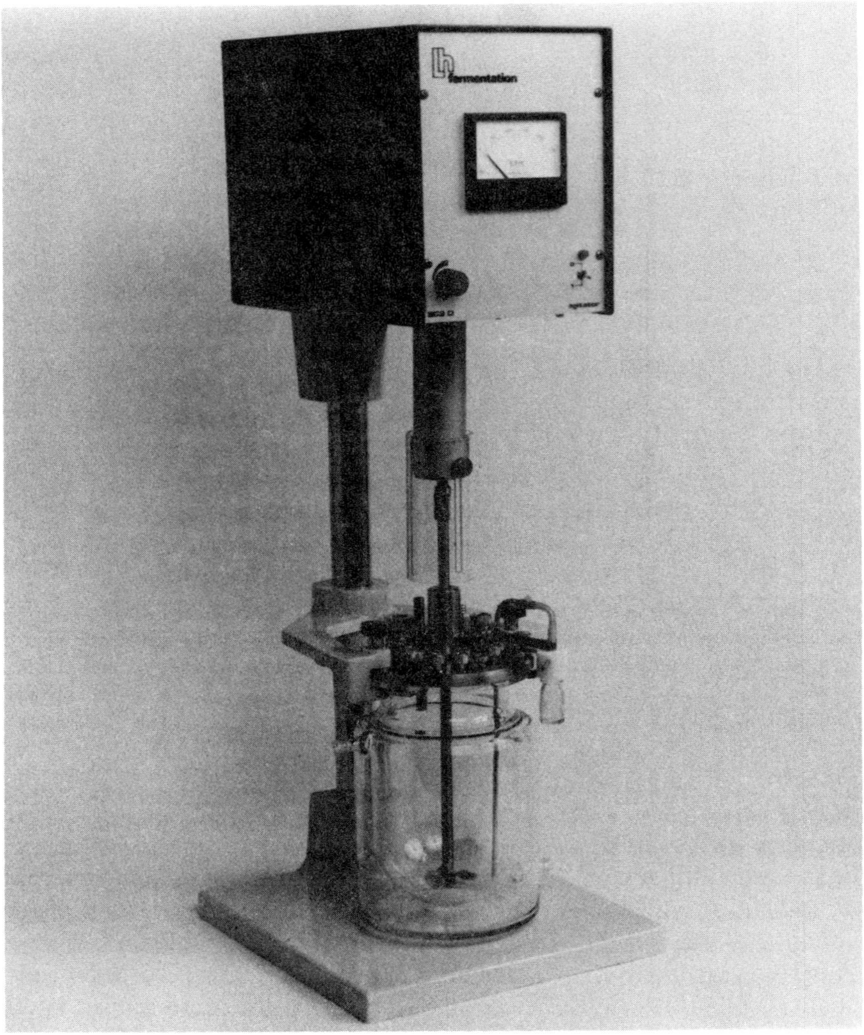

Figure 3. Typical laboratory scale stirred tank fermenter. (Courtesy of LH Fermentation, Stake Pages, UK.)

4.2. Continuous Systems

During most of the lag phase and the stationary phase of a batch culture no cells are produced. Continuous culture systems have been devised so that after an initial adaptation period the *Saccharomyces* cells can grow at high specific growth rates for long periods of time without a stationary phase occurring.

A simple continuous culture system, a continuous stirred tank reactor (CSTR) as shown diagramatically in Fig. 4, would comprise a fermentation vessels containing a constant volume (V) of culture medium. A continuous supply of fresh culture medium would be added at a constant flow rate (F) and fermented liquor would be removed from the fermentation vessel at the same flow rate. Cells would leave the vessel with the spent medium. The dilution rate (D) of such a system is defined as

$$D = \frac{F}{V} \tag{10}$$

$$\text{and the mean liquid residence time} = \frac{1}{D} = \frac{V}{F} \tag{11}$$

The overall accumulation of cells in the fermentation vessel (per unit time) is given as the difference between the rate of growth and the rate of removal of the cells, i.e.,

$$\frac{dx}{dt} = \mu x - Dx \tag{12}$$

After adapting to the conditions in the fermenter, the growth rate of the yeast cells remains constant, and the system is said to be at a steady state until there is a change in the environmental conditions. Since under steady-state conditions the cell concentration remains constant, $\frac{dx}{dt} = 0$ and therefore $\mu = D$. As the dilution rate is increased, the specific growth rate of the yeast cells will increase until D reaches its critical value D_c at which $\mu = \mu_{max}$. If the dilution rate is further increased above D_c

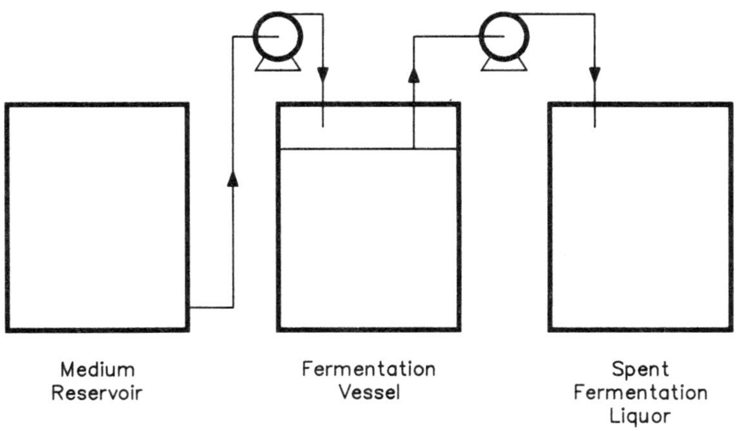

Figure 4. A simple continuous culture system.

the yeast cells will be removed from the fermenter at a faster rate than they can be produced and hence all cells will eventually be washed out of the vessel.

At steady state there is a relationship between the substrate concentration and the yeast cell concentration. Assuming the rate of substrate uptake $\frac{dS}{dt}$ is proportional to the rate of growth of the cells, then

$$-\frac{dS}{dt} = \frac{1}{Y}\frac{dx}{dt} \tag{13}$$

where Y, the yield coefficient, is the ratio of the dry weight of cells produced to the weight of substrate consumed over a given time period. The substrate concentration in the feed reservoir (S_R) will be higher than in the fermentation vessel (S) as the yeast cells consume the substrate for growth. The rate of change of the substrate concentration in the fermentation vessel is the difference between the input, the output, and the rate of consumption:

$$\frac{dS}{dt} = \mathrm{D}\,\mathrm{S_R} - D\,S - \frac{1}{Y}\frac{dx}{dt} \tag{14}$$

Substituting from equation (8)

$$\frac{dS}{dt} = \mathrm{D}\,\mathrm{S_R} - D\,S - \frac{\mu x}{Y} \tag{15}$$

and substituting for μ from equation (9)

$$\frac{\mathrm{dS}}{\mathrm{dt}} = \mathrm{D}\,(S_R - S) - \frac{\mu_{max} x}{Y}\,\frac{\mathrm{S}}{\mathrm{S} + \mathrm{K_m}} \tag{16}$$

and

$$\frac{dx}{dt} = \mu_{max} x\,\frac{S}{S + K_m} - D\,x \tag{17}$$

The behavior of single-cell organisms in continuous culture is defined by equations (16) and (17).

Under steady-state conditions there is no change in the number of cells or of the limiting substrate concentration in the fermenter.

Hence

$$\frac{dx}{dt} = \frac{dS}{dt} = 0 \tag{18}$$

Equations (16) and (17) will then become, respectively,

$$D\,(S_R - S) = \frac{\mu_{max}\,x}{Y}\,\frac{S}{S + K_m} \tag{19}$$

and

$$D = \mu_{max}\,\frac{S}{S + K_m} \tag{20}$$

where x is the steady-state yeast cell concentration and S is the steady-state limiting substrate concentration in the fermenter. Equations (19) and (20) can be simplified by substitution to

$$S_R - S = \frac{x}{Y} \tag{21}$$

hence

$$x = Y\,(S_R - S) \tag{22}$$

and

$$S = K_m\,\frac{D}{\mu_{max} - D} \tag{23}$$

At steady state the production of cells per unit volume of culture medium is Dx and from equations (22) and (23)

$$Dx = DY\,[S_R - K_m\,\frac{D}{\mu_{max} - D}] \tag{24}$$

Cell production will increase with dilution rate up to a maximum value D_{max} which can be determined by differentiating equation (24) with respect to D, i.e.,

$$\frac{d(Dx)}{dD} = Y\,[S_R + K_m - \frac{K_m\,\mu_{max}^2}{(\mu_{max} - D)^2}\,] \tag{25}$$

and $D = D_{max}$ when $\frac{d(Dx)}{dD} = 0$, so:

at $$\frac{d(Dx)}{dD} = 0,\ S_R + K_m = \frac{K_m\,\mu_{max}^2}{(\mu_{max} - D_{max})^2} \tag{26}$$

or

$$\mu_{max} - D_{max} = \left[\frac{K_m\,\mu_{max}^2}{S_R + K_m}\right]^{0.5} \tag{27}$$

and

$$D_{\max} = \mu_{\max}\left[1 - \frac{K_m}{S_R + K_m}\right]^{0.5} \tag{28}$$

If S_R is large $\left[\frac{K_m}{K_m + S_R}\right]^{0.5}$ will be small and therefore $D_{\max} = \mu_{\max}$ and hence $= D_c$.

4.3. Practical Continuous Culture Systems

The simplest type of vessel for a continuous culture fermentation, the CSTR, is almost identical in configuration to that described for batch cultures (Fig. 3), differing only in that for a continuous system an inlet port and outlet port for the flow of liquid media are needed. Improvements on this basic system have been made for the continuous production of food and fodder yeasts at industrial scale by changes in the stirring and aeration system (Hospodka, 1966). Examples of these are shown in Fig. 5.

Figure 5a shows the Vogelbusch aeration system. Air flows down the hollow central stirring shaft and out into the culture medium through small holes, 0.5–1.0 mm in diameter, in the stirring paddle. Figure 5b shows the system of Scholler and Seidel, based on the principle of the airlift pump. Fermentation medium is circulated through tubes arranged outside the fermentation vessel. The liquid and air enter the vessel through ceramic aeration candles at the bottom of the tubes and are either recycled through the tubes or removed from the fermenter via a central overflow cylinder. Figure 5c shows the Phrix aeration system, which consists of three stirrer paddles mounted on a central shaft. Air enters under, and is dispersed by, the bottom paddle, which is made from perforated plate. The middle paddle is for breaking up any large air bubbles and preventing further bubble coalescence, while the top paddle defoams the culture medium by a suction action as it rotates. Figure 5d shows the Lafrancois system, which comprises two inlet tubes, one for air and a narrower one for liquid medium, which extend almost to the bottom of the fermenter. The ends of these tubes are contained in a cooling tube to aid circulation.

These systems were devised during or just after World War II, when there was an increased demand for food and fodder yeast. Baker's yeast has only been produced by continuous fermentation during the past 25 years. The first successful plant was located at the Distillers Co., Dovercourt, England, and consisted of six fermentation vessels, each with a working volume of 40 m^3, connected in series as shown in Fig. 6. The first vessel was filled with liquid medium and inoculated from the

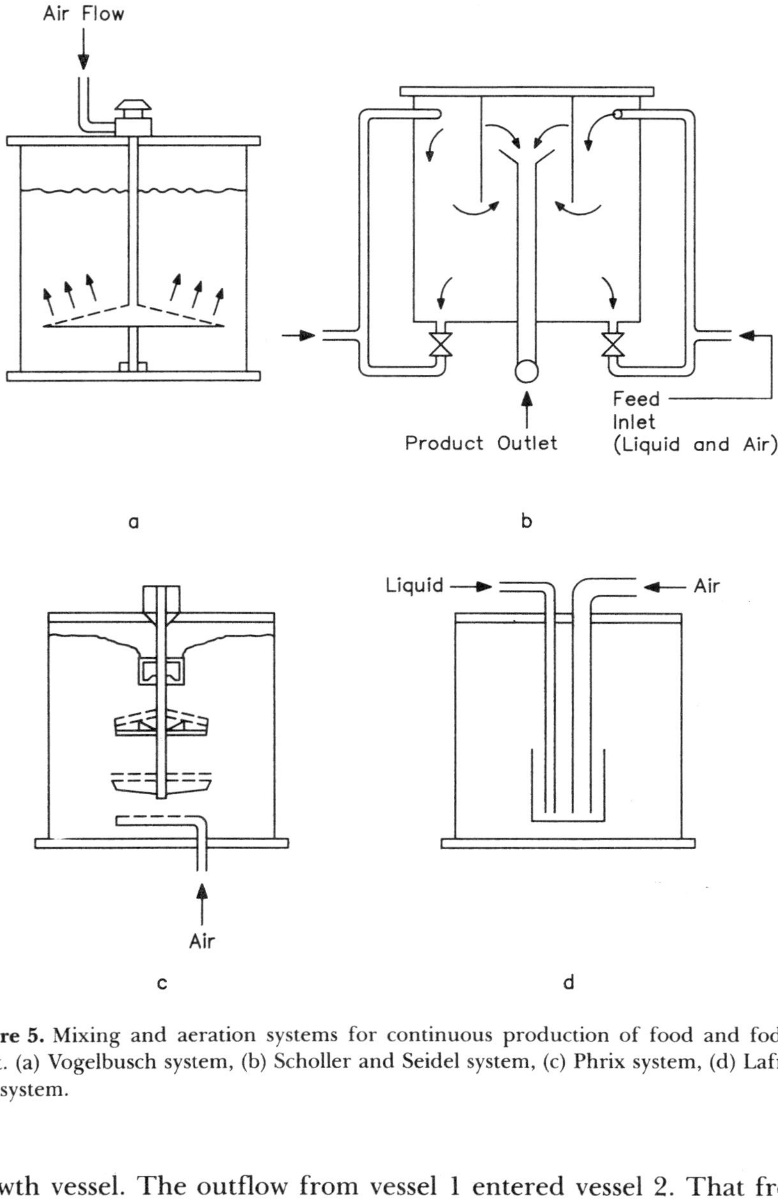

Figure 5. Mixing and aeration systems for continuous production of food and fodder yeast. (a) Vogelbusch system, (b) Scholler and Seidel system, (c) Phrix system, (d) Lafrancois system.

growth vessel. The outflow from vessel 1 entered vessel 2. That from vessel 2 entered vessel 3, and so on in sequence. When the yeast culture had grown in all six tanks, usually after approximately 27 hr, the operation became continuous. Liquid media was pumped into the first five vessels, which were fully aerated. The 6th vessel was only gently aerated, which allowed the yeast to mature before being separated and dried for use. Continuous operation of this system varied from 50 to 80 hr, and

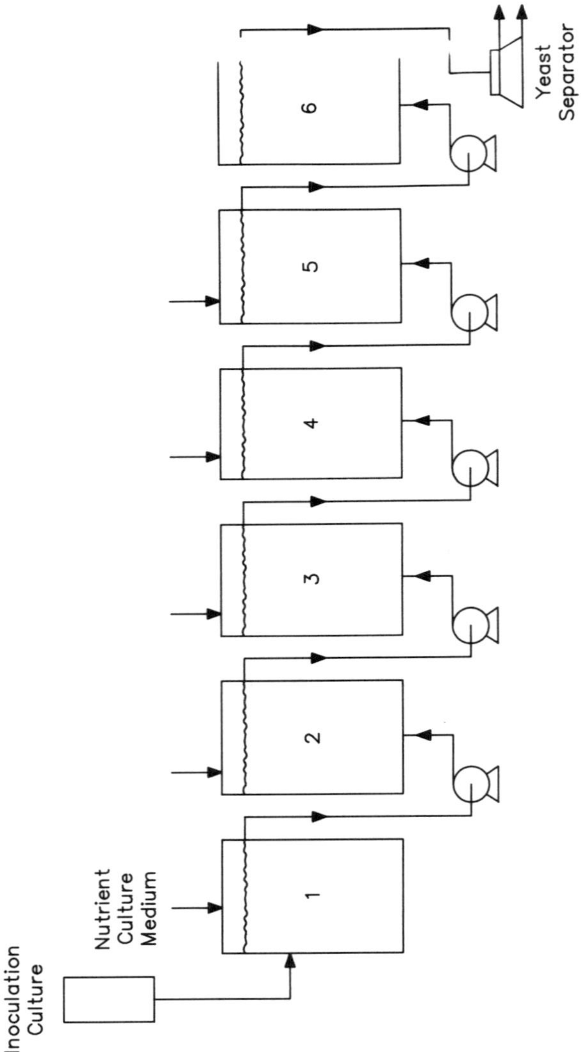

Figure 6. Distiller's system for continuous baker's yeast production.

the baker's yeast productivity was reported to be 33% higher than in batch systems.

An alternative smaller continuous production plant for baker's yeast proved to be successful in Russia. This, the Plevato system, comprised only two connected fermentation vessels. The first, with a volume of 30 m^3, was fully aerated and had a complete nutrient medium added continuously, while the second was only 10 m^3 in volume, only mildly aerated, and no nutrients were added. Eighty-five percent of the total biomass formed was produced in the first vessel. Continuous operation lasted up to 140 hr and gave a 30–35% increase in *S. cerevisiae* productivity over a batch system.

During the 1960s the brewing industry developed tower fermentation systems for the continuous production of beer. This type of fermenter is, in essence, a vertical cylindrical tower with a conical bottom; one such, developed by the APV Company Ltd. (Royston, 1966), is shown schematically in Fig. 7. A settling zone containing a yeast separator is included above the tower section and in principle allows the yeast to settle and return to the tower while clear beer is removed from the fermenter. Culture medium (wort) and air enter at the base of the tower. For this system to work it is essential that the yeast strain selected is highly flocculent and will settle back into the tower section, otherwise it

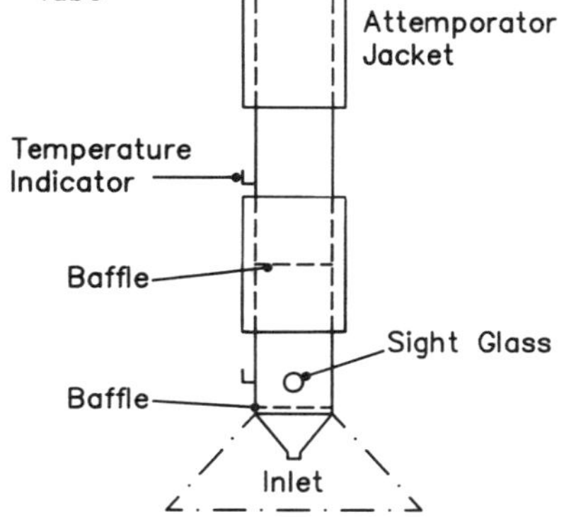

Figure 7. APV tower fermenter.

would be washed out with the fermentation effluent. Initially only one strain of *S. cerevisiae* was found to be suitable, but later, tests were developed for yeast flocculence under tower fermentation conditions and several more suitable *Saccharomyces* strains were identified (Greenshields and Smith, 1971). Wort with an initial sugar concentration of 5–8% has successfully been converted to beer with an alcohol content of 2–4%, using such a system (Prince and Barford, 1982).

Attempts to produce *S. cerevisiae* by either batch or continuous operation of a tower fermenter system resulted in low yields, but harvesting of the yeast was facilitated by its flocculent nature (Greenshields and Smith, 1971).

More recently, as interest in power alcohol production has increased, the performance of the tower fermenter with high sugar concentrations has been studied. Quantitative conversion of 20% glucose to ethanol by flocculent strains of *S. cerevisiae* and *S. diastaticus* has been shown to be possible at low dilution rates (Prince and Barford, 1982).

With the increased interest in ethanol production and in order to obtain higher ethanol productivities many innovative fermentation systems have been developed since the mid-1970s.

Cysewski and Wilke (1977,1978) designed a fermentation system for the production of ethanol by *S. cerevisiae* which incorporated a cell recycle system whereby cells washed out of the fermentation vessel are separated from the effluent liquor by sedimentation or centrifugation and returned to give increased cell densities and hence higher ethanol productivities. Ethanol inhibition effects were eliminated by running the fermenter under reduced pressure, allowing the alcohol to be boiled off from the fermentation liquor at a sufficiently low-temperature (35°C) so as not to affect the yeast.

More conventional continuous stirred tank reactors with external cell settling devices have also been considered for the production of ethanol (Ghose and Tyagi, 1979a; Fricker and Witts, 1981). Another system, where the fermentation broth is agitated by recycled gas and incorporating an external separator for the *Saccharomyces* cells, has been operated by Bu'Lock and Comberbach (1981).

Ultrafiltration offers an alternative method for cell retention within a fermenter. The rotofermenter as proposed by Margaritis and Wilke (1978) is a pressurized vessel with an internal rotating filter which retained *S. cerevisiae* cells. A similar system with a static filter has been used by Chem and Zall (1982) with *S. fragi.*

The economics of operating all fermentation systems for ethanol production needs serious consideration; recycling of yeast cells which settle under gravity appears to have a cost advantage over systems employing centrifuges or filters.

5. MONITORING THE GROWTH OF *SACCHAROMYCES*

A number of techniques are available to monitor the growth of a *Saccharomyces* culture. All require a well-mixed, representative sample of the culture broth, and the most appropriate should be chosen for a particular application.

Cell counting, the most direct method, involves the use of a counting chamber (either a hemocytometer or a Petroff–Hauser slide). The slide is etched with a grid of squares of known size and is recessed to a specified depth, typically 0.1 mm. The number of cells per square is determined by microscopic observation and the concentration of cells is found by dividing by the square area times depth. The viability of the cells can be determined by staining 1 part liquid culture with 9 parts methylene blue solution (Lee *et al.*, 1981). Viable cells will remain colorless and nonviable cells will take up the blue dye. The technique is quick but is inaccurate if cells are flocculated.

Similarly, colonies may be counted. A sample is spread over a nutrient agar and incubated before counting the number of colonies formed, either by naked eye (from a plate) or microscopically (from an agar-coated slide). Early detection of colonies can be aided by the use of fluorescent dyes added to the culture. Cell viability may be estimated from the microscopic examination by comparing the initial number of cells present with the number of colonies formed. The technique is again inaccurate if cells are flocculated, since only colony-forming units (not necessarily single cells) are counted.

Automatic cell counting can be achieved by the use of a Coulter counter, in which cells suspended in an electrolyte solution pass through an orifice and the change in current is recorded. Size as well as number of particles is recorded, enabling flocculated cells to be taken into account.

The most common method of monitoring cell growth is by measurement of cell mass. This can be achieved by filtering or centrifuging and drying a sample of cell suspension and then weighing to give the dry weight. Alternatively, by measuring the optical density, at 600 nm, of the sample and comparing this against a cell-free sample, then referring to a standard, an almost instant measure of cell mass can be obtained. The turbidity of a *Saccharomyces* culture can also be used to give a measure of the cell mass.

In some alcoholic fermentations, carried out at industrial scale, online analysis for the evolution of carbon dioxide gas is used to check the fermentation rates and hence give an indirect measure of the amount of active cell mass present. NADH, upon UV radiation at 366 nm, emits fluorescence at about 400 nm. Since the concentration of NADH is directly related to biomass density, flouresence measurements provide another possibility for determining cell mass *in situ*, using on-line probes.

6. CELL SEPARATION TECHNIQUES

Saccharomyces cells are generally separated from the fermented liquor at the end of a fermentation. Flocculent yeasts can be allowed to settle under gravity, but this method is too slow for fine suspensions of nonflocculent cells. Filtration or centrifugation is more commonly employed for separating yeast cells from the bulk liquor.

In the industrial processing of baker's and brewer's yeast the cells are initially separated from the liquor by centrifugation. The yeast concentrate obtained is then washed, recentrifuged, and filtered either through cloth in presses or in rotary vacuum filters to obtain pressed yeast with a dry matter content of 28–30% wt./vol. (Burrows, 1970).

7. IMMOBILIZED CELL SYSTEMS

The immobilization of whole cells within a fermentation vessel is not a new idea but has only been applied to the *Saccharomyces* for alcoholic fermentations during the past 10 years. Immobilization techniques are aimed at the aggregation of cells to form some sizable structure which can be retained within a fermenter. There are several advantages in using an immobilized cell system over a free cell system for the production of ethanol.

- A higher fermenter yeast cell concentration can be achieved than with free cell batch or continuous systems.
- With free cells the dilution rate is limited by the growth rate of the cells, but as immobilized yeast cells are retained within the fermenter, dilution rates can be optimized for maximum ethanol productivity.
- Cell recycling is unnecessary, and since fewer yeast cells leave the fermenter with the product, ethanol extraction is easier than with free cell fermentations.
- A further advantage of high cell densities and high dilution rates is that the risk of contamination may be greatly reduced.

In a comparison of an immobilized cell reactor (ICR) and a CSTR for ethanol production by *S. cerevisiae* (Sitton *et al.*, 1980), it was found that the immobilized cell reactor could be operated at dilution rates six times higher than the washout rate of the CSTR and that the ethanol productivity of the ICR was better than that of the CSTR by a factor of nine.

Methods available for the immobilization of *Saccharomyces* cells can be categorized as either passive immobilization, which involves the natural growth of the yeasts onto or within a support, or active immobilization, which requires some chemical process to immobilize pregrown yeast cells.

7.1. Passive Immobilization

7.1.1. Films on Solid Supports

Saccharomyces cells have been immobilized as films on a wide range of inert solid supports, including PVC flake (Ghose and Bandyopadhyay, 1980), porous brackets (Ghose and Bandyopadhyay, 1980), Raschig rings (Ghose and Bandyopadhyay, 1980; Sitton and Gaddy, 1980), gelatin-coated Raschig rings (Sitton and Gaddy, 1980), ceramic spheres (Moo-Young *et al.*, 1980), wood chips (Moo-Young *et al.*, 1980; Gencer and Mutharasan, 1981), spherical mullite (Fricker and Witts, 1981), and brick (Minier and Gomas, 1982).

However, in some cases the *Saccharomyces* cells were sloughed off the supports at superficial liquid velocities above a critical value and hence were washed out of the fermenter.

7.1.2. Biomass Support Particles

Some *Saccharomyces* species are not normally flocculent or film forming and hence cannot be immobilized passively onto solid supports. An alternative immobilization technique, which can be applied to all *Saccharomyces* yeasts and which takes advantage of the sloughing off of overgrowth, is given by biomass support particles (BSPs) (Atkinson *et al.*, 1978,1979).

BSPs are open, porous structures which are available in a selection of sizes and shapes and which have contiguous internal voids accounting for up to 97% of the volume of the particle. Yeast cells can grow normally and be retained within the structure and any external growth can be sheared off.

BSPs constructed from knitted crushed stainless steel wire and flexible reticulated polyurethane foam have been used to immobilize strains of *S. cerevisiae* and *S. uvarum* for ethanol production (Black *et al.*, 1984). A fermenter containing BSPs and fermentation medium can be inoculated with a cell suspension, and after an initial batch growth period during which the cells become immobilized, continuous operation can be maintained for long periods.

7.2. Active Immobilization

7.2.1. Adsorption to a Charged Surface

This technique depends on the *Saccharomyces* cell wall surface containing constituents providing the necessary ionic sites for attachment to a charged support and generally requires more expensive supports such as ion exchange resins.

7.2.2. Covalent Bonding

The attachment of *Saccharomyces* cells to a solid support by covalent bonds is limited in that only a monolayer of yeast can be formed on the support and the chemical reagents required to assist the formation of covalent linkages are generally toxic and hence can cause a loss of viability in the yeast cells.

7.2.3. Entrapment within Gels and Resins

This technique is the most widely used method of *Saccharomyces* cell immobilization and all *Saccharomyces* species can be immobilized in this way. A wide range of gels and resins are available, including calcium alginate (Kierstan and Bucke, 1977; White and Portno, 1978; Williams and Munnecke, 1981; Linko and Linko, 1981; Cho and Choi, 1981), kappa carrageenan (Wada *et al.,* 1979,1980,1981), agar (Holcberg and Margalith, 1981), polyacrylamide (Holcberg and Margalith, 1981), gelatin (Sivaraman *et al.,* 1982), and epoxy resins (Klein and Kressdorf, 1982).

The size and shape of the gels and resins may vary according to the method of production. Calcium alginate is generally produced as spherical beads whereas other gels may be extruded and cut to size or produced as a sheet and then cut to the required shape.

In general, a concentrated suspension of cells is mixed with one or more of the constituents of the gel prior to gel formation. If a complete nutrient feed is produced, the yeast cells can continue to grow after immobilization, and this has been shown to produce higher yeast concentrations near the gel surface where there are less diffusion limitations (Cho and Choi, 1981; Wada *et al.,* 1979).

The strength of gels and resins can be varied by changing the ratio of their constituents. However, calcium alginate can be weakened in the presence of phosphate ions, and polyacrylamide and some epoxy resins can have a toxic effect on *Saccharomyces;* so care is needed during the immobilization procedure with these. The most common problem encountered with gel-immobilized yeast is the evolution of carbon dioxide gas during fermentation. The gel can be disrupted or caused to float if the gas remains trapped.

7.3. Fermenters for Immobilized *Saccharomyces* Cells

CSTRs and tower fermenters are generally not suitable for immobilized *Saccharomyces* cells. A range of fermenters have been developed specifically for their required application and these include the following.

7.3.1. Packed Beds

These vary in size and shape but all contain a retaining volume in which the immobilized cells are packed. Fermentation medium is pumped through this volume and may be recycled. In general, liquid mixing is poor and the large amounts of CO_2 generated during fermentation make these fermenters susceptible to gas flooding (the densely packed particles restrict the movement of the gas phase through the bed, resulting in a buildup of gas which displaces the liquid medium; Cho and Choi, 1981); hence sampling and parameter control are difficult.

7.3.2. Sheet Reactors

These are similar to packed beds but contain parallel plates of gel or support material over which the fermentation broth can be pumped. Liquid mixing is improved and evolved gases can be removed more easily than in a packed bed (Larsson and Mosbach, 1979).

7.3.3. Fluidized Beds

Particles containing immobilized *Saccharomyces* with a density greater than the fermentation medium can be fluidized by the upflow of the fermentation medium which is usually recycled. In order to achieve adequate particle movement, the bed is generally expanded by around 20%. Gel beads may be weighted, for example, by the inclusion of magnetite (Larsson and Mosbach, 1979), in order that the density may be increased sufficiently for the beads to be fluidized. A fluidized bed fermenter used with stainless steel BSPs is shown in Fig. 8.

7.3.4. Gas Mixed Reactors

Particles of immobilized biomass with a density similar to that of the fermentation medium can be circulated by the introduction of air or recycled gases at the base of the fermenter. Liquid recycling is not necessary. Better mixing occurs when gas is introduced only over a segment of

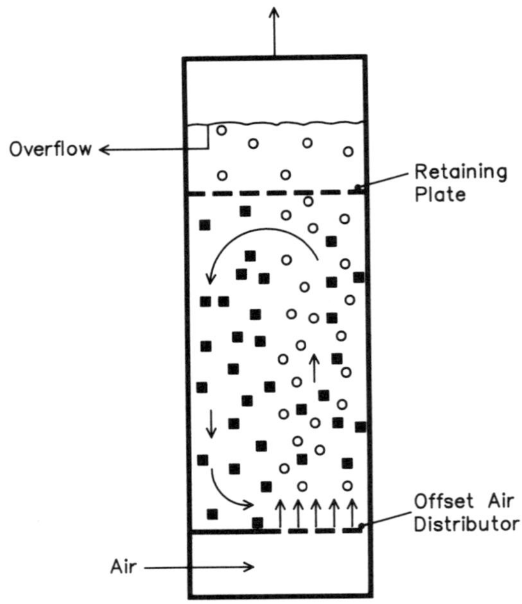

Figure 8. Fluidized bed reactor for use with dense particles of immobilized yeast cells.

the fermenter base and when a bed expansion of around 20% is permitted (Black *et al.*, 1984). A circulating bed fermenter used with foam BSPs is shown in Fig. 9.

In a comparison of a packed bed and a fluidized bed reactor each containing the same loading of *S. cerevisiae* immobilized in calcium alginate gel beads (Cho and Choi, 1981), it was found that over a range of dilution rates the ethanol productivity obtained from the fluidized bed fermenter was double that of the packed bed fermenter. However, obtaining very high ethanol productivities is not always desirable if there is a corresponding decrease in ethanol concentration, increasing the costs of downstream processing.

8. DOWNSTREAM PROCESSING

The products of *Saccharomyces* fermentation are yeast and alcohol. Processing methods for yeast and alcoholic beverages are well established, although fermentations for fuel alcohol production are much newer processes in competition with established production methods

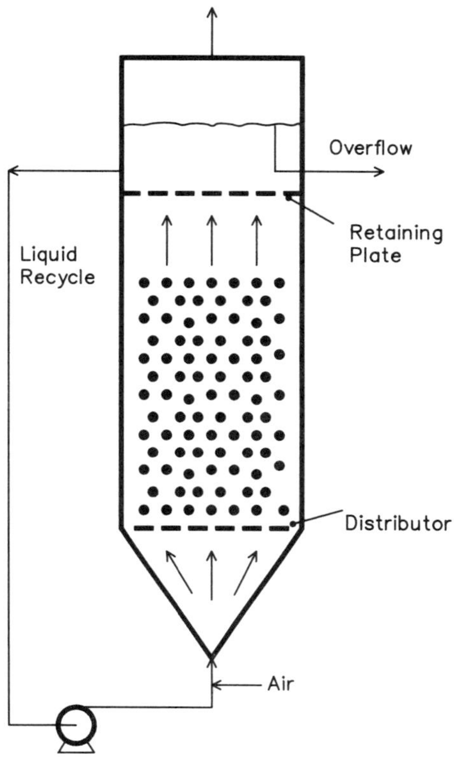

Figure 9. Circulating bed reactor for use with immobilized yeast cells where the particles have a similar density to the liquid medium.

(see Chapter 7). The downstream processing costs of such fermentations will affect the economic viability of the process.

8.1. Pressed Yeast

Traditionally yeast is processed as pressed yeast. After separation of the cells from the broth, as described previously, the yeast is removed from the presses or filter and passes to a hopper. Water and an emulisifer may be added to assist extrusion and packing, and the yeast is either granulated or compressed into blocks and packed in waxed papers or plastic containers. For storage the yeast must be kept below 10°C to prevent metabolization of storage carbohydrate (Burrows, 1970).

8.2. Dried Yeast

The commercial production of dried yeast has the advantage of a cost reduction in storage and transport but is generally lower in activity than pressed yeast. Pressed yeast is extruded through a screen to give continuous threads which are chopped and dried in a rotary drum drier or by tunnel drying. Dry air at between 35°C and 90°C is passed over the yeast but the surface temperature is kept low by evaporative cooling; this process takes between 6 and 24 hr. Loss of activity occurs at the end of the drying period (usually between 6 and 24 hr) when the cells are no longer protected by the evaporative cooling. Where high levels of cell activity are to be retained, drying agents such as methyl cellulose can be used to absorb final moisture.

Spray drying is sometimes used but requires higher temperatures which cause biological damage to the yeast unless protective agents are used.

8.3. Yeast Extract

The most common method of producing yeast extract is to plasmolyze pressed yeast with salt, heat to around 65°C for 40 hr, and then concentrate the extract. Other plasmolyzing agents can be used to reduce saltiness. An alternative method is by acid hydrolysis of yeast under pressure and at high temperatures (Pyke, 1957).

8.4. Alcoholic Beverages

Fermented liquor usually contains some *Saccharomyces* cells, and hence in order to prevent further fermentation, the liquid may be filtered, heat-treated, or pasteurized or a chemical antiseptic may be added. For spirits various methods of distillation are available depending on the requirements of the final products. Beers, ciders, and wines may be clarified by racking, filtration, or centrifugation prior to bottling or kegging. Carbon dioxide may be introduced to some alcoholic beverages at the bottling stage.

8.5. Ethanol

Ethanol is normally required at concentrations about the 95% wt./vol. which can be achieved by distillation. Very pure ethanol can be obtained by extractive distillation and azeotropic columns can be included in multiple-column stills for absolute alcohol (99.7%). However, the ethanol content of fermented liquor from a *Saccharomyces* fermentation is rarely above 10% and hence the distillation step can be costly. Research into energy-efficient alternative methods of alcohol recovery is currently in progress, and it is clear that the development of such economic downstream processing methods is essential if *Saccharomyces* fermentations are to provide alternative fuels for the future.

REFERENCES

Atkin, L., Schultz, S., and Frey, C. N. 1946. *Yeast Fermentation in Enzymes and Their Role in Wheat Technology*, Interscience, New York.

Atkinson, B., Black, G. M., Lewis, P. J. S., and Pinches, A. 1978. Growth of biological material, British Patent No. 2006 181 B.

Atkinson, B., Black, G. M., Lewis, P. J. S., and Pinches, A. 1979. Biological particles of given size, shape and density for use in biological reactors, *Biotech. Bioeng.* **21**:193–200.

Barford, J. P., Jeffery, P. M., and Hall, R. J. 1980. The Crabtree effect in *Saccharomyces cerevisiae*—Primary control method or transient? *Adv. Biotechnol.* **1**:255–274.

Black, G. M., Webb, C., Matthews, T., and Atkinson, B. 1984. Practical reactor systems for yeast cell immobilisation using biomass support particles, *Biotech. Bioeng.* **26**:134–141.

Brown, S. W., Oliver, S. G., Harrison, D. E. F., and Righelato, R. C. 1981. Ethanol inhibition of yeast growth and fermentation differences in the magnitude and complexity of the effect, *Eur. J. Appl. Microbiol. Biotechnol.* **11**:151–155.

Bu'Lock, J. D., and Comberbach, D. M. 1981. A practical system for high productivity ethanol fermentations, *Second European Congress on Biotechnology, Abstracts of Communications,* Society of Chemical Industry, London, p. 204.

Burrows, S. 1970. Bakers' Yeasts in: *The Yeasts,* Vol. 3 (A. H. Rose and J. S. Harrison, eds.), Academic Press, New York, p. 349.

Chen, H. C., and Zall, R. R. 1982. Continuous fermentations of whey into alcohol using an attached film expanded bed reactor, *Process Biochem.* **17**:20–25.

Chen, S. L., and Gutmanis, F. 1976. Carbon dioxide inhibition of yeast growth in biomass production, *Biotech. Bioeng.* **18**:1462.

Cho, G. H., and Choi, C. Y. 1981. Continuous ethanol production by immobilised yeast in a fluidized reactor, *Biotech. Lett.* **3**:667–671.

Cysewski, G. R., and Wilke, C. R. 1977. Rapid ethanol fermentations using vacuum and cell recycle, *Biotech. Bioeng.* **19**:1125–1143.

Cysewski, G. R., and Wilke, C. R. 1978. Process design and economic studies of alternative fermentation methods for the production of ethanol, *Biotech. Bioeng.* **20**:1421–1444.

de Deken, R. H. 1966. The Crabtree effect: A regulatory system in yeast, *J. Gen. Microbiol.* **44**:149–156.

Del Rosario, E. J., Lee, K. J., and Rogers, P. L. 1979. Kinetics of alcohol fermentation at high yeast levels, *Biotech. Bioeng.* **21**:1477–1482.

Eroshin, V. K., Utkin, I. S., Ladynichev, S. V., Samoylov, V. V., Kushinnikov, V. D., and Skryabin, G. K. 1976. Influence of pH and temperature on the substrate yield coefficient of yeast growth in a chemostat, *Biotech. Bioeng.* **18**:289–295.

Fiechter, A., Fuhrmann, G. F., and Kappeli, C. 1981. Regulation of glucose metabolism in growing yeast cells, *Adv. Microbial Physiol.* **22**:123–183.

Fricker, R., and Witts, W. S. 1981. The alcon process for continuous production of ethanol by fermentation, *Second European Congress on Biotechnology, Abstracts of Communications,* Society of Chemical Industry, London, p. 275.

Gencer, M. A., and Mutharasan, R. 1981. Ethanol fermentation in a yeast immobilised column fermenter, *Adv. Biotechnol.* **1**:627–633.

Ghose, T. K., and Bandyopadhyay, K. K. 1980. Rapid ethanol fermentation in immobilised yeast cell reactor, *Biotech. Bioeng.* **22**:1489–1496.

Ghose, T. K., and Tyagi, R. D. 1979a. Rapid ethanol fermentation of cellulose hydrolysate. I. Batch versus continuous systems, *Biotech. Bioeng.* **21**:1387–1400.

Ghose, T. K., and Tyagi, R. D. 1979b. Rapid ethanol fermentation of cellulose hydrolysate. II. Product and substrate inhibition and optimisation of fermenter design, *Biotech. Bioeng.* **21**:1401–1420.

Gordon, P. A., and Stewart, P. R. 1972. Effect of lipid status on cytoplasmic and mytochondrial protein synthesis in anaerobic cultures of Saccharomyces cerevisiae, *J. Gen. Mirobiol.* **72:**231.

Gray, W. D. 1941. Studies on the alcohol tolerance of yeasts, *J. Bacteriol.* **42:**561–574.

Greenshields, R. N., and Smith, E. L. 1971. Tower fermentation systems and their applications, *Chem. Engineer* **249:**182–190.

Harding, S. A., and Kirsop, B. H. 1979. Relative significance of oxygen and other nutrients as fermentation regulators in *Saccharomyces cerevisiae, J. Inst. Brew.* **85:**174.

Hoggan, J. 1977. Aspects of fermentation in conical vessels, *J. Inst. Brew.* **83:**133–138.

Holcberg, I. B., and Margalith, P. 1981. Alcoholic fermentation by immobilised yeast at high sugar concentrations, *Eur. J. Appl. Microbiol. Biotechnol.* **13:**133–140.

Hospodka, J. 1966. Industrial Application of Continuous Fermentation, in: *Theoretical and Methodological Basis of Continuous Culture of Microorganisms* (I. Malek and Z. Fenel, eds.), Academic Press, New York, p. 535.

Jones, M., Pragnell, M. J., and Pierce, J. S. 1969. Adsorption of amino acids by yeasts from a semi-defined medium simulating wort, *J. Inst. Brew.* **75:**520–536.

Jones, R. P., Pamment, N., and Greenfield, P. F. 1981. Alcohol fermentation by yeasts—The effect of environmental and other variables, *Process Biochem.* **16:**42–49.

Kierstan, M., and Bucke, C. 1977. The immobilisation of microbial cells, subcellular organelles and enzymes in calcium alginate gels, *Biotech. Bioeng.* **19:**387–397.

Kirsop, B. H., and Brown, M. L. 1972. Some effects of wort compositions on rate and extent of fermentation by brewing yeasts, *J. Inst. Brew.* **78:**51.

Klein, J., and Kressdorf, B. 1982. Immobilisation of living whole cells in an epoxy matrix, *Biotech. Lett.* **4:**375–380.

Krouwel, P. G., and Braber, L. 1979. Ethanol production by yeast at supraoptimal temperatures, *Biotech. Lett.* **1:**403.

Kunkee, R. E., and Ough, C. S. 1966. Multiplication and fermentation of *Saccharomyces cerevisiae* under carbon dioxide pressure in wine, *Appl. Microbiol.* **14:**643.

Larsson, P. O., and Mosbach, K. 1979. Alcohol production by magnetic immobilised yeast, *Biotech. Lett.* **1:**501–506.

Lee, S. S., Robinson, F. M., and Wang, H. Y. 1981. Rapid determination of yeast viability, *Biotech. Bioeng. Symp.* **11:**641–649.

Leibowitz, J., and Hestrin, S. 1939. The direct fermentation of maltose by yeast, *Enzymologia* **6:**15–16.

Lewis, M. J., and Wildenradt, H. L. 1969. Sulphur in brewing, *Brewers Digest* **44:**88.

Linko, Y. Y., and Linko, P. 1981. Continuous ethanol production by immobilised yeast reactor, *Biotech. Lett.* **3:**21–26.

Margaritis, A., and Wilke, C. R. 1978. The rotorfermentor. II. Application to ethanol fermentation, *Biotech. Bioeng.* **20:**727–753.

Markham, E., and Byrne, W. J. 1968. Uptake storage and utilisation of phosphate by yeast. 3. Behaviour of phosphate starved yeast, *J. Inst. Brew.* **44:**374.

Martiny, S. C. 1972. *Analysis and Simulation of Biochemical Systems,* Vol. 25, Fed. Eur. Biochem. Soc. Meet [Proc.], p. 387–397.

Menzinsky, G. 1943. Application of microorganisms to sugar analysis. I. Quick microbilogical procedures for the determination of fermentable sugar in the sulfite waste liquor, *Biochem. Z.* **314:**312–326.

Minier, M., and Goma, G. 1982. Ethanol production by extractive fermentation, *Biotechnol. Bioeng.* **24:**1565–1579.

Moo-Young, M., Lamptey, J., and Robinson, C. W. 1980. Immobilisation of yeast cells on various supports for ethanol production, *Biotech. Lett.* **2:**541–548.

Nagodawithana, T. W., Castellano, C., and Steinkraus, K. H. 1974. The effect of dissolved oxygen, temperature, initial cell count and sugar concentration on the viability of *S. cerevisiae* in rapid fermentations, *Appl. Microbiol.* **28:** 383–391.

Nes, W. R., Sekula, B. C., Nes, W. D., and Adler, J. L. 1978. The functional importance of structural features of ergosterol in yeasts, *J. Biol. Chem.* **253:**6218–6225.

Power, D. M., and Challinor, S. W. 1969. Effects of Inositol deficiency on chemical composition of yeast cell wall, *J. Gen. Microbiol.* **55:**169–176.

Prince, I. G., and Barford, J. P. 1982. Continuous tower fermentation for power ethanol production, *Biotech. Lett.* **4:**263–268.

Pyke, M. 1957. Industrial Production in: *Yeasts* (W. Roman, ed.), Dr. W. Junk Publishers, Holland, pp. 69–78.

Ranganathan, B., and Bhat, J. V. 1958. Ethanol tolerance of some yeasts, *J. Indian Inst. Sci.* **40:**105–110.

Rhodes, A., and Fletcher, P. L. 1966. *Principles of Industrial Microbiology*, Pergamon Press, Oxford.

Rickard, P. A. D., and Hogan, C. B. J. 1978. Effects of glucose on the activty and synthesis of fermentative and respiratory pathways of *Saccharomyces* sp., *Biotech. Bioeng.* **20:**1105–1110.

Rose, A. 1976. *Chemical Microbiology*, 3rd ed., Plenum Press, New York.

Rothstein, A. 1961. Interrelationships between the ion transporting systems of the yeast cell in: *Membrane Transport and Metabolism* (A. Kleinzeller and A. Kotyk, eds.), Academic Press, Prague, pp. 270–284.

Royston, M. G. 1966. Tower fermentation of beer, *Process Biochem.* **1:**215–221.

Schultz, A. S., and McManus, D. K. 1950. Amino acids and inorganic sulphur as sulphure source for the growth of yeasts, *Arch. Biochem.* **25:**401–409.

Schultz, A. S., and Pomper, S. 1948. Amino acids as a nitrogen source for growth of yeasts, *Arch. Biochem.* **19:**184–192.

Schultz, A. S., Atkin, L., and Frey, C. N. 1940. Influence of oxygen on the fermentation of maltose and galactose, *J. Am. Chem. Soc.* **62:**2271–2272.

Sitton, O. C., and Gaddy, J. L. 1980. Ethanol production in an immobilised cell reactor, *Biotech. Bioeng.* **22:**1735–1748.

Sitton, O. C., Manrider, G. C., Book, N. L., and Gaddy, J. L. 1980. Comparison of immobilised cell reactor and CSTR for ethanol production, *Biotech. Bioeng. Symp.* **10:**213–235.

Sivaraman, H., Seetarama Rao, B., Pundle, A. U., and Sivaraman, C. 1982. Continuous ethanol produciton by yeast cells immobilised in open pore gelatin matrix, *Biotech. Lett.* **4:**359–364.

Smith, R. H. 1951. A study of the role of inositol in the nutrition of *Nematospora gossypii* and *Saccharomyces carlsbergensis*, *J. Gen. Microbiol.* **5:**772–780.

Sobotka, H., Holzman, M., and Reiner, M. 1936. Selective fermentation. III. Fermentation of hexose-pentose mixtures, *Biochem. J.* **30:**933–940.

Thompson, E. D., and Parks, L. W. 1974. Effects of altered sterol composition on growth characteristics of *Saccharomyces cerevisiae*, *J. Bacteriol.* **120:**779–784.

Thorne, R. S. W. 1944. Growth of yeasts in binary mixtures of nitrogen nutrients, *J. Inst. Brew.* **50:**186.

Travassos, L. R., and Cury, A. 1971. Thermophilic enteric yeasts, *Annu. Rev. Microbiol.* **25:**49–74.

Wada, M., Kato, J., and Chibata, I. 1979. A new immobilisation of microbial cells. Immobilised growing cells using carrageenan gel and their properties, *Eur. J. Appl. Microbiol. Biotechnol.* **8:**241–247.

Wada, M., Kato, J., and Chibata, I. 1980. Continuous production of ethanol using immobilized growing yeast cells, *Eur. J. Appl. Microbiol. Biotechnol.* **10:**275–287.

Wada, M., Kato, J., and Chibata, I. 1981. Continuous production of ethanol in high concentration using immobilised growing yeast cells, *Eur. J. Appl. Microbiol. Biotechnol.* **11:**67–71.

Walsh, R. M., and Martin, P. A. 1977. Growth of *Saccharomyces cerevisiae* and *Saccharomyces uvarum* in a temperature gradient incubator, *J. Inst. Brew.* **83:**169–172.

Walter, F. G. 1940. *The Manufacture of Compressed Yeast,* Chapman and Hall, London.

White, F. H., and Portno, A. D. 1978. Continuous fermentation by immobilised brewers yeast, *J. Inst. Brew.* **84:**228.

White, J., and Munns, D. J. 1950. Nutrilites and the production of pressed yeast, *J. Inst. Brew.* **56:**194–202.

White, J., and Munns, D. J. 1951. Influence of temperature on yeast growth and fermentation, *J. Inst. Brew.* **57:**280–284.

Williams, D., and Munnecke, D. M. 1981. The production of ethanol by immobilized yeast cells, *Biotech. Bioeng.* **23:**1813–1825.

Williams, R. J., Eakin, R. E., and Snell, E. E. 1940. Relationship of inositol thiamin biotin pantothenic acid and vitamin B_6 to growth of yeasts, *J. Am. Chem. Soc.* **62:**1204–1207.

Biochemical Techniques 9

MICHAEL F. TUITE and STEPHEN G. OLIVER

1. INTRODUCTION

A plethora of biochemical techniques have been developed to study fundamental cellular processes of yeasts of the *Saccharomyces* genera. All of these generally require efficient disruption of the yeast cell followed either by purification of particular classes of macromolecules (e.g., proteins, nucleic acids, ribonucleoprotein complexes) or by partial fractionation of the lysate to yield subfractions that are particularly enriched in some biosynthetic activity, be it a given enzyme or a multicomponent process such as protein synthesis. In this chapter we describe the basic strategies that have been developed for *Saccharomyces* species (especially *S. cerevisiae*) for both cell disruption and fractionation, together with consideration of how one radiolabels specific macromolecules *in vivo*. A detailed account of strategies for subcellular fractionation of yeast cell components—mitochondria, membranes, vacuoles, and so forth—can be found in Chapter 2.

2. CELL DISRUPTION

The cell wall of *S. cerevisiae* presents a daunting barrier to the biochemist and molecular biologist alike. It comprises a thick, complex mixture of proteins and polysaccharides (Ballou, 1982). The outer surface is a tight aggregation of mannoproteins lying among the long, thin polysaccharides glucan and chitin and is largely impenetrable to nucleic acids, antibiotics, and other macromolecules. It is not readily susceptible to disruption by classical methods such as sonication and freeze–thawing, and therefore other strategies have been devised that allow for the

MICHAEL F. TUITE • Biological Laboratory, University of Kent, Canterbury, Kent CT2 7NJ, England. STEPHEN G. OLIVER • Manchester Biotechnology Centre, University of Manchester, Institute for Science and Technology, Manchester M60 1QD, England effective breakage of yeast cells. These strategies can be subdivided into those employing whole cells and those employing protoplasts.

2.1. Whole Cell Disruption

Traditionally yeast enzymes such as invertase have been extracted from whole cells by prolonged periods of autolysis, a process of "self-digestion" that relies on endogenous enzyme activities, particularly proteases, to disrupt the cell and to solubilize at least some of the yeast protein (Peppler, 1979). In the laboratory the use of autolysis is best avoided—particularly if one is setting out to purify a single polypeptide or trying to generate an active cell-free lysate, since most cellular enzymes are inactivated by this technique due to proteolysis or thermal inactivation.

A number of alternative, but no less effective, mechanical methods for disrupting yeast cells and releasing their proteins intact have been devised. The French Press has been frequently and widely used, as has the Eaton Press (for details, see Hughes *et al.*, 1971). More recently, methods that involve adding small mesh glass beads to a yeast cell slurry followed by vigorous stirring or shaking have come into favor. This approach was originally devised for large-scale yeast cell disruption in specially designed pieces of apparatus such as the Braun homogenizer (e.g., Schatz, 1967) and the Nossal shaker (e.g., Gregory *et al.*, 1967). More recently the method has been adapted for small-scale disruption using a vortex mixer (Mills, 1974) or by carefully choreographed hand shaking (Land *et al.*, 1977) to agitate the glass bead/yeast slurry. The size of the glass beads can be critical for achieving efficient breakage, and usually acid-washed glass beads with a diameter of 0.4–0.45 mm have proven to be most effective. Glass bead disruption can be used in preparing DNA, RNA, or protein and, in addition, has proven capable of generating active cell-free translation systems (see Section 6.2).

In all cell disruption techniques, steps must be taken to minimize the likelihood of proteolysis, and this generally entails carrying out the procedures at low temperatures (+2°C to +4°C) and in the presence of nonspecific protease inhibitors such as phenylmethyl sulfonyl fluoride (PMSF) and leupeptin (Pringle, 1975). The chances of avoiding proteolysis in an extract can be further increased by employing a strain of yeast deficient in one or more of the now well-characterized proteinases (Achstetter and Wolf, 1985). In particular, strains carrying the *pep4*, *prb1*, and *prc1* mutations are already widely in use (e.g., Rothblatt and Meyer, 1986).

2.2. Protoplast Lysis

A protoplast can be simply defined as an osmotically sensitive spherical body formed from a fungal cell without regard to completeness of removal of all cell wall material (Brunner *et al.,* 1958). This term is frequently used interchangeably with the term spheroplast although the latter should only be applied to protoplasts completely devoid of cell wall material.

A variety of lytic enzyme preparations are effective in producing spheroplasts from *S. cerevisiae* provided an osmotic stabilizer such as sorbitol, mannitol, or $MgSO_4$ is present to protect the osmotically fragile spheroplast. Only recently have many of these enzymes become commercially available (Table I). The majority of these are also effective against other *Saccharomyces* species and other yeast genera, e.g., *Candida.* Many of the favored lytic enzymes originate from microbial sources, with the exception of the widely used preparations from the gut of the snail *Helix pomatia* (Eddy and Williamson, 1957). These enzyme preparations contain a large number of, often undefined, enzymatic activities, although by and large they contain a β-1,3-glucanase activity necessary to degrade the rigid glucan layer in the cell wall, together with proteases, chitinases, and deoxyribonucleases (DNases). It has been suggested that the protease activity may be a necessary component of the lytic activity of these enzyme preparations (Villanueva *et al.,* 1973; Scott and Schekman, 1980).

A number of factors affect the efficiency of protoplast formation, some of which are controllable. Perhaps the most frustrating problem is the resistance of a number of laboratory strains to most protoplasting enzymes even though closely related strains are not. While this implicates some genetic component in susceptibility to protoplasting, the physiological state of the cell can also profoundly affect susceptibility to these enzymes. Particularly critical is the growth phase from which one harvests the cells; exponentially growing cells are highly susceptible, but as cells enter stationary phase they become relatively resistant (Eddy and

Table I. Commercially Available Enzymes Used for Preparing Protoplasts of *S. cerevisiae*

Enzyme	Source	Supplier
Novozym 234	*Trichoderma* spp.	Novo Enzyme Products Ltd.
Lyticase	*Arthrobacter luteus*	Sigma
Zymolyase 20T[a] Zymolyase 100T[a]	*Arthrobacter luteus*	ICN Biomedicals Ltd.
Glusulase	*Helix pomatia*	New England Nuclear (E.I. du Pont de Nemours and Co.)
Protoplast-forming enzyme	*Helix pomatia*	Boehringer Mannheim
Helicase	*Helix pomatia*	L'Industrie Biologique Francaise

[a]20T = 20,000 units/g; 100T = 100,000 units/g.

Williamson, 1957), probably as a consequence of changes in cell wall structure. Some enzyme preparations do, however, show an ability to attack the cell walls of stationary phase cells—zymolyase and lyticase are particularly effective in this respect (Kitamura and Yamamoto, 1972).

Yeast cells can be rendered more susceptible to protoplasting enzymes by the addition of a mercapto compound such as dithiothreitol, β-mercaptoethanol, or thioglycollate (Sommer and Lewis, 1971), which presumably destroys disulfide bonds in cell wall mannoproteins making the glucan layer more accessible to the β-1,3-glucanase activity. Such treatment is particularly important for preparing protoplasts from stationary phase cells (Kuo and Yamamoto, 1975) and for preparing protoplasts in the cold.

Prior to lysis of the protoplasts, it is generally necessary to remove all traces of the lytic enzyme since many of the enzyme preparations are contaminated with nucleases and proteases. This can be achieved by multiple washes with an isotonic buffer, based on either sorbitol or mannitol.

Effective disruption of protoplasts can be achieved by mechanical means, e.g., Dounce homogenizer, glass beads, cycles of freeze–thaw, or by lysis with a nonisotonic buffer containing a detergent such as SDS or Briij. The efficacy of the latter procedures depends on the degree of protoplasting achieved, since nonprotoplasted cells remain largely intact in the presence of low concentrations of detergents.

The strategies employed in the purification of specific enzymes and/or structural proteins from the yeast lysate are generally those applied for other cells lysates. The reader is referred to a number of excellent texts on the general subject of protein purification, e.g., Scopes (1982).

3. RADIOACTIVE LABELING OF MACROMOLECULES

The ability to radioactively label macromolecules to a high specific activity is a key biochemical tool not only for studying the rate of synthesis of a particular macromolecular species, but also as an accurate qualitative and quantitative estimate of different species within a particular group of macromolecules. The addition of radiolabeled precursors to growing cells, for brief periods (i.e., pulse labeling) or for longer periods (often many generations), is the most effective means of achieving this for studies on RNA, DNA, and protein synthesis. Such studies by and large use whole cells, but spheroplasts, incubated in an osmotically stabilized medium, can incorporate radiolabeled precursors into RNA, DNA, and protein for several hours, often at a rate similar to that observed for whole cells in an identical osmotically stabilized medium (Hutchison and Hartwell, 1967).

3.1. RNA

RNA, which represents greater than 98% of the total nucleic acids in *S. cerevisiae,* can be labeled *in vivo* using either [^{3}H]- or [^{14}C]-purines or pyrimidines. Alternatively, [^{32}P]orthophosphate can be used (Rubin, 1975), but in this case DNA will also be labeled (see Section 3.2). For routine analysis of RNA synthesis, pulse labeling with 2–5 μCi/ml [^{3}H]-adenine or [^{3}H]-uracil has proven most effective and labeling can be achieved in a rich medium such as YEPD (Yeast Extract, Peptone, Dextrose) (Sogin *et al.,* 1972), although much higher specific activity labeling can be achieved in a minimal medium using a mutant defective in adenine or uracil biosynthesis, respectively. As in labeling all macromolecules, there are serious problems in labeling RNA with precursors because of the time needed for equilibration of internal and external precursor pool sizes. A further complication is that the rate of uptake of precursors can be drastically altered by changes in cell physiology brought about, for example, by amino acid starvation.

A number of yeast RNA species are methylated, e.g., ribosomal RNAs and mRNAs at their 5′ methylated cap ($m^{7}G$), and these can be labeled *in vivo* with [methyl-^{3}H]-methionine provided the labeling medium is supplemented with sufficient guanine and adenine (approximately 2 mM each) to prevent labeling of the purine ring via the one-carbon pool (Retel *et al.,* 1969; Sriptai *et al.,* 1976).

To label RNA to a high specific activity with [^{32}P]orthophosphate requires a low-phosphate medium. Such a medium can be prepared from a rich medium (such as YEPD) by selective precipitation of inorganic phosphate using $MgSO_4$ (Rubin, 1975), and RNA with a specific activity in excess of 10^{6} cpm/μg can be obtained using [^{32}P]orthophosphate added at 0.1–0.5 mCi/ml of growing cells.

3.2. DNA

To specifically label DNA in yeast using precursors is very difficult because, unlike *Escherichia coli,* yeast cannot efficiently incorporate either thymine or thymidine into DNA. The reason for this is that *S. cerevisiae,* like many fungi, does not possess the enzyme responsible for phosphorylating thymidine monophosphate (TMP), i.e., thymidine kinase (Grivell and Jackson, 1968), thus preventing *S. cerevisiae* from utilizing either thymine or thymidine as a precursor for DNA synthesis.

While certain strains can incorporate small amounts of exogenous dTMP into DNA, the efficiency of incorporation is very low. To achieve efficient incorporation requires the use of mutants (called *tup* mutants) that can efficiently incorporate dTMP into DNA, and these can be readily selected for in most strains after mutagenesis and using inhibitors of endogenous folate synthesis such as aminopterin and *p*-aminobenzoic acid (Wickner, 1975). While the use of *tup* mutants permits the routine labeling of DNA in physiological studies, there are still a number of problems: for example, mitochondrial DNA seems to be preferentially labeled (Cryer *et al.*, 1973), although this problem can, of course, be eliminated by use of a mitochrondria-less (*petite*) mutant.

3.3. Proteins

To label yeast proteins to a high specific activity *in vivo* requires that the radioactive precursor (i.e., an amino acid) is rapidly incorporated both into new polypeptide chains and into the majority of protein species. The choice of amino acid as the precursor is critical in this respect since the intracellular pools of amino acids can differ in molar concentrations by as much as 250-fold; e.g., in nitrogen-rich medium the intracellular concentration of serine was reported to be about 5000 μmoles/10 g cells, whereas cysteine was only present at around 20 μmoles/10 g cells (Chan and Cossins, 1976). Clearly, the larger the intracellular pool, the longer it takes to reach isotopic equilibrium; for example, for the relatively abundant amino acid lysine it takes 70–80 min to reach isotopic equilibrium, whereas methionine, present at relatively low intracellular concentrations, takes less than 30 sec (Gross and Pogo, 1974). Care must also be taken in the choice of medium since intracellular amino acid levels can vary considerably depending on the media composition (Chan and Cossins, 1976). An appropriate labeling medium is a defined medium, based on yeast nitrogen base and buffered to pH 5.2 with a sodium succinate–sodium hydroxide buffer. To increase the efficiency of labeling, the cold amino acid precursor can be reduced or eliminated from such a medium, or an appropriate mutant unable to synthesize *de novo* the supplied radiolabeled amino acid can be used.

For routine monitoring of the rate of protein synthesis *in vivo*, pulse labeling with a [^{3}H]-amino acid is appropriate using, for example, [^{3}H]-leucine or [^{3}H]-phenylalanine, both of which have low endogenous pool sizes. To rapidly terminate the incorporation of the radioactive precursor, the simultaneous addition of the potent protein synthesis inhibitor cycloheximide (to 50 μg/ml) and crushed ice to the culture has proven very effective in the authors' hands. To label proteins for one-dimensional and two-dimensional polyacrylamide gel analysis, methionine is widely used because of its availability as [^{35}S]-methionine. This fact, coupled with the rapidity of isotopic equilibration for methionine, means that a 10-min pulse of a 25-ml culture with 15–25 μCi of [^{35}S]-methionine (specific activity > 1000 Ci/mmole) can generate sufficient sample for several gel analyses. Addition of histidine or tyrosine to the labeling medium can significantly enhance methonine incorporation, although the reason for this is unclear (Miller *et al.*, 1979).

The danger in using methionine in this context is that methionine is used comparatively rarely in yeast proteins and the *N*-terminal methionine is generally cleaved by an amino peptidase. An alternative option is to use a high-specific-activity [^{14}C]-amino acid mixture to generate a polypeptide electrophoretic pattern that is more quantitative and a truer reflection of the total number and relative abundance of proteins synthesized by yeast under various physiological conditions. An electrophoretic comparison of total yeast proteins labeled under identical conditions with either [^{14}C]-amino acid mixture or with [^{35}S]-methionine is shown in Fig. 1, and a number of differences can be noted.

If one wants to specifically label only mitochondrial proteins, i.e., proteins synthesized on mitochondrial ribosomes, this can be achieved by pulse-labeling cells for 1 hr in the presence of an antibiotic such as cycloheximide that specifically inhibits protein synthesis on cytoplasmic ribosomes, but not mitochondrial ribosomes (Douglas *et al.*, 1979). Labeling can be achieved with [^{35}S]-methionine, although a higher specific activity labeling with $^{35}SO_4^{2-}$ requires that the strain is not a methionine auxotroph and Mg^{2+} and NH_4^+ components of the growth medium are in the form of chloride salts.

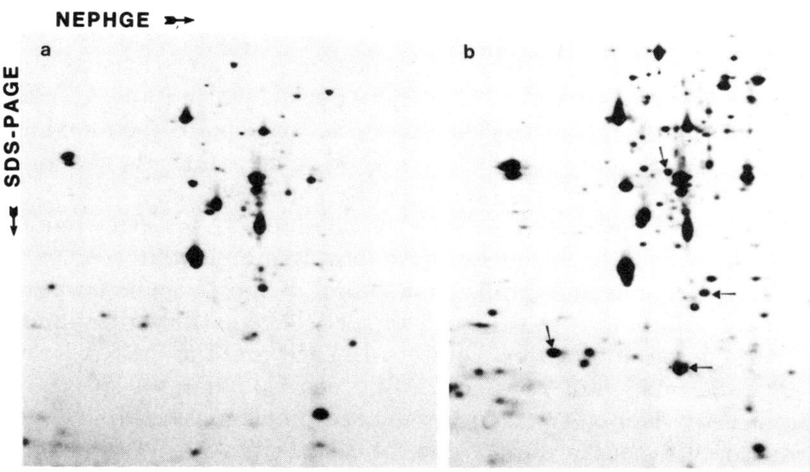

Figure 1. Two-dimensional gel electrophoresis of yeast proteins synthesized *in vivo.* (a) Total soluble proteins labeled with [^{35}S]-methionine. (b) Total soluble proteins labeled with [^{14}C]-amino acid mixture. The first dimension was by nonequilibrium pH gradient gel electrophoresis (NEPHGE), the second dimension by SDS-polyacrylamide gel electrophoresis (SDS-PAGE). Shown are autoradiographs exposed for 2 days (a) and 14 days (b). The arrows in b indicate proteins not detectable by [^{35}S]-methionine labeling. (Photographs courtesy of Dr. Ian Fitch, University of Kent.)

4. RNA PREPARATION

4.1. Differential Extraction Techniques

It is possible to select for a particular class of RNA molecule by use of the different extraction techniques described in this chapter. However, it is relatively simple to separate the different classes of yeast RNA from a total cell extract by using CF11 cellulose chromatography (Franklin, 1966). The CF11 cellulose (Whatman) column should be equilibrated with STE buffer (0.1 M NaC1, 1 mM EDTA, 0.05 M Tris, pH 6.85) containing 1% β-mercaptoethanol and then washed with STE containing 35% (vol./vol.) ethanol. The RNA sample is resuspended in 0.1 column volume of STE plus 35% ethanol and loaded onto the column. The different RNA species may then be eluted in a stepwise manner: (1) STE plus 35% ethanol—tRNA; (2) STE plus 15% ethanol—rRNA; (3) STE without ethanol—dsRNA. The progress of the stepwise process can be monitored by recording the OD_{260} of the eluate.

4.2. Messenger RNA

Messenger RNA (mRNA) represents between 1 and 5% of total cell RNA in *S. cerevisiae* and comprises a complex heterogeneous collection of single-stranded RNAs with sizes generally in the range 500–5000 nucleotides. Like most eukaryotic mRNAs, *S. cerevisiae* mRNAs show a certain degree of posttranscriptional modification; a methylated 5′ terminus (or cap; Sripati *et al.*, 1976) and a polyadenylate [poly(A)] extension of approximately 50 nucleotides at the 3′ terminus (McLaughlin *et al.*, 1973) are the most obvious. Few yeast cellular mRNAs contain introns, which is in contrast to higher eukaryotic mRNAs.

In studies with *S. cerevisiae* the isolation of mRNA is usually required for one of four reasons: (1) for *in vitro* translation studies (see Section 6.2); (2) for Northern hybridization studies of specific gene transcripts (3) for mapping the 5′ and 3′ ends of specific gene transcripts; and (4) for the generation of cDNA libraries. In each case the mRNA isolated needs to be intact, but not necessarily free of other RNA species. For example, translation of yeast mRNAs in a homologous cell-free lysate is much more efficient in the presence of excess rRNA or tRNA (Gasior *et al.*, 1979a; Tuite and Plesset, 1986), and the sensitivity of Northern hydridization methods allows for the detection of even minor mRNAs in total cellular RNA.

mRNAs are highly susceptible to degradation by endogenous ribonucleases, and in *S. cerevisiae* they have half-lives ranging from 3.5 to 70 min, with an average of 22 min (Chia and McLaughlin, 1979; Koch and Friesen, 1979). Isolation of these RNAs therefore presents the experimenter with a range of problems, not least being the inactivation of ribonucleases. Generally, ribonuclease degradation can be avoided by "standard" precautions: use of baked glassware, use of diethylpyrocarbonate-treated buffers, and inclusion of ribonuclease inhibitors. In addition, yeast strains deficient in one or more ribonuclease activities are available; e.g., the $RNase_3^-$ mutant (Littlewood *et al.*, 1971) is deficient in a potassium-activated ribonuclease found associated with 40S ribosomal subunits in wild-type strains (Baan *et al.*, 1981).

A range of extraction protocols can be used for the isolation of intact mRNA. These generally include extraction of either glass bead-disrupted whole cells or spheroplasts (e.g., Sripati and Warner, 1978) with organic solvents such as phenol or phenol:chloroform:isoamyl alcohol mixtures. Total polysomal RNA, extracted from pelleted polysomes, has also proven to be an effective source of translatable mRNA for *in vitro* studies (Gasior *et al.*, 1979a). More recently, a chaotropic disruption method, based on the ability of high concentrations of guanidinium thiocyanate and other guanidinium salts to dissociate ribonucleoprotein complexes and denature proteins, has been employed in conjunction with glass bead disruption to obtain intact mRNA (Feinberg and McLaughlin, 1988). This method has proven of value in obtaining intact mRNAs from mammalian tissues with high ribonuclease content (Chirgwin *et al.*, 1979) and may be the method of choice for cDNA work, and for 5′ and 3′ end mapping studies.

Should pure mRNA be required, it can be isolated from total RNA preparations using the standard affinity chromatography method that exploits the ability of the 3′ poly(A) tail of mRNA to bind to oligo (dT) cellulose or poly(U) sepharose at high salt concentrations (Holland *et al.*, 1977; Sripati and Warner, 1978). However, because the length of the poly(A) tail in *S. cerevisae* mRNAs is somewhat shorter than found in higher eukaryotic mRNAs (50-as opposed to 200-adenylic acid residues), the "flowthrough" RNA has to be recycled several times through the affinity column to ensure maximal yields. The ability to isolate mRNA from *S. cerevisiae* in this way becomes increasingly difficult as batch-grown cells enter stationary phase since this is accompanied by a shortening of the poly(A) tail (Sogin and Saunders, 1980).

4.3. Transfer RNA

The low molecular weight of transfer RNA (tRNA) species greatly simplifies their purification from yeast. Phenol extraction of whole yeast cells is sufficient to release tRNA (Monier *et al.*, 1960) in a relatively pure form. Figure 2 shows absorbance scans of both 2.6% and 12% polyacrylamide gels on which nucleic extracts from yeast have been run and demonstrates that tRNA is the only low-molecular-weight species ex-

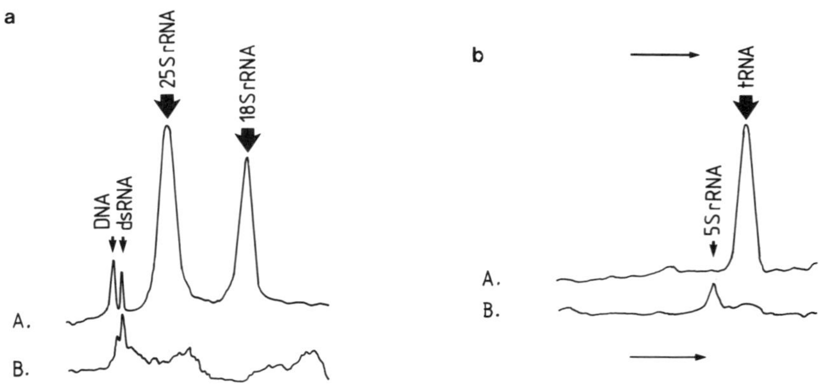

Figure 2. Electrophoretic analysis of different yeast RNA preparations. Shown are absorbance scans (A_{260} nm) of ethidium bromide stained gels; (a) RNA separated on a 12% polyacrylamide gel; (b) RNA separated on a 2.6% polyacrylamide gel. Sample A was prepared by phenol extraction of a total cell lysate; sample B was prepared by phenol extraction of whole cells. (Adapted from Clare, 1982.)

tracted from whole cells and that subsequent breakage of the cells releases no more tRNA. The whole cell extraction procedure may therefore be used for quantitative recovery of tRNA from yeast. The only RNA species that contaminates such preparations is double-stranded RNA (dsRNA; see Section 4.5) and this accounts for less than 2.5% of the total nucleic acid content of the extract. Such small amounts of dsRNA do not appear to interfere with the use of tRNA in *in vitro* translation systems or the determination of tRNA charging levels.

Determination of the proportion of molecules of any tRNA species that are charged with their cognate amino acid may be performed on extracts prepared in the manner described earlier. The principle of the charging assay is outlined in Fig. 3. The tRNA extract, containing both charged and uncharged species, is divided into two fractions. One fraction is chemically deacylated by mild alkaline hydrolysis (Yang and Novelli, 1971) while the other is treated with periodate. The oxidizing agent reacts with the ribosyl hydroxyl groups of the terminal adenosine residues of uncharged tRNA such that they can no longer esterify amino acids (Schmidt, 1968). The oxidized preparation is then subjected to mild alkaline hydrolysis to remove amino acids from the charged species. Both fractions are then used in separate charging reactions. Incorporation into the first fraction gives a measure of the total amount of tRNA (charged plus uncharged) of a given species present in the extract. Incorporation into the second fraction measures the charged species alone. The ratio of counts incorporated thus provides the proportion of charged tRNA molecules in the extract.

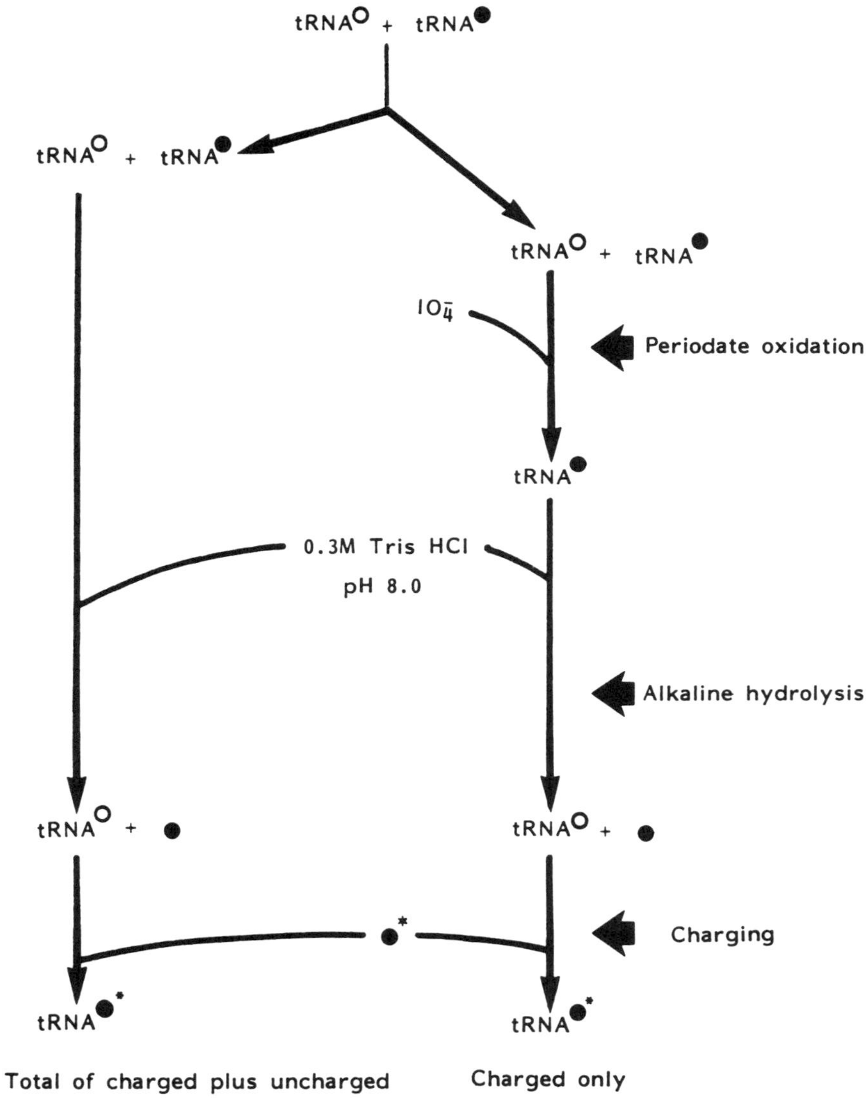

Figure 3. Protocol used for determination of *in vivo* tRNA charging levels. tRNA•, charged tRNA; tRNA°, uncharged tRNA; ○, amino acid; ●* radiolabeled amino acid.

Charging reactions (Hampel *et al.,* 1979) may be conveniently performed in drops of the appropriate cocktail on the surface of a stretched sheet of Saran Wrap ("Cling Film"). The aminoacyl–tRNA synthetase preparation which catalyzes the reaction may be prepared by a method involving differential centrifugation, the precipitation of nucleic acids with plyethylenediamine (Polymin P), and two ammonium sulphate precipitations followed by ion exchange chromatography and molecular sieve chromatography (Hartwell and McLaughlin, 1968; Von der Haar, 1979; Faulhammer and Cramer, 1977). However, a rapid purification method has been devised (Clare, 1982; Clare and Oliver, 1982) involving two differential centrifugation steps which on their own are sufficient to remove endogenous tRNAs. The final dialysis step is important since, in the absence of ammonium sulfate precipitation, it is the only step that frees the synthetase preparation from endogenous amino acids. Synthetase preparations may be stored in 50% glycerol at −80°C with little or no loss of activity.

4.4. Ribosomal RNA

Yeast, as a fast-growing microorganism, is an abundant source of ribosomal RNA (rRNA), which may readily be isolated from broken cells using standard phenol extraction methods (e.g., Kirby, 1965). While the large 35S rRNA precursor molecule can be isolated from disrupted cells (Van den Bos and Planta, 1971), it is probably better prepared from lysed spheroplasts (Udem and Warner, 1972). Of the two small molecules of the four mature species, 5.8S rRNA is tightly hydrogen-bonded to the 5′ end of the 25S species. It must therefore be released by either heat or formamide denaturation before it can be identified. A convenient method of releasing the 5.8S species is simply to freeze and thaw the total RNA extract two or three times. Rubin (1975) has demonstrated that both small mature rRNA species, 5.8S and 5S, may be prepared, in quantitative amounts, from intact yeast cells by treatment with 20 mM EDTA, 0.1 M Na acetate, and 1% SDS (pH 5) at 33°C. The use of phenol extraction to release low-molecular-weight RNAs from intact yeast cells was dealt with in Section 4.3.

4.5. Double-Stranded RNA

Almost all laboratory strains and many wild-type (nonlaboratory) strains of *S. cerevisiae* contain one or more species of dsRNA encapsidated within viruslike particles designated ScV particles (Mitchell and Bevan, 1987, a review). The major species of dsRNA, designated L, is approximately 4.7 kb in length, while a second group of smaller dsRNAs, designated M, are associated with the killer phenotype originally described by Bevan and co-workers. M dsRNAs vary in size between 1.9 and 2.7 kb.

Two general strategies are available for purifying dsRNAs from *S. cerevisiae*. In the first, ScV particles are partially purified on a CsCl gradient followed by final purification by sucrose gradient centrifugation (Oliver *et al.*, 1977). The dsRNA is subsequently extracted from the purified particles using simple phenol extraction methods (Kandel, 1979). In the second strategy, the dsRNA species are differentially precipitated from total RNA samples using LiCl (Diaz-Ruiz and Kaper, 1978). The latter method involves first precipitating rRNA and mRNA with 2 M LiCl and then precipitating the dsRNA from the 2 M LiCl supernatant with 4 M LiCl, leaving small single-stranded RNAs (particularly tRNA) and DNA in the supernatant (Fig. 4). This strategy works for both L and M dsRNAs and is relatively quick and simple. Should further purification of the dsRNA be necessary, it can be achieved using CF-11 chromatography (Franklin, 1966) or by preparative electrophoresis.

One useful property of dsRNA is that it is very resistant to ribonucleases; for example, it is insensitive to ribonuclease A at high, but not low, ionic strength (Wickner and Leibowitz, 1976). This property can also serve as a simple confirmation that the isolated material is indeed dsRNA.

5. DNA PREPARATION

5.1. Chromosomal DNA

An ability to isolate intact, high-molecular-weight chromosomal DNA is an obvious prerequisite if one is to study the yeast genome by classical or recombinant DNA techniques. Both lengthy large-scale and more rapid miniscale preparative techniques have been developed for *S. cerevisiae* that generate chromosomal DNA of varying yields and degrees of purity, but which is suitable for a range of analytical purposes, including restriction enzyme digestion. To avoid contamination with mitochondrial DNA, [*rho*°] petite mutants that totally lack that species may be employed.

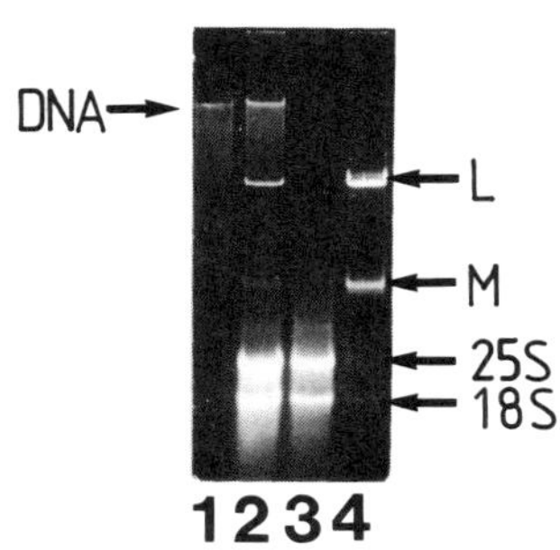

Figure 4. Fractionation of yeast RNAs using differential precipitation with lithium chloride. Lane 1, 4 M LiCl supernatant; Lane 2, total RNA; Lane 3, 2 M LiCl precipitate; Lane 4, 2 M LiCl supernatant. The strain used contained both L and M double-stranded RNAs. The position of the 18S and 25S ribosomal RNAs is also indicated.

The method first described by Cryer *et al.* (1975) has been widely adopted as the method of choice, and yields up to 0.25 mg DNA/g cells can be achieved. The initial step invariably involves removal of the cell wall by treating cells with an appropriate lytic enzyme (see Section 2.2). While diethyl oxydiformate (i.e. diethyl pyrocarbonate) can be used to inactivate any nucleases associated with the enzyme preparation (Davis *et al.*, 1980), this compound unfortunately also chemically reacts with the DNA, often rendering it useless for cloning and transformation experiments. To ensure that the chromosomal DNA remains intact, the spheroplasts are gently lysed by SDS in the presence of proteinase K, the latter ensuring that partial proteolysis of the lysate occurs. Deproteinization and extraction of the lipids from the lysate is achieved by mixing with an appropriate organic solvent (usually chloroform:isoamylalcohol in a ratio of 24:1), and the high-molecular-weight DNA is either "spooled" or pelleted by centrifugation from the resulting aqueous phase following ethanol precipitation. Contaminating RNA and polysaccharide often remains at this stage and can be removed by treatment with DNase-free RNase A. Deproteinization and separation of the DNA from the digestion products can be achieved by selective isopropanol precipitation. In addition, either before or (more generally) after the RNase/isopropanol step, further purification can be achieved by employing an isopycnic cesium chloride gradient, using either ethidium bromide (Radloff *et al.*, 1967) or 4′,6-diamidino-2-phenylindole (DAPI) (Williamson and Fennell, 1975) to enhance the separation of chromosomal (buoyant density 1.699 g/cm^3) and mitochrondrial DNA (buoyant density 1.683 g/cm^3). Additionally the ribosomal DNA-containing heavy satellite DNA band (buoyant density 1.704 g/cm^3) can usually also be resolved in these gradients.

A number of more rapid miniprocedures employing small cultures (5–50 ml) have been devised to accommodate the need to process more than one strain at a time, for example, when studying integrated plasmids (see Chapter 5). Some are adaptations of the large-scale method described earlier (e.g., Davis *et al.*, 1980; Winston *et al.*, 1986) while others use harsh cell disruption techniques to avoid the need to prepare spheroplasts (e.g., Specht *et al.*, 1982; Weeks *et al.*, 1986), although in the latter cases a CsCl gradient is almost certainly required to produce DNA of sufficient purify for restriction enzyme digestion. A more rapid and efficient protocol has recently been devised that involves lysing spheroplasts at 65°C in the presence of the chaotropic agent guanidine hydrochloride (GuHCl) followed by deproteination with proteinase K and pheno-chloroform extraction (Holm *et al.*, 1986). The yields with this protocol can reach 75% of the theoretical yield, a value reported to be much higher than with other miniprocedures (Holm *et al.*, 1986). We have also found the latter method to be the most reproducible mini-procedure for a wide variety of laboratory strains (M. F. Tuite *et al.*, unpublished). A slightly different approach is taken when preparing chromosomal DNA for analysis by pulsed-field gradient gel electrophoresis (Schwartz and Cantor, 1984) or orthogonal-field-alternation gel electrophoresis (Carle and Olson, 1984) where it is critical that chromosomes are isolated intact (see Section 7).

5.2. Mitochondrial DNA

Mitochondrial DNA (mtDNA) represents 10–15% of the total DNA of *S. cerevisiae,* depending on carbon source and growth rate, and a number of protocols have been devised to isolate this fraction in high yield and free of chromosomal DNA. These protocols do *not* employ isolated mitochondria since this is a difficult task and isolated mitochondria are invariably damaged and contaminated with chromosomal DNA. The strategy is to isolate total DNA from a crude mitochondrial lysate and rely on the difference in the A + T base content between nuclear and mitochondrial DNA (61% and 83%, respectively). This difference generates a buoyant density difference that can be exaggerated by the use of the fluorescent, DNA-binding agents DAPI (Williamson and Fennell, 1975) and bisbenzimide (Hoechst 33258; Hudspeth *et al.,* 1980). Advantages to using bisbenzimide include greater separation between chromosomal and mtDNA and the fact that there is no need to remove the dye from DNA preparations prior to restriction enzyme digestion and transformation (Hudspeth *et al.,* 1980). Mitochondrial DNA prepared by this strategy can be in excess of 99.5% pure.

5.3. Plasmid DNA

It is more practicable to propagate and manipulate yeast recombinant plasmids in *E. coli* since the majority of such plasmids are constructed to allow for both selection and replication in this bacterium (see Chapter 5), and the methodology is both simpler and more reproducible for producing high yields of pure covalently closed circular DNA (cccDNA). Techniques do exist, however, for extracting native cccDNA molecules from yeast, the most successful of which is that described by Devenish and Newlon (1982). In this case spheroplasts are first lysed in alkaline conditions (pH 12.45) to denature linear DNA (but not cccDNA) and then subjected to a high salt-phenol extraction to remove single-stranded DNA (ssDNA), the cccDNA being recovered by ethanol precipitation. This technique, originally devised as a means of purifying a circular derivative of chromosome III (the "ring chromosome" Strathern *et al.,* 1979), can also be used for general plasmid isolation from yeast (Devenish and Newlon, 1982).

If one is interested in obtaining plasmid DNA from yeast without destroying its chromatin structure, Shalitin and Vishlizky (1984) have described a technique that employs nondetergent lysis of purified nuclei coupled with a metrizamide [2-(3-acetamido-5-*N*-methylacetamido-2,4,6-triiodo-benzamido)-2-deoxy-D-glucose] gradient. This technique allows for the purification of intact 2-μm "minichromosomes," free of cellular chromatin and ribosomal protein contamination, a problem often associated with techniques based on separation in sucrose gradients (Shalitin *et al.*, 1983).

6. *IN VITRO* SYSTEMS

6.1. *In Vitro* Transcription Systems

Test tube systems that can accurately and efficiently transcribe cloned yeast genes are an essential component of the armory of techniques required for a complete understanding of gene expression. The subject has a rather checkered history. Initial enthusiasm following the purification of the three RNA polymerase complexes from yeast was tempered by the finding that such pure preparations usually failed to transcribe accurately *in vitro*. A retreat was made into the use of isolated nuclei and of whole cell lysates, which were often combined with techniques for selective isolation of transcripts that had been initiated in the cell-free system. Recently, identification of specific proteins that act as transcription activating factors has led to a more complete understanding of the components required for *in vitro* transcription and permitted the design of more defined systems.

Sawadago *et al.* (1981) demonstrated that purified RNA polymerase I (or A) together with the transcription factor "P_{37}" was able to selectively transcribe a cloned rRNA gene in a plasmid vector. However, it was found that the *in vitro* transcript was initiated at least 200 bp upstream of the start site which had been identified *in vivo*. Swanson and Holland (1983) obtained similar results with a cell-free system in which a yeast lysate was subjected to $(NH_4)_2SO_4$ fractionation and ultracentrifugation. A correspondence between the *in vivo* and *in vitro* sites of initiation of synthesis of the ^{35}S pre-rRNA was only obtained using less defined systems. Since a considerable amount of "run-off" synthesis occurs in these crude systems, it is essential to have some means of selecting, for further analysis, only those transcripts which have been initiated *in vitro*. This is achieved by including γ-SH derivatives of ribonucleotide triphosphates in the reaction mix and purifying molecules that carry a 5′ thiol group (and hence were initiated *in vitro*) by Hg-agarose chromatography. This approach was first exploited by Lohr and Ide (1983) in a system based on isolated nuclei. Both RNA polymerase I (or A) and III (or C) activities may be detected in these nuclear preparations, and these enzymes accurately initiate the synthesis of ^{35}S pre-rRNA and of 5S rRNA, respectively. Klootwijk *et al.* (1984) employed the 5′-thiol selection system to identify transcripts synthesized by highly concentrated whole cell lysates. These workers included two ribonuclease inhibitors, vanadyl ribonucleotide complex and RNAsin, in their reaction mixes and were able to demonstrate the activity of all three yeast RNA polymerase complexes. Accurate initiation of the synthesis of the ^{35}S rRNA precursor molecule from the start site defined *in vivo* was recorded.

The development of cell-free systems that faithfully transcribe protein-encoding genes via the activity of RNA polymerase II (or B) has been difficult and efficient methods have been developed only recently. Initial studies using the purified polymerase (Carnevali *et al.*, 1982) demonstrated that initiation occurred *in vitro*, as could be recognized by the formation of "ternary transcription complexes," but that elongation was defective. A fully functional system has now been developed by Lue and Kornberg (1987). The method involves isolation of yeast nuclei and their subsequent lysis, the lysate being fractionated using two $(NH_4)_2SO_4$ precipitation steps. Three different protease inhibitors [phenylmethyl sulfonyl fluoride (PMSF), pepstatin, and leupeptin] are included in the buffers for both spheroplast and nuclear lysis. However, the most important innovation appears to be the incorporation of Ficoll into the spheroplast lysis buffer. This polymer is essential as it permits the nuclei to retain protein transcription factors that would otherwise be lost in the supernatant when the organelles are collected by centrifugation.

Nuclear extracts formed the basis of the first RNA polymerase III (or C) *in vitro* transcription system developed by Klekamp and Weil (1982). However, they found that the yields were too low, and the preparation method too tedious, for routine use. These workers then developed a system based on a 100,000-g supernatant (S100) of a whole cell lysate prepared by glass bead breakage. The S100 is subsequently fractionated using $(NH_4)_2SO_4$ precipitation, residual nucleic acids being removed by ion exchange chromatography. This system has formed the basis for most subsequent protocols for the cell-free synthesis of transfer RNA. Recent developments have come from Geiduschek's laboratory (Kassovetis *et al.*, 1989). These workers have developed systems in which either multiple or single rounds of transcription may be studied. In the former case, a cell extract of the Klekamp and Weil (1982) type is fractionated using DEAE-sephadex chromatography. Such extracts contain not only RNA polymerase III (or C), but also the transcription factor

TFIIIC and will carry out multiple rounds of transcription on a plasmid template in the presence of all four ribonucleotide triphosphates. Alternatively purified RNA polymerase III, prepared using the heparin-agarose methodology of Hammond and Holland (1983), may be used together with affinity-purified TFIIIC. This is the most defined yeast system yet developed for accurate *in vitro* transcription. In the single-round protocol, the initial incubation is performed in the absence of GTP to obtain stable ternary transcription complexes. GTP is then added back to the reaction mix in the presence of heparin. This permits elongation of the nascent RNA chain but prevents any reinitiation or transcriptional processing.

6.2. *In Vitro* Translation Systems

The ability to translate natural mRNAs in a cell-free lysate requires that all three functional steps of the process of translation are occurring *in vitro,* namely, initiation, elongation, and termination. On the other hand, artificial templates such as poly(U) require only a functional elongation system since the initiation upon this template is "forced" by using supraoptimal levels of Mg^{2+} ions. Both kinds of system are powerful tools for studying mRNA-coding potential, fidelity of translation, and the basic mechanisms of protein synthesis.

A poly(U)-driven system was developed for *S. cerevisiae* by straightforward extrapolation of the techniques used successfully with other organisms (see Sissons, 1978), yet attempts at establishing an mRNA-dependent cell-free translation system were numerous, but singularly unsuccessful, until 1979, when Gasior *et al.* (1979a) finally reported the successful establishment of a cell-free system able to initiate translation upon natural mRNAs. Since this first report the system has been optimized for a wide range of mRNAs and has been used in a number of studies (reviewed in Tuite and Plesset, 1986). In addition, a number of modifications have been made, largely in the cell disruption protocols, that allow researchers to reproducibly generate highly active systems from a range of yeast strains.

6.2.1. Systems for Translating Synthetic Templates

Translation of a polynucleotide can be achieved in the absence of a natural initiation mechanism, but generally requires the presence of supraoptimal Mg^{2+} concentrations. Two types of systems have been developed for *S. cerevisiae.* The first of these is analogous to the *E. coli* S30 systems and simply consists of a crude S30 lysate, prepared from either disrupted whole cells or lysed spheroplasts (Sissons, 1978) and subjected to either dialysis or gel filtration chromatography to lower the amount of endogenous amino acids. The second type of system comprises highly purified ribosomes and translation factors, particularly those required for elongation in the yeast, namely, elongation factors EF-1, EF-2, and EF-3. This highly fractionated system has provided important insights into the mechanism of elongation in yeast (e.g., the existence of a novel elongation factor, EF-3; Skogerson and Wakatama, 1976) and the basis of ribosome-mediated antibiotic resistance (e.g., resistance to the trichothecene antibiotic trichodermin is expressed in the 60S subunit; Schindler *et al.*, 1974). Nevertheless, it is far more difficult to prepare than the S30 system and more prone to failure because of the increased chance of damaging labile translation components during their lengthy purification.

The yeast S30 system, capable of efficiently elongating on synthetic templates such as poly(U) and poly(U,C), has been useful in studying the mechanism of action of inhibitors of translation (Vazquez and Jimenez, 1980), the fidelity of protein synthesis (e.g., Masurekar *et al.*, 1981), and the ionic requirements for elongation (Sissons, 1978), yet it is unable to initiate translation upon exogenously supplied natural mRNA species. It has been demonstrated, however, that the system can translate endogenous yeast mRNAs already bound to ribosomes prior to cell disruption (Gallis and Young, 1975).

6.2.2. Systems for Translating Natural mRNAs

A yeast cell-free system capable of carrying out all phases of protein synthesis—initiation, elongation, and termination—using exogenously supplied natural mRNAs was first described by Gasior *et al.* (1979a). This system basically consists of an S100 lysate from which polysomes (but not monosomes or ribosomal subunits) have been removed by differential ultracentrifugation and which has been subjected to gel filtration chromatography to remove low-molecular-weight components such as amino acids and nucleoside triphosphates. An unusual characteristic of the system is that it will only initiate translation at temperatures of 20°C or less, a characteristic unique to yeast. The lysate can be effectively rendered mRNA-dependent by treating with micrococcal nuclease (Pelham and Jackson, 1976). Detailed studies of the ionic requirements and cell disruption strategies (Tuite *et al.*, 1980; Hofbauer *et al.*, 1982; Tuite and Plesset, 1986) have allowed the system to be optimized for the efficient and faithful translation of both homologous mRNAs and a wide range of heterologous mRNAs, be they viral, plant, or mammalian in origin (Tuite and Plesset, 1986).

The Gasior system (Gasior *et al.*, 1979a) has already proven useful in studying the basic mechanism of translational initiation in yeast (Gasior *et al.*, 1979b), in analyzing yeast nonsense suppressors and their modifiers (Tuite *et al.,* 1981, 1983), in identifying the defects in conditional lethal protein synthesis mutants (review, Moldave and McLaughlin, 1988), and in examining the role of translational control in the yeast heat shock response (Plesset *et al.,* 1982).

Since the initial report by Gasior *et al.* (1979a), a number of other yeast cell-free systems have been reported that are also capable of translating exogenous mRNAs (Szczesna and Filipowicz, 1980; Hofbauer *et al.,* 1982; Kreutzfeldt and Lochmann, 1983; Chanda and Kung, 1983; Hussain and Leibowitz, 1986). While all of these systems show the same temperature requirement for initiation as reported by Gasior *et al.* (1979a), their preparation is generally less time-consuming, employing S30 rather than S100 lysates and using whole cells rather than spheroplasts to generate the lysate. A number of them suffer from having high endogenous mRNA activity, although the use of micrococcal nuclease (see above) can reduce this problem. One problem with the Gasior system has been that, because efficient spheroplast formation is an absolute necessity, and not all laboratory strains can be efficiently converted to spheroplasts, there is a lack of strain-to-strain reproducibility. Employing whole cell disruption methods such as glass bead lysis (Hofbauer *et al.,* 1982; Tuite and Plesset, 1986; Hussain and Leibowitz, 1986) or the French Press (Kreutzfeldt and Lochmann, 1983) allows one to overcome this problem, although it is found to be essential that a protease inhibitor such as PMSF be included during cell disruption to ensure the high levels of translational activity of the Gasior system (Chanda and Kung, 1983; Hussain and Leibowitz, 1986).

6.2.3. Systems that Posttranslationally Process Secretory Proteins

Cell-free translation systems have been widely used to study the first step in the secretion of proteins from eukaryotic cells, namely, translocation of the secretory polypeptide across the rough endoplasmic reticulum (RER) membrane. Microsomes are a source of translocation competent membranes. With the increasing use of yeast as a model organism for studying the secretory process, the need for a yeast cell-free system that is capable not only of translating natural mRNAs, but also of translocating a secretory protein across microsomal membranes, became important. Such a system has now been developed (Rothblatt and Meyer, 1986; Waters and Blobel, 1986) and this was achieved by including a microsomal membrane fraction from yeast that contains the activities necessary for translocation, glycosylation, and cleavage of the signal sequence from the precursors of such yeast secretory proteins as invertase and mating factor α. The microsome-containing membrane fraction is prepared from gently lysed spheroplasts.

Rothblatt and Meyer (1986) reported that active yeast microsome membranes failed to process the prepro-α-factor secretory protein in a heterologous cell-free translation system, namely, wheat germ, nor would pancreatic microsomes process prepro-α-factor in the yeast lysate, pointing to an inability of yeast translation complexes to interact properly with membranes from higher eukaryotes, and *vice versa*. The homologous yeast-processing system, however, will prove of great value in dissection of the yeast secretion pathway, which has already been so effectively defined genetically, and in studying the role of the signal recognition particle (SRP) and its receptor.

6.2.4. Systems that Translate Mitochondrial mRNAs

Utilization of a different genetic code by mitochondria, both in yeast and in higher eukaryotes, means that mitochondrial mRNAs are generally not translatable in cytoplasmically derived cell-free systems such as those described in Section 6.2.2. The major problem in yeast is that the UGA codon is recognized as specifying tryptophan in mitochondria. While supplementing a wheat germ cell-free system with yeast UGA suppressor tRNA allows the faithful translation of yeast mitochondrial mRNAs (de Ronde *et al.*, 1980), the ideal solution is to prepare an *in vitro* translation system composed entirely of yeast mitochondrial translation components. Two alternative approaches have been taken in the development of such systems. In the first, yeast mitochondria are purified, fractionated into the various necessary components, and then reconstituted (Pfisterer and Buetow, 1981). Such a reconstituted system has proven useful in the definition of the components, and to some extent the mechanism, of mitochondrial protein synthesis, but, like its cytoplasmic counterpart (see Section 6.2.2), the system is difficult and time-consuming to prepare. The second approach has been the development of an *in vitro* mitochondrial protein synthesis system which makes use of *intact* yeast mitochondria (e.g., McKee and Poyton, 1984). Such a system is able to synthesize authentic yeast mitochondrial translation products and can initiate new polypeptide chains (McKee *et al.*, 1984).

6.3. *In Vitro* DNA Synthesis

The development of a cell-free synthetic system is essential for elucidating the detailed biochemistry of macromolecular synthesis. DNA synthesis is no exception to this rule, as has been demonstrated by studies on bacteria and on animal viruses (see review by Kornberg, 1980). The first attempts to generate *in vitro* systems for yeast DNA synthesis, like those for bacteria, involved the use of permeabilized cells. Hereford and Hartwell (1971) permeabilized yeast using the nonionic detergent "Brij 58" and an incubation cocktail which included the four deoxynucleotide triphosphates, ATP and a regenerating system, β-mercaptoethanol, and Mg^{2+} ions. Using this system, they were able to show that DNA synthesis in permeabilized cells carrying the *cdc8* mutation was temperature-sensitive. Banks (1973) demonstrated that only mitochondrial DNA is synthesized in Brij-permeabilized yeast cells, although this has been contested by Kuo and Campbell (1982), who claimed that DNA-zero petite mutants were still active in such a system.

Permeabilized cell systems appear to offer little opportunity for experiments involving reconstruction of the DNA replication complex. Nevertheless, having demonstrated that thymidylate kinase was the *CDC8* gene product, Jong *et al.* (1984) were able to complement the *cdc8* temperature-sensitive lesion by adding purified enzyme to permeabilized cells. Thymidylate kinase may be a special case, however, since it is not clear whether the protein must interact directly with the replication complex or whether complementation can be effected by precursor molecules traveling in both directions across the permeabilized cell membranes. More complex reconstruction experiments demand a truly cell-free DNA synthesis system.

Three such systems are currently in use, developed in the laboratories of Jazwinski (Jazwinski and Edelman, 1982), Sugino (Kojo *et al.*, 1978), and Campbell (Celniker and Campbell, 1982), respectively. All three involve the preparation of protoplasts using lytic enzyme and their lysis in the presence (Jazwinski; Campbell) or absence (Sugino) of detergent. A high-speed supernatant is then prepared and fractionated by ammonium sulfate precipitation. The resuspended precipitate may then be used without further fractionation (Sugino) or subjected to gel filtration (Jazwinski) or ion exchange chromatography (Campbell). The incubation cocktail employed includes the four deoxyribonucleoside triphosphates, the four ribonucleotide triphosphates, glycerol, and dithiothreitol. The Jazwinski method includes EDTA and an ATP regenerating system, whereas both Campbell and Sugino omit these but include spermidine. The Sugino cocktail also used NAD^+. There have been various claims and counterclaims about the merits of these three systems and they have been critically evaluated by Jong and Scott (1985).

The incorporation of radiolabeled triphosphate in all three cases is dependent on exogenously added DNA and, it is claimed, is absolutely specific for DNA containing yeast origins of replication and capable of the *de novo* initiation of DNA synthesis. Jong and Scott (1985) claim that the Jazwinski system, at least as originally described (Jazwinski and Edelman, 1982), is not specific for yeast DNA. These workers have further suggested that, when recombinant plasmids prepared from bacteria are used as templates, only elongation rather than initiation is observed since oligonucleotide primers are copurified with the plasmid molecules (Jong and Scott, 1985). The alkaline denaturation of the plasmids, to remove these primers, followed by renaturation was said to prevent DNA synthesis in a modified Sugino system. Both Campbell (Celniker and Campbell, 1982) and Jazwinski (Jazwinski and Edelman, 1984; Jazwinski, 1987) have reported that alkali pretreatment of template molecules does not prevent DNA synthesis in their systems. This prepriming effect may account for the very high levels of DNA synthesis found in the Sugino system (Koko *et al.*, 1978).

The replication activity from such yeast extracts may be isolated as high-molecular-weight complex of ca. 2×10^6 Da (Jazwinski and Edelman, 1982). A number of enzyme activities that might be expected to be found in a "replisome" (Kornberg, 1980) are associated with this complex. These include DNA polymerase I, DNA ligase, DNA primase, and topoisomerase II (Jazwinski and Edelman, 1984). The large size of the replication complex means that its association with DNA may be visualized by electron microscopy. Protein "knobs" are found associated with portions of the 2-μm molecule which have been shown (Newlon *et al.*, 1981) to function as *in vivo* origins (Jazwinski and Edelman, 1982). Similarly, these knobs are also found associated with restriction fragments of YRp plasmids, which are known to contain a functional *ARS* element (Jazwinski *et al.*, 1983).

Plasmid DNA treated with any of the three types of extract has been found, by electron microscopy, to contain replication bubbles with appropriately positioned centers (Jazwinski and Edelman, 1979; Kojo *et al.*, 1978; Celniker and Campbell, 1982). In the case of the Campbell system such analysis demonstrated the existence of "bubbles within bubbles," indicating that reinitiation events had occurred (Celniker and Campbell, 1982). Reinitiation is prohibited in eukaryotic nuclear DNA *in vivo,* and it is not clear whether this phenomenon reflects an intrinsic defect of the cell-free system or whether it is an artifact produced by the inclusion of the chain-terminating analog cytosine-β-D-arabinoside-5′-triphosphate (ara CTP) in the reaction mix. Celniker and Campbell (1982) added this analog in order to enrich the system for replication complexes. Its use enabled them to perform labeling experiments to confirm the site of the replication origin within the plasmids used and also to demonstrate that replication was bidirectional. The latter was only obtained in extracts made from exponential phase cells; when stationary phase cells were used, then replication appeared to be unidirectional.

The combination of an effective cell-free DNA replication system and the many temperature-sensitive mutants of yeast available which have blocks in DNA synthesis is a powerful tool in the analysis of the biochemistry of eukaryotic DNA replication. However, doubts still remain as to the accuracy with which the currently employed cell-free system represents normal DNA replication. The availability of improved methods of analyzing the *in vivo* situation, such as two-dimensional gel electrophoresis (Brewer and Fangman, 1987; Huberman *et al.*, 1987), has tended to shift attention away from *in vitro* techniques, at least for the present.

6.4. Two-Dimensional Gel Electrophoresis in the Analysis of DNA Replication

The position, in an electrophoresis gel, of any DNA sequence for which a suitable restriction fragment or oligonucleotide probe exists may be accurately determined by the Southern blotting procedure (Southern, 1975). This opens the way to the identification of those sequences which act as origins of yeast chromosomal DNA replication. To achieve this it is necessary to be able to select or enrich for replication intermediates in DNA prepared from cultures of growing yeast cells and to utilize gel electrophoresis to resolve those intermediates from the bulk of the chromosomal DNA. Two methods, which both exploit two-dimensional agarose gel electrophoresis (Brewer and Fangman, 1987; Huberman *et al.*, 1987), have been developed to achieve this.

In the Brewer and Fangman (1987) approach, DNA preparations enriched for replication intermediates are obtained by using synchronized cultures. The synchrony technique employed involves holding cells in G1 by use of a combined α-factor/*cdc7* block (Fangman *et al.*, 1983) and then releasing the inhibition by proteolytically degrading the pheromone and shifting to the permissive temperature.

The two-dimensional agarose gel electrophoresis methods employed separate molecules in the first phase according to their mass and in the second phase according to their shape (Bell and Byers, 1983). The first dimension is run in 0.4% agarose at 1 V/cm whereas the second dimension is run in 1% agarose and 5 V/cm in the presence of ethidium bromide. Nonlinear DNA molecules, such as the "Y" and "eye" forms expected of replication intermediates, are known to migrate anomalously in agarose gels as compared to linear molecules of equal mass (Oppenheim, 1981). The increase in voltage and agarose concentration in the second dimension exaggerates these differences (Bell and Byers, 1983). Figure 5 diagrams possible replication intermediates and shows the way they are predicted to migrate in the two-dimensional gel system.

The approach of Huberman *et al.* (1987) differs from that of Brewer and Fangman (1987) in both the method of DNA preparation and the electrophoretic conditions employed. Synchronized cultures are not used, but DNA preparations from random cultures are enriched for replication intermediates using a biochemical method. Molecules containing single-stranded regions, which will include the replicating forms, are selectively adsorbed onto benzoylated naphthoylated DEAE-cellulose (BND-cellulose; Gamper *et al.*, 1985). The first electrophoresis dimension employs 0.5% agarose at 40–45 V under neutral conditions

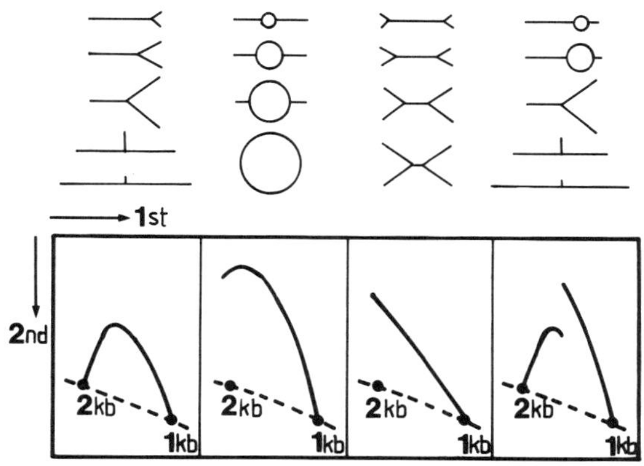

Figure 5. Two-dimensional gel electrophoresis in the analysis of DNA replication. (Upper panel) The types of replication intermediate that arise depending on the placement of the replication origin on an arbitrary 1-kb restriction fragment. (Lower panel) The predicted migration of the four types of replication intermediate in two-dimensional agarose gels (for details, see text). The dashed lines mark the locations of linear molecules of various sizes. (Adapted from Brewer and Fangman, 1987.)

in the presence of ethidium bromide, whereas the second dimension uses 1.2% agarose gels run at 12 or 17 V under alkaline conditions. The second, denaturing, phase separates simple linear DNA fragments into their two single strands. Molecules containing a replication fork, on the other hand, will produce four strands on denaturation: the two parental strands and two, smaller, nascent strands. The linear molecules and parental strands migrate as an arc in the two-dimensional system, with the smaller nascent strands running ahead of this major arc. Closed and open circular molecules also migrate ahead of the major arc, and this complicates the analysis of plasmid replication unless all molecules are linearized using an appropriate restriction cut. Single strand nicks in the DNA also complicate the patterns, leading to vertical streaking in the second dimension. In all, the data obtained from the Huberman *et al.* (1987) method is not as clear or as easily interpreted as that provided by the Brewer and Fangman (1987) technique.

7. PULSED FIELD GEL ELECTROPHORESIS OF YEAST CHROMOSOMES

Conventional agarose gel electrophoresis techniques fail to resolve DNA molecules much greater than 20 kb in size. While large molecules will enter the gel, different sized species do not migrate at significantly different rates (Fisher and Dingman, 1971) and so are not resolved from one another. Undigested DNA extracts from yeast or *E. coli* commonly produce a chromosomal smear near the top of the gel. A solution to this problem has been found in pulsed field gel electrophoresis, in which DNA molecules are subjected to an electric field which periodically changes its orientation. The time it takes molecules to realign to the new field direction is proportional to their size, and molecules up to 1 megabase (mb) in size can be resolved one from another. Fortunately for yeast researchers, the size of yeast chromosomes (240 kb–1.5 mb) falls within these limits and thus pulsed field techniques may be used to provide a karyotype of *S. cerevisiae* in a way never achieved in a convincing manner by either light or electron microscopy. Moreover, the technique of Southern blotting may be used to locate any cloned gene or other DNA fragment to a particular chromosome and so an alternative to conventional mapping procedures by meiotic segregation analysis (see Chapter 4) is provided. Furthermore, individual chromosomes may be isolated from these agarose gels for restriction analysis or the construction of gene banks.

A large number of different types of apparatus have been designed to carry out pulsed field gel electrophoresis and have been applied successfully to the resolution of yeast chromosomes. Unfortunately, our theoretical understanding of the technique is poor and so a large amount of empiricism is involved. A short discussion of the theory will be provided before describing the main types of apparatus and the methods of sample preparation.

7.1. Theoretical Background

All theories on pulsed gel electrophoresis assume that large DNA molecules move through the gel matrix by a process of reptation. This term, which literally means "snakelike," implies that the molecules move through the gel and on, with their long axis parallel to the electric field. Indeed, large circular molecules, such as circular derivatives of yeast chromosome III (Strathern *et al.,* 1979), fail to enter the gel. Small circular molecules, such as the 2-μm plasmid, show a quite different rate of migration relative to their size than the large linear chromosomes.

The pore size of a 1.5% agarose gel is much smaller than the coiled length of a DNA molecule greater than 30 kb in size (Lumpkin and Zimm, 1982; Serwer and Hayes, 1986). Thus, when the orientation of the electric field through the gel is changed, the molecule is lying across many of the gel pores, which will allow it to move in the new direction of electrophoresis. The molecule must therefore reorientate before it can move forward again, and the time taken for this reorientation is proportional to its size (Schwartz and Cantor, 1984). In a pulsed field gel, according to this theory, large linear DNA molecules are resolved from one another on the basis of their reorientation time.

An alternative view has been presented by Southern *et al.* (1987). Their theory is again based on reptatory movement and postulates that the leading end of the molecule can never cross the path of the rest of the DNA trailing behind it. In the simplest form of the theory, when the direction of the electric field changes only one end or the other is able to lead the molecule in the new direction. With a field change angle of 90° either end may lead off in the new direction. When the new field is at less than 90° to the previous one, the old leading end continues to lead. But when the change in field orientation is greater than 90°, it is the old trailing end that becomes the leader. In support of their theory, Southern *et al.* (1987) demonstrated that field change angles of <90° provided no better resolution of large linear molecules than conventional gel electrophoresis, while angles of >90° produced increasing resolution of large DNA species. The critical importance of the angle of field reorientation was also noted by Chu *et al.* (1986).

While the heuristic properties of the second model are appealing, they do not explain how molecules are resolved when the field change angle is 180°. Such a situation obtains in the field inversion gel electrophoresis (FIGE) technique (Carle *et al.*, 1986) when the direction of electrophoresis is reversed periodically. In such a situation the old trailing ends should always be chosen to be the new leader, but the new trailing ends would all be in the same position relative to the new field. The success of FIGE implies that there is some physical difference between the two ends of the molecule, and Carle *et al.* (1986) talk of DNA adopting a "wedgelike" configuration. The physical basis of this is not clear and an equivalent result would obtain if there was, for instance, a charge gradient along the length of the molecule. We have far to go in understanding this new tool of pulsed field gradient gel electrophoresis (PFGGE); however, our theoretical inadequacy does not prevent us from exploiting the technique.

7.2. Pulsed Field Gradient Gel Electrophoresis

This was the original pulsed field technique developed by Schwartz and Cantor (1984) and may now be considered obsolete. The apparatus employs arrays of small electrodes perpendicular to one another (Fig. 6) to give a field reorientation angle of 90°. Usually, all available electrodes were used in one orientation, to create a uniform field, but only a single anode and all of the cathodes in the second orientation. This produced a gradient of field strength in the second direction and resulted in bands that had migrated the same distance in adjacent tracks lying on a curved diagonal across the gel. This makes track-to-track comparisons difficult,

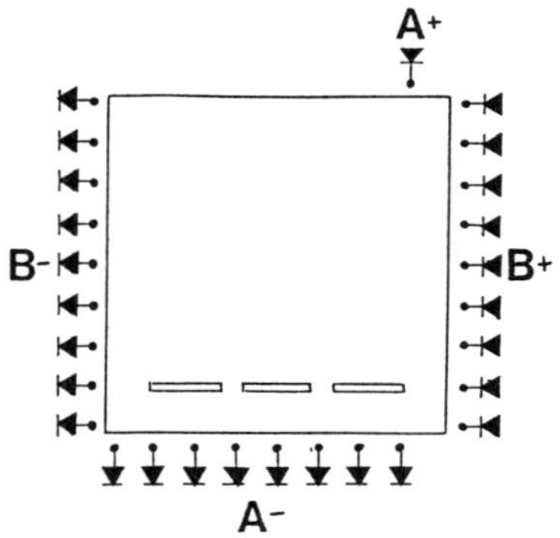

Figure 6. Geometries for pulsed field gradient gel electrophoresis (PFGGE). A and B represent the two alternative electric fields used in the system. Their polarities are indicated by plus (+) and minus (−) signs, the electrodes employed being shown as either heavy lines or filled circles. The heavy arrows (▼) represent diodes. Three sample slots, indicated by open rectangles, are also shown.

so this method was quickly replaced by the orthogonal-field-alternation gel electrophoresis (OFAGE) system of Carle and Olson (1984).

7.3. Orthogonal-Field-Alternation Gel Electrophoresis

The OFAGE system uses two sets of continuous electrodes placed at right angles to one another. In the usual configuration (Fig. 7) the anodes are much smaller than, and are placed near to one end of, the cathodes. This means that the field change angle changes continuously across the gel and produces a field gradient analogous to that obtained with the discontinuous electrode system of PFGGE. It is unlikely that this gradient is essential to the resolving power of the system (Chu *et al.*, 1986; Southern *et al.*, 1987) and causes a distortion of the track path, particularly for tracks distal to the center of the gel (see Fig. 8). However, even when cathode and anode are of equal length in both sets of electrodes, similar distortion is observed (Carle and Olson, 1984). This is because the resistance of the platinum electrodes is less than that of the running buffer, so an induced current is produced in the two inactive electrodes. This results in an inhomogeneous field and distorts the mi-

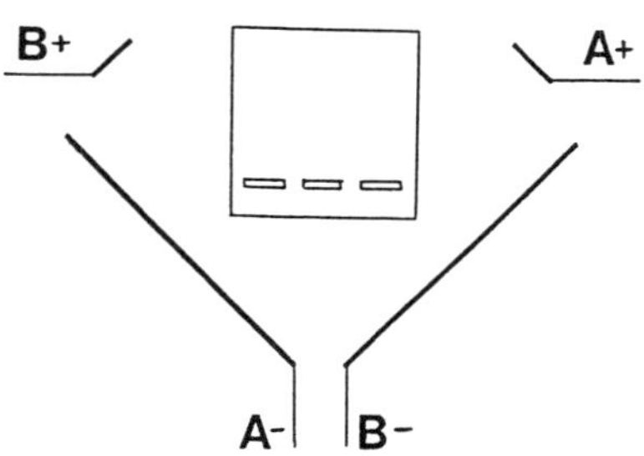

Figure 7. Geometries for orthogonal-field-alternation gel electrophoresis (OFAGE). For details of symbols see legend to Fig. 6.

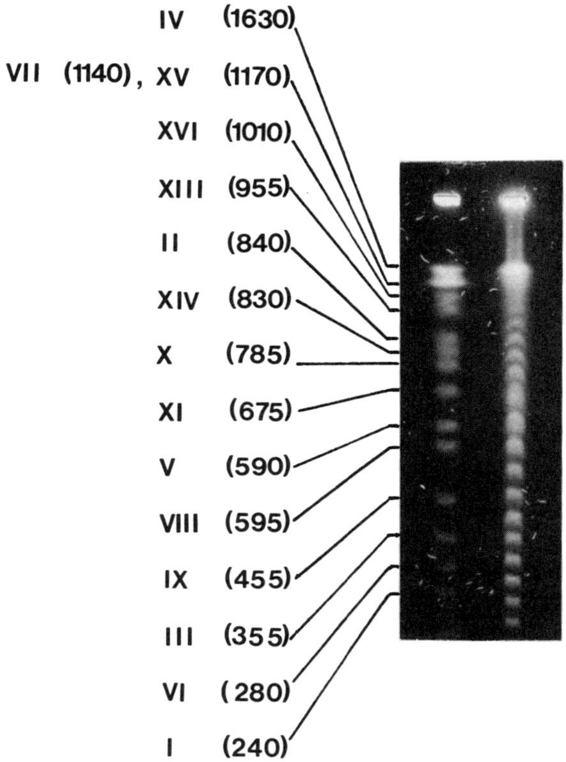

Figure 8. Separation of yeast chromosomes by CHEF pulsed field gel electrophoresis. The left track presents the electrophoretic karyotype of *S. cerevisiae* with the bands corresponding to the different chromosomes identified together with their estimated size (in kb). The right track shows a "ladder" of multimers of phage lambda DNA which were used as size standards. (The help of Mike Musialowski and Nadia Danhash in the preparation of this figure is gratefully acknowledged.)

gration pathway. While OFAGE is a great improvement in PFGGE, it is still difficult to make accurate track-to-track comparisons for a large number of samples. Various novel electrode configurations have been employed to obtain homogeneous electric field and obviate track distortions.

7.4. Contour-Clamped Homogeneous Electric Field

The reasoning behind the electrode configuration in the contour-clamped homogeneous electric field (CHEF) system (Fig. 9) can be described as follows: The induced current in the inactive electrodes which distorts the field in the OFAGE system could be obviated if the electrodes

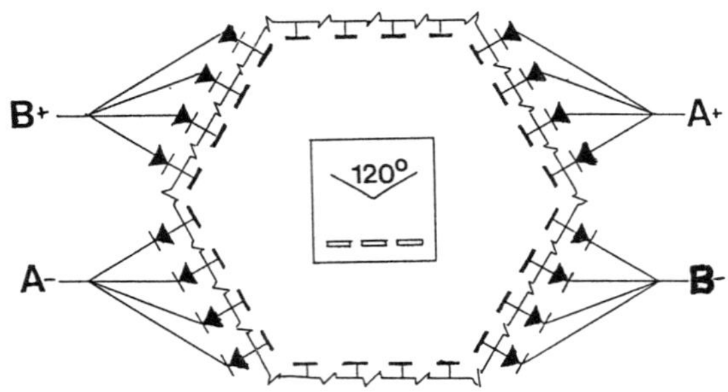

Figure 9. Geometries for CHEF electrophoresis. For details of symbols see legend to Fig. 6.

were infinitely long. This is clearly impracticable, but the situation may be approximated by using multiple electrodes separated by resistors that "clamp" the potential of each individual electrode to the value it would have reached in an infinitely long continuous wire. In the system developed by Chu *et al.* (1986) a hexagonal array of electrodes is employed (Fig. 9). The CHEF system is currently the method of choice for yeast chromosome separations. By the use of low field strengths and low agarose concentrations, Vollrath and Davis (1987) have succeeded in resolving DNA molecules of greater than 5 mb in size and in producing electrophoretic karyotypes of *Candida albicans* and *Schizosaccharomyces pombe,* yeast species that have much larger chromosome than *S. cerevisiae.*

7.5. Vertical Pulsed-Field Gradient Gel Electrophoresis and Constant Electric Field with Rotating Gel Platform

Two other solutions to the field distortion problem of OFAGE have been found. In vertical pulsed-field gradient gel electrophoresis (VPFGGR) (Gardiner *et al.,* 1986), lateral distortions are removed by alternating the field through the thickness of a vertical gel rather than along the length of a horizontal one. In the method of Southern *et al.* (1987) a conventional horizontal gel system with just one set of electrodes of equal length is employed. However, the gel itself is rotated through an angle of 110° with respect to the electric field. Good results for the separation of yeast chromosomes have been published for both systems.

7.6. Field Inversion Gel Electrophoresis

A radically different system of pulsed gel electrophoresis has been developed by Carle *et al.* (1986). This employs a conventional horizontal gel electrophoresis system but periodically inverts the electric field such that the DNA molecules migrate toward the anode in a "two steps for-

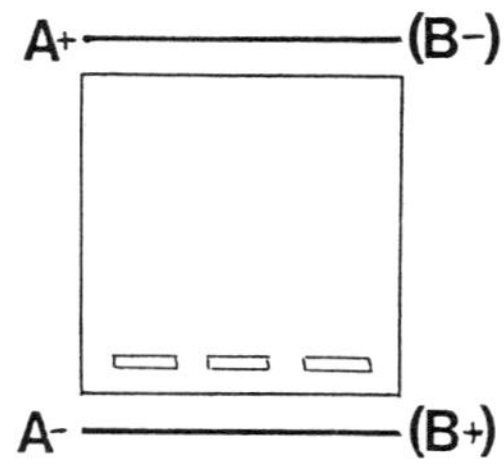

Figure 10. Geometries for FIGE. For details of symbols see legend to Fig. 6.

ward, one step back" fashion (Fig. 10). Thus the field reorientation angle is a maximum 180°. The system has the merit of simplicity for its apparatus but has the procedural complication that the migration of DNA molecules within a certain size range is length-independent at any given switching interval. Thus molecules of quite disparate size can unexpectedly comigrate. This problem may be largely overcome by ramping the switch interval such that it increases linearly as the run progresses. Such ramped control of switching may easily be achieved by use of a very simple microcomputer. The FIGE and OFAGE techniques may be combined to produce a two-dimensional separation of yeast chromosomes (Carle *et al.*, 1986).

7.7. General Considerations

All of the systems discussed here are available commercially, but OFAGE, CHEF, and FIGE systems may readily be constructed in-house. Any standard power pack that can maintain 350 V at 350 mA (milliamps) may be used and cyclic time delay relays may be purchased cheaply from most suppliers of electronic and electrical components. The cooling of the running buffer is very important and a temperature of 12–14°C must be maintained. A suitable heat exchanger and a recirculating pump must be supplied. Recirculation of the buffer is most effectively maintained using a peristaltic pump and wide-bore silicone tubing.

7.8. Sample Preparation

Systems of sample preparation that avoid shear are essential if intact chromosomes are to be obtained. This is achieved by embedding yeast cells in slabs or beads of agarose, removing their cell walls by enzyme treatment, and then lysing the resultant spheroplasts *in situ*. The slab system was originally introduced by Schwartz and Cantor (1984) and is the most widely used procedure since it is simple and reliable. The production of agarose beads involves suspending the cells in molten agarose and then shaking with paraffin oil to form an emulsion. The main merit of the bead system is that lower enzyme concentrations are required for spheroplast formation or for the digestion of DNA with rare-cutting restriction endonucleases.

REFERENCES

Achstetter, T., and Wolf, D. H. 1985. Proteinases, proteolysis and biological control in the yeast *Saccharomyces cerevisiae, Yeast* **1:**139–158.

Baan, R. A., Keller, P. B., and Dahlberg, A. E. 1981. Isolation of eukaryotic initiation factor 2 from yeast *Saccharomyces cerevisiae, J. Biol. Chem.* **256:**1063–1066.

Ballou, C. E. 1982. Yeast cell wall and cell surface, in: *The Molecular Biology of the Yeast* Saccharomyces: *Metabolism and Gene Expression* (J. N. Strathern, E. W. Jones, and J. R. Broach, eds.), Cold Spring Harbor Laboratory, Cold Spring Harbor, NY, pp. 335–360.

Banks, G. R. 1973. Mitochondrial DNA synthesis in permeable cells, *Nature* **245:**196–199.

Bell, L., and Byers, B. 1983. Separation of branched from linear DNA by two-dimensional gel electrophoresis, *Anal. Biochem.* **130:**527–535.

Brewer, B. J., and Fangman, W. L. 1987. The localization of replication origins on *ARS* plasmids in *S. cerevisiae, Cell* **51:**463–471.

Brunner, S., Dark, F. A., Gerhardt, P., Jeynes, M. H., Kandler, O., Kellenberger, E., Klienberger-Nobel, E., McQuillen, K., Rubio-Huertos, M., Salton, M. R. J., Strange, R. E., Tomcsik, J., and Weibull, C. 1958. Bacterial protoplasts, *Nature* **181:**1713–1715.

Carle, G. F., and Olson, M. V. 1984. Separation of chromosomal DNA molecules from yeast by orthogonal-field-alternation gel electrophoresis, *Nucleic Acids Res.* **12:**5647–5664.

Carle, G. F., Frank, M., and Olson, M. V. 1986. Electrophoretic separations of large DNA molecules by periodic inversion of the electric field, *Science* **232:**65–68.

Carnevali, F., Caserta, M., and DiMauro, E. 1982. *In vitro* transcription by purified RNA polymerase II. Coarse promoter mapping on homologous cloned genes, *Nucleic Acids Res.* **10:**3195–3209.

Celniker, S. E., and Campbell, J. L. 1982. Yeast DNA replication *in vitro:* Initiation and elongation events mimic *in vivo* processes, *Cell* **51:**203–210.

Chan, P. Y., and Cossins, F. A. 1976. General properties and regulation of arginine transporting systems in *Saccharomyces cerevisiae, Plant Cell Physiol.* **17:**341–349.

Chanda, P. K., and Kung, H-F. 1983. *In vitro* synthesis of biologically active human leukocyte inteferon in a RNA-dependent system from *Saccharomyces cerevisiae, Proc. Natl. Acad. Sci. USA* **80:**2569–2573.

Chia, L-L., and McLaughlin, C. S. 1979. The half-life of mRNA in *Saccharomyces cerevisiae, Mol. Gen. Genet.* **170:**137–144.

Chirgwin, J. M., Przybyla, A. E., MacDonald, R. J., and Rutter, W. J. 1979. Isolation of biologically active ribonucleic acid from sources enriched in ribonuclease, *Biochemistry* **18:**5294–5299.

Chu, G., Vollrath, D., and Davis, R. W. 1986. Separation of large DNA molecules by contour-clamped homogeneous electric fields, *Science* **234:**1582–1585.

Clare, J. J. 1982. Studies on the regulation of RNA synthesis in the yeast, *Saccharomyces cerevisiae.* Ph.D. thesis, University of Kent at Canterbury.

Clare, J. J., and Oliver, S. G. 1982. The regulation of RNA synthesis in yeast. V. tRNA charging studies, *Mol. Gen. Genet.* **188:**86–102.

Cryer, D. R., Goldthwaite, C. D., Zinker, S., Lam, K.-B., Storm, E., Hirschberg, R., Blamire, J., Finkelstein, D. B., and Marmur, J. 1973. Studies on nuclear and mitochondrial DNA of *Saccharomyces cerevisiae, Cold Spring Harb. Symp. Quant. Biol.* **38:**17–29.

Cryer, D. R., Eccleshall, R., and Marmur, J. 1975. Isolation of yeast DNA, *Methods Cell Biol.* **12:**39–44.

Davis, R. W., Thomas, M., Cameron, J., St. John, T. P., Scherer, S., and Padgett, R. A. 1980. Rapid DNA isolations for enzymatic and hybridisation analysis, *Methods Enzymol.* **65:**404–411.

de Ronde, A., van Loon, A. P. G. M., and Grivell, L. A. 1980. *In vitro* suppression of UGA codons in a mitochrondrial mRNA, *Nature* **287:**361–363.

Devenish, R. J., and Newlon, C. S. 1982. Isolation and characterisation of yeast ring chromosome III by a method applicable to other circular DNAs. *Gene* **18:**277–288.

Diaz-Ruiz, J. R., and Kaper, J. M. 1978. Isolation of viral double-stranded RNAs using LiCl fractionation procedure, *Prep Biochem.* **8:**1–17.

Douglas, M., Finkelstein, D., and Butow, R. A. 1979. Analysis of products of mitochondrial protein synthesis in yeast: Genetic and biochemical aspects, *Methods Enzymol.* **56:**58–66.

Eddy, A. A., and Williamson, D. H. 1957. A method of isolating protoplasts from yeast, *Nature* **179:**1252–1253.

Fangman, W. L., Hice, R. H., and Chlebowicz-Sledziewska, E. 1983. *ARS* replication during the yeast S-phase, *Cell* **32:**831–838.

Faulhammer, H. G., and Cramer, F. 1977. Tyrosyl-tRNA synthetase from Baker's yeast: Rapid isolation by affinity elution, molecular weight of the enzyme and determination of sulfhydryl groups, *Eur. J. Biochem.* **75:**1664–1671.

Feinberg, B., and McLaughlin, C. S. 1988. Isolation of yeast mRNA and *in vitro* translation in a yeast cell-free system, in: *Yeast: A Practical Approach* (I. Campbell and J. H. Duffus, eds.), IRL Press, Oxford, pp. 147–161.

Fisher, M. P., and Dingman, C. W. 1971. Role of mo' cular conformation in determining the electrophoretic properties of polynucleotides in agarose-acrylamide composite gels, *Biochemistry* **10:**1895–1899.

Franklin, R. M. 1966. Purification and properties of the replicative intermediate of the RNA bacteriophase R17, *Proc. Natl. Acad. Sci. USA* **55:**1504–1511.

Gallis, B. M., and Young, E. T. 1975. Endogenous messenger RNA-directed polypeptide chain elongation in a cell-free system from the yeast *Saccharomyces cerevisiae, J. Bacteriol.* **122:**719–726.

Gamper, H., Lehman, H., Piette, J., and Hearst, J. 1985. Purification of DNA using benzoylated naphthoylated DEAE-cellulose, *Gene* **4:**157–164.

Gardiner, K., Laas, W., and Patterson, D. 1986. Fractionation of large mammalian DNA restriction fragments using vertical pulsed-field gel electrophoresis, *Somatic Cell Genet.* **12:**185–195.

Gasior, E., Herrera, F., Sadnik, I., McLaughlin, C. S., and Moldave, K. 1979a. The preparation and characterisation of a cell-free system from *Saccharomyces cerevisiae* that translates natural messenger ribonucleic acid, *J. Biol. Chem.* **254:**3965–3969.

Gasior, E., Herrera, F., McLaughlin, C. S., and Moldave, K. 1979b. The analysis of intermediary reactions involved in protein synthesis, in a cell-free extract of *Saccharomyces cerevisiae* that translates natural messenger ribonucleic acid, *J. Biol. Chem.* **254:**3970–3976.

Gregory, D. W., Polakis, E. S., and Bartley, W. 1967. The isolation and purification of yeast mitochondria, in: *Methods in Enzymology,* Vol. 12 (L. Grossman and K. Moldave, eds.), Academic Press, New York, pp. 469–475.

Grivell, A. R., and Jackson, J. F. 1968. Thymidine kinase: Evidence for its absence from *Neurospora crassa* and some other microorganisms and the relevance of this to the specific labelling of DNA, *J. Gen. Microbiol.* **54:**307–312.

Gross, K. J., and Pogo, O. 1974. Control mechanism of ribonucleic acid synthesis in eukaryotes, *J. Biol. Chem.* **249:**568–576.

Hammond, C. I., and Holland, M. J. 1983. Purification of yeast RNA polymerase using heparin agarose affinity chromatography, *J. Biol. Chem.* **258:**3230–3241.

Hampel, A. E., Enger, M. D., and Ritler, P. E. 1979. Aminoacyl-tRNA synthetases from cultured CHO cells, *Methods Enzymol.* **59:**229–234.

Hartwell, L. H., and McLaughlin, C. S. 1968. Temperature-sensitive mutants of yeast exhibiting a rapid inhibition of protein synthesis, *J. Bacteriol.* **96:**1664–1671.

Hereford, L. M., and Hartwell, L. H. 1971. Defective DNA synthesis in permeabilized yeast mutants, *Nature* **234:**171–172.

Hofbauer, R., Fessl, F., Hamilton, B., and Ruis, H. 1982. Preparation of an mRNA-dependent cell-free translation system from whole cells of *Saccharomyces cerevisiae, Eur. J. Biochem.* **122:**199–203.

Holland, M. J., Hager, G. L., and Rutter, W. J. 1977. Characterisation of purified poly(adenylic acid)-containing messenger RNA from *Saccharomyces cerevisiae, Biochemistry* **16:**8–16.

Holm, C., Meeks-Wagner, D. W., Fangman, F. L., and Botstein, D. 1986. A rapid, efficient method for isolating DNA from yeast, *Gene* **42:**169–173.

Huberman, J. A., Spotila, L. D., Nawotka, K. A., El-Assouli, S. M., and Davis, L. R. 1987. The *in vivo* replication origin of the yeast 2 μm plasmid, *Cell* **51:**473–481.

Hudspeth, M. E. S., Shumard, D. S., Tatti, K. M., and Grossman, L. I. 1980. Rapid purification of yeast mitochondrial DNA in high yield, *Biochim. Biophys. Acta* **610:**221–228.

Hughes, D. E., Wimpenny, J. W. T., and Lloyd, D. 1971. The disintegration of micro-organisms, in: *Methods in Microbiology,* Vol. 5B (J-R. Norris and D. W. Ribbons, eds.), Academic Press, New York, pp. 1–54.

Hussain, I., and Leibowitz, M. J. 1986. Translation of homologous and heterologous mRNAs in a yeast cell-free system, *Gene* **46:**13–23.

Hutchison, H. T., and Hartwell, L. H. 1967. Macromolecule synthesis in yeast sphaeroplasts, *J. Bacteriol.* **94:**1697–1705.

Jazwinski, S. M. 1987. Participation of ATP in the binding of a yeast replicative complex to DNA, *Biochem. J.* **246:**213–219.

Jazwinski, S. M., and Edelman, G. M. 1982. Protein complexes from active replication fractions associate *in vitro* with the replication origins of yeast—2μm DNA plasmid, *Proc. Nat. Acad. Sci. USA* **79:**3428–3432.

Jazwinski, S. M., and Edelman, G. M. 1984. Evidence for the participation of a multiprotein complex in yeast DNA replication *in vitro, J. Biol. Chem.* **259:**6852–6857.

Jazwinski, S. M., Niedzwiecka, A., and Edelman, G. M. 1983. *In vitro* association of a replication complex with a yeast chromosomal replicator, *J. Biol. Chem.* **258:**2754–2757.

Jong, A. Y. S., and Scott, J. F. 1985. DNA synthesis in yeast cell-free extracts dependent on recombinant plasmid purified from *Escherichia coli, Nucleic Acids Res.* **13:**2943–2949.

Jong, A. Y. S., Kuo, C-L., and Campbell, J. L. 1984. The *CDC8* gene of yeast encodes thymidylate kinase, *J. Biol. Chem.* **259:**11052–11059.

Kandel, J. 1979. Isolation and detection of double-stranded RNA from fungi, *Methods Enzymol.* **60:**549–554.

Kassovetis, G. A., Riggs, D. L., Negri, R., Nguayen, L., and Geiduschek, E. P. 1989. Transcription factor III B generates extended DNA interactions in RNA polymerase III transcription complexes on tRNA genes, *Mol. Cell Biol.* **9:**2551–2566.

Kirby, K. S. 1965. Isolation and characterization of ribosomal ribonucleic acid, *Biochem. J.* **96:**266–269.

Kitamura, K., and Yamamoti, Y. 1972. Purification and properties of an enzyme, zymolyase, which lyses viable yeast cells, *Arch. Biochem. Biophys.* **153:**403–406.

Klekamp, M. S., and Weil, P. A. 1982. Specific transcription of homologous class III genes in yeast soluble cell-free extracts, *J. Biol. Chem.* **257:**8432–8441.

Klootwijk, J., Verbeet, M.Ph., Veldmann, G. M., deRegt, V. C. H. F., van Heerikhuizen, H., Bogerd, J., and Planta, R. J. 1984. The *in vivo* and *in vitro* initiation site for transcription of the RNA operon of *Saccharomyces carlsbergensis, Nucleic Acids Res.* **12:**1377–1390.

Koch, H., and Friesen, J. D. 1979. Individual messenger RNA half lives in *Saccharomyces cerevisiae, Mol. Gen. Genet.* **170:**129–135.

Kojo, H., Greenberg, B. D., and Sugino, A., 1978, Yeast 2 μm plasmid replication *in vitro:* Origin and direction, *Proc. Natl. Acad. Sci. USA* **78:**7261–7265.

Kornberg, A. 1980. *DNA Replication,* W. H. Freeman, San Francisco.

Kreutzfeldt, C., and Lochmann, E-R. 1983. Preparation of a cell-free extract from yeast that is active in protein synthesis, *FEMS Microbiol. Lett.* **16:**179–182.

Kuo, C-L., and Campbell, J. L. 1982. Purification of the *cdc8* protein of *Saccharomyces cerevisiae* by complementation in an aphidocolin-sensitive *in vitro* DNA replication system, *Proc. Natl. Acad. Sci. USA* **79:**4243–4247.

Kuo, S-C., and Yamamoto, S. 1975. Preparation and growth of yeast protoplasts, *Methods Cell Biol.* **11:**169–183.

Lang, B., Burger, G., Doxiadis, I., Thomas, D. Y., Bandlow, W., and Kaudewitz, F. 1977. A simple method for the large scale preparation of mitochondria from microorganisms, *Anal. Biochem.* **77:**110–121.

Littlewood, R., Shaffer, B., and Davies, J. 1971. Mutants of *Saccharomyces cerevisiae* with reduced levels of ribonuclease activity, *Genetics* **68:**s39.

Lohr, D., and Ide, G. I. 1983. *In vitro* initiation and termination of ribosomal RNA transcription in isolated yeast nuclei, *J. Biol. Chem.* **258:**4668–4671.

Lue, N. F., and Kornberg, R. D. 1987. Accurate initiation at RNA polymerase II promoters in extracts from *Saccharomyces cerevisiae, Proc. Natl. Acad. Sci. USA* **84:**8839–8843.

Lumpkin, O. J., and Zimm, B. H. 1982. Mobility of DNA in gel electrophoresis, *Biopolymers* **11:**2315–2316.

Masurekar, N., Palmer, E., Ono, B-I., Wilhelm, J. M., and Sherman, F. 1981. Misreading of the ribosomal suppressor *SUP46* is due to an altered 40S subunit in yeast, *J. Mol. Biol.* **147:**381–390.

McKee, E. E., and Poyton, R. O. 1984. Mitochondrial gene expression in *Saccharomyces cerevisiae.* I. Optimal conditions for protein synthesis in isolated mitochondria, *J. Biol. Chem.* **259:**9320–9331.

McKee, E. E., McEwen, J. E., and Poyton, R. O. 1984. Mitochondrial gene expression in *Saccharomyces cerevisiae.* II. Fidelity of translation in isolated mitochondria from wild type and respiratory-deficient mutant cells, *J. Biol. Chem.* **259:**9332–9336.

McLaughlin, C. S., Warner, J. R., Edmonds, M., Nakazato, H., and Vaughan, M. H. 1973. Polyadenylic acid sequences in yeast messenger RNA, *J. Biol. Chem.* **248:**1466–1471.

Miller, M. J., Xuong, N., and Geiduschek, E. P. 1979. A response of protein synthesis to a temperature shift in the yeast *Saccharomyces cerevisiae, Proc. Natl. Acad. Sci. USA* **76:**5222–5225.

Mills, D. 1974. Isolation of polyribosomes from yeast during sporulation and vegative growth, *Appl. Microbiol.* **27:**944–948.

Mitchell, D. J., and Bevan, E. A. 1987. dsRNA killer systems in yeast, in: *Yeast Biotechnology* (D. R. Berry, I. Russell, and G. G. Stewart, eds.), Allen and Unwin, London, pp. 104–155.

Moldave, K., and McLaughlin, C. S. 1988. The analysis of temperature-sensitive mutants of *Saccharomyces cerevisiae* altered in components required for protein synthesis, in: *Genetics of Translation: New Approaches* (M. F. Tuite, M. Picard, and M. Bolotin-Fukuhara, eds.), Springer-Verlag, Berlin, pp. 271–282.

Monier, R., Stephenson, M. L., and Zamechnik, P. C. 1960. The preparation and some properties of a low molecular weight ribonucleic acid from baker's yeast, *Biochim. Biophys. Acta* **43:**1–8.

Newlon, C. S., Devenish, R. J., Suci, P. A., and Roffi, C. J. 1981. Replication of small chromosomal DNAs in yeast, *ICN-UCLA Symp. Mol. Biol.* **22:**501–516.

Oliver, S. G., McCready, S. J., Holm, C., Sutherland, P., McLaughlin, C. S., and Cox, B. S. 1977. Biochemical and physiological studies of the yeast virus-like particle, *J. Bacteriol.* **130:**1303–1309.

Oppenheim, A. 1981. Separation of closed circular DNA from linear DNA by electrophoresis in two dimensions in agarose gels, *Nucleic Acids Res.* **9:**6805–6812.

Pelham, H. R. B., and Jackson, R. J. 1976. An efficient mRNA-dependent translation system from reticulocyte lysates, *Eur. J. Biochem.* **67:**247–256.

Peppler, H. J. 1979. Production of yeasts and yeast products, in: *Microbial Technology*, 2nd ed., Vol. 1 (H. J. Peppler and D. Perlman, eds.), Academic Press, New York, pp. 157–185.

Pfisterer, J., and Buetow, D. E. 1981. *In vitro* reconstitution of the mitochondrial translation system of yeast, *Proc. Natl. Acad. Sci. USA* **78:**4917–4921.

Plesset, J., Foy, J. J., Chia, L. L., and McLaughlin, C. S. 1982. Heat shock in *Saccharomyces cerevisiae:* Quantitation of transcriptional and translational effects, in: *Transcription/Translational Regulation of Gene Expression* (M. Grunberg-Manago and B. Safer, eds.), Elsevier, New York, pp. 495–514.

Pringle, J. R. 1975. Methods for avoiding proteolytic artefacts in studies of enzymes and other proteins from yeasts, *Methods Cell Biol.* **12:**149–184.

Radloff, R., Bauer, W., and Vinograd, J. 1967. A dye-buoyant density method for the detection and isolation of closed circular duplex DNA: The closed circular DNA in HeLa cells, *Proc. Natl. Acad. Sci. USA* **57:**1514–1521.

Retel, J., Van den Bos, R. C., and Planta, R. J. 1969. Characteristics of the methylation *in vivo* of ribosomal RNA in yeast, *Biochim. Biophys. Acta* **195:**370–380.

Rothblatt, J. A., and Meyer, D. I. 1986. Secretion in yeast: Reconstitution of the translocation and glycosylation of α-factor and invertase in a homologous cell-free system, *Cell* **44:**619–628.

Rubin, G. M. 1975. Preparation of RNA and ribosomes from yeast, *Methods Cell Biol.* **12:**45–64.

Sawadago, M., Sentenac, A., and Fromageot, P. 1981. *In vitro* transcription of cloned yeast ribosomal DNA by yeast RNA polymerase A, *Biochem. Biophys. Res. Commun.* **101:**250–257.

Schantz, G. 1967. Stable phosphorylating submitochondrial particles from baker's yeast, *Methods Enzymol.* **10:**197–202.

Schindler, D., Grant, P. G., and Davies, J. E. 1974. Trichodermin resistance-mutation affecting eukaryotic ribosomes, *Nature* **248:**535–536.

Schmidt, G. 1968. Periodate oxidation of ribonucleic acids and their derivatives, *Methods Enzymol.* **12B:**230–235.

Schwartz, D. C., and Cantor, C. R. 1984. Separation of yeast chromosome-sized DNAs by pulse field gradient gel electrophoresis, *Cell* **37:**67–75.

Scopes, R. K. 1982. *Protein Purification: Principles and Practice,* Springer, New York.

Scott, J. H., and Schekman, R. 1980. Lyticase: Endoglucanase and protease activities that act together in yeast cell lysis, *J. Bacteriol.* **142:**414–423.

Serwer, P., and Hayes, S. J. 1986. Exclusion of spheres by agarose gels during agarose gel electrophoresis: Dependence on the sphere's radius and the gel's concentration, *Anal. Biochem.* **158:**72–78.

Shalitin, C., and Vishlizky, A. 1984. An improved isolation procedure for yeast two-micrometer minichromosomes, *Current Genet.* **9:**107–111.

Shalitin, C., Pan, C. J., and Davie, J. R. 1983. Isolation of 2μm minichromosomes from *Saccharomyces cerevisiae* using shallow metrizamide gradients, *Exp. Mycol.* **7:**175–181.

Sissons, C. H. 1978. Methods for yeast protein synthesis in a cell-free system, *Methods Cell Biol.* **20:**83–99.

Skogerson, L., and Wakatama, E. 1976. A ribosome-dependent GTPase from yeast distinct from elongation factor 2, *Proc. Natl. Acad. Sci. USA* **73**:73–76.

Sogin, S. J., and Saunders, C. A. 1980. Fluctuation in polyadenylate size and content in exponential- and stationary-phase cells of *Saccharomyces cerevisiae, J. Bacteriol.* **144**:74–81.

Sogin, S. J., Haber, J. E., and Halvorson, H. O. 1972. Relationship between sporulation-specific 20S ribonucleic acid and ribosomal ribonucleic acid processing in *Saccharomyces cerevisiae, J. Bacteriol.* **112**:806–814.

Sommer, A., and Lewis, M. J. 1971. Effect of dithiothreitol on yeast: Sphaeroplast formation and invertase release, *J. Gen. Microbiol.* **68**:327–335.

Southern, E. M. 1975. Detection of specific sequences among DNA fragments separated by gel electrophoresis, *J. Mol. Biol.* **98**:503–517.

Southern, E. M., Anand, R., and Fletcher, D. S. 1987. A model for the separation of large DNA molecules by crossed field gel electrophoresis, *Nucleic Acids Res.* **15**:5925–5943.

Specht, C. A., Di Russo, C. C., Novotny, C. P., and Ullrich, R. C. 1982. A method for extracting high molecular weight DNA from fungi, *Anal. Biochem.* **119**:158–163.

Sripati, C. E., and Warner, J. R. 1978. Isolation, characterisation and translation of mRNA from yeast, *Methods Cell Biol.* **20**:61–81.

Sripati, C. E., Groner, Y., and Warner, J. R. 1976. Methylated blocked 5′termini of yeast mRNA, *J. Biol. Chem.* **251**:2898–2904.

Strathern, J. N., Newlon, C. S., Herskowitz, I., and Hicks, J. B. 1979. Isolation of a circular derivative of yeast chromosome. III. Implications for the mechanism of mating type interconversion, *Cell* **18**:309–319.

Swanson, M. E., and Holland, M. J. 1983. RNA polymerase I-dependent selective transcription of yeast ribosomal DNA, *J. Biol. Chem.* **258**:3242–3250.

Szczesna, E., and Filipowicz, W. 1980. Faithful and efficient translation of viral and cellular eukaryotic mRNAs in a cell-free S-27 extract of *Saccharomyces cerevisiae, Biochem. Biophys. Res. Commun,* **92**:563–569.

Tuite, M. F., and Plesset, J. 1986. mRNA-dependent yeast cell-free translation systems: Theory and practice, *Yeast* **2**:35–52.

Tuite, M. F., Plesset, J., Moldave, K., and McLaughlin, C. S. 1980. Faithful and efficient translation of homologous and heterologous mRNAs in an mRNA-dependent cell-free system from *Saccharomyces cerevisiae, J. Biol. Chem.* **255**:8761–8766.

Tuite, M. F., Cox, B. S., and McLaughlin, C. S. 1981. An homologous *in vitro* assay for yeast nonsense suppressors, *J. Biol. Chem.* **256**:7298–7304.

Tuite, M. F., Cox, B. S., and McLaughlin, C. S. 1983. *In vitro* nonsense suppression in [*psi*$^+$] and [*psi*$^-$] cell-free lysates of *Saccharomyces cerevisiae, Proc. Natl. Acad. Sci. USA* **80**:2824–2828.

Udem, S. A., and Warner, J. R. 1972. Ribosomal RNA synthesis in *Saccharomyces cerevisiae, J. Mol. Biol.* **65**:227–242.

Van den Bos, R. C., and Planta, R. J. 1971. Studies on the role of rapidly labelled 20S RNA in the biosynthesis of ribosomal RNA in yeast, *Biochim. Biophys. Acta* **247**:175–180.

Vazquez, D., and Jimenez, A. 1980. Antibiotic inhibitors of translation in eukaryotes, in: *Ribosomes: Structure, Function and Genetics* (G. Chambliss, G. R. Graven, J. E. Davis, K. Davis, L. Kahan, and M. Nomura, eds.), University Park Press, Baltimore, pp. 847–869.

Villanueva, J. R., Gacto, M., and Sierva, S. M. 1973. Enzymatic composition of a lytic system from *Micromonospora chalcea, in: Yeast, Mold and Plant Protoplasts* (J. R. Villanueva, I. Garcia-Aeha, S. Gascon, and F. Urburu, eds.), Acaemic Press, New York, pp. 3–24.

Vollrath, D., and Davis, R. W. 1987. Resolution of DNA molecules greater than 5 megabases by contour-clamped homogeneous electric fields, *Nucleic Acids Res.* **15**:7865–7876.

Von der Haar, F. 1979. Purification of aminoacyl-tRNA synthetases, *Methods Enzymol.* **59:**257–267.

Waters, M. G., and Blobel, G. 1986. Secretory protein translocation in a yeast cell-free system can occur posttranslationally and require ATP hydrolysis, *J. Cell Biol.* **102:**1543–1550.

Weeks, D. P., Beerman, N., and Griffith, O. M. 1986. A small-scale five hour procedure for isolating multiple samples of CsC1-purified DNA: Application to isolations from mammalian, insect, higher plant, algal, yeast and bacterial sources, *Anal. Biochem.* **152:**376–385.

Wickner, R. B. 1975. Mutants of *Saccharomyces cerevisiae* that incorporate deoxythymidine 5′-monophosphate into DNA *in vivo, Methods Cell Biol.* **11:**295–302.

Wickner, R. B., and Leibowitz, M. J. 1976. Chromosomal genes essential for replication of a dsRNA plasmid of *Saccharomyces cerevisiae:* The killer character of yeast, *J. Mol. Biol.* **105:**427–434.

Williamson, D. H., and Fennell, D. J. 1975. The use of fluorescent DNA-binding agent for detecting and separating yeast mitochondrial DNA, *Methods Cell Biol.* **11:**335–351.

Winston, F., Chumley, F., and Fink, G. R. 1983. Eviction and transplacement of mutant genes in yeast, *Methods Enzymol.* **101:**211–227.

Yang, W. K., and Novelli, G. D. 1971. Analysis of isoaccepting tRNAs in mammalian tissues and cells, *Methods Enzymol.* **20:**44–55.

Index